Spon's
Civil Engineering and Highway Works Price Book

2023

Ebook Set up and Use Instructions

Use of the ebook is subject to the single user licence at the back of this book.

Electronic access to your price book is now provided as an ebook on the VitalSource® Bookshelf platform. You can access it online or offline on your PC/Mac, smartphone or tablet. You can browse and search the content across all the books you've got, make notes and highlights and share these notes with other users.

Setting up

1. Create a VitalSource Bookshelf account at https://online.vitalsource.com/user/new (or log into your existing account if you already have one).

2. Retrieve the code by scratching off the security-protected label inside the front cover of this book. Log in to Bookshelf and click the **Redeem Codes link in the Tools** menu at the top right of the screen. Enter the code in the **Redeem code** box and press **Redeem**. Once the code has been redeemed your Spon's Price Book will download and appear in your library. N.B. the code in the scratch-off panel can only be used once.

When you have created a Bookshelf account and redeemed the code you will be able to access the ebook online or offline on your smartphone, tablet or PC/Mac. Your notes and highlights will automatically stay in sync no matter where you make them.

Use ONLINE

1. Log in to your Bookshelf account at https://online.vitalsource.com).
2. Double-click on the title in your **library** to open the ebook.

Use OFFLINE

Download BookShelf to your PC, Mac, iOS device, Android device or Kindle Fire, and log in to your Bookshelf account to access your ebook, as follows:

On your PC/Mac
Go to https://support.vitalsource.com/hc/en-us/categories/360001056774-Bookshelf#360002383594 and follow the instructions to download the free VitalSource Bookshelf app to your PC or Mac. Double-click the VitalSource Bookshelf icon that appears on your desktop and log into your Bookshelf account. Select **All Titles** from the menu on the left - you should see your price book on your Bookshelf. If your Price Book does not appear, select **Update Booklist** from the **Account** menu. Double-click the price book to open it.

On your iPhone/iPod Touch/iPad
Download the free Bookshelf App available via the iTunes App Store. Open the Bookshelf app and log into your Bookshelf account. Select **All Titles** – you should see your price book on your Bookshelf. Select the price book to open it. You can find more information at https://support.vitalsource.com/hc/en-us/articles/360014105113-iOS

On your Android™ smartphone or tablet
Download the free Bookshelf App available via Google Play. Open the Bookshelf app and log into your Bookshelf account. You should see your price book on your Bookshelf. Select the price book to open it. You can find more information at https://support.vitalsource.com/hc/en-us/articles/360014007294-Android

On your Kindle Fire
Download the free Bookshelf App from Amazon. Open the Bookshelf app and log into your Bookshelf account. Select All Titles – you should see your price book on your Bookshelf. Select the price book to open it. You can find more information at https://support.vitalsource.com/hc/en-us/articles/360014107953-Kindle-Fire

Support

If you have any questions about downloading Bookshelf, creating your account, or accessing and using your ebook edition, please visit http://support.vitalsource.com/

For questions or comments on content, please contact us on sponsonline@tandf.co.uk

---

Free Updates

with three easy steps...

1. Register today on www.pricebooks.co.uk/updates
2. We'll alert you by email when new updates are posted on our website
3. Then go to www.pricebooks.co.uk/updates and download the update.

All four Spon Price Books – *Architects' and Builders'*, *Civil Engineering and Highway Works*, *External Works and Landscape* and *Mechanical and Electrical Services* – are supported by an updating service. Two or three updates are loaded on our website during the year, typically in November, February and May. Each gives details of changes in prices of materials, wage rates and other significant items, with regional price level adjustments for Northern Ireland, Scotland and Wales and regions of England. The updates terminate with the publication of the next annual edition.

As a purchaser of a Spon Price Book you are entitled to this updating service for this 2023 edition – free of charge. Simply register via the website www.pricebooks.co.uk/updates and we will send you an email when each update becomes available.

If you haven't got internet access or if you've some questions about the updates please write to us at Spon Price Book Updates, Spon Press Marketing Department, 4 Park Square, Milton Park, Abingdon, Oxfordshire, OX14 4RN.

Find out more about Spon books
Visit www.pricebooks.co.uk for more details.

Spon's Civil Engineering and Highway Works Price Book

Edited by

AECOM

2023

Thirty-seventh edition

CRC Press
Taylor & Francis Group

First edition 1984
Thirty-seventh edition published 2023
by CRC Press
4 Park Square, Milton Park, Abingdon, Oxon, OX14 4RN

and by CRC Press
Taylor & Francis, Broken Sound Parkway, NW, Suite 300, Boca Raton, FL 33487

CRC Press is an imprint of the Taylor & Francis Group, an informa business

British Library Cataloguing in Publication Data
A catalogue record for this book is available from the British Library

ISBN: 978-1-032-33175-1
Ebook: 978-1-003-31890-3

ISSN: 0957-171X

DOI: 10.1201/9781003318903

Typeset in Arial by Taylor & Francis Books
Printed and bound by CPI Group (UK) Ltd, Croydon, CR0 4YY

Contents

PART 8: DAYWORK

PART 9: PROFESSIONAL FEES

PART 10: OUTPUTS

Preface to the Thirty-Seventh Edition

At the time of writing the construction industry continues to experience retarded growth in both turnover and margin because of the impacts of Brexit, the COVID 19 pandemic and the Russia-Ukraine war, however only part of these impacts has been included in this book. UK inflation (based on the Consumer Prices Index, CPI) jumped to a 40-year high of 9% in the 12 months to April 2022 as food and energy prices spiralled. Raw materials, supplier and manufactured goods have in recent months shown moderate to high price increases against the published list price. There has been reasonable workload in the infrastructure following the construction of HS2 Project, modernisations of London Underground and Network rail lines, Thames Tideway Tunnel, and other highway projects like Lower Thames Crossing. The start of the site delivery of HS2 project has driven some skilled labour demands in the infrastructure projects somehow to the limits. This is due to the scale of labour, plant and material requirements. The resultant from increased spending announcements on HS2 and other infrastructure projects will continue to increase growth expectation in 2022/2023 even though this has been dwindled by the impact of Covid 19. Levels of expenditure in infrastructure are expected to rise over the next two to three years to sustain the economy and that may impact on the prices of labour, material, and plant. On the other side we have seen decrease in labour resource due to Brexit and this scarcity of labour resource has resulted to increase in labour costs.

The Russian Ukraine war has exacerbated an already turbulent market, hit by Brexit, Covid−19, rising energy cost and spiralling inflationary pressure. It has caused significant uncertainty about the viability of both current and future UK construction projects, particularly in terms of the availability and cost of critical material stell, such as steel. Energy price increases are another issue, particularly given the European countries' high dependency on Russian gas and oil. British Steel lifts prices by 'record 25% in facing of the soaring manufacturing costs, which has been the highest one-off increase, according to one industry expert. The price rise underlines the pressure European steelmakers are under because of surging electricity prices following Russia's invasion of Ukraine. Cement and aggregates prices have increased significantly resulting to high output rates as are bituminous products which form a significant part of infrastructure material costs. Labour rates have also increased in 2022 and expected to continue with due to impact of Brexit and COVID−19. Skilled labour will continue to be scarce and will push overall costs higher. Failures through insolvency are reducing competition with bidders being more selective over which projects they commit resources to, seeking to ensure higher success rates and increased profitability.

Under these market conditions the prices in the various parts of this publication can only be taken as a guide to actual costs. Any sustained upturn may give rise to marked increases in selling prices as manufacturers and specialists seek to recover margins lost in the downturn.

London, the North and the Southeast all remain active and construction activities should continue in the next few years across the UK with mainly HS2 projects plus other projects like Lower Thames Crossing, road schemes, Northumberland line and Thames Tideway Tunnelling leading to increased expenditure. Network Rail continues its ongoing renewals programme. This is expected to generate significant activity throughout 2022/2023. The impact of HS2 is expected to affect performance levels as its now on-site delivery stage. A pick on aviation will also have a resource demand impact on the infrastructure capital investment expenditure.

For the 2023 edition, we have undertaken a general update of all prices up to May/June 2022 in consultation with leading manufacturers, suppliers and specialist contractors and included revisions as necessary. Our efforts have been directed at reviewing, revising and consolidating the scope, range and detail of information to help the user to compare or adjust any unit costs with reference to allocated resources or outputs.

The rates, prices and outputs included in the Resources and Unit Cost calculations, including allowances for wastage, normal productivity, and efficiency, are based on medium sized Civil Engineering schemes of about £50–£300 million in value, with no acute access or ground condition problems. However, they are equally applicable, with little or no adjustment, to a wide ranges of construction projects from £2–£250 million. Where suitable, tables of multipliers have been given to enable easy adjustment to outputs or costs for varying work conditions.

As with all attempts to provide price guidance on a general basis, this must be loaded with caveats. In applying the rates to any specific project, the user must take into account the general nature of the project, i.e., matters such as scale, site difficulties, locale, tender climate etc. This book aims at providing as much information as possible about the nature of the rate so as to assist the user to adapt it if necessary. With safety a foremost objective on every project, an added element of cost is incurred as represented by the preliminaries percentage adjustments. Each project experiences differing levels of risk and as such the adjustments should be made relative to each situation.

This edition continues to provide the reader with cost guidance at a number of levels, varying from the more general functional costs shown in Part 5, through the detailed unit costs in Parts 6 and 7 which relate respectively to the CESMM4 and the Highways Method of Measurement bills of quantities formats, down to the detailed resource costing given in Part 4 supplemented by the further advice on output factors in Part 10.

The outputs used in this work have been compiled in detail from the editors' wide-ranging experience and are based almost exclusively on studies and records derived from a number of recent Civil Engineering schemes. This information is constantly being re-appraised to ensure consistency with current practice. A number of prices and outputs are based upon detailed specialist advice and acknowledgements to the main contributors are included within the book.

The current market is showing signs of volatility in both labour and materials in the forthcoming period, however, to monitor this and maintain accuracy levels readers should use the free price book update service, which advises of any significant changes to the published information over the period covered by this edition. The Update is posted free every three months on the publishers' website, until the publication of the next annual Price Book, to those readers who have registered with the publisher. Details of how to register can be found at the end of the book.

Whilst all efforts are made to ensure the accuracy of the data and information used in the book, neither the editors nor the publishers can in any way accept liability for loss of any kind resulting from the use made by any person of such information.

COVID–19 Pandemic

At the time the book data was being collated, the global pandemic is not yet over and there is considerable uncertainty about the path that the pandemic will not take in the UK. As a result of the vaccine rollout, the Government was able to ease restrictions in the England through the first half of 2021. There are a range of possible futures for the course of the pandemic and its extraordinary impacts continue in many industry sectors nationally and globally. The pandemic brought many once-in-a-lifetime changes to everyday life and that will become the 'new normal' for people around the world.

The expansive disruption to populations, economies and businesses introduced by the public health crisis will continue, as governments across the world are taking measures to manage the post COVID–19 situations. These mitigation measures directly address the public health risks and implications from COVID–19, but an economic demand and supply shock is also a consequence of the actions.

With all this in mind we believe that UK construction supply chain firms will face an array of risks during 2022 as the fallout from the crisis continue to evolve. Due to the evolving and still uncertain outlook, for the purposes of forming this baseline scenario on market drivers, the following assumptions have been used:

- The domestic restrictions are removed, while safer behaviours are encouraged through public health advice.
- Construction sites begin reopening in the 2022 Q1, steadily increasing in number over the upcoming years.

- Coronavirus infection transfer risk is mitigated by ongoing facing covering and social distancing measures in force across the population. It should be acknowledged that further lockdowns could take place if the transfer rate of infection begins to increase once again due to the new variant being introduced.
- Social distancing rules are no longer required on construction sites, hence negative impacts on site productivity are considered as minimum.
- Construction output across 2021 was severely impacted because of site lockdowns, furloughs and decisions made by businesses to pause their construction activity. In the medium-term (6–12 months) output improves from restarted construction activity but is still materially impacted though to a lesser extent than 2021.
- Underlying construction demand into the medium-term remains broadly in place, although some projects are deferred or held because of funding challenges or client-side decisions.
- Risks and impacts from the potential for the coronavirus public health crisis developing into a financial crisis.

AECOM Ltd
Aldgate Tower
2 Leman Street
London
E1 8FA

Acknowledgements

The Editors wish to record their appreciation of the assistance given by many individuals and organizations in the compilation of this edition.

Materials suppliers and subcontractors who have contributed this year include:

Abacus Lighting Ltd
Lighting and Street Furniture
Tel: 01623 511111
Fax: 01623 552133
Website: www.abacuslighting.com
Email: sales@abacuslighting.com

ACO Technologies Plc
Drainage Systems
Tel: 01462 816666
Fax: 01462 815895
Website: www.aco.co.uk
Email: technologies@aco.co.uk

Aggregate Industries
Kerbs, Edgings and Pavings
Tel: 01530 510066
Website: www.aggregate.com

Ainscough Crane Hire Limited
Crane Hire
Tel: 0800 272 637
Website: http://www.ainscough.co.uk/
Email: info@ainscough.co.uk

Alumet Systems (UK) Ltd
Fixings
Tel: 01926 811 677
Website: http://www.alumet.co.uk/
Email: enquiries@alumet.co.uk

Amec Foster Wheeler Plc (Wood Plc)
Tel: 020 7429 7500
Fax: 020 7429 7550
Website: http://www.woodplc.com/

Anaco Trading Ltd
Reinforcing Bar
Tel: 01832 272109
Fax: 01832 275759
Website: www.anacotrading.com

Andrews Sykes
Dewatering
Tel: 0800 211 611

ArcelorMittal Commercial UK Ltd
Steel Sheet Piling
Tel: 0121 713 6674
Mob: 07772 324 595
Website: http://sheetpiling.arcelormittal.com/
Email: david.baxter@arcelormittal.com

Asset International Structures
Steel Culverts
Tel: 01633 637505
Website: www.assetint.co.uk
Email: info@assetint.co.uk

AWBS LTD
Walling
Tel: 01865 326500
Website: http://www.awbsltd.com/
Email: office@awbsltd.com

Bachy Soletanche Ltd
Diaphragm Walling
Tel: 01704 895686
Fax: 01704 895581
Website: www.bacsol.co.uk

Balfour Beatty Engineering Services
Subcontractor Piles
Tel: 0207 216 6800
Website: http://www.balfourbeatty.com/

Barhale Plc
Pipe Jacking
Tel: 01922 707 700
Fax: 01922 721 808
Website: http://www.barhale.co.uk/
Email info@barhale.co.uk

Bison Manufacturing Limited
Flooring
Tel: 01283 817500
Fax: 01283 220563
Website: http://www.bison.co.uk/
Email: concrete@bison.co.uk

BRC Reinforcement
Masonry Reinforcement and Accessories
Tel: 02074 741 800
Fax: 02038 762 909
Website: http://www.brc.ltd.uk
Email: london.sales@brc.ltd.uk

Bromborough Paints
Paint
Tel: 0151 334 1237
Fax: 0151 334 0086
Website: http://www.bromboroughpaints.co.uk/
Email: sales@bromboroughpaints.co.uk

Broxap Ltd
Street Furniture
Tel: 0844 800 4085 / 01782 564411
Fax: 01782 565 357/562 546
Website: http://www.broxap.com/
Email: enquiries@broxap.com

Buildbase Civils & Lintels (Grafton Group Plc)
Drainage Materials
Tel: 01322 446200
Website: https: //www.civilsandlintels.co.uk/
Email: Erith@civilsandlintels.co.uk

Capital Demolition
Demolition and Clearance
Tel: 01932 355 737
Fax: 01932 340 244
Website: www.capitaldemolition.co.uk
Email: d.read@capitaldemolition.freeserve.co.uk

CCL Stressing Systems
Bridge Bearings
Tel: 0113 270 1221
Website: http://www.cclint.com/
Email: enquiries@cclint.com

Celsa Steel (UK) Ltd
Bar Reinforcement
Tel: 0292 035 1800
Fax: 0292 035 1801
Website: www.celsauk.com
Email: salesuk@celsauk.com

Cementation Foundations Skanska Ltd
Piling
Tel: 01923 776666
Website: http://www.skanska.co.uk/services/cementation-piling-and-foundations

Cogne UK LTD
Steel Works
Tel: 0114 221 2020
Fax: 0114221 3030
Website: http://www.cogne.co.uk/

Craig West Turf Ltd
Landscaping
Tel: 01179 982 2255
Website: http://cwtls.co.uk/
Email: chris@cwtls.co.uk or adam@cwtls.co.uk

Craigton Industries
Precast Concrete Manholes
Tel: 0141 956 6585 or 2311
Fax: 0141 956 3757
Website: http://craigtonindustries.co.uk/
Email: sales@craigtonindustries.co.uk

Darlaston Builders Merchants
Walling
Tel: 0121 526 2449
Website: http://www.dbmdiy.com/

Dorsey Construction Materials Ltd
Water Stops
Tel: 0118 975 3377
Fax: 0118 975 3393
Website: http://www.dorseyconstructionmaterials.co.uk/home/
Email: sales@dorseys.co.uk

Don & Low Ltd
Geotextiles
Tel: 01307 452 200
Fax: 01307 452 300
Website: www.donlow.co.uk

Dulux
Paint
Tel: 0333 222 7070
Website: https: //www.duluxtradepaintexpert.co.uk/
Email: duluxtrade.advice@akzonobel.com

Euromix Concrete
Ready Mix Concrete
Tel: 01245 464545
Website: http://www.euromixconcrete.com/

Ecochoice Ltd
Timber Piles
Tel: 0345 638 1340
Fax: 0845 642 1340
Website: www.ecochoice.co.uk
Email: info@ ecochoice.co.uk

EPCO PLASTICS
Drains
Tel: 0113 249 1155
Website: https: //www.epco-plastics.com/
Email: sales@epco-plastics.com

Everyvalve Ltd
Valves and fittings
Tel: 01707 642018
Fax: 01707 646340
Website: http://www.everyvalve.com/
Email: sales@everyvalve.com

Expamet
General Building Materials
Tel: 0845 5172761
Website: www.expamet.co.uk
Email: info@expament.co.uk

Fischer Fixings (UK) Ltd
Fixings
Tel: 01491 827900
Website: www.fischer.co.uk
Email: sales@fischer.co.uk

Forterra Building Products Ltd
Bricks
Tel: 01604 707600
Website: http://forterra.co.uk/

Fosroc Ltd
Waterproofing and Expansion Joints
Tel: 01827 262222
Fax: 01827 262444
Website: www.fosrocuk.com
Email: enquiryuk@fosroc.com

Freyssinet Ltd
Reinforced Earth Systems
Tel: 01952 201901
Fax: 01952 201753
Website: www.freyssinet.co.uk
Email: info@freyssinet.co.uk

George Walker Ltd
Timber and Fencing
Tel: 0116 260 8330
Fax: 0116 269 7450
Website: http://www.george-walker.co.uk/
Email: sales@george-walker.co.uk

George Lines Merchants
Paving
Tel: 01753 685354
Fax: 01753 686031
Website: http://www.georgelines.co.uk/
Email: sales@georgelines.co.uk

GCP Applied Technologies
Waterproofing/Expansion Joints
Tel: 01480 478421
Fax: 01753 490000
Website: https: //gcpat.uk/en-gb
Email: info-gb@gcpat.com

Geolabs
Ground Investigation
Tel: 01923 892 190
Fax: 01923 892 191
Website: http://geolabs.co.uk/
Email: admin@geolabs.co.uk

Greenham Trading Ltd
Contractors Site Equipment
Tel: 0845 300 6672
Fax: 0208 568 8423
Website: www.greenham.com
Email: sales@greenham.com

Griffiths Signs
Signage
Tel: 01633 895566
Fax: 01633 895577
Website: http://www.griffiths-signs.co.uk/
Email: sales@griffiths-signs.co.uk

Hanson UK
Cement
Tel: 01628 774100
Website: http://www.hanson.co.uk/en
Email: enquiries@hanson.com

Hill & Smith Ltd
Safety Barrier Systems
Tel: 01902 499400
Website: https: //www.hill-smith.co.uk/
Email: sales@hill-smith.co.uk

HSS Hire
Access Staging and Towers
Tel: 0161 7494090
Fax: 0161 749 4401

Website: http://www.hss.com/
Email: customercaredirector@hss.com

IKO Roofing Specification Division
Asphalt Products
Tel: 01257 255 771
Fax: 01257 252514
Website: http://www.ikogroup.co.uk/
Email: info@ikogroup.co.uk

Inform UK Ltd
Formwork and Accessories
Tel: 01223 862230
Fax: 01223 440246
Website: http://www.informuk.co.uk/
Email: enquiries@informuk.co.uk

Jacksons Fencing Ltd
Fencing
Tel: 01233 750393 / 0800 408 2234
Fax: 01233 750403
Website: www.jackson-fencing.co.uk
Email: sales@jacksons-fencing.co.uk

James Latham Plc
Timber Merchants
Tel: 01442 849100
Fax: 01442 267241
Website: www.lathamtimber.co.uk
Email: marketing@lathams.co.uk

John Bell Pipeline
Carbon Steel Drainage
Tel: 01224 714514
Website: http://www.jbpipeline.co.uk
Email: sales@jbpipeline.co.uk

K B Rebar Ltd
Steel Reinforcement
Tel: 0161 790 8635
Fax: 0161 799 7083
Website: www.kbrebar.co.uk
Email: sales@kbrebar.co.uk

Keyline Builders Merchants Ltd
Waterproofing
Tel: 0141 777 8979
Website: http://www.keyline.co.uk/
Email: customerservice@keyline.co.uk

L&R Roadlines Ltd
Road Markings
Tel: +44 (0)151 355 4100 / 356 2222
Website: http://www.lrroadlines.co.uk/
Email: info@hitexinternational.com

M & T Pipeline Supplies Ltd
Drains
Tel: 0141 774 8300
Fax: 0141 774 7300
Website: http://www.mtpipe.co.uk/ Email: info@mtpipe.co.uk

Maccaferri Ltd
River and Sea Gabions
Tel: 01865 770555
Website: https: //www.maccaferri.com/uk/
Email: technical.uk@maccaferri.com

Marley Plumbing & Drainage Ltd
UPVC Drainage and Rainwater Systems
Tel: 01622 858888
Website: http: //www.marleyplumbinganddrainage.com/
Email: customerservice@marleypd.com

Marshalls Mono Ltd
Kerbs, Edgings and Pavings
Tel: 01422 312000
Website: www.marshalls.co.uk
Email: customeradvice@marshalls.co.uk

Milton Precast
Pre-cast Concrete Solutions
Tel: 01795 425191
Fax: 01795 420360
Website: http://www.miltonprecast.co.uk/
Email: sales@miltonprecast.com

Mone Bros
Geotech
Tel: 0113 252 3636
Website: http://www.monebros.co.uk/index.htm
Email: info@monebros.co.uk

Morelock Signs
Traffic Signs
Tel: 01902 637 575
Website: www.morelock.co.uk

Moveright International Ltd
Abnormal Load Specialists
Tel: +44 (0)1675 475590

Fax: +44 (0)1675 475591
Website: http://moverightinternational.com/
Email: andrew@moverightinternational.com

Naylor Drainage Ltd
Clayware Pipes and Fittings
Tel: 01226 790591
Fax: 01226 790531
Website: www.naylor.co.uk
Email: drainage@naylor.co.uk

Pearce Signs Ltd
Gantry Signs
Tel: 01159 409620
Website: http://www.pearcesigns.com/
Email: info@pearcesigns.com

PHI Group Ltd
Retaining Walls, Gabions
Tel: 01242 707600
Website: www.phigroup.co.uk
Email: southern@phigroup.co.uk

Pipestock Limited
Conduits
Tel: 01794 835835
Fax: 0845 634 1056 (local rate)
Website: http://www.pipestock.com/
Email: info@pipestock.com

Platipus Anchors Ltd
Earth Anchors
Tel: 01737 762300
Website: www.platipus-anchors.com
Email: info@platipus-anchors.com

Redman Fisher Ltd
Flooring and Metal Work
Tel: +44 (0) 1952 685110
Fax: +44 (0) 1952 685117
Website: www.redmanfisher.co.uk
Email: sales@redmanfisher.co.uk

Remmers (UK) Limited
Waterproofing
Tel: 01293 594 010
Fax: 01293 594 037
Website: http://www.remmers.co.uk/
Email: sales@remmers.co.uk

RIW Limited
Waterproofing
Tel: 01344 397777 / 7788
Website: http://www.riw.co.uk/
Email: enquiries@riw.co.uk

Roger-Bullivant Ltd
Pilling and Foundations
Tel: 0845 838 1801
Website: www.roger-bullivant.co.uk
Email: info@roger-bullivant.co.uk

ROM Ltd
Formwork, Reinforcement and Accessories
Tel: 01543 421600
Website: www.rom.co.uk
Email: sales@rom.co.uk / sales@rom-tech.co.uk

Saint-Gobain PAM UK
Cast Iron Soil, Water and Rainwater Pipes and Fittings.
Tel: 0115 9305000
Website: www.saint-gobain-pam.co.uk
Email: sales.uk.pam@saint-gobain.com

Stocksigns Ltd
Road Signs and Posts
Tel: 01737 774 072
Website: www.stocksigns.co.uk
Email: sales@stocksigns.co.uk

Structural Soil Ltd
Ground Investigation
Tel: 0117 947 1000
Website: www.soils.co.uk
Email: ask@soils.co.uk

Sundeala Limited
Flexcell Board
Tel: 01453 708689
Website: http://www.sundeala.co.uk/
Email: enquiries@sundeala.co.uk

Tarmac Trading Limited
Prestressed Concrete Beams and Blockwork
Tel: 0121 787 5001
Website: http: //www.tarmac.com/
Email: contractinghqreception@tarmac.com

Tata Steel Ltd
Steel Products
Tel: 020 7717 4444
Fax: 020 7717 4455
Website: http://www.tatasteeleurope.com/
Email: feedback@tatasteeleurope.com

Tensar International
Soil Reinforcement
Tel: 01254 262431
Fax: 01254 266868
Website: www.tensar.co.uk
Email: info@tensar-international.com

Travis Perkins plc
Builders Merchants
Tel: 01604 752424
Website: www.travisperkinsplc.co.uk
Email: customerservice@travisperkins.co.uk

Unistrut
Concrete Inserts
Tel: 0121 580 6300
Fax: 0121 580 6370
Website: http://www.unistrut.co.uk/

Varley and Gulliver Ltd
Bridge Parapets
Tel: 0121 773 2441
Website: www.v-and-g.co.uk
Email: sales@v-and-g.co.uk

Vibro Menard
Ground Consolidation
Tel: 01704 891039
Website: http://www.vibromenard.co.uk/
Email: sales@vibromenard.co.uk

Wavin (OSMA)
Plastic Piping and Drainage
Tel: 01709 856 300
Website: https: //www.wavin.com/en-gb/

Wavin Plastics Ltd
UPVC Drain Pipes and Fittings
Tel: 01249 766600
Website: https: //www.wavin.com/en-gb/

Wells Spiral Tubes Ltd
Steel Culverts
Tel: 01535 664231
Fax: 01535 664235
Website: www.wells-spiral.co.uk
Email: sales@wells-spiral.co.uk

Winn & Coales (Denso) Ltd
Anti-corrosion and Sealing Products
Tel: 0208 670 7511
Fax: 0208 761 2456
Website: www.denso.net
Email: mail@denso.net

AECOM
A303 Stonehenge
(Amesbury to Berwick Down)
major improvements
Delivering a better world

PART 1

General

Greg Watts and Norman Watts
R
CPD in the Built Environment

Purpose and Content of the Book

For many years the Editors have compiled a price book for use in the building industry with, more recently, companion volumes for use in connection with mechanical and electrical engineering services contracts and landscaping work. All the price books take their reliability from established practice (within these sectors of the building industry) of pricing work by the application of unit rates to quantities measured from the designer's drawings. This practice remains valid because most construction work can be carried out under similar circumstances regardless of site location. A comparatively low proportion of contract value is subject to the risks that attend upon work below ground level and, once the building envelope is watertight, most trades can proceed without serious disruption from the weather.

This is not, however, the general method of pricing Civil Engineering Work. The volume of work below ground, increased exposure to weather and the tremendous variety of projects, in terms of type, complexity and scale, makes the straightforward use of unit rates less reliable. So, whilst in building work, similar or identical measured items attract a fairly broad range of prices, the range is much greater in Civil Engineering Bills. This uncertainty is compounded by the lower number of bill items generated under Civil Engineering Methods of Measurement, so that the precise nature of the work is less apparent from the bill descriptions and the statistical effect of 'swings and roundabouts' has less scope to average out extremes of pricing.

To prepare an estimate for a Civil Engineering project, it is necessary to be cognisant of the method to be adopted in executing the work, able to draw up a detailed programme and then cost out the resources necessary to execute the chosen method. Because the first two parts of this process are in the control of the Contractor's planner, there has been a tendency to postpone detailed estimating until the tendering stage itself, with the employer relying up to that point upon an estimate prepared with a high degree of pricing tolerance, use of Optimum Bias.

The result has been a growing pressure on the part of project sponsors for an improvement in budgetary advice, so that a decision to commit expenditure to a particular project is taken on firmer grounds. The absence of a detailed pricing method during the pre-contract phase inhibits the accurate costing of alternative designs and necessitates regular cost reviews to ensure that the design is being developed within the employer's budget.

This book therefore seeks to draw together the information appropriate to two methods of pricing: the cost of resources for use where an operational plan has been outlined, and unit rates for use where quantities can be taken from available drawings.

To take some note of the range of unit rates that might apply to an item, the rates themselves are in some cases related to working method – for example by identifying the different types of plant that would suit varying circumstances. Nonetheless, it would be folly to propose that all types of Civil Engineering and Highways work could be covered by a price book such as this. The Editors have therefore had in mind the type and scale of work commissioned by a local authority, a public corporation or a large private company.

This does embrace the great majority of work undertaken by the industry each year. Although almost all projects will have individual features that require careful attention in pricing, there will be some projects that are so specialist that they will not conform to standard pricing information at all. The user is expected to contact Specialists directly in such circumstances.

For most projects, within the range of work covered, this book should provide a firm foundation of cost information upon which a job-specific estimate can be built.

PURPOSE AND CONTENT OF THE BOOK

The contents of the book are therefore set out in a form that permits the user to follow the estimating process through in a structured way, as follows:

Part 1: General

The balance of this section describes in narrative form the work stages normally followed in a Contractor's office from receipt of the tender documents through to the submission of the tender.

It also includes a generalised introduction to the user of the principles and current legislation affecting the application of The Aggregates Levy, Capital Allowances, VAT, Landfill Tax and Land Remediation within the construction industry.

Part 2: Oncosts and Profit

Having produced an estimate for the predicted cost of the work, being the sum of the preliminaries and the measured work, the estimate must be converted to a tender by the application of any adjustment made by management (which follows the Management Appraisal described later in this part of the book) and by additions for financing charges, head office overheads and profit. These additions are discussed in this section and also included is a worked example of an indicative tender summary.

A Checklist of items to be priced with Preliminaries and General Items (or Method Related Charges) is included within this section as a 'Worked Example'.

Part 3: Cost and Tender Price Indices

The cost and tender price indices included in this part of the book provide a basis for updating historical cost or price information, by presenting changes in the indices since 2010. Caution must be taken when applying these indices as individual price fluctuations outside the general trend may have significant effect on contract cost.

Part 4: Resources

This deals with the basic cost of resources, so that a resource-based system of estimating can be adopted where it is possible to develop an outline programme and method statement. Reference to this section will also assist the user to make adjustments to unit rates where different labour or material costs are thought to apply and to calculate analogous rates for work based on the hypothetical examples given. It is stressed that all of the costs given in this section exclude the items costed within the preliminaries, financing charges, head office overheads and profit. The materials and plant costs as shown are gross, with no deduction of discount.

Part 5: Approximate Estimating

The prices in this section have been assembled from a number of sources, including the relevant items in the unit costs section and recovered data from recent projects. They are intended to give broad price guides or to assist in comparison exercises. This section is to be utilized in conjunction with Parts 6 and 7 to enable the user to incorporate within the estimate, items more normally associated with Building Work and/or Civil Engineering and which do not fall readily under recognized methods of measurement for either Building and/or Civil Engineering Work. Due to the diversity of items that fall under such a definition, (because of specification differences), the format for this section is structured to incorporate a range of items to allow the production of the estimate for such items prior to detailed design information being available.

Building Prices per Square Metre

This section provides an indicative range of values against specific building and engineering facilities exclusive of preliminaries, specialist equipment and professional fees, but does include for overheads and profit.

This section provides an indicative range of values against general civil engineering works exclusive of preliminaries, specialist equipment and professional fees, but does include for overheads and profit.

PURPOSE AND CONTENT OF THE BOOK

Approximate Estimating Rates – General Building Works

This section provides an indicative range of values against general building works exclusive of preliminaries, specialist equipment and professional fees, but does include for overheads and profit.

Parts 6 and 7: Unit Costs

These sections are structured around methods of measurement for Civil Engineering Work and Highway Works and give 'trade by trade' unit rates for those circumstances where the application of unit rates to measured quantities is possible and practical. Again, it is stressed that the rates exclude the items costed within preliminaries, financing charges, head office overheads and profit. Both materials and plant costs are adjusted to allow a normal level of discount, with allowances for materials wastage and plant usage factors. Guidance prices are included for work normally executed by specialists, though there will be some projects that are so specialized that they will not conform to standard pricing information at all. The user is expected to contact Specialists directly in such circumstances.

Part 8: Daywork

This section includes details of the CECA day works schedule and supplementary charges and advice on costing excluded items.

Part 9: Professional Fees

These contain reference to a standard suite of agreements relating to a scale of professional charges/ fees and conditions of appointment for Consulting Engineers and Quantity Surveyors.

Part 10: Outputs

Scheduled here are various types of operations and the outputs expected of them for mechanical and hand operated equipment. Also listed are outputs per man hour for various trades found in Civil Engineering.

Part 11: Useful Addresses for Further Information

This section provides useful contact addresses in the UK construction industry.

Part 12: Tables and Memoranda

These include conversion tables, formulae, and a series of reference tables structured around trade headings.

Market Update

Brexit, the coronavirus pandemic and the Ukraine war have, together, generated exceptional inflationary pressure in the UK construction sector. These factors have proved to be the ultimate collective test of globalised supply chains and just in time delivery models – and vulnerabilities have been exposed. In the past 12 months we have seen overall construction cost inflation hit historic highs and uncertainty around the near-term inflationary outlook is heightened.

Prices presented in this book have a base date of Q1 2022 – meaning indirect impacts of the Ukraine war have largely been excluded. Immediate, direct impacts of the Ukraine war have been factored into the rates but indirect impacts are likely to be significant and will take time to feed through the supply chain.

It's an exceptional, rapidly evolving time, and pricing decisions require a greater degree of judgement.

This article provides a summary of current inflationary drivers and identifies the most prominent risk factors as at July 2022. Updates, reporting on market developments, will be issued throughout the lifespan of this book to provide guidance on adjustments it would be sensible to make, and any factors that should be taken into consideration when interpreting the material presented.

Inflationary trends and drivers

Construction materials prices are under significant pressure as energy price inflation feeds through. Supply of some core materials and components is still restricted due to manufacturing capacity not yet returning to pre-pandemic levels and the impact of the war on Ukrainian and Russian supply.

Recent data tracking factory gate price movements suggests inflationary pressure on manufactured product prices is extremely strong, and accelerating, as lagged impacts from supply chain disruption and high energy prices feed through as stocks are replenished and supply agreements renegotiated.

In May 2022, construction materials price inflation reported by BEIS, rose at an annual rate of 27%. The list of materials with factory gate inflation exceeding 20% in May is long, with energy intensive steel and concrete reinforcing bars increasing at the highest rate.

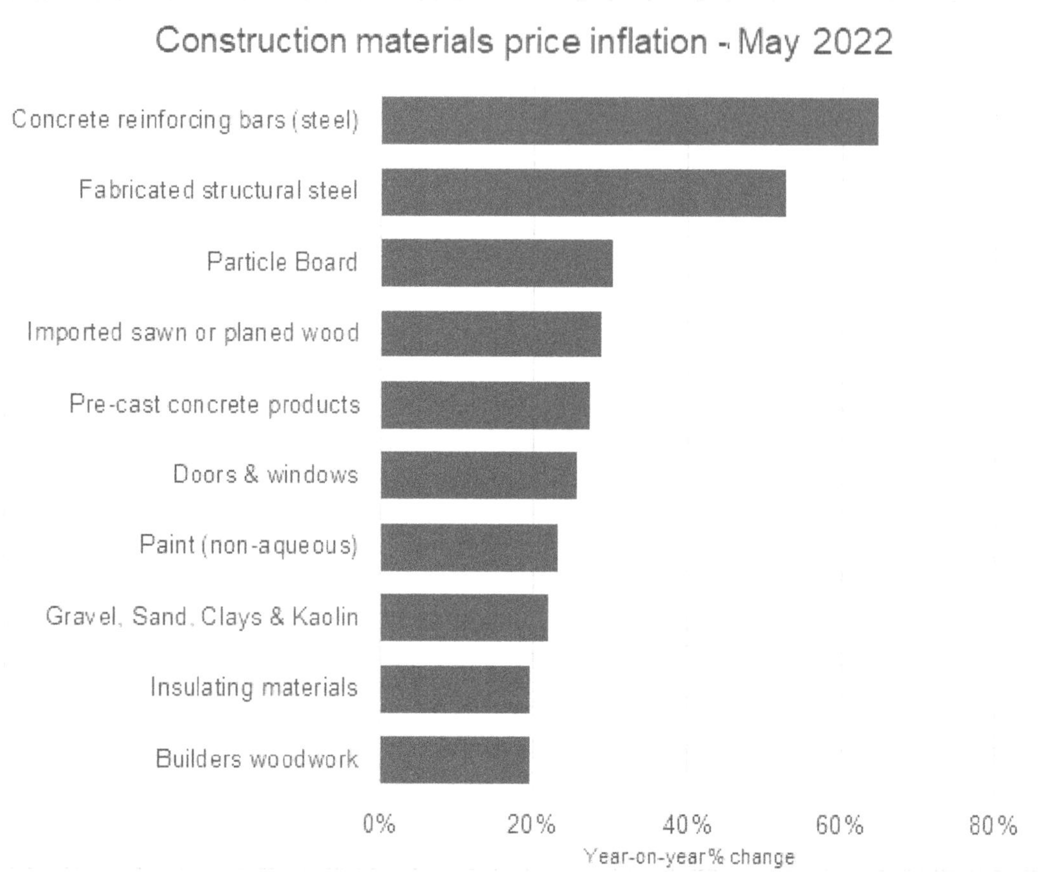

Source: BEIS

Price pressures on raw materials have eased slightly, although absolute levels remain very high relative to 2019. Oil prices have exceeded $100 per barrel for the past three months, around 65% up on levels a year ago. European gas prices have fallen back from their March peak but are still around 230% higher year-on-year. Lagged impacts from this exceptional inflationary price shock will impact prices of energy-intensive construction products for several quarters as manufacturers' supply agreements come up for renegotiation. Furthermore, wholesale gas prices may have softened slightly but the situation remains precarious and significant upside risks still exist.

Pressure on construction wage rates is adding to inflationary woes as workers face the biggest squeeze on real income in decades. With vacancies for construction jobs at record highs, an aging workforce, a covid-induced increase in the early retirement rate and restrictions on economic migration, pressure on wages will persist. Currently annual construction sector wage inflation falls in the 5% to 7% range.

The inflationary outlook

AECOM's latest forecast suggests building cost inflation, incorporating anticipated trends in labour, plant and materials, will average 11% during 2022 and 10% in 2023. Tender price inflation is forecast to average 8.7% during 2022, reducing to 6.1% in 2023.

This forecast assumes the UK construction sector will avoid recession and that industry demand will continue to grow at a modest pace. Against this backdrop, demand will be such that pipeline visibility is sufficiently stable to enable cost increases to be passed through the supply chain.

A chart presenting AECOM's latest building and tender price forecasts is presented in Building Cost Indices; Tender Price Indices and Location Factors.

Risks

Risks to the outlook are numerous and broad. From uncertainty around the domestic policy agenda, and heightened geopolitical tension, to the potential impact of the generational reduction in real household disposable income, rising interest rates on economic activity and, ultimately, UK construction demand, the outlook is exceptionally uncertain.

Energy prices continue to pose a risk to the inflationary outlook. Proposed strike action by Norwegian gas workers recently threatened to stop supply to the UK. Action was averted but it highlights the extreme vulnerability of supply stability. With prices still at extremely high levels by historical standards, a further supply-side shock would have serious ramifications.

Insolvencies in the construction sector increased sharply in the first quarter of 2022. Construction contractors and specialists will continue to face financial headwinds over the next 12 months as inflation remains stubbornly high. Delivering on fixed price contracts agreed before the Ukraine war is challenging and maintaining cash flow is essential.

Plans to replace EU CE testing certification for construction products sold in the UK with a domestic testing regime have been relaxed slightly in response to industry warnings. New testing requirements will still take effect in January 2023, but CE marked products imported into the UK by the end of 2022 will no longer need to be retested. While this is welcomed by manufacturers, fundamental concerns remain about limited testing capacity, along with the ability to hamper product supply.

From a demand perspective, risks are also exceptionally high. Orders remain healthy and projects are progressing but household and business confidence has plummeted. With the Help to Buy scheme – which supported close to 40% of all newly built homes in the private market between 2013 and 2021 – ending just as mortgage interest rates are rising and household disposable income is reducing, headwinds in the house building sector are strong.

Outline of the Tendering and Estimating Process

This section of the book outlines the nature and purpose of Civil Engineering estimating and provides background information for users. It comprises an outline of the estimating and tendering process with supporting notes and commentaries on particular aspects. Some worked examples on tender preparation referred to in this part are included at the end of **Part 2**.

It must be emphasized that the main purpose of this book is to aid the estimating process. Thus, it is concerned more with the predicted cost of Civil Engineering work than with the prices in a bill of quantities. To ensure the correct interpretation of the information provided it is important to distinguish clearly between estimating and tendering; the following definitions are followed throughout.

The Estimate is the prediction of the cost of a project to the Contractor. The Tender is the price submitted by the Contractor to the Employer.

The tender is based on the estimate with adjustments being made after review by management; these include allowances for risk, overheads, profit and finance charges. As discussed later in this section, prices inserted against individual items in a bill of quantities may not necessarily reflect the true cost of the work so described due to the view taken by the Contractor on the risks and financial aspects involved in executing the work.

Whilst projects are now constructed using many different forms of contract the core estimating process falls into two main divisions namely 'Design & Construct' and 'Construct only'. The following list summarizes the activities involved in the preparation of a process for a typical design and construct Civil Engineering project where the client issues detail drawings, specifications and Bills describing the extent of the works to be priced.

Due to the size and complex nature of civil and highway projects, the delivery process is usually divided into eight distinct stages. Network rail, for example, uses PACE stages (replacing GRIP stages from January 2021) while other bodies may use different terminology like Gateway or RIBA stages. The overall approach is product rather than process driven, and within each stage an agreed set of products are delivered. Formal stage gate reviews are held at varying points within the gateway stages. The stage gate review process examines a project at critical stages in its lifecycle to provide assurance that it can successfully progress to the next stage. In this pre contract stage of the project some of the activities involved are site visits, obtaining current prices, assessing programme and method statement, design of temporary works, verification of estimates, tender and management appraisal, among other things.

Typical pre- and post contract stages of the project are outlined below and the diagram that follows illustrates the relationships between the stages.

Output definition – Initial Client Brief

2. Feasibility
3. Option selection
4. Single option development
5. Detailed design
6. Construction test and commission
7. Scheme hand back
8. Project close out

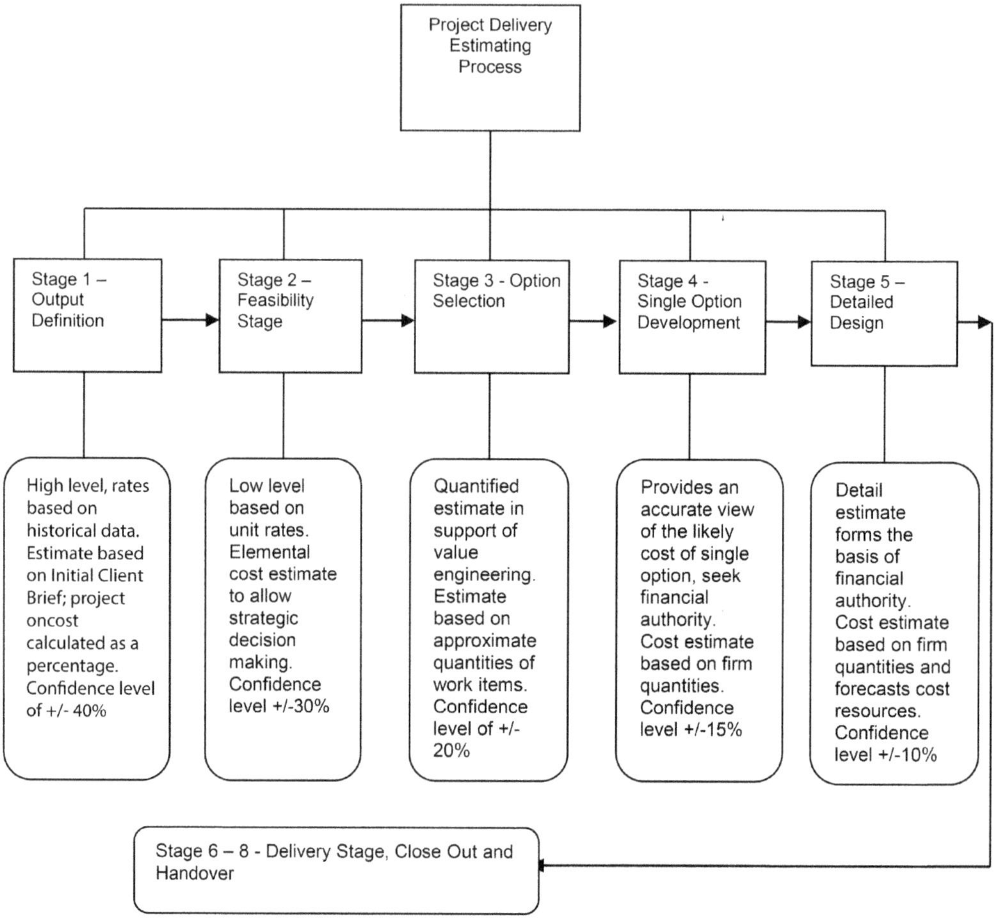

The following are some notes on factors and processes to consider in delivering the project and its definition.

1. An overall appraisal of the project is made including any variations to the standard contract form, insurance provisions and any other unusual or onerous requirement.
2. Material requirements are abstracted from the tender documents and prices are obtained from suppliers. Details of work normally subcontracted are abstracted, together with relevant extracts from the tender documents and prices are obtained from subcontractors. The site of the works and the surrounding area is visited and studied. Local information is obtained on factors affecting the execution of the contract.
3. A programme and detailed method of working for the execution of the contract is prepared, to include details of plant requirements, temporary works, unmeasured work, appropriate supervisory staff, etc.
4. Designs are made for temporary works and other features left to be designed by the Contractor, and quantities are taken off for costing.
5. Major quantities given in the tender documents are checked.
6. The cost estimate for the project is prepared by costing out all of the resources identified by stages 2 to 6. A more detailed report is made on the conditions of contract, financial requirements, etc. and an assessment of risk/opportunity is prepared.
7. The Tender documents are priced.
8. Management reviews the estimate, evaluates the risks and makes allowances for overheads, profit and finance.

9. Adjust spread of oncosts and profit.
10. The tender is completed and submitted.

1. INITIAL APPRAISAL

The purpose of the initial overall appraisal is to highlight any high value areas or any particular problems which may require specialist attention; it can also identify possible alternative methods of construction or temporary works.

Other points to be considered:

- the location and type of project and its suitability to the tenderer's particular expertise
- the size of project, its financing requirement, the proportion of annual turnover it would represent and the availability of resources
- the size of the tender list and the nature of the competition
- the identity of the employer and his professional consultants
- the adequacy of the tender documents
- an initial appraisal of risk and opportunity

Corporate governance requires that the directors of companies are aware of all the liabilities inherent in the contract being sought. Few contracts are offered on totally un-amended standard forms and specifications and it is the estimator's duty to ensure that a report is prepared to advise on the precise terms and conditions being offered and their potential implications.

It is essential for the estimator to study the contract documents issued with the enquiry or made available for inspection and to note those parts which will affect pricing or involve the Contractor in liabilities which must be evaluated and drawn to the attention of management. The following comments are indicative only.

Conditions of Contract

For Civil Engineering work these are normally, but not exclusively, based on ECC (NEC4) standard forms. However these forms are rarely offered without addition and amendment and it is imperative that the full implications are understood and directors informed. Any required bonds, guarantees and warranties must be identified, reported and included in subcontract enquiries where appropriate. Insurance requirements and excesses must be checked against company policies.

Bill of Quantities

Where a Bill is provided it serves three purposes: first and foremost it must be prepared with the objective of providing the estimator with as accurate a picture of the project as possible, so as to provide a proper basis for pricing. Second, it should enable the employer to compare tenders on an equal basis and third it will be used to evaluate the work executed for payment purposes. Individual items in the Bill do not necessarily describe in detail the work to be carried out under a particular item; reference must be made to the specification and the drawings to ascertain the full scope of the work involved.

The method of preparing the Bill may be based on the 'Civil Engineering Standard Method of Measurement' issued by the Institution of Civil Engineers or the 'Method of Measurement for Highway Works' issued by the Highways Agency, but some employing authorities have evolved their own methods and it is important for the estimator to study the Bill and its preambles to ensure that rates and prices are comprehensive.

In all cases the quantities given in the Bill are not a guarantee and the drawings usually have precedence. The estimator must understand whether he is pricing a re-measurable or fixed price contract and make due allowance.

Specification

This gives a detailed description of the workmanship, finish and materials to be used in the construction of the work. It may also give completion periods for sections and/or the whole of the work together with details of the requirements for the employer and/or the Consulting Engineer in connection with their supervision on site.

Water Authority and Highway Works in particular are based around a standard specification. However even standard specifications will have contract specific appendices and tables. The estimator must take due note of these requirements and ensure that this information is issued and taken into account by all potential subcontractors and material suppliers.

Drawings

These give details of the work to be carried out and must be read in conjunction with the specification. It is important for the estimator to study the notes and descriptions given on the drawings as these amplify the specification.

Should the estimator discover any conflict between the various documents, it is important to have such discrepancies clarified by the Employer or Engineer prior to submission of any offer.

2. OBTAINING PRICES

(a) Materials

When pricing materials, the following points must be noted:

- checks must be made to ensure that the quality of materials to be supplied meets with the requirements of the specification. If necessary, samples should be obtained and tested for compliance
- checks must be made to ensure that the rates of delivery and fabrication periods can meet the demands of the programme. It is sometimes necessary to use more than one supplier, with differing prices, to ensure a sufficient flow of materials
- tests should be carried out to ascertain allowances to be made for operations such as compaction of soils and aggregates. Records of past contracts using similar materials can give this information, providing such records are accurate and reliable

(b) Subcontractors

It is common practice among Civil Engineering Contractors to subcontract a significant proportion of their work.

Subcontracted work can represent the bulk of the value of measured work.

When utilizing subcontractors' prices it is extremely important to ensure that the rates given cover the full extent of the work described in the main contract, and that the subcontractors quotation allows for meeting the main Tender programme and methods of working.

Unless an exclusive relationship has been entered into prior to the tendering process it is likely that the same subcontractor will submit prices to a number of competing tenderers. It is important for the estimator to ensure that the price offered represents the method intended and is not a generic sum which is subject to variation if successful.

3. SITE VISIT

Factors to check during the site visit include, but are not limited to:

- access
- limitations of working space
- existing overhead and underground services

- nearby public transport routes
- availability of services – water, gas, electricity, telephone, etc.
- availability of labour and subcontractors
- availability of materials – particularly aggregates and fill materials and location of nearest tipping facility
- nature of the ground, including inspection of trial bores/pits if dug below ground water level
- presence of other Contractors working on or adjacent to the site.

4. METHOD STATEMENT AND PROGRAMME

As previously stated, whilst an estimate is being prepared it is necessary that a detailed method of working and a programme for the execution of the works is drawn up; the latter can take the form of a bar chart or, for large and more complex projects, may be prepared on more sophisticated computerised platforms. Compliance with the employer's target completion dates is, of course, essential. The method of working will depend on this programme in so far as the type and size of plant and the gang sizes to be used must be capable of achieving the output necessary to meet the programmed times. Allowance must be made for delays due to adverse weather, other hazards particular to the site and the requirements of the specification, particularly with regard to periods to be allowed for service diversions and other employer's requirements.

A method statement is prepared in conjunction with the programme, setting out the resources required, outputs to be achieved and the requirements in respect of temporary works, etc.

At the same time, separate bar charts may be produced giving:

- plant requirements
- staff and site supervision requirements
- labour requirements

These programmes and method statements will form the basis of the actual contract programme should the tender be accepted. They will also enable the Contractor to assure himself that he has available or can gain access to the necessary resources in plant, labour, materials and supervision to carry out the work should he be awarded the contract.

5. DESIGN OF TEMPORARY WORKS, ETC.

Normally the period of time allowed for tendering is relatively short and therefore it is important that those aspects requiring design work are recognized as early as possible in the tender period. Design can be carried out either by the Contractor using his own engineers or by utilizing the services of a Consulting Engineer. There are three aspects of design to be considered:

(a) Temporary works to the contractor's requirements to enable the works to be constructed

Design of temporary works covers the design of those parts of the work for which the contractor accepts full responsibility in respect of stability and safety during construction. Such parts include support structures, cofferdams, temporary bridges, jetties and river diversions, special shuttering, scaffolding, haul roads and hard standings, compounds, traffic management etc. Design must be in sufficient detail to enable materials, quantities and work content to be assessed and priced by the estimator. In designing such work, it is important that adequate attention is given to working platforms and access for labour and plant and also to ease of dismantling and re-erection for further uses without damage.

It should be noted that many specialist subcontractors will provide a design service when submitting quotations. For example, scaffolding contractors will design suitable support work for soffit shuttering, etc.

(b) Specific items of the permanent works to meet a performance specification set out by the client or the Consulting Engineer

It is common practice for certain parts of the work to be specified by means of a performance specification. For example, concrete is specified by strength only, piles by load carrying capacity, etc. It is then left to the contractor to use those materials; workmanship and design which he feels are most suited to the particular site and conditions. In many cases such design will be carried out by specialist suppliers or subcontractors.

(c) Alternative designs for sections of the permanent works where the Contractor's experience leads him to consider that a more economical design could be used

It is possible for a Contractor to use his expertise and experience to design and submit alternative proposals and prices for complete sections of the permanent work without, of course, altering the basic requirements of the original design. Examples would be foundations in difficult ground conditions, bridge superstructures, use of precast in place of in situ concrete, etc. Such designs may be carried out by the Contractor's own staff or in conjunction with Consulting Engineers or specialist contractors.

Obviously this is only done when the Contractor can offer considerable savings in cost and/or a reduced construction period. It is necessary to include a price for the original design in the tender, but the decision to submit a keen tender may be underpinned by the hope of sharing such savings with the Employer.

In all cases, the Contractor's designs and calculations must be checked and approved by the Employer's Consulting Engineer, but such checks and approvals do not relieve the Contractor of his responsibilities for the safety and stability of the work.

Where any design work is undertaken it is important to ensure that adequate design insurance is maintained.

6. QUANTITIES CHECK

Working within obvious time constraints, the estimating team will endeavour to complete a quantities check on at least the major and high price quantities of the Bill, as this could affect pricing strategy, for example, in pricing provisional items. Any major discrepancies noted should be referred to the management appraisal agenda.

7. COST ESTIMATE

At this stage the estimator draws together the information and prepares a cost estimate made up of:

- preliminaries and general items.
- temporary works.
- labour, costed by reference to the method statement, with appropriate allowances for labour oncosts (non-productive overtime, travelling time/fares, subsistence, guaranteed time, bonus, Employer's Liability Insurance, training, statutory payments, etc.).
- material costs taken from price lists or suppliers' quotations, with appropriate allowances for waste.
- plant, whether owned or hired, with appropriate allowances for transport to/from site, erection/dismantling, maintenance, fuel and operation. Heavy static plant (batching plant, tower cranes, etc.) will normally be priced with general items; the remainder will normally be allocated to the unit rates.)
- sublet work, as quoted by specialist subcontractors, with appropriate allowances for all attendances required to be provided by the main contractor.

At the same time, a preliminary assessment of risk/opportunity will be made for consideration with the Management Appraisal (see Section 9 below). On major tenders a formal quantified risk assessment (QRA) will be undertaken – the estimator will be expected to evaluate the effects of the risks for inclusion in the calculations. This will include a look at:

- weather conditions – costs not recoverable by claims under the Conditions of Contract, ground becoming unsuitable for working due to the effect of weather, etc.
- flooding – liability of site to flooding and the consequent costs.
- suitability of materials – particular risk can arise if prices are based on the use of borrow pits or quarries and inadequate investigation has been carried out or development is refused by the local authorities.
- reliability of subcontractors – failure of a subcontractor to perform can result in higher costs through delays to other operations and employing an alternative subcontractor at higher cost.
- non-recoverable costs – such as excesses on insurance claims.

- estimator's ability – e.g. outputs allowed. This can only be gauged from experience.
- cost increase allowances for fixed price contract.
- terms and conditions contained in the contract documents.
- ability to meet specification requirements for the prices allowed.
- availability of adequate and suitable labour.

Risk is, of course, balanced by opportunity and consideration needs to be given to areas for which particular expertise possessed by the Contractor will lead to a price advantage against other tenderers.

8. PRICING THE TENDER

Once the cost estimate is complete, the estimator prices the items in the Bill. The rate to be entered against the items at this stage should be the correct rate for doing the job, whatever the quantity shown.

Where the overall operation covers a number of differing Bill items, the estimator will allocate the cost to the various items in reasonable proportions; the majority of the work is priced in this manner. The remaining items are normally priced by unit rate calculation.

Resources schedules, based on the programme and giving details of plant, labour and staff, perform an important role in enabling the estimator to check that he has included the total cost of resources for the period they are required on site. It is not unusual for an item of plant to be used intermittently for more than one operation. A reconciliation of the total time for which the cost has been included in the estimate against the total time the item is needed on site as shown on the programme, gives a period of non-productive time. The cost of this is normally included in site oncosts and preliminaries. A similar situation can arise in the cases of skilled labour, craftsmen and plant drivers.

Having priced the Bill at cost, there will remain a sum to be spread over the Bill items. The way in which this is done depends on the view taken by the Contractor of the project, for example:

- sums can be put against the listed general items in the Preliminaries Bill.
- a fixed 'Adjustment Item' can be included in the Bill for the convenience of the estimator. This can be used for the adjustment made following the Management Appraisal, and for taking advantage of any late but favourable quotations received from subcontractors or suppliers.
- the balance can be spread over the rates by equal percentage on all items, or by unequal percentages to assist in financing the contract or to take advantage of possible contract variations or expected quantity changes.

The Contractor will normally assess the financial advantages to be gained from submitting his bid in this manner and possibly enabling him to submit a more competitive offer.

After completing the pricing of all aspects of the tender, the total costs are summarized and profit, risk, etc., added to arrive at the total value of the Tender.

Finally, reasonable forecasts of cash flow and finance requirements are essential for the successful result of the project. Preliminary assessments may have been made for the information of management, but contract cash flow and the amount of investment required to finance the work can now be estimated by plotting income against expenditure using the programme of work and the priced Bill of Quantities. Payment in arrears and retentions, both from the Employer and to the suppliers and subcontractors, must be taken into account.

It is unlikely that sufficient time will be available during the tender period to produce such information accurately, but an approximate figure, for use as a guide for finance requirements, can be assessed.

A worked example is set out in Part 2: On Costs and Profit.

9. MANAGEMENT APPRAISAL

Clearly, as far as the detail of the tender build-up is concerned, management must rely upon its established tendering procedures and upon the experience and skill of its estimators. However, the comprehensive review of tenders prior to submission is an onerous duty and the estimator should look upon the process as an opportunity to demonstrate his skill. The Management Appraisal will include a review of:

- the major quantities.
- the programme and method statement.
- plant usage.
- major suppliers and/or subcontractors, and discounts.
- the nature of the competition.
- risk and opportunity.
- contract conditions, including in particular the level of damages for late completion, the minimum amount of certificates, retention and bonding requirements.
- cash flow and finance.
- margin for head office overheads and profit.
- the weighting and spreading of the cost estimate over the measured items in the Bill.

10. SUBMISSION OF TENDER

On completion of the tender, the documents are read over, comp checked and then despatched to the employer in accordance with the conditions set down in the invitation letter.

A complete copy of the tender as submitted should be retained by the contractor. Drawings on which the offer has been based should be clearly marked 'Tender Copy', their numbers recorded, and the drawings filed for future reference. These documents will then form the basis for price variations should the design be amended during the currency of the contract.

The contractor may wish to qualify his offer to clarify the basis of his price. Normally such qualifications are included in a letter accompanying the tender. Legally the Form of Tender constitutes the offer and it is important that reference to such a letter is made on the Form of Tender to ensure that it forms part of the offer. Wording such as 'and our letter dated..., reference...' should be added prior to quoting the Tender Sum.

Before any qualifications are quoted, careful note must be taken of the 'instructions to tenderers', as qualifications, or at least qualifications submitted without a conforming tender, may be forbidden.

Research & Development (R&D) Tax Relief

Introduction

Research and Development (R&D) tax relief, or credit, is a Corporation Tax benefit that can either reduce a company's tax bill or, in some circumstances, provide a cash sum. It is based on the company's expenditure on R&D. R&D tax relief/credits are aimed at encouraging expenditure by companies on innovative activities that advance the overall knowledge or capability in the fields of science and technology. The tax credits consist of additional tax relief beyond the amount of the expenditure incurred on qualifying R&D activities and are given against Corporation Tax. In some situations, where the company is making losses, the tax credits can be surrendered for a cash payment.

For there to be R&D for the purpose of the tax relief, a company must be undertaking activity as part of a specific project that seeks to make an advance in science or technology. It is necessary to be able to state what the intended advance is, and to show how, through the resolution of scientific or technological uncertainty, the activity seeks to achieve this.

HMRC has confirmed that qualifying activities include those typically undertaken in the construction and property sector, such as engineering design and developing new construction techniques.

Definition of R&D for Tax Purposes

The primary legislation is contained in Part 13 of the Corporation Tax Act 2009 (CTA 2009) and covers two separate schemes:

> R&D expenditure credits (s.1040 A), commonly referred to as RDEC are available to large companies.
> Additional tax relief for small or medium sized enterprises (s.1044) is commonly referred to as the SME scheme.

R&D is defined in section 1138 of the Corporation Tax Act 2010 as 'activities that fall to be treated as research and development in accordance with generally accepted accounting practice.' This is subject to regulations made by HM Treasury on the definition of R&D, which refer to the detailed guidelines published by the Secretary of State for Business, Energy and Industrial Strategy (BEIS). The BEIS guidelines define R&D for the purpose of obtaining relief as follows:

- R&D for tax purposes takes place when a project seeks to achieve an advance in science or technology. The activities which directly contribute to achieving this advance in science or technology through the resolution of scientific or technological uncertainty are R&D. Certain qualifying indirect activities related to the project are also R&D. Activities other than qualifying indirect activities which do not directly contribute to the resolution of the project's scientific or technological uncertainty are not R&D.

Guidance is provided by HM Revenue & Customs (HMRC) on what constitutes R&D for the purposes of the tax relief and can be found within the Corporate Intangibles R&D Manual at CIRD81900. The work that qualifies for R&D tax relief must be part of a specific project to make an advance in science or technology. It cannot be an advance within arts, humanities, a social science such as economics or a theoretical field such as pure maths. The project must relate to the company's trade.

Advance in Science or Technology

An advance in science or technology means an advance in overall knowledge or capability in a field of science or technology (not a company's own state of knowledge or capability alone). This includes the adaptation of knowledge or capability from another field of science or technology in order to make such an advance where this adaptation was not readily deducible. A process, material, device, product, service, or source of knowledge does not become an advance in science or technology simply because science or technology is used in its creation.

A project which seeks to either,

- Extend overall knowledge or capability in a field of science or technology; or
- Create a process, material, device, product, or service which incorporates or represents an increase in overall knowledge or capability in a field of science or technology; or
- Make an appreciable improvement (for example, changing or adapting the scientific or technological characteristics of something to the point where it is 'better' than the original and beyond minor or routine upgrading) to an existing process, material, device, product, or service through scientific or technological changes; or
- Use science or technology to duplicate the effect of an existing process, material, device, product, or service in a new or appreciably improved way (for example, a product which has exactly the same performance characteristics as existing models, but is built in a fundamentally different manner) will therefore be R&D.

Even if the advance in science or technology sought by a project is not achieved or not fully realised, R&D still takes place and the project could qualify for relief. Moreover, if a particular advance in science or technology has already been made or attempted but details are not readily available (for example, if it is a trade secret), work to achieve such an advance can still be an advance in science or technology.

Scientific or Technological Uncertainty

Scientific or technological uncertainty exists when knowledge of whether something is scientifically possible or technologically feasible, or how to achieve it in practice, is not readily available or deducible by a competent professional (an individual who is knowledgeable about the relevant scientific and technological principles involved, aware of the current state of knowledge and has accumulated experience) working in the field. This includes system uncertainty, or a scientific or technological uncertainty that results from the complexity of a system rather than uncertainty about how its individual components behave. Scientific or technological uncertainty will often arise from turning something that has already been established as scientifically feasible into a cost-effective, reliable, and reproducible process, material, device, product, or service.

Uncertainties that can readily be resolved by a competent professional working in the field are not scientific or technological uncertainties. Similarly, improvements, optimizations and fine-tuning which do not materially affect the underlying science or technology do not constitute work to resolve scientific or technological uncertainty.

Qualifying Indirect Activity

These are activities which form part of a project but do not directly contribute to the resolution of the scientific or technological uncertainty.

This includes the following:

- Scientific and technical information services, insofar as they are conducted for the purpose of R&D support (such as the preparation of the original report of R&D findings);
- ndirect supporting activities such as maintenance, security, administration and clerical activities, and finance and personnel activities, insofar as undertaken for R&D;
- Ancillary activities essential to the undertaking of R&D (for example, taking on and paying staff, leasing laboratories, and maintaining R&D equipment including computers used for R&D purposes);
- Training required to directly support an R&D project;
- Research by students and researchers carried out at universities;

- Research (including related data collection) to devise new scientific or technological testing, survey, or sampling methods, where this research is not R&D in its own right; and
- Feasibility studies to inform the strategic direction of a specific R&D activity.

Start and End of R&D for Tax Purposes

R&D begins when work to resolve the scientific or technological uncertainty starts, and ends when that uncertainty is resolved or work to resolve it ceases. For example, R&D could end once a working prototype or a material, device, product, or process is ready for testing, production or if the project does not continue. However, R&D can take place even after production starts. For example, if any new problems arise involving scientific or technological uncertainty after the product has been put into production or into use, then the R&D process may start again.

Types of R&D Tax Relief

There are different types of R&D relief, depending on the size of a company and if the project has been subcontracted to the company or not.

The two principal tax reliefs available to companies undertaking R&D in the UK are as follows:

- SME R&D Tax Relief.
- Research and Development Expenditure Credit (RDEC).

SME R&D Tax Relief

An SME for R&D tax relief purposes is a company or organization with less than 500 staff and either of the following:

- Annual turnover not exceeding €100 million.
- Balance sheet not exceeding €86 million.

A company may need to take into account its own data, a proportion of a partner enterprise's data or the data of a linked enterprise. A company may not be considered an SME if it is part of a larger enterprise that would fail these tests if taken as a whole.

SME R&D Tax relief allows companies to:

- Deduct an extra 130% of their qualifying costs from their yearly profit, as well as the normal 100% deduction, to make a total 230% deduction. For profit making SMEs, R&D tax credits reduce the Corporation Tax bill and businesses can receive up to £24.70 for every £100 spent on qualifying R&D activities.
- Claim a tax credit if the company is loss making, worth up to 14.5% of the surrenderable loss. Loss-making SMEs can receive up to £33.35 for every £100 spent on qualifying R&D activities.

A company can only claim R&D tax relief as an SME if the company is a going concern and not in administration or liquidation when making the claim.

In some circumstances, an SME company may be allowed to claim only the RDEC in respect of particular expenditure (for example, for work subcontracted to it or that is otherwise subsidized due to a grant or subsidy). The SME scheme is a notifiable State Aid, and a company can't get the SME Tax relief if it is receiving any other notifiable State Aids for the same R&D project. If an SME company has received a grant which is notifiable State Aid, for an R&D project, it cannot claim R&D tax relief under the SME scheme, but eligible expenditure will qualify under the RDEC scheme.

The UK government has recently announced that SME R&D tax credit claims will be subject to a cap in order to deter abuse of the scheme. This measure limits the amount of payable R&D tax credit which an SME can claim to £20,000 plus 300% of its total Pay As You Earn (PAYE) and National Insurance contributions (NICs) liability for the period. The measure will have effect for accounting periods beginning on or after 1 April 2021.

The measure amends CTA 2009 Part 13 Chapter 2. It adds a new condition at section 1058 (which sets the amount of payable tax credit) limiting the amount that can be claimed to the level of the new cap, which is defined as £20,000 plus three times the company's relevant expenditure on workers.

A company is however exempt from the cap if:

- Its employees are creating, preparing to create or managing Intellectual Property (IP) and
- It does not spend more than 15% of its qualifying R&D expenditure on subcontracting R&D to, or the provision of externally provided workers by, connected persons.

RDEC

Large companies (a company which does not meet the SME test for R&D tax relief purposes) can claim RDEC for working on R&D projects. It can also be claimed by SMEs who have been subcontracted to do R&D work by a large company or who have received a grant or subsidy for their R&D project.

The RDEC is an 'above the line' tax credit equal to 13% of qualifying R&D costs. RDEC is a taxable credit. This means that large companies receive up to £10.53 for every £100 spent on qualifying R&D activities. For profit making companies, the credit discharges the Corporation Tax liability that the company would have to pay. Companies with no Corporation Tax liability will benefit from the RDEC either through a cash payment or a reduction of tax or other duties due. Companies in groups can also surrender the RDEC against another group companies' Corporation Tax liability.

Categories of Qualifying Expenditure

- Staffing costs – This includes salaries, wages, Class 1 NICs, pension fund contributions and certain reimbursed expenses for staff directly and actively engaged in the R&D project. Support staff costs, for example administrative or clerical staff, do not qualify, except when they relate to qualifying indirect activities. These can be activities like maintenance, clerical, administrative and security work.
- Externally provided workers – These are the staff costs paid to an external agency for staff who are directly and actively engaged in the R&D project. These are not employees and subcontractors. Relief is usually given on 65% of the payments made to the staff provider. Special rules apply if the company and staff provider are connected or elect to be connected.
    - Consumable items
    - Cost of items that are directly employed and consumed in qualifying R&D projects. These include materials and the proportion of water, fuel and power of any kind consumed in the R&D process.
    - From 1 April 2015, the costs of materials incorporated in products that are sold are not eligible for relief.
    - Where a prototype is created to test the R&D being undertaken, the design, construction and testing costs will normally be qualifying expenses. However, if the prototype is also built with a view to selling the prototype itself (for example, the construction of a bespoke machine), HMRC considers that to be production and outside the R&D scheme, even if R&D was undertaken to create the prototype. In this case, the R&D expenditure and production costs will need to be split. For example, the construction costs and materials consumed would not be qualifying expenses, but design, modelling and testing costs could still qualify for R&D tax relief.
- Software – A company may claim for the cost of software that is directly employed in the R&D activity. Where software is only partly employed in direct R&D, an appropriate apportionment should be made.
- Subcontracted expenditure
    - SME Scheme: A company can generally claim for 65% of the payments made to unconnected parties. The subcontracted work may be further subcontracted to any third party. Special rules apply where the parties are connected or elect to be connected. Under the SME Scheme, the subcontractor does not need to be a UK resident and there is no requirement for the subcontracted R&D to be performed in the UK.
    - RDEC Scheme: R&D expenditure subcontracted to other persons is generally not allowable unless it is directly undertaken by a charity, higher education institute, scientific research organization, health service body, or by an individual or a partnership of individuals.

Qualifying expenditure must be allowable as a deduction in calculating the profits of the trade. Capital expenditure is therefore excluded; it may however qualify for R&D allowances.

Costs that Cannot be Claimed for R&D Tax Relief

Companies cannot receive R&D tax relief for the following:

- The production and distribution of goods and services.
- Capital expenditure under either the SME tax relief or RDEC scheme. However, a 100% Research and Development Allowance may be due on capital assets, such as plant, machinery and buildings used for R&D activity.
- The cost of land.
- Cost of patents and trademarks.
- Rent or rates.

Examples of Qualifying R&D activities in the Construction Industry

Design

- Integrating new or improved technology into buildings.
- Redesigning to improve performance, reliability, quality, safety and life-cycle costs.
- Designing prefabricated components for off-site assembly of large and complex systems.
- Developing innovative sustainable solutions to achieve net zero carbon.
- Developing advanced Buildings Information Management (BIM) capabilities.

On site

- Solving technical problems that arise during or after construction.
- Developing new techniques to remove contaminants and remediate difficult ground conditions.
- Complex diversions of utilities such as gas, water and electricity.
- Developing new, more efficient construction techniques.
- Developing bespoke temporary solutions on challenging sites.

Research and Development Allowances

Research and Development Allowance (RDA) gives relief for capital expenditure incurred on R&D by a business. RDA is only due if the research and development expenditure is related to the trade being carried on or about to be carried on. Qualifying RDA expenditure is capital expenditure that a business incurs for providing facilities for carrying out R&D, for example, a laboratory in which to carry on R&D. RDA is only available to traders and is not available to a person carrying on a profession or vocation. The rate of RDA is 100% of the qualifying expenditure. However, if a disposal value is brought into account on expenditure in a chargeable period in which an RDA is made, the RDA is 100% of the expenditure less the disposal value. RDA is made in the chargeable period in which the qualifying expenditure was incurred unless it was incurred before the trade began, in which case RDA is made in the chargeable period in which trade began.

How to Claim R&D Tax relief from HMRC

A company can make a claim for R&D tax relief up to two years after the end of the accounting period it relates to. To make a claim, a company must enter their total qualifying expenditure on the full Company Tax Return form, CT600, to HMRC. If claiming under the SME scheme, a company enters the enhanced expenditure figure into the CT600, and the payable tax credit figure if applicable. If claiming under the RDEC scheme, a company must enter both the total qualifying expenditure and RDEC figures into the CT600. After submitting their full CT600 form, a claimant company can use the HMRC online service to support their R&D tax relief claim and send any additional details, as required.

Contact

For further information or assistance, please contact one of the Fiscal Incentives team:

help.r&d@aecom.com

Capital Allowances

Introduction

Capital Allowances provide tax relief by prescribing a statutory rate of depreciation for tax purposes in place of that used for accounting purposes. They are utilized by government to provide an incentive to invest in capital equipment, including assets within commercial property, by allowing the majority of taxpayers a deduction from taxable profits for certain types of capital expenditure, thereby reducing, or deferring, tax liabilities.

The capital allowances most commonly applicable to real estate are those given for plant and machinery in all buildings, other than residential dwellings, except for common areas and Structures and Buildings Allowances (SBA) for the residual construction spend on non-residential property.

Enterprise Zone Allowances are also available for capital expenditure within designated areas only where there is a focus on high value manufacturing. Enhanced rates of allowances were also available on certain types of energy and water saving plant and machinery assets, until their withdrawal from April 2020.

Freeports, introduced in 2020 are specifically designed to incentivize investment and job creation by including lower customs duties, simplified planning rules and tax breaks, such as an enhanced level of capital allowances, suspension of SDLT, business rates, customs and NIC obligations for a five-year period until 2026.

The Act

The primary legislation is contained in the Capital Allowances Act 2001. Major changes to the system were introduced in 2008, 2014, 2018 and 2020 affecting the treatment of tax relief to include plant and machinery allowances, SBA and the incentivized tax relief for the newly introduced Freeports.

Plant and Machinery

Qualifying expenditure on plant and machinery is allocated to either the main pool, or special rate pool, which provide for different rates of recovery and these are set out later.

Various legislative changes and case law precedents in recent years have introduced major changes to the availability of Capital Allowances on property expenditure. The Capital Allowances Act 2001 excludes expenditure on the provision of a building from qualifying for plant and machinery, with prescribed exceptions.

List A in Section 21 of the 2001 Act sets out those assets treated as parts of buildings:

- *Walls, floors, ceilings, doors, gates, shutters, windows and stairs.*
- *Mains services, and systems, for water, electricity and gas.*
- *Waste disposal systems.*
- *Sewerage and drainage systems.*
- *Shafts, or other structures, in which lifts, hoists, escalators and moving walkways are installed.*
- *Fire safety systems.*

Similarly, List B in Section 22 identifies excluded structures and other assets.

Both sections are, however, subject to Section 23. This section sets out expenditure, which, although being part of a building, may still be expenditure on the provision of Plant and Machinery.

List C in Section 23 is reproduced below.

Sections 21 and 22 do not affect the question whether expenditure on any item in List C is expenditure on the provision of Plant or Machinery.

1. Machinery (including devices for providing motive power) not within any other item in this list.

2. Gas and sewerage systems provided mainly –
    a. to meet the particular requirements of the qualifying activity, or
    b. to serve particular plant or machinery used for the purposes of the qualifying activity.
3. Omitted
4. Manufacturing or processing equipment; storage equipment (including cold rooms); display equipment; and counters, checkouts and similar equipment.
5. Cookers, washing machines, dishwashers, refrigerators and similar equipment; washbasins, sinks, baths, showers, sanitary ware and similar equipment; and furniture and furnishings.
6. Hoists.
7. Sound insulation provided mainly to meet the particular requirements of the qualifying activity.
8. Computer, telecommunication and surveillance systems (including their wiring or other links).
9. Refrigeration or cooling equipment.
10. Fire alarm systems; sprinkler and other equipment for extinguishing or containing fires.
11. Burglar alarm systems.
12. Strong rooms in bank or building society premises; safes.
13. Partition walls, where moveable and intended to be moved in the course of the qualifying activity.
14. Decorative assets provided for the enjoyment of the public in hotel, restaurant or similar trades.
15. Advertising hoardings; signs, displays and similar assets.
16. Swimming pools (including diving boards, slides and structures on which such boards or slides are mounted).
17. Any glasshouse constructed so that the required environment (namely, air, heat, light, irrigation and temperature) for the growing of plants is provided automatically by means of devices forming an integral part of its structure.
18. Cold stores.
19. Caravans provided mainly for holiday lettings.
20. Buildings provided for testing aircraft engines run within the buildings.
21. Moveable buildings intended to be moved in the course of the qualifying activity.
22. The alteration of land for the purpose only of installing Plant or Machinery.
23. The provision of dry docks.
24. The provision of any jetty or similar structure provided mainly to carry Plant or Machinery.
25. The provision of pipelines or underground ducts or tunnels with a primary purpose of carrying utility conduits.
26. The provision of towers to support floodlights.
27. The provision of –
    a. any reservoir incorporated into a water treatment works, or
    b. any service reservoir of treated water for supply within any housing estate or other particular locality.
28. The provision of –
    a. silos provided for temporary storage, or
    b. storage tanks.
29. The provision of slurry pits or silage clamps.
30. The provision of fish tanks or fish ponds.
31. The provision of rails, sleepers and ballast for a railway or tramway.
32. The provision of structures and other assets for providing the setting for any ride at an amusement park or exhibition.
33. The provision of fixed zoo cages.

Case Law

The fact that an item appears in List C does not automatically mean that it will qualify for capital allowances. It only means that it may potentially qualify.

Guidance about what can qualify as plant is found in case law dating back to 1887 (*Yarmouth v France*). The case of *Wimpy International Ltd and Associated Restaurants Ltd v Warland* in the late 1980s is one of the most important case law references for determining what can qualify as plant.

The Judge in that case applied three tests when considering whether, or not, an item is plant.

1. Is the item stock in trade? If the answer is yes, then the item is not plant.

2. Is the item used for carrying on the business? In order to pass the business use test the item must be employed in carrying on the business; it is not enough for the asset to be simply used in the business. For example, product display lighting in a retail store may be plant but general lighting in a warehouse would fail the test. *(Please note, this case law relates to the pre-Integral Feature Legislation, which introduced lighting as an eligible asset.)*
3. Is the item the business premises, or part of the business premises? An item cannot be plant if it fails the premises test, i.e. if the business use is the premises itself, or part of the premises, or a place in which the business is conducted. The meaning of 'part of the premises' in this context should not be confused with real property law. HMRC's internal manuals suggest there are four general factors to be considered, each of which is a question of fact and degree:
    - Does the item appear visually to retain a separate identity?
    - With what degree of permanence has it been attached to the building?
    - To what extent is the structure complete without it?
    - To what extent is it intended to be permanent, or alternatively, is it likely to be replaced within a short period?

Certain assets will qualify as plant in most cases. However, many others need to be considered on a case-by-case basis. For example, decorative assets in a hotel restaurant may be plant, but similar assets in an office reception area may be ineligible.

Main Pool Plant and Machinery

Capital Allowances on main pool plant and machinery are currently given in the form of writing down allowances at the current rate of 18% per annum on a reducing balance basis. For every £100 of qualifying expenditure £18 is claimable in year 1, £14.76 in year 2 and so on until either all of the allowances have been claimed, or the asset is sold.

Special Rate Pool Plant and Machinery

Capital Allowances on special rate pool plant and machinery are currently given in the form of writing down allowances at the current rate of 6% per annum on a reducing balance basis. For every £100 of qualifying expenditure £6 is claimable in year 1, £5.64 in year 2 and so on until either all of the allowances have been claimed, or the asset is sold.

The special rate pool specifically covers expenditure on integral features, thermal insulation, solar panels and long life assets.

Integral Features

The category of qualifying expenditure on 'integral features' was introduced with effect from April 2008. The following items are integral features:

- An electrical system (including a lighting system)
- A cold water system
- A space or water heating system, a powered system of ventilation, air cooling or air purification, and any floor or ceiling comprised in such a system
- A lift, an escalator or a moving walkway
- External solar shading

Thermal Insulation

The installation of thermal insulation to buildings is allowable where it is provided to an existing building owned or occupied by the taxpayer. The allowance is not available for expenditure incurred on dwellings.

Long-Life Assets

A long-life asset is defined as plant and machinery that has an expected useful economic life of at least 25 years. The useful economic life is taken as the period from first use until it is likely to cease to be used as a fixed asset of any business. It is important to note that this is likely to be a shorter period than an asset's physical life.

Plant and machinery provided for use in a building used wholly, or mainly, as a dwelling house, showroom, hotel, office, retail shop, or similar premises, or for purposes ancillary to such use, cannot be classified as a long-life asset.

In contrast certain plant and machinery assets in buildings such as factories, cinemas, hospitals etc. could potentially be treated as long-life assets.

Temporary Super-deduction and Enhanced First Year Allowance for Plant and Machinery Assets

Following Budget 2021, temporary special first year allowances were introduced for new and unused plant and machinery to stimulate investment. The relief will only apply to companies within the charge to corporation tax for expenditure incurred from 1 April 2021 up to and including 31 March 2023. This applies to expenditure contracted after 3 March 2021.

This temporary first year allowance will apply as follows:

- 130% 'super-deduction' for main pool plant and machinery (currently 18% per annum on a reducing balance basis); and
- 50% first year allowance for special rate plant and machinery.

In addition, unincorporated businesses will not benefit because qualifying expenditure must be incurred by a company within the charge to corporation tax.

Refurbishment Schemes

Building refurbishment projects will typically be a mixture of capital costs and revenue expenses, unless the works are so extensive that they are more appropriately classified as a redevelopment. A straightforward repair or a 'like for like' replacement of part of an asset would be a revenue expense, meaning that the entire amount can be deducted from taxable profits in the same year.

Where capital expenditure is incurred which is incidental to the installation of plant or machinery then Section 25 of the Capital Allowances Act 2001 allows it to be treated as part of the expenditure on the qualifying item. Incidental expenditure will often include parts of the building that would be otherwise disallowed, as shown in the Lists reproduced above. For example, the cost of forming a lift shaft inside an existing building would be deemed to be part of the expenditure on the provision of the new lift.

The extent of the application of section 25 was reviewed for the first time by the Special Commissioners in the case of JD Wetherspoon. The key areas of expenditure considered were overheads and preliminaries where it was held that such costs could be allocated on a pro-rata basis; decorative timber panelling which was found to be part of the premises and so ineligible for allowances; toilet lighting which was considered to provide an attractive ambience and qualified for allowances; and incidental building alterations of which enclosing walls to toilets and kitchens and floor finishes did not qualify except for tiled splash backs, bespoke toilet cubicles and drainage did qualify, along with the related sanitary fittings and kitchen equipment.

Structures and Building Allowances

The 2018 Autumn Budget introduced the Structures and Buildings Allowance (SBA) for expenditure incurred on non-residential structures and buildings.

Initially, the SBA was available at a flat rate of 2% per annum over a 50-year period, but this increased to 3% per annum over a 33⅓ year period from April 2020. This relates to expenditure incurred on new commercial structures and buildings, including costs for new conversions and renovations, applicable where contracts, for the main construction works, were entered into on, or after 29 October 2018. A temporary 10% rate (over ten years) applies for freeports between 1 April 2021 and 30 September 2026, as noted earlier.

The SBA extends to qualifying expenditure incurred under capital contributions – for example, a landlord contributing towards a tenant's fitting out costs.

The relief will be available when the building, or structure, first comes into qualifying use and is available for both UK and overseas assets, provided the business is within the charge to UK tax. This relief is not applicable for expenditure on assets used for residential purposes but does include nursing and care homes.

The SBA expenditure will not qualify for the Annual Investment Allowance (AIA).

The relief will cease to be available if the building, or structure, is brought into residential use, or if it is demolished. However, SBA continues to be available for periods of temporary disuse.

If the building, or structure, is sold, the new owner would benefit from the unutilized residue of the SBA for the remaining period, via an Allowances Statement in accordance with s170IA CAA2001.

The costs of construction will include only the net direct costs related to the construction expenditure of the building, or structure, after any discounts, refunds, or other adjustments. This will include demolition costs, or any land alterations necessary for construction and direct costs required to bring the building, or structure, into existence. Excluded from the above expenditure are indirect costs not associated with the physical construction, such as finance, planning, legal and marketing costs, landscaping etc.

SBA qualifying expenditure will also be applicable to purchases of new and second-hand buildings, or structures. The basis of claim will be dependent upon the vendor's holding structure and whether the property has been used, or unused, at the time of acquisition, as well as being entitled to claim this tax relief.

Where a building, or structure, is acquired and subsequently altered, or renovated, this will trigger a new SBA qualification period. Similar additional expenditure on existing properties will also trigger new SBA streams of qualifying expenditure.

It is important to note that expenditure on plant and machinery does not qualify for SBA. A purchaser will be unable to reclassify expenditure on plant and machinery previously claimed as SBA to plant and machinery at a later date.

Expenditure on qualifying land remediation will also not qualify for SBA.

Annual Investment Allowance

The AIA is available to all businesses of any size and allows a deduction for the whole AIA of qualifying expenditure on plant and machinery, including integral features and long-life assets. The AIA rates have fluctuated over the years:

- 1 April 2014 to 31 December 2015 – £500,000
- 1 January 2016 to 31 December 2018 – £200,000
- Currently, the AIA rate from 1 January 2019 to 31 March 2023 is £1,000,000.

For accounting periods less, or greater than, 12 months, or if claiming in periods where the rates have changed, time apportionment rules will apply to calculate hybrid rates applicable to the period of claim.

Enterprise Zones

Enterprise zones benefit from a number of reliefs, including a 100 per cent first year allowance for new and unused non-leased plant and machinery assets, where there is a focus on high-value manufacturing.

Freeports

Following a government bid process launched late in 2020, the first eight freeports in England were confirmed as follows:

- East Midlands Airport
- Felixstowe and Harwich
- Humber
- Liverpool City
- Plymouth and South Devon
- Solent
- Teesside
- Thames

Further sites are expected to follow including those in devolved administrations in Scotland, Wales and Northern Ireland.

Dedicated 'tax sites' in freeports will benefit from generous tax allowances to 30 September 2026 including:

- An enhanced 10% rate of SBA, meaning that the relief can be claimed over 10 years, rather than the current 33⅓ years;
- An enhanced capital allowance of 100% for plant and machinery allowances, providing a full deduction, rather than the current rates of 18% (MP) and 6% (SRP) on a reducing balance basis. The allowance is only available to occupiers and not property owners or developers leasing premises;
- Full relief from Stamp Duty Land Tax (SDLT) on land and property purchased for a qualifying commercial purpose;
- Full Business Rates relief for new businesses and existing businesses that expand operations;
- Relief against employer National Insurance contributions, subject to Parliamentary process and approval.

Other Capital Allowances

Other types of allowances include those available for capital expenditure on Mineral Extraction, Research and Development, Know-How, Patents, Dredging and Assured Tenancy.

International Tax Depreciation

The UK is not the only tax regime offering investors, owners and occupiers valuable incentives to invest in plant and machinery and environmentally friendly equipment. Ireland, Australia, Malaysia, Hong Kong and Singapore also have capital allowances regimes which are broadly similar to the UK and provide comparable levels of tax relief to businesses.

Many other overseas countries have tax depreciation regimes based on accounting treatment, instead of capital allowances. Some use a systematic basis over the useful life of the asset and others have prescribed methods spreading the cost over a statutory period, not always equating to the asset's useful life. Some regimes have prescribed statutory rates, whilst others have rates which have become acceptable to the tax authorities through practice.

Contact

For further information or assistance, please contact one of the Fiscal Incentives team:

help.capitalallowances@aecom.com

Value Added Tax

Introduction

Value Added Tax (VAT) is a tax on the consumption of goods and services. The UK introduced a domestic VAT regime when it joined the European Community in 1973. The principal source of European law in relation to VAT is Council Directive 2006/112/EC, a recast of Directive 77/388/EEC, which is currently restated and consolidated in the UK through the VAT Act 1994 and various Statutory Instruments, as amended by subsequent Finance Acts.

VAT Notice 708: Buildings and construction (December 2020) provides HMRC's interpretation of the VAT law in connection with construction works, however, the UK VAT legislation should always be referred to in conjunction with the publication. Recent VAT tribunals and court decisions since the date of this publication will affect the application of the VAT law in certain instances. The Notice is available on HM Revenue & Customs website at www.hmrc.gov.uk.

The Scope of VAT

VAT is payable on:

- Supplies of goods and services made in the UK
- By a taxable person
- In the course or furtherance of business; and
- Which are not specifically exempted or zero-rated.

Rates of VAT

There are three rates of VAT:

- A standard rate, currently 20%;
- A reduced rate, currently 5%; and
- A zero rate of 0%.

Additionally some supplies are exempt from VAT and others are considered outside the scope of VAT.

Recovery of VAT

When a taxpayer makes taxable supplies he must account for VAT, known as output VAT, at the appropriate rate of either 20%, 5% or 0%. Any VAT due then has to be declared and submitted on a VAT submission to HM Revenue & Customs and will normally be charged to the taxpayer's customers.

As a VAT registered person, the taxpayer is entitled to reclaim from HM Revenue & Customs, commonly referred to as input VAT, the VAT incurred on their purchases and expenses directly related to its business activities in respect of standard-rated, reduced-rated and zero-rated supplies. A taxable person cannot, however, reclaim VAT that relates to any non-business activities (but see below) or, depending on the amount of exempt supplies they made, input VAT may be restricted or not recoverable.

At predetermined intervals the taxpayer will pay to HM Revenue & Customs the excess of VAT collected over the VAT they can reclaim. However, if the VAT reclaimed is more than the VAT collected, the taxpayer, who will be in a net repayment position, can reclaim the difference from HM Revenue & Customs.

Example

X Ltd constructs a block of flats. It sells long leases to buyers for a premium. X Ltd has constructed a new building designed as a dwelling and will have granted a long lease. This first sale of a long lease is a VAT zero-rated supply. This means any VAT incurred in connection with the development, which X Ltd will have properly paid (e.g. payments for consultants and certain preliminary services) will be recoverable. For reasons detailed below, the contractor employed by X Ltd will have charged VAT on his construction services at the zero rate of VAT.

Use for Business and Non-Business Activities

Where a supply relates partly to business use and partly to non-business use, then the basic rule is that it must be apportioned on a fair and reasonable basis so that only the business element is potentially recoverable. In some cases, VAT on land, buildings and certain construction services, purchased for both business and non-business use, could be recovered in full by applying what is known as 'Lennartz' accounting, to reclaim VAT relating to the non-business use and account for VAT on the non-business use over a maximum period of 10 years. Following an ECJ case restricting the scope of this approach, its application to immovable property was removed completely in January 2011 by HMRC (business brief 53/10), when UK VAT law was amended to comply with EU Directive 2009/162/EU.

Taxable Persons

A taxable person is an individual, firm, company, etc., who is required to be registered for VAT. A person who makes taxable supplies above certain turnover limits is compulsorily required to be VAT registered. The current registration limit (since 1 April 2017), known as the VAT threshold, is £85,000. If the threshold is exceeded in any 12 month rolling period, or there is an expectation that the value of the taxable supplies in a single 30 day period, or goods are received into the UK from the EU worth more than the £85,000, then registration for UK VAT is compulsory.

A person who makes taxable supplies below the limit is still entitled to be registered on a voluntary basis if they wish, for example, in order to recover input VAT incurred in relation to those taxable supplies, however output VAT will then become due on the sales and must be accounted for.

VAT Exempt Supplies

Where a supply is exempt from VAT this means that no output VAT is payable – but equally the person making the exempt supply cannot normally recover any of the input VAT on their own costs relating to that exempt supply.

Generally commercial property transactions such as leasing of land and buildings are exempt unless a landlord chooses to standard-rate its interest in the property by a applying for an option to tax. This means that VAT is added to rental income and also that VAT incurred, on say, an expensive refurbishment, is recoverable.

Supplies outside the scope of VAT

Supplies are outside the scope of VAT if they are:

- Made by someone who is not a taxable person
- Made outside the UK; or
- Not made in the course or furtherance of business

In course or furtherance of business

VAT must be accounted for on all taxable supplies made in the course or furtherance of business, with the corresponding recovery of VAT on expenditure incurred.

If a taxpayer also carries out non-business activities, then VAT incurred in relation to such supplies is generally not recoverable.

In VAT terms, business means any activity continuously performed which is mainly concerned with making supplies for a consideration. This includes:

- Anyone carrying on a trade, vocation or profession;
- The provision of membership benefits by clubs, associations and similar bodies in return for a subscription or other consideration; and
- Admission to premises for a charge.

It may also include the activities of other bodies including charities and non-profit making organizations.

Examples of non-business activities are:

- Providing free services or information;
- Maintaining some museums or particular historic sites;
- Publishing religious or political views.

Construction Services

In general the provision of construction services by a contractor will be VAT standard rated at 20%, however, there are a number of exceptions for construction services provided in relation to certain relevant residential properties and charitable buildings.

The supply of building materials is VAT standard rated at 20%, however, where these materials are supplied and installed as part of the construction services, the VAT liability of those materials follows that of the construction services supplied.

Zero-rated construction services

The following construction services are VAT zero-rated, including the supply of related building materials.

The construction of new dwellings

The supply of services in the course of the construction of a new building designed for use as a dwelling or number of dwellings is zero-rated, other than the services of an architect, surveyor or any other person acting as a consultant or in a supervisory capacity.

The following basic conditions must ALL be satisfied in order for the works to qualify for zero-rating:

1. A qualifying building has been, is being, or will be constructed;
2. Services are made 'in the course of the construction' of that building;
3. Where necessary, you hold a valid certificate; and
4. Your services are not specifically excluded from zero-rating.

The construction of a new building for 'relevant residential or charitable' use

The supply of services in the course of the construction of a building designed for use as a relevant residential Purpose (RRP), or relevant charitable purpose (RCP), is zero-rated, other than the services of an architect, surveyor or any other person acting as a consultant or in a supervisory capacity.

A 'relevant residential' use building means:

1. A home or other institution providing residential accommodation for children;
2. A home or other institution providing residential accommodation with personal care for persons in need of personal care by reason of old age, disablement, past or present dependence on alcohol or drugs or past or present mental disorder;
3. A hospice;
4. Residential accommodation for students or school pupils;
5. Residential accommodation for members of any of the armed forces;

6. A monastery, nunnery, or similar establishment; or
7. An institution which is the sole or main residence of at least 90% of its residents.

A 'relevant residential' purpose building does not include use as a hospital, a prison, or similar institution, or as a hotel, inn, or similar establishment.

A 'relevant charitable' purpose means use by a charity in either, or both of the following ways:

1. Otherwise than in the course or furtherance of a business; or
2. As a village hall, or similarly in providing social or recreational facilities for a local community.

Non qualifying use, which is not expected to exceed 10% of the time the building is normally available for use, can be ignored. The calculation of business use can be based on time, floor area, or head count, subject to approval being acquired from HM Revenue & Customs.

The construction services can only be zero-rated if a certificate is given by the end user to the contractor carrying out the works, confirming that the building is to be used for a qualifying purpose, i.e. for a 'relevant residential or charitable' purpose. It follows that such services can only be zero-rated when supplied to the end user and, unlike supplies relating to dwellings, supplies by subcontractors cannot be zero-rated.

The construction of an annexe used for a 'relevant charitable' purpose

Construction services provided in the course of construction of an annexe for use entirely, or partly for a 'relevant charitable' purpose, can be zero-rated.

In order to qualify, the annexe must:

1. Be capable of functioning independently from the existing building;
2. Have its own main entrance; and
3. Be covered by a qualifying use certificate.

The conversion of a non-residential building into dwellings, or the conversion of a building from non-residential use to 'relevant residential' use, where the supply is to a 'relevant' housing association

The supply to a 'relevant' housing association in the course of conversion of a non-residential building, or non-residential part of a building, into:

1. A new eligible dwelling designed as a dwelling, or number of dwellings; or
2. A building, or part of a building, for use solely for a relevant residential purpose.

Any services related to the conversion, other than the services of an architect, surveyor or any person acting as a consultant or in a supervisory capacity, are zero-rated.

A 'relevant' housing association is defined as:

1. A private registered provider of social housing;
2. A registered social landlord within the meaning of Part I of the Housing Act 1996 (Welsh registered social landlords);
3. A registered social landlord within the meaning of the Housing (Scotland) Act 2001 (Scottish registered social landlords), or
4. A registered housing association within the meaning of Part II of the Housing (Northern Ireland) Order 1992 (Northern Irish registered housing associations).

If the building is to be used for a 'relevant residential' purpose, the housing association should issue a qualifying use certificate to the contractor completing the works. Subcontractors' services that are not made directly to a relevant housing association are standard-rated.

The development of a residential caravan park

The supply in the course of the construction of any civil engineering work 'necessary for' the development of a permanent park for residential caravans, and any services related to the construction, other than the services of an architect, surveyor or any person acting as a consultant or in a supervisory capacity, are zero-rated when a new permanent park is being developed, the civil engineering works are necessary for the development of the park and the services are not specifically excluded from zero-rating. This includes access roads, paths, drainage, sewerage and the installation of mains water, power and gas supplies.

Certain building alterations for 'disabled' persons

Certain goods and services supplied to a 'disabled' person, or a charity making these items and services available to 'disabled' persons, can be zero-rated. The recipient of these goods or services needs to give the supplier an appropriate written declaration that they are entitled to benefit from zero rating.

The following services (amongst others) are zero-rated:

1. The installation of specialist lifts and hoists and their repair and maintenance;
2. The construction of ramps, widening doorways or passageways including any preparatory work and making good work;
3. The provision, extension and adaptation of a bathroom, washroom or lavatory; and
4. Emergency alarm call systems.

Sale of Reconstructed Buildings

A protected building is not to be regarded as substantially reconstructed unless, when the reconstruction is completed, the reconstructed building incorporates no more of the original building than the external walls, together with other external features of architectural or historical interest.

DIY Builders and Converters

Private individuals who decide to construct their own home are able to reclaim VAT they pay on goods they use to construct their home by use of a special refund mechanism made by way of an application to HM Revenue & Customs. This also applies to services provided in the conversion of an existing non-residential building to form a new dwelling.

The scheme is meant to ensure that private individuals do not suffer the burden of VAT if they decide to construct their own home.

Charities may also qualify for a refund on the purchase of materials incorporated into a building used for non-business purposes where they provide their own free labour for the construction of a 'relevant charitable' use building.

Installation of energy saving materials (ESM) to domestic properties

At Spring Statement 2022, it was announced that the supply and installation of certain energy saving materials including insulation, draught stripping, central heating, hot water controls and solar panels in a residential building, would benefit from a zero rate until 31 March 2027. From 1 April 2027, ESM will revert to the reduced rate.

The scope of ESM has also been extended to add back wind and water turbines, which had previously been excluded.

Reduced-rated construction services

The following construction services are subject to the reduced rate of VAT of 5%, including the supply of related building materials.

Conversion – changing the number of dwellings

In order to qualify for the 5% rate, there must be a different number of 'single household dwellings' within a building than there were before commencement of the conversion works. A 'single household dwelling' is defined as a dwelling that is designed for occupation by a single household.

These conversions can be from 'relevant residential' purpose buildings, non-residential buildings and houses in multiple occupation.

A house in multiple occupation conversion

This relates to construction services provided in the course of converting a 'single household dwelling', a number of 'single household dwellings', a non-residential building or a 'relevant residential' purpose building into a house for multiple occupation, such as bed sit accommodation.

A special residential conversion

A special residential conversion involves the conversion of a 'single household dwelling', a house in multiple occupation, or a non-residential building into a 'relevant residential' purpose building, such as student accommodation or a care home.

Renovation of derelict dwellings

The provision of renovation services in connection with a dwelling or 'relevant residential' purpose building that has been empty for two or more years prior to the date of commencement of construction works can be carried out at a reduced rate of VAT of 5%.

Installation of energy saving materials

As stated above, ESM installed in domestic properties are subject to a zero rate until 31 March 2027, after which they will revert to the reduced rate of VAT.

Buildings that are used by charities for non-business purposes, and/or as village halls, are not eligible for the reduced rate for the supply of energy saving materials.

Grant-funded installation of heating equipment or connection of a gas supply

The grant-funded supply and installation of heating appliances, connection of a mains gas supply, supply, installation, maintenance and repair of central heating systems, and supply and installation of renewable source heating systems, to qualifying persons. A qualifying person is someone aged 60 or over or is in receipt of various specified benefits.

Grant-funded installation of security goods

The grant-funded supply and installation of security goods to a qualifying person.

Housing alterations for the elderly

Certain home adaptations that support the needs of elderly people are reduced rated.

Building Contracts

Design and build contracts

If a contractor provides a design and build service relating to works to which the reduced or zero rate of VAT is applicable, then any design costs incurred by the contractor will follow the VAT liability of the principal supply of construction services.

Management contracts

A management contractor acts as a main contractor for VAT purposes and the VAT liability of his services will follow that of the construction services provided. If the management contractor only provides advice without engaging trade contractors, his services will be VAT standard rated.

Construction Management and Project Management

The project manager or construction manager is appointed by the client to plan, manage and co-ordinate a construction project. This will involve establishing competitive bids for all the elements of the work and the appointment of trade contractors. The trade contractors are engaged directly by the client for their services. The VAT liability of the trade contractors will be determined by the nature of the construction services they provide and the building being constructed.

The fees of the construction manager or project manager will be VAT standard rated. If the construction manager also provides some construction services, these works may be zero or reduced rated if the works qualify.

Liquidated and Ascertained Damages

Liquidated damages are outside of the scope of VAT as compensation. The employer should not reduce the VAT amount due on a payment under a building contract on account of a deduction of damages. In contrast, an agreed reduction in the contract price will reduce the VAT amount.

Similarly, in certain circumstances, HM Revenue & Customs may agree that a claim by a contractor under a JCT, or other form of contract, is also compensation payment and outside the scope of VAT.

Reverse Charge for Building and Construction Services

A VAT reverse charge was introduced on 1 March 2021 for certain building and construction services. This measure has been introduced to combat missing trader fraud in the construction industry, removing the opportunity for fraudsters to charge VAT and then go missing, before paying it over to the HMRC.

For certain supplies of construction services, the customer will now be liable to account to HMRC for the VAT in respect of those purchases rather than the supplier. This reverse charge will apply through the supply chain where payments are required to be reported through the Construction Industry Scheme (CIS) up to the point where the customer receiving the supply is no longer a business that makes supplies of specified services (end users).

The introduction of a reverse charge does not change the liability of the supply of the specified services: it is the way in which the VAT on those supplies is accounted for. Rather than the supplier charging and accounting for the VAT, the recipient of those supplies accounts for the VAT instead of paying the VAT amount to its supplier. It will be able to reclaim that VAT amount as input tax, subject to the normal rules. The supplier will need to issue a VAT invoice that indicates the supplies are subject to the reverse charge.

The types of construction services covered by the reverse charge are based on the definition of 'construction operations' used in CIS under section 74 of the Finance Act 2004.

As well as excluding supplies of specified services to end users, the reverse charge does not capture supplies of specified services where the supplier and customer are connected in a particular way, and for supplies between landlords and tenants.

Contact

For further information or assistance, please contact one of the Fiscal Incentives team:

help.vat@aecom.com

The Aggregates Levy

The Aggregates Levy came into operation on 1 April 2002 in the UK, except for Northern Ireland where it was phased in over five years from 2003.

It was introduced to ensure that the external costs associated with the exploitation of aggregates are reflected in the price of aggregate, and to encourage the use of recycled aggregate. There continues to be strong evidence that the levy is achieving its environmental objectives, with sales of primary aggregate down and production of recycled aggregate up. The Government expects that the rates of the levy will at least keep pace with inflation over time, although it accepts that the levy is still bedding in.

The rate of the levy remains at £2.00 per tonne from 1 April 2022 and is levied on anyone considered to be responsible for commercially exploiting 'virgin' aggregates in the UK and should naturally be passed by price increase to the ultimate user.

All materials falling within the definition of 'Aggregates' are subject to the levy unless specifically exempted.

It does not apply to clay, soil, vegetable or other organic matter.

The intention is that it will:

- Encourage the use of alternative materials that would otherwise be disposed of to landfill sites
- Promote development of new recycling processes, such as using waste tyres and glass
- Promote greater efficiency in the use of virgin aggregates
- Reduce noise and vibration, dust and other emissions to air, visual intrusion, loss of amenity and damage to wildlife habitats

Definitions

'Aggregates' means any rock, gravel or sand which is extracted or dredged in the UK for aggregates use. It includes whatever substances are for the time being incorporated in it or naturally occur mixed with it.

'Exploitation' is defined as involving any one or a combination of any of the following:

- Being removed from its original site, a connected site which is registered under the same name as the originating site or a site where it had been intended to apply an exempt process to it, but this process was not applied
- Becoming subject to a contract or other agreement to supply to any person
- Being used for construction purposes
- Being mixed with any material or substance other than water, except in permitted circumstances

The definition of 'aggregate being used for construction purposes' is when it is:

- Used as material or support in the construction or improvement of any structure
- Mixed with anything as part of a process of producing mortar, concrete, tarmacadam, coated roadstone or any similar construction material

Incidence

It is a tax on primary aggregates production – i.e. 'virgin' aggregates won from a source and used in a location within the UK territorial boundaries (land or sea). The tax is not levied on aggregates which are exported or on aggregates imported from outside the UK territorial boundaries.

It is levied at the point of sale.

Exemption from tax

An 'aggregate' is exempt from the levy if it is:

- Material which has previously been used for construction purposes
- Aggregate that has already been subject to a charge to the Aggregates Levy
- Aggregate which was previously removed from its originating site before the start date of the levy
- Aggregate which is moved between sites under the same Aggregates Levy Registration
- Aggregate which is removed to a registered site to have an exempt process applied to it
- Aggregate which is removed to any premises where china clay or ball clay will be extracted from the aggregate
- Aggregate which is being returned to the land from which it was won provided that it is not mixed with any material other than water
- Aggregate won from a farm land or forest where used on that farm or forest
- Rock which has not been subjected to an industrial crushing process
- Aggregate won by being removed from the ground on the site of any building or proposed building in the course of excavations carried out in connection with the modification or erection of the building and exclusively for the purpose of laying foundations or of laying any pipe or cable
- Aggregate won by being removed from the bed of any river, canal or watercourse or channel in or approach to any port or harbour (natural or artificial), in the course of carrying out any dredging exclusively for the purpose of creating, restoring, improving or maintaining that body of water
- Aggregate won by being removed from the ground along the line of any highway or proposed highway in the course of excavations for improving, maintaining or constructing the highway otherwise than purely to extract the aggregate
- Drill cuttings from petroleum operations on land and on the seabed
- Aggregate resulting from works carried out in exercise of powers under the New Road and Street Works Act 1991, the Roads (Northern Ireland) Order 1993 or the Street Works (Northern Ireland) Order 1995
- Aggregate removed for the purpose of cutting of rock to produce dimension stone, or the production of lime or cement from limestone.

Aggregate arising as a waste material during the processing of the following industrial minerals:

- anhydrite
- ball clay
- barites
- calcite
- china clay
- clay, coal, lignite and slate
- feldspar
- flint
- fluorspar
- fuller's earth
- gems and semi-precious stones
- gypsum
- any metal or the ore of any metal
- muscovite
- perlite
- potash
- pumice

- rock phosphates
- sodium chloride
- talc
- vermiculite
- spoil from the separation of the above industrial minerals from other rock after extraction
- material that is mainly but not wholly the spoil, waste or other by-product of any industrial combustion process or the smelting or refining of metal.

Anything that consists 'wholly or mainly' of the following is exempt from the levy (note that 'wholly' is defined as 100% but 'mainly' as more than 50%, thus exempting any contained aggregates amounting to less than 50% of the original volumes:

- clay, soil, vegetable or other organic matter
- drill cuttings from oil exploration in UK waters
- material arising from utility works, if carried out under the New Roads and Street Works Act 1991

However, when ground that is more than half clay is mixed with any substance (for example, cement or lime) for the purpose of creating a firm base for construction, the clay becomes liable to Aggregates Levy because it has been mixed with another substance for the purpose of construction.

Anything that consists completely of the following substances is exempt from the levy:

- Spoil, waste or other by-products from any industrial combustion process or the smelting or refining of metal – for example, industrial slag, pulverised fuel ash and used foundry sand. If the material consists completely of these substances at the time it is produced it is exempt from the levy, regardless of any subsequent mixing
- Aggregate necessarily arising from the footprint of any building for the purpose of laying its foundations, pipes or cables. It must be lawfully extracted within the terms of any planning consent
- Aggregate necessarily arising from navigation dredging
- Aggregate necessarily arising from the ground in the course of excavations to improve, maintain or construct a highway or a proposed highway.
- Aggregate necessarily arising from the ground in the course of excavations to improve, maintain or construct a railway, monorail or tramway.

Relief from the levy either in the form of credit or repayment is obtainable where:

- it is subsequently exported from the UK in the form of aggregate
- it is used in an exempt process
- it is used in a prescribed industrial or agricultural process
- it is waste aggregate disposed of by dumping or otherwise, e.g. sent to landfill or returned to the originating site

An exemption for aggregate obtained as a by-product of railway, tramway and monorail improvement, maintenance and construction was introduced in 2007.

Discounts

Water which is added to the aggregate after the aggregate has been won (washing, dust dampening etc.) may be discounted from the tax calculations. There are two accepted options by which the added water content can be calculated.

The first is to use HMRC's standard added water percentage discounts listed below: .

- washed sand 7%
- washed gravel 3.5%
- washed rock/aggregate 4%

Alternatively a more exact percentage can be agreed for dust dampening of aggregates.

Whichever option is adopted, it must be agreed in writing in advance with HMRC.

Impact

The British Aggregates Association suggested that the additional cost imposed by quarries is more likely to be in the order of £3.40 per tonne on mainstream products, applying an above average rate on these in order that by-products and low grade waste products can be held at competitive rates, as well as making some allowance for administration and increased finance charges.

With many gravel aggregates costing in the region of £20.00 per tonne, there is a significant impact on construction costs.

Avoidance

An alternative to using new aggregates in filling operations is to crush and screen rubble which may become available during the process of demolition and site clearance as well as removal of obstacles during the excavation processes.

More detailed information can be found on the HMRC website (www.hmrc.gov.uk) in Notice AGL1 Aggregates Levy published 1 April 2014 (updated 9 February 2017).

Contact

For further information or assistance, please contact one of the Fiscal Incentives team:

help.fiscalincentives@aecom.com

Land Remediation

The purpose of this section is to review the general background of ground contamination, the cost implications of current legislation and to consider the various remedial measures and to present helpful guidance on the cost of Land Remediation.

It must be emphasized that the cost advice given is an average and that costs can vary considerably from contract to contract depending on individual Contractors, site conditions, type and extent of contamination, methods of working and various other factors as diverse as difficulty of site access and distance from approved tips.

We have structured this Unit Cost section to cover as many aspects of Land Remediation works as possible.

The introduction of the Landfill Directive in July 2004 has had a considerable impact on the cost of Remediation works in general and particularly on the practice of Dig and Dump. The number of Landfill sites licensed to accept Hazardous Waste has drastically reduced and inevitably this has led to increased costs.

Market forces will determine future increases in cost resulting from the introduction of the Landfill Directive and the cost guidance given within this section will require review in light of these factors.

Statutory Framework

In July 1999 new contaminated land provisions, contained in Part IIa of the Environmental Protection Act 1990 were introduced. Primary objectives of the measures included a legal definition of Contaminated Land and a framework for identifying liability, underpinned by a 'polluter pays' principle meaning that remediation should be paid for by the party (or parties) responsible for the contamination. A secondary, and indirect, objective of Part IIA is to provide the legislative context for remediation carried out as part of development activity which is controlled through the planning system. This is the domain where other related objectives, such as encouraging the recycling of brownfield land, are relevant.

Under the Act action to remediate land is required only where there are unacceptable actual or 'significant possibility of significant harm' to health, controlled waters or the environment. Only Local Authorities have the power to determine a site as Contaminated Land and enforce remediation. Sites that have been polluted from previous land use may not need remediating until the land use is changed; this is referred to as 'land affected by contamination'. This is a risk-based assessment on the site specifics in the context of future end uses. As part of planning controls, the aim is to ensure that a site is incapable of meeting the legal definition of Contaminated Land post-development activity. In addition, it may be necessary to take action only where there are appropriate, cost-effective remediation processes that take the use of the site into account.

The Environment Act 1995 amended the Environment Protection Act 1990 by introducing a new regime designed to deal with the remediation of sites which have been seriously contaminated by historic activities. The regime became operational on 1 April 2000. Local authorities and/or the Environment Agency regulate seriously contaminated sites which are known as 'special sites'. The risks involved in the purchase of potentially contaminated sites are high, particularly considering that a transaction can result in the transfer of liability for historic contamination from the vendor to the purchaser.

The contaminated land provisions of the Environmental Protection Act 1990 are only one element of a series of statutory measures dealing with pollution and land remediation that have been and are to be introduced. Others include:

- Groundwater regulations, including pollution prevention measures
- An integrated prevention and control regime for pollution
- Sections of the Water Resources Act 1991, which deals with works notices for site controls, restoration and clean up

April 2012 saw the first revision of the accompanying Part IIa Statutory Guidance. This has introduced a new categorisation scheme for assessing sites under Part IIA. Category 1 is land which definitely is Contaminated Land and Category 4 is for land which definitely is not Contaminated Land. This is intended to assist prioritisation of sites which pose the greatest risk. Still included in the statutory guidance are matters of inspection, definition, remediation, apportionment of liabilities and recovery of costs of remediation. The measures are to be applied in accordance with the following criteria:

- the planning system
- the standard of remediation should relate to the present use
- the costs of remediation should be reasonable in relation to the seriousness of the potential harm
- the proposals should be practical in relation to the availability of remediation technology, impact of site constraints and the effectiveness of the proposed clean-up method

Liability for the costs of remediation rests with either the party that 'caused or knowingly permitted' contamination, or with the current owners or occupiers of the land.

Apportionment of liability, where shared, is determined by the local authority. Although owners or occupiers become liable only if the polluter cannot be identified, the liability for contamination is commonly passed on when land is sold.

If neither the polluter nor owner can be found, the clean up is funded from public resources.

The ability to forecast the extent and cost of remedial measures is essential for both parties, so that they can be accurately reflected in the price of the land.

At the end of March 2012, the National Planning Policy Framework replaced relevant planning guidance relating to remediation, most significantly PPS 23 Planning and Pollution Control. This has been replaced by a need to investigate and assess land contamination, which must be carried out by a competent person.

The EU Landfill Directive

The Landfill (England and Wales) Regulations 2002 came into force on 15 June 2002 followed by Amendments in 2004 and 2005. These new regulations implement the Landfill Directive (Council Directive 1999/31/EC), which aims to prevent, or to reduce as far as possible, the negative environmental effects of landfill. These regulations have had a major impact on waste regulation and the waste management industry in the UK.

The Scottish Executive implements the Landfill Directive through the Landfill (Scotland) Regulations 2003 and subsequent 2013 Amendments, and the Northern Ireland Assembly implements the Directive through the Landfill Regulations (Northern Ireland) 2003, the Waste and Emissions Trading Act 2003 and the Landfill Allowances Scheme (Northern Ireland) Regulations 2004 (as amended).

In summary, the Directive requires that:

- Sites are to be classified into one of three categories: hazardous, non-hazardous or inert, according to the type of waste they will receive
- Higher engineering and operating standards will be followed
- Biodegradable waste will be progressively diverted away from landfills
- Certain hazardous and other wastes, including liquids, explosive waste and tyres will be prohibited from landfills
- Pretreatment of wastes prior to landfilling will become a requirement

On 15 July 2004 the co-disposal of hazardous and non-hazardous waste in the same landfill site ended and in July 2005 new waste acceptance criteria (WAC) were introduced which also prevents the disposal of materials contaminated by coal tar.

The effect of this Directive has been to dramatically reduce the hazardous disposal capacity post July 2004, resulting in a **SIGNIFICANT** increase in remediating costs. This has significantly increased travelling distance and cost for disposal to landfill. The increase in operating expenses incurred by the landfill operators has also resulted in higher tipping costs.

However, there is now a growing number of opportunities to dispose of hazardous waste to other facilities such as soil treatment centres, often associated with registered landfills potentially eliminating landfill tax. Equally, improvements in on-site treatment technologies are helping to reduce the costs of disposal by reducing the hazardous properties of materials going offsite.

All hazardous materials designated for disposal off-site are subject to WAC tests. Samples of these materials are taken from site to laboratories in order to classify the nature of the contaminants. These tests, which cost approximately £200 each, have resulted in increased costs for site investigations and as the results may take up to 3 weeks this can have a detrimental effect on programme.

As from 1 July 2008 the WAC derogations which have allowed oil contaminated wastes to be disposed in landfills with other inert substances were withdrawn. As a result the cost of disposing oil contaminated solids has increased.

There has been a marked slowdown in brownfield development in the UK with higher remediation costs, longer clean-up programmes and a lack of viable treatment options for some wastes.

The UK Government established the Hazardous Waste Forum in December 2002 to bring together key stakeholders to advise on the way forward on the management of hazardous waste.

Effect on Disposal Costs

Although most landfills are reluctant to commit to future tipping prices, tipping costs have generally stabilized. However, there are significant geographical variances, with landfill tip costs in the North of England typically being less than their counterparts in the Southern regions.

For most projects to remain viable there is an increasing need to treat soil in situ by bioremediation, soil washing or other alternative long-term remediation measures. Waste untreatable on-site such as coal tar remains a problem. Development costs and programmes need to reflect this change in methodology.

Types of Hazardous Waste

- Sludges, acids and contaminated wastes from the oil and gas industry
- Acids and toxic chemicals from chemical and electronics industries
- Pesticides from the agrochemical industry
- Solvents, dyes and sludges from leather and textile industries
- Hazardous compounds from metal industries
- Oil, oil filters and brake fluids from vehicles and machines
- Mercury-contaminated waste from crematoria
- Explosives from old ammunition, fireworks and airbags
- Lead, nickel, cadmium and mercury from batteries
- Asbestos from the building industry
- Amalgam from dentists
- Veterinary medicines

[Source: Sepa]

Foam insulation materials containing ODP (Ozone Depletant Potential) are also considered as hazardous waste under the EC Regulation 2037/2000.

Land Remediation Techniques

There are two principal approaches to remediation – dealing with the contamination in situ or ex situ. The selection of the approach will be influenced by factors such as: initial and long term cost, timeframe for remediation, types of contamination present, depth and distribution of contamination, the existing and planned topography, adjacent land uses, patterns of surface drainage, the location of existing on-site services, depth of excavation necessary for foundations and below-ground services, environmental impact and safety, interaction with geotechnical performance, prospects for future changes in land use and long-term monitoring and maintenance of in situ treatment.

On most sites, contamination can be restricted to the top couple of metres, although gasholder foundations for example can go down 10 to 15 metres. Underground structures can interfere with the normal water regime and trap water pockets.

There could be a problem if contaminants get into fissures in bedrock.

In Situ Techniques

A range of in situ techniques is available for dealing with contaminants, including:

- Clean cover – a layer of clean soil is used to segregate contamination from receptor. This technique is best suited to sites with widely dispersed contamination. Costs will vary according to the need for barrier layers to prevent migration of the contaminant.
- On-site encapsulation – the physical containment of contaminants using barriers such as slurry trench cut-off walls. The cost of on-site encapsulation varies in relation to the type and extent of barriers required, the costs of which range from £50/m² to more than £175/m².

There are also in situ techniques for treating more specific contaminants, including:

- Bioremediation – for removal of oily, organic contaminants through natural digestion by micro-organisms. Most bioremediation is ex situ, i.e. it is dug out and then treated on site in bio-piles. The process can be slow, taking up to three years depending upon the scale of the problem, but is particularly effective for the long-term improvement of a site, prior to a change of use.
- Phytoremediation – the use of plants that mitigate the environmental problem without the need to excavate the contaminant material and dispose of it elsewhere. Phytoremediation consists of mitigating pollutant concentrations in contaminated soils, water or air, with plants able to contain, degrade or eliminate metals, pesticides, solvents, explosives, crude oil and its derivatives and various other contaminants from the media that contain them.
- Vacuum extraction – involving the extraction of volatile organic compounds (e.g. benzene) from soil and groundwater by vacuum.
- Thermal treatment – the incineration of contaminated soils on site. Thermal processes use heat to increase the volatility to burn, decompose, destroy or melt the contaminants. Cleaning soil with thermal methods may take only a few months to several years.
- Stabilization – cement or lime, is used to physically or chemically bind oily or metal contaminants to prevent leaching or migration. Stabilization can be used in both in situ and ex situ conditions.
- Aeration – if the ground contamination is highly volatile, e.g. fuel oils, then the ground can be ploughed and rotovated to allow the substance to vaporize.
- Air sparging – the injection of contaminant-free air into the sub-surface enabling a phase transfer of hydrocarbons from a dissolved state to a vapour phase.
- Chemical oxidization – the injection of reactive chemical oxidants directly into the soil for the rapid destruction of contaminants.
- Pumping – to remove liquid contaminants from boreholes or excavations. Contaminated water can be pumped into holding tanks and allowed to settle; testing may well prove it to be suitable for discharging into the foul sewer subject to payment of a discharge fee to the local authority. It may be necessary to process the water through an approved water treatment system to render it suitable for discharge.

Ex Situ Techniques

Removal for landfill disposal has, historically, been the most common and cost-effective approach to remediation in the UK, providing a broad spectrum solution by dealing with all contaminants. As mentioned above, the implementation of the Landfill Directive has resulted in other techniques becoming more competitive for the disposal of hazardous waste.

If used in combination with material-handling techniques such as soil washing, the volume of material disposed at landfill sites can be significantly reduced. The disadvantages of these techniques include the fact that the contamination is not destroyed, there are risks of pollution during excavation and transfer; road haulage may also cause a local nuisance. Ex situ techniques include:

- Soil washing – involving the separation of a contaminated soil fraction or oily residue through a washing process. This also involves the excavation of the material for washing ex situ. The de-watered contaminant still requires disposal to landfill. In order to be cost effective, 70–90% of soil mass needs to be recovered. It will involve constructing a hard area for the washing, intercepting the now-contaminated water and taking it away in tankers.
- Thermal treatment – the incineration of contaminated soils ex situ. The uncontaminated soil residue can be recycled. By-products of incineration can create air pollution and exhaust air treatment may be necessary.

Soil treatment centres are now beginning to be established. These use a combination of treatment technologies to maximize the potential recovery of soils and aggregates and render them suitable for disposal to the landfill. The technologies include:

- Physico-chemical treatment – a method which uses the difference in grain size and density of the materials to separate the different fractions by means of screens, hydrocyclones and upstream classification.
- Bioremediation – the aerobic biodegradation of contaminants by naturally occurring micro-organisms placed into stockpiles/windrows.
- Stabilization/solidification – a cement or chemical stabilization unit capable of immobilizing persistent leachable components.

Tax Relief for Remediation of Contaminated Land

The Finance Act 2001 included provisions that allow companies (but not individuals or partnerships) to claim tax relief on capital and revenue expenditure on the 'remediation of contaminated land' in the United Kingdom. The relief is available for expenditure incurred on or after 11 May 2001. From April 2020, non-resident companies are within the scope of UK Corporation Tax and can also claim relief.

From 1 April 2009 there was an increase in the scope of costs that qualify for Land Remediation Relief where they are incurred on long-term derelict land. The list includes costs that the Treasury believe to be primarily responsible for causing dereliction, such as additional costs for removing building foundations and machine bases. However, while there is provision for the list to be extended, the additional condition for the site to have remained derelict since 1998 is likely to render this relief redundant in all but a handful of cases. The other positive change was the fact that Japanese Knotweed removal and treatment (on-site only) now qualified for the relief under the existing legislation, thereby allowing companies to make retrospective claims.

A company is able to claim an additional 50% deduction for 'qualifying land remediation expenditure' allowed as a deduction in computing taxable profits, and may elect for the same treatment to be applied to qualifying capital expenditure.

With Landfill Tax Exemption (LTE) now phased out, Land Remediation Relief (LRR) for contaminated and derelict land is the Government's primary tool to create incentives for brownfield development. LRR is available to companies engaged in land remediation that are not responsible for the original contamination.

Over 7 million tonnes of waste each year were being exempted from Landfill Tax in England alone, so this change could have a major impact on the remediation industry. The modified LRR scheme, which provides Corporation Tax relief on any costs incurred on qualifying land remediation expenditure, is in the long run designed to yield benefits roughly equal to those lost through the withdrawal of LTE, although there is some doubt about this stated equity in reality.

However, with much remediation undertaken by polluters or public authorities, who cannot benefit from tax relief benefits, the change could result in a net withdrawal of Treasury support to a vital sector. Lobbying and consultation continues to ensure the Treasury maintains its support for remediation.

While there are no financial penalties for not carrying out remediation, a steep escalator affected the rate of landfill tax for waste material other than inert or inactive wastes, which has increased to £98.60/tonne from 1 April 2022. This means that for schemes where there is no alternative to dig and dump and no pre-existing LTE, the cost of remediation has risen to prohibitive levels should contaminated material be disposed off-site to licensed landfills.

Looking forward, tax-relief benefits under LRR could provide a significant cash contribution to remediation. Careful planning is the key to ensure that maximum benefits are realised, with actions taken at the points of purchase, formation of JV arrangements, procurement of the works and formulation of the Final Account (including apportionment of risk premium) all influencing the final value of the claim agreed with HM Revenue & Customs.

The Relief

Qualifying expenditure may be deducted at 150% of the actual amount expended in computing profits for the year in which it is incurred.

For example, a property trading company may buy contaminated land for redevelopment and incurs £250,000 on qualifying land remediation expenditure that is an allowable for tax purposes. It can claim an additional deduction of £125,000, making a total deduction of £375,000. Similarly, a company incurring qualifying capital expenditure on a fixed asset of the business is able to claim the same deduction provided it makes the relevant election within 2 years.

What is Remediation?

Land remediation is defined as the doing of works including preparatory activities such as condition surveys, to the land in question, any controlled waters affected by the land, or adjoining or adjacent land for the purpose of:

- Preventing or minimizing, or remedying or mitigating the effects of, any relevant harm, or any pollution of controlled waters, by reason of which the land is in a contaminated state.

Definitions

Contaminated land is defined as land that, because of substances on or under it, is in such a condition that relevant harm is or has the significant possibility of relevant harm being caused to:

- The health of living organisms
- Ecological systems
- Quality of controlled waters
- Property

Relevant harm is defined as meaning:

- death of living organisms or significant injury or damage to living organisms,
- significant pollution of controlled waters,
- a significant adverse impact on the ecosystem, or
- structural or other significant damage to buildings or other structures or interference with buildings or other structures that significantly compromises their use

Land includes buildings on the land, and expenditure on asbestos removal is expected to qualify for this tax relief. It should be noted that the definition is not the same as that used in the Environmental Protection Act Part IIA.

Sites with a nuclear licence are specifically excluded.

Conditions

To be entitled to claim LRR, the general conditions for all sites, which must all be met, are:

- Must be a company (non-resident companies can also claim from April 2020)
- Must be land in the United Kingdom
- Must acquire an interest in the land
- Must not be the polluter or have a relevant connection to the polluter
- Must not be in receipt of a subsidy
- Must not also qualify for Capital Allowances (particular to capital expenditure only)

Additional conditions introduced since 1 April 2009:

- The interest in land must be major – freehold or leasehold longer than 7 years
- Must not be obligated to carry out remediation under a statutory notice

Additional conditions particular to derelict land:

- Must not be in or have been in productive use at any time since at least 1 April 1998
- Must not be able to be in productive use without the removal of buildings or other structures

In order for expenditure to become qualifying, it must relate to substances present at the point of acquisition.

Furthermore, it must be demonstrated that the expenditure would not have been incurred had those substances not been present.

Contact

For further information or assistance, please contact one of the Fiscal Incentives team:

help.landremediationrelief@aecom.com

PART 2

On Costs and Profit

On Costs and Profit

In Part 1 of this book, it is stressed that the cost information given in Parts 4 to 7 leads to a cost estimate that requires further adjustment before it is submitted as a tender. This part deals with those adjustments and includes a worked example of a Tender Summary.

RISK/OPPORTUNITY

The factors to be taken into account when gauging the possibility of the Estimator's prediction of cost being inadequate or excessive are given in Part 2. Clearly it is considered in parallel with profit and it is not possible to give any indicative guidance on the level of adjustment that might result. For the purpose of a preliminary estimate, it is suggested that no adjustment is made to the costs generated by the other parts of this book.

At the same time as making a general appraisal of risk/opportunity, management will look at major quantities and may suggest amendments to the unit rates proposed.

HEAD OFFICE OVERHEADS

An addition needs to be made to the net estimate to cover all costs incurred in operating the central services provided by head office. Apart from general management and accountancy, this will normally include the departments dealing with:

- Estimating and commercial management
- Planning and design
- Purchasing
- Surveying
- Insurance
- Wages and bonus
- Site safety

The appropriate addition varies with the extent of services provided centrally, rather than on site, and with size of organization, but a range of 5% to 10% on turnover would cover most circumstances.

Some organizations would include finance costs with head office overheads, as a general charge to the company, but for the purposes of this book finance costs are treated separately (see below).

PROFIT

Obviously, the level of profit is governed by the degree of competition applicable to the job – which is in turn a function of the industry's current workload. Again, the appropriate addition is highly variable, but for the purposes of a preliminary estimate an addition of 5% to 15% onto net turnover is suggested.

FINANCE COSTS – ASSESSMENT OF CONTRACT INVESTMENT

The following procedure may be followed to give an indication of the average amount of capital investment required to finance the contract. It must be emphasized that this method will not give an accurate investment as this can only be done by preparing a detailed cash flow related to the programme for the contract. The example is based on the same theoretical contract used for the worked example in Part 2 and should be read in conjunction with the Tender Summary that follows.

The average monthly income must first be assessed. This is done by deducting from the Tender total the contingency items and those items for which immediate payment is necessary.

	£	£
Tender total (excluding finance charges)		10,396,313
Deduct		
Sub-contractors	2,000,000	
Prime cost sums	100,000	
Employer's contingencies	245,500	2,345,500
Amount to be financed	£	8,050,813

The average monthly income is this sum (£8,050,813) divided by the contract period (12 months), which is £670,901.

The average contract investment may now be calculated as follows:

	£	£
Plant and equipment to be purchased		90,000
Non time related		
Contractor	240,000	
Employer	8,000	
Other services, charges and fees	NIL	
Subtotal £	248,000	
Take 50% as an average [1]		124,000
Stores and unfixed materials on site		20,000
Work done but not paid for		
2½ months at £670,901 (see table above) [2]	1,677,253	
Less retention at 5% [3]	-83,863	1,593,390
Retention (5% with limit of 3%)		
Average retention [4] 3% of £ 8,050,813 (see table above)		241,524
Subtotal £		2,068,914
Deduct		
Advance payment by client	NIL	

	£	£
Plant and equipment to be purchased		90,000
Non time related		
Contractor	240,000	
Employer	8,000	
Other services, charges and fees	NIL	
Subtotal £	248,000	
Take 50% as an average [1]		124,000
Stores and unfixed materials on site		20,000
Work done but not paid for		
2½ months at £670,901 (see table above) [2]	1,677,253	
Less retention at 5% [3]	-83,863	1,593,390
Retention (5% with limit of 3%)		
Average retention [4] 3% of £ 8,050,813 (see table above)		241,524
Subtotal £		2,068,914
Deduct		
Advance payment by client	NIL	
Bill loading [5]	180,000	
Creditors (suppliers)	500,000	
Average contract investment	£	1,388,914

The interest charges that must be added to the Tender price (or absorbed from profit if capital needs to be borrowed) are therefore:

£ 1,388,914 × say 3.5% [6] × 1 year = say – £ 48,612

Notes

1. These non-times related oncosts and services are incurred as lump sums during the contract and, therefore, only 50% of such costs are taken for investment purposes.
2. This period depends on the terms of payment set out in the contract.
3. Retention is deducted as full retention is taken into account later.
4. Average retention will depend on the retention condition set out in the contract, taking into account any partial completion dates.
5. The contractor assesses here any financial advantage he may obtain by varying his items.
6. Assumed Rate – current market varies greatly.

VALUE ADDED TAX

All of the figures quoted in this book exclude value added tax, which the conditions of contract normally make the subject of a separate invoicing procedure between the contractor and the employer.

Value Added Tax will be chargeable at the standard rate, currently 20%, on supplies of services in the course of:

7. The construction of a non-domestic building
8. The construction or demolition of a civil engineering work
9. The demolition of any building, and
10. The approved alteration of a non-domestic protected building

PART 3

Cost Indices and Location Factors

Cost Indices and Location Factors

The purpose of this part is to present historic changes in civil engineering costs and tender prices. It gives published and constructed indices and diagrammatic comparisons between building and civil engineering costs and tender prices and the retail price index and will provide a basis for updating historical cost or tender price information.

INTRODUCTION

It is important to distinguish between costs and tender prices. Civil engineering costs are the costs incurred by a contractor in the course of their business. Civil engineering tender prices are the prices for which a contractor undertakes work, i.e. cost to client. Tender prices will be based on contractors' costs but will also take into account market considerations such as the availability of labour and materials and the prevailing workload for civil engineering contractors. This can mean that in a period when work is scarce tender prices may fall as costs are rising, while when there is plenty of work prices will tend to increase at a faster rate than costs. This section comprises published civil engineering cost indices, road construction cost indices and a constructed Civil Engineering Cost Index and comparisons of these with building cost and tender indices and with the retail price index.

A CONSTRUCTED COST INDEX

The table below shows relative cost movement for contractors since 2020.

Constructed Civil Engineering Cost Index base: 2015 = 100

Year	First quarter	Second quarter	Third quarter	Fourth quarter	Annual Average	Annual Change
2020	110.6	108.0	109.1	110.3	109.5	−1.6%
2021	113.1 (P)	113.9 (F)	114.6	115.4	114.2	4.3%
2022	116.1	116.9	117.6	118.3	117.2	2.6%
2023	119.1	119.8	120.6	121.3	120.2	2.6%

Source: AECOM

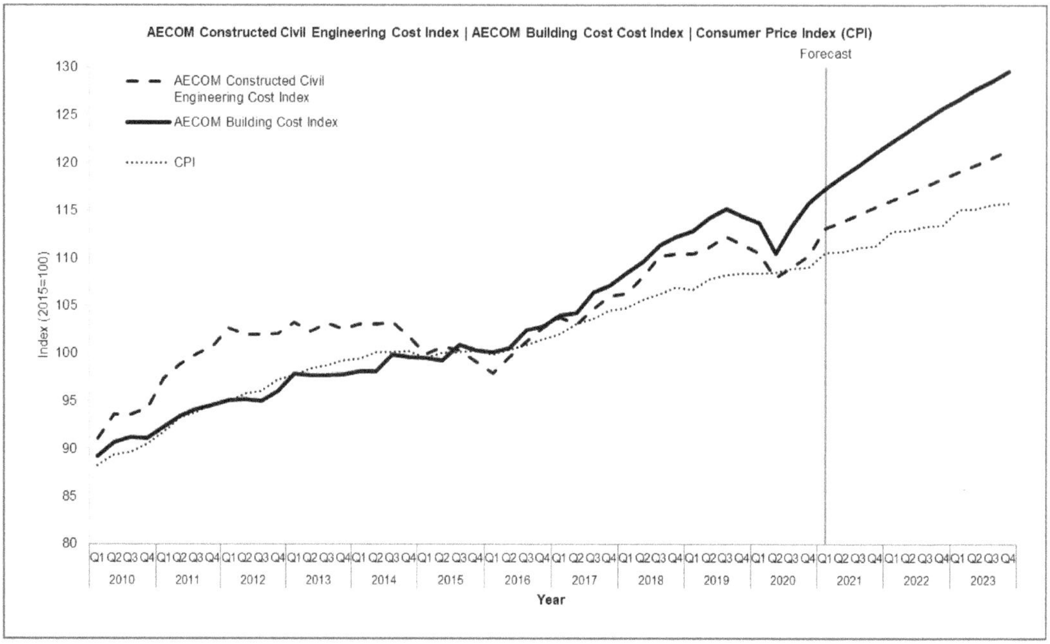

The chart illustrates the relative movement of the Constructed Civil Engineering Cost Index, AECOM Building Cost Index and the Consumer Price Index since 2010.

THE ROAD CONSTRUCTION COST INDEX

Civil Engineering work generally does not lend itself easily to the preparation of tender price indices in the same way as building work. The index is intended to indicate the movement in cost prices for road construction contracts. It is based on priced rates contained in accepted tenders for Road Construction, Motorway Widening and Major Maintenance Schemes.

Road Construction Cost Index – Base: 2015 = 100

Year	First quarter	Second quarter	Third quarter	Fourth quarter	Annual Average	Annual Change
2019	110.3	110.7	111.5	110.3	110.7	2.6%
2020	110.7	109.0	109.4	109.4	109.6	−0.9%

Source: BEIS/ONS

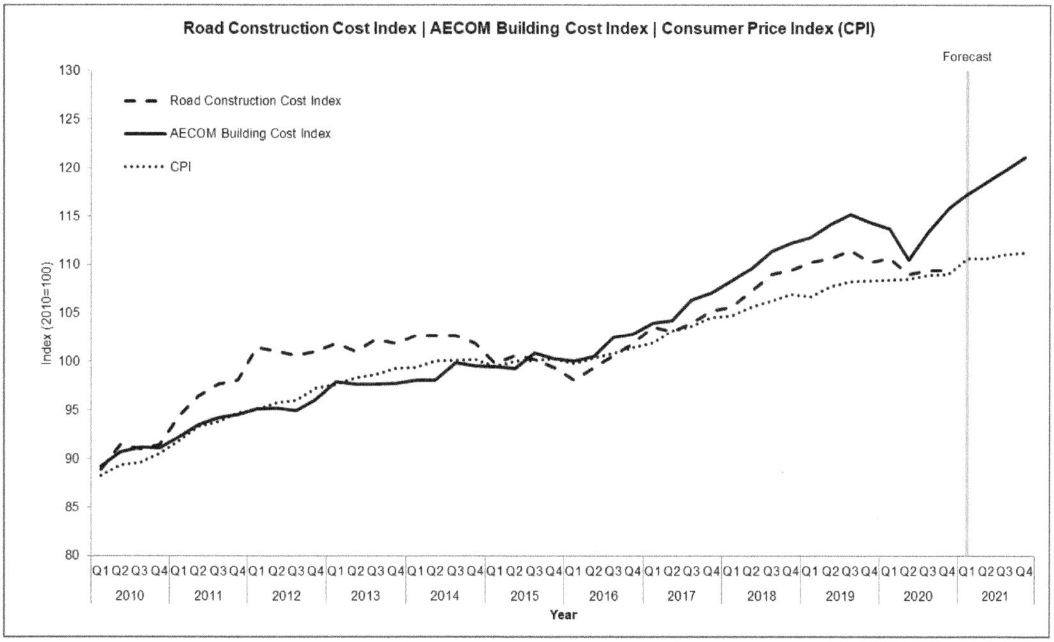

The chart illustrates the relative movement of the Road Construction Cost Index; the AECOM Building Cost Index and the Consumer Price Index since 2010.

Over time, price differentials can change depending on regional workloads and local hot spots. *Spon's Updates* together with the *Market Forecast* feature in *Building* magazine will keep readers informed of the latest developments and changes as they occur.

REGIONAL VARIATIONS

As well as being aware of inflationary trends when preparing an estimate, it is also important to establish the appropriate price level for the project location.

The regional variation factors shown in the following table are based on our forecast of price differentials in each of the economic planning regions at the time the book was published. Historically the adjustments do not vary very much year-on-year; however, there can be wider variations within any region, particularly between urban and rural locations. The following table shows suggested adjustments required to cost plans or estimates prepared using *Spon's Civil Engineering and Highway Works Price Book 2022*. Please note: these adjustments need to be applied to the total cost plan value, not to individual element items.

Region	Factor
Outer London	1.00
Inner London	1.05
South East	0.96
East of England	0.93
South West	0.90
North West	0.88
East Midlands	0.88
Scotland	0.88
West Midlands	0.87
Yorkshire & Humberside	0.86
Wales	0.88
North East	0.83
Northern Ireland	0.76

PART 4
Resources

This part comprises sections on labour, materials and plant for civil engineering work. These resources form the basis of the unit costs in Parts 6 and 7 and are given so that users of the book may:

- Calculate rates for work similar to, but differing in detail from, the unit costs given
- Compare the costs given here with those used in their own organization
- Calculate the effects of changes in wage rates, material prices, etc.

Adjustments should be made to the rates shown to allow for time, location, local conditions, site constraints and any other factors likely to affect the cost of the specific scheme.

BASIS OF THIS SECTION

CIJC Basic Rates of Pay

Copies of the Working Rule Agreement may be obtained from:

Construction Industry Joint Council,
55 Tufton Street,
London,
SW1P 3QL

Rates of Pay (Rule WR.1)

The Construction Industry Joint Council (CIJC) has completed negotiations with trade union GMB and Unite the Union and reached an agreement on an increase in all rates of 2.5%, which took effect on Monday 28 June 2021.

The following basic rates of pay are effective from 28 June 2021	Rate per hour (£)
Craft Rate	12.99
Skill Rate 1	12.36
Skill Rate 2	11.91
Skill Rate 3	11.14
Skill Rate 4	10.53
General operative	9.77

The following table shows the hourly labour rates used for Spon's Civil Engineering and Railway Rates 2023. This year's rates are calculated on the rounded value of a 40 man working team and from historical civil engineering and railway projects for the past year and therefore typical average rates currently used. These rates are higher than the normal labour rates due to a number of logistic and safety factors, highly skilled requirements and control standards. The above rates however have been used on standard and general activities.

Labourer/General Operative	£22.80
Skill Rate 4	£26.30
Skill Rate 3	£29.17
Skill Rate 2	£35.46
Skill Rate 1	£49.49
Skilled Operative/UG	£50.46

Bonus (Rule WR.2)

The Working Rule leaves it open to employers and employees to agree a bonus scheme based on measured output and productivity for any operation or operations on a job.

Working Hours (Rule WR.3)

Normal working hours are unchanged at 39 hour week – Monday to Thursday at 8 hours per day and Friday at 7 hours. The working hours for operatives working shifts are 8 hours per weekday, 40 hours per week.

Rest/Meal Breaks (Rule WR.3.1)

Times fixed by the employer, not to exceed 1 hour per day in aggregate, including a meal break of not less than half an hour.

Overtime Rates (Rule WR.4)

During the period of Monday to Friday, the first 4 hours after normal working day is paid at time and a half, after 4 hours double time shall be paid. Saturday is paid at time and a half until completion of the first four hours. Remainder of Saturday and all Sunday paid at double time.

Daily Fare and Travelling Allowances (Rule WR.5)

These have increased by 3%. This applies only to distances from home to the site of between 10 m and 46 m. Again Employers have individual agreements with typical ranges of £1.05 to £9.05 for Travelling Allowance (taxed) and £4.43 to £17.20 for Fare Allowance (not taxed). For the one way distance of 10 m used in the following example we have used a rate of £4.78.

Rotary Shift Working (Rule WR.6)

This relates to the situation where more than one shift of minimum 8 hours is worked on a job in a 24 hour period, and the operative rotates between the shifts either in the same or different pay weeks.

The basic rate shall be the operative's normal hourly rate plus 14%. Overtime beyond the 8 hour shift shall be at time and a half for the first 4 hours at normal rate plus 14% thereafter double normal rate.

Night Work (Rule WR.7)

Providing night work is carried out by a separate gang from those working during daytime, an addition of 25% shall be paid on top of the normal hourly rate.

Overtime payments: during the period of Monday to Friday, the first 4 hours after normal working day is paid at time and a half plus the 25% addition of normal hourly rate, after 4 hours double time shall be paid. All hours worked on Saturday and all Sunday paid at double time.

Tide Work (Rule WR.9)

Where the operative is also employed on other work during the day, additional time beyond the normal working day shall be paid in accordance with the rules for overtime payments.

Where the operative is solely involved in work governed by tidal conditions, he shall be paid a minimum of 6 hours at normal rates for each tide. Payment for hours worked in excess of 8 over two tides shall be calculated proportionately.

$$\frac{(\text{Total Hours Worked} - 8\ \text{hours}) \times \text{Total Hours Worked}}{8\ \text{hours}}$$

Work done after 4pm Saturday and during Sunday shall be at double time. Operatives are guaranteed 8 hours at ordinary rate for any time worked between 4pm and midnight on Saturdays and 16 hours for two tides worked on a Sunday.

Tunnel Work (Rule WR.10)

The first part of a shift equivalent to the length of the normal working day shall be paid at the appropriate normal rate, the first four hours thereafter at time and a half and thereafter at double time. In the case of shifts on a Saturday, the first 4 hours are at time and a half, thereafter at double time. All shifts on a Sunday at double time.

Saturday shifts extending into Sunday are at double time. Sunday shifts extending into Monday, time after midnight: first 4 hours at time and a half, thereafter at double time.

Annual Holidays Allowance (Rules WR.18 and 21)

Annual and Public Holiday Pay is included in accordance with the Building & Civil Engineering Benefit Schemes (B&CE) Template Scheme. Allowances are calculated on total weekly earnings inclusive of overtime and bonus payments and the labour cost calculation assumes 22 days annual and 8 days public holidays.

Easy Build Pension Contributions (Rule WR 21.3)

The minimum employer contribution is £7.50 per week. Where the operative contributes between £5.01 and £10.00 per week the employer will increase the minimum contribution to match that of the operative up to a maximum of £10.00 per week. The calculations below assume 1 in 10 employees make contributions of £10.00 per week.

LABOUR COSTS

CALCULATION OF LABOUR COSTS (This is just an example.)

This section sets out a method used in calculating all-in labour rates used within this book. The calculations are based on the wage rates, plus rates and other conditions of the Working Rule Agreement (WRA); important points are discussed below. The calculations can be used as a model to enable the user to make adjustments to suit specific job conditions in respect of plus rate, bonus, subsistence and overhead allowances together with working hours etc. to produce alternative all-in rates which can be substituted for those printed.

All-in labour costs are calculated on the pages following for six categories of labour reflecting the different classifications set out in the WRA.

AVERAGE WORKING WEEK has been calculated to an equal balance between winter and summer (46.20 working weeks per year).

SUBSISTENCE PAYMENTS are included for key men only: for a stated percentage of the workforce.

TRAVELLING ALLOWANCES are based on rate payable for a journey of 10 m at £4.78 per day. These allowances are adjusted to cater for operatives receiving Subsistence Payments.

BONUS PAYMENTS reflect current average payments in the industry.

WORKING HOURS & NON-PRODUCTIVE OVERTIME are calculated thus:

SUMMER (hours worked 8.00 am–6.00pm, half hour for lunch)

							Total	Deductions		Effective
	Mon	Tue	Wed	Thu	Fri	Sat	Paid Hours	Lost & Wet time	Paid Breaks	Total Hours
Total hrs	9.50	9.50	9.50	9.50	9.50	4.00	51.50	0.50	1.00	**50.00**
Overtime	0.75	0.75	0.75	0.75	1.25	2.00	6.25			
TOTAL PAID:							**57.75**	EFFECTIVE		**50.00**

WINTER (hours worked 7.30 am–4.30pm, half hour for lunch)

							Total	Deductions:		Effective
	Mon	Tue	Wed	Thu	Fri	Sat	Paid Hours	Lost & Wet time	Paid Breaks	Total Hours
Total hrs	8.50	8.50	8.50	8.50	7.50	4.00	45.50	1.50	-	**44.00**
Overtime	0.25	0.25	0.25	0.25	0.25	2.00	3.25	-	-	-
TOTAL PAID							**48.75**	EFFECTIVE		**44.00**

AVERAGE TOTAL HOURS PAID	(57.75 + 48.75) / 2	= 53.25 hours per week
AVERAGE EFFECTIVE HOURS WORKED	(50.00 + 44.00) / 2	= 47.00 hours per week

Payment details (assumed effective from June 2021)	general operative (£)	skill rate 4 (£)	skill rate 3 (£)	skill rate 2 (£)	skill rate 1 (£)	craft rate (£)
Basic rate per hour	9.79	10.55	11.17	11.94	12.39	13.02
53.25 Hours Paid @ Total Rate	521.43	561.92	594.75	635.78	659.86	693.23
Weekly Bonus Allowance	38.38	38.38	76.54	64.47	115.41	153.97
TOTAL WEEKLY EARNINGS	**559.81**	**600.30**	**671.29**	**700.25**	**775.26**	**847.20**
Travelling Allowances (say, 10 m per day, i.e. £4.78 × 6 days)						
General, skill rate 4, skill rate 3 (100% of time)	28.67	28.67	28.67	-	-	-
Skill rate 2, skill rate 1 (80% of time)	-	-	-	22.93	22.93	-
craft rate (905 of time)	-	-	-	-	-	25.80
Subsistence allowance £39.68 (Average 6.66 nights + 7.00% to cover periodic travel)						
General, skill rate 4, skill rate 3 (0% allowance)						
Skill rate 2, skill rate 1 (20% of time)				44.12	44.12	
Craft rate (10% of time)						22.06
TOTAL WAGES	**588.47**	**628.96**	**699.96**	**767.30**	**842.32**	**895.06**
National Insurance contributions – (13.8% above Earnings Threshold except Fares)	49.29	54.88	64.68	68.68	79.03	88.95
Annual Holiday Allowance – 4.4 weeks @ Total Weekly Earnings over 46 weeks	56.04	59.90	66.66	73.08	80.22	85.24
Public Holidays with Pay – 1.6 weeks @ Total Weekly Earnings over 46.2 weeks	22.93	24.51	27.27	29.89	32.82	34.87
Easy Build pension contribution	7.50	7.50	7.50	7.50	7.50	7.50
CITB Levy (0.50% Wage Bill)	3.34	3.57	3.97	4.35	4.78	5.08
Allowance for Employer's liability & third-party insurances, safety officer's time	**719.01**	**770.12**	**859.74**	**940.05**	**1034.75**	**1103.67**
QA policy/inspection and all other costs & overheads:	18.19	19.48	21.75	23.77	26.17	27.92
TOTAL WEEKLY COST (47 HOURS)	**745.77**	**798.80**	**891.79**	**974.57**	**1072.83**	**1144.62**
COST PER HOUR	**15.87**	**17.00**	**18.97**	**20.74**	**22.83**	**24.35**
Plant Operators rates: Addition to Cost per Hour for Rule WR.11 – plant servicing time (6 hrs)	1.47	1.58	1.67	1.79	1.85	1.95
COST PER HOUR (include plant)	**17.33**	**18.57**	**20.65**	**22.52**	**24.68**	**26.30**

LABOUR CATEGORIES

Schedule 1 to the WRA lists specified work establishing entitlement to the Skilled Operative Pay Rate 4, 3, 2, 1 or the Craft Rate as follows:

General Operative

Unskilled general labour

Skilled Operative Rate 4

supervision	Gangers + trade charge hands
plant	Contractors plant mechanics mate; Greaser
transport	Dumper <7t; Agricultural Tractor (towing use); Road going motor vehicle <10t; Loco driver
scaffolding	Trainee scaffolder
drilling	Attendee on drilling
explosives	Attendee on shot firer
piling	General skilled piling operative
tunnels	Tunnel Miner's mates; Operatives driving headings over 2 m in length from the entrance (in with drain, cable & mains laying)
excavation	Banksmen for crane/hoist/derrick; Attendee at loading/tipping; Drag shovel; Trenching machine (multi bucket) <30hp; Power roller <4t; Timberman's attendee
coal	Opencast coal washeries and screening plants
tools	Compressor/generator operator; Power driven tools (breakers, tamping machines etc.); Power-driven pumps; air compressors 10 KW+; Power driven winches
concrete	Concrete leveller/vibrator operator/screeder + surface finisher, concrete placer; Mixer < 21/14 or 400 litres; Pumps/booms operator
linesmen	Linesmen-erector's mate
timber	Carpentry 1 st year trainee
pipes	Pipe layers preparing beds and laying pipes <300 mm dia.; Pipe jointers, stoneware or concrete pipes; pipe jointers flexible or lead joints <300 mm dia.
paving	Rolled asphalt, tar and/or bitumen surfacing: Mixing Platform Chargehand, Chipper or Tamperman; Paviors; Rammerman, kerb & paving jointer
dry lining	Trainee dry liners
cranes	Forklift truck <3t; Crane < 5t; Tower crane <2t; Mobile crane/hoists/fork-lifts <5t; Power driven hoist/crane

Skilled Operative Rate 3

transport	Road going motor vehicle 10t+
drilling	Drilling operator
explosives	Explosives/shot firer
piling	Piling ganger/chargehand; Pile frame winch driver
tunnels	Tunnel Miner (working at face/operating drifter type machine)
excavation	Tractors (wheeled/tracked with/without equipment) <100hp; Excavator <0.6 m^3 bucket; Trenching machine (multi bucket) 30–70hp; Dumper 7–16t; Power roller 4t+; Timberman
concrete	Mixer 400–1500 litres; Mobile concrete pump with/without concrete placing boom; Hydraulic jacks & other tensioning devices in post-tensioning and/or pre-stressing

formwork	Formwork carpenter 2nd year trainee
masonry	Face pitching or dry walling
linesmen	Linesmen-erectors 2nd grade
steelwork	Steelwork fixing simple work; Plate layer
pipes	Pipe jointers flexible or lead joints (300–535 mm dia.)
paving	Rolled asphalt, tar and/or bitumen surfacing: Raker, Power roller 4 t+, mechanical spreader operator/leveller
dry lining	Certified dry liners
cranes	Cranes with grabs fitted; Crane 3–10t; Mobile crane/hoist/fork-lift 4–10t; Overhead/gantry crane <10t; Power-driven derrick <20t;Tower crane 2–10t; Forklift 3t+

Skilled Operative Rate 2

plant	Maintenance mechanic; Tyre fitter on heavy earthmover
transport	Dumper 16–60t
scaffolding	Scaffolder < 2 years scaffolding experience and < 1 year as Basic Scaffolder
piling	Rotary or specialist mobile piling rig driver
tunnels	Face tunnelling machine
excavation	Tractors (wheeled/tracked with/without equipment) 100–400hp; Excavator 0.6–3.85 m^3 bucket; Trenching machine (multi bucket) 70hp +; Motorized scraper; Motor grader
concrete	Mixer 1500 litres+; Mixer, mobile self-loading and batching <2500 litres
linesmen	Linesmen-erectors 1 st grade
welding	Gas or electric arc welder up to normal standards
pipes	Pipe jointers flexible or lead joints over 535 mm dia.
cranes	Mobile cranes/hoists/fork-lifts 5–10t; Power-driven derrick 20t+, with grab <20t; Tower crane 10–20t

Skilled Operative Rate 1

plant	Contractors plant mechanic
transport	LGV driver; Lorry driver Class C+E licence; Dumper 60–125t
excavation	Excavator 3.85–7.65 m^3; Tractors (wheeled/tracked with/without equipment) 400–650hp
concrete	
linesmen	
steelwork	Steelwork assembly, erection and fixing of steel framed construction
welding	Electric arc welder up to highest standards for structural fabrication & simple pressure vessels (air receivers including CO_2 processes)
drilling	Drilling rig operator
cranes	Power-driven derrick with grab 20t +; Tower crane 20t+

Craft Operative

transport	Dumper 125t+
scaffolding	Scaffolder a least 2 years scaffolding experience and at least 1 year as Basic and Advanced Scaffolder
excavation	Excavator 7.65t + (see WR 1.2.2 above); Tractors (wheeled/tracked with/without equipment) 650hp+
concrete	Reinforcement bender & fixer
formwork	Formwork carpenter
welding	Electric arc welder capable of all welding processes on all weldable materials including working on own initiative from drawings
cranes	Crane 10t+; Mobile cranes/hoists/fork-lifts 10t +

PLANT COSTS

Introduction

The information, rates and prices included in this section are calculated examples of actual owning and operating costs of a range of construction plant and equipment. To an extent they will serve as a guide to prevailing commercial plant hire rates, but be aware that many factors will influence actual hire rates. For example, rates could be lower because of long term hire, use of older or second hand machines, low market demands/loss leaders, easy tasks or sites near to the depot. Rates could be higher due to short term hire, high production demands, low utilization factors, specialist operations, restricted working and restricted accessibility problems, high profit or overhead demands and finance fluctuations. The use factors will mean a hired machine may not be used productively for each hour on site, it is therefore up to the estimator to allow for reasonable outputs in his unit rate calculations.

These and other considerations must be borne in mind if using these costs as a comparison to hire rates, especially operated plant.

All rates quoted EXCLUDE the following:

- Cost of drivers/operators
- VAT

Contractor Owned Plant

Plant owned by a Contractor generally falls under two headings:

- Small plant and tools which are the subject of a direct charge to contracts and for estimating purposes are normally allowed for as a percentage of the labour cost in site oncosts (see Preliminaries and General Items), although many items are shown in this section for information.
- Power driven plant and major non-mechanical plant such as steel trestling, scaffolding, gantries etc. Such plant is normally charged to the contract on a rental basis except in the case of purpose-made or special plant bought specifically for a particular operation; this latter is normally charged in full to contracts and allowance made for disposal on completion (often at scrap value).

A very wide range of plant is readily available from plant hire companies; it is not usually economical for Contractors to own plant unless they can ensure at least 75 to 80% utilization factor based on the Contractor's normal working hours. Where a Contractor does own plant, however, it is essential that he maintains reasonably accurate records of the working hours and detailed costs of maintenance and repairs in order that he may estimate the charges to be made for each item of plant. For this reason, where a Contractor owns a large quantity of plant, it is normal for a separate plant hire department or company to be formed.

This department or company should be financially self-supporting and may hire plant to other Contractors when it is not needed. Maintenance of contractor owned plant can be carried out on site; it is not necessary to maintain a centrally based repair workshop. It is, however, desirable to have some storage facilities available for plant when not in use. The cost of owning plant and hence the rental charges to be made to contracts must take into account:

- Capital cost
- Depreciation charges
- Maintenance and repairs
- Cost of finance
- Insurances and licences
- Administration, head office, depot and other overhead charges

Calculated examples of hourly owning costs of a range of plant and equipment.

PLANT ITEMS	A UTILIZATION FACTOR %	B P.A. HOURS	C PLANT	D LIFE Hours	E PURCHASE PRICE £ *	F RESALE PRICE %	G RESALE PRICE	H LOSS IN VALUE	I DEPREC.	J MAINTEN.	K FINANCE @ 3.5%	L INSURANCE £/HR	M ADMIN £/HR	N TOTAL COST £/HR
ACCESS PLATFORMS														
Boom Lift, 230 kg load 15.8 m height	65	1,365	10	13,650	39,375	20	7,875	31,500	2.31	2.81	1.01	0.58	1.34	8.04
Boom Lift, 450 kg load 18.6 m height	65	1,365	10	13,650	61,950	20	12,390	49,560	3.63	4.41	1.59	0.86	2.10	12.58
ASPHALT PAVERS														
Wheeled paver, 6.4 m wide, 79.2 m/min	65	1,365	7	9,555	84,525	27.50	23,244	61,281	6.41	6.02	2.17	1.18	3.16	18.93
Wheeled paver, 7.9 m wide, 90.5 m/min	65	1,365	7	9,555	135,450	27.50	37,249	98,201	10.28	9.64	3.47	1.89	5.06	30.34
COMPRESSORS (PORTABLE)														
7.3 m^3/min, electric motor	85	1,785	9	16,065	13,125	20	2,625	10,500	0.65	0.71	0.26	0.14	0.35	2.12
17 m^3/min, potable	85	1,785	9	16,065	52,500	20	10,500	42,000	2.61	2.86	1.03	0.56	1.41	8.47
CRANES														
16 t mobile	60	1,260	12	15,120	203,175	25	50,794	152,381	10.08	15.67	5.64	3.07	6.89	41.35
35 t mobile	60	1,260	15	18,900	225,750	27.50	62,081	163,669	8.66	17.40	6.27	3.41	7.15	42.89
GENERATORS														
10 kVA, water cooled	85	1,785	5	8,925	6,825	15	1,024	5,801	0.65	0.37	0.13	0.07	0.24	1.47
10 kVA, air cooled	85	1,785	5	8,925	5,775	15	866	4,909	0.55	0.32	0.11	0.06	0.21	1.25
EXCAVATORS														
Tracked backhoe 0.7 m^3 max bucket	80	1,680	9	15,120	90,300	27.50	24,833	65,468	4.33	5.22	1.88	1.02	2.49	14.94
Tracked backhoe 4.5 m^3 max bucket	75	1,575	9	14,175	425,250	27.50	116,944	308,306	21.75	26.22	9.45	5.14	12.51	75.08
JCB 3CX	85	1,785	6	10,710	54,075	27.50	14,871	39,204	3.66	2.95	1.06	0.58	1.65	9.90

PLANT ITEMS	A UTILIZATION FACTOR %	B P.A. HOURS	C PLANT	D LIFE Hours	E PURCHASE PRICE £ *	F RESALE PRICE %	G RESALE PRICE	H LOSS IN VALUE	I DEPREC.	J MAINTEN.	K FINANCE @ 3.5%	L INSURANCE £/HR	M ADMIN £/HR	N TOTAL COST £/HR
ROLLERS														
Double Drum 757 kg. 650 mm wide	85	1,785	6	10,710	17062.50	27.50	4,692	12,370	1.16	0.93	0.33	0.18	0.52	3.12
Double Drum 1040 kg. 750 mm wide	85	1,785	6	10,710	22,575	27.50	6,208	16,367	1.53	1.22	0.44	0.24	0.69	4.12
Aticulated tandem 1.6 t. 800 mm wide	85	1,785	6	10,710	34,125	27.50	9,384	24,741	2.31	1.86	0.67	0.36	1.04	6.23
CRAWLER LOADERS														
20 t weight, 2.0 m^3 bucket	80	1,680	9	15,120	157,500	27.50	43,313	114,188	7.55	9.11	3.28	1.79	4.35	26.08
TIPPERS														
10 t capacity	90	1,890	6	11,340	50,925	22	11,204	39,722	3.50	2.62	0.94	0.51	1.52	9.09
20 t capacity	90	1,890	6	11,340	79012.50	22	17,383	61,630	5.43	4.06	1.46	0.80	2.35	14.11
DUMP TRUCKS														
Articulated dump 39 t, 24 m^3 capacity	75	1,575	10	15,750	330,750	30	99,225	231,525	14.70	20.40	7.35	4	9.29	55.74

NOTES
* The purchase price is illustrative only and will require a quotation from the relevant manufacturer.
* No allowance is included for Road Fund Licence on road-going vehicles.

Notes – example plant ownership hourly costs

The above costs would be updated annually or bi-annually, the rental rate being revised to ensure complete recovery of the costs associated with the item of plant and return of capital to enable the machine to be replaced at the end of its life with the Contractor. The purchase price must also be adjusted to ensure recovery of the replacement cost and not the original cost of purchase.

Driver's wages and costs should be charged direct to site wages.

Column A: The utilization factors used in conjunction with the period of ownership is the percentage of time (per annum) that an item of plant can be expected to be used productively on a site or job and therefore is a very important influence of hourly costs

(Note: Utilization factors are not the same as site utilization rates – see notes regarding Fuel Consumption)

Economical owning periods/use factors:	Life (years)	Usage (%)
Hydraulic Excavators, large	9	75
Hydraulic Excavators, medium	7	76
Hydraulic Excavators, mini	6	77
Dozers/Scrapers	10	78
Loaders/Shovels	9	79
Mobile Cranes	12	80
Crawler Cranes	15	81
Dump Trucks	8	82
Dumpers	6	83
Rollers/Compaction	6	84
Compressors/Generators	7–9	85
Diesel engine road vehicles	6–7	86
Petrol engine road vehicles	5–6	87

- Column B: The result of applying the utilization factor to 2,100 hours per annum.
- Columns C and D: The period over which the Contractor owns the plant and during which time the items of plant can be maintained at reasonable efficiency will vary according to the Contractor's experience, the type of plant under consideration and the work on which is it employed. The owning period of an item of plant can vary considerably on this basis. The data we have used in our calculations varies considerably (see the tables shown above and on the next page).
- Column E: The list price being quoted by dealers, against which should be allowed a level of discount suited to the availability of the plant and the bargaining power between the purchaser and the agent.
- Column J: Maintenance costs cover major overhauls and replacement costs for wear items only, excluding insurable damage and will vary with each type of plant. This does not include day to day general servicing on site.
- Column K: Finance interest charges are taken on the average base rate over the previous 12 months plus 3%.
- Column L: Insurance premiums have been taken as 2% on the replacement value of the plant (purchase price).
- Column M: Administration, head office, depot and other overhead charges have been taken at +20%.
- Consumables should be charged direct to site costs. The costs included with this section are based upon manufacturers' data and are used in good faith. (For outline estimates fuel consumption can be taken as approximately 0.15–0.20 litres per kW per working hour; lube oils, filters, grease etc., can be taken as 2% to 10% of the fuel cost, depending upon the working conditions of the machine and its attachments). See below for details.

Please note that the costs in this section have been reviewed for recent editions, based on the latest available advice and data from manufacturers and Plant Hire firms, the basis of calculation, including retention period of the plant, has been changed in a number of cases, providing a realistic overall owning and operating cost of the machine.

These examples are for machine costs only and do not include for operator costs or profit element or for transport costs to and from site.

Fuel consumption is based on the following site utilization rates. (Site utilization being the percentage of time that the machine is operating at its average fuel consumption during a working day).

Typical utilization factors:	Utilization factor	Typical utilization factors:	Utilization factor
Excavators	0.75	Compressors, mixers, generators	
Dozers/Scrapers	0.80	Tractors	0.75
Loaders/Shovels/Graders	0.75	Hoists, etc.	0.65
Mobile Cranes	0.50	Drills and Saws	0.90
Crawler Cranes	0.25	Rollers/Compaction Plant	0.75
Dump Trucks (not Tippers)	0.75	Piling/Asphalt Equipment	0.65
Dumpers	0.80	All other items	1.00

Approximate fuel consumption in litres per hour and percentage addition for consumables (examples only) based on the above site utilization rates, assuming that reasonably new and well maintained plant would be used.

		Fuel	consumables			Fuel	consumables
		l/hr	% of fuel			l/hr	% of fuel
Tractor	116 hp	9.0	2.0%	Rollers	BW 90 AD	2.0	7.5%
Paver	Bitelli BB650	16.0	5.0%		BW 120 AD	4.4	7.5%
Compactor plate	188 kg	0.9	3.0%		BW6 (towed)	7.6	7.5%
Compressor	3.5 m³/min	5.9	3.0%	Dozers	CAT D6N	21.0	6.0%
	10 m³/min	18.0	3.0%		CAT D8T	38.0	3.0%
	Tractor mounted	9.0	5.0%	Loaders	CAT 953C	18.0	4.0%
Mixer	5/3.5	1.2	2.0%	Scrapers	CAT 621G	41.5	3.0%
Mobile crane	15t	5.1	10.0%		CAT 657E	125.0	6.0%
Mobile crane	40t	6.3	10.0%	Skidsteer	Bobcat 553	3.4	5.0%
Generator	4 kVA	1.6	3.0%	Graders	CAT 140H	19.0	5.0%
	25 kVA	4.9	5.0%	Dumpers	2 tonne	3.0	3.0%
Backacter	11.5 t	9.5	7.5%		Volvo A25D	17.0	3.0%
	19 t	14.0	7.5%		36 tonne	25.0	5.0%
	30 t	24.0	7.5%	Pumps	75 mm 65 m³/hr	1.1	2.0%
Backhoes	JCB 3CX	7.5	6.0%		150 mm 360 m³/hr	4.2	3.0%

Note that the above consumption figures represent 'medium' plant operation, i.e. the plant would not be operating at full throttle and working conditions such as the soil and grades are average. A study of manufacturers' figures has indicated that 'high' or 'low' usage can affect consumption on average by ± 25%; 'high' usage would involve full loads, continuous working at full throttle with difficult ground and adverse grades, whereas 'low' usage would involve more intermittent working with perhaps considerable idling periods, more easily worked ground and easier grades

FUEL COSTS

Average fuel costs used in build-up of rates in this edition (May 2022 Base)

Petrol Unleaded	173.02	Pence per litre
Petrol Unleaded Super	184.80	Pence per litre
Fuel Oil, DERV	182.58	Pence per litre
Fuel Oil Gas Oil	179.99	Pence per litre
Lubrication Oil (15/40)	15653	Pence per 5 litres
Main Electric Power	28.38	Pence per kwh
Propane (47 kg Refillable Cylinder)	178.25	Pence per kg

Also included here are allowances for Transmission Oils, Hydraulic Oil, Filters, Grease, etc.

PART 5

Approximate Estimating

Preliminaries

This part deals with that portion of Civil Engineering costs not, or only indirectly, related to the actual quantity of work being carried out. It comprises a definition of Method Related Charges, a checklist of items to be accounted for on a typical Civil Engineering contract and a worked example illustrating how the various items on the checklist can be dealt with.

Approximate Estimating Rates – Ancillary Building Works and General Building Works

Prices given in this section, which is generally arranged in elemental order, also include for all the incidental items and labours which are normally measured separately in Bills of Quantities. They include overheads and profit but do not include for preliminaries.

Whilst every effort is made to ensure the accuracy of these figures, they have been prepared for approximate estimating purposes only and on no account should they be used for the preparation of tenders.

Unless otherwise described units denoted as m² refer to appropriate unit areas (rather than gross internal floor areas).

As elsewhere in this edition prices do not include Value Added Tax, which should be applied at the current rate, together with consultants' fees.

Preliminaries and General Items

GENERAL REQUIREMENTS AND METHOD RELATED CHARGES

This part deals with that portion of Civil Engineering cost not, or only indirectly, related to the actual quantity of work being carried out. It comprises a definition Related Charges, a checklist of items to be accounted for in a typical Civil Engineering contract and a worked example illustrating how the various items on the checklist can be dealt with.

The concept of METHOD RELATED CHARGES can be summarized as follows:

In commissioning Civil Engineering work, the Employer buys the materials left behind, but only hires from the Contractor the men and machines which manipulate them and the management skills to manipulate them effectively. It is logical to assess their values in the same terms as the origin of their costs. It is illogical not to do so if the Employer is to retain the right at any time to vary what is left behind and if the financial uncertainties affecting Employer and Contractors are to be minimized.

Tenderers have the option to define a group of bill items and insert charges against them to cover those expected costs which are not proportional to the quantities of Permanent Works. To distinguish these items they are called Method Related Charges. They are themselves divided into charges for recurrent or time related cost elements, such as maintaining site facilities or operating major plant, and charges for elements which are neither recurrent nor directly related to quantities, such as setting up, bringing plant to site and Temporary Works.

Another hope expressed with the introduction of Method Related Charges was that they should accurately reflect the work described in the item and that they should not, as had become the practice with some of the vague general items frequently included in Civil Engineering Bills, be used as a home for lump sum tender adjustments quite unrelated to the item. Where cost information is given in the worked example presented at the end of this part of the book, therefore, it must be stressed that only direct and relevant costs are quoted.

Where no detailed information is available, it is suggested that when preparing a preliminary estimate an addition of between 15% and 35% of net contract value is made to cover Contractor's Site Oncosts, both time and non time related. The higher percentages' should apply to specialist works such as railways, marine and tunnelling.

CHECKLIST OF ITEMS

The following checklist is representative but not exhaustive. It lists and describes the major preliminary and general items which are included, implicitly or explicitly, in a typical Civil Engineering contract and, where appropriate, gives an indication of how they might be costed. Generally contract documents give detailed requirements for the facilities and equipment to be provided for the Employer and for the Engineer's representative and Bills of Quantities produced in conformity with CESMM4 Class A provide items against which these may be priced; no such items are provided for Contractor's site oncosts or, usually, for temporary works and general purpose plant. For completeness a checklist of both types of item is given here under the following main headings:

PRELIMINARIES AND GENERAL ITEMS

Contractor's site oncosts	time related non time related
Employer's and consultants' site requirements	time related non time related
Site requirements	time related non time related
Other services, charges and fees	
Temporary works (other than those included in unit costs)	
General purpose plant (other than that included in unit costs)	

CONTRACTOR'S SITE ONCOSTS – TIME RELATED

Site staff salaries

All non-productive supervisory staff on site including: agent, sub-agent, engineers, general foremen, non-productive section foremen, clerks, typists, timekeepers, checkers, quantity surveyors, cost engineers, security guards, etc. Cost includes salaries, subsistence allowance, National Health Insurance and Pension Scheme contributions, etc. Average cost approximately 3% to 5% of contract value.

Site staff expenses

Travelling, hotel and other incidental expenses incurred by staff. Average cost approximately 1% of staff salaries.

Attendant labour

Chainmen, storemen, drivers for staff vehicles, watchmen, cleaners, etc.

General yard labour

Labour employed on loading and offloading stores, general site cleaning, removal of rubbish, etc.

Plant maintenance

Fitters, electricians, and assistants engaged on general plant maintenance on site. This excludes drivers and banksmen who are provided for specifically in the Unit Costs Sections.

Site transport for staff and general use

Vehicles provided for use of staff and others including running costs, licence and insurance and maintenance if not carried out by site fitters.

Transport for labour to and from site

Buses or coaches provided for transporting employees to and from site including cost of drivers and running costs, etc., or charges by coach hire company for providing this service.

Contractor's office rental

This includes:

- rental charges for provision of offices for Contractor's staff
- main office
- section offices

PRELIMINARIES AND GENERAL ITEMS

- timekeepers, checkers and security
- laboratory, etc.
- an allowance of approximately 8 m^2 per staff member should be made

Contractor's site huts

Rental charges for stores and other general-use site huts

Canteen and welfare huts

Rental charges for canteen and huts for other welfare facilities required under Rule XVI of the Working Rule Agreement

Rates

Chargeable by local authorities on any site, temporary buildings or quarry

General office expenditure

Provision of postage, stationery and other consumables for general office use

Telecommunications

Rental charges and charges for calls

Furniture and equipment

Rental charges for office furniture and equipment including photocopiers, calculators, personal computers and laser printers, etc.

Surveying equipment

Rental

Canteen and welfare equipment

Rental charges for canteen and other welfare equipment

Radio communication equipment

Rental

Testing and laboratory equipment

Rental

Lighting and heating for offices and huts

Electricity, gas or other charges in connection with lighting and heating site offices and hutting

Site lighting electrical consumption

Electricity charges in connection with general external site lighting

Water consumption

Water rates and charges

Canteen operation

Labour, consumables and subsidy costs in operating site canteens

PRELIMINARIES AND GENERAL ITEMS

Carpenter's shop equipment

Rental of building and mechanical equipment

Fitter's shop equipment

Rental of building and equipment

Small tools

Provision of small tools and equipment for general use on site. Average cost 5% of total labour cost

Personal protective equipment

Provision of protective clothing for labour including boots, safety helmets, etc. Average cost 2% of total labour cost

Traffic control

Hire and operation of traffic lights

Road lighting

Hire and operation of road lighting and traffic warning lights

Cleaning vehicles

Equipment and labour cleaning vehicles before entering public roads

Cleaning roads and footpaths

Equipment and labour cleaning public roads and footpaths

Progress photographs for Contractor's records

Cost of taking and processing photographs to demonstrate progress

CONTRACTOR'S SITE ONCOSTS – NON TIME RELATED

Rent of additional land

For Contractor's use for erection of huts, storage of soil and other materials, etc.

Staff removal expenses

Costs of staff moving house to new location. Generally only applies on longer-term contracts.

Erection of offices including drainage, paths, etc.

Construction of foundations, drainage, footpaths and parking areas, erection of huts, installation of electric wiring, in situ fittings and decorating, etc.

Dismantle offices and restore site on completion

Dismantling and taking away huts and furniture, disconnecting and removing services, removing temporary foundations etc. and re-instating ground surface to condition prevailing before construction.

Erection of general site huts

Construction of foundations, drainage, footpaths and parking areas, erection of huts, installation of electric wiring, in situ fittings and decorating, etc.

PRELIMINARIES AND GENERAL ITEMS

Dismantle general site huts

Dismantling and taking away huts and furniture, disconnecting and removing services, removing temporary foundations etc. and re-instating ground surface to condition prevailing before construction

Erection of canteen and welfare huts

Construction of foundations, drainage, footpaths and parking areas, erection of huts, installation of electric wiring, in situ fittings and decorating, etc.

Dismantle canteen and welfare huts

Dismantling and taking away huts and furniture, disconnecting and removing services, removing temporary foundations etc. and re-instating ground surface to condition prevailing before construction

Caravan site construction and clearance

Construction of site for employees' caravans including provision for water, electricity and drainage, and subsequently clear away and restore site on completion, allow credit for any charges to be levied

Telecommunications

Charges for initial installation and removal

Furniture and equipment

Purchase costs of furniture and equipment, allow for residual sale value

Survey equipment

Purchase costs of survey equipment including pegs, profiles, paint, etc., for setting out

Canteen and welfare equipment

Purchase costs of equipment

Testing and laboratory equipment

Purchase cost of equipment

Radio communication

Installation costs

Electrical connection and installation

Initial charges for connections to mains supply

Electrical connection site plant

Connection to site mains supply and final disconnection and removal

Electrical connection site lighting

Connection to site mains supply and final disconnection and removal

Water supply

Installation on site and connection charges

Haulage plant

Cost of transport of plant and equipment to and from site

PRELIMINARIES AND GENERAL ITEMS

Progress photographs

Depot loading and unloading charges

Carpenter's shop

Erection of building, installation of equipment including electrical installation, etc. Dismantle and clear away on completion

Fitter's shop

Erection of building, installation of equipment including electrical installation, etc. Dismantle and clear away on completion

Stores compound

Erect and dismantle stores compound

Notice boards and signs

Supply, erect and remove Contractor's signboards, traffic control signs, etc.

Insurances

Payment of premiums for all Contractor's insurance obligations (see separate section on insurances and bond below)

Bond

Charges for provision of bond (see separate section on insurances and bond below)

Plant erection

Cost of erection of Contractor's plant on site including foundations, hard standings, drainage, etc.

Plant dismantling

Cost of removal of Contractor's plant on site including foundations, hard standings, drainage, etc.

Clear site on completion including removal of rubbish and reinstatement

EMPLOYER'S AND CONSULTANTS' SITE REQUIREMENTS – TIME RELATED

Office and other huts

Rental of office accommodation, sub offices, laboratory, etc.

Office and site attendant labour

Office cleaning, chainmen, laboratory assistants, etc.

Site transport

Rental for vehicles for use of client and engineer

Telecommunications

Rental and cost of calls (if to be borne by Contractor)

PRELIMINARIES AND GENERAL ITEMS

Furniture and equipment

Rental of office furniture and equipment

Survey equipment

Rental of surveying equipment

Testing and laboratory equipment

Rental of testing and laboratory equipment

Radio communication equipment

Rental and maintenance

Office lighting and heating

Cost of heating and lighting all offices and huts

Office consumables

Cost of office consumables to be provided by the Contractor

EMPLOYER'S AND CONSULTANTS' SITE REQUIREMENTS – NON TIME RELATED

Erection of huts and offices

Client's and Engineer's offices and other huts including foundations, pathways, parking area, electrical installation and drainage, etc.

Dismantling huts and offices

Restoration of site on completion

Telecommunications installation charges

Furniture and equipment purchase

Purchase cost for furniture and equipment

Survey equipment purchase

Testing and laboratory equipment purchase

Radio communication equipment installation

Progress photographs

Cost of professional photographer and supplying prints as required

OTHER SERVICES, CHARGES AND FEES

Design fees for alternative designs for permanent works

Design and drawing office costs and charges for preparing alternative designs and specifications and bill of quantities for alternative designs for permanent works

Design and design office charges for temporary works

Design and drawing office costs and charges for preparing designs and drawings for temporary works

PRELIMINARIES AND GENERAL ITEMS

Preparation of bending schedules

Drawing office charges for preparation of bending schedules

Fees to local authorities

Legal advice and fees

Fees and charges from legal adviser

Fencing

Traffic diversions

Lighting

Traffic signs

Traffic control

Footpath diversions

Stream or river diversion

Cofferdam installation

Cofferdam removal

Support works

Jetties

Bridges

De-watering

General pumping

Including construction of collecting sumps, etc.

Site access roads and maintenance

Scaffolding

GENERAL PURPOSE PLANT – OTHER THAN THAT INCLUDED IN UNIT COSTS

Lorries and dumpers for general transport around site

Tractors and trailers

Crane age for general use

Compressed air plant

PRELIMINARIES AND GENERAL ITEMS

Pumps

Bowsers for fuelling plant

Bowsers for water supply

Non productive time for plant on site

Obtained by comparing plant requirements as on programme with the plant time included in the build up of bill rates

Note: For all items of plant listed above the cost of drivers and other attendants must be allowed for together with consumables and other operating costs

INSURANCES AND BOND

Contractors are legally required to insure against liability which may be incurred when employees are injured at work and when individuals are injured by owners' vehicles. There is also a statutory requirement for certain types of machinery to be inspected at regular intervals. In addition to legal requirements, companies insure against possible loss due to fire, explosion, fraud, liability incurred as a result of damage to the property of others and through serious injury to individuals.

Certain risks are excluded from insurers' policies; these include war, revolution, etc., contamination by radioactivity and risks which arise from bad management.

Generally insurance companies take into account the claims record of a Contractor when assessing premiums payable on a particular contract or policy. Premiums are related to the risks involved and, on large Civil Engineering contracts, the insurers will require full details of the work, the methods of construction, plant used and risks involved due to flood, ground conditions, etc. Insurance companies or brokers should be consulted before submitting a tender for major Civil Engineering work.

The following gives an outline of the items to be allowed for in a tender.

Employer's liability insurance

This provides indemnity to the Employer against legal liability for death of or injury to employees sustained in the course of their employment. The cost is normally allowed for in the build-up of the 'all-in' labour rate as a percentage addition to the gross cost. This will vary, depending on the Contractor's record. An allowance of 2% has been made in this book.

Vehicle insurance

This can cover individual vehicles or fleets. The cost is normally covered in the rate charged to the contract for the use of the vehicles.

All risks insurance

This provides for loss of or damage to permanent and/or temporary works being executed on the contract. It also covers plant, materials, etc. The cost is allowed for in the tender as a percentage of the total contract value. This will vary depending on the Contractor's record and type of contract undertaken and also on the value of excesses included in the policy.

Public liability insurance

This provides indemnity against legal liability which arises out of business activities resulting in bodily injury to any person (other than employees), loss of or damage to property (not owned or under the control of the company), obstruction, trespass or the like.

PRELIMINARIES AND GENERAL ITEMS

Such insurance can be extended to include labour-only subcontractors and self-employed persons if required. The cost is generally included with head office overheads.

Professional indemnity insurance

This provides against liability arising out of claims made against the conduct and execution of the business. This covers such items as design liability, etc. The cost may be with head office overheads where such insurance is considered desirable.

Loss of money insurance

This covers loss of money and other negotiable items and loss or damage to safes and strong rooms as a result of theft. It is necessary to cover cash in transit for wages, etc. The cost is included with head office overheads.

Fidelity guarantee insurance

This covers loss by reason of any act of fraud or dishonesty committed by employees. The cost is included with head office overheads.

Other insurances

Other insurances which may be carried by a Contractor include fire insurance on his permanent premises and contents, consequential loss insurance in relation to his permanent premises, personal accident insurance on a 24 hour per day basis for employees.

Contract bond

Where the contract calls for a bond to be provided, this is normally given by either banks or insurance companies. The total value of these guarantees available to any company is limited, depending on the goodwill and assets of the company. It will also affect the borrowing facilities available to the company and therefore, to some extent, can restrict his trading. An allowance should be made at a rate of 1½% per annum on the amount of the bond for the construction period plus a rate of ½% per annum for the maintenance period.

PRELIMINARIES AND GENERAL ITEMS

WORKED EXAMPLE – PROGRAMME OF ACTIVITIES, STAFFING AND PLANT

The example is of a contract for the construction of an Airport extension. The contract includes concrete surfaced aprons/runways, surface water drainage, construction of two concrete structures, minor accommodation works and two culverts. The conditions of contract are ICE Standard conditions; the contract period is 12 months; and the approximate value is £10.5 million.

It is assumed that the main Contractor will sublet bulk earthworks and landscape, fencing, concrete surfacing, signs and lighting as well as waterproofing to structures. It is also taken that all materials are obtained off site (including concrete).

The worked example demonstrates a method of assessing preliminary costs and is based on the programme below together with an assessment of general purpose plant (a plant reconciliation is also given).

CONTRACTOR'S SITE ONCOSTS – TIME RELATED

			£	£
Site staff salaries (see programme)				
Agent	54 wks at	£1,895	102,330	
Senior Engineer	54 wks at	£1,745	94,230	
Engineers	76 wks at	£1,320	100,320	
General Foreman	52 wks at	£1,020	53,040	
Office Manager/Cost Clerk	52 wks at	£870	45,240	
Timekeeper/Storeman/Checker	44 wks at	£610	26,840	
Admin support	52 wks at	£460	23,920	
Security Guard	52 wks at	£510	26,520	
Quantity Surveyor	52 wks at	£1,566	81,432	
Fitter	42 wks at	£850	35,700	589,572
Site staff expenses (1% staff salaries)				5,896
Attendant labour				
Chainman	128 wks at	£580	74,240	
Driver	52 wks at	£620	32,240	
Office Cleaner (part-time)	52 wks at	£170	8,840	115,320
General yard labour				
(Part-time involvement in loading and offloading, clearing site rubbish etc.)				
1 ganger	10 wks at	£620	6,200	
4 labourers	40 wks at	£590	23,600	29,800
Plant maintenance (Contractor's own plant)				
(Fitter included in site staff salaries above)				
Fitter's mate	32 wks at	£580	18,560	18,560
Site transport for staff and general use				
QS/Agent's cars	54 wks at	£160	8,640	
Engineers' cars (contribution)	130 wks at	£80	10,400	
Land Rover or similar SWB	54 wks at	£390	21,060	40,100

PRELIMINARIES AND GENERAL ITEMS

				£	£
Site transport – Labour	48	wks at	£610	29,280	29,280
Transport for labour to and from site					NIL
Contractor's office rental					
Mobile offices (10 staff × 8 m²) = 80 m²	52	wks at	£205	10,660	
Section offices (2 nr at 10 m²) = 20 m²	52	wks at	£50	2,600	13,260
Contractor's site huts					
Stores hut, 22 m²	52	wks at	£56	2,912	2,912
Canteen and welfare huts					
Canteen 70 m² (assume 70 men)	52	wks at	£180	9,360	
Washroom 30 m²	52	wks at	£70	3,640	
Staff toilets	52	wks at	£180	9,360	
Site toilets	52	wks at	£240	12,480	34,840
Rates					0
General office expenditure etc.					
Postage, stationery and other consumables	52	wks at	£130	6,760	
Telephone/fax calls/e-mail and rental	52	wks at	£140	7,280	
Furniture and equipment rental	52	wks at	£70	3,640	
Personal computers, laser printers, scanners rental	52	wks at	£140	7,280	
Surveying equipment rental	52	wks at	£60	3,120	
Canteen and welfare equipment rental	52	wks at	£80	4,160	
Photocopier rental	52	wks at	£120	6,240	
Testing equipment rental	52	wks at	£50	2,600	
Lighting and heating offices and huts (200 m²)	52	wks at	£125	6,500	47,580
Water consumption					
2,250,000 litres at £2.75 per 5,000 litres	450	units at	£2.75	1,238	1,238
Small tools					
1% on labour costs of say £1,400,000					14,000
Protective clothing					
0.5% on labour costs of say £1,400,000					7,000
Cleaning vehicles					
Cleaning roads					
Towed sweeper (tractors elsewhere)	52	wks at	£150	7,800	
Brushes			say,	1,000	
Labour (skill rate 4)	52	wks at	£638	33,176	41,976
Progress photographs			say,		1,000
Total Contractor's site oncosts – Time related					**£927,520**

PRELIMINARIES AND GENERAL ITEMS

CONTRACTOR'S SITE ONCOSTS – NON TIME RELATED

			£	£
Erect and dismantle offices				
Mobile	108 m² at	£9.00	972	
Site works		say,	1,000	
Toilets		say,	300	
Wiring, water, etc.		say,	750	3,022
Erect and dismantle other buildings				
Stores and welfare	130 m² at	£16.00	2,080	
Site works		say,	1,000	
Toilets		say,	700	3,780
Telephone installation				500
Survey equipment and setting out				
Purchase cost including pegs, profiles, paint ranging rods, etc.				1,200
Canteen and welfare equipment				
Purchase cost less residual value				1,500
Electrical installation				2,500
Water supply				
Connection charges			1,000	
Site installation			1,000	2,000
Transport of plant and equipment				3,500
Stores compound and huts				1,000
Sign boards and traffic signs				1,000
Insurances (dependent on Contractor's policy and record)				
Contractor's all risks 2.5% on £7,800,000			195,000	
Allow for excesses			20,000	215,000
General site clearance				3,500
Total Contractor's site oncosts – Non time related				**£238,502**

PRELIMINARIES AND GENERAL ITEMS

EMPLOYER'S AND CONSULTANTS' SITE REQUIREMENTS – TIME RELATED

(details of requirements will be defined in the contract documents)

Offices (50 m²)	52 wks at	£110	5,720
Site attendant labour (man weeks)	150 wks at	£580	87,000
Site transport (2 Land Rovers or similar)	52 wks at	£780	40,560
Telephones and calls	52 wks at	£120	6,240
Furniture and equipment	52 wks at	£50	2,600
Survey equipment	52 wks at	£60	3,120
Office heating and lighting (50 m²)	52 wks at	£45	2,340
Office consumables (provided by Contractor)	52 wks at	£50	2,600
Total employer's and consultants' requirements – time related			**£150,180**

EMPLOYER'S AND CONSULTANTS' SITE REQUIREMENTS – NON TIME RELATED

(details of requirements will be defined in the contract documents)

Erection and dismantling of huts and offices (50 m²)	50 m² at	£9.00	450
Site works, toilets, etc.			1,000
Telephone installation			500
Electrical installation			1,000
Furniture and equipment Purchase cost less residual value			750
Progress photographs	100 sets at	£40	4,000
Total employer's and consultants' requirements – non time related			**£7,700**

OTHER SERVICES, CHARGES AND FEES

Not applicable to this cost model.

TEMPORARY WORKS – OTHER THAN THOSE INCLUDED IN UNIT COSTS

Temporary Fencing				
1200 mm chestnut fencing				
Materials	5,000 m at	£6.94	34,700	
Labour	500 hrs at	£12.70	6,350	41,050
Traffic diversions				
Structure No. 1			8,000	
Structure No. 2			6,500	14,500
Footpath diversion				5,000
Stream diversion				15,500
Site access roads	400 m at	£120.00		48,000
Total temporary works				**£124,050**

PRELIMINARIES AND GENERAL ITEMS

GENERAL PURPOSE PLANT – OTHER THAN THAT INCLUDED IN UNIT COSTS

Description				Labour	Plant	Fuel etc.	Total
Wheeled tractor							
Hire charge	52	wks at	£580		30,160		30,160
driver (skill rate 4)	52	wks at	£580	30,160			30,160
consumables	52	wks at	£80			4,160	4,160
Trailer							
Hire charge	52	wks at	£25		1,300		1,300
10 t Crawler Crane							
Hire charge	40	wks at	£1,000		40,000		40,000
driver (skill rate 3)	40	wks at	£713	28,520			28,520
consumables	40	wks at	£45			1,800	1,800
14.5 t hydraulic backacter							
Hire charge	6	wks at	£1,010		6,060		6,060
Driver (skill rate 3)	6	wks at	£713	4,278			4,278
Banksman (skill rate 4)	6	wks at	£639	3,834			3,834
consumables	6	wks at	£90			540	540
Concrete vibrators (two)							
Hire charge in total	24	wks at	£100		2,400		2,400
D6 Dozer or similar							
Hire charge	4	wks at	£1,550		6,200		6,200
driver (skill rate 2)	4	wks at	£796	3,184			3,184
consumables	4	wks at	£195			780	780
Towed roller BW6 or similar							
Hire charge	6	wks at	£350		2,100		2,100
consumables	6	wks at	£60			360	360
Loading shovel Cat 939 or similar							
Hire charge	16	wks at	£1,100		17,600		17,600
driver (skill rate 2)	16	wks at	£796	12,736			12,736
consumables	16	wks at	£95			1,520	1,520
Compressor 22.1 m³/min (silenced)							
Hire charge	12	wks at	£492		5,904		5,904
consumables	12	wks at	£410			4,920	4,920
Plate compactor (180 kg)							
Hire charge	12	wks at	£50		600		600
consumables	12	wks at	£15			180	180
75 mm 750 l/min pumps							
Hire charge	25	wks at	£79		1,975		1,975
consumables	25	wks at	£15			375	375
Total costs			**£**	**82,712**	**114,299**	**14,635**	**211,646**

PRELIMINARIES AND GENERAL ITEMS

SUMMARY OF PRELIMINARIES AND GENERAL ITEMS

Contractor's site oncosts – Time related	927,520
Contractor's site oncosts – Non time related	238,502
Employer's and consultants' requirements on site – Time related	150,180
Employer's and consultants' requirements on site – Non Time related	7,700
Other services, charges and fees	N/A
Temporary works not included in unit costs	124,050
General purpose plant and plant not included in unit costs	211,646
Total of Preliminaries and General Items	**£1,659,598**

BUILDING PRICES PER SQUARE METRE

Item	Unit	Range £		
UNICLASS D1 UTILITIES, CIVIL ENGINEERING FACILITIES				
Car parking				
surface car parking	m^2	96.00	to	120.00
surface car parking; landscaped	m^2	120.00	to	160.00
Multi-storey car parks				
flat slab	m^2	650.00	to	820.00
Underground car parks				
partially underground under buildings; naturally ventilated	m^2	730.00	to	930.00
completely underground under buildings with mechanical ventilation	m^2	1000.00	to	1275.00
completely underground with landscaped roof and mechanical ventilation	m^2	1200.00	to	1550.00
Transport facilities				
railway stations	m^2	4200.00	to	5300.00
bus and coach stations	m^2	2950.00	to	3750.00
bus garages	m^2	1100.00	to	1425.00
petrol stations	m^2	3650.00	to	4650.00
Vehicle showrooms with workshops, garages etc.				
up to 2,000 m^2	m^2	1425.00	to	1825.00
over 2,000 m^2	m^2	1225.00	to	1575.00
Vehicle showrooms without workshops, garages etc.				
up to 2,000 m^2	m^2	1425.00	to	1825.00
Vehicle repair and maintenance buildings				
up to 500 m^2	m^2	2125.00	to	2700.00
over 500 m^2 up to 2000 m^2	m^2	1425.00	to	1825.00
car wash buildings	m^2	1225.00	to	1575.00
Airport facilities (excluding aprons)				
airport terminals	m^2	3750.00	to	4800.00
airport piers/satellites	m^2	3950.00	to	5100.00
Airport campus facilities				
cargo handling bases	m^2	950.00	to	1225.00
distribution centres	m^2	530.00	to	680.00
hangars (type C and D facilities)	m^2	1650.00	to	2075.00
TV, radio and video studios	m^2	2700.00	to	3500.00
Telephone exchanges	m^2	1425.00	to	1825.00
Telephone engineering centres	m^2	970.00	to	1250.00
Branch post offices	m^2	1425.00	to	1825.00
Postal delivery offices/sorting offices	m^2	1425.00	to	1825.00
Electricity substations	m^2	2050.00	to	2600.00
Garages, domestic	m^2	880.00	to	1125.00
UNICLASS D2 INDUSTRIAL FACILITIES				
B1 Light industrial/offices buildings				
economical shell, and core with heating only	m^2	870.00	to	1100.00
medium shell and core with heating and ventilation	m^2	1225.00	to	1575.00
high quality shell and core with air conditioning	m^2	1925.00	to	2450.00
developers Category A fit-out	m^2	700.00	to	900.00
tenants Category B fit-out	m^2	465.00	to	590.00
Agricultural storage	m^2	740.00	to	940.00

BUILDING PRICES PER SQUARE METRE

Item	Unit	Range £		
UNICLASS D2 INDUSTRIAL FACILITIES – cont				
Factories				
for letting (incoming services only)	m^2	750.00	to	960.00
for letting (including lighting, power and heating)	m^2	960.00	to	1225.00
nursery units (including lighting, power and heating)	m^2	960.00	to	1225.00
workshops	m^2	890.00	to	1125.00
maintenance/motor transport workshops	m^2	1150.00	to	1475.00
owner occupation for light industrial use	m^2	1225.00	to	1575.00
owner occupation for heavy industrial use	m^2	2050.00	to	2600.00
Factory/office buildings high technology production				
for letting (shell and core only)	m^2	810.00	to	1025.00
for owner occupation (controlled environment fully finished)	m^2	1075.00	to	1375.00
High technology laboratory workshop centres, air-conditioned	m^2	3500.00	to	4450.00
Warehouse and distribution centres				
high bay (10–15 m high) for owner occupation (no heating) up to 10,000 m^2	m^2	390.00	to	500.00
high bay (10–15 m high) for owner occupation (no heating) over 10,000 m^2 up to 20,000 m^2	m^2	295.00	to	375.00
high bay (16–24 m high) for owner occupation (no heating) over 10,000 m^2 up to 20,000 m^2	m^2	430.00	to	550.00
high bay (16–24 m high) for owner occupation (no heating) over 20,000 m^2	m^2	355.00	to	455.00
Fit-out cold stores, refrigerated stores inside warehouse	m^2	780.00	to	1000.00
Industrial buildings				
Shell with heating to office areas only				
500–1,000 m^2	m^2	640.00	to	810.00
1,000–2,000 m^2	m^2	540.00	to	690.00
greater than 2,000 m^2	m^2	590.00	to	750.00
Unit including services to production area				
500–1,000 m^2	m^2	880.00	to	1125.00
1,000–2,000 m^2	m^2	800.00	to	1025.00
greater than 2,000 m^2	m^2	800.00	to	1025.00
UNICLASS D3 ADMINISTRATIVE, COMMERCIAL, PROTECTIVE SERVICES FACILITIES				
Embassies	m^2	2600.00	to	3300.00
County courts	m^2	3150.00	to	4000.00
Magistrates courts	m^2	1700.00	to	2175.00
Civic offices				
non air-conditioned	m^2	1700.00	to	2175.00
fully air-conditioned	m^2	2050.00	to	2600.00
Probation/registrar offices	m^2	1275.00	to	1650.00
Offices for owner occupation				
low rise, medium quality; air-conditioned	m^2	1850.00	to	2350.00
medium rise, medium quality; air-conditioned	m^2	2325.00	to	3000.00
Offices – City and West End				
To developers Cat A finish only				
high quality, speculative 8–20 storeys, air-conditioned	m^2	2600.00	to	3300.00
high rise, air-conditioned, iconic speculative towers	m^2	3950.00	to	5100.00

BUILDING PRICES PER SQUARE METRE

Item	Unit	Range £		
Business park offices				
Generally up to 3 storeys only				
functional non air-conditioned less than 2,000 m^2	m^2	1500.00	to	1925.00
functional non air-conditioned more than 2,000 m^2	m^2	1575.00	to	2000.00
medium quality non air-conditioned less than 2,000 m^2	m^2	1650.00	to	2075.00
medium quality non air-conditioned more than 2,000 m^2	m^2	1750.00	to	2225.00
medium quality air-conditioned less than 2,000 m^2	m^2	1925.00	to	2450.00
medium quality air-conditioned more than 2,000 m^2	m^2	2050.00	to	2600.00
good quality – naturally ventilated to meet BCO specification (exposed soffits, solar shading) less than 2,000 m^2	m^2	1850.00	to	2350.00
good quality – naturally ventilated to meet BCO specification (exposed soffits, solar shading) more than 2,000 m^2	m^2	1925.00	to	2450.00
high quality air-conditioned less than 2,000 m^2	m^2	2325.00	to	3000.00
high quality air-conditioned more than 2,000 m^2	m^2	2375.00	to	3050.00
Large trading floors in medium sized offices	m^2	3750.00	to	4800.00
Two storey ancillary office accommodation to warehouses/factories	m^2	1500.00	to	1925.00
Fitting out offices (NIA – Net Internal Area)				
City and West End				
basic fitting out including carpets, decorations, partitions and services	m^2	680.00	to	870.00
good quality fitting out including carpets, decorations, partitions and services	m^2	750.00	to	960.00
high quality fitting out including carpets, decorations, partitions, ceilings, furniture, air conditioning and electrical services	m^2	820.00	to	1050.00
reception areas	m^2	3400.00	to	4400.00
conference suites	m^2	3300.00	to	4200.00
meeting areas	m^2	4100.00	to	5200.00
back of house/storage	m^2	750.00	to	960.00
sub-equipment room	m^2	2175.00	to	2800.00
kitchen	m^2	4800.00	to	6100.00
Out-of town (South East England)				
basic fitting out including carpets, decorations, partitions and services	m^2	550.00	to	700.00
good quality fitting out including carpets, decorations, partitions and services	m^2	620.00	to	790.00
high quality fitting out including carpets, decorations, partitions, ceilings, furniture, air conditioning and electrical services	m^2	750.00	to	960.00
reception areas	m^2	2050.00	to	2600.00
conference suites	m^2	2800.00	to	3600.00
meeting areas	m^2	2700.00	to	3500.00
back of house/storage	m^2	620.00	to	790.00
sub-equipment room	m^2	2175.00	to	2800.00
kitchen	m^2	3150.00	to	4000.00
Office refurbishment (including developers finish – (Gross Internal Floor Area – GIFA; Central London)				
minor refurbishment	m^2	640.00	to	820.00
medium refurbishment	m^2	1375.00	to	1750.00
major refurbishment	m^2	2125.00	to	2700.00
Banks and Building Societies				
banks; local	m^2	1975.00	to	2500.00
banks; city centre/head office	m^2	2800.00	to	3600.00
building society branches	m^2	1925.00	to	2450.00
building society; refurbishment	m^2	1225.00	to	1575.00

BUILDING PRICES PER SQUARE METRE

Item	Unit	Range £		
UNICLASS D3 ADMINISTRATIVE, COMMERCIAL, PROTECTIVE SERVICES FACILITIES – cont				
Shop shells				
small	m^2	830.00	to	1050.00
large, including department stores and supermarkets	m^2	760.00	to	970.00
Fitting out shell for small shop (including shop fittings)				
simple store	m^2	790.00	to	1000.00
designer/fashion store	m^2	1575.00	to	2000.00
Fitting out shell for department store or supermarket	m^2	2250.00	to	2900.00
Retail Warehouses				
shell	m^2	570.00	to	730.00
fitting out, including all display and refrigeration units, check outs and IT systems	m^2	340.00	to	440.00
Hypermarkets/Supermarkets				
shell	m^2	690.00	to	880.00
supermarket fit-out	m^2	1375.00	to	1750.00
hypermarket fit-out	m^2	960.00	to	1225.00
Shopping Centres				
Malls, including fit-out				
comfort cooled	m^2	4400.00	to	5600.00
air-conditioned	m^2	5600.00	to	7200.00
food court	m^2	4850.00	to	6200.00
factory outlet centre – enclosed	m^2	3950.00	to	5100.00
factory outlet centre – open	m^2	790.00	to	1000.00
anchor tenants; capped off services	m^2	1275.00	to	1650.00
medium/small units; capped off services	m^2	1225.00	to	1575.00
centre management	m^2	2800.00	to	3600.00
enclosed surface level service yard	m^2	1975.00	to	2500.00
landlords back of house and service corridors	m^2	1975.00	to	2500.00
Refurbishment				
mall; limited scope	m^2	1425.00	to	1825.00
mall; comprehensive	m^2	2050.00	to	2600.00
Rescue/aid facilities				
ambulance stations	m^2	1225.00	to	1575.00
ambulance control centres	m^2	1925.00	to	2450.00
fire stations	m^2	2050.00	to	2600.00
Police stations	m^2	1975.00	to	2500.00
Prisons	m^2	2450.00	to	3150.00
UNICLASS D4 MEDICAL, HEALTH AND WELFARE FACILITIES				
District hospitals	m^2	2275.00	to	2900.00
refurbishment	m^2	1650.00	to	2075.00
Hospices	m^2	1850.00	to	2350.00
Private hospitals	m^2	2275.00	to	2900.00
Pharmacies	m^2	2450.00	to	3150.00
Hospital laboratories	m^2	2700.00	to	3500.00
Ward blocks	m^2	2050.00	to	2600.00
refurbishment	m^2	1225.00	to	1575.00
Geriatric units	m^2	2050.00	to	2600.00
Psychiatric units	m^2	1925.00	to	2450.00

BUILDING PRICES PER SQUARE METRE

Item	Unit	Range £		
Psycho-geriatric units	m²	1850.00	to	2350.00
Maternity units	m²	2050.00	to	2600.00
Operating theatres	m²	2700.00	to	3500.00
Outpatients/casualty units	m²	2050.00	to	2600.00
Hospital teaching centres	m²	1775.00	to	2275.00
Health centres	m²	1650.00	to	2075.00
Welfare centres	m²	1775.00	to	2275.00
Day centres	m²	1650.00	to	2075.00
Group practice surgeries	m²	1850.00	to	2350.00
Homes for the mentally handicapped	m²	1500.00	to	1925.00
Homes for the physically handicapped	m²	1700.00	to	2175.00
Geriatric day hospital	m²	1925.00	to	2450.00
Accommodation for the elderly				
residential homes	m²	1225.00	to	1575.00
nursing homes	m²	1650.00	to	2075.00
Homes for the aged	m²	1925.00	to	2450.00
refurbishment	m²	1025.00	to	1300.00
Observation and assessment units	m²	1375.00	to	1750.00
Primary Health Care				
doctors surgery – basic	m²	1425.00	to	1825.00
doctors surgery/medical centre	m²	1700.00	to	2175.00
Hospitals				
diagnostic and treatment centres	m²	3500.00	to	4450.00
acute services hospitals	m²	3350.00	to	4300.00
radiotherapy and ocology units	m²	3500.00	to	4450.00
community hospitals	m²	2800.00	to	3600.00
trauma unit	m²	2600.00	to	3300.00
UNICLASS D5 RECREATIONAL FACILITIES				
Public houses	m²	2050.00	to	2600.00
Dining blocks and canteens	m²	1650.00	to	2075.00
Restaurants	m²	1850.00	to	2350.00
Community centres	m²	1775.00	to	2275.00
General purpose halls	m²	1500.00	to	1925.00
Visitors' centres	m²	2450.00	to	3150.00
Youth clubs	m²	2050.00	to	2600.00
Arts and drama centres	m²	2175.00	to	2800.00
Theatres, including seating and stage equipment				
large – over 500 seats	m²	4650.00	to	5900.00
studio/workshop – less than 500 seats	m²	3100.00	to	3900.00
refurbishment	m²	2450.00	to	3150.00
Concert halls, including seats and stage equipment	m²	3500.00	to	4450.00
Cinema				
shell	m²	1075.00	to	1375.00
multiplex; shell only	m²	2175.00	to	2800.00
fitting out, including all equipment, air-conditioned	m²	1375.00	to	1750.00
Exhibition centres	m²	2050.00	to	2600.00

BUILDING PRICES PER SQUARE METRE

Item	Unit	Range £		
UNICLASS D5 RECREATIONAL FACILITIES – cont				
Swimming pools				
international standard	m^2	4650.00	to	5900.00
local authority standard	m^2	3100.00	to	3900.00
school standard	m^2	1425.00	to	1825.00
leisure pools, including wave making equipment	m^2	3600.00	to	4600.00
Ice rinks	m^2	1650.00	to	2075.00
Rifle ranges	m^2	1375.00	to	1750.00
Leisure centres				
dry	m^2	1925.00	to	2450.00
extension to hotels; shell and fitout, including small pool	m^2	2700.00	to	3500.00
wet and dry	m^2	2900.00	to	3700.00
Sports halls including changing rooms	m^2	1275.00	to	1650.00
School gymnasiums	m^2	1225.00	to	1575.00
Squash courts	m^2	1275.00	to	1650.00
Indoor bowls halls	m^2	920.00	to	1175.00
Bowls pavilions	m^2	1125.00	to	1450.00
Health and Fitness Clubs	m^2	1925.00	to	2450.00
Sports pavilions	m^2	1375.00	to	1750.00
changing only	m^2	1575.00	to	2000.00
social and changing	m^2	1575.00	to	2000.00
Clubhouses	m^2	1275.00	to	1650.00
UNICLASS D6 RELIGIOUS FACILITIES				
Temples, mosques, synagogues	m^2	1775.00	to	2275.00
Churches	m^2	1650.00	to	2075.00
Mission halls, meeting houses	m^2	1775.00	to	2275.00
Convents	m^2	1925.00	to	2450.00
Crematoria	m^2	2175.00	to	2800.00
Mortuaries	m^2	3000.00	to	3850.00
UNICLASS D7 EDUCATION, SCIENTIFIC AND INFORMATION FACILITIES				
Nursery schools	m^2	1650.00	to	2075.00
Primary/junior schools	m^2	1700.00	to	2175.00
Secondary/middle schools	m^2	2050.00	to	2600.00
Secondary schools and further education colleges				
classrooms	m^2	1375.00	to	1750.00
laboratories	m^2	1500.00	to	1925.00
craft design and technology	m^2	1500.00	to	1925.00
music	m^2	1700.00	to	2175.00
Extensions to schools				
classrooms	m^2	1650.00	to	2075.00
laboratories	m^2	1700.00	to	2175.00
Sixth form colleges	m^2	1650.00	to	2075.00
Special schools	m^2	1375.00	to	1750.00
Training colleges	m^2	1375.00	to	1750.00
Management training centres	m^2	1775.00	to	2275.00

BUILDING PRICES PER SQUARE METRE

Item	Unit	Range £		
Universities				
arts buildings	m^2	1500.00	to	1925.00
science buildings	m^2	1850.00	to	2350.00
College/University Libraries	m^2	1500.00	to	1925.00
Laboratories and offices, low level servicing	m^2	2900.00	to	3700.00
Computer buildings	m^2	2450.00	to	3150.00
Museums and Art Galleries				
national standard museum	m^2	6600.00	to	8500.00
national standard independent specialist museum, excluding fit-out	m^2	4650.00	to	5900.00
regional, including full air conditioning	m^2	3950.00	to	5100.00
local, including full air conditioning	m^2	2950.00	to	3750.00
conversion of existing warehouse to regional standard museum	m^2	1850.00	to	2350.00
conversion of existing warehouse to local standard museum	m^2	1575.00	to	2000.00
Galleries				
international standard art gallery	m^2	4000.00	to	5200.00
national standard art gallery	m^2	3100.00	to	3900.00
independent commercial art gallery	m^2	1700.00	to	2175.00
Arts and drama centre	m^2	1650.00	to	2075.00
Learning resource centre				
economical	m^2	1500.00	to	1925.00
high quality	m^2	1925.00	to	2450.00
Libraries				
regional; 5,000 m^2	m^2	3400.00	to	4400.00
national; 15,000 m^2	m^2	4450.00	to	5700.00
international; 20,000 m^2	m^2	5800.00	to	7400.00
Conference centres	m^2	2375.00	to	3050.00
UNICLASS D8 RESIDENTIAL FACILITIES				
Local Authority and Housing Association schemes				
Housing Association Developments (Code for Sustainable Homes Level 3)				
Bungalows				
semi-detached	m^2	1325.00	to	1675.00
terraced	m^2	1250.00	to	1575.00
Two storey housing				
detached	m^2	1325.00	to	1675.00
semi-detached	m^2	1250.00	to	1575.00
terraced	m^2	1100.00	to	1400.00
Three storey housing				
semi-detached	m^2	1325.00	to	1675.00
terraced	m^2	1100.00	to	1400.00
Apartments/flats				
low rise	m^2	1325.00	to	1675.00
medium rise	m^2	1625.00	to	2075.00
Sheltered housing with wardens accommodation	m^2	1250.00	to	1575.00
Private developments				
single detached houses	m^2	1775.00	to	2250.00
two and three storey houses	m^2	1825.00	to	2350.00

BUILDING PRICES PER SQUARE METRE

Item	Unit	Range £		
UNICLASS D8 RESIDENTIAL FACILITIES – cont				
Private developments – cont				
Apartments/flats generally				
standard quality; 3–5 storeys	m^2	2200.00	to	2800.00
warehouse/office conversion to apartments	m^2	2425.00	to	3100.00
high quality apartments in residential tower – Inner London	m^2	2900.00	to	3750.00
Hotels (including fittings, furniture and equipment)				
luxury city-centre with conference and wet leisure facilities	m^2	3900.00	to	5000.00
business town centre with conference and wet leisure facilities	m^2	2850.00	to	3650.00
mid-range with conference and leisure facilities	m^2	2225.00	to	2800.00
budget city-centre with dining and bar facilities	m^2	1950.00	to	2475.00
budget roadside excluding dining facilities	m^2	1675.00	to	2175.00
Hotel accommodation facilities (excluding fittings, furniture and equipment)				
bedroom areas	m^2	1175.00	to	1475.00
front of house and reception	m^2	1475.00	to	1925.00
restaurant areas	m^2	1750.00	to	2250.00
bar areas	m^2	1575.00	to	1975.00
function rooms/conference facilities	m^2	1475.00	to	1925.00
Students residences				
large budget schemes with en-suite accommodation	m^2	1475.00	to	1925.00
smaller schemes (40–100 units) with mid range specifications	m^2	1825.00	to	2325.00
smaller high quality courtyard schemes, college style	m^2	2475.00	to	3150.00

GENERAL BUILDING WORKS

Item	Unit	Range £		
1 SUBSTRUCTURE				
1.1 SUBSTRUCTURE				
1.1.1 Standard Foundations				
Foundations and floor slab composite				
Strip or trenchfill foundations with masonry up to 150 mm above floor level only; blinded hardcore bed; slab insulation; reinforced ground bearing slab 200 mm thick. To suit residential and small commercial developments with good ground bearing capacity				
shallow foundations up to 1.00 m deep	m^2	195.00	to	235.00
shallow foundations up to 1.50 m deep	m^2	225.00	to	275.00
Foundations in poor ground; mini piles; typically 300 mm dia.; 15 m long; 175 mm thick reinforced concrete slab; for single storey commercial type development				
minipiles to building columns only; 1 per column	m^2	155.00	to	190.00
minipiles to columns and perimeter ground floor beam; piles at 2.00 m centres to ground beams	m^2	300.00	to	365.00
minipiles to entire building at 2.00 m × 2.00 m grid	m^2	330.00	to	405.00
Raft foundations				
simple reinforced concrete raft on poorer ground for development up to two storey high	m^2	175.00	to	210.00
Trench fill foundations				
Machine excavation, disposal, plain in situ concrete 20.00 N/mm² – 20 mm aggregate (1:2:4) trench fill, 300 mm high cavity masonry in cement mortar (1:3), pitch polymer damp roof course				
With 3 courses of common brick outer skin up to dpc level (PC bricks @ £380 / 1000)				
600 mm × 1000 mm deep	m	125.00	to	150.00
600 mm × 1500 mm deep	m	170.00	to	210.00
With 3 courses of facing bricks brick outer skin up to dpc level (PC bricks @ £500 / 1000)				
600 mm × 1000 mm deep	m	130.00	to	155.00
600 mm × 1500 mm deep	m	175.00	to	210.00
With 3 courses of facing bricks brick outer skin up to dpc level (PC bricks @ £700 / 1000)				
600 mm × 1000 mm deep	m	130.00	to	160.00
600 mm × 1500 mm deep	m	180.00	to	215.00
Strip foundations				
Excavate trench 600 mm wide, partial backfill, partial disposal, earthwork support (risk item), compact base of trench, plain in situ concrete 20.00 N/mm² – 20 mm aggregate (1:2:4), 250 mm thick, 3 courses of common brick up to dpc level, cavity brickwork/blockwork in cement mortar (1:3), pitch polymer damp-proof course machine excavation				
With 3 courses of common brick outer skin up to dpc level (PC bricks @ £380 / 1000)				
600 mm × 1000 mm deep	m	140.00	to	165.00
600 mm × 1500 mm deep	m	180.00	to	220.00

GENERAL BUILDING WORKS

Item	Unit	Range £		
1.1 SUBSTRUCTURE – cont				
Strip foundations – cont				
With 3 courses of facing brick outer skin up to dpc level (PC bricks @ £500 / 1000)				
600 mm × 1000 mm deep	m	140.00	to	170.00
600 mm × 1500 mm deep	m	190.00	to	230.00
With 3 courses of facing brick outer skin up to dpc level (PC bricks @ £700 / 1000)				
600 mm × 1000 mm deep	m	150.00	to	180.00
600 mm × 1500 mm deep	m	195.00	to	235.00
Column bases				
Excavate pit in firm ground by machine, partial backfill, partial disposal, support, compact base of pit; in situ concrete 25.00 N/mm²; reinforced at 50 kg/m³				
rates per base for base size				
600 mm × 600 mm × 300 mm; 1000 mm deep pit	nr	430.00	to	520.00
900 mm × 900 mm × 450 mm; 1250 mm deep pit	nr	520.00	to	630.00
1500 mm × 1500 mm × 600 mm; 1500 mm deep pit	nr	860.00	to	1050.00
2700 mm × 2700 mm × 1000 mm; 1500 mm deep pit	nr	2475.00	to	3000.00
rates per m³ for base size				
600 mm × 600 mm × 300 mm; 1000 mm deep pit	m^3	3850.00	to	4650.00
900 mm × 900 mm × 450 mm; 1250 mm deep pit	m^3	1450.00	to	1750.00
1500 mm × 1500 mm × 600 mm; 1500 mm deep pit	m^3	650.00	to	790.00
2700 mm × 2700 mm × 1000 mm; 1500 mm deep pit	m^3	350.00	to	430.00
extra for reinforcement at 75 kg/m³ concrete, base size	m^3	82.00	to	99.00
extra for reinforcement at 100 kg/m³ concrete, base size	m^3	100.00	to	120.00
1.1.2 SPECIALIST FOUNDATIONS				
Piling				
Enabling Works				
excavate to form piling mat; supply and lay imported hardcore – recycled brick and similar to form piling mat	m^3	11.00	to	13.40
provision of plant (1nr rig); including bringing to and removing from site; maintenance, erection and dismantling at each pile position	item	19500.00	to	23500.00
Supply and install concrete Continuous Flight Auger (CFA) piles; cart away inactive spoil; measure total length of pile				
450 mm dia. reinforced concrete CFA piles	m	53.00	to	64.00
600 mm dia. reinforced concrete CFA piles	m	84.00	to	100.00
750 mm dia. reinforced concrete CFA piles	m	130.00	to	160.00
900 mm dia. reinforced concrete CFA piles	m	190.00	to	230.00
Pile testing				
using tension piles as reaction; typical loading test 1000 kN to 2000 kN	item	8000.00	to	9700.00
integrity testing; minimum 20 per visit	nr	16.00	to	19.30
Secant wall piling, 750 mm dia. piles' including site mobilization and demobilization; Measure total area of piling (total length × total depth)				
supply and install secant wall piling	m^2	275.00	to	330.00

GENERAL BUILDING WORKS

Item	Unit	Range £		
Steel sheet piling; Measure total area of piling (total length × total depth) interlocking steel sheet piling to excavation perimeter; Corus LX or similar; extraction on completion	m^2	210.00	to	250.00
Pile caps				
Excavate pit in firm ground by machine, partial backfill, partial disposal, support, compaction, cut off pile and prepare reinforcement, formwork; reinforced in situ concrete cap 25 N/mm²; reinforcement at 50 kg/m³; rates per cap for cap size				
900 mm × 900 mm × 1000 mm; 1–2 piles	nr	760.00	to	920.00
2100 mm × 2100 mm × 1000 mm; 2–3 piles	nr	1700.00	to	2075.00
2700 mm × 2700 mm × 1500 mm; 3–5 piles	nr	3400.00	to	4150.00
Excavate pit in firm ground by machine, partial backfill, partial disposal, support, compaction, cut off pile and prepare reinforcement, formwork; reinforced in situ concrete cap 25 N/mm²; reinforcement at 50 kg/m³; rates per m^3 for typical cap size				
up to 900 mm × 900 mm × 1000 mm; 1–2 piles	m^3	910.00	to	1075.00
up to 2100 mm × 2100 mm × 1000 mm; 2–3 piles	m^3	390.00	to	475.00
up to 2700 mm × 2700 mm × 1500 mm; 3–5 piles	m^3	315.00	to	380.00
extra for reinforcement at 75 kg/m³ concrete	m^3	80.00	to	97.00
extra for reinforcement at 100 kg/m³ concrete	m^3	100.00	to	120.00
extra for alternative strength concrete C30 (30.00 N/mm²)	m^3	1.45	to	1.75
extra for alternative strength concrete C40 (40.00 N/mm²)	m^3	7.30	to	8.80
Concrete ground beams				
Reinforced in situ concrete ground beams; bar reinforcement; formwork				
300 mm × 300 mm, reinforcement at 180 kg/m³	m	53.00	to	64.00
450 mm × 450 mm, reinforcement at 200 kg/m³	m	100.00	to	120.00
450 mm × 600 mm, reinforcement at 270 kg/m³	m	145.00	to	180.00
Precast concrete ground beams; square ends and dowelled. Supplied and installed in lengths to span between stanchion bases; concrete strength C50; maximum UDL of 4.5 kN/m; in situ work at beam ends not included.				
350 mm × 425 mm	m	93.00	to	110.00
350 mm × 875 mm	m	220.00	to	270.00
350 mm × 1075 mm	m	300.00	to	365.00
450 mm × 400 mm	m	110.00	to	130.00
Concrete lift pits				
Excavate and disposal; reinforced concrete floor and walls; bitumen tanking as necessary				
1.65 m × 1.81 m × 1.60 m deep pit – 8 person lift / 630 kg	nr	2700.00	to	3250.00
1.80 m × 2.50 m × 1.60 m deep pit −13 person lift / 1000 kg	nr	3300.00	to	4000.00
3.00 m × 2.50 m × 1.60 m deep pit – 21 person lift / 1700 kg	nr	3900.00	to	4700.00

GENERAL BUILDING WORKS

Item	Unit	Range £		
1.1 SUBSTRUCTURE – cont				
Underpinning				
In stages not exceeding 1500 mm long from one side of existing wall and foundation, excavate preliminary trench by machine and underpinning pit by hand, partial backfill, partial disposal, earthwork support (open boarded), cutting away projecting foundations, prepare underside of existing, compact base of pit, plain in situ concrete 20.00 N/mm² – 20 mm aggregate (1:2:4), formwork, brickwork in cement mortar (1:3), pitch polymer damp-proof course, wedge and pin to underside of existing with slates. Commencing at 1.00 m below ground level with common bricks, depth of underpinning				
900 mm high, one brick wall	m²	400.00	to	490.00
1500 mm high, one brick wall	m²	580.00	to	710.00
extra for commencing 2.00 m below ground level	m²	79.00	to	96.00
extra for commencing 3.00 m below ground level	m²	155.00	to	190.00
extra for commencing 4.00 m below ground level	m²	220.00	to	260.00
1.1.3 LOWEST FLOOR CONSTRUCTION				
Ground floor construction				
Mechanical excavation to reduce levels, disposal, level and compact, hardcore bed blinded with sand, 1200 gauge polythene damp-proof membrane, in situ concrete 20.00 N/mm² – 20 mm aggregate (1:2:4)				
150 mm thick concrete slab with 1 layer of A195 fabric reinforcement	m²	99.00	to	120.00
200 mm thick concrete slab with 1 layer of A252 fabric reinforcement	m²	110.00	to	130.00
250 mm thick concrete slab with 1 layer of A393 fabric reinforcement	m²	110.00	to	130.00
extra for every additional 50 mm thick concrete	m²	7.60	to	9.20
extra for alternative strength concrete C30 (30.00 N/mm²)	m³	1.45	to	1.75
extra for alternative strength concrete C40 (40.00 N/mm²)	m³	7.30	to	8.80
extra for reinforcement at 25 kg/m³ concrete	m³	30.00	to	36.50
extra for reinforcement at 50 kg/m³ concrete	m³	50.00	to	61.00
extra for reinforcement at 75 kg/m³ concrete	m³	80.00	to	97.00
extra for reinforcement at 100 kg/m³ concrete	m³	100.00	to	120.00
Warehouse ground floor				
Steel fibre reinforced floor slab placed using large pour construction techniques providing a finish floor flatness complying with FM2 special +/- 15 mm from datum. NOTE: Excavation, subbase and damp-proof membrane not included				
nominal 200 mm thick in situ concrete floor slab, concrete grade C40, reinforced with steel fibres, surface power floated and cured with a spray application of curing and hardening agent	m²	33.00	to	40.00
Suspended ground floor				
Beam and block flooring				
suspended floor with 150 mm deep precast concrete beams and infill blocks	m²	60.00	to	73.00
suspended floor with 225 mm deep precast concrete beams and infill blocks	m²	120.00	to	145.00

GENERAL BUILDING WORKS

Item	Unit	Range £		
Board or slab insulation				
Kingspan Thermafloor TF70 (Thermal conductivity 0.022 W/mK) rigid urethane floor insulation for solid concrete and suspended ground floors				
50 mm thick	m^2	12.10	to	14.70
75 mm thick	m^2	16.10	to	19.40
100 mm thick	m^2	19.10	to	23.00
125 mm thick	m^2	25.00	to	30.00
150 mm thick	m^2	29.00	to	35.00
Styrofoam Floormate 500 (Thermal conductivity 0.033 W/mK) extruded polystyrene foam or other equal and approved				
50 mm thick	m^2	18.30	to	22.00
75 mm thick	m^2	22.00	to	26.50
100 mm thick	m^2	26.00	to	32.00
1.1.5 Basements				
Reinforced concrete basement floors; NOTE: excluding bulk excavation costs and hardcore fill				
200 mm thick waterproof concrete with 2 × layers of A252 mesh reinforcement	m^2	77.00	to	93.00
Reinforced concrete basement walls; NOTE: excluding bulk excavation costs and hardcore fill				
200 mm thick waterproof reinforced concrete	m^2	370.00	to	450.00

GENERAL BUILDING WORKS

Item	Unit	Range £		
2 SUPERSTRUCTURES				
2.1 FRAME				
2.1.1 Frame and Floor Composite				
Concrete frame; flat slab reinforced concrete floors up to 250 mm thick				
Suspended slab; no coverings or finishes				
up to six storeys	m^2	150.00	to	180.00
six to twelve storeys	m^2	160.00	to	195.00
thirteen to eighteen storeys	m^2	190.00	to	230.00
Steel frame; composite beam and slab floors				
Suspended slab; permanent steel shuttering with 130 mm thick concrete; no coverings or finishes				
up to six storeys	m^2	150.00	to	180.00
seven to twelve storeys	m^2	175.00	to	210.00
thirteen to eighteen storeys	m^2	190.00	to	230.00
Concrete frame				
Reinforced in situ concrete columns, bar reinforcement, formwork. Generally all formwork assumes four uses				
Reinforcement rate 180 kg/m³; column size				
225 mm × 225 mm	m	89.00	to	110.00
300 mm × 600 mm	m	195.00	to	235.00
450 mm × 900 mm	m	350.00	to	430.00
Reinforcement rate 240 kg/m³; column size				
225 mm × 225 mm	m	98.00	to	120.00
300 mm × 600 mm	m	210.00	to	255.00
450 mm × 900 mm	m	390.00	to	470.00
In situ concrete casing to steel column, formwork; column size				
225 mm × 225 mm	m	80.00	to	97.00
300 mm × 600 mm	m	185.00	to	225.00
450 mm × 900 mm	m	320.00	to	385.00
Reinforced in situ concrete beams, bar reinforcement, formwork. Generally all formwork assumes four uses				
Reinforcement rate 200 kg/m³; beam size				
225 mm × 450 mm	m	115.00	to	140.00
300 mm × 600 mm	m	185.00	to	220.00
450 mm × 600 mm	m	235.00	to	280.00
600 mm × 600 mm	m	280.00	to	340.00
Reinforcement rate 240 kg/m³; beam size				
225 mm × 450 mm	m	120.00	to	150.00
300 mm × 600 mm	m	190.00	to	230.00
450 mm × 600 mm	m	240.00	to	295.00
600 mm × 600 mm	m	300.00	to	365.00

GENERAL BUILDING WORKS

Item	Unit	Range £		
In situ concrete casing to steel beams, formwork; beam size				
225 mm × 450 mm	m	105.00	to	130.00
300 mm × 600 mm	m	160.00	to	190.00
450 mm × 600 mm	m	210.00	to	250.00
600 mm × 600 mm	m	250.00	to	300.00
Other floor and frame constructions				
reinforced concrete cantilevered balcony; up to 1.50 m wide × 1.20 deep	nr	2800.00	to	3400.00
reinforced concrete cantilevered walkways; up to 1.00 m wide	m^2	200.00	to	240.00
reinforced concrete walkways and supporting frame; up to 1.00 m wide × 2.50 m high	m^2	260.00	to	315.00
Steel frame				
Fabricated steelwork erected on site with bolted connections, primed				
smaller sections not exceeding 40 kg/m	tonne	2175.00	to	2650.00
universal beams; grade S275	tonne	2075.00	to	2500.00
universal beams; grade S355	tonne	2175.00	to	2600.00
universal columns; grade S275	tonne	2075.00	to	2500.00
universal columns; grade S355	tonne	2175.00	to	2600.00
composite columns	tonne	2050.00	to	2475.00
hollow section circular	tonne	2350.00	to	2850.00
hollow section square or rectangular	tonne	2350.00	to	2850.00
cellular beams (FABSEC)	tonne	2350.00	to	2850.00
lattice beams	tonne	2900.00	to	3500.00
roof trusses	tonne	3000.00	to	3650.00
Steel finishes				
grit blast and one coat zinc chromate primer	m^2	8.60	to	10.40
touch up primer and one coat of two pack epoxy zinc phosphate primer	m^2	5.40	to	6.50
blast cleaning	m^2	2.80	to	3.45
galvanizing	m^2	19.20	to	23.00
galvanizing	tonne	315.00	to	380.00
Other floor and frame constructions				
steel space deck on steel frame, unprotected	m^2	56.00	to	68.00
18.00 m high bay warehouse; steel propped portal frame, cold rolled purlin sections, primed surface treatment only, excluding decorations and protection	m^2	115.00	to	140.00
columns and beams to mansard roof, 60 min fire protection	m^2	180.00	to	220.00
feature columns and beams to glazed atrium roof unprotected	m^2	190.00	to	230.00

GENERAL BUILDING WORKS

Item	Unit	Range £		
2.1 FRAME – cont				
Fire protection to steelwork				
Sprayed mineral fibre; gross surface area				
60 minute protection	m^2	17.60	to	21.00
90 minute protection	m^2	31.00	to	38.00
Sprayed vermiculite cement; gross surface area				
60 minute protection	m^2	20.00	to	24.00
90 minute protection	m^2	29.50	to	36.00
Supply and fit fire-resistant boarding to steel columns and beams; noggins, brackets and angles, intumescent paste. Beamclad or similar; measure board area				
30 minute protection for concealed applications	m^2	29.00	to	35.00
60 minute protection for concealed applications	m^2	64.00	to	78.00
30 minute protection left exposed for decoration	m^2	32.00	to	38.50
60 minute protection left exposed for decoration	m^2	77.00	to	93.00
Intumescent fire protection coating/decoration to exposed steelwork; Gross surface area (m^2) or per tonne; On site application, spray applied				
30 minute protection per m^2	m^2	12.20	to	14.80
30 minute protection per tonne	tonne	310.00	to	375.00
60 minute protection per m^2	m^2	15.50	to	18.70
60 minute protection per tonne	tonne	390.00	to	475.00
Intumescent fire protection coating/decoration to exposed steelwork; Gross surface area (m^2) or per tonne; Off site application, spray applied				
60 minute protection per m^2	m^2	24.00	to	29.00
60 minute protection per tonne	tonne	620.00	to	750.00
90 minute protection per m^2	m^2	44.00	to	53.00
90 minute protection per tonne	tonne	1125.00	to	1350.00
120 minute protection per m^2	m^2	75.00	to	91.00
120 minute protection per tonne	tonne	1900.00	to	2300.00
2.2 UPPER FLOORS				
2.2.1 Floors				
Composite steel deck flooring				
Composite steel and concrete upper floors (Note: all floor thicknesses are nominal); A142 mesh reinforcement				
TATA Slimdek SD225 steel decking 1.25 mm thick; 130 mm reinforced concrete topping	m^2	97.00	to	120.00
Re-entrant type steel deck 0.90 mm thick; 150 mm reinforced concrete	m^2	82.00	to	99.00
Re-entrant type steel deck 60 mm deep, 1.20 mm thick; 200 mm reinforced concrete	m^2	88.00	to	105.00
Trapezoidal steel decking 0.90 mm thick; 150 mm reinforced concrete topping	m^2	84.00	to	100.00
Trapezoidal steel decking 1.20 mm thick; 150 mm reinforced concrete topping	m^2	88.00	to	105.00
Composite precast concrete and in situ concrete upper floors				
Omnidec flooring system comprising of 50 mm thick precast concrete deck (as permanent shuttering); 210 mm polystyrene void formers; concrete topping and reinforcement	m^2	120.00	to	140.00

GENERAL BUILDING WORKS

Item	Unit	Range £		
Precast concrete suspended floors; 1200 mm wide suspended slab; 75 mm thick screed; no coverings or finishes				
3.00 m span; 150 mm thick planks; 5.00 kN/m^2 loading	m^2	76.00	to	88.00
6.00 m span; 150 mm thick planks; 5.00 kN/m^2 loading	m^2	78.00	to	90.00
7.50 m span; 200 mm thick planks; 5.00 kN/m^2 loading	m^2	82.00	to	95.00
9.00 m span; 250 mm thick planks; 5.00 kN/m^2 loading	m^2	86.00	to	99.00
12.00 m span; 350 mm thick planks; 5.00 kN/m^2 loading	m^2	90.00	to	105.00
3.00 m span; 150 mm thick planks; 8.50 kN/m^2 loading	m^2	75.00	to	86.00
6.00 m span; 200 mm thick planks; 8.50 kN/m^2 loading	m^2	77.00	to	89.00
7.50 m span; 250 mm thick planks; 8.50 kN/m^2 loading	m^2	86.00	to	99.00
3.00 m span; 150 mm thick planks; 12.50 kN/m^2 loading	m^2	78.00	to	90.00
6.00 m span; 250 mm thick planks; 12.50 kN/m^2 loading	m^2	86.00	to	99.00
Post-tensioned concrete upper floors				
reinforced post-tensioned suspended concrete slab 150–225 mm thick, 40 N/mm^2, reinforcement 60 kg/m^3, formwork	m^2	110.00	to	135.00
Softwood floors				
Joisted floor; plasterboard ceiling; skim; emulsion; t&g chipboard, sheet vinyl flooring and painted softwood skirtings	m^2	68.00	to	89.00
2.3 ROOF				
2.3.1 Roof Structure				
Flat roof decking; Structure only				
Softwood				
comprising roof joists; 100 mm × 50 mm wall plates; herringbone strutting; 18 mm thick external quality plywood boarding	m^2	55.00	to	67.00
Metal decking				
galvanized steel roof decking; insulation; three layer felt roofing and chippings; 0.70 mm thick steel decking (U-value = 0.25 W/m^2 K)	m^2	75.00	to	91.00
aluminium roof decking; three layer felt roofing and chippings; 0.90 mm thick aluminium decking (U-value = 0.25 W/m^2 K)	m^2	8.50	to	10.30
Concrete				
precast concrete suspended slab with sand: cement screed over	m^2	82.00	to	99.00
reinforced concrete slabs; on steel permanent steel shuttering; 150 mm reinforced concrete topping	m^2	84.00	to	100.00
Screeds/decks to receive roof coverings				
50 mm thick cement and sand screed	m^2	18.10	to	22.00
75 mm thick lightweight bituminous screed and vapour barrier	m^2	63.00	to	77.00
Softwood trussed pitched roofs; Structure only				
Timber roof trusses and bracings only; plan area				
comprising 75 mm × 50 mm roof trusses at 600 mm centres (measured on plan)	m^2	36.00	to	44.00
comprising 100 mm × 38 mm roof trusses at 600 mm centres (measured on plan)	m^2	43.00	to	52.00
Mansard type roof comprising 100 mm × 50 mm roof trusses at 600 mm centres; 70° pitch	m^2	44.00	to	53.00
forming small dormers up to 1.50 m^2	nr	740.00	to	890.00

GENERAL BUILDING WORKS

Item	Unit	Range £		
2.3 ROOF – cont				
2.3.2 Roof Coverings				
Timber trusses with tile coverings				
Timber roof trusses; insulation; roof coverings; PVC rainwater goods; plasterboard; skim and emulsion to ceilings (U-value = 0.25 W/m^2 K); roof plan area				
concrete interlocking tile coverings	m^2	190.00	to	230.00
clay plain tile coverings	m^2	195.00	to	240.00
clay pantile coverings	m^2	210.00	to	255.00
natural Welsh slate coverings	m^2	260.00	to	320.00
reconstructed stone coverings	m^2	250.00	to	300.00
Timber dormer roof trusses; insulation; roof coverings; PVC rainwater goods; plasterboard; skim and emulsion to ceilings (U-value = 0.25 W/m^2 K)				
concrete interlocking tile coverings	m^2	220.00	to	270.00
clay pantile coverings	m^2	210.00	to	250.00
plain clay tile coverings	m^2	220.00	to	265.00
natural slate coverings	m^2	330.00	to	400.00
reconstructed stone coverings	m^2	280.00	to	340.00
Comparative tiling and slating finishes; including underfelt, battening, eaves courses and ridges				
concrete troughed or bold roll interlocking tiles; sloping	m^2	67.00	to	81.00
Tudor clay pantiles; sloping	m^2	81.00	to	98.00
blue composition (cement fibre) slates; sloping	m^2	110.00	to	130.00
Spanish slates; sloping	m^2	130.00	to	160.00
Welsh slates; sloping	m^2	220.00	to	270.00
reconstructed stone slates; random slates; sloping	m^2	170.00	to	205.00
Steel trussed with metal sheet cladding				
Steel roof trusses and beams; thermal and acoustic insulation (U-value = 0.25 W/m^2 K)				
aluminium profiled composite cladding	m^2	295.00	to	360.00
copper roofing on boarding	m^2	325.00	to	395.00
Sheet roof claddings				
Fibre cement sheet profiled cladding				
Profile 6; single skin; natural grey finish	m^2	26.00	to	32.00
P61 Insulated System; natural grey finish; metal inner lining panel (U-value = 0.25 W/m^2 K)	m^2	50.00	to	60.00
extra for coloured fibre cement sheeting	m^2	1.10	to	1.35
double skin GRP translucent roof sheets	m^2	55.00	to	67.00
triple skin GRP translucent roof sheets	m^2	62.00	to	75.00
Steel PVF2 coated galvanized trapezoidal profile cladding fixed to steel purlins (not included); for roof pitches greater than 4°				
single skin trapezoidal sheeting only	m^2	16.60	to	20.00
composite insulated roofing system; 80 mm overall panel thickness (U-value = 0.25 W/m^2 K)	m^2	44.00	to	53.00
composite insulated roofing system; 115 mm overall panel thickness (U-value = 0.18 W/m^2 K)	m^2	46.00	to	56.00

GENERAL BUILDING WORKS

Item	Unit	Range £		
composite insulated roofing system; 150 mm overall panel thickness (U-value = 0.14 W/m² K)	m²	62.00	to	75.00
For very shallow roof pitches				
standing seam joints composite insulated roofing system; 90 mm overall panel thickness (U-value = 0.25 W/m² K)	m²	71.00	to	86.00
standing seam joints composite insulated roofing system; 110 mm overall panel thickness (U-value = 0.20 W/m² K)	m²	77.00	to	93.00
standing seam joints composite insulated roofing system; 125 mm overall panel thickness (U-value = 0.18 W/m² K)	m²	80.00	to	97.00
Aluminium roofing; standing seam				
Kalzip standard natural aluminium; 0.9 mm thick; 180 mm glassfibre insulation; vapour control layer; liner sheets (U-value = 0.25 W/m² K)	m²	65.00	to	79.00
Copper roofing				
copper roofing with standing seam joints; insulation breather membrane or vapour barrier (U-value = 0.25 W/m² K)	m²	190.00	to	230.00
Stainless steel				
Terne-coated stainless steel roofing on and including Metmatt underlay (U-value = 0.25 W/m² K)	m²	160.00	to	200.00
Zinc; standing seam				
Terne-coated zinc roofing on and including Metmatt underlay (U-value = 0.25 W/m² K)	m²	115.00	to	140.00
Lead				
Roof coverings in welded seam construction including Geotec underlay fixed to prevent lifting and distortion	m²	165.00	to	200.00
Flat Roofing Systems				
Includes insulation and vapour control barrier; excludes decking (U-value = 0.25 W/m² K)				
single layer polymer roofing membrane	m²	75.00	to	91.00
single layer polymer roofing membrane with tapered insulation	m²	145.00	to	180.00
20 mm thick polymer modified asphalt roofing including felt underlay	m²	100.00	to	125.00
high performance bitumen felt roofing system	m²	99.00	to	120.00
high performance polymer modified bitumen membrane	m²	115.00	to	140.00
Kingspan KS1000TD composite Single Ply roof panels for roof pitches greater than 0.7° (after deflection); 1.5 mm Single Ply External Covering, Internal Coating Bright White Polyester (steel)				
71 mm thick (U-value = 0.25)	m²	58.00	to	70.00
100 mm thick (U-value = 0.18)	m²	59.00	to	71.00
120 mm thick (U-value = 0.15)	m²	68.00	to	82.00
Edges to felt flat roofs; softwood splayed fillet				
280 mm × 25 mm painted softwood fascia; no gutter aluminium edge trim	m	48.00	to	58.00
Edges to flat roofs; code 4 lead drip dresses into gutter; 230 mm × 25 mm painted softwood fascia; buildup				
170 mm uPVC gutter	m	110.00	to	135.00
100 mm aluminium gutter	m	135.00	to	165.00
100 mm cast iron gutter; decorated	m	120.00	to	145.00

GENERAL BUILDING WORKS

Item	Unit	Range £		
2.3 ROOF – cont				
2.3.3 Specialist Roof Systems				
Landscaped roofs				
Polyester-based elastomeric bitumen waterproofing and vapour equalization layer, copper lined bitumen membrane root barrier and waterproofing layer, separation and slip layers, protection layer, 50 mm thick drainage board, filter fleece, insulation board, Sedum vegetation blanket				
intensive (high maintenance – may include trees and shrubs require deeper substrate layers, are generally limited to flat roofs)	m^2	200.00	to	240.00
extensive (low maintenance – herbs, grasses, mosses and drought tolerant succulents such as Sedum)	m^2	170.00	to	205.00
Ethylene tetrafluoroethylene (ETFE) systems				
Multiple layered ETFE inflated cushions supported by a lightweight aluminium or steel structure	m^2	880.00	to	1050.00
2.3.4 Roof Drainage				
Gutters				
100 mm uPVC gutter	m	30.00	to	36.00
170 mm uPVC gutter	m	49.50	to	60.00
100 mm cast iron gutter; decorated	m	47.00	to	57.00
150 mm cast iron gutter; decorated	m	76.00	to	92.00
Rainwater downpipes pipes; fixed to backgrounds; including offsets and shoes				
68 mm dia. uPVC	m	28.00	to	34.00
110 mm dia. uPVC	m	49.00	to	60.00
75 mm dia. cast iron; decorated	m	71.00	to	86.00
100 mm dia. cats iron; decorated	m	92.00	to	110.00
2.3.5 Rooflights/Patent Glazing and Glazed Roofs				
Rooflights				
individual polycarbonate rooflights; rectangular; fixed light	m^2	520.00	to	630.00
individual polycarbonate rooflights; rectangular; manual opening	m^2	720.00	to	870.00
individual polycarbonate rooflights; rectangular; electric opening	m^2	1075.00	to	1325.00
Velux style rooflights to traditional roof construction (tiles\slates)	m^2	570.00	to	690.00
Patent glazing; including flashings, standard aluminium alloy bars; Georgian wired or laminated glazing; fixed lights				
single glazed	m^2	205.00	to	250.00
low-e clear toughened and laminated double glazed units	m^2	540.00	to	650.00
2.3.6 Roof Features				
Flat roof feature				
roof walkways 600 mm × 600 mm × 50 mm precast concrete slabs on support system; pedestrian access	m^2	56.00	to	68.00
solar reflective paint	m^2	5.30	to	6.40
limestone chipping finish	m^2	5.70	to	6.90
grip tiles in hot bitumen	m^2	70.00	to	85.00

GENERAL BUILDING WORKS

Item	Unit	Range £		
Insulation				
Glass fibre roll; Crown Loft Roll 44 (Thermal conductivity 0.044 W/mK) or other equal; laid loose				
100 mm thick	m^2	2.80	to	3.40
150 mm thick	m^2	3.70	to	4.45
200 mm thick	m^2	4.65	to	5.60
Glass fibre quilt; Isover Modular roll (Thermal conductivity 0.043 W/mK) or other equal and approved; laid loose				
100 mm thick	m^2	2.80	to	3.40
150 mm thick	m^2	3.60	to	4.40
200 mm thick	m^2	5.40	to	6.50
Crown Rafter Roll 32 (Thermal conductivity 0.032 W/mK) glass fibre flanged building roll; pinned vertically or to slope between timber framing				
50 mm thick	m^2	5.80	to	7.05
75 mm thick	m^2	7.55	to	9.15
100 mm thick	m^2	9.20	to	11.20
Thermafleece EcoRoll (0.039W/mK)				
50 mm thick	m^2	6.75	to	8.20
75 mm thick	m^2	8.60	to	10.40
100 mm thick	m^2	10.60	to	12.90
140 mm thick	m^2	14.40	to	17.50
2.4 STAIRS AND RAMPS				
2.4.1 Stair/Ramp Structures				
Reinforced concrete construction				
Escape staircase; granolithic finish; mild steel balustrades and handrails				
3.00 m rise; dogleg	nr	6600.00	to	8000.00
plus or minus for each 300 mm variation in storey height	nr	630.00	to	770.00
Staircase; terrazzo finish; mild steel balustrades and handrails; plastered soffit; balustrades and staircase soffit decorated				
3.00 m rise; dogleg	nr	9900.00	to	12000.00
plus or minus for each 300 mm variation in storey height	nr	960.00	to	1175.00
Staircase; terrazzo finish; stainless steel balustrades and handrails; plastered and decorated soffit				
3.00 m rise; dogleg	nr	12000.00	to	14000.00
plus or minus for each 300 mm variation in storey height	nr	1150.00	to	1400.00
Staircase; high quality finishes; stainless steel and glass balustrades; plastered and decorated soffit				
3.00 m rise; dogleg	nr	1875.00	to	2275.00
plus or minus for each 300 mm variation in storey height	nr	2100.00	to	2550.00
Metal construction				
Steel access/fire ladder				
3.00 m high	nr	760.00	to	920.00
4.00 m high; epoxide finished	nr	1150.00	to	1375.00

GENERAL BUILDING WORKS

Item	Unit	Range £		
2.4 STAIRS AND RAMPS – cont				
Metal construction – cont				
Light duty metal staircase; galvanized finish; perforated treads; no risers; balustrades and handrails; decorated				
3.00 m rise; straight; 900 mm wide	nr	4400.00	to	5300.00
plus or minus for each 300 mm variation in storey height	nr	350.00	to	425.00
Light duty circular metal staircase; galvanized finish; perforated treads; no risers; balustrades and handrails; decorated				
3.00 m rise; straight; 1548 mm dia.	nr	4850.00	to	5900.00
plus or minus for each 300 mm variation in storey height	nr	415.00	to	500.00
Heavy duty cast iron staircase; perforated treads; no risers; balustrades and hand rails; decorated				
3.00 m rise; straight	nr	5500.00	to	6600.00
plus or minus for each 300 mm variation in storey height	nr	570.00	to	690.00
3.00 m rise; spiral; 1548 mm dia.	nr	6300.00	to	7700.00
plus or minus for each 300 mm variation in storey height	nr	610.00	to	740.00
Feature metal staircase; galvanized finish perforated treads; no risers; decorated				
3.00 m rise; spiral balustrades and handrails	nr	7200.00	to	8700.00
3.00 m rise; dogleg; hardwood balustrades and handrails	nr	8400.00	to	10000.00
3.00 m rise; dogleg; stainless steel balustrades and handrails	nr	11000.00	to	13000.00
plus or minus for each 300 mm variation in storey height	nr	710.00	to	860.00
galvanized steel catwalk; nylon coated balustrading 450 mm wide	m	400.00	to	480.00
Timber construction				
Softwood staircase; softwood balustrades and hardwood handrail; plasterboard; skim and emulsion to soffit				
2.60 m rise; standard; straight flight	nr	1125.00	to	1375.00
2.60 m rise; standard; top three treads winding	nr	1325.00	to	1575.00
2.60 m rise; standard; dogleg	nr	1450.00	to	1750.00
Oak staircase; balustrades and handrails; plasterboard; skim and emulsion to soffit				
2.60 m rise; purpose-made; dogleg	nr	9600.00	to	11500.00
plus or minus for each 300 mm variation in storey height	nr	1275.00	to	1550.00
2.4.2 Stair/Ramp Finishes				
Finishes to treads and risers; including nosings etc.				
vinyl or rubber	m	60.00	to	73.00
carpet (PC sum £22/m²)	m	93.00	to	110.00
2.4.3 Stair/Ramp Balustrades and Handrails				
Wall handrails				
softwood handrail and brackets	m	84.00	to	100.00
hardwood handrail and brackets	m	115.00	to	140.00
mild steel handrail and brackets	m	155.00	to	190.00
stainless steel handrail and brackets	m	195.00	to	235.00
Balustrading and handrails				
mild steel balustrade and steel or timber handrail	m	305.00	to	370.00
balustrade and handrail with metal infill panels	m	385.00	to	465.00
balustrade and handrail with glass infill panels	m	425.00	to	520.00

GENERAL BUILDING WORKS

Item	Unit	Range £		
stainless steel balustrade and handrail	m	510.00	to	610.00
stainless steel and structural glass balustrade	m	890.00	to	1075.00
2.5 EXTERNAL WALLS				
2.5.1 External Enclosing Walls above Ground Level				
Brick/block walling; solid walls				
Common brick solid walls; bricks PC £380.00/1000				
half brick thick	m^2	58.00	to	70.00
one brick thick	m^2	110.00	to	130.00
one and a half brick thick	m^2	160.00	to	190.00
two brick thick	m^2	205.00	to	250.00
extra for fair face per side	m^2	2.55	to	3.10
Engineering brick walls; class B; bricks PC £590.00/1000				
half brick thick	m^2	65.00	to	78.00
one brick thick	m^2	120.00	to	145.00
one and a half brick thick	m^2	170.00	to	210.00
two brick thick	m^2	230.00	to	275.00
Facing brick walls; sand faced facings; bricks PC £500.00/1000				
half brick thick; pointed one side	m^2	74.00	to	89.00
one brick thick; pointed both sides	m^2	140.00	to	170.00
Facing bricks solid walls; handmade facings; bricks PC £700.00/1000				
half brick thick; fair face one side	m^2	89.00	to	110.00
one brick thick; fair face both sides	m^2	170.00	to	210.00
add or deduct for each variation of £10.00/1000 in PC value				
half brick thick	m^2	0.75	to	0.90
one brick thick	m^2	1.50	to	1.80
one and a half brick thick	m^2	2.20	to	2.70
two brick thick	m^2	3.00	to	3.60
Aerated lightweight block walls				
100 mm thick	m^2	38.00	to	46.00
140 mm thick	m^2	51.00	to	62.00
215 mm thick	m^2	97.00	to	120.00
Dense aggregate block walls				
100 mm thick	m^2	42.00	to	51.00
140 mm thick	m^2	51.00	to	62.00
Coloured dense aggregate masonry block walls; Lignacite or similar				
100 mm thick single skin, weathered face, self coloured	m^2	87.00	to	105.00
100 mm thick single skin, polished face	m^2	135.00	to	160.00
Brick/block walling; cavity walls				
Cavity wall; facing brick outer skin; insulation; with plaster on standard weight block inner skin; emulsion (U-value = 0.30 $W/m^2 K$)				
machine-made facings; PC £550.00/1000	m^2	140.00	to	170.00
machine-made facings; PC £700.00/1000	m^2	160.00	to	190.00
self finished masonry block, weathered finish outer skin and paint grade block inner skins; fair faced both sides	m^2	140.00	to	170.00
self finished masonry block, polished finish outer skin and paint grade block inner skins; fair faced both sides	m^2	185.00	to	225.00

GENERAL BUILDING WORKS

Item	Unit	Range £		
2.5 EXTERNAL WALLS – cont				
Brick/block walling – cont				
Cavity wall; rendered block outer skin; insulation; with plaster on standard weight block inner skin; emulsion (U-value = 0.30 W/m^2 K)				
block outer skin; insulation; lightweight block inner skin outer block rendered	m^2	170.00	to	210.00
Cavity wall; facing brick outer skin; insulation; plasterboard on stud inner skin; emulsion (U-value = 0.30 W/m^2 K)				
machine-made facings; PC £550.00/1000	m^2	110.00	to	130.00
machine-made facings; PC £700.00/1000	m^2	125.00	to	150.00
Reinforced concrete walling				
In situ reinforced concrete 25.00 N/mm^2; 15 kg/m^2 reinforcement; formwork both sides				
150 mm thick	m^2	140.00	to	165.00
225 mm thick	m^2	155.00	to	190.00
300 mm thick	m^2	170.00	to	205.00
Panelled walling				
Precast concrete panels; including insulation; lining and fixings generally 7.5 m × 0.15 thick × storey height (U-value = 0.30 W/m^2 K)				
standard panels	m^2	270.00	to	325.00
standard panels; exposed aggregate finish	m^2	290.00	to	350.00
reconstructed stone faced panels	m^2	51.00	to	62.00
brick clad panels (P.C. £350.00 / 1000 for bricks)	m^2	43.00	to	52.00
natural stone faced panels (Portland Stone or similar)	m^2	580.00	to	700.00
marble or granite faced panels	m^2	750.00	to	910.00
Kingspan TEK cladding panel; 142 mm thick (15 mm thick OSB board and 112 mm thick rigid eurathane core) fixed to frame; metal edge trim flashings	m^2	385.00	to	465.00
Sheet claddings				
Cement fibre profiled cladding				
Profile 6; single skin; natural grey finish	m^2	27.00	to	33.00
P61 Insulated System; natural grey finish; metal inner lining panel (U-value = 0.30 W/m^2 K)	m^2	50.00	to	61.00
P61 Insulated System; natural grey finish; 100 mm lightweight concrete block (U-value = 0.30 W/m^2 K)	m^2	99.00	to	120.00
P61 Insulated System; natural grey finish; 12 mm self finish plasterboard lining (U-value = 0.30 W/m^2 K)	m^2	69.00	to	84.00
extra for coloured fibre cement sheeting	m^2	0.80	to	1.00
Metal profiled cladding				
Standard trapazoidal profile				
composite insulated roofing system; 60 mm overall panel thickness (U-value = 0.35 W/m^2 K)	m^2	50.00	to	61.00
composite insulated roofing system; 80 mm overall panel thickness (U-value = 0.26 W/m^2 K)	m^2	53.00	to	64.00
composite insulated roofing system; 100 mm overall panel thickness (U-value = 0.20 W/m^2 K)	m^2	56.00	to	68.00

GENERAL BUILDING WORKS

Item	Unit	Range £		
Micro-rib profile				
composite insulated roofing system; micro-rib panel; 60 mm overall panel thickness (U-value = 0.35 W/m^2 K)	m^2	140.00	to	170.00
composite insulated roofing system; micro-rib panel; 80 mm overall panel thickness (U-value = 0.26 W/m^2 K)	m^2	150.00	to	180.00
composite insulated roofing system; micro-rib panel; 100 mm overall panel thickness (U-value = 0.20 W/m^2 K)	m^2	160.00	to	200.00
Flat panel				
composite insulated roofing system; micro-rib panel; 60 mm overall panel thickness (U-value = 0.35 W/m^2 K)	m^2	225.00	to	270.00
composite insulated roofing system; micro-rib panel; 80 mm overall panel thickness (U-value = 0.26 W/m^2 K)	m^2	215.00	to	260.00
composite insulated roofing system; micro-rib panel; 100 mm overall panel thickness (U-value = 0.20 W/m^2 K)	m^2	250.00	to	305.00
Glazed walling				
Curtain walling				
stick curtain walling with double glazed units, aluminium structural framing and spandrel rails. Standard colour powder coated capped	m^2	540.00	to	650.00
unitized curtain walling system with double glazed units, aluminium structural framing and spandrel rails. Standard colour powder coated	m^2	1125.00	to	1350.00
unitized curtain walling bespoke project specific system with double glazed units, aluminium structural framing and spandrel rails. Standard colour powder coated.	m^2	1125.00	to	1350.00
visual mock-ups for project specific curtain walling solutions	item	36000.00	to	43500.00
Other systems				
lift surround of double glazed or laminated glass with aluminium or stainless steel framing	m^2	1000.00	to	1200.00
Other cladding systems				
Tiles (clay/slate/glass/ceramic)				
machine-made clay tiles; including battens	m^2	33.00	to	40.00
best handmade sand faced tiles; including battens	m^2	45.00	to	55.00
concrete plain tiles; including battens	m^2	47.00	to	57.00
natural slates; including battens	m^2	94.00	to	110.00
20 mm × 20 mm thick mosaic glass or ceramic; in common colours; fixed on prepared surface	m^2	120.00	to	150.00
Steel				
vitreous enamelled insulated steel sandwich panel system; with insulation board on inner face	m^2	210.00	to	255.00
Formalux sandwich panel system; with coloured lining tray; on steel cladding rails	m^2	240.00	to	295.00
Comparative external finishes				
Comparative concrete wall finishes				
wrought formwork one side including rubbing down	m^2	6.20	to	7.45
shotblasting to expose aggregate	m^2	7.50	to	9.05
bush hammering to expose aggregate	m^2	19.30	to	23.00

GENERAL BUILDING WORKS

Item	Unit	Range £		
2.5 EXTERNAL WALLS – cont				
Comparative external finishes – cont				
Comparative in situ finishes				
two coats Sandtex Matt cement paint to render	m^2	6.55	to	7.95
13 mm thick cement and sand plain face rendering	m^2	24.50	to	30.00
ready-mixed self-coloured acrylic resin render on blockwork	m^2	65.00	to	79.00
three coat Tyrolean rendering; including backing	m^2	47.00	to	57.00
Stotherm Lamella; 6 mm thick work in 2 coats	m^2	91.00	to	110.00
Insulation				
Crown Dritherm Cavity Slab 37 (Thermal conductivity 0.037 W/mK) glass fibre batt or other equal; as full or partial cavity fill; including cutting and fitting around wall ties and retaining discs				
50 mm thick	m^2	4.25	to	5.15
75 mm thick	m^2	5.85	to	7.10
100 mm thick	m^2	6.70	to	8.10
Crown Dritherm Cavity Slab 34 (Thermal conductivity 0.034 W/mK) glass fibre batt or other equal; as full or partial cavity fill; including cutting and fitting around wall ties and retaining discs				
65 mm thick	m^2	4.45	to	5.40
75 mm thick	m^2	4.95	to	6.00
85 mm thick	m^2	6.30	to	7.65
100 mm thick	m^2	6.45	to	7.80
Crown Dritherm Cavity Slab 32 (Thermal conductivity 0.032 W/mK) glass fibre batt or other equal; as full or partial cavity fill; including cutting and fitting around wall ties and retaining discs				
65 mm thick	m^2	5.10	to	6.20
75 mm thick	m^2	5.80	to	7.00
85 mm thick	m^2	6.20	to	7.50
100 mm thick	m^2	6.95	to	8.40
Crown Frametherm Roll 40 (Thermal conductivity 0.040 W/mK) glass fibre semi-rigid or rigid batt or other equal; pinned vertically in timber frame construction				
90 mm thick	m^2	6.10	to	7.35
140 mm thick	m^2	8.15	to	9.85
Kay-Cel (Thermal conductivity 0.033 W/mK) expanded polystyrene board standard grade SD/N or other equal; fixed with adhesive				
25 mm thick	m^2	6.45	to	7.80
50 mm thick	m^2	9.30	to	11.30
75 mm thick	m^2	12.10	to	14.70
100 mm thick	m^2	14.70	to	17.80
Kingspan Kooltherm K8 (Thermal conductivity 0.022 W/mK) zero ODP rigid urethene insulation board or other equal; as partial cavity fill; including cutting and fitting around wall ties and retaining discs				
40 mm thick	m^2	15.90	to	19.30
50 mm thick	m^2	15.90	to	19.30
60 mm thick	m^2	18.40	to	22.00
75 mm thick	m^2	18.20	to	22.00

GENERAL BUILDING WORKS

Item	Unit	Range £		
Kingspan Thermawall TW50 (Thermal conductivity 0.022 W/mK) zero ODP rigid urethene insulation board or other equal and approved; as partial cavity fill; including cutting and fitting around wall ties and retaining discs				
25 mm thick	m^2	12.30	to	14.90
50 mm thick	m^2	14.90	to	18.10
75 mm thick	m^2	20.00	to	24.50
100 mm thick	m^2	25.00	to	30.50
Thermafleece TF35 high density wool insulating batts (0.035 W/mK); 60% British wool, 30% recycled polyester and 10% polyester binder with a high recycled content				
50 mm thick	m^2	15.10	to	18.20
75 mm thick	m^2	19.20	to	23.00
2.5.3 Solar/Rain Screen				
Rainscreen				
25 mm thick tongued and grooved tanalized softwood boarding; including timber battens	m^2	49.50	to	60.00
timber shingles, Western Red Cedar, preservative treated in random widths; not including any sub-frame or battens	m^2	64.00	to	77.00
25 mm thick tongued and grooved Western Red Cedar boarding including timber battens	m^2	56.00	to	68.00
25 mm thick Western Red Cedar tongued and grooved wall cladding on and including treated softwood battens on breather membrane, 10 mm Eternit Blueclad board and 50 mm insulation board; the whole fixed to Metsec frame system; including sealing all joints etc.	m^2	130.00	to	160.00
Sucupira Preta timber boarding 19 mm thick × 75 mm wide slats with chamfered open joints fixed to 38 mm × 50 mm softwood battens with stainless steel screws on 100 mm Kingspan K2 insulations and 28 mm WBP plywoodd fxed to 50 mm × 1000 softwood studs; complete with breather membrane, aluminium cills and Sucupira Petra batten at external joints	m^2	310.00	to	380.00
Trespa single skin rainscreen cladding, 8 mm thick panel, with secondary support/frame system; adhesive fixed panels; open joints; 100 mm Kingspan K15 insulation; aluminium sub frame fixed to masonry/ concrete	m^2	260.00	to	315.00
Terracotta rainscreen cladding; aluminium support rails; anti-graffiti coating	m^2	445.00	to	540.00
Corium brick tiles in metal tray system; standard colour; including all necessary angle trim on main support system; comprising of polythene vapour check; 75 mm × 50 mm timber studs; 50 mm thick rigid insulation with taped joints; 38 mm × 50 mm timber counter battens	m^2	340.00	to	410.00
Solar shading				
fixed aluminium Brise Soleil including uni-strut supports; 300 mm deep	m	180.00	to	220.00

GENERAL BUILDING WORKS

Item	Unit	Range £		
2.6 WINDOWS AND EXTERNAL DOORS				
2.6.1 External Windows				
Softwood windows (U-value = 1.6 W/m² K)				
Standard windows				
painted; double glazed; up to 1.50 m²	m²	475.00	to	570.00
painted; double glazed; over 1.50 m², up to 3.20 m²	m²	355.00	to	430.00
Purpose-made windows				
painted; double glazed; up to 1.50 m²	m²	650.00	to	780.00
painted; double glazed; over 1.50 m²	m²	580.00	to	700.00
Hardwood windows (U-value = 1.4 W/m² K)				
Standard windows; stained				
double glazed	m²	1050.00	to	1275.00
Purpose-made windows; stained				
double glazed	m²	1250.00	to	1525.00
Steel windows (U-value = 1.6 W/m² K)				
Standard windows				
double glazed; powder coated	m²	600.00	to	720.00
Purpose-made windows				
double glazed; powder coated	m²	830.00	to	1000.00
uPVC windows				
Windows; standard ironmongery; sills and factory glazed with low E 24 mm double glazing				
standard with low E 24 mm double glazing	m²	230.00	to	275.00
WER A rating	m²	230.00	to	280.00
WER C rating	m²	230.00	to	275.00
Secured by Design accreditation	m²	230.00	to	280.00
extra for colour finish to uPVC	m²	60.00	to	73.00
Composite aluminium/timber windows; U value = 1.5 W/m² K				
Purpose-made windows; stainless steel ironmongery; Velfac System 200 or similar				
fixed windows up to 1.50 m²	m²	290.00	to	350.00
fixed windows over 1.50 m² up to 4.00 m²	m²	260.00	to	310.00
outward opening pivot windows up to 1.50 m²	m²	700.00	to	850.00
outward opening pivot windows over 1.50 m² up to 4.00 m²	m²	310.00	to	375.00
round porthole window 1.40 m dia.	nr	1025.00	to	1250.00
round porthole window 1.80 m dia.	nr	155.00	to	190.00
round porthole window 2.60 m dia.	nr	3000.00	to	3600.00
purpose-made entrance screens and doors double glazed	m²	1250.00	to	1500.00

GENERAL BUILDING WORKS

Item	Unit	Range £		
2.6.2 External Doors				
Standard external softwood doors and hardwood frames; doors painted; including ironmongery				
Matchboarded, framed, ledged and braced door, 838 mm × 1981 mm	nr	580.00	to	700.00
flush door; cellular core; plywood faced; 838 mm × 1981 mm	nr	590.00	to	720.00
Heavy duty solid flush door				
single leaf	nr	1250.00	to	1500.00
double leaf	nr	2050.00	to	2475.00
single leaf; emergency fire exit	nr	1700.00	to	2050.00
double leaf; emergency fire exit	nr	2475.00	to	3000.00
Steel external doors				
Standard doors				
single external steel door, including frame, ironmongery, powder coated finish	nr	1025.00	to	1250.00
single external steel security door, including frame, ironmongery, powder coated finish	nr	2050.00	to	2475.00
double external steel security door, including frame, ironmongery, powder coated finish	nr	3050.00	to	3700.00
Overhead doors				
electric operation standard lift, 42 mm thick insulated sandwich panels	m^2	230.00	to	280.00
rapid lift fabric door, external, electric operation	m^2	990.00	to	1175.00
Dock shelters				
Curtain mechanical shelter; extruded aluminium frame; spring loaded-frame connected with parallel bracing arms; 2 side curtains, one top curtain; double-layered high quality polyester, coated on both sides, longitudinal strength: approx. 750 N, transverse strength	n	1275.00	to	1475.00
Inflatable mechanical shelter; hot dipped galvanized surface treatment, polyester painted, top bag with polyester fabric panels; side bags with polyester fabric panels, strong blower for straight movement of top and side bags; side section with steel guards; colour from standard range	n	4000.00	to	4650.00

GENERAL BUILDING WORKS

Item	Unit	Range £		
2.6 WINDOWS AND EXTERNAL DOORS – cont				
uPVC external doors				
Entrance doors; residential standard; PVCu frame; brass furniture (spyhole/security chain/letter plate/draught excluder/multipoint locking)				
overall 900 × 2100 mm half glazed	nr	490.00	to	590.00
overall 900 × 2100 mm half glazed; WER A rated	nr	495.00	to	600.00
overall 900 × 2100 mm half glazed; WER C rated	nr	490.00	to	590.00
overall 900 × 2100 mm half glazed; coloured	nr	560.00	to	670.00
Composite aluminium/timber entrance screens and doors; U-value = 1.5 W/m^2 K				
Purpose-made doors				
glazed single personnel door; stainless steel ironmongery; Velfac System100 or similar	nr	2150.00	to	2600.00
glazed double personnel door; stainless steel ironmongery; Velfac System 100 or similar	pair	3300.00	to	4000.00
automatic revolving door; 2000 mm dia.; clear laminated glazing; 4nr wings; glazed curved walls	nr	39000.00	to	47000.00
automatic sliding door; bi-parting	m^2	7600.00	to	9200.00
Stainless steel entrance screens and doors				
Purpose-made screen; double glazed				
with manual doors	m^2	2500.00	to	3050.00
with automatic doors	m^2	2850.00	to	3450.00
Shop fronts, shutters and grilles				
Purpose-made screen				
temporary timber shop fronts	m^2	84.00	to	100.00
hardwood and glass; including high enclosed window beds	m^2	930.00	to	1125.00
flat facade; glass in aluminium framing; manual centre doors only	m^2	800.00	to	970.00
grilles or shutters	m^2	475.00	to	580.00
fire shutters; powers operated	m^2	460.00	to	560.00

GENERAL BUILDING WORKS

Item	Unit	Range £		
Automatic Glazed Entrance Doors				
Automatic glazed sliding doors in aluminium; polyester powder coated; linear sliding doors; inner pivoted pocket screens; glazed with safety units in accordance with BS. ASSA ABLOY SL510 P RC2 Anti-Burglar Sliding Door System complete with 100 mm operator providing automated operation of the sliding door leaves				
Double bi-parting; opening 2.00 m × 2.30 m	nr	9400.00	to	12000.00
Automatic revolving door, framed. ASSA ABLOY automatic revolving door system providing hands free operation and escape functionality. Modern modular design comprising four collapsible door leaves with electro-mechanical breakout, framed curved glazed outer walls, radially segmented ceiling and semi-circular dustcovers, complete with operator and imbedded digital control				
3 or 4 wing 2100 mm dia., 2200 mm internal height	nr	28000.00	to	35500.00
3 or 4 wing 3000 mm dia., 2200 mm internal height	nr	30000.00	to	37500.00
Automatic revolving door, structural glass. ASSA ABLOY automatic glass revolving door system providing hands free operation and Hi-Tech functionality. Modern modular design comprising four door leaves, curved glazed outer walls, glass ceiling/canopy, complete with in-floor mounted operator and imbedded digital control				
3 or 4 wing 1800 mm dia., 2200 mm internal height	nr	41000.00	to	52000.00
3 or 4 wing 3000 mm dia., 2200 mm internal height	nr	44000.00	to	56000.00
2.7 INTERNAL WALLS AND PARTITIONS				
2.7.1 Internal Walls and Partitions				
Frame and panel partitions				
Timber stud partitions				
structure only comprising 100 mm × 38 mm softwood studs at 600 mm centres; head and sole plates	m^2	25.00	to	30.50
softwood stud comprising 100 mm × 38 mm softwood studs at 600 mm centres; head and sole plates; 12.5 mm thick plasterboard each side; tape and fill joints; emulsion finish	m^2	62.00	to	75.00
Metal stud and board partitions; height range from 2.40 m to 3.30 m				
73 mm partition; 48 mm studs and channels; one layer of 12.5 mm Gyproc Wallboard each side; joints filled with joint filler and joint tape; emulsion paint finish; softwood skirtings with gloss finish	m^2	48.00	to	58.00
102 mm partition; 70 mm studs and channels; one layer of 15 mm Gyproc Wallboard each side; joints filled with joint filler and joint tape; emulsion paint finish; softwood skirtings with gloss finish	m^2	51.00	to	62.00
102 mm thick partition; 70 mm steel studs at 600 mm centres generally; 1 layer 15 mm Fireline board each side; joints filled with joint filler and joint tape; emulsion paint finish; softwood skirtings with gloss finish	m^2	60.00	to	72.00

GENERAL BUILDING WORKS

Item	Unit	Range £		
2.7 INTERNAL WALLS AND PARTITIONS – cont				
Frame and panel partitions – cont				
Metal stud and board partitions – cont				
130 mm thick partition; 70 mm steel studs at 600 mm centres generally; 2 layers 15 mm Fireline board each side; joints filled with joint filler and joint tape; emulsion paint finish; softwood skirtings with gloss finish	m^2	71.00	to	86.00
102 mm thick partition; 70 mm steel studs at 600 mm centres generally; 1 layer 15 mm Soundbloc board each side; joints filled with joint filler and joint tape; emulsion paint finish; softwood skirtings with gloss finish	m^2	60.00	to	72.00
130 mm thick partition; 70 mm steel studs at 600 mm centres generally; 2 layers 15 mm Soundbloc board each side; joints filled with joint filler and joint tape; emulsion paint finish; softwood skirtings with gloss finish	m^2	79.00	to	96.00
Alternative board finishes				
12 mm plywood boarding	m^2	11.00	to	13.30
one coat Thistle board finish 3 mm thick work to walls; plasterboard base	m^2	7.20	to	8.70
extra for curved work	%	15.00	to	20.00
Acoustic insulation to partitions				
Mineral fibre quilt; Isover Acoustic Partition Roll (APR 1200) or other equal; pinned vertically to timber or plasterboard				
25 mm thick	m^2	2.70	to	3.25
50 mm thick	m^2	3.50	to	4.20
Brick partitions				
common brick half brick thick wall; bricks PC £380.00/1000	m^2	56.00	to	67.00
Block partitions				
aerated/lightweight block partitions				
100 mm thick	m^2	34.00	to	41.00
140 mm thick	m^2	46.50	to	56.00
200 / 215 mm thick	m^2	59.00	to	71.00
dense aggregate block walls				
100 mm thick	m^2	42.00	to	51.00
140 mm thick	m^2	51.00	to	62.00
extra for fair face (rate per side)	m^2	1.00	to	1.20
extra for plaster and emulsion (rate per side)	m^2	21.00	to	25.50
extra for curved work	%	15.00	to	20.00
Concrete partitions				
Reinforced concrete walls; C25 strength; standard finish; reinforced at 100 kg/m^3				
150 mm thick	m^2	170.00	to	200.00
225 mm thick	m^2	190.00	to	230.00
300 mm thick	m^2	210.00	to	255.00
extra for plaster and emulsion (rate per side)	m^2	21.00	to	25.50
extra for curved work	%	20.00	to	30.00

GENERAL BUILDING WORKS

Item	Unit	Range £		
Glass				
Hollow glass block walling; cement mortar joints; reinforced with 6 mm dia. stainless steel rods; pointed both sides with mastic or other equal and approved				
240 mm × 240 mm × 80 mm clear blocks	m^2	360.00	to	440.00
240 mm × 240 mm × 80 mm craft blocks	m^2	360.00	to	440.00
190 mm × 190 mm × 100 mm glass blocks; 30 minute fire-rated	m^2	400.00	to	485.00
190 mm × 190 mm × 100 mm glass blocks; 60 minute fire-rated	m^2	660.00	to	800.00
2.7.2 Stair/Ramp Balustrades and Handrails				
Wall handrails				
softwood handrail and brackets	m	84.00	to	100.00
hardwood handrail and brackets	m	115.00	to	140.00
mild steel handrail and brackets	m	155.00	to	190.00
stainless steel handrail and brackets	m	195.00	to	235.00
Balustrading and handrails				
mild steel balustrade and steel or timber handrail	m	305.00	to	370.00
balustrade and handrail with metal infill panels	m	385.00	to	465.00
balustrade and handrail with glass infill panels	m	425.00	to	520.00
stainless steel balustrade and handrail	m	510.00	to	610.00
stainless steel and structural glass balustrade	m	890.00	to	1075.00
2.7.3 Moveable Room Dividers				
Demountable/folding partitions				
Demountable partitioning; aluminium framing; veneer finish doors				
medium quality; 46 mm thick panels factory finish vinyl faced	m^2	380.00	to	460.00
high quality; 46 mm thick panels factory finish vinyl faced	m^2	560.00	to	680.00
Demountable aluminium/steel partitioning and doors				
high quality; folding	m^2	445.00	to	540.00
high quality; sliding	m^2	950.00	to	1150.00
Demountable fire partitions				
enamelled steel; half hour	m^2	680.00	to	820.00
stainless steel; half hour	m^2	1075.00	to	1300.00
soundproof partitions; hardwood doors luxury veneered	m^2	310.00	to	370.00
Aluminium internal patent glazing				
single glazed laminated	m^2	205.00	to	250.00
double glazed; 1 layer toughened and 1 layer laminated glass	m^2	280.00	to	340.00
Stainless steel glazed manual doors and screens				
high quality; to inner lobby of malls	m^2	900.00	to	1075.00
Acoustic folding partition; headtrack suspension and bracing; aluminium framed with high density particle board panel with additional acoustic insulation; melamine laminate finish; acoustic seals. Nominal weight approximately 55 kg per m^2.				
sound reduction 48 db (Rw)	m^2	520.00	to	630.00
sound reduction 55 db (Rw)	m^2	600.00	to	730.00

GENERAL BUILDING WORKS

Item	Unit	Range £		
2.7 INTERNAL WALLS AND PARTITIONS – cont				
2.7.4 Cubicles				
Framed panel cubicles				
Changing and WC cubicles; high pressure laminate faced mdf; proprietary system				
standard quality WC cubicle partition sets; aluminium framing; melamine face chipboard dividing panels and doors; ironmongery; small range (up to 5 cubicles); standard cubicle set; (rate per cubicle)	nr	490.00	to	600.00
medium quality WC cubicle partition sets; stainless steel framing; real wood veneer face chipboard diiding panels and doors; ironmongery; small range (up to 6 cubicles); standard cubicle set; (rate per cubicle)	nr	950.00	to	1150.00
high end specification flush fronted system floor to ceiling 44 mm doors, no visible fixings; real wood veneer or HPL or high gloss paint and lacquer; cubicle set; 800 mm × 1500 mm × 2400 mm high per cubicle, with satin finished stainless steel ironmongery; 30 mm high pressure laminated (HPL) chipboard divisions and 44 mm solid cored real wood veneered doors and pilasters; small range (up to 6 cubicles); standard cubicle set (rate per cubicle)	nr	2500.00	to	3050.00
IPS duct panel systems; melamine finish chipboard; softwood timber subframe				
2.70 m high IPS back panelling system; to accommodate wash hand basins or urinals; access hatch; frame support	m^2	140.00	to	170.00
2.8 INTERNAL DOORS				
2.8.1 Internal Doors				
The following rates include for the supply and hang of doors, complete with all frames, architrave, typical medium standard ironmongery set and appropriate finish.				
Standard doors				
Standard doors; cellular core; softwood; softwood architrave; aluminium ironmongery (latch only)				
single leaf; moulded panel; gloss paint finish	nr	310.00	to	380.00
single leaf; Sapele veneered finish	nr	350.00	to	425.00
Purpose-made doors				
Softwood panelled; softwood lining; softwood architrave; aluminium ironmongery (latch only); brass or stainless ironmongery (latch only); painting and polishing				
single leaf; four panels; mouldings	nr	480.00	to	580.00
double leaf; four panels; mouldings	nr	930.00	to	1125.00
Hardwood panelled; hardwood lining; hardwood architrave; aluminium ironmongery (latch only); brass or stainless ironmongery (latch only); painting and polishing				
single leaf; four panelled doors; mouldings	nr	970.00	to	1175.00
double leaf; four panelled doors; mouldings	nr	1875.00	to	2275.00

GENERAL BUILDING WORKS

Item	Unit	Range £		
Fire doors				
Standard fire doors; cellular core; softwood lining; softwood architrave; aluminuim ironmongery (lockable, self-closure); painting or polishing;				
single leaf; Oak veneered; 30 min fire resistance; polished	nr	510.00	to	620.00
double leaf; Oak veneered; 30 min fire resistance; polished	nr	1175.00	to	1425.00
single leaf; Oak veneered; 60 min fire resistance; polished	nr	790.00	to	960.00
double leaf; Oak veneered; 60 min fire resistance; polished	nr	1450.00	to	1775.00
Ironmongery sets				
Stainless steel ironmongery; euro locks; push plates; kick plates; signage; closures; standard sets				
office door; non locking; fire-rated	nr	340.00	to	410.00
office/store; lockable; fire-rated	nr	385.00	to	465.00
classroom door; lockable; fire-rated	nr	510.00	to	610.00
maintenance/plant room door; lockable; fire-rated	nr	360.00	to	440.00
standard bathroom door (unisex)	nr	300.00	to	360.00
accessible toilet door	nr	170.00	to	210.00
fire escape door	nr	1850.00	to	2250.00
Overhead doors				
single skin; manual	m²	170.00	to	200.00
single skin; electric	m²	305.00	to	370.00
electric operation standard lift, 42 mm thick insulated sandwich panels	m²	230.00	to	280.00
rapid lift fabric door, internal, electric operation	m²	990.00	to	1175.00
3.1 WALL FINISHES				
3.1.1 Wall Finishes				
Internal wall area (unless otherwise described)				
In situ wall finishes				
Comparative finishes				
one mist and two coats emulsion paint	m²	4.20	to	5.10
two coats of lightweight plaster	m²	15.00	to	18.10
two coats of lightweight plaster with emulsion finish	m²	19.20	to	23.00
two coat sand cement render and emulsion finish	m²	27.00	to	32.50
plaster and vinyl wallpaper coverings	m²	25.00	to	30.00
polished plaster system; Armourcoat or similar; 11 mm thick first coat and 2 mm thick finishing coat with a polished finish	m²	120.00	to	150.00
Rigid tile/panel/board finishes				
Timber boarding/panelling; on and including battens; plugged to wall				
12 mm thick softwood boarding	m²	41.00	to	50.00
hardwood panelling; t&g & v-jointed	m²	125.00	to	150.00
Ceramic wall tiles; including backing				
economical quality	m²	46.00	to	55.00
medium to high quality	m²	56.00	to	68.00

GENERAL BUILDING WORKS

Item	Unit	Range £		
3.1 WALL FINISHES – cont				
Rigid tile/panel/board finishes – cont				
Porcelain; including backing				
Porcelain mosaic tiling; walls and floors	m²	88.00	to	105.00
Marble				
Roman Travertine marble wall linings 20 mm thick; polished	m²	320.00	to	385.00
Granite				
Dakota mahogany granite cladding 20 mm thick; polished finish; jointed and pointed in coloured mortar	m²	420.00	to	510.00
Dakota mahogany granite cladding 40 mm thick; polished finish; jointed and pointed in coloured mortar	m²	700.00	to	840.00
Comparative woodwork finishes				
Knot; one coat primer; two undercoats; gloss on wood surfaces; number of coats:				
two coats gloss	m²	10.60	to	12.90
three coats gloss	m²	11.70	to	14.20
three coats gloss; small girth ne 300 mm	m	7.85	to	9.50
Polyurethane lacquer				
two coats	m²	7.95	to	9.60
three coats	m²	8.70	to	10.60
Flame-retardant paint				
Unitherm or similar; two coats	m²	19.30	to	23.00
Polish				
wax polish; seal	m²	15.40	to	18.70
wax polish; stain and body in	m²	19.00	to	23.00
French polish; stain and body in	m²	28.00	to	34.00
Other wall finishes				
Wall coverings				
decorated paper backed vinyl wall paper	m²	9.90	to	12.00
PVC wall linings – Altro or similar; standard satins finish	m²	65.00	to	78.00
3.2 FLOOR FINISHES				
3.2.1 Floor Finishes				
Internal floor area (unless otherwise described)				
In situ screed and floor finishes; laid level; over 300 mm wide				
Cement and sand (1:3) screeds; steel trowelled				
50 mm thick	m²	18.10	to	22.00
75 mm thick	m²	23.50	to	28.00
100 mm thick	m²	25.50	to	31.00
Flowing screeds				
Latex screeds; self colour; 3 mm to 5 mm thick	m²	11.10	to	13.40
Lafarge Gyvlon flowing screed 50 mm thick	m²	12.20	to	14.80
Treads, steps and the like (small areas)	m²	34.00	to	41.50

GENERAL BUILDING WORKS

Item	Unit	Range £		
Granolithic; laid on green concrete				
20 mm thick	m²	25.50	to	31.00
38 mm thick	m²	29.50	to	36.00
Resin floor finish				
Altrotect; 2 coat application nominally 350–500 micron thick	m²	17.10	to	21.00
AltroFlow EP; 3 part solvent-system; up to 3 mm thick	m²	37.00	to	44.50
Atro Screed (TB Screed; Quartz; Multiscreed), 3 mm to 4 mm thick	m²	48.00	to	58.00
Altrocrete heavy duty polyurethane screed, 6 mm to 8 mm screed	m²	54.00	to	65.00
Sheet/board flooring				
Chipboard				
18 mm to 22 mm thick chipboard flooring; t&g joints	m²	17.70	to	21.50
Softwood				
22 mm thick wrought softwood flooring; 150 mm wide; t&g joints;	m²	31.00	to	37.00
softwood skirting, gloss paint finish	m	17.40	to	21.00
mdf skirting, gloss paint finish	m	16.60	to	20.00
Hardwood				
Wrought hardwood t&g strip flooring; polished; including fillets	m²	115.00	to	140.00
Hardwood skirting, stained finish	m	22.00	to	27.00
Sprung floors				
Taraflex Combisport 85 System with Taraflex Sports M Plus; t&g plywood 22 mm thick on softwood battens and crumb rubber cradles	m²	180.00	to	215.00
Sprung composition block flooring (sports), court markings, sanding and sealing	m²	115.00	to	140.00
Rigid tile/slab finishes (includes skirtings; excludes screeds)				
Quarry tile flooring	m²	73.00	to	88.00
Glazed ceramic tiled flooring				
standard plain tiles	m²	49.00	to	59.00
anti slip tiles	m²	53.00	to	64.00
designer tiles	m²	110.00	to	130.00
Terrazzo tile flooring 28 mm thick polished	m²	55.00	to	66.00
York stone 50 mm thick paving	m²	180.00	to	215.00
Slate tiles, smooth; straight cut	m²	78.00	to	95.00
Portland stone paving	m²	290.00	to	350.00
Roman Travertine marble paving; polished	m²	280.00	to	340.00
Granite paving 20 mm thick paving	m²	430.00	to	520.00
Parquet/wood block wrought hardwood block floorings; 25 mm thick; polished; t&g joints	m²	150.00	to	180.00
Flexible tiling; welded sheet or butt joint tiles; adhesive fixing				
vinyl floor tiling; 2.00 mm thick	m²	16.00	to	19.40
vinyl safety flooring; 2.00–2.50 mm thick	m²	50.00	to	60.00
vinyl safety flooring; 3.5 mm thick heavy duty	m²	58.00	to	70.00
linoleum tile flooring; 333 mm × 333 mm × 3.20 mm tiles	m²	41.00	to	50.00
linoleum sheet flooring; 2.00 mm thick	m²	32.00	to	39.00
rubber studded tile flooring; 500 mm × 500 mm × 2.50 mm thick	m²	45.00	to	54.00
Carpet tiles; including underlay, edge grippers				
heavy domestic duty; to floors; PC Sum £38 / m²	m²	65.00	to	79.00
heavy domestic duty; to treads and risers; PC Sum £38 / m²	m	51.00	to	62.00
heavy contract duty; Forbo Flotex HD; PC Sum £25 / m²	m²	48.50	to	59.00

GENERAL BUILDING WORKS

Item	Unit	Range £		
3.2 FLOOR FINISHES – cont				
Rigid tile/slab finishes (includes skirtings; excludes screeds) – cont				
Entrance matting				
door entrance and circulation matting; soil and water removal	m²	59.00	to	72.00
Entrance matting and matwell				
18 mm thick Gradus Topguard barrier matting with aluminium frame	m²	400.00	to	485.00
18 mm thick Gradus Topguard barrier matting with stainless steel frame	m²	340.00	to	415.00
18 mm thick Gradus Topguard barrier matting with polished brass frame	m²	480.00	to	580.00
3.2.2 Raised Access Floors				
Raised access floors: including 600 mm × 600 mm steel encased particle boards on height adjustable pedestals < 300 mm				
light grade duty	m²	50.00	to	60.00
medium grade duty	m²	50.00	to	60.00
heavy grade duty	m²	67.00	to	81.00
battened raft chipboard floor with sound insulation fixed to battens; medium quality carpeting	m²	100.00	to	125.00
Common floor coverings bonded to access floor panels				
anti static vinyl	m²	31.00	to	38.00
heavy duty fully flexible vinyl tiles	m²	26.00	to	31.50
needle punch carpet	m²	19.00	to	23.00
3.3 CEILING FINISHES				
3.3.1 Finishes to Ceiling				
In situ finishes				
Decoration only to soffits; one mist and two coats emulsion paint				
to exposed steelwork (surface area)	m²	6.45	to	7.80
to concrete soffits (surface area)	m²	4.70	to	5.70
to plaster/plasterboard	m²	4.30	to	5.20
Plaster to soffits				
plaster skim coat to plasterboard ceilings	m²	7.45	to	9.00
lightweight plaster to concrete	m²	19.80	to	24.00
Plasterboard to soffits				
9 mm Gyproc board and skim coat	m²	30.50	to	37.00
12.50 mm Gyproc board and skim coat	m²	31.00	to	37.50
Other board finishes; with fire-resisting properties; excluding decoration				
12.50 mm thick Gyproc Fireline board	m²	23.50	to	28.50
15 mm thick Gyproc Fireline board	m²	24.50	to	29.50
12 mm thick Supalux	m²	63.00	to	77.00
Specialist plasters; to soffits				
sprayed acoustic plaster; self-finished	m²	86.00	to	100.00
rendering; Tyrolean finish	m²	86.00	to	100.00
Other ceiling finishes				
softwood timber t&g boarding	m	52.00	to	69.00

GENERAL BUILDING WORKS

Item	Unit	Range £		
Comparative ceiling finishes				
Emulsion paint to plaster				
two coats	m²	4.80	to	5.80
one mist and two coats	m²	5.20	to	6.30
Gloss paint to timber or metal				
primer and two coats	m²	7.10	to	8.60
primer and three coats	m²	10.30	to	12.40
Other coatings				
Artex plastic compound one coat; textured to plasterboard	m²	8.10	to	9.80
3.3.3 Demountable Suspended Ceilings				
Armstrong suspended ceiling; assume large rooms over 250 m²				
mineral fibre; basic range; Cortega, Tatra, Academy; exposed grid	m²	22.50	to	29.50
mineral fibre; medium quality; Corline; exposed grid	m²	32.50	to	43.00
mineral fibre; medium quality; Corline; concealed grid	m²	48.50	to	64.00
mineral fibre; medium quality; Corline; silhouette grid	m²	61.00	to	80.00
mineral fibre; Specific; Bioguard; exposed grid	m²	27.00	to	35.50
mineral fibre; Specific; Bioguard; clean room grid	m²	42.50	to	56.00
mineral fibre; Specific; Hygiene, Cleanroom; exposed grid	m²	58.00	to	77.00
metal open cell; Cellio Global White; exposed grid	m²	55.00	to	72.00
metal open cell; Cellio Black; exposed grid	m²	66.00	to	87.00
wood veneer board; Microlook 8; plain; exposed grid	m²	72.00	to	95.00
wood veneer; plain; exposed grid	m²	99.00	to	130.00
wood veneers; perforated; exposed grid	m²	170.00	to	225.00
wood veneers; plain; concealed grid	m²	160.00	to	215.00
wood veneers; perforated; concealed grid	m²	225.00	to	295.00
Other suspended ceilings				
perforated aluminium ceiling tiles 600 mm × 600 mm	m²	37.50	to	45.50
perforated aluminium ceiling tiles 1200 mm × 600 mm	m²	45.00	to	55.00
metal linear strip; Dampa/Luxalon	m²	57.00	to	69.00
metal linear strip micro perforated acoustic ceiling with Rockwool acoustic infill	m²	70.00	to	85.00
metal tray	m²	60.00	to	72.00
egg-crate	m²	92.00	to	110.00
open grid; Formalux/Dimension	m²	120.00	to	145.00
Integrated ceilings				
coffered; with steel surfaces	m²	160.00	to	190.00
acoustic suspended ceilings on anti vibration mountings	m²	73.00	to	88.00
4.1 FITTINGS, FURNISHINGS AND EQUIPMENT				
4.1.1 General Fittings, Furnishings and Equipment				
Office furniture and equipment				
There is a large quality variation for office furniture and we have assumed a medium level, even so prices will vary between suppliers				
Reception desk				
straight counter; 3500 mm long; 2 person	nr	2375.00	to	3000.00
curved counter; 3500 mm long; 2 person	nr	5700.00	to	7200.00
curved counter; 3500 mm long; 2 person; real wood veneer finish	nr	11500.00	to	14500.00

GENERAL BUILDING WORKS

Item	Unit	Range £		
4.1 FITTINGS, FURNISHINGS AND EQUIPMENT – cont				
Office furniture and equipment – cont				
Furniture and equipment to general office area; standard off the shelf specification				
workstation; 2000 mm long desk; drawer unit; task chair	nr	910.00	to	1150.00
4.1.2 Domestic Kitchen Fittings and Equipment				
Residential fittings (volume housing)				
Kitchen fittings for residential units (not including white goods). NB quality and quantity of units can varies enormously. Always obtain costs from your preferred supplier. The following assume medium standard units from a large manufacturer and includes all units, worktops, stainless steel sink and taps, not including white goods				
one person flat	nr	2350.00	to	2950.00
two person flat/house	nr	2600.00	to	3250.00
three person house	nr	4300.00	to	5400.00
four person house; includes utility area	nr	8000.00	to	10000.00
five person house; includes utility area	nr	11000.00	to	14000.00
Hotel Bathroom Pods				
Fully fitted out, finished and furnished bathroom pods; installed; suitable for business class hotels				
standard pod (4.50 m plan area)	nr	5700.00	to	7200.00
accessible pod (4.50 m plan area)	nr	7700.00	to	9800.00
4.1.3 Special Purpose Fittings, Furnishings and Equipment				
Window blinds				
Louvre blind				
89 mm louvres, manual chain operation, fixed to masonry	m²	61.00	to	77.00
127 mm louvres, manual chain operation, fixed to masonry	m²	53.00	to	67.00
Roller blind				
fabric blinds, roller type, manual chain operation, fixed to masonry	m²	47.00	to	59.00
solar black out blinds, roller type, manual chain operation, fixed to masonry	m²	81.00	to	100.00
Fire curtains				
Electrically operated automatic 2 hr fire curtains to form a virtually continuous barrier against both fire and smoke; size in excess of 4 m^2 (minimum size 800 mm)	m²	1725.00	to	2175.00

GENERAL BUILDING WORKS

Item	Unit	Range £		
Dock levellers and shelters				
ASSA ABLOY Entrance systems				
curtain mechanical shelter; 2 side curtains, one top curtain	nr	1350.00	to	1675.00
teledock leveller; electro-hydraulic operation with two lifting cylinders; one lip ram cylinder; movable telescopic lip; Control for door and leveller from one communal control panel, including interlocking of leveller/door	nr	4000.00	to	5000.00
inflatable mechanical shelter; top bag with polyester fabric panels, 1000 mm extension; side bags with polyester fabric panels, 650 mm extension, strong blower for straight movement of top and side bags; side section with steel guards; colour from standard range	nr	4200.00	to	5300.00
Stand-alone Load House; a complete loading unit attached externally to building; modular steel skin/insulated cladding; two wall elements, one roof element	nr	6700.00	to	8500.00
5 SERVICES				
5.1 SANITARY INSTALLATIONS				
5.1.1 Sanitary Appliances				
Please refer to Spon's Mechanical and Electrical Services Price Book for a more comprehensive range of rates				
Rates are for gross internal floor area – GIFA unless otherwise described				
shopping malls and the like (not tenant fit-out works)	m²	1.10	to	1.50
office building – multi-storey	m²	9.90	to	13.80
office building – business park	m²	8.70	to	12.10
performing arts building (medium specification)	m²	16.00	to	22.00
sports hall	m²	12.70	to	17.60
hotel	m²	42.50	to	59.00
hospital; private	m²	25.00	to	35.00
school; secondary	m²	12.20	to	16.90
residential; multi-storey tower	m²	0.65	to	0.90
supermarket	m²	1.60	to	2.25
Comparative sanitary fittings/sundries				
Note: Material prices vary considerably, the following composite rates are based on average prices for mid priced fittings:				
Individual appliances (including fittings)				
low level WC's; vitreous china pan and cistern; black plastic seat; low pressure ball valve; plastic flush pipe; fixing brackets	nr	350.00	to	450.00
bowl type wall urinal; white glazed vitreous china flushing cistern; chromium plated flush pipes and spreaders; fixing brackets	nr	395.00	to	510.00
sink; glazed fireclay; chromium plated waste; plug and chain	nr	600.00	to	760.00
sink; stainless steel; chromium plated waste; plug and chain; single drainer; double bowl (bowl and half)	nr	375.00	to	480.00
sink; stainless steel; chromium plated waste; plug and chain; double drainer; double bowl (bowl and half)	nr	400.00	to	510.00
bath; reinforced acrylic; chromium plated taps; overflow; waste; chain and plug; P trap and overflow connections	nr	360.00	to	460.00
bath; enamelled steel; chromium plated taps; overflow; waste; chain and plug; P trap and overflow connections	nr	760.00	to	970.00
shower tray; glazed fireclay; chromium plated waste; riser pipe; rose and mixing valve	nr	640.00	to	820.00

GENERAL BUILDING WORKS

Item	Unit	Range £		
5.4 WATER INSTALLATIONS				
5.4.1 Mains Water Supply				
Please refer to Spon's Mechanical and Electrical Services Price Book for a more comprehensive range of rates				
Hot and cold water installations; mains supply; hot and cold water distribution. Gross internal floor area (unless described other wise)				
shopping malls and the like (not tenant fit-out works)	m²	13.60	to	18.90
office building – multi-storey	m²	19.90	to	27.50
office building – business park	m²	14.90	to	20.50
performing arts building (medium specification)	m²	23.00	to	32.50
sports hall	m²	17.70	to	24.50
hotel	m²	50.00	to	69.00
hospital; private	m²	60.00	to	83.00
school; secondary; potable and non potable to labs, art rooms	m²	46.00	to	64.00
residential; multi-storey tower	m²	56.00	to	78.00
supermarket	m²	31.00	to	43.00
distribution centre	m²	2.15	to	3.00
5.6 SPACE HEATING AND AIR CONDITIONING				
5.6.1 Space Heating and Air Conditioning				
Please refer to Spon's Mechanical and Electrical Services Price Book for a more comprehensive range of rates				
Gross internal floor area (unless described otherwise)				
shopping malls and the like (not tenant fit-out works); LTHW, air conditioning, ventilation	m²	88.00	to	115.00
office building – multi-storey; LTHW, ductwork, chilled water, ductwork	m²	120.00	to	160.00
office building – business park; LTHW, ductwork, chilled water, ductwork	m²	140.00	to	185.00
performing arts building (medium specification); LTHW, ductwork, chilled water, ductwork	m²	245.00	to	325.00
sports hall; warm air to main hall, LTHW radiators to ancillary	m²	43.50	to	57.00
hotel; air conditioning	m²	165.00	to	215.00
hospital; private; LPHW	m²	245.00	to	325.00
school; secondary; LTHW, cooling to ICT server room	m²	135.00	to	180.00
residential; multi-storey tower; LTHW to each apartment	m²	74.00	to	97.00
supermarket	m²	18.50	to	24.00
distribution centre; LTHW to offices, displacement system to warehouse	m²	26.00	to	34.00
5.7 VENTILATION SYSTEMS				
5.7.1 Ventilation Systems				
Please refer to Spon's Mechanical and Electrical Services Price Book for a more comprehensive range of rates				
Gross internal area (unless otherwise described)				
shopping malls and the like (not tenant fit-out works)	m²	42.50	to	56.00
office building – multi-storey; toilet and kitchen areas only	m²	7.30	to	9.55
office building – business park; toilet and kitchen areas only	m²	8.50	to	11.20

GENERAL BUILDING WORKS

Item	Unit	Range £		
performing arts building (medium specification); toilet and kitchen areas, workshop	m²	16.90	to	22.00
sports hall	m²	15.80	to	21.00
hotel; general toilet extraction, bathrooms, kitchens	m²	49.00	to	65.00
hospital; private; toilet and kitchen areas only	m²	9.05	to	11.90
school; secondary; toilet and kitchen areas, science labs	m²	20.50	to	27.00
residential; multi-storey tower	m²	42.50	to	56.00
supermarket	m²	6.15	to	8.10
distribution centre; smoke extract system	m²	6.15	to	8.10
5.8 ELECTRICAL INSTALLATIONS				
5.8.1 Electric Mains and Sub-mains Distribution				
Please refer to Spon's Mechanical and Electrical Services Price Book for a more comprehensive range of rates				
Including LV distribution, HV distribution, lighting, small power. Gross internal floor area (unless described other wise)				
shopping malls and the like (not tenant fit-out works)	m²	130.00	to	180.00
office building – multi-storey	m²	110.00	to	150.00
office building – business park	m²	58.00	to	81.00
performing arts building (medium specification)	m²	210.00	to	290.00
sports hall	m²	64.00	to	89.00
hotel	m²	200.00	to	280.00
hospital; private	m²	235.00	to	325.00
school; secondary	m²	140.00	to	195.00
residential; multi-storey tower	m²	93.00	to	130.00
supermarket	m²	75.00	to	105.00
distribution centre	m²	70.00	to	97.00
Comparative fittings/rates per point				
Consumer control unit; 63–100 Amp 230 volt; switched and insulated; RCDB protection. Gross internal floor area (unless described other wise)	nr	340.00	to	425.00
Fittings; excluding lamps or light fittings				
lighting point; PVC cables	nr	50.00	to	64.00
lighting point; PVC cables in screwed conduits	nr	57.00	to	72.00
lighting point; MICC cables	nr	75.00	to	95.00
Switch socket outlet; PVC cables				
single	nr	63.00	to	80.00
double	nr	83.00	to	105.00
Switch socket outlet; PVC cables in screwed conduit				
single	nr	94.00	to	120.00
double	nr	110.00	to	140.00
Switch socket outlet; MICC cables				
single	nr	91.00	to	115.00
double	nr	105.00	to	130.00
Other power outlets				
Immersion heater point (excluding heater)	nr	115.00	to	145.00
Cooker point; including control unit	nr	190.00	to	240.00

GENERAL BUILDING WORKS

Item	Unit	Range £		
5.9 FUEL INSTALLATIONS				
5.9.2 Fuel Distribution Systems				
Please refer to Spon's Mechanical and Electrical Services Price Book for a more comprehensive range of rates				
Gas mains service to plantroom. Gross internal floor area				
shopping mall/supermarket	m²	3.65	to	4.60
warehouse/distribution centre	m²	1.00	to	1.30
office/hotel	m²	1.70	to	2.15
5.10 LIFT AND CONVEYOR INSTALLATIONS				
5.10.1 Lifts and Enclosed Hoists				
Please refer to Spon's Mechanical and Electrical Services Price Book for a more comprehensive range of rates				
Passenger lifts				
Passenger lifts (standard brushed stainless steel finish; 2 panel centre opening doors)				
8-person; 4 stops; 1.0 m/s speed	nr	88000.00	to	110000.00
8-person; 4 stops; 1.6 m/s speed	nr	93000.00	to	110000.00
13-person; 6 stops; 1.0 m/s speed	nr	110000.00	to	135000.00
13-person; 6 stops; 1.6 m/s speed	nr	120000.00	to	145000.00
13-person; 6 stops; 2.0 m/s speed	nr	120000.00	to	145000.00
21-person; 8 stops; 1.0 m/s speed	nr	140000.00	to	170000.00
21-person; 8 stops; 1.6 m/s speed	nr	160000.00	to	190000.00
21-person; 8 stops; 2.0 m/s speed	nr	160000.00	to	190000.00
extra for enhanced finishes; mirror; carpet	nr	4350.00	to	5300.00
extra for lift car LCD TV	nr	9400.00	to	11500.00
extra for intelligent group control; 5 cars; 11 stops	nr	45000.00	to	55000.00
Special installations				
wall climber lift; 10-person; 0.50 m/sec; 10 levels	nr	440000.00	to	540000.00
disabled platform lift single wheelchair; 400 kg; 4 stops; 0.16 m/s	nr	11000.00	to	13000.00
Non-passenger lifts				
Goods lifts; prime coated internal finish				
2000 kg load; 4 stops; 1.0 m/s speed	nr	110000.00	to	135000.00
2000 kg load; 4 stops; 1.6 m/s speed	nr	120000.00	to	145000.00
2500 kg load; 4 stops; 1.0 m/s speed	nr	120000.00	to	145000.00
2000 kg load; 8 stops; 1.0 m/s speed	nr	150000.00	to	185000.00
2000 kg load; 8 stops; 1.6 m/s speed	nr	160000.00	to	190000.00
Other goods lifts				
Hoist	nr	30500.00	to	37000.00
Kitchen service hoist 50 kg; 2 levels	nr	13000.00	to	15500.00
5.10.2 Escalators				
30° escalator; 0.50 m/sec; enamelled steel glass balustrades				
3.50 m rise; 800 mm step width	nr	97000.00	to	115000.00
4.60 m rise; 800 mm step width	nr	105000.00	to	125000.00
5.20 m rise; 800 mm step width	nr	110000.00	to	130000.00
6.00 m rise; 800 mm step width	nr	120000.00	to	145000.00
extra for enhanced finish; enamelled finish; glass balustrade	nr	13000.00	to	15000.00

GENERAL BUILDING WORKS

Item	Unit	Range £		
5.10.6 Dock Levellers and Scissor Lifts				
Electro-hydraulic operation with two lifting cylinders; integrated frame to suit various pit applications; Control for door and leveller from one communal control pane				
2500 mm long dock leveller with swing lip	nr	3450.00	to	4200.00
2500 mm long dock leveller with telescope lip	nr	3700.00	to	4450.00
5.11 FIRE AND LIGHTNING PROTECTION				
5.11.1 Fire Fighting Systems				
Please refer to Spon's Mechanical and Electrical Services Price Book for a more comprehensive range of rates				
Gross internal floor area (unless described otherwise); Including lightning protection, sprinklers unless stated otherwise				
shopping malls and the like (not tenant fit-out works)	m²	16.30	to	23.00
office building – multi-storey	m²	35.00	to	49.00
office building – business park	m²	5.80	to	8.05
performing arts building (medium specification); lightning protection only	m²	1.75	to	2.45
sports hall; lightning protection only	m²	3.50	to	4.85
hospital; private; lightning protection only	m²	2.30	to	3.20
school; secondary; lightning protection only	m²	2.20	to	3.05
hotel	m²	40.00	to	56.00
residential; multi-storey tower	m²	58.00	to	81.00
supermarket; lightning protection only	m²	1.15	to	1.60
distribution centre	m²	45.00	to	63.00
5.12 COMMUNICATION AND SECURITY INSTALLATIONS				
5.12.1 Communication, Security and Control Systems				
Please refer to Spon's Mechanical and Electrical Services Price Book for a more comprehensive range of rates				
Gross internal floor area (unless described otherwise); Including fire alarm, public address system, security installation unless stated otherwise				
shopping malls and the like (not tenant fit-out works)	m²	48.00	to	63.00
office building – multi-storey	m²	34.50	to	45.00
office building – business park; fire alarm and wireways only	m²	16.60	to	22.00
performing arts building (medium specification)	m²	105.00	to	140.00
sports hall	m²	69.00	to	91.00
hotel	m²	115.00	to	150.00
hospital; private	m²	89.00	to	115.00
school; secondary	m²	67.00	to	88.00
residential; multi-storey tower	m²	67.00	to	88.00
supermarket	m²	20.50	to	27.00
distribution centre	m²	24.50	to	32.00

GENERAL BUILDING WORKS

Item	Unit	Range £		
5.13 SPECIAL INSTALLATIONS				
5.13.3 Specialist Mechanical Installations				
Window cleaning equipment				
twin track	m	190.00	to	255.00
manual trolley/cradle	nr	13000.00	to	17500.00
automatic trolley/cradle	nr	27000.00	to	35500.00
Sauna				
2.20 m × 2.20 m internal Finnish sauna; benching; heater; made of sauna grade Aspen or similar	nr	10000.00	to	14000.00
Jacuzzi Installation				
2.20 m × 2.20 m × 0.95 m; 5 adults; 99 jets; lights; pump	nr	14000.00	to	18000.00
5.13.4 Specialist Electrical/Electronic Installations				
Solar PV systems				
Residential solar water heating including collectors; dual coil cylinders; pump; controller (NB excludes any grant allowance)	m²	950.00	to	1250.00
Solar power including 4 kWp monocrystalline solar modules (12 × 185 Wp) on roof mounting kit; certified inverter; DC and AC isolation switches and connection to the grid and certification	nr	8200.00	to	11000.00
5.14 BUILDERS WORK IN CONNECTION WITH SERVICES				
5.14.1 Builders Work in Connection with Services				
Please refer to Spon's Mechanical and Electrical Services Price Book for a more comprehensive range of rates				
Warehouses, sports halls and shopping malls. Gross internal floor area				
main supplies, lighting and power to landlord areas	m²	3.90	to	4.40
central heating and electrical installation	m²	11.80	to	13.40
central heating, electrical and lift installation	m²	13.40	to	15.20
air conditioning, electrical and ventilation installations	m²	30.00	to	34.00
Offices and hotels. Gross internal floor area				
main supplies, lighting and power to landlord areas	m²	11.40	to	15.20
central heating and electrical installation	m²	16.60	to	22.00
central heating, electrical and lift installation	m²	19.70	to	26.00
air conditioning, electrical and ventilation installations	m²	36.00	to	48.50
8.2 ROADS, PATHS, PAVINGS AND SURFACINGS				
8.2.1 Roads, Paths and Pavings				
Paved areas				
Gravel paving rolled to falls and chambers paving on subbase; including excavation	m²	17.00	to	21.50
Resin bound paving				
16 mm–24 mm deep of natural gravel	m²	73.00	to	92.00
16 mm–24 mm deep of crushed rock	m²	74.00	to	94.00
16 mm–24 mm deep of marble chips	m²	80.00	to	100.00

GENERAL BUILDING WORKS

Item	Unit	Range £		
Tarmacadam paving				
two layers; limestone or igneous chipping finish paving on subbase; including excavation and type 1 subbase	m^2	94.00	to	120.00
Slab paving				
precast concrete paving slabs on subbase; including excavation	m^2	67.00	to	84.00
precast concrete tactile paving slabs on subbase; including excavation	m^2	120.00	to	150.00
York stone slab paving on subbase; including excavation	m^2	160.00	to	200.00
imitation York stone slab paving on subbase; including excavation	m^2	97.00	to	120.00
Brick/Block/Setts paving				
brick paviors on subbase; including excavation	m^2	86.00	to	110.00
precast concrete block paviors to footways including excavation; 150 mm hardcore subbase with dry sand joints	m^2	100.00	to	130.00
granite setts on 100 mm thick concrete subbase; including excavation; pointing mortar joints	m^2	180.00	to	225.00
cobblestone paving cobblestones on 100 mm thick concrete subbase; including excavation; pointing mortar joints	m^2	185.00	to	235.00
Reinforced grass construction				
Grasscrete or similar on 200 mm type 1 subbase; topsoil spread across units and grass seeded upon completion	m^2	81.00	to	100.00
Car parking alternatives				
Surface parking; include drains, kerbs, lighting				
surface level parking	m^2	105.00	to	130.00
surface car parking with landscape areas	m^2	135.00	to	170.00
Rates per car				
surface level parking	car	2325.00	to	2900.00
surface car parking with landscape areas	car	3000.00	to	3800.00
Other surfacing options				
permeable concrete block paving; 80 mm thick Formpave Aquaflow or equal; maximum gradient 5%; laid on 50 mm sand/gravel; subbase 365 mm thick clean crushed angular non plastic aggregate, geotextiles filter layer to top and bottom of subbases	m^2	59.00	to	75.00
8.2.2 Special Surfacing and Pavings Sports Grounds				
Pitch plateau construction and pitch drainage not included				
Hockey and support soccer training (no studs)				
Sand filled polypropelene synthetic artificial pitch to FIH national standard. Notts Pad XC FIH shockpad; sport needlepunched polypropelene sand filled synthetic multi sport carpet; white lining; 3 m high perimeter fence	m^2	33.00	to	43.00
FIFA One Star Football and IRB reg 22 Rugby Union				
Lano Rugby Max 60 mm monofilament laid on shockpad on a dynamic base infill with 32 kg/m^2 sand and 11 kg/m^2 SBR rubber – IRB Clause 22 compliant; supply and spread to required infill rates, 2EW sand and 0. 5–2 mm SBR Rubber. Sand (32 kg/m^2)and rubber (11 kg/m^2) spread and brushed into surface; white lining; 3.5 m high perimeter fence	m^2	27.00	to	36.00
Tennis and Netball Court				
Lay final layer on existing subbase of 50 mm porous stone followed by 2 coats of porous bitumen macadam. 40 mm binder course; 14 mm aggregate and 20 mm surface course with 6 mm aggregate. Laid to fine level tolerances and laser leveled, +/- 6 mm tolerance over 3 m; white lining; 3 m high perimeter fencing	m^2	43.50	to	57.00

GENERAL BUILDING WORKS

Item	Unit	Range £		
8.2 ROADS, PATHS, PAVINGS AND SURFACINGS – cont				
8.2.2 Special Surfacing and Pavings Sports Grounds – cont				
Natural Grass Winter Sports				
Strip topsoil and level existing surface (cut and fill); laser grade formation and proof roll. Load and spread topsoil, grade and consolidate; cultivate and grade topsoil. Supply and spread 25 mm approved sand and mix into topsoil by power harrow; apply pre-seeding fertilizer and sow winter rye sports grass	m^2	9.80	to	12.90
8.3 SOFT LANDSCAPING, PLANTING AND IRRIGATION SYSTEMS				
8.3.1 Seeding and Turfing				
Please refer to Spon's External Works and Landscape Price Book for a more comprehensive range of rates				
Surface treatment				
spread and lightly consolidate top soil from spoil heap 150 mm thick; by machine	m^2	2.40	to	2.90
spread and lightly consolidate top soil from spoil heap 150 mm thick; by hand	m^2	8.20	to	9.90
Plant supply, planting, maintenance and 12 months guarantee				
seeded areas	m^2	5.20	to	6.35
turfed areas	m^2	6.90	to	8.35
8.3.2 External Planting				
Planted areas (per m^2 of planted area)				
herbaceous plants	m^2	6.20	to	7.55
climbing plants	m^2	9.85	to	11.90
general planting	m^2	24.00	to	29.00
woodland	m^2	35.50	to	43.00
shrubbed planting	m^2	60.00	to	73.00
dense planting	m^2	61.00	to	74.00
shrubbed area including allowance for small trees	m^2	78.00	to	94.00
Trees				
light standard bare root tree (PC £10)	nr	67.00	to	81.00
standard root balled tree (PC 25)	nr	79.00	to	96.00
heavy standard root ball tree (PC £40)	nr	145.00	to	180.00
semi mature root balled tree (PC £130)	nr	470.00	to	570.00
Parklands				
NOTE: Work on parklands will involve different techniques of earth shifting and cultivation. The following rates include for normal surface excavation, they include for the provision of any land drainage.				
Parklands, including cultivating ground, applying fertilizer, etc. and seeding with parks type grass	ha	23000.00	to	28000.00
Lakes including excavation average 1.0 m deep, laying 1000 micron sheet with welded joints; spreading top soil evenly on top 200 mm deep				
regular shaped lake	1000 m^2	26500.00	to	32000.00

GENERAL BUILDING WORKS

Item	Unit	Range £		
8.4 FENCING, RAILINGS AND WALLS				
8.4.1 Fencing and Rails				
Chain link fencing; plastic coated				
1.20 m high	m	23.00	to	28.00
1.80 m high	m	30.00	to	37.00
Timber fencing				
1.20 m high chestnut pale facing	m	23.00	to	28.00
1.80 m high cross-boarded fencing	m	31.50	to	38.00
8.4.2 Walls and Screens				
Screen walls; one brick thick; including foundations etc.				
1.80 m high facing brick screen wall	m	325.00	to	390.00
1.80 m high smooth faced masonry block boundary wall	m	200.00	to	245.00
8.4.3 Retaining Walls				
Crib retaining walls				
Permacrib timber crib retaining walls on concrete foundation; granular infill material; measure face area m^2. Excludes backfill material to the rear of the wall				
wall heights up to 2.0 m	m^2	200.00	to	245.00
wall heights up to 3.0 m	m^2	215.00	to	260.00
wall heights up to 4.0 m	m^2	230.00	to	280.00
wall heights up to 5.0 m	m^2	240.00	to	290.00
wall heights up to 6.0 m	m^2	250.00	to	305.00
Andacrib concrete crib retaining walls on concrete foundation; granular infill material; measure face area m^2. Excludes backfill material to the rear of the wall				
wall heights up to 2.0 m	m^2	215.00	to	260.00
wall heights up to 3.0 m	m^2	230.00	to	280.00
wall heights up to 4.0 m	m^2	240.00	to	290.00
wall heights up to 5.0 m	m^2	260.00	to	310.00
wall heights up to 6.0 m	m^2	270.00	to	325.00
Textomur green faced reinforced soil system; geogrid/geotextile reinforcement and compaction of reinforced soil mass.				
wall heights up to 3.0 m	m^2	145.00	to	175.00
wall heights up to 4.0 m	m^2	150.00	to	180.00
wall heights up to 5.0 m	m^2	160.00	to	190.00
wall heights up to 6.0 m	m^2	165.00	to	200.00
Titan/Geolock vertical modular block reinforced soil system; steel ladder. Geogrid reinforcement and soil fill material				
wall heights up to 3.0 m	m^2	240.00	to	290.00
wall heights up to 4.0 m	m^2	255.00	to	310.00
wall heights up to 5.0 m	m^2	265.00	to	320.00
wall heights up to 6.0 m	m^2	275.00	to	335.00

GENERAL BUILDING WORKS

Item	Unit	Range £		
8.4 FENCING, RAILINGS AND WALLS – cont				
8.4.4 Barriers and Guardrails				
Post and rail fencing				
open metal post and rail fencing 1.00 m high	m	165.00	to	200.00
galvanized steel post and rail fencing 2.00 m high	m	190.00	to	235.00
Guard rails				
steel guard rails and vehicle barriers	m	66.00	to	80.00
Bollards and barriers				
parking bollards precast concrete or steel	nr	180.00	to	215.00
vehicle control barrier; manual pole	nr	1000.00	to	1200.00
8.5 EXTERNAL FIXTURES				
8.5.1 Site/Street Furniture and Equipment				
Please refer to Spon's External Works and Landscape Price Book for a more comprehensive range of rates				
Street furniture				
Roadsigns				
reflected traffic signs 0.25 m² area on steel post	nr	140.00	to	165.00
internally illuminated traffic signs; dependent on area	nr	220.00	to	270.00
externally illuminated traffic signs; dependent on area	nr	860.00	to	1050.00
Benches; bolted to ground				
benches – hardwood or precast concrete	nr	1125.00	to	1375.00
Litter bins; bolted to ground				
precast concrete	nr	205.00	to	250.00
hardwood slatted	nr	210.00	to	255.00
cast iron	nr	420.00	to	510.00
large aluminium	nr	640.00	to	770.00
Bus stops including basic shelter	nr	2650.00	to	3200.00
Pillar box	nr	640.00	to	770.00
Galvanized steel cycle stand	nr	52.00	to	63.00
Galvanized steel flag staff	nr	1250.00	to	1525.00
Playground equipment				
Modern swings with flat rubber safety seats: four seats; two bays	nr	1575.00	to	1925.00
Stainless steel slide, 3.40 m long	nr	1850.00	to	2250.00
Climbing frame – igloo type 3.20 m × 3.75 m on plan × 2.00 m high	nr	1225.00	to	1475.00
See-saw comprising timber plank on sealed ball bearings 3960 mm × 230 mm × 70 mm thick	nr	105.00	to	130.00
Wicksteed Tumbleguard type safety surfacing around play equipment	m²	13.80	to	16.70
Bark particles type safety surfacing 150 mm thick on hardcore bed	nr	13.80	to	16.70

GENERAL BUILDING WORKS

Item	Unit	Range £		
8.6 EXTERNAL DRAINAGE				
8.6.1 Surface Water and Foul Water Drainage				
Overall £/m² of drained area allowances				
site drainage (per m² of paved area)	m²	25.00	to	32.00
building storm water drainage (per m² of gross internal floor area)	m²	20.50	to	26.00
Machine excavation, grade bottom, earthwork support, laying and jointing pipes and accessories, backfill and compact, disposal of surplus soil. uPVC pipes and fittings, lip seal coupling joints				
up to 1.50 m deep; nominal size				
100 mm dia. pipe	m	56.00	to	70.00
160 mm dia. pipe	m	69.00	to	87.00
over 1.50 m not exceeding 3.00 m deep; nominal size				
100 mm dia. pipe	m	96.00	to	120.00
160 mm dia. pipe	m	110.00	to	140.00
Machine excavation, grade bottom, earthwork support, laying and jointing pipes and accessories, backfill and compact, disposal of surplus soil. uPVC Ultra-Rib ribbed pipes and fittings, sealed ring push fit joints				
up to 1.50 m deep; nominal size				
150 mm dia. pipe	m	57.00	to	72.00
225 mm dia. pipe	m	75.00	to	95.00
300 mm dia. pipe	m	89.00	to	110.00
over 1.50 m not exceeding 3.00 m deep; nominal size				
150 mm dia. pipe	m	97.00	to	120.00
225 mm dia. pipe	m	120.00	to	150.00
300 mm dia. pipe	m	140.00	to	175.00
Machine excavation, grade bottom, earthwork support, laying and jointing pipes and accessories, backfill and compact, disposal of surplus soil. Vitrified clay pipes and fittings, Hepsleve; shingle bed and surround				
up to 1.00 m deep; nominal size				
150 mm dia. pipe	m	76.00	to	97.00
225 mm dia. pipe	m	100.00	to	130.00
300 mm dia. pipe	m	185.00	to	230.00
over 1.50 m not exceeding 3.00 m deep; nominal size				
150 mm dia. pipe	m	115.00	to	150.00
225 mm dia. pipe	m	145.00	to	180.00
300 mm dia. pipe	m	220.00	to	280.00

GENERAL BUILDING WORKS

Item	Unit	Range £		
8.6 EXTERNAL DRAINAGE – cont				
Machine excavation, grade bottom, earthwork support, laying and jointing pipes and accessories, backfill and compact, disposal of surplus soil . Class M tested concrete centrifugally spun pipes and fittings, flexible joints; concrete bed and surround				
up to 3.00 m deep; nominal size				
300 mm dia. pipe	m	180.00	to	230.00
525 mm dia. pipe	m	280.00	to	355.00
900 mm dia. pipe	m	500.00	to	640.00
1200 mm dia. pipe	m	730.00	to	920.00
Machine excavation, grade bottom, earthwork support, laying and jointing pipes and accessories, backfill and compact, disposal of surplus soil . Cast iron Timesaver drain pipes and fittings, mechanical coupling joints				
up to 1.50 m deep; nominal size				
100 mm dia. pipe	m	110.00	to	140.00
150 mm dia. pipe	m	160.00	to	200.00
over 1.50 m not exceeding 3.00 m deep; nominal size				
100 mm dia. pipe	m	150.00	to	190.00
150 mm dia. pipe	m	200.00	to	250.00
Brick manholes				
Excavate pit in firm ground, partial backfill, partial disposal, earthwork support, compact base of pit, 150 mm plain in situ concrete 20.00 N/mm² – 20 mm aggregate (1:2:4) base, formwork, one brick wall of engineering bricks in cement mortar (1:3) finished fair face, vitrified clay channels, plain in situ concrete 25.00 N/mm² – 20 mm aggregate (1:2:4) cover and reducing slabs, fabric reinforcement, formwork step irons, medium duty cover and frame; Internal size of manhole				
600 mm × 450 mm; cover to invert				
not exceeding 1.00 m deep	nr	480.00	to	580.00
over 1.00 m not exceeding 1.50 m deep	nr	640.00	to	780.00
over 1.50 m not exceeding 2.00 m deep	nr	800.00	to	960.00
900 mm × 600 mm; cover to invert				
not exceeding 1.00 m deep	nr	630.00	to	760.00
over 1.00 m not exceeding 1.50 m	nr	860.00	to	1050.00
over 1.50 m not exceeding 2.00 m	nr	1075.00	to	1325.00
900 mm × 900 mm; cover to invert				
not exceeding 1.00 m deep	nr	750.00	to	910.00
over 1.00 m not exceeding 1.50 m	nr	1000.00	to	1225.00
over 1.50 m not exceeding 2.00 m	nr	1275.00	to	1550.00
1200 × 1800 mm; cover to invert				
not exceeding 1.00 m deep	nr	1475.00	to	1775.00
over 1.00 m not exceeding 1.50 m deep	nr	1925.00	to	2350.00
over 1.50 m not exceeding 2.00 m deep	nr	2375.00	to	2900.00

GENERAL BUILDING WORKS

Item	Unit	Range £		
Concrete manholes				
Excavate pit in firm ground, disposal, earthwork support, compact base of pit, plain in situ concrete 20.00 N/mm² – 20 mm aggregate (1:2:4) base, formwork, reinforced precast concrete chamber and shaft rings, taper pieces and cover slabs bedded jointed and pointed in cement; mortar (1:3) weak mix concrete filling to working space, vitrified clay channels, plain in situ concrete 25.00 N/mm² – 20 mm aggregate (1:1:5:3) benchings, step irons, medium duty cover and frame; depth from cover to invert; Internal dia. of manhole				
900 mm dia.; cover to invert				
over 1.00 m not exceeding 1.50 m deep	nr	770.00	to	930.00
1050 mm dia.; cover to invert				
over 1.00 m not exceeding 1.50 m deep	nr	840.00	to	1025.00
over 1.50 m not exceeding 2.00 m deep	nr	960.00	to	1150.00
over 2.00 m not exceeding 3.00 m deep	nr	1175.00	to	1425.00
1500 mm dia.; cover to invert				
over 1.50 m not exceeding 2.00 m deep	nr	1600.00	to	1950.00
over 2.00 m not exceeding 3.00 m deep	nr	2050.00	to	2475.00
over 3.00 m not exceeding 4.00 m deep	nr	2500.00	to	3000.00
1800 mm dia.; cover to invert				
over 2.00 m not exceeding 3.00 m deep	nr	2750.00	to	3300.00
over 3.00 m not exceeding 4.00 m deep	nr	3400.00	to	4100.00
2100 mm dia.; cover to invert				
over 2.00 m not exceeding 3.00 m deep	nr	4250.00	to	5100.00
over 3.00 m not exceeding 4.00 m deep	nr	5200.00	to	6300.00
Polypropylene inspection chambers				
475 mm dia. PPIC inspection chamber including all excavations; earthwork support; cart away surplus spoil; concrete bed and surround; lightweight cover and frame				
600 mm deep	nr	405.00	to	490.00
900 mm deep	nr	465.00	to	560.00
8.6.2 Ancillary Drainage Systems				
Septic tanks				
Excavate for, supply and install Klargester glass fibre septic tank, complete with lockable cover				
2800 litre capacity	nr	3700.00	to	3700.00
3800 litre capacity	nr	4500.00	to	4500.00
4600 litre capacity	nr	5300.00	to	5300.00
Urban and landscape drainage				
Exacavate for and lay oil separators, complete with lockable cover				
Excavate for, supply and lay 1000 litre polyethelyne by pass oil separator	nr	3800.00	to	3800.00
Excavate for, supply and lay 1000 litre polyethelyne full retention oil separator	nr	3900.00	to	3900.00

GENERAL BUILDING WORKS

Item	Unit	Range £		
8.6 EXTERNAL DRAINAGE – cont				
8.6.4 Land Drainage				
NOTE: If land drainage is required on a project, the propensity of the land to flood will decide the spacing of the land drains. Costs include for excavation and backfilling of trenches and laying agricultural clay drain pipes with 75 mm dia. lateral runs average 600 mm deep, and 100 mm dia. mains runs average 750 mm deep.				
land drainage to parkland with laterals at 30 m centres and main runs at 100 m centres	ha	9100.00	to	10500.00
Rainwater harvesting				
FP McCann Ltd Easi-Rain domestic and commercial rainwater harvesting tank system designed to reduce demands on the mains network by providing a sustainable water collection and distribution system ideal for non-potable uses such as toilet flushing, washing and landscape or garden watering. Including excavations				
1500 litre tank	nr	2475.00	to	2475.00
7500 litre tank	nr	3400.00	to	3400.00
Soakaways: Stormwater management				
Soakaway crates; heavyweight (60 tonne) polypropylene high-void box units; including excavation and backfilling as required. Note final surfacing not included	m^3	190.00	to	230.00
Soakaway crates; lightweight (20 tonne) polypropylene high-void box units; including excavation and backfilling as required. Note final surfacing not included	m^3	170.00	to	205.00
8.7 EXTERNAL SERVICES				
8.7.1 Water Mains Supply				
Water main; all laid in trenches including excavation and backfill with excavated material				
up to 50 mm dia. MDPE pipe	m	115.00	to	140.00
Service connection charges				
The privatization of telephone, water, gas and electricity has complicated the assessment of service connection charges. Typically, service connection charges will include the actual cost of the direct connection plus an assessment of distribution costs from the main. The latter cost is difficult to estimate as it depends on the type of scheme and the distance from the mains. In addition, service charges are complicated by discounts that may be offered. For instance, the electricity boards will charge less for housing connections if the house is all electric. However, typical charges for a reasonably sized housing estate might be as follows:				
Water and sewerage connections				
water connections; water main up to 2 m from property	house	560.00	to	680.00
water infrastructure charges for new properties	house	355.00	to	430.00
waste and water sewerage infrastructure charges for new properties	house	710.00	to	860.00

GENERAL BUILDING WORKS

Item	Unit	Range £		
8.7.2 Electricity Mains Supply				
Electric main; all laid in trenches including excavation and backfill with excavated material				
600/1000 volt cables. Two core 25 mm dia. cable including 100 mm dia. clayware duct	m	125.00	to	150.00
Service connection charges				
all electric housing	house	2500.00	to	3050.00
pre-packaged sub-station housing	nr	35000.00	to	43000.00
8.7.5 Gas Mains Supply				
Gas main; all laid in trenches including excavation and backfill with excavated material				
150 mm dia. gas pipe	m	350.00	to	420.00
Service connection charges				
Gas				
gas connection to house	house	325.00	to	390.00
governing station	nr	19000.00	to	23000.00
8.7.6 Telecommunication and other Communication System Connections				
Telephone duct; all laid in trenches including excavation and backfill with excavated material				
100 mm dia. uPVC duct	m	82.00	to	99.00
BT approved communication boxes, including excavations and bedding, precast reinforced concrete box with cast iron cover (medium grade)				
Communication box 1200 mm long × 600 mm wide × 895 mm high	nr	415.00	to	480.00
Communication box 910 mm long × 890 mm wide × 1000 mm high	nr	480.00	to	560.00
Telephone connections charges	house	200.00	to	240.00
8.7.9 External/Street Lighting Systems				
Lighting				
lighting to pedestrian areas an estate roads on 4.00 m–6.00 m columns with up to 70 W lamps	nr	27.50	to	33.00
lighting to main roads;10.00 m–12.00 m columns with 250 W lamps	nr	600.00	to	730.00
lighting to main roads; 12.00 m–15.00 m columns with 400 W high pressure sodium lighting	nr	750.00	to	910.00
8.8 ANCILLARY BUILDINGS AND STRUCTURES				
Multiparking Systems/Stack Parkers				
Fully automatic systems				
integrated robotic parking system using robotic car transporter to store vehicles	car	24000.00	to	30500.00
Semi automatic systems				
integrated parking system, transverse and vertical positioning; semi automatic parking achieving 17 spaces in a 6 car width × 3 car height grid	car	13500.00	to	17000.00

GENERAL BUILDING WORKS

Item	Unit	Range £		
8.8 ANCILLARY BUILDINGS AND STRUCTURES – cont				
Multiparking Systems/Stack Parkers – cont				
Integrated stacker systems				
integrated parking system, vertical positioning only; double width, double height pit stacker achieving 4 spaces with each car stacker	car	7100.00	to	8900.00
integrated parking system, vertical positioning only; triple stacker achieving 3 spaces with each car stacker, generally 1 below ground and 2 above ground	car	14000.00	to	18000.00
integrated parking system, vertical positioning only; triple height double width stacker, achieving 6 spaces with each car stacker, generally 2 below ground and 4 above ground	car	10500.00	to	13000.00
9 MAIN CONTRACTOR'S PRELIMINARIES				
9.2 MAIN CONTRACTOR'S PRELIMINARIES				
9.2.2 Site Establishment				
Typical rates for site accommodation; Delivery not included; minimum hire periods generally apply				
Office cabins – 24 ft × 9 ft (20 m²)	week	30.00	to	37.50
Office cabins – 32 ft × 10 ft (30 m²)	week	38.00	to	48.00
Fire-rated cabins – 24 ft × 9 ft (20 m²)	week	63.00	to	80.00
Fire-rated cabins – 32 ft × 10 ft (30 m²)	week	86.00	to	110.00
Meeting room – 24 ft × 9 ft (20 m²)	week	30.00	to	37.50
Meeting room – 32 ft × 9 ft (30 m²)	week	38.00	to	48.00
Mess cabins (incl furniture)	week	35.50	to	45.00
Drying rooms (incl furniture)	week	40.00	to	51.00
Safestore – 20ft × 8ft (15 m²)	week	20.50	to	26.00
Container with padlock – 20ft × 8ft (15 m²)	week	11.50	to	14.50
Toilets – 13 ft × 9 ft	week	46.00	to	58.00
Signs and notices	week	57.00	to	72.00
Fire extinguishers	week	9.20	to	11.60
Kitchen with cooker, fridge, sink, water heater; 32 ft × 10 ft	week	170.00	to	220.00
Mess room with wash basin, water heater, seating; 16 ft × 7 ft	week	100.00	to	130.00
Toilets – fitted to mains; three pan unit	week	180.00	to	230.00
Toilets – fitted to mains; four pan unit	week	220.00	to	275.00
Haulage to and from site; site offices, storage sheds, toilets	item	5100.00	to	6400.00
9.2.7 Mechanical Plant				
Power and lighting; delivery not included; minimum hire periods generally apply				
3 Kva transformer	week	11.10	to	14.00
5 Kva transformer	week	18.40	to	23.00
10 Kva transformer	week	37.00	to	47.00
4 Way distribution box; 110v	week	9.40	to	11.90
14 m extension cable	week	4.50	to	5.70
25 m extension cable	week	11.90	to	15.00
2 × 500 W floodlights	week	13.90	to	17.50
2 × 500 W floodlights on stands	week	17.30	to	22.00

GENERAL BUILDING WORKS

Item	Unit	Range £		
Generators; delivery not included; minimum hire periods generally apply				
1 Kva 240v portable (low noise)	week	43.00	to	55.00
2 Kva dual volt, portable petrol	week	30.00	to	38.00
5 Kva silenced	week	84.00	to	105.00
7.5 Kva silenced	week	125.00	to	160.00
15 Kva silenced	week	180.00	to	230.00
40 Kva silenced	week	270.00	to	340.00
60 Kva silenced	week	280.00	to	355.00
108 Kva silenced	week	350.00	to	440.00
165 Kva silenced	week	430.00	to	550.00
Mixers and small plant; delivery not included; minimum hire periods generally apply				
small mixer; 5–3 1/2 diesel	week	40.50	to	51.00
mixer; 7–5 diesel	week	51.00	to	64.00
compresser	week	105.00	to	130.00
small dehumidifers	week	27.50	to	35.00
large dehumidifers	week	74.00	to	93.00
turbo dryer	week	46.00	to	58.00
gas jet air heater; 260,000 Btu	week	55.00	to	70.00
9.2.8 Temporary Works				
Roadways				
formation of temporary roads to building perimeter comprising of geotextile membrane and 300 mm MOT type 1; reduce level; spoil to heap on site within 25 m of excavation	m^2	24.50	to	30.00
installation of wheel wash facility and maintenance	nr	3050.00	to	3700.00

ANCILLARY BUILDING WORKS

Item	Unit	Range £		
WATER AND TREATMENT FACILITIES				
Reinforced concrete tanks				
excavation, fill, structural work, valves, penstocks, pipework	m^3	820.00	to	990.00
Water facilities				
fire ponds and lagoons	m^3	710.00	to	860.00
reservoirs	m^3	730.00	to	870.00
Reinforced concrete river weirs				
quay and wave walls including trmporary dams, caissons, pumping, anchorages	m^2	650.00	to	840.00
Reinforced concrete dams up to 12 m high				
arch dam, including excavation anchorages and structural work only, per m^3 of structure	m^3	2075.00	to	2700.00
flat slab dam, including excavation anchorages and structural work only, per m^3 of structure	m^3	1650.00	to	2150.00
Earth dams				
rock fill with concrete core, per m^3 of structure	m^3	740.00	to	960.00
hydraulic fill embankment dam, per m^3 of structure	m^3	360.00	to	465.00
EARTH RETENTION AND STABILISATION				
The following prices have been assembled from the relevant items in the unit costs section. They are intended to give a broad overall price or help in comparisons between a number of basic construction procedures				
These approximate estimates are for construction only. They do not include for preliminaries, design/supervisioncosts, land purchase or OH&P etc.				
Prices in this section are based on the same information and outputs as used in the unit costs section. Costs per m^2 or m^3 of completed structure				
Reinforced in situ concrete retaining wall				
including excavation; reinforcement; formwork; expansion joints; granular backfill and 100 mm land drain; profiled formwork finish to one side typical retaining wall, allowing for profiling finishes				
1.0 m high	m^2	680.00	to	820.00
3.0 m high	m^2	600.00	to	720.00
6.0 m high	m^2	700.00	to	850.00
9.0 m high	m^2	800.00	to	960.00
Precast concrete block, earth retaining wall				
including granular fill; earth anchors and proprietary units or concrete panels, strips, fixings and accessories (Reinforced Earth)				
1.5 m high	m^2	510.00	to	620.00
3.0 m high	m^2	470.00	to	560.00
6.0 m high	m^2	510.00	to	620.00
Precast, reinforced concrete unit retaining wall (in situ foundation):				
including excavation and fill; reinforced concrete foundation; pre-cast concrete units, joints				
1.0 m high	m^2	620.00	to	750.00
2.0 m high	m^2	640.00	to	760.00
3.0 m high	m^2	670.00	to	800.00

ANCILLARY BUILDING WORKS

Item	Unit	Range £		
Precast concrete crib wall				
including excavation; stabilization and foundation work				
up to 1.0 m high	m²	415.00	to	500.00
up to 1.5 m high	m²	400.00	to	480.00
up to 2.5 m high	m²	400.00	to	480.00
up to 4.0 m high	m²	370.00	to	450.00
Timber crib walling				
including preparation; excluding anchoring				
up to 1.5 m high	m²	200.00	to	240.00
up to 3.7 m high	m²	265.00	to	320.00
up to 5.9 m high	m²	315.00	to	380.00
up to 7.4 m high	m²	370.00	to	440.00
Rock gabions				
including preparation; excluding anchoring				
1.0 m thick	m²	110.00	to	130.00
BRIDGEWORK				
ROAD BRIDGES				
per m² of deck maximum span between piers or abutments; include for the works described to the bridge decks and abutments, but exclude any approach works				
Reinforced in situ concrete viaduct				
including excavation; reinforcement; formwork; concrete; bearings; expansion joints; deck waterproofing; deck finishings; P1 parapet				
span: 15 m	m²	3300.00	to	5600.00
span: 20 m	m²	3100.00	to	5300.00
span: 25+ m	m²	3050.00	to	5100.00
Reinforced concrete bridge with precast beams				
including excavation; reinforcement; formwork; concrete; bearings; expansion joints; deck waterproofing; deck finishings; P1 parapet				
span: 12 m	m²	2900.00	to	4850.00
span: 17 m	m²	2900.00	to	4950.00
span: 22 m	m²	2850.00	to	4800.00
span: 30+ m	m²	2700.00	to	4600.00
Reinforced concrete bridge with pre-fabricated steel beams				
including excavation; reinforcement; formwork; concrete; bearings expansion joints; deck waterproofing; deck finishings; P1 parapet				
span: 20 m	m²	3650.00	to	6400.00
span: 30 m	m²	3850.00	to	6700.00
span: 40 m	m²	3750.00	to	6600.00

ANCILLARY BUILDING WORKS

Item	Unit	Range £		
BRIDGEWORK – cont				
FOOTBRIDGES				
per m² of deck maximum span between piers or abutments; include for the works described to the bridge decks and abutments, but exclude any approach works				
Reinforced in situ concrete with precast beams				
widths up to 6 m of deck; including excavation; reinforcement; formwork; concrete; bearings expansion joints; deck waterproofing; deck finishings; P6 parapet; maximum span between piers or abutments per m² of deck				
span: 5 m	m²	3950.00	to	6900.00
span: 10 m	m²	3850.00	to	6700.00
span: 20+ m	m²	3750.00	to	6600.00
Structural steel bridge with concrete foundations				
width up to 4 m wide per m² of deck				
span: 10 m	m²	3000.00	to	4600.00
span: 12 m	m²	2950.00	to	4600.00
span: 16 m	m²	2900.00	to	4550.00
span: 20 m	m²	2800.00	to	4400.00
Timber footbridge (stress graded with concrete piers) per m² of deck				
width up to 2 m wide per m² of deck				
span: 12 m	m²	1625.00	to	2275.00
span: 18 m	m²	1625.00	to	2275.00
TUNNELLING FIELD COSTS				
The following have been extracted from costs recorded on historical projects and are indicative of the costs expected in constructing larger dia. tunnels. All up as constructed cost based on 60 m² to 90 m² tunnels				
The following exclude the cost of land, wayleaves, portals, vent shafts, etc., and are an indication of the order of magnitude expected but actual costs are dependent on ground conditions, location and ease of access, etc.				
Tunnels; 75 m² face				
lower	m³	350.00	to	420.00
median	m³	630.00	to	750.00
higher	m³	760.00	to	910.00
lower	km	26230000.00	to	31480000.00
median	km	44550000.00	to	53460000.00
upper	km	56810000.00	to	68170000.00

ANCILLARY BUILDING WORKS

Item	Unit	Range £		
HIGHWAY WORKS				
The following prices are the approximate costs per metre run of roadway, and are based on information from a number of sources including engineers estimates, tenders, final account values etc. on a large number of highways contracts. Motorway and All Purpose Road prices include for earthworks, structures, drainage, pavements, line markings, reflective studs, footways signs, lighting, motorway communications, fencing and barrier works as well as allowance for accommodation works, statutory undertakings and landscaping as appropriate to the type and location of the carriageway. Motorway and All Purpose Road prices do NOT include for the cost of associated features such as side roads, interchanges, underbridges, overbridges, culverts, sub-ways, gantries and retaining walls. These are shown separately beside the cost range for each road type, based on statistical frequency norms.				
MOTORWAYS				
The following costs are based on a 850 mm construction comprising 40 mm wearing course, 60 mm base course, 250 mm road base, 200 mm subbase and 350 mm capping layer; central reserve incorporating two 0.7 m wide hardstrips; no footpaths or cycle paths included				
Rural motorways				
Grassed central reserve and outer berms; no kerbs or edgings; assumption that 30% of excavated material is unsuitable for filling; costs allow for forming embankments for 50% of highway length average 4.70 m high and 50% of length in cuttings average 3.90 m deep; accommodation fencing each side; allowance of 25% of length having crash barriers and 20% of length having lighting				
dual two lane (D2M_R); 25.20 m overall width; each carriageway 7.30 m with 3.30 m hard shoulder; 4.00 m central reserve	m	4900.00	to	5900.00
dual three lane (D3M_R); 32.60 m overall width; each carriageway 11.00 m with 3.30 m hard shoulder; 4.00 m central reserve	m	6200.00	to	7400.00
dual four lane (D4M_R); 39.80 m overall width; each carriageway 14.60 m with 3.30 m hard shoulder; 4.00 m central reserve	m	7300.00	to	8800.00
Urban motorways				
Hard paved central reserve and outer berms; precast concrete kerbs; assumption that 30% of excavated material is unsuitable for filling; costs allow for forming embankments for 50% of highway length average 3.20 m high and 50% of length in cuttings average 1.60 m accommodation fencing each side; allowance of 25% of length having crash barriers and 20% of length having lighting dual two lane (D2M_U); 23.10 m overall width; each carriageway				
dual two lane (D2M_U); 23.10 m overall width; each carriageway 7.30 m with 2.75 m hard shoulder; 3.00 m central reserve	m	4700.00	to	5600.00
dual three lane (D3M_U); 30.50 m overall width; each carriageway 11.00 m with 2.75 m hard shoulder; 3.00 m central reserve	m	4800.00	to	5800.00
dual four lane (D4M_U); 37.70 m overall width; each carriageway 14.60 m with 2.75 m hard shoulder; 3.00 m central reserve	m	7000.00	to	8400.00

ANCILLARY BUILDING WORKS

Item	Unit	Range £		
HIGHWAY WORKS – cont				
ALL PURPOSE ROADS				
The following costs are based on a 800 mm construction comprising 40 mm wearing course, 60 mm base course, 200 mm road base, 150 mm subbase and 350 mm capping layer; no footpaths or cycle paths included				
Rural all purpose road				
Grassed central reserve; no kerbs or edgings; assumption that 30% of excavated material is unsuitable for filling; costs allow for forming embankments for 50% of highway length average 4.00 m high and 50% of length in cuttings average 3.75 m deep; allowance of 25% of length having crash barriers and 20% of length having lighting				
dual two lane (D2AP_R); 18.60 m overall width; each carriageway 7.30 m; 4.00 m central reserve	m	3100.00	to	4000.00
dual three lane (D3AP_R); 26.00 m overall width; each carriageway 11.00 m ; 4.00 m central reserve	m	3950.00	to	5200.00
Urban all purpose roads				
hard paved central reserve; precast concrete kerbs; assumption that 30% of excavated material is unsuitable for filling; costs allow for forming embankments for 50% of highway length in average 2.2 m high and 50% of length in cuttings average 1.62 m deep; allowance of 25% of length having crash barriers and 20% of length having lighting				
dual two lane (D2AP_U); 23.10 m overall width; each carriageway 7.30 m with 2.75 m hard shoulder; 3.00 m central reserve	m	4300.00	to	4700.00
dual three lane (D3AP_U); 30.50 m overall width; each carriageway 11.00 m with 2.75 m hard shoulder; 3.00 m central reserve	m	5600.00	to	6100.00
Other roads				
Rural all purpose roads				
single carriageway all-purpose road (carriageway is 7.3 m wide)	m	1550.00	to	1875.00
wide single carriageway all-purpose road (carriageway is 10.0 m wide)	m	1950.00	to	2350.00
Rural link roads				
two lane link road (carriageway is 7.3 m wide)	m	1450.00	to	1750.00
single lane link road (carriageway is 3.7 m wide)	m	880.00	to	1050.00
Rural motorway or dual carriageway slip roads				
single carriageway slip road (carriageway is 5.0 m wide)	m	1050.00	to	1275.00
Urban motorway or dual carriageway slip roads				
single carriageway slip road (carriageway is 6.0 m wide)	m	1325.00	to	1575.00
Wide single carriageway all-purpose road with footway each side				
carriageway is 10.0 m wide each footway is 3.0 m wide	m	2025.00	to	2425.00
nominal 3.0 m cycle track to all-purpose roads (one side only)	m	245.00	to	295.00
Urban link roads				
two lane link road (carriageway is 7.3 m wide)	m	1575.00	to	1900.00
single lane link road (carriageway is 3.7 m wide)	m	960.00	to	1150.00

ANCILLARY BUILDING WORKS

Item	Unit	Range £		
The following are approximate costs for the installation of incidental road works to serve as part of the development of infrastructure for rousing, retail or industrial development.				
Type A Construction = medium duty carriageway consisting of 100 mm surfacing, roadbase up to 115 mm subbase of 150 mm				
turning or passing bay = 35 m² overall area; suitable for cars vans	nr	6600.00	to	8000.00
turning or passing bay = 100 m² overall area; suitable for semi trailer	nr	1075.00	to	1325.00
parking lay-by = 200 m² overall area	nr	2075.00	to	2500.00
Type B Construction = heavy duty carriageway consisting of reinforced concrete slab 225 mm thick subbase 130 mm thick capping layer 280 mm thick				
turning or passing bay = 35 m² overall area; suitable for cars vans	nr	9500.00	to	11500.00
turning or passing bay = 100 m² overall area; suitable for semi trailer	nr	24000.00	to	29000.00
bus lay-by = 40 m² overall area	nr	9500.00	to	11000.00
parking lay-by = 200 m² overall area	nr	46500.00	to	56000.00
Footpaths and crossings				
vehicle crossing verge/footway/central reserve	m²	14.60	to	17.60
footway construction (bit-mac plus edgings)	m²	81.00	to	97.00
Road crossings				
Major roads; costs include road markings beacons lights signs advance danger signs etc.				
4 way traffic signal installation	nr	96000.00	to	105000.00
puffin crossing	nr	71000.00	to	78000.00
pelican crossing	nr	32000.00	to	35000.00
zebra crossing	nr	8700.00	to	9500.00
pedestrian guard railing	m	255.00	to	280.00
Estate roads; costs include road markings beacons lights signs advance danger signs etc.				
4 way traffic signal installation	nr	56000.00	to	62000.00
zebra crossing	nr	6700.00	to	7400.00
puffin crossing	nr	65000.00	to	71000.00
pelican crossing	nr	24000.00	to	26500.00
pedestrian guard railing	m	170.00	to	190.00
UNDERPASSES				
Major roads				
Provision of underpasses to new major roads constructed as part of a road building programme				
pedestrian underpass 3 m wide × 2.5 m high	m	6800.00	to	7600.00
vehicle underpass 7 m wide × 5 m high	m	30000.00	to	34000.00
vehicle underpass 14 m wide × 5 m high	m	70000.00	to	79000.00
Estate roads				
Provision of underpasses to new estate roads constructed as part of a road building programme				
pedestrian underpass 3 m wide × 2.5 m high	m	5700.00	to	6300.00
vehicle underpass 7 m wide × 5 m high	m	23500.00	to	26000.00
vehicle underpass 14 m wide × 5 m high	m	57000.00	to	62000.00

ANCILLARY BUILDING WORKS

Item	Unit	Range £		
ACCESS SCAFFOLDING				
The following are guideline costs for the hire and erection of proprietary scaffold systems (tube and coupling). Costs are very general and are based on a minimum area of 360 m² at 1.80 m deep. Erection of scaffolding system is based on 3 men erecting a 16 m² bay in about 1 hour and dismantling the same in 20 minutes. Note: Although scaffolding is essentially plant hire, allowance must be made for inevitable loss and damage to fittings for consumables used and for maintenance during hire periods				
approximate hire (supply and fix) of patent scaffold per 4 week, cost dependent upon length of hire quality of system number of toe boards handrails etc.	m^2	9.05	to	10.80
approximate hire, supply and fix of mobile access towers per 4 week hire	m^2	31.00	to	37.00
additional costs of Pole ladder access per 4.0 m high per 4 week	nr	9.65	to	11.60
additional cost of stair towers extra over the cost of scaffold system per 2 m rise per 4 week	m^2	97.00	to	115.00
additional cost of hoarding around base perimeter using multi-use ply sheeting per 4 week period	m^2	2.50	to	3.00
additional cost of polythene debris netting, no reuse	m^2	1.30	to	1.60
additional cost of Monaflex 'T' plus weather-proof sheeting including anchors and straps, based on 3 uses	m^2	2.20	to	2.70
TEMPORARY WORKS				
Bailey bridges				
Installation of temporary Bailey bridges (including temporary concrete abutments; erection maintenance; dismantling)				
span up to 10 m	nr	39000.00	to	42500.00
span 15 m	nr	43000.00	to	47000.00
span 20 m	nr	55000.00	to	60000.00
span 25 m	nr	60000.00	to	66000.00
Cofferdams				
Installation of cofferdams; based on driven steel sections with recovery value, including all plant for installation and dismantling; loss of materials; pumping and maintenance excluding excavation and disposal of material – backfilling on completion. Cost range based on 12 weeks installation on soft-medium ground conditions				
Depth of drive up to 5 m dia. or side length				
up to 2 m	nr	5600.00	to	6200.00
up to 10 m	nr	28000.00	to	30500.00
up to 20 m	nr	52000.00	to	58000.00
Depth of drive 5 to 10 m dia. or side length				
up to 2 m	nr	9600.00	to	10500.00
up to 10 m	nr	45000.00	to	50000.00
up to 20 m	nr	88000.00	to	96000.00
Depth of drive 10 to 15 m dia. or side length				
up to 2 m	nr	13500.00	to	15000.00
up to 10 m	nr	59000.00	to	64000.00
up to 20 m	nr	97000.00	to	110000.00

ANCILLARY BUILDING WORKS

Item	Unit	Range £		
Depth of drive 15 to 20 m dia. or side length				
up to 2 m	nr	18000.00	to	19500.00
up to 10 m	nr	52000.00	to	57000.00
up to 20 m	nr	130000.00	to	140000.00
extra for				
medium-hard ground	%	25.00	to	30.00
hard, but not rock	%	40.00	to	50.00
Earthwork support				
The following are comparative costs for earthwork and trench support based on the hire of modular systems (trench box hydraulic) with an allowance for consumable materials (maximum 5 day hire allowance). for larger excavations requiring Earthwork support refer to cofferdam estimates; or sheet piling within unit cost sections				
earthwork support not exceeding 1 m deep distance between opposing faces not exceeding 2 m	m^2	10.00	to	11.00
earthwork support not exceeding 1 m deep distance between opposing faces 2–4 m	m^2	15.00	to	16.50
earthwork support not exceeding 2 m deep distance between opposing faces not exceeding 2 m	m^2	17.60	to	19.30
earthwork support not exceeding 2 m deep distance between opposing faces 2–4 m	m^2	31.00	to	34.50
earthwork support not exceeding 4 m deep distance between opposing faces not exceeding 2 m	m^2	45.00	to	50.00
earthwork support not exceeding 4 m deep distance between opposing face 2–4 m	m^2	59.00	to	65.00
The following are approximate weekly hire costs for a range of basic support equipment used on site				
Steel sheet piling				
AU or AZ series section	t/wk	51.00	to	56.00
Trench sheeting				
standard overlapping sheets	m^2/wk	1.95	to	2.10
interlocking type	m^2/wk	6.55	to	7.20
heavy duty overlappingsheets				
6 mm thick	m^2/wk	8.75	to	9.65
8 mm thick	m^2/wk	8.60	to	9.45
driving cap	nr/wk	10.80	to	11.80
extraction clamp	nr/wk	27.00	to	29.50
trench struts				
No 0; 0.3 to 0.40 m	nr/wk	1.10	to	1.20
No 1; 0.5 to 0.75 m	nr/wk	1.10	to	1.20
No 2; 0.7 to 1.14 m	nr/wk	1.10	to	1.20
No 3; 1.0 to 1.75 m	nr/wk	1.10	to	1.20

PART 6

Unit Costs – Civil Engineering Works

INTRODUCTORY NOTES

The Unit Costs in this part represent the net cost to the Contractor of executing the work on site and not the prices which would be entered in a tender Bill of Quantities.

It must be emphasized that the unit rates are averages calculated on unit outputs for typical site conditions. Costs can vary considerably from contract to contract depending on individual Contractors, site conditions, methods of working and various other factors. Reference should be made to Part 1 for a general discussion on Civil Engineering Estimating.

Guidance prices are included for work normally executed by specialists, with a brief description where necessary of the assumptions upon which the costs have been based. Should the actual circumstances differ, it would be prudent to obtain check prices from the specialists concerned, on the basis of actual/likely quantity of the work, nature of site conditions, geographical location, time constraints, etc.

The method of measurement adopted in this section is the **CESMM4**, *subject to variances where this has been felt to produce more helpful price guidance.*

We have structured this Unit Costs section to cover as many aspects of Civil Engineering works as possible.

The Gang hours column shows the output per measured unit in actual time, not the total labour hours; thus for an item involving a gang of 5 men each for 0.3 hours, the total labour hours would normally be 1.5, whereas the Gang hours shown would be 0.3.

This section is structured to provide the User with adequate background information on how the rates have been calculated, so as to allow them to be readily adjusted to suit other conditions to the example presented:

- Alternative gang structures as well as the effect of varying bonus levels, travelling costs etc.
- Other types of plant or else different running costs from the medium usage presumed
- Other types of materials or else different discount/waste allowances from the levels presumed

GUIDANCE NOTES

Generally

Adjustments should be made to the rates shown for time, location, local conditions, site constraints and any other factors likely to affect the costs of a specific scheme.

Method of Measurement

Although this part of the book is primarily based on **CESMM 4**, the specific rules have been varied from in cases where it has been felt that an alternative presentation would be of value to the book's main purpose of providing guidance on prices. This is especially so with a number of specialist contractors but also in the cases of work where a more detailed presentation will enable the user to allow for ancillary items.

Materials Cost

Materials cost within the rates have been calculated using the 'list prices' contained in Part 4: Resources, adjusted to allow for delivery charges (if any) and a 'reasonable' level of discount obtainable by the contractor. This will vary very much depending on the contractor's standing, the potential size of the order and the supplier's eagerness and will vary also between raw traded goods such as timber which will attract a low discount of perhaps 3%, if at all, and manufactured goods where the room for bargaining is much greater and can reach levels of 30% to 40%. High demand for a product at the time of pricing can dramatically reduce the potential discount, as can the world economy in the case of imported goods such as timber and copper. Allowance has also been made for wastage on site (generally 2½% to 5%), dependent upon the risk of damage, the actual level should take account of the nature of the material and its method of storage and distribution about the site.

Labour Cost

The composition of the labour and type of plant is generally stated at the beginning of each section, more detailed information on the calculation of the labour rates is given in Part 4: Resources. In addition is a summary of labour grades and responsibilities extracted from the Working Rule Agreement. Within Parts 6 and 7 each section is prefaced by a detailed build-up of the labour gang assumed for each type of work. This should allow the user to see the cost impact of a different gang as well as different levels of bonus payments, allowances for skills and conditions, travelling allowances, etc. The user should be aware that the output constants are based on the gangs shown and would also need to be changed.

Plant Cost

A rate build-up of suitable plant is generally stated at the beginning of each section, with more detailed information on alternative machines and their average fuel costs being given in Part 4: Resources. Within Parts 6 and 7 each section is prefaced by a detailed build-up of plant assumed for each type of work. This should allow the user to see the cost impact of using alternative plant as well as different levels of usage. The user should be aware that the output constants are based on the plant shown and would also need to be changed.

Outputs

The user is directed to Part 10: Outputs which contains a selection of output constants, in particular a chart of haulage times for various capacities of Tippers.

CLASS A: GENERAL ITEMS

Item	Gang hours	Labour £	Plant £	Material £	Unit	**Total rate £**
03 METHOD RELATED CHARGES						
Note: Refer also to the example calculation of Preliminaries and also to Oncosts and Profit						
CONTRACTUAL REQUIREMENTS						
Performance bond						
The cost of the bond will relate to the nature and degree of difficulty and risk inherent in the type of works intended, the perceived ability and determination of the Contractor to complete them, his financial status and whether he has a proven track record in this field with the provider.						
Insurance of the Works						
Refer to the discussion of the matter in Preliminaries						
Third party insurance						
Refer to the discussion of the matter in Preliminaries						
SPECIFIED REQUIREMENTS						
General						
This section entails the listing of services and facilities over and above the Permanent Works which the Contractor would be instructed to provide in the Contract Documents.						
Accommodation for the Engineer's Staff						
Refer to Resources – Plant section for a list of accommodation types.						
Services for the Engineer's staff						
Transport vehicles						
4 WD utility short wheelbase		–	1125.71	–	week	**1125.71**
4 WD long wheelbase		–	990.07	–	week	**990.07**
(for other vehicles, refer to Resources – Plant)						
Telephones (allow for connection charges and usage of telephones required for use by the Engineer's Staff)						
Equipment for use by the Engineer's staff						
Allow for equipment specifically required; entailing Office Equipment, Laboratory Equipment and Surveying Equipment.						
Attendance upon the Engineer's staff						
Driver	40.00	976.32	–	–	week	**976.32**
Chainmen	40.00	767.66	–	–	week	**767.66**
Laboratory assistants	40.00	1052.35	–	–	week	**1052.35**

CLASS A: GENERAL ITEMS

Item	Gang hours	Labour £	Plant £	Material £	Unit	Total rate £
03 METHOD RELATED CHARGES – cont						
Temporary Works						
Temporary Works relate to other work which the contractor may need to carry out due to his construction method. These are highly specific to the particular project and envisaged construction method. Examples are:						
* access roads and hardstandings/ bases for plant and accommodation as well as to the assembly/ working area generally						
* constructing ramps for access to low excavations						
* steel sheet, diaphragm wall or secant pile cofferdam walling to large excavations subject to strong forces/water penetration						
* bridges						
* temporary support works and decking						
The need for these items should be carefully considered and reference made to the other sections of the book for guidance on what costs should be set against the design assumptions made. Extensive works could well call for the involvement of a contractor's temporary works engineer for realistic advice.						
METHOD RELATED CHARGES						
Accommodation and buildings						
Offices; establishment and removal; Fixed Charge						
80 m^2 mobile unit (10 staff × 8 m^2)		–	2098.17	–	item	**2098.17**
10 m^2 section units; two		–	899.51	–	item	**899.51**
Offices; maintaining; Time Related Charge						
80 m^2 mobile unit (10 staff × 8 m^2)		–	539.70	–	week	**539.70**
10 m^2 section units; two		–	299.83	–	week	**299.83**
Stores; establishment and removal; Fixed Charge						
22 m^2 section unit		–	719.60	–	item	**719.60**
Stores; maintaining; Time Related Charge						
22 m^2 section unit		–	119.95	–	item	**119.95**
Canteens and messrooms; establishment and removal; Fixed Charge						
70 m^2 mobile unit (70 men)		–	1799.00	–	item	**1799.00**
Canteens and messrooms; maintaining; Time-Related Charge						
70 m^2 mobile unit (70 men)		–	509.71	–	week	**509.71**
Plant						
General purpose plant not included in Unit Costs: – Transport						
wheeled tractor including driver	40.00	784.08	1640.39	–	week	**2424.47**
trailer		–	19.01	–	week	**19.01**

CLASS A: GENERAL ITEMS

Item	Gang hours	Labour £	Plant £	Material £	Unit	**Total rate £**
Tractor						
hire charge		–	1131.91	–	week	**1131.91**
driver (skill rate 4)	40.00	1051.92	–	–	week	**1051.92**
fuel and consumables		–	12400.75	–	week	**12400.75**
Cranes						
10t Crane	40.00	1166.83	2720.80	–	week	**3887.63**
Miscellaneous						
sawbench		–	73.56	–	week	**73.56**
concrete vibrator		–	5978.46	–	week	**5978.46**
75 mm 750 l/min pump		–	223.24	–	week	**223.24**
compressor; 22.10 m^3/min, 216 kW, four tool		–	5253.34	–	week	**5253.34**
plate compactor; 180 kg		–	160.04	–	week	**160.04**
towed roller; BW6		–	1571.95	–	week	**1571.95**
Excavators etc.						
hydraulic backacter; 14.5 tonne; driver + banksman	80.00	1974.67	2575.24	–	week	**4549.92**
bulldozer; D6; driver	40.00	1418.26	4339.50	–	week	**5757.75**
loading shovel; CAT 939; driver	40.00	1166.83	2217.10	–	week	**3383.93**
Temporary Works						
Supervision and labour						
Supervision for the duration of construction; Time Related Charge						
Agent	40.00	2815.78	–	–	week	**2815.78**
Senior Engineer	40.00	2418.34	–	–	week	**2418.34**
Engineers	40.00	1820.02	–	–	week	**1820.02**
General Foreman	40.00	2248.56	–	–	week	**2248.56**
Administration for the duration of construction; Time Related Charge						
Office Manager/Cost Clerk	40.00	1280.02	–	–	week	**1280.02**
Timekeeper/Storeman/Checker	40.00	767.66	–	–	week	**767.66**
Typist/Telephonist	40.00	1043.28	–	–	week	**1043.28**
Security Guard	40.00	727.49	–	–	week	**727.49**
Quantity Surveyor	40.00	1791.94	–	–	week	**1791.94**
Labour teams for the duration of construction; Time Related Charge						
General yard labour (part time); loading and offloading, clearing site rubbish etc.; ganger	40.00	807.84	–	–	week	**807.84**
General yard labour (part time); loading and offloading, clearing site rubbish etc.; four unskilled operatives	160.00	3231.36	–	–	week	**3231.36**
Maintenance of ontractor's own plant; Fitter	40.00	975.89	–	–	week	**975.89**
Maintenance of contractor's own plant; Fitter's Mate	40.00	767.66	–	–	week	**767.66**

CLASS A: GENERAL ITEMS

Item	Gang hours	Labour £	Plant £	Material £	Unit	Total rate £
03 METHOD RELATED CHARGES – cont						
DE-WATERING						
The following unit costs are for dewatering pervious ground only and are for sets of equipment comprising:						
1 nr diesel driven pump (WP 150/60 or similar) complete with allowance of £50 for fuel		–	–	–	day	**119.74**
50 m of 150 mm dia. header pipe		–	–	–	day	**22.04**
35 nr of disposable well points		–	–	–	day	**441.48**
18 m of delivery pipe		–	–	–	day	**10.07**
1 nr diesel driven standby pump		–	–	–	day	**69.30**
1 nr jetting pump with hoses (for installation of wellpoints only)		–	–	–	-	**119.74**
attendant labour and plant (2 hrs per day) inclusive of small dumper and bowser)		–	–	–	-	**40.59**
Guide price for single set of equipment comprising pump, 150 mm dia. header pipe, 35 nr well points, delivery pipes and attendant labour and plant						
Bring to site equipment and remove upon completion		–	–	–	item	**5686.45**
Installation costs						
hire of jetting pump with hoses; 1 day		–	–	–	item	**119.74**
purchase of well points; 35 nr		–	–	–	item	**441.48**
labour and plant; 10 hours		–	–	–	item	**405.93**
Operating costs						
hire of pump, header pipe, delivery pipe and standby pump complete with fuel etc. and 2 hours attendant labour and plant		–	–	–	day	**302.34**

CLASS B: GROUND INVESTIGATION

Item	Gang hours	Labour £	Plant £	Material £	Unit	**Total rate £**
RESOURCES – LABOUR						
Labour Gang Build-up Rate						
Trial hole gang						
1 ganger or chargehand (skill rate 4)		21.42	–	–		
1 skilled operative (skill rate 4)		20.20	–	–		
2 unskilled operatives (general)		39.20	–	–		
1 plant operator (skill rate 3)		29.17	–	–		
Total Gang Rate/Hour		**109.99**	–	–		
RESOURCES – PLANT						
Plant Resources Build-up Rate						
Trial holes						
8 tonne wheeled backacter		–	51.09	–		
3 tonne dumper		–	17.43	–		
3.7 m^3/min compressor, 2 tool		–	18.06	–		
two 2.4 m^3/min road breakers		–	13.00	–		
extra 50 ft / 15 m hose		–	0.46	–		
plate compactors; vibrating compaction; plate 180 kg/600 mm		–	4.00	–		
Total Rate/Hour		–	**104.04**	–		
01 TRIAL PITS AND TRENCHES						
The following costs assume the use of mechanical plant and excavating and backfilling on the same day						
Trial holes measured by number						
Excavating trial hole; plan size 1.0 × 2.0 m; supports, backfilling						
ne 1.0 m deep	0.24	26.40	63.45	–	nr	**89.84**
1.0–2.0 m deep	0.47	51.69	82.31	–	nr	**134.00**
over 2.0 m deep	0.53	58.29	100.92	–	nr	**159.22**
Excavating trial hole in rock or similar; plan size 1.0 × 2.0 m; supports, backfilling						
ne 1.0 m deep	0.28	30.80	68.70	–	nr	**99.50**
1.0–2.0 m deep	0.51	56.09	99.02	–	nr	**155.12**
over 2.0 m deep	0.58	63.79	118.26	–	nr	**182.05**
Trial holes measured by depth						
Excavating trial hole; plan size 1.0 × 2.0 m; supports, backfilling						
ne 1.0 m deep	0.24	26.40	64.90	–	m	**91.30**
1.0–2.0 m deep	0.53	58.29	92.51	–	m	**150.80**
2.0–3.0 m deep	0.59	64.89	98.12	–	m	**163.01**
3.0–5.0 m deep	0.65	71.49	129.16	–	m	**200.66**

CLASS B: GROUND INVESTIGATION

Item	Gang hours	Labour £	Plant £	Material £	Unit	Total rate £
01 TRIAL PITS AND TRENCHES – cont						
Trial holes measured by depth – cont						
Excavating trial hole in rock or similar; plan size 1.0 × 2.0 m; supports, backfilling						
ne 1.0 m deep	2.95	324.46	448.77	–	m	**773.24**
1.0–2.0 m deep	3.65	401.45	683.58	–	m	**1085.03**
2.0–3.0 m deep	3.95	434.45	712.18	–	m	**1146.63**
3.0–5.0 m deep	4.50	494.94	764.41	–	m	**1259.35**
Sundries in trial holes						
Removal of obstructions from trial holes irrespective of depth	1.00	109.99	263.40	–	hr	**373.39**
Pumping; maximum depth 4.0 m minimum 750 litres per hour	0.06	2.97	27.47	–	hr	**30.44**
02 LIGHT CABLE PERCUSSION BOREHOLES						
General						
The following costs are based on using Specialist Contractors and are for guidance only.						
Establishment of standard plant and equipment and removal on completion		–	–	–	item	**14332.50**
Number; 150 mm nominal dia. at base		–	–	–	nr	**74.74**
Depth; 150 mm nominal dia. of base						
in holes of maximum depth not exceeding 5 m		–	–	–	m	**29.89**
in holes of maximum depth 5–10 m		–	–	–	m	**29.89**
in holes of maximum depth 10–20 m		–	–	–	m	**37.36**
in holes of maximum depth 20–30 m		–	–	–	m	**47.82**
Depth backfilled; imported pulverised fuel ash		–	–	–	m	**7.74**
Depth backfilled; imported gravel		–	–	–	m	**10.82**
Depth backfilled; bentonite grout		–	–	–	m	**15.48**
Chiselling to prove rock or to penetrate obstructions		–	–	–	hour	**74.76**
Standing time of rig and crew		–	–	–	hour	**74.76**
03 ROTARY DRILLED BOREHOLES						
The following costs are based on using Specialist Contractors and are for guidance only.						
Establishment of standard plant and equipment and removal on completion		–	–	–	item	**35831.25**
Setting up at each borehole position		–	–	–	nr	**14.95**
Depth without core recovery; nominal minimum core dia. 100 mm						
ne 5.0 m deep		–	–	–	m	**83.25**
5–10 m deep		–	–	–	m	**83.25**
10–20 m deep		–	–	–	m	**92.13**
20–30 m deep		–	–	–	m	**92.13**

CLASS B: GROUND INVESTIGATION

Item	Gang hours	Labour £	Plant £	Material £	Unit	**Total rate £**
Depth with core recovery; nominal minimum core dia. 75 mm						
ne 5.0 m deep		–	–	–	m	**136.53**
5–10 m deep		–	–	–	m	**137.64**
10–20 m deep		–	–	–	m	**160.95**
20–30 m deep		–	–	–	m	**219.78**
Depth cased; semi-rigid plastic core barrel liner		–	–	–	m	**13.32**
04 SAMPLES						
General						
From the surface or from trial pits and trenches						
undisturbed soft material; minimum 200 mm cube		–	–	–	nr	**24.07**
disturbed soft material; minimum 5 kg		–	–	–	nr	**3.77**
rock; minimum 5 kg		–	–	–	nr	**15.01**
groundwater; minimum 1 l		–	–	–	nr	**6.67**
From boreholes						
open tube; 100 mm dia. × 450 mm long		–	–	–	nr	**23.72**
disturbed; minimum 5 kg		–	–	–	nr	**3.67**
groundwater; minimum 1 l		–	–	–	nr	**5.75**
stationary piston		–	–	–	nr	**5.84**
Swedish foil		–	–	–	nr	**37.04**
Delft		–	–	–	nr	**37.04**
Bishop sand		–	–	–	nr	**5.92**
06 INSTRUMENTAL OBSERVATIONS						
General						
Pressure head						
standpipe; 75 mm dia. HDPE pipe		–	–	–	m	**60.48**
piezometer		–	–	–	m	**96.12**
install protective cover		–	–	–	nr	**156.60**
readings		–	–	–	nr	**64.80**
07 LABORATORY TESTS						
General						
Classification						
moisture content		–	–	–	nr	**6.50**
specific gravity		–	–	–	nr	**9.37**
particle size analysis by sieve		–	–	–	nr	**39.07**
particle size analysis by pipette or hydrometer		–	–	–	nr	**52.09**
Chemical content						
organic matter		–	–	–	nr	**41.68**
sulphate		–	–	–	nr	**28.63**
pH value		–	–	–	nr	**9.12**
contaminants; Comprehensive		–	–	–	nr	**325.66**
contaminants; Abbreviated		–	–	–	nr	**227.95**
contaminants; Mini, Screening		–	–	–	nr	**188.89**
contaminants; nitrogen herbicides		–	–	–	nr	**169.35**
contaminants; organophosphorus pesticides		–	–	–	nr	**143.29**
contaminants; organochlorine pesticides		–	–	–	nr	**143.29**

CLASS B: GROUND INVESTIGATION

Item	Gang hours	Labour £	Plant £	Material £	Unit	Total rate £
07 LABORATORY TESTS – cont						
General – cont						
Compaction						
standard		–	–	–	nr	**169.34**
heavy		–	–	–	nr	**169.35**
vibratory		–	–	–	nr	**188.89**
Permeability						
falling head		–	–	–	nr	**78.16**
Soil strength						
quick undrained triaxial; set of three 38 mm dia. specimens		–	–	–	nr	**39.07**
shear box; peak only; size of shearbox 100 × 100 mm		–	–	–	nr	**48.84**
California bearing ratio; typical		–	–	–	nr	**57.32**
Rock strength						
point load test; minimum 5 kg sample		–	–	–	nr	**84.17**
08 PROFESSIONAL SERVICES						
General						
Technician		–	–	–	hr	**48.82**
Technician engineer		–	–	–	hr	**66.96**
Engineer or geologist						
graduate		–	–	–	hr	**66.96**
chartered		–	–	–	hr	**83.03**
principal or consultant		–	–	–	hr	**111.24**
Visits to the site						
technician		–	–	–	nr	**48.82**
technician engineer/graduate engineer or geologist		–	–	–	nr	**66.96**
chartered engineer		–	–	–	nr	**83.03**
principal or consultant		–	–	–	nr	**111.24**
Overnight stays in connection with visits to the site						
technician		–	–	–	nr	**93.31**
technician engineer/graduate engineer or geologist		–	–	–	nr	**92.88**
chartered engineer/senior geologist		–	–	–	nr	**120.96**
principal or consultant		–	–	–	nr	**120.96**

CLASS C: GEOTECHNICAL AND OTHER SPECIALIST PROCESSES

Item	Gang hours	Labour £	Plant £	Material £	Unit	**Total rate £**
01 DRILLING FOR GROUT HOLES						
Note: The processes referred to in this Section are generally carried out by Specialist Contractors and the costs are therefore an indication of the probable costs based on good site conditions.						
The following unit costs are based on drilling 100 grout holes on a clear site with reasonable access						
Establishment of standard drilling plant and equipment and removal on completion		–	–	–	item	**13072.50**
Rate per hole		–	–	–	nr	**130.73**
Standing time		–	–	–	hour	**299.03**
Multiple water pressure tests		–	–	–	nr	**9.32**
Drilling through material other than rock or artificial hard material						
vertically downwards						
depth ne 5 m		–	–	–	m	**25.41**
depth 5–10 m		–	–	–	m	**29.42**
depth 10–20 m		–	–	–	m	**34.85**
depth 20–30 m		–	–	–	m	**42.64**
downwards at an angle 0–45° to the vertical						
depth ne 5 m		–	–	–	m	**25.41**
depth 5–10 m		–	–	–	m	**29.42**
depth 10–20 m		–	–	–	m	**34.85**
depth 20–30 m		–	–	–	m	**42.64**
horizontally or downwards at an angle less than 45° to the horizontal						
depth ne 5 m		–	–	–	m	**25.41**
depth 5–10 m		–	–	–	m	**29.42**
depth 10–20 m		–	–	–	m	**34.85**
depth 20–30 m		–	–	–	m	**42.64**
upwards at an angle 0–45° to the horizontal						
depth ne 5 m		–	–	–	m	**41.07**
depth 5–10 m		–	–	–	m	**49.95**
depth 10–20 m		–	–	–	m	**53.28**
depth 20–30 m		–	–	–	m	**56.83**
upwards at an angle less than 45° to the vertical						
depth ne 5 m		–	–	–	m	**41.07**
depth 5–10 m		–	–	–	m	**49.95**
depth 10–20 m		–	–	–	m	**53.28**
depth 20–30 m		–	–	–	m	**56.83**
Drilling through rock or artificial hard material						
Vertically downwards						
depth ne 5 m		–	–	–	m	**27.22**
depth 5–10 m		–	–	–	m	**33.41**
depth 10–20 m		–	–	–	m	**40.66**
depth 20–30 m		–	–	–	m	**50.83**

CLASS C: GEOTECHNICAL AND OTHER SPECIALIST PROCESSES

Item	Gang hours	Labour £	Plant £	Material £	Unit	Total rate £
01 DRILLING FOR GROUT HOLES – cont						
Drilling through rock or artificial hard material – cont						
Downwards at an angle 0–45° to the vertical						
depth ne 5 m		–	–	–	m	**27.22**
depth 5–10 m		–	–	–	m	**33.41**
depth 10–20 m		–	–	–	m	**40.66**
depth 20–30 m		–	–	–	m	**50.83**
Horizontally or downwards at an angle less than 45° to the horizontal						
depth ne 5 m		–	–	–	m	**27.22**
depth 5–10 m		–	–	–	m	**33.41**
depth 10–20 m		–	–	–	m	**40.66**
depth 20–30 m		–	–	–	m	**50.83**
Upwards at an angle 0–45° to the horizontal						
depth ne 5 m		–	–	–	m	**29.97**
depth 5–10 m		–	–	–	m	**56.83**
depth 10–20 m		–	–	–	m	**64.60**
depth 20–30 m		–	–	–	m	**72.48**
Upwards at an angle less than 45° to the horizontal						
depth ne 5 m		–	–	–	m	**29.97**
depth 5–10 m		–	–	–	m	**56.83**
depth 10–20 m		–	–	–	m	**64.60**
depth 20–30 m		–	–	–	m	**72.48**
04 GROUT MATERIALS AND INJECTION						
The following unit costs are based on drilling 100 grout holes on a clear site with reasonable access						
Materials						
ordinary portland cement		–	–	–	t	**169.05**
sulphate-resistant cement		–	–	–	t	**172.12**
cement grout		–	–	–	t	**249.88**
pulverised fuel ash		–	–	–	t	**24.23**
sand		–	–	–	t	**31.78**
pea gravel		–	–	–	t	**31.78**
bentonite (2:1)		–	–	–	t	**218.25**
Injection						
establishment of standard injection plant and removal on completion		–	–	–	item	**15855.00**
standing time		–	–	–	hr	**299.03**
number of injections		–	–	–	nr	**92.13**
neat cement grout		–	–	–	t	**156.60**
cement/PFA grout		–	–	–	t	**96.28**

CLASS C: GEOTECHNICAL AND OTHER SPECIALIST PROCESSES

Item	Gang hours	Labour £	Plant £	Material £	Unit	**Total rate £**
05 DIAPHRAGM WALLS						
Note: Diaphragm walls are the construction of vertical walls, cast in place in a trench excavation. They can be formed in reinforced concrete to provide structural elements for temporary or permanent retaining walls. Wall thicknesses of 500 mm to 1.50 m and up to 40 m deep may be constructed. Special equipment such as the Hydrofraise can construct walls up to 100 m deep. Restricted urban sites will significantly increase the costs.						
The following costs are based on constructing a diaphragm wall with an excavated volume of 4000 m³ using standard equipment Typical progress would be up to 500 m per week.						
Establishment of standard plant and equipment including bentonite storage tanks and removal on completion		–	–	–	item	**437500.00**
Standing time		–	–	–	hour	**1196.14**
Excavation, disposal of soil and placing of concrete		–	–	–	m^3	**568.10**
Provide and place reinforcement cages		–	–	–	tonne	**968.30**
Excavate/chisel in hard material/rock		–	–	–	hour	**1258.10**
Waterproofed joints		–	–	–	m	**8.40**
Guide walls (twin)		–	–	–	m	**661.50**
06 GROUND REINFORCEMENT						
Note: Ground anchorages consist of the installation of a cable or solid bar tendon fixed in the ground by grouting and tensioned to exceed the working load to be carried. Ground anchors may be of a permanent or temporary nature and can be used in conjunction with diaphragm walls or sheet piling to eliminate the use of strutting etc.						
The following costs are based on the installation of 50 nr ground anchors.						
Establishment of standard plant and equipment and removal on completion		–	–	–	item	**15855.00**
Standing time		–	–	–	hour	**299.03**
Ground anchorages; temporary or permanent						
15.0 m maximum depth; in rock, alluvial or clay; 0–50 t load		–	–	–	m	**53.98**
15.0 m maximum depth; in rock or alluvial; 50–90 t load		–	–	–	m	**76.25**
15.0 m maximum depth; in rock only; 90–150 t load		–	–	–	m	**156.87**

CLASS C: GEOTECHNICAL AND OTHER SPECIALIST PROCESSES

Item	Gang hours	Labour £	Plant £	Material £	Unit	Total rate £
06 GROUND REINFORCEMENT – cont						
The following costs are based on the installation of 50 nr ground anchors – cont						
Temporary tendons						
in rock, alluvial or clay; 0–50 t load		–	–	–	m	**42.18**
in rock or alluvial; 50–90 t load		–	–	–	m	**61.05**
in rock only; 90–150 t load		–	–	–	m	**81.03**
Permanent tendons						
in rock, alluvial or clay; 0–50 t load		–	–	–	m	**42.18**
in rock or alluvial; 50–90 t load		–	–	–	m	**61.05**
in rock only; 90–150 t load		–	–	–	m	**129.65**
07 SAND, BAND AND WICK DRAINS						
Note: Vertical drains are a technique by which the rate of consolidation of fine grained soils can be considerably increased by the installation of vertical drainage paths commonly in the form of columns formed by a high-quality plastic material encased in a filter sleeve. Columns of sand are rarely used in this country these days.						
Band drains are generally 100 mm wide and 3–5 mm thick. Water is extracted through the drain from the soft soils when the surface is surcharged. The rate of consolidation is dependent on the drain spacing and the height of surcharge.						
Drains are usually quickly installed up to depths of 25 m by special lances either pulled or vibrated into the ground. typical drain spacing would be one per 1–2 m with the rate of installation varying between 1,500 to 6,000 m per day depending on ground conditions and depths.						
The following costs are based on the installation of 2,000 nr vertical band drains to a depth of 12 m						
Establishment of standard plant and equipment and removal on completion		–	–	–	item	**9513.00**
Standing time		–	–	–	hour	**299.03**
Set up installation equipment at each drain position		–	–	–	nr	**2.41**
Install drains maximum depth 10–15 m		–	–	–	m	**2.25**
Additional costs in pre-drilling through hard upper strata at each drain position:						
establishment of standard drilling plant and equipment and removal on completion		–	–	–	item	**3105.90**
set up at each drain position		–	–	–	nr	**7.53**
drilling for vertical band drains up to a maximum depth of 3 m		–	–	–	m	**16.34**

CLASS C: GEOTECHNICAL AND OTHER SPECIALIST PROCESSES

Item	Gang hours	Labour £	Plant £	Material £	Unit	Total rate £
08 VIBRO-FLOTATION AND DYNAMIC COMPACTION						
VIBRO-FLOTATION						
There are many considerations regarding the soil types to be treated, whether cohesive or non-cohesive, made-up or natural ground, which influence the choice of wet or dry processes, pure densification or stone column techniques with added granular backfill. It is therefore possible only to give indicative costs; a Specialist Contractor should be consulted for more accurate costs for a particular site.						
Plant Mobilization						
Vibro-flotation is a method of considerably increasing the ground bearing pressure and consists of a specifically designed powerful poker vibrator penetrating vertically into the ground hydraulically. Air and water jets may be used to assist penetration. In cohesive soils a hole is formed into which granular backfill is placed and compacted by the poker, forming a dense stone column. In natural sands and gravels the existing loose deposits may be compacted without the addition of extra material other than making up levels after settlement resulting from compaction. They are three main types of Compact, Replacement and Displacement method, the cost has considered replacement method of flotation.						
Testing of conditions after consolidation can be static or dynamic penetration tests, plate bearing tests or zone bearing tests. A frequently adopted specification calls for plate bearing tests at 1 per 1000 stone columns. Allowable bearing pressures of up to 400 kN/m^2 by the installation of stone columns in made or natural ground.						
establishment of standard plant and equipment and removal on completion		–	–	–	m	**6340.32**
standing time		–	385.15	–	m	**385.15**
Construct stone columns to a depth ne 4 m						
dry formed	1.00	60.46	–	–	m	**60.46**
water jet formed		–	–	–	m	**428.47**
Plate bearing test						
ne 11 t or 2 hour duration		–	–	–	nr	**679.32**
zone specification testing		–	–	–	nr	**13586.40**
Vibro-flotation						
Depth not exceeding 5 m	0.60	15.01	74.45	–	m	**2219.56**
Depth: 5 m–10 m	1.50	37.54	38.95	–	m	**2759.91**
Depth: 10 m–20 m	1.50	37.54	71.87	–	m	**3853.99**
Depth: 30 m–40 m	3.00	75.07	77.89	–	m	**6198.02**

CLASS C: GEOTECHNICAL AND OTHER SPECIALIST PROCESSES

Item	Gang hours	Labour £	Plant £	Material £	Unit	**Total rate £**
08 VIBRO-FLOTATION AND DYNAMIC COMPACTION – cont						
DYNAMIC COMPACTION						
Ground consolidation by dynamic compaction is a technique which involves the dropping of a steel or concrete pounder several times in each location on a grid pattern that covers the whole site. For ground compaction up to 10 m, a 15 t pounder from a free fall of 20 m would be typical. Several passes over the site are normally required to achieve full compaction. The process is recommended for naturally cohesive soils and is usually uneconomic for areas of less than 4,000 m^2 for sites with granular or mixed granular cohesive soils and 6,000 m for a site with weak cohesive soils. The main considerations to be taken into account when using this method of consolidation are:						
* sufficient area to be viable						
* proximity and condition of adjacent property and services						
* need for blanket layer of granular material for a working surface and as backfill to offset induced settlement						
* water table level						
The final bearing capacity and settlement criteria that can be achieved depends on the nature of the material being compacted. Allowable bearing capacity may be increased by up to twice the pre-treated value for the same settlement. Control testing can be by crater volume measurements, site levelling between passes, penetration tests or plate loading tests.						
The following range of costs are average based on treating an area of about 10,000 m^2 for a 5–6 m compaction depth. Typical progress would be 1,500–2,000 m^2 per week						
Establishment of standard plant and equipment and removal on completion.		–	–	–	item	**16275.00**
Ground treatment		–	–	–	m^2	**2.55**
Laying free-draining granular blanket layer as both working surface and backfill material (300 mm thickness required of filter material)		–	–	–	m^2	**12.42**
Control testing including levelling, piezometers and penetrameter testing		–	–	–	m^2	**5.46**
Kentledge load test		–	–	–	nr	**11655.00**

CLASS C: GEOTECHNICAL AND OTHER SPECIALIST PROCESSES

Item	Gang hours	Labour £	Plant £	Material £	Unit	**Total rate £**
CONSOLIDATION OF ABANDONED MINE WORKINGS						
The following costs assume using Specialist Contractors to undertake the works						
Transport plant, labour and all equipment to and from site (max. 50 miles)		–	–	–	item	**554.78**
Drilling bore holes						
move to each seperate bore position; erect equipment; dismantle prior to next move		–	–	–	nr	**53.88**
drill 50 mm dia. bore holes		–	–	–	m	**14.97**
drill 100 mm dia. bore holes for pea gravel injection		–	–	–	m	**26.20**
extra for casing, when required		–	–	–	m	**22.43**
standing time for drilling rig and crew		–	–	–	hr	**149.69**
Grouting drilled bore holes						
connecting grout lines		–	–	–	m	**27.22**
injection of grout		–	–	–	t	**108.91**
add for pea gravel injection		–	–	–	t	**128.02**
standing time for grouting rig and crew		–	–	–	hour	**229.37**
Provide materials for grouting						
ordinary portland cement		–	–	–	t	**169.05**
sulphate-resistant cement		–	–	–	t	**172.12**
pulverised fuel ash (PFA)		–	–	–	t	**24.23**
sand		–	–	–	t	**31.78**
pea gravel		–	–	–	t	**264.87**
bentonite (2:1)		–	–	–	t	**218.25**
cement grout		–	–	–	t	**249.88**
Capping to old shafts or similar; reinforced concrete grade C20P, 20 mm aggregate; thickness						
ne 150 mm		–	–	–	m^3	**254.12**
150–300 mm		–	–	–	m^3	**246.88**
300–500 mm		–	–	–	m^3	**239.60**
over 500 mm		–	–	–	m^3	**196.04**
Mild steel bars BS4449; supplied in bent and cut lengths						
6 mm nominal size		–	–	–	t	**1394.53**
8 mm nominal size		–	–	–	t	**1394.55**
10 mm nominal size		–	–	–	t	**1394.55**
12 mm nominal size		–	–	–	t	**1394.55**
16 mm nominal size		–	–	–	t	**1394.55**
20 mm nominal size		–	–	–	t	**1394.55**
High yield steel bars BS4449 or 4461; supplied in bent and cut lengths						
6 mm nominal size		–	–	–	t	**1394.53**
8 mm nominal size		–	–	–	t	**1394.55**
10 mm nominal size		–	–	–	t	**1394.55**
12 mm nominal size		–	–	–	t	**1394.55**
16 mm nominal size		–	–	–	t	**1394.55**
20 mm nominal size		–	–	–	t	**1394.55**

CLASS D: DEMOLITION AND SITE CLEARANCE

Item	Gang hours	Labour £	Plant £	Material £	Unit	Total rate £
01 SITE CLEARANCE						
The rates for site clearance include for all sundry items, small trees (i.e. under 500 mm dia.), hedges etc., but exclude items that are measured separately; examples of which are given in this section						
Clear site vegetation						
generally		–	–	–	ha	**1592.00**
invasive plant species		–	–	–	ha	**4387.93**
areas below tidal level		–	–	–	ha	**5363.05**
DRILLING AND BLASTING IN ROCK						
The cost of blasting is controlled by the number of holes and the length of drilling required to achieve the tolerances and degree of shatter required, e.g. line drilling to trenches, depth of drilling to control horizontal overbreak.						
Drilling with rotary percussion drills						
105–110 mm dia.; hard rock		–	–	–	m	**21.06**
105–110 mm dia.; sandstone		–	–	–	m	**10.60**
125 mm dia.; hard rock		–	–	–	m	**16.61**
125 mm dia.; sandstone		–	–	–	m	**15.03**
165 mm dia.; hard rock		–	–	–	m	**17.19**
165 mm dia.; sandstone		–	–	–	m	**19.04**
Drilling and blasting in open cut for bulk excavation excluding cost of excavation or trimming						
hard rock		–	–	–	m	**4.33**
sandstone		–	–	–	m	**6.55**
Drilling and blasting for quarry operations with face height exceeding 10 m						
hard rock		–	–	–	m	**4.55**
sandstone		–	–	–	m	**4.55**
Drilling and blasting in trenches excluding cost of excavation or trimming						
trench width ne 1.0 m		–	–	–	m	**36.15**
trench width 1.0–1.5 m		–	–	–	m	**31.60**
trench width over 1.5 m		–	–	–	m	**26.56**
Secondary blasting to boulders						
pop shooting		–	–	–	m	**11.39**
plaster shooting		–	–	–	m	**6.08**

CLASS D: DEMOLITION AND SITE CLEARANCE

Item	Gang hours	Labour £	Plant £	Material £	Unit	**Total rate £**
02 TREES						
The following rates are based on removing a minimum of 100 trees, generally in a group. Cutting down a single tree on a site would be many times these costs						
Remove trees						
girth 500 mm–1 m		–	–	–	nr	**58.83**
girth 1–2 m		–	–	–	nr	**99.90**
girth 2–3 m		–	–	–	nr	**346.32**
girth 3–5 m		–	–	–	nr	**1254.30**
girth 7 m		–	–	–	nr	**1758.24**
03 STUMPS						
Remove stumps and cart away						
Clearance of stumps						
dia. 150–500 mm		–	–	–	nr	**59.94**
dia. 500 mm–1 m		–	–	–	nr	**113.22**
dia. 2 m		–	–	–	nr	**170.94**
Clearance of stumps; backfilling holes with topsoil from site						
dia. 150–500 mm		–	–	–	nr	**72.82**
dia. 500 mm–1 m		–	–	–	nr	**160.95**
dia. 2 m		–	–	–	nr	**346.32**
Clearance of stumps; backfilling holes with imported hardcore						
dia. 150–500 mm		–	–	–	nr	**84.80**
dia. 500 mm–1 m		–	–	–	nr	**294.93**
dia. 2 m		–	–	–	nr	**971.25**
04 BUILDINGS						
The following rates are based on assuming a non urban location where structure does not take up a significant area of the site						
Demolish building to ground level and dispose off site						
brickwork with timber floor and roof		–	–	–	m^3	**9.20**
brickwork with concrete floor and roof		–	–	–	m^3	**15.35**
masonry with timber floor and roof		–	–	–	m^3	**12.03**
reinforced concrete frame with brick infill		–	–	–	m^3	**16.04**
steel frame with brick cladding		–	–	–	m^3	**8.71**
steel frame with sheet cladding		–	–	–	m^3	**8.29**
timber		–	–	–	m^3	**7.44**
Demolish buildings with asbestos linings to ground level and dispose off site						
brick with concrete floor and roof		–	–	–	m^3	**37.33**
reinforced concrete frame with brick infill		–	–	–	m^3	**38.85**
steel frame with brick cladding		–	–	–	m^3	**21.25**
steel frame with sheet cladding		–	–	–	m^3	**22.07**

CLASS D: DEMOLITION AND SITE CLEARANCE

Item	Gang hours	Labour £	Plant £	Material £	Unit	**Total rate £**
05 OTHER STRUCTURES						
The following rates are based on assuming a non urban location where structure does not take up a significant area of the site						
Demolish walls to ground level and dispose off site						
reinforced concrete wall	–	–	–		m^3	**205.82**
brick or masonry wall	–	–	–		m^3	**95.26**
brick or masonry retaining wall	–	–	–		m^3	**116.42**
06 PIPELINES						
Removal of redundant services						
electric cable; LV	–	–	–		m	**3.66**
75 mm dia. water main; low pressure	–	–	–		m	**7.33**
150 mm dia. gas main; low pressure	–	–	–		m	**8.79**
earthenware ducts; one way	–	–	–		m	**3.66**
earthenware ducts; two way	–	–	–		m	**7.34**
100 or 150 mm dia. sewer or drain	–	–	–		m	**15.77**
225 mm dia. sewer or drain	–	–	–		m	**22.93**
300 mm dia. sewer or drain	–	–	–		m	**25.81**
450 mm dia. sewer or drain	–	–	–		m	**31.52**
750 mm dia. sewer or drain	–	–	–		m	**40.09**
extra for breaking up concrete surround	–	–	–		m	**40.09**
Grouting redundant drains or sewers						
100 mm dia.	–	–	–		m	**8.79**
150 mm dia.	–	–	–		m	**21.49**
225 mm dia.	–	–	–		m	**50.15**
manhole chambers	–	–	–		m^3	**1217.88**

CLASS E: EARTHWORKS

Item	Gang hours	Labour £	Plant £	Material £	Unit	**Total rate £**
NOTES						
Ground conditions						
The following unit costs for 'excavation in material other than topsoil, rock or artificial hard material' are based on excavation in firm sand and gravel soils. For alternative types of soil, multiply the following rates by:						
Scrapers						
Stiff clay × 1.5						
Chalk × 2.5						
Soft rock × 3.5						
Broken rock × 3.7						
Tractor dozers and loaders						
Stiff clay × 2.0						
Chalk × 3.0						
Soft rock × 2.5						
Broken rock × 2.5						
Backacter (minimum bucket size 0.5 m^3)						
Stiff clay × 1.7						
Chalk × 2.0						
Soft rock × 2.0						
Broken rock × 1.7						
Basis of disposal rates						
All pricing and estimating for disposal is based on the volume of solid material excavated and rates for disposal should be adjusted by the following factors for bulkage. Multiply the rates by:						
Sand bulkage × 1.10						
Gravel bulkage × 1.20						
Compacted soil bulkage × 1.30						
Compacted subbase, suitable fill etc. bulkage × 1.30						
Stiff clay bulkage × 1.20						
See also Part 12: Tables and Memoranda						
Basis of rates generally						
To provide an overall cost comparison, rates, prices and outputs have been based on a medium sized Civil Engineering project of £ 10–12 million, location neither in city centre nor excessively remote, with no abnormal ground conditions that would affect the stated output and consistency of work produced. The rates are optimum rates and assume continuous output with no delays caused by other operations or works.						

CLASS E: EARTHWORKS

Item	Gang hours	Labour £	Plant £	Material £	Unit	**Total rate £**
RESOURCES – LABOUR						
Labour Gang Build-up Rates						
Excavation for cuttings gang						
1 plant operator (skill rate 1) – 33% of time		16.33	–	–		
1 plant operator (skill rate 2) – 66% of time		23.40	–	–		
1 banksman (skill rate 4)		20.20	–	–		
Total Gang Rate/Hour		**59.93**	–	–		
Excavation for foundations gang						
1 plant operator (skill rate 3) – 33% of time		9.63	–	–		
1 plant operator (skill rate 2) – 66% of time		23.40	–	–		
1 banksman (skill rate 4)		20.20	–	–		
Total Gang Rate/Hour		**53.22**	–	–		
General excavation gang						
1 plant operator (skill rate 3)		29.17	–	–		
1 plant operator (skill rate 3) – 25% of time		7.29	–	–		
1 banksman (skill rate 4)		20.20	–	–		
Total Gang Rate/Hour		**56.66**	–	–		
Filling gang						
1 plant operator (skill rate 4)		26.30	–	–		
2 unskilled operatives (general)		39.20	–	–		
Total Gang Rate/Hour		**65.50**	–	–		
Treatment of filled surfaces gang						
1 plant operator (skill rate 2)		35.46	–	–		
Total Gang Rate/Hour		**35.46**	–	–		
Geotextiles (light sheets) gang						
1 ganger/chargehand (skill rate 4) – 20% of time		4.28	–	–		
2 unskilled operatives (general)		39.20	–	–		
Total Gang Rate/Hour		**43.49**	–	–		
Geotextiles (medium sheets) gang						
1 ganger/chargehand (skill rate 4) – 20% of time		4.28	–	–		
3 unskilled operatives (general)		58.81	–	–		
Total Gang Rate/Hour		**63.09**	–	–		
Geotextiles (heavy sheets) gang						
1 ganger/chargehand (skill rate 4) – 20% of time		4.28	–	–		
2 unskilled operatives (general)		39.20	–	–		
1 plant operator (skill rate 3)		29.17	–	–		
Total Gang Rate/Hour		**72.66**	–	–		
Horticultural works gang						
1 skilled operative (skill rate 4)		20.20	–	–		
1 unskilled operative (general)		19.60	–	–		
Total Gang Rate/Hour		**39.80**	–	–		
RESOURCES – PLANT						
Plant Resources Build-up Rate						
Excavation for foundations						
hydraulic Backacter – 21 tonne (33% of time)		–	28.22	–		
hydraulic Backacter – 14.5 tonne (33% of time)		–	21.46	–		
1000 kg hydraulic breaker (33% of time)		–	6.90	–		
tractor loader – 0.80 m^3 (33% of time)		–	18.47	–		
Total Rate/Hour		–	**75.05**	–		

CLASS E: EARTHWORKS

Item	Gang hours	Labour £	Plant £	Material £	Unit	**Total rate £**
Filling						
1.5 m^3 tractor loader		–	93.72	–		
6t vibratory roller		–	39.30	–		
pedestrian Roller, Bomag BW35		–	8.34	–		
Total Rate/Hour		–	**141.36**	–		
Treatment of filled surfaces						
1.5 m^3 tractor loader		–	93.72	–		
pedestrian Roller, Bomag BW35		–	8.34	–		
Total Rate/Hour		–	**102.06**	–		
Geotextiles (heavy sheets)						
1.5 m^3 tractor loader		–	93.72	–		
Total Rate/Hour		–	**93.72**	–		
01 EXCAVATION BY DREDGING						
Note: Dredging can be carried out by land based machines or by floating plant. The cost of the former can be assessed by reference to the excavation costs of the various types of plant given below, suitably adjusted to take account of the type of material to be excavated, depth of water and method of disposal. The cost of the latter is governed by many factors which affect the rates and leads to wide variations.						
For reliable estimates it is advisable to seek the advice of a Specialist Contractor. The prices included here are for some typical dredging situations and are shown for a cost comparison and EXCLUDE initial mobilization charges which can range widely between £3,000 and £10,000 depending on plant, travelling distance etc. Some clients schedule operations for when the plant is passing so as to avoid the large travelling cost.						
Of the factors affecting the cost of floating plant, the matter of working hours is by far one of the most important. The customary practice in the dredging industry is to work 24 hours per day, 7 days per week. Local constraints, particularly noise restrictions, will have a significant impact.						
Other major factors affecting the cost of floating plant are:						
* type of material to be dredged						
* depth of water						
* depth of cut						
* tidal range						
* disposal location						
* size and type of plant required						
* current location of plant						
* method of disposal of dredged material						

CLASS E: EARTHWORKS

Item	Gang hours	Labour £	Plant £	Material £	Unit	Total rate £
01 EXCAVATION BY DREDGING – cont						
In tidal locations, creating new channels on approaches to quays and similar locations or within dock systems						
Backhoe dredger loading material onto two hopper barges with bottom dumping facility maximum water depth 15 m, distance to disposal site less than 20 miles						
approximate daily cost of backhoe dredger and two hopper barges £12,500; average production 100 m^3/hr, 1,500 m^3/day; locate, load, deposit and relocate		–	–	–	m^3	**10.72**
For general bed lowering or in maintaining shipping channels in rivers, estuaries or deltas						
Trailer suction hopper dredger excavating non-cohesive sands, grits or silts; hopper capacity 2000 m^3, capable of dredging to depths of 25 m with ability either to dump at disposal site or pump ashore for reclamation. Costs totally dependent on nature of material excavated and method of disposal.						
locate, load, deposit and relocate		–	–	–	tonne	**4.34**
Harbour bed control						
Maintenance dredging of this nature would in most cases be carried out by trailing suction hopper dredger as detailed above, at similar rates. The majority of the present generation of trailers have the ability to dump at sea or discharge ashore.						
Floating craft using diesel driven suction method with a 750 mm dia. flexible pipe for a maximum distance of up to 5000 m from point of suction to point of discharge using a booster (standing alone, the cutter suction craft should be able to pump up to 2000 m). Maximum height of lift 10 m.						
average pumping capacity of silt/sand type materials containing maximum 30% volume of solids would be about 8,000 m^3/day based on 24 hour working, Daily cost (hire basis) in the region of £20,000 including all floating equipment and discharge pipes, maintenance and all labour and plant to service but excluding mobilization/initial set-up and demobilisation costs (minimum £10,000).		–	–	–	m^3	**4.00**

CLASS E: EARTHWORKS

Item	Gang hours	Labour £	Plant £	Material £	Unit	**Total rate £**
For use in lakes, canals, rivers, industrial lagoons and from silted locations in dock systems						
Floating craft using diesel driven suction method with a 200 mm dia. flexible pipe maximum distance from point of suction to point of discharge 1500 m. Maximum dredge depth 5 m.						
average pumping output 30 m^3 per hour and approximate average costs (excluding mobilization and demobilization costs ranging between £3,000and £6,000 a piece)	–	–	–		m^3	**7.36**
02 EXCAVATION FOR CUTTINGS						
The following unit costs are based on backacter and tractor loader machines						
Excavate topsoil						
maximum depth ne 0.25 m	0.03	1.80	2.29	–	m^3	**4.09**
Excavate material other than topsoil, rock or artificial hard material						
ne 0.25 m maximum depth	0.03	1.80	2.29	–	m^3	**4.09**
0.25–0.5 m maximum depth	0.03	1.80	2.29	–	m^3	**4.09**
0.5–1.0 m maximum depth	0.04	2.40	2.97	–	m^3	**5.37**
1.0–2.0 m maximum depth	0.06	3.60	4.51	–	m^3	**8.10**
2.0–5.0 m maximum depth	0.10	5.99	7.47	–	m^3	**13.47**
5.0–10.0 m maximum depth	0.20	11.99	15.05	–	m^3	**27.04**
10.0–15.0 m maximum depth	0.29	17.38	21.77	–	m^3	**39.15**
The following unit costs are based on backacter machines fitted with hydraulic breakers and tractor loader machines						
Excavate rock (medium hard)						
ne 0.25 m maximum depth	0.31	18.58	23.28	–	m^3	**41.86**
0.25–0.5 m maximum depth	0.43	25.77	32.30	–	m^3	**58.06**
0.5–1.0 m maximum depth	0.58	34.76	43.52	–	m^3	**78.27**
1.0–2.0 m maximum depth	0.80	47.94	60.17	–	m^3	**108.11**
Excavate unreinforced concrete exposed at the commencing surface						
ne 0.25 m maximum depth	0.57	34.16	42.84	–	m^3	**77.00**
0.25–0.5 m maximum depth	0.60	35.96	45.05	–	m^3	**81.01**
0.5 −1.0 m maximum depth	0.67	40.15	50.32	–	m^3	**90.47**
1.0–2.0 m maximum depth	0.70	41.95	52.53	–	m^3	**94.48**
Excavate reinforced concrete exposed at the commencing surface						
ne 0.25 m maximum depth	0.90	53.93	67.58	–	m^3	**121.51**
0.25–0.5 m maximum depth	0.90	53.93	67.58	–	m^3	**121.51**
0.5–1.0 m maximum depth	0.95	56.93	71.33	–	m^3	**128.26**
1.0–2.0 m maximum depth	1.08	64.72	81.10	–	m^3	**145.82**

CLASS E: EARTHWORKS

Item	Gang hours	Labour £	Plant £	Material £	Unit	Total rate £
02 EXCAVATION FOR CUTTINGS – cont						
The following unit costs are based on backacter machines fitted with hydraulic breakers and tractor loader machines – cont						
Excavate unreinforced concrete not exposed at the commencing surface						
ne 0.25 m maximum depth	0.65	38.95	48.80	–	m^3	**87.75**
0.25–0.5 m maximum depth	0.67	40.15	50.32	–	m^3	**90.47**
0.5 –1.0 m maximum depth	0.69	41.35	51.85	–	m^3	**93.20**
1.0–2.0 m maximum depth	0.72	43.15	54.06	–	m^3	**97.21**
03 EXCAVATION FOR FOUNDATIONS						
The following unit costs are based on the use of backacter machines						
Excavate topsoil						
maximum depth ne 0.25 m	0.03	1.60	2.29	–	m^3	**3.89**
Excavate material other than topsoil, rock or artificial hard material						
0.25–0.5 m deep	0.05	2.66	3.81	–	m^3	**6.47**
0.5–1.0 m deep	0.06	3.19	4.51	–	m^3	**7.70**
1.0–2.0 m deep	0.07	3.73	5.20	–	m^3	**8.93**
2.0–5.0 m deep	0.12	6.39	9.01	–	m^3	**15.40**
5.0–10.0 m deep	0.23	12.24	17.32	–	m^3	**29.57**
The following unit costs are based on backacter machines fitted with hydraulic breakers						
Excavate unreinforced concrete exposed at the commencing surface						
ne 0.25 m maximum depth	0.60	31.93	45.05	–	m^3	**76.99**
0.25–0.5 m maximum depth	0.66	35.13	49.56	–	m^3	**84.69**
0.5–1.0 m maximum depth	0.72	38.32	54.06	–	m^3	**92.39**
1.0–2.0 m maximum depth	0.80	42.58	60.17	–	m^3	**102.75**
Excavate reinforced concrete exposed at the commencing surface						
ne 0.25 m maximum depth	0.94	50.03	70.49	–	m^3	**120.52**
0.25–0.5 m maximum depth	0.98	52.16	73.68	–	m^3	**125.84**
0.5–1.0 m maximum depth	1.00	53.22	75.05	–	m^3	**128.28**
1.0–2.0 m maximum depth	1.04	55.35	78.19	–	m^3	**133.54**
Excavate tarmacadam exposed at the commencing surface						
ne 0.25 m maximum depth	0.30	15.97	22.53	–	m^3	**38.49**
0.25–0.5 m maximum depth	0.34	18.10	25.43	–	m^3	**43.53**
0.5–1.0 m maximum depth	0.37	19.69	27.73	–	m^3	**47.42**
1.0–2.0 m maximum depth	0.40	21.29	29.94	–	m^3	**51.23**
Excavate unreinforced concrete not exposed at the commencing surface						
ne 0.25 m maximum depth	0.70	37.26	52.47	–	m^3	**89.72**
0.25–0.5 m maximum depth	0.85	45.24	63.77	–	m^3	**109.01**
0.5 –1.0 m maximum depth	0.96	51.09	72.09	–	m^3	**123.18**
1.0–2.0 m maximum depth	1.02	54.29	76.59	–	m^3	**130.88**

CLASS E: EARTHWORKS

Item	Gang hours	Labour £	Plant £	Material £	Unit	**Total rate £**
04 GENERAL EXCAVATION						
The following unit costs are based on backacter and tractor loader machines						
Excavate topsoil						
maximum depth ne 0.25 m	0.03	1.70	2.29	–	m^3	**3.99**
Excavate material other than topsoil, rock or artificial hard material						
ne 0.25 m maximum depth	0.03	1.70	2.20	–	m^3	**3.90**
0.25–0.5 m maximum depth	0.03	1.70	2.29	–	m^3	**3.99**
0.5–1.0 m maximum depth	0.04	2.27	2.97	–	m^3	**5.24**
1.0–2.0 m maximum depth	0.06	3.40	4.51	–	m^3	**7.91**
2.0–5.0 m maximum depth	0.10	5.67	7.47	–	m^3	**13.14**
5.0–10.0 m maximum depth	0.20	11.33	15.05	–	m^3	**26.38**
10.0–15.0 m maximum depth	0.29	16.43	21.77	–	m^3	**38.20**
The following unit costs are based on backacter machines fitted with hydraulic breakers and tractor loader machines						
Excavate rock (medium hard)						
ne 0.25 m maximum depth	0.31	17.56	23.28	–	m^3	**40.85**
0.25–0.5 m maximum depth	0.43	24.36	32.30	–	m^3	**56.66**
0.5–1.0 m maximum depth	0.58	32.86	43.52	–	m^3	**76.38**
1.0–2.0 m maximum depth	0.80	45.33	60.17	–	m^3	**105.49**
Excavate unreinforced concrete exposed at the commencing surface						
ne 0.25 m maximum depth	0.57	32.30	42.84	–	m^3	**75.14**
0.25–0.5 m maximum depth	0.60	34.00	45.05	–	m^3	**79.05**
0.5 –1.0 m maximum depth	0.67	37.96	50.32	–	m^3	**88.28**
1.0–2.0 m maximum depth	0.70	39.66	52.53	–	m^3	**92.19**
Excavate reinforced concrete exposed at the commencing surface						
ne 0.25 m maximum depth	0.90	50.99	67.58	–	m^3	**118.57**
0.25–0.5 m maximum depth	0.90	50.99	67.58	–	m^3	**118.57**
0.5–1.0 m maximum depth	0.95	53.83	71.33	–	m^3	**125.15**
1.0–2.0 m maximum depth	1.08	61.19	81.10	–	m^3	**142.29**
Excavate unreinforced concrete not exposed at the commencing surface						
ne 0.25 m maximum depth	0.65	36.83	48.80	–	m^3	**85.63**
0.25–0.5 m maximum depth	0.67	37.96	50.32	–	m^3	**88.28**
0.5 –1.0 m maximum depth	0.69	39.10	51.85	–	m^3	**90.95**
1.0–2.0 m maximum depth	0.72	40.79	54.06	–	m^3	**94.86**

CLASS E: EARTHWORKS

Item	Gang hours	Labour £	Plant £	Material £	Unit	**Total rate £**
04 GENERAL EXCAVATION – cont						
EARTH MOVING						
Note: The cost of earth moving and other associated works is dependent on matching the overall quantities and production rate called for by the programme of works with the most appropriate plant and assessing the most suitable version of that plant which will:						
* deal with the site conditions (e.g. type of ground, type of excavation, length of haul, prevailing weather, etc.)						
* comply with the specification requirements (e.g. compaction, separation of materials, surface tolerances, etc.)						
* complete the work economically (e.g. provide surface tolerances which will avoid undue expense of imported materials)						
Labour costs are based on a plant operative skill rate 3 unless otherwise stated						
Comparative costs of excavation equipment						
The following are comparative costs using various types of excavation equipment and include loading into transport. All costs assume 50 minutes productive work in every hour and adequate disposal transport being available to obviate any delay.						
Dragline (for excavations requiring a long reach and long discharge, mainly in clearing streams and rivers)						
bucket capacity ne 1.15 m^3	0.06	2.13	5.05	–	m^3	**7.18**
bucket capacity ne 2.00 m^3	0.03	1.06	4.50	–	m^3	**5.56**
bucket capacity ne 3.00 m^3	0.02	0.71	4.11	–	m^3	**4.81**
Hydraulic backacter (for all types of excavation and loading, including trenches, breaking hard ground, etc.)						
bucket capacity ne 0.40 m^3	0.05	1.77	2.78	–	m^3	**4.55**
bucket capacity ne 1.00 m^3	0.02	0.71	1.69	–	m^3	**2.40**
bucket capacity ne 1.60 m^3	0.01	0.35	1.72	–	m^3	**2.07**
Hydraulic face shovel (predominantly for excavating cuttings and embankments over 2 m high requiring high output)						
bucket capacity ne 1.50 m^3	0.04	1.42	4.83	–	m^3	**6.25**
bucket capacity ne 2.60 m^3	0.02	0.71	3.27	–	m^3	**3.98**
bucket capacity ne 3.40 m^3	0.01	0.35	2.36	–	m^3	**2.71**
Tractor loader (for loading, carrying, placing materials, spreading and levelling, some site clearance operations and reducing levels)						
bucket capacity ne 0.80 m^3	0.02	0.71	1.11	–	m^3	**1.82**
bucket capacity ne 1.50 m^3	0.01	0.35	0.94	–	m^3	**1.29**
bucket capacity ne 3.00 m^3	0.01	0.25	1.27	–	m^3	**1.52**

CLASS E: EARTHWORKS

Item	Gang hours	Labour £	Plant £	Material £	Unit	**Total rate £**
Multipurpose wheeled loader/backhoe (versatile machine for small to medium excavations, trenches, loading, carrying and back filling)						
bucket capacity ne 0.76 m^3	0.08	2.84	2.87	–	m^3	**5.70**
bucket capacity ne 1.00 m^3	0.06	2.13	2.82	–	m^3	**4.95**
Comparative costs of transportation equipment						
The following are comparative costs for using various types of transportation equipment to transport excavated loose material. The capacity of the transport must be suitable for the output of the loading machine. The cost will vary depending on the number of transport units required to meet the output of the loading unit and the distance to be travelled.						
Loading loose material into transport by wheeled loader						
ne 2.1 m^3 capacity	0.02	0.71	1.54	–	m^3	**2.25**
ne 5.4 m^3 capacity	0.01	0.35	2.29	–	m^3	**2.65**
ne 10.5 m^3 capacity	0.01	0.25	2.14	–	m^3	**2.39**
Transport material within site by dump truck (rear dump)						
ne 24 m^3 heaped capacity, distance travelled ne 0.5 km	0.03	1.06	5.28	–	m^3	**6.35**
add per 0.5 km additional distance	0.01	0.35	1.76	–	m^3/0.5 km	**2.12**
ne 57 m^3 heaped capacity, distance travelled ne 0.5 km	0.03	1.48	6.07	–	m^3	**7.56**
add per 0.5 km additional distance	0.03	1.48	3.04	–	m^3/0.5 km	**4.52**
Transport material within site by dump truck (articulated)						
ne 32 t payload, distance travelled ne 0.5 km	0.04	1.98	4.83	–	m^3	**6.81**
Add per 0.5 km additional distance	0.02	0.99	2.42	–	m^3/0.5 km	**3.41**
Transport material within or off site by tipping lorry						
ne 10 t payload, distance travelled ne 1 km	0.05	1.46	3.82	–	m^3	**5.28**
add per 1 km additional distance	0.03	0.73	1.91	–	m^3/km	**2.64**
10–15 t payload, distance travelled ne 1 km	0.04	1.17	3.74	–	m^3	**4.91**
add per 1 km additional distance	0.02	0.58	1.87	–	m^3/km	**2.45**
15–25 t payload, distance travelled ne 1 km	0.03	1.06	3.22	–	m^3	**4.29**
add per 1 km additional distance	0.02	0.53	1.61	–	m^3/km	**2.14**
Comparative costs of earth moving equipment						
The following are comparative costs using various types of earth moving equipment to include excavation, transport, spreading and levelling.						
Bulldozer up to 74 KW (CAT D4H sized machine used for site strip, reducing levels or grading and spreading materials over smaller sites)						
average push one way 10 m	0.03	0.88	1.97	–	m^3	**2.85**
average push one way 30 m	0.08	2.33	5.26	–	m^3	**7.59**
average push one way 50 m	0.14	4.08	9.20	–	m^3	**13.28**
average push one way 100 m	0.29	8.46	19.05	–	m^3	**27.51**

CLASS E: EARTHWORKS

Item	Gang hours	Labour £	Plant £	Material £	Unit	Total rate £
04 GENERAL EXCAVATION – cont						
Comparative costs of earth moving equipment – cont						
Bulldozer up to 104 KW (CAT D6E sized machine for reducing levels, excavating to greater depths at steeper inclines, grading surfaces, small cut and fill operations; maximum push 100 m)						
average push one way 10 m	0.03	1.06	3.25	–	m^3	**4.32**
average push one way 30 m	0.07	2.48	7.59	–	m^3	**10.08**
average push one way 50 m	0.09	3.19	9.76	–	m^3	**12.96**
average push one way 100 m	0.19	6.74	20.61	–	m^3	**27.35**
Bull or Angle Dozer up to 212 KW (CAT D8N sized machine for high output, ripping and excavating by reducing levels at steeper inclines or in harder material than with D6E, larger cut and fill operations used in conjunction with towed or S.P. scrapers. Spreading and grading over large areas; maximum push 100 m)						
average push one way 10 m	0.03	1.06	6.00	–	m^3	**7.06**
average push one way 30 m	0.06	2.13	12.00	–	m^3	**14.13**
average push one way 50 m	0.07	2.48	14.00	–	m^3	**16.48**
average push one way 100 m	0.11	3.90	22.00	–	m^3	**25.90**
Motorized scraper, 15 m^3 capacity (for excavating larger volumes over large haul lengths, excavating to reduce levels and also levelling ground, grading large sites, moving and tipping material including hard material – used in open cast sites)						
average haul one way 500 m	–	0.14	0.81	–	m^3	**0.95**
average haul one way 1,000 m	0.01	0.21	1.22	–	m^3	**1.43**
average haul one way 2,000 m	0.01	0.28	1.63	–	m^3	**1.91**
average haul one way 3,000 m	0.01	0.35	2.03	–	m^3	**2.39**
Twin engined motorized scraper, 16 m^3 capacity						
average haul one way 500 m	–	0.11	0.83	–	m^3	**0.94**
average haul one way 1,000 m	–	0.14	1.11	–	m^3	**1.25**
average haul one way 2,000 m	0.01	0.18	1.39	–	m^3	**1.56**
average haul one way 3,000 m	0.01	0.25	1.94	–	m^3	**2.19**
Twin engined motorized scraper, 34 m^3 capacity						
average haul one way 500 m	–	0.07	1.13	–	m^3	**1.20**
average haul one way 1,000 m	–	0.11	1.70	–	m^3	**1.81**
average haul one way 2,000 m	–	0.11	1.70	–	m^3	**1.81**
average haul one way 3,000 m	–	0.14	2.27	–	m^3	**2.41**
Excavation by hand						
Desirable for work around live services or in areas of highly restricted access.						
Excavate and load into skip or dumper bucket						
loose material	2.02	39.60	–	–	m^3	**39.60**
compacted soil or clay	3.15	61.75	–	–	m^3	**61.75**
mass concrete or sandstone	3.00	58.81	–	–	m^3	**58.81**
broken rock	2.95	57.83	–	–	m^3	**57.83**
existing subbase or pipe surrounds	3.12	61.16	–	–	m^3	**61.16**

CLASS E: EARTHWORKS

Item	Gang hours	Labour £	Plant £	Material £	Unit	**Total rate £**
Excavate by hand using pneumatic equipment						
Excavate below ground using 1.80 m^3/min single tool compressor and pneumatic breaker and load material into skip or dumper bucket						
rock (medium drill)	0.50	9.80	30.04	–	m^3	**39.84**
brickwork or mass concrete	2.76	54.10	38.69	–	m^3	**92.79**
reinforced concrete	3.87	75.86	52.11	–	m^3	**127.97**
asphalt in carriageways	1.15	22.54	21.01	–	m^3	**43.56**
Comparative prices for ancillary equipment						
Excavate using 6 tonne to break out (JCB 3CX and Montalbert 125 breaker)						
medium hard rock	0.43	12.54	28.73	–	m^3	**41.27**
brickwork or mass concrete	0.54	15.75	35.00	–	m^3	**50.75**
reinforced concrete	0.64	18.67	41.48	–	m^3	**60.15**
Load material into skip or dumper						
Load material into skip or dumper bucket using a 11.5 tonne crawler backacter with 0.80 m^3 rock bucket						
medium hard rock	0.07	2.04	4.16	–	m^3	**6.20**
brickwork or mass concrete	0.08	2.33	4.75	–	m^3	**7.08**
reinforced concrete	0.09	2.63	5.33	–	m^3	**7.96**
05 EXCAVATION ANCILLARIES						
The following unit costs are for various machines appropriate to the work						
Trimming topsoil; using D4H dozer						
horizontal	0.04	0.99	1.31	–	m^2	**2.30**
10–45° to horizontal	0.05	1.23	1.64	–	m^2	**2.88**
Trimming material other than topsoil, rock or artificial hard material; using D4H dozer, tractor loader or motor grader average rate						
horizontal	0.04	0.99	1.78	–	m^2	**2.77**
10–45° to horizontal	0.04	0.99	1.78	–	m^2	**2.77**
45–90° to horizontal	0.06	1.48	2.66	–	m^2	**4.15**
Trimming rock; using D6E dozer						
horizontal	0.08	1.97	43.40	–	m^2	**45.37**
10–45° to horizontal	0.80	19.75	43.40	–	m^2	**63.14**
45–90° to horizontal	1.06	26.16	57.50	–	m^2	**83.66**
vertical	1.06	26.16	57.50	–	m^2	**83.66**
Preparation of topsoil; using D4H dozer, tractor loader or motor grader average rate						
horizontal	0.06	1.48	2.39	–	m^2	**3.87**
10–45° to horizontal	0.06	1.48	2.39	–	m^2	**3.87**
Preparation of material other than rock or artificial hard material; using D6E dozer, tractor loader or motor grader average rate						
horizontal	0.06	1.48	3.09	–	m^2	**4.57**
10–45° to horizontal	0.06	1.48	3.09	–	m^2	**4.57**
45–90° to horizontal	0.06	1.48	3.09	–	m^2	**4.57**

CLASS E: EARTHWORKS

Item	Gang hours	Labour £	Plant £	Material £	Unit	Total rate £
05 EXCAVATION ANCILLARIES – cont						
The following unit costs are for various machines appropriate to the work – cont						
Preparation of rock; using D6E dozer						
horizontal	0.80	19.75	43.40	–	m^2	**63.14**
10–45° to horizontal	0.60	14.81	32.55	–	m^2	**47.36**
45–90° to horizontal	0.60	14.81	32.55	–	m^2	**47.36**
The following unit costs for disposal are based on using a 22.5 t ADT for site work and 20 t tipper for off-site work. The distances used in the calculation are quoted to assist estimating, although this goes beyond the specific requirements of CESMM4						
Disposal of excavated topsoil						
storage on site; 100 m maximum distance	0.05	1.77	5.05	–	m^3	**6.82**
removal; 5 km distance	0.11	3.90	13.03	–	m^3	**16.93**
removal; 15 km distance	0.21	7.45	24.88	–	m^3	**32.32**
Disposal of excavated earth other than rock or artificial hard material						
storage on site; 100 m maximum distance; using 22.5 t ADT	0.05	1.77	5.05	–	m^3	**6.82**
removal; 5 km distance; using 20 t tipper	0.12	4.25	14.22	–	m^3	**18.47**
removal; 15 km distance; using 20 t tipper	0.22	7.80	26.06	–	m^3	**33.86**
Disposal of excavated rock or artificial hard material						
storage on site; 100 m maximum distance; using 22.5 t ADT	0.06	2.13	6.06	–	m^3	**8.19**
removal; 5 km distance; using 20 t tipper	0.13	4.61	15.40	–	m^3	**20.01**
removal; 15 km distance; using 20 t tipper	0.23	8.16	27.25	–	m^3	**35.40**
Add to the above rates where tipping charges apply:						
non-hazardous waste		–	–	–	m^3	**57.07**
hazardous waste		–	–	–	m^3	**171.16**
special waste		–	–	–	m^3	**180.69**
contaminated liquid		–	–	–	m^3	**228.21**
contaminated sludge		–	–	–	m^3	**285.31**
Add to the above rates where Landfill Tax applies:						
inactive or inert waste material		–	–	–	m^3	**4.73**
other taxable waste material		–	–	–	m^3	**147.90**
The following unit costs for double handling are based on using a 1.5 m^3 tractor loader and 22.5 t ADT. A range of distances is listed to assist in estimating, although this goes beyond the specific requirements of CESMM4						
Double handling of excavated topsoil; using 1.5 m^3 tractor loader and 22.5 t ADT						
300 m average distance moved	0.08	2.84	7.79	–	m^3	**10.63**
600 m average distance moved	0.10	3.55	9.74	–	m^3	**13.28**
1000 m average distance moved	0.12	4.25	11.69	–	m^3	**15.94**

CLASS E: EARTHWORKS

Item	Gang hours	Labour £	Plant £	Material £	Unit	**Total rate £**
Double handling of excavated earth other than rock or artificial hard material; using 1.5 m^3 tractor loader and 22.5 t ADT						
300 m average distance moved	0.08	2.84	7.79	–	m^3	**10.63**
600 m average distance moved	0.10	3.55	9.74	–	m^3	**13.28**
1000 m average distance moved	0.12	4.25	11.69	–	m^3	**15.94**
Double handling of rock or artificial hard material; using 1.5 m^3 tractor loader and 22.5 t ADT						
300 m average distance moved	0.16	5.67	15.58	–	m^3	**21.25**
600 m average distance moved	0.18	6.38	17.53	–	m^3	**23.91**
1000 m average distance moved	0.20	7.09	19.48	–	m^3	**26.57**
The following unit rates for excavation below Final Surface are based on using a 16 t backacter machine						
Excavation of material below the Final Surface and replacement of with:						
granular fill	1.20	29.62	40.87	38.52	m^3	**109.00**
concrete Grade C7.5P	0.60	14.81	20.43	118.04	m^3	**153.28**
Miscellaneous						
Timber supports left in	0.70	17.28	–	12.08	m^2	**29.36**
Metal supports left in	1.54	38.17	–	17.00	m^2	**55.17**
06 FILLING						
Excavated topsoil; DfT specified type 5 A						
Filling						
to structures	0.07	4.68	10.11	–	m^3	**14.79**
embankments	0.03	1.64	3.53	–	m^3	**5.17**
general	0.02	1.38	2.97	–	m^3	**4.34**
150 mm thick	0.01	0.46	0.99	–	m^2	**1.45**
250 mm thick	0.01	0.79	1.70	–	m^2	**2.48**
400 mm thick	0.02	1.11	2.40	–	m^2	**3.52**
600 mm thick	0.02	1.57	3.39	–	m^2	**4.96**
Imported topsoil DfT specified type 5B;						
Filling						
embankments	0.03	1.64	3.53	32.07	m^3	**37.24**
general	0.02	1.38	2.97	32.07	m^3	**36.41**
150 mm thick	0.01	0.46	0.99	4.81	m^2	**6.26**
250 mm thick	0.01	0.79	1.70	8.02	m^2	**10.50**
400 mm thick	0.02	1.11	2.40	12.83	m^2	**16.34**
600 mm thick	0.02	1.57	3.39	19.24	m^2	**24.21**

CLASS E: EARTHWORKS

Item	Gang hours	Labour £	Plant £	Material £	Unit	Total rate £
06 FILLING – cont						
Non-selected excavated material other than topsoil or rock						
Filling						
to structures	0.04	2.36	5.09	–	m^3	**7.45**
embankments	0.02	0.98	2.12	–	m^3	**3.10**
general	0.01	0.72	1.56	–	m^3	**2.28**
150 mm thick	–	0.26	0.57	–	m^2	**0.83**
250 mm thick	0.01	0.39	0.85	–	m^2	**1.24**
400 mm thick	0.01	0.59	1.27	–	m^2	**1.86**
600 mm thick	0.01	0.79	1.70	–	m^2	**2.48**
Selected excavated material other than topsoil or rock						
Filling						
to structures	0.04	2.69	5.80	–	m^3	**8.48**
embankments	0.01	0.92	1.98	–	m^3	**2.90**
general	0.01	0.79	1.70	–	m^3	**2.48**
150 mm thick	–	0.26	0.57	–	m^2	**0.83**
250 mm thick	0.01	0.46	0.99	–	m^2	**1.45**
400 mm thick	0.01	0.66	1.41	–	m^2	**2.07**
600 mm thick	0.01	0.85	1.84	–	m^2	**2.69**
Imported natural material other than topsoil or rock; subsoil						
Filling						
to structures	0.04	2.36	5.09	33.38	m^3	**40.83**
embankments	0.02	0.98	2.12	33.38	m^3	**36.48**
general	0.02	0.98	2.12	33.38	m^3	**36.48**
150 mm thick	0.01	0.59	1.27	5.01	m^2	**6.87**
250 mm thick	0.01	0.85	1.84	8.34	m^2	**11.03**
400 mm thick	0.02	0.98	2.12	13.35	m^2	**16.46**
600 mm thick	0.02	1.11	2.40	20.03	m^2	**23.54**
Imported natural material other than topsoil or rock; granular graded material						
Filling						
to structures	0.04	2.36	5.09	40.32	m^3	**47.77**
embankments	0.02	1.11	2.40	40.32	m^3	**43.83**
general	0.02	1.11	2.40	40.32	m^3	**43.83**
150 mm thick	0.01	0.79	1.70	6.05	m^2	**8.53**
250 mm thick	0.02	1.11	2.40	10.08	m^2	**13.60**
400 mm thick	0.02	1.31	2.83	16.13	m^2	**20.26**
600 mm thick	0.02	1.44	3.11	24.19	m^2	**28.74**

CLASS E: EARTHWORKS

Item	Gang hours	Labour £	Plant £	Material £	Unit	Total rate £
Imported natural material other than topsoil or rock; granular selected material						
Filling						
to structures	0.04	2.36	5.09	34.73	m³	**42.17**
embankments	0.02	1.11	2.40	34.73	m³	**38.24**
general	0.02	1.11	2.40	34.73	m³	**38.24**
150 mm thick	0.01	0.79	1.70	5.21	m²	**7.69**
250 mm thick	0.02	1.11	2.40	8.68	m²	**12.20**
400 mm thick	0.02	1.31	2.83	13.89	m²	**18.03**
600 mm thick	0.02	1.44	3.11	20.84	m²	**25.39**
Excavated rock						
Filling						
to structures	0.04	2.62	5.65	–	m³	**8.27**
embankments	0.05	3.28	7.07	–	m³	**10.34**
general	0.05	3.28	7.07	–	m³	**10.34**
150 mm thick	0.02	1.11	2.40	–	m²	**3.52**
250 mm thick	0.03	1.70	3.68	–	m²	**5.38**
400 mm thick	0.04	2.62	5.65	–	m²	**8.27**
600 mm thick	0.06	3.60	7.77	–	m²	**11.38**
Imported rock						
Filling						
to structures	0.04	2.62	5.65	76.49	m³	**84.76**
embankments	0.02	1.57	3.39	76.49	m³	**81.45**
general	0.02	1.31	2.83	76.49	m³	**80.62**
150 mm thick	0.01	0.79	1.70	11.47	m²	**13.96**
250 mm thick	0.02	1.31	2.83	19.12	m²	**23.26**
400 mm thick	0.03	1.97	4.24	30.59	m²	**36.80**
600 mm thick	0.03	1.97	4.24	45.89	m²	**52.10**
High energy impaction general fill						
embankments	0.03	1.64	3.53	89.08	m³	**94.25**
general	0.02	1.38	2.97	44.54	m³	**48.88**
150 mm thick	0.01	0.46	0.99	8.91	m²	**10.36**
250 mm thick	0.01	0.79	1.70	13.36	m²	**15.84**
400 mm thick	0.02	1.11	2.40	14.48	m²	**17.99**
600 mm thick	0.02	1.57	3.39	22.27	m²	**27.23**
07 FILLING ANCILLARIES						
Trimming of filled surfaces						
Topsoil						
horizontal	0.02	0.74	2.14	–	m²	**2.89**
inclined at an angle of 10–45° to horizontal	0.02	0.74	2.14	–	m²	**2.89**
inclined at an angle of 45–90° to horizontal	0.03	0.96	2.76	–	m²	**3.71**
Material other than topsoil, rock or artificial hard material						
horizontal	0.02	0.74	2.14	–	m²	**2.89**
inclined at an angle of 10–45° to horizontal	0.02	0.74	2.14	–	m²	**2.89**
inclined at an angle of 45–90° to horizontal	0.03	0.96	2.76	–	m²	**3.71**

CLASS E: EARTHWORKS

Item	Gang hours	Labour £	Plant £	Material £	Unit	Total rate £
07 FILLING ANCILLARIES – cont						
Trimming of filled surfaces – cont						
Rock						
horizontal	0.51	18.08	52.05	–	m^2	**70.13**
inclined at an angle of 10–45° to horizontal	0.52	18.44	53.07	–	m^2	**71.51**
inclined at an angle of 45–90° to horizontal	0.70	24.82	71.44	–	m^2	**96.26**
Preparation of filled surfaces						
Topsoil						
horizontal	0.03	1.06	3.06	–	m^2	**4.13**
inclined at an angle of 10–45° to horizontal	0.03	1.06	3.06	–	m^2	**4.13**
inclined at an angle of 45–90° to horizontal	0.04	1.31	3.78	–	m^2	**5.09**
Material other than topsoil, rock or artificial hard material						
horizontal	0.03	1.06	3.06	–	m^2	**4.13**
inclined at an angle of 10–45° to horizontal	0.03	1.06	3.06	–	m^2	**4.13**
inclined at an angle of 45–90° to horizontal	0.04	1.31	3.78	–	m^2	**5.09**
Rock						
horizontal	0.33	11.70	33.68	–	m^2	**45.38**
inclined at an angle of 10–45° to horizontal	0.33	11.70	33.68	–	m^2	**45.38**
inclined at an angle of 45–90° to horizontal	0.52	18.44	53.07	–	m^2	**71.51**
GEOTEXTILES						
The geotextile products mentioned below are not specifically confined to the individual uses stated but are examples of one of many scenarios to which they may be applied. Conversely, the scenarios are not limited to the geotextile used as an example.						
Geotextiles; stabilization applications for reinforcement of granular subbases, capping layers and railway ballast placed over weak and variable soils						
For use over weak soils with moderate traffic intensities e.g. car parks, light access roads; Tensar SS20 Polypropylene Geogrid						
horizontal	0.04	1.83	–	2.19	m^2	**4.02**
inclined at an angle of 10 to 45° to the horizontal	0.05	2.30	–	2.19	m^2	**4.50**
For use over weak soils with high traffic intensities and/or high axle loadings; Tensar SS30 Polypropylene Geogrid						
horizontal	0.05	2.84	–	2.24	m^2	**5.08**
inclined at an angle of 10 to 45° to the horizontal	0.06	3.53	–	2.24	m^2	**5.77**
For use over very weak soils e.g. alluvium, marsh or peat or firmer soil subject to exceptionally high axle loadings;Tensar SS40 Polypropylene Geogrid						
horizontal	0.05	3.27	4.22	3.93	m^2	**11.42**
inclined at an angle of 10 to 45° to the horizontal	0.06	4.07	5.25	3.93	m^2	**13.25**

CLASS E: EARTHWORKS

Item	Gang hours	Labour £	Plant £	Material £	Unit	**Total rate £**
For trafficked areas where fill comprises aggregate exceeding 100 mm; Tensar SSLA20 Polypropylene Geogrid						
horizontal	0.04	1.83	–	4.06	m^2	**5.89**
inclined at an angle 10–45° to the horizontal	0.05	2.30	–	4.06	m^2	**6.37**
Stabilization and separation of granular fill from soft sub grade to prevent intermixing: Terram 1000						
horizontal	0.05	3.22	–	0.68	m^2	**3.90**
inclined at an angle of 10 to 45° to the horizontal	0.06	4.04	–	0.68	m^2	**4.72**
Stabilization and separation of granular fill from soft sub grade to prevent intermixing: Terram 2000						
horizontal	0.04	3.05	3.94	1.62	m^2	**8.61**
inclined at an angle of 10 to 45° to the horizontal	0.05	3.85	4.97	1.62	m^2	**10.44**
Geotextiles; reinforcement applications for asphalt pavements						
For roads, hardstandings and airfield pavements; Tensar AR-G composite comprising Tensar AR−1 grid bonded to a geotextile, laid within asphalt						
horizontal	0.05	2.84	–	3.66	m^2	**6.50**
inclined at an angle of 10 to 45° to the horizontal	0.06	3.53	–	3.66	m^2	**7.19**
Geotextiles; slope reinforcement and embankment support; for use where soils can only withstand limited shear stresses, therefore steep slopes require external support						
Paragrid 30/155; 330 g/m^2						
horizontal	0.04	1.83	–	2.28	m^2	**4.11**
inclined at an angle of 10–45° to the horizontal	0.05	2.30	–	2.28	m^2	**4.59**
Paragrid 100/255; 330 g/m^2						
horizontal	0.04	1.83	–	1.40	m^2	**3.23**
Paralink 200 s; 1120 g/m^2						
horizontal	0.05	3.92	5.06	4.07	m^2	**13.06**
inclined at an angle of 10–45° to the horizontal	0.07	4.94	6.37	4.07	m^2	**15.38**
Paralink 600 s; 2040 g/m^2						
horizontal	0.06	4.58	5.90	8.96	m^2	**19.44**
inclined at an angle of 10–45° to the horizontal	0.08	5.74	7.40	8.96	m^2	**22.10**
Terram grid 30/30						
horizontal	0.06	4.14	5.34	3.02	m^2	**12.51**
inclined at an angle of 10–45° to the horizontal	0.07	5.16	6.65	3.02	m^2	**14.83**
Geotextiles; scour and erosion protection						
For use where erosion protection is required to the surface of a slope once its geotechnical stability has been achieved, and to allow grass establishment; Tensar Mat Polyethylene mesh; fixed with Tensar pegs						
horizontal	0.04	2.59	–	5.60	m^2	**8.19**
inclined at an angle of 10–45° to the horizontal	0.05	3.22	–	5.60	m^2	**8.82**

CLASS E: EARTHWORKS

Item	Gang hours	Labour £	Plant £	Material £	Unit	**Total rate £**
07 FILLING ANCILLARIES – cont						
Geotextiles – cont						
For use where hydraulic action exists, such as coastline protection from pressures exerted by waves, currents and tides; Typar SF56						
horizontal	0.05	3.71	4.78	0.60	m^2	**9.09**
inclined at an angle of 10–45° to the horizontal	0.06	4.65	6.00	0.60	m^2	**11.25**
For protection against puncturing to reservoir liner; Typar SF563						
horizontal	0.05	3.71	4.78	0.60	m^2	**9.09**
inclined at an angle of 10–45° to the horizontal	0.06	4.65	6.00	0.60	m^2	**11.25**
Geotextiles; temporary parking areas						
For reinforcement of grassed areas subject to wear from excessive pedestrian and light motor vehicle traffic; Netlon CE131 high density polyethylene geogrid						
horizontal	0.04	1.91	–	4.67	m^2	**6.59**
Geotextiles; landscaping applications						
For prevention of weed growth in planted areas by incorporating a geotextile over top soil and below mulch or gravel; Typar SF20						
horizontal	0.08	4.73	–	0.25	m^2	**4.98**
inclined at an angle of 10–45° to the horizontal	0.09	5.93	–	0.25	m^2	**6.18**
For root growth control; prevention of lateral spread of roots and mixing of road base and humus; Typar SF20						
horizontal	0.08	4.73	–	0.25	m^2	**4.98**
inclined at an angle of 10–45° to the horizontal	0.09	5.93	–	0.25	m^2	**6.18**
Geotextiles; drainage applications						
For clean installation of pipe support material and to prevent silting of the drainage pipe and minimizing differential settlement; Typar SF10						
horizontal	0.04	2.59	–	0.40	m^2	**2.98**
inclined at an angle of 10–45° to the horizontal	0.05	3.22	–	0.40	m^2	**3.62**
For wrapping to prevent clogging of drainage pipes surrounded by fine soil; Typar SF10						
sheeting	0.08	5.05	–	0.51	m^2	**5.56**
For wrapping to prevent clogging of drainage pipes surrounded by fine soil; Terram 1000						
sheeting	0.08	5.05	–	0.74	m^2	**5.78**
For vertical structure drainage to sub-surface walls, roofs and foundations; Filtram 1B1						
sheeting	0.08	5.05	–	4.55	m^2	**9.60**
For waterproofing to tunnels, buried structures, etc. where the membrane is buried, forming part of the drainage system; Filtram 1BZ						
sheeting	0.08	5.05	–	8.35	m^2	**13.40**

CLASS E: EARTHWORKS

Item	Gang hours	Labour £	Plant £	Material £	Unit	**Total rate £**
Geotextiles; roofing insulation and protection						
Protection of waterproofing membrane from physical damage and puncturing; Typar SF56						
sheeting	0.05	3.71	4.78	0.60	m^2	**9.09**
Internal reinforcement of in situ spread waterproof bitumen emulsion; Typar SF10						
sheeting	0.07	4.16	–	0.40	m^2	**4.56**

CLASS F: IN SITU CONCRETE

Item	Gang hours	Labour £	Plant £	Material £	Unit	**Total rate £**
NOTES						
The unit rates in this Section are based on net measurements and appropriate adjustments should be made for unmeasured excess (e.g. additional blinding thickness as a result of ground conditions).						
The unit rates for the provision of concrete are based on ready mixed concrete in full loads delivered to site within 5 miles (8 km) of the concrete mixing plant and include an allowance for wastage prior to placing.						
This section assumes optimum outputs of an efficiently controlled pour with no delays caused by out of sequence working and no abnormal conditions that would affect continuity of work.						
RESOURCES – LABOUR						
Labour Gang Build-up Rates						
Concrete gang						
1 ganger or chargehand (skill rate 4)		21.42	–	–		
2 skilled operatives (skill rate 4)		40.39	–	–		
4 unskilled operatives (general)		78.41	–	–		
1 plant operator (skill rate 3) – 25% of time		7.29	–	–		
Total Gang Rate/Hour		**147.51**	–	–		
RESOURCES – MATERIAL						
Material Resources Build-up Rate						
The following costs do not reflect in the rates generally shown later and should be considered separately						
delivery to site for each additional mile from the concrete plant further than 5 miles (8 km)		–	–	2.77	m^3	**2.77**
mix design, per trial mix		–	–	309.44	mix	**309.44**
part loads, per m^3 below full load		–	–	34.72	m^3	**34.72**
waiting time (in excess of 6 mins/m^3 'norm' discharge time)		–	–	107.36	hr	**107.36**
making and testing concrete cube	0.69	13.49	–	–	nr	**13.49**
pumping from ready mix truck to point of placing at the rate of 25 m^3 / hour	0.11	2.06	8.47	–	m^3	**10.52**
pumping from ready mix truck to point of placing at the rate of 45 m^3 / hour	0.05	0.99	8.92	–	m^3	**9.91**
RESOURCES – PLANT						
Plant Resources Build-up Rate						
Concrete						
10t Crane (50% of time)		–	34.01	–		
1.00 m^3 concrete skip (50% of time)		–	2.14	–		
11.30 m^3/min compressor, 2 tool		–	63.39	–		
four 54 mm poker vibrators		–	7.48	–		
Total Rate/Hour		–	**107.02**	–		

CLASS F: IN SITU CONCRETE

Item	Gang hours	Labour £	Plant £	Material £	Unit	**Total rate £**
01 DESIGNED CONCRETE						
Designed Concrete						
Strength						
C12/15		–	–	123.66	m^3	**123.66**
C16/20		–	–	125.18	m^3	**125.18**
C20/25		–	–	126.79	m^3	**126.79**
C25/30		–	–	128.33	m^3	**128.33**
C32/40		–	–	129.91	m^3	**129.91**
RCC8/10		–	–	139.53	m^3	**139.53**
RC25/30		–	–	141.05	m^3	**141.05**
RC30/37		–	–	142.66	m^3	**142.66**
RC32/40		–	–	144.21	m^3	**144.21**
RC 40/50		–	–	145.79	m^3	**145.79**
Designed mix; sulphate-resisting cement to BS 4027						
Grade C7.5						
10 mm aggregate		–	–	134.11	m^3	**134.11**
20 mm aggregate		–	–	123.40	m^3	**123.40**
40 mm aggregate		–	–	121.79	m^3	**121.79**
Grade C10						
10 mm aggregate		–	–	134.36	m^3	**134.36**
20 mm aggregate		–	–	125.95	m^3	**125.95**
40 mm aggregate		–	–	129.32	m^3	**129.32**
Grade C15						
10 mm aggregate		–	–	122.79	m^3	**122.79**
20 mm aggregate		–	–	131.10	m^3	**131.10**
40 mm aggregate		–	–	129.58	m^3	**129.58**
Grade C20						
10 mm aggregate		–	–	127.87	m^3	**127.87**
20 mm aggregate		–	–	131.68	m^3	**131.68**
40 mm aggregate		–	–	129.96	m^3	**129.96**
Grade C25						
10 mm aggregate		–	–	138.78	m^3	**138.78**
20 mm aggregate		–	–	134.77	m^3	**134.77**
40 mm aggregate		–	–	138.81	m^3	**138.81**
Grade C30						
10 mm aggregate		–	–	147.98	m^3	**147.98**
20 mm aggregate		–	–	151.54	m^3	**151.54**
40 mm aggregate		–	–	149.53	m^3	**149.53**
02 DESIGNATED CONCRETE						
Designated Concrete						
Grade C7.5						
10 mm aggregate		–	–	118.60	m^3	**118.60**
20 mm aggregate		–	–	107.89	m^3	**107.89**
40 mm aggregate		–	–	106.27	m^3	**106.27**

CLASS F: IN SITU CONCRETE

Item	Gang hours	Labour £	Plant £	Material £	Unit	**Total rate £**
02 DESIGNATED CONCRETE – cont						
Designated Concrete – cont						
Grade C10						
10 mm aggregate		–	–	118.60	m^3	**118.60**
20 mm aggregate		–	–	110.19	m^3	**110.19**
40 mm aggregate		–	–	113.56	m^3	**113.56**
Grade C15						
10 mm aggregate		–	–	106.77	m^3	**106.77**
20 mm aggregate		–	–	115.08	m^3	**115.08**
40 mm aggregate		–	–	113.56	m^3	**113.56**
Grade C20						
10 mm aggregate		–	–	111.70	m^3	**111.70**
20 mm aggregate		–	–	115.51	m^3	**115.51**
40 mm aggregate		–	–	113.79	m^3	**113.79**
Grade C25						
10 mm aggregate		–	–	122.07	m^3	**122.07**
20 mm aggregate		–	–	118.06	m^3	**118.06**
40 mm aggregate		–	–	122.10	m^3	**122.10**
Grade C30						
10 mm aggregate		–	–	131.01	m^3	**131.01**
20 mm aggregate		–	–	134.57	m^3	**134.57**
40 mm aggregate		–	–	132.56	m^3	**132.56**
Grade C40						
10 mm aggregate		–	–	144.67	m^3	**144.67**
20 mm aggregate		–	–	142.49	m^3	**142.49**
Grade C50						
10 mm aggregate		–	–	147.55	m^3	**147.55**
20 mm aggregate		–	–	145.33	m^3	**145.33**
Grade C60						
10 mm aggregate		–	–	150.51	m^3	**150.51**
20 mm aggregate		–	–	148.22	m^3	**148.22**
03 STANDARDISED PRESCRIBED CONCRETE						
Prescribed mix; cement to BS EN 197–1						
Grade C7.5						
10 mm aggregate		–	–	112.27	m^3	**112.27**
20 mm aggregate		–	–	110.58	m^3	**110.58**
40 mm aggregate		–	–	108.92	m^3	**108.92**
Grade C10						
10 mm aggregate		–	–	114.28	m^3	**114.28**
20 mm aggregate		–	–	112.54	m^3	**112.54**
40 mm aggregate		–	–	110.89	m^3	**110.89**
Grade C15						
10 mm aggregate		–	–	114.91	m^3	**114.91**
20 mm aggregate		–	–	116.95	m^3	**116.95**
40 mm aggregate		–	–	111.48	m^3	**111.48**

CLASS F: IN SITU CONCRETE

Item	Gang hours	Labour £	Plant £	Material £	Unit	**Total rate £**
Grade C20						
10 mm aggregate		–	–	120.20	m^3	**120.20**
20 mm aggregate		–	–	118.39	m^3	**118.39**
40 mm aggregate		–	–	116.63	m^3	**116.63**
Grade C25						
10 mm aggregate		–	–	122.07	m^3	**122.07**
20 mm aggregate		–	–	122.07	m^3	**122.07**
40 mm aggregate		–	–	122.07	m^3	**122.07**
Grade C30						
10 mm aggregate		–	–	140.03	m^3	**140.03**
20 mm aggregate		–	–	137.95	m^3	**137.95**
40 mm aggregate		–	–	150.07	m^3	**150.07**
Grade C40						
10 mm aggregate		–	–	147.70	m^3	**147.70**
20 mm aggregate		–	–	146.06	m^3	**146.06**
Grade C50						
10 mm aggregate		–	–	151.24	m^3	**151.24**
20 mm aggregate		–	–	148.97	m^3	**148.97**
Grade C60						
10 mm aggregate		–	–	154.28	m^3	**154.28**
20 mm aggregate		–	–	151.96	m^3	**151.96**
Prescribed mix; sulphate-resisting cement to BS 4027						
Grade C7.5						
10 mm aggregate		–	–	127.78	m^3	**127.78**
20 mm aggregate		–	–	126.09	m^3	**126.09**
40 mm aggregate		–	–	124.43	m^3	**124.43**
Grade C10						
10 mm aggregate		–	–	130.04	m^3	**130.04**
20 mm aggregate		–	–	128.30	m^3	**128.30**
40 mm aggregate		–	–	126.65	m^3	**126.65**
Grade C15						
10 mm aggregate		–	–	130.93	m^3	**130.93**
20 mm aggregate		–	–	132.97	m^3	**132.97**
40 mm aggregate		–	–	127.50	m^3	**127.50**
Grade C20						
10 mm aggregate		–	–	136.38	m^3	**136.38**
20 mm aggregate		–	–	134.57	m^3	**134.57**
40 mm aggregate		–	–	132.81	m^3	**132.81**
Grade C25						
10 mm aggregate		–	–	138.78	m^3	**138.78**
20 mm aggregate		–	–	138.78	m^3	**138.78**
40 mm aggregate		–	–	138.78	m^3	**138.78**
Grade C30						
10 mm aggregate		–	–	157.00	m^3	**157.00**
20 mm aggregate		–	–	154.93	m^3	**154.93**
40 mm aggregate		–	–	167.04	m^3	**167.04**

CLASS F: IN SITU CONCRETE

Item	Gang hours	Labour £	Plant £	Material £	Unit	Total rate £
06 PLACING OF CONCRETE; MASS						
Blinding; thickness						
ne 150 mm	0.18	26.55	1.14	–	m^3	**27.69**
150–300 mm	0.16	23.60	3.29	–	m^3	**26.89**
300–500 mm	0.14	20.65	5.42	–	m^3	**26.07**
exceeding 500 mm	0.12	17.70	7.57	–	m^3	**25.27**
Bases, footings, pile caps and ground slabs; thickness						
ne 150 mm	0.20	29.50	21.40	–	m^3	**50.91**
150–300 mm	0.17	25.08	18.26	–	m^3	**43.34**
300–500 mm	0.15	22.13	15.98	–	m^3	**38.11**
exceeding 500 mm	0.14	20.65	14.99	–	m^3	**35.64**
ADD to the above for placing against an excavated surface	0.03	3.69	2.64	–	m^3	**6.32**
Walls; thickness						
ne 150 mm	0.21	30.98	22.54	–	m^3	**53.52**
150–300 mm	0.15	22.13	15.99	–	m^3	**38.12**
300–500 mm	0.13	19.18	13.98	–	m^3	**33.16**
exceeding 500 mm	0.12	17.70	12.84	–	m^3	**30.54**
ADD to the above for placing against an excavated surface	0.03	4.43	3.29	–	m^3	**7.71**
Other concrete forms						
plinth 1000 × 1000 × 600 mm	0.33	48.68	35.39	–	m^3	**84.06**
plinth 1500 × 1500 × 750 mm	0.25	36.88	26.83	–	m^3	**63.71**
plinth 2000 × 2000 × 600 mm	0.20	29.50	21.40	–	m^3	**50.91**
surround to precast concrete manhole chambers 200 mm thick	0.29	42.78	30.96	–	m^3	**73.74**
07 PLACING OF CONCRETE; REINFORCED						
Bases, footings, pile caps and ground slabs; thickness						
ne 150 mm	0.21	30.98	22.54	–	m^3	**53.52**
150–300 mm	0.18	26.55	19.26	–	m^3	**45.81**
300–500 mm	0.16	23.60	17.12	–	m^3	**40.73**
exceeding 500 mm	0.15	22.13	15.98	–	m^3	**38.11**
Suspended slabs; thickness						
ne 150 mm	0.27	39.83	28.97	–	m^3	**68.80**
150–300 mm	0.21	30.98	22.54	–	m^3	**53.52**
300–500 mm	0.19	28.03	20.41	–	m^3	**48.44**
exceeding 500 mm	0.19	28.03	20.41	–	m^3	**48.44**
Walls; thickness						
ne 150 mm	0.29	42.78	30.96	–	m^3	**73.74**
150–300 mm	0.22	32.45	23.56	–	m^3	**56.01**
300–500 mm	0.20	29.50	21.40	–	m^3	**50.91**
exceeding 500 mm	0.20	29.50	21.40	–	m^3	**50.91**
Columns and piers; cross-sectional area						
ne 0.03 m^2	0.50	73.75	53.52	–	m^3	**127.27**
0.03–0.10 m^2	0.40	59.00	42.83	–	m^3	**101.83**
0.10–0.25 m^2	0.35	51.63	37.39	–	m^3	**89.02**
0.25–1.00 m^2	0.35	51.63	37.39	–	m^3	**89.02**
exceeding 1 m^2	0.28	41.30	29.97	–	m^3	**71.27**

CLASS F: IN SITU CONCRETE

Item	Gang hours	Labour £	Plant £	Material £	Unit	**Total rate £**
Beams; cross-sectional area						
ne 0.03 m^2	0.50	73.75	53.52	–	m^3	**127.27**
0.03–0.10 m^2	0.40	59.00	42.83	–	m^3	**101.83**
0.10–0.25 m^2	0.35	51.63	37.39	–	m^3	**89.02**
0.25–1.00 m^2	0.35	51.63	37.39	–	m^3	**89.02**
exceeding 1 m^2	0.28	41.30	29.97	–	m^3	**71.27**
Casings to metal sections; cross-sectional area						
ne 0.03 m^2	0.47	69.33	50.38	–	m^3	**119.71**
0.03–0.10 m^2	0.47	69.33	50.38	–	m^3	**119.71**
0.10–0.25 m^2	0.40	59.00	42.81	–	m^3	**101.81**
0.25–1.00 m^2	0.40	59.00	42.81	–	m^3	**101.81**
exceeding 1 m^2	0.35	51.63	37.39	–	m^3	**89.02**
08 PLACING OF CONCRETE; PRESTRESSED						
Suspended slabs; thickness						
ne 150 mm	0.28	41.30	29.97	–	m^3	**71.27**
150–300 mm	0.22	32.45	23.55	–	m^3	**56.01**
300–500 mm	0.20	29.50	21.40	–	m^3	**50.91**
exceeding 500 mm	0.19	28.03	20.41	–	m^3	**48.44**
Beams; cross-sectional area						
ne 0.03 m^2	0.50	73.75	53.52	–	m^3	**127.27**
0.03–0.10 m^2	0.40	59.00	42.83	–	m^3	**101.83**
0.10–0.25 m^2	0.35	51.63	37.39	–	m^3	**89.02**
0.25–1.00 m^2	0.35	51.63	37.39	–	m^3	**89.02**
exceeding 1 m^2	0.28	41.30	29.97	–	m^3	**71.27**

CLASS G: CONCRETE ANCILLARIES

Item	Gang hours	Labour £	Plant £	Material £	Unit	Total rate £
RESOURCES – LABOUR						
Labour Resources Build-up Rate						
Formwork gang – small areas						
1 joiner (craftsman)		26.14	–	–		
1 unskilled operative (general)		19.60	–	–		
Total Gang Rate/Hour		**45.74**	–	–		
Formwork gang – large areas						
1 foreman joiner (craftsman)		26.14	–	–		
2 joiners (craftsman)		45.60	–	–		
1 unskilled operative (general)		19.60	–	–		
Total Gang Rate/Hour		**91.34**	–	–		
Reinforcement gang						
1 foreman steel fixer (craftsman)		26.14	–	–		
4 steel fixers (craftsman)		91.20	–	–		
1 unskilled operative (general)		19.60	–	–		
Total Gang Rate/Hour		**136.93**	–	–		
Reinforcement – on-site bending/baling gang						
1 steel fixer (craftsman)		22.80	–	–		
1 unskilled operative (general)		19.60	–	–		
Total Gang Rate/Hour		**42.40**	–	–		
Joints gang						
1 ganger/chargehand (skill rate 4)		21.42	–	–		
1 skilled operative (skill rate 4)		20.20	–	–		
1 unskilled operative (general)		19.60	–	–		
Total Gang Rate/Hour		**61.21**	–	–		
Concrete accessories gang						
1 ganger/chargehand (skill rate 4)		21.42	–	–		
1 skilled operative (skill rate 4)		20.20	–	–		
1 unskilled operative (general)		19.60	–	–		
Total Gang Rate/Hour		**61.21**	–	–		
RESOURCES – MATERIAL						
Material Resources Build-up Rate						
Reinforcement; supply only ex works materials rates						
Grade 500C deformed reinforcing bars						
Reinforcement; high tensile steel bars; straight; T12		–	–	952.07	tonne	**952.07**
Bars; stainless steel;straight;lengths; 16 mm nominal size-alloy 1.4301						
Stainless steel bars; straight		–	–	5213.68	tonne	**5213.68**
Stainless steel bars; bent		–	–	5213.68	tonne	**5213.68**
Welded fabric to BS 4483 in sheets 4.80 × 2.40						
A98		–	–	4.83	m^2	**4.83**
A142		–	–	4.45	m^2	**4.45**
A193		–	–	6.03	m^2	**6.03**

CLASS G: CONCRETE ANCILLARIES

Item	Gang hours	Labour £	Plant £	Material £	Unit	**Total rate £**
A252		–	–	6.68	m²	**6.68**
A393		–	–	9.37	m²	**9.37**
B196		–	–	10.58	m²	**10.58**
B283		–	–	8.21	m²	**8.21**
B385		–	–	9.11	m²	**9.11**
B503		–	–	12.76	m²	**12.76**
B785		–	–	16.49	m²	**16.49**
B1131		–	–	21.70	m²	**21.70**
C283		–	–	5.20	m²	**5.20**
C385		–	–	6.83	m²	**6.83**
C363		–	–	10.86	m²	**10.86**
C503		–	–	8.66	m²	**8.66**
C785		–	–	12.32	m²	**12.32**
D49		–	–	1.23	m²	**1.23**
D98		–	–	2.47	m²	**2.47**
Reinforcement couplers, Lenton type						
Type A; one bar twistable						
12 mm dia.		–	–	4.91	nr	**4.91**
16 mm dia.		–	–	6.25	nr	**6.25**
20 mm dia.		–	–	11.42	nr	**11.42**
25 mm dia.		–	–	17.80	nr	**17.80**
32 mm dia.		–	–	24.14	nr	**24.14**
40 mm dia.		–	–	34.84	nr	**34.84**
Type B; no bar twistable						
12 mm dia.		–	–	29.49	nr	**29.49**
16 mm dia.		–	–	34.63	nr	**34.63**
20 mm dia.		–	–	37.05	nr	**37.05**
25 mm dia.		–	–	43.80	nr	**43.80**
32 mm dia.		–	–	59.00	nr	**59.00**
40 mm dia.		–	–	89.43	nr	**89.43**
extra for threaded bar at factory						
12 mm dia.		–	–	3.28	nr	**3.28**
16 mm dia.		–	–	3.64	nr	**3.64**
20 mm dia.		–	–	4.31	nr	**4.31**
25 mm dia.		–	–	4.91	nr	**4.91**
32 mm dia.		–	–	6.54	nr	**6.54**
40 mm dia.		–	–	8.43	nr	**8.43**
RESOURCES – PLANT						
Plant Resources Build-up Rate						
Formwork – small areas						
20t crawler crane – 50% of time		–	24.26	–		
22" dia. saw bench		–	4.03	–		
allowance for small power tools		–	0.96	–		
Total Rate/Hour		–	**29.25**	–		

CLASS G: CONCRETE ANCILLARIES

Item	Gang hours	Labour £	Plant £	Material £	Unit	Total rate £
RESOURCES – PLANT – cont						
Plant Resources Build-up Rate – cont						
Formwork – large areas						
20t crawler crane – 25% of time		–	12.13	–		
22" dia. saw bench		–	4.03	–		
allowance for small power tools		–	0.96	–		
Total Rate/Hour		–	**17.12**	–		
Reinforcement						
30 t crawler crane – 25% of time		–	12.72	–		
bar cropper		–	4.83	–		
small power tools		–	1.92	–		
support acrows, tirfors, kentledge, etc.		–	0.89	–		
Total Rate/Hour		–	**20.36**	–		
01 FORMWORK; ROUGH FINISH						
Formwork						
Formwork materials include for shutter, bracing, ties, support, kentledge and all consumables.						
The following unit costs do not include for formwork outside the payline and are based on an optimum of a minimum 8 uses with 10% per use towards the cost of repairs/replacement of components damaged during disassembly						
ADD to formwork material costs generally depending on the number of uses:						
Nr of uses / % Addition / % Waste						
1 / + 90 to 170 / +7						
2 / + 50 to 80 / +7						
3 / + 15 to 30 / +6						
6 / + 5 to 10 / +6						
8 / No change / +5						
10 / − 5 to 7 / +5						
Plane horizontal, width						
ne 0.1 m	0.10	4.57	2.93	5.25	m	**12.75**
0.1–0.20 m	0.18	8.23	5.27	5.46	m	**18.96**
0.2–0.40 m	0.40	36.53	6.85	9.74	m^2	**53.12**
0.4–1.22 m	0.40	36.53	6.85	9.74	m^2	**53.12**
exceeding 1.22 m	0.38	34.71	6.51	9.74	m^2	**50.95**
Plane sloping, width						
ne 0.1 m	0.11	5.03	3.22	5.25	m	**13.51**
0.1- 0.20 m	0.20	9.15	5.85	5.46	m	**20.46**
0.2–0.40 m	0.43	39.27	7.33	13.34	m^2	**59.95**
0.4–1.22 m	0.43	39.27	7.33	13.34	m^2	**59.95**
exceeding 1.22 m	0.38	34.71	6.51	13.34	m^2	**54.55**
Plane battered, width						
ne 0.1 m	0.12	5.49	3.51	6.15	m	**15.15**
0.1- 0.20 m	0.20	9.15	5.85	6.36	m	**21.36**
0.2–0.40 m	0.43	39.27	7.49	15.14	m^2	**61.90**
0.4–1.22 m	0.43	39.27	7.49	15.14	m^2	**61.90**
exceeding 1.22 m	0.40	36.53	6.85	15.14	m^2	**58.52**

CLASS G: CONCRETE ANCILLARIES

Item	Gang hours	Labour £	Plant £	Material £	Unit	**Total rate £**
Plane vertical, width						
ne 0.1 m	0.12	5.49	3.51	6.15	m	**15.15**
0.1- 0.20 m	0.19	8.69	5.56	6.36	m	**20.61**
0.2–0.40 m	0.70	63.93	11.98	13.34	m^2	**89.26**
0.4–1.22 m	0.51	46.58	8.61	13.34	m^2	**68.53**
exceeding 1.22 m	0.47	42.93	8.17	13.34	m^2	**64.44**
Curved to one radius in one plane, 0.5 m radius, width						
ne 0.1 m	0.19	8.69	5.56	7.96	m	**22.21**
0.1- 0.20 m	0.25	11.43	7.32	8.16	m	**26.91**
0.2–0.40 m	0.90	82.20	15.41	15.14	m^2	**112.75**
0.4–1.22 m	0.72	65.76	12.33	15.14	m^2	**93.23**
exceeding 1.22 m	0.65	59.37	11.01	15.14	m^2	**85.52**
Curved to one radius in one plane, 2 m radius, width						
ne 0.1 m	0.18	8.23	5.27	7.96	m	**21.45**
0.1- 0.20 m	0.25	11.43	6.44	8.16	m	**26.03**
0.2–0.40 m	0.90	82.20	14.38	15.14	m^2	**111.72**
0.4–1.22 m	0.70	63.93	11.30	15.14	m^2	**90.37**
exceeding 1.22 m	0.52	47.49	8.90	15.14	m^2	**71.54**
For voids						
small void; depth ne 0.5 m	0.12	5.49	–	3.11	nr	**8.60**
small void; depth 0.5–1.0 m	0.25	11.43	–	5.98	nr	**17.42**
small void; depth ne 1.0–2.0 m	0.30	13.72	–	11.04	nr	**24.76**
large void; depth ne 0.5 m	0.30	13.72	–	9.30	nr	**23.02**
large void; depth 0.5–1.0 m	0.50	22.87	–	19.28	nr	**42.15**
large void; depth 1.0–2.0 m	0.90	41.16	–	39.14	nr	**80.31**
For concrete components of constant cross-section						
beams; 200 × 200 mm	0.70	32.02	14.04	8.95	m	**55.00**
beams; 500 × 500 mm	0.85	38.88	16.09	13.19	m	**68.16**
beams; 500 × 800 mm	0.90	41.16	19.60	14.92	m	**75.68**
columns; 200 × 200 mm	0.70	32.02	16.09	9.34	m	**57.45**
columns; 300 × 300 mm	0.80	36.59	16.09	12.16	m	**64.84**
columns; 300 × 500 mm	0.80	36.59	18.14	13.62	m	**68.35**
to walls; 1.0 m high thickness 250 mm	1.20	109.60	18.83	15.31	m	**143.75**
to walls; 1.5 m high thickness 300 mm	1.60	146.14	25.68	20.89	m	**192.71**
box culvert; 2 × 2 m internally and wall thickness 300 mm	4.70	429.28	77.04	115.19	m	**621.51**
projections (100 mm deep)	0.10	4.57	2.93	6.22	m	**13.72**
intrusions (100 mm deep)	0.10	4.57	2.93	6.22	m	**13.72**
Allowance for additional craneage and rub up where required						
ADD to items measures linear	0.08	3.66	1.17	0.04	m	**4.87**
ADD to items measures m^2	0.24	10.98	3.51	0.12	m^2	**14.60**

CLASS G: CONCRETE ANCILLARIES

Item	Gang hours	Labour £	Plant £	Material £	Unit	Total rate £
02 FORMWORK; FAIR FINISH						
Formwork						
Plane horizontal, width						
ne 0.1 m	0.10	4.57	2.93	7.55	m	**15.05**
0.1- 0.20 m	0.18	8.23	5.27	9.04	m	**22.54**
0.2–0.40 m	0.40	36.53	6.85	26.44	m^2	**69.82**
0.4–1.22 m	0.40	36.53	6.85	26.44	m^2	**69.82**
exceeding 1.22 m	0.38	34.71	6.51	26.44	m^2	**67.65**
Plane sloping, width						
ne 0.1 m	0.11	5.03	3.22	7.55	m	**15.81**
0.1- 0.20 m	0.20	9.15	3.42	9.04	m	**21.61**
0.2–0.40 m	0.43	39.27	20.03	34.36	m^2	**93.67**
0.4–1.22 m	0.43	39.27	20.03	34.36	m^2	**93.67**
exceeding 1.22 m	0.38	34.71	6.51	34.36	m^2	**75.58**
Plane battered, width						
ne 0.1 m	0.12	5.49	3.51	9.54	m	**18.53**
0.1- 0.20 m	0.20	9.15	5.85	11.02	m	**26.02**
0.2–0.40 m	0.43	39.27	7.37	38.33	m^2	**84.97**
0.4–1.22 m	0.43	39.27	7.37	38.33	m^2	**84.97**
exceeding 1.22 m	0.40	36.53	6.85	38.33	m^2	**81.71**
Plane vertical, width						
ne 0.1 m	0.12	5.49	3.51	9.54	m	**18.53**
0.1- 0.20 m	0.20	9.15	5.85	11.02	m	**26.02**
0.2–0.40 m	0.72	65.76	12.33	34.36	m^2	**112.45**
0.4–1.22 m	0.53	48.41	9.08	34.36	m^2	**91.85**
exceeding 1.22 m	0.48	43.84	8.22	34.36	m^2	**86.42**
Curved to one radius in one plane, 0.5 m radius, width						
ne 0.1 m	0.20	9.15	5.85	13.50	m	**28.50**
0.1- 0.20 m	0.25	11.43	7.32	14.98	m	**33.73**
0.2–0.40 m	0.90	82.20	15.41	38.33	m^2	**135.94**
0.4–1.22 m	0.72	65.76	12.33	38.33	m^2	**116.41**
exceeding 1.22 m	0.65	59.37	11.13	38.33	m^2	**108.83**
Curved to one radius in one plane, 2.0 m radius, width						
ne 0.1 m	0.18	8.23	5.27	13.50	m	**27.00**
0.1- 0.20 m	0.22	10.06	6.44	14.98	m	**31.48**
0.2–0.40 m	0.84	76.72	14.38	38.33	m^2	**129.43**
0.4–1.22 m	0.66	60.28	11.30	38.33	m^2	**109.91**
exceeding 1.22 m	0.52	47.49	8.90	38.33	m^2	**94.72**
For voids, using void former						
small void; depth ne 0.5 m	0.07	3.20	–	17.10	nr	**20.30**
small void; depth 0.5–1.0 m	0.12	5.49	–	33.39	nr	**38.88**
small void; depth ne 1.0–2.0 m	0.16	7.32	–	63.55	nr	**70.87**
large void; depth ne 0.5 m	0.14	6.40	–	37.93	nr	**44.33**
large void; depth 0.5–1.0 m	0.33	15.09	–	72.83	nr	**87.92**
large void; depth 1.0–2.0 m	0.58	26.53	–	126.54	nr	**153.07**

CLASS G: CONCRETE ANCILLARIES

Item	Gang hours	Labour £	Plant £	Material £	Unit	**Total rate £**
For concrete components of constant cross-section						
beams; 200 × 200 mm	0.49	22.41	14.34	20.70	m	**57.45**
beams; 500 × 500 mm	0.57	26.07	16.67	37.58	m	**80.32**
beams; 500 × 800 mm	0.69	31.56	20.18	45.85	m	**97.59**
columns; 200 × 200 mm	0.57	26.07	16.67	23.57	m	**66.31**
columns; 300 × 300 mm	0.57	26.07	16.67	33.27	m	**76.01**
columns; 300 × 500 mm	0.64	29.27	18.72	41.00	m	**88.99**
to walls; 1.0 m high thickness 250 mm	1.20	109.60	20.54	48.72	m	**178.87**
to walls; 1.5 m high thickness 300 mm	1.60	146.14	27.39	71.00	m	**244.53**
box culvert; 2 × 2 m internally and wall thickness 300 mm	4.60	420.14	78.75	354.72	m	**853.61**
projections (100 mm deep)	0.10	4.57	2.93	11.02	m	**18.52**
intrusions (100 mm deep)	0.10	4.57	2.93	11.02	m	**18.52**
Allowance for additional craneage and rub up where required						
ADD to items measures linear	0.28	12.81	4.10	0.05	m	**16.95**
ADD to items measures m^2	0.20	9.15	3.51	0.15	m^2	**12.81**
03 FORMWORK; EXTRA SMOOTH FINISH						
Formwork						
Plane horizontal, width						
ne 0.1 m	0.10	4.57	2.93	8.00	m	**15.50**
0.1- 0.20 m	0.18	8.23	5.27	10.56	m	**24.06**
0.2–0.40 m	0.40	36.53	6.85	34.20	m^2	**77.58**
0.4–1.22 m	0.40	36.53	6.85	34.20	m^2	**77.58**
exceeding 1.22 m	0.38	34.71	6.50	34.20	m^2	**75.40**
Plane sloping, width						
ne 0.1 m	0.11	5.03	3.22	8.39	m	**16.64**
0.1- 0.20 m	0.20	9.15	5.85	10.56	m	**25.56**
0.2–0.40 m	0.43	39.27	7.39	42.12	m^2	**88.79**
0.4–1.22 m	0.43	39.27	7.39	42.12	m^2	**88.79**
exceeding 1.22 m	0.38	34.71	6.51	42.12	m^2	**83.34**
Plane battered, width						
ne 0.1 m	0.12	5.49	3.51	10.37	m	**19.36**
0.1- 0.20 m	0.20	9.15	5.85	12.54	m	**27.54**
0.2–0.40 m	0.43	39.27	7.39	46.08	m^2	**92.75**
0.4–1.22 m	0.43	39.27	7.39	46.08	m^2	**92.75**
exceeding 1.22 m	0.40	36.53	6.85	46.08	m^2	**89.47**
Plane vertical, width						
ne 0.1 m	0.12	10.96	3.51	10.37	m	**24.84**
0.1- 0.20 m	0.20	18.27	5.85	12.54	m	**36.66**
0.2–0.40 m	0.74	67.59	12.67	42.12	m^2	**122.38**
0.4–1.22 m	0.53	48.41	9.10	42.12	m^2	**99.63**
exceeding 1.22 m	0.48	43.84	8.22	42.12	m^2	**94.18**
Curved to one radius in one plane, 0.5 m radius, width						
ne 0.1 m	0.20	9.15	5.85	14.33	m	**29.33**
0.1- 0.20 m	0.25	11.43	7.30	12.54	m	**31.27**
0.2–0.40 m	0.90	82.20	15.41	46.08	m^2	**143.69**
0.4–1.22 m	0.72	65.76	12.33	46.08	m^2	**124.17**
exceeding 1.22 m	0.65	59.37	11.16	46.08	m^2	**116.61**

CLASS G: CONCRETE ANCILLARIES

Item	Gang hours	Labour £	Plant £	Material £	Unit	Total rate £
03 FORMWORK; EXTRA SMOOTH FINISH – cont						
Formwork – cont						
Curved to one radius in one plane, 1.0 m radius, width						
ne 0.1 m	0.18	8.23	5.27	14.33	m	**27.83**
0.1- 0.20 m	0.22	10.06	6.44	12.54	m	**29.04**
0.2–0.40 m	0.84	76.72	24.09	46.08	m^2	**146.89**
0.4–1.22 m	0.66	60.28	11.30	46.08	m^2	**117.66**
exceeding 1.22 m	0.52	47.49	8.90	46.08	m^2	**102.48**
For voids, using void former						
small void; depth ne 0.5 m	0.07	3.20	–	15.17	nr	**18.37**
small void; depth 0.5–1.0 m	0.12	5.49	–	29.78	nr	**35.27**
small void; depth ne 1.0–2.0 m	0.16	7.32	–	46.21	nr	**53.53**
large void; depth ne 0.5 m	0.14	6.40	–	37.30	nr	**43.71**
large void; depth 0.5–1.0 m	0.33	15.09	–	64.53	nr	**79.63**
large void; depth 1.0–2.0 m	0.58	26.53	–	102.25	nr	**128.78**
For concrete components of constant cross-section						
beams; 200 × 200 mm	0.50	22.87	14.63	25.35	m	**62.85**
beams; 500 × 500 mm	0.57	26.07	9.73	49.22	m	**85.02**
beams; 500 × 800 mm	0.69	31.56	11.83	59.82	m	**103.21**
columns; 200 × 200 mm	0.57	26.07	9.73	29.78	m	**65.58**
columns; 300 × 300 mm	0.57	26.07	9.73	42.58	m	**78.38**
columns; 300 × 500 mm	0.64	29.27	10.96	53.41	m	**93.64**
to walls; 1.0 m high thickness 250 mm	1.20	109.60	20.54	64.24	m	**194.39**
to walls; 1.5 m high thickness 300 mm	1.60	146.14	27.39	94.28	m	**267.81**
box culvert; 2 × 2 m internally and wall thickness 300 mm	4.60	420.14	78.75	437.10	nr	**936.00**
projections (100 mm deep)	0.10	4.57	2.93	12.54	m	**20.04**
intrusions (100 mm deep)	0.10	4.57	2.93	12.54	m	**20.04**
Allowance for additional craneage and rub up where required						
ADD to items measures linear	0.28	12.81	4.10	0.05	m	**16.95**
ADD to items measures m^2	0.20	9.15	3.51	0.20	m^2	**12.86**
05 REINFORCEMENT						
Supply and fit reinforcement materials include for bars, tying wire, spacers, couplers and steel supports for bottom layer reinforcement (stools, chairs and risers).						
Plain round mild steel bars to BS 4449						
Bars; supplied in straight lengths						
6 mm nominal size	8.00	1095.47	162.87	844.66	tonne	**2103.00**
8 mm nominal size	6.74	922.93	137.22	844.66	tonne	**1904.81**
10 mm nominal size	6.74	922.93	137.22	844.63	tonne	**1904.78**
12 mm nominal size	6.74	922.93	137.22	844.63	tonne	**1904.78**
16 mm nominal size	6.15	842.14	125.22	844.63	tonne	**1811.99**
20 mm nominal size	4.44	607.98	90.40	844.63	tonne	**1543.01**
25 mm nominal size	4.44	607.98	90.40	844.63	tonne	**1543.01**
32 mm nominal size	4.44	607.98	90.40	844.63	tonne	**1543.01**
40 mm nominal size	4.44	607.98	90.40	844.63	tonne	**1543.01**

CLASS G: CONCRETE ANCILLARIES

Item	Gang hours	Labour £	Plant £	Material £	Unit	**Total rate £**
Bars; supplied in bent and cut lengths						
6 mm nominal size	8.00	1095.47	162.87	952.67	tonne	**2211.01**
8 mm nominal size	6.74	922.93	137.22	952.67	tonne	**2012.82**
10 mm nominal size	6.74	922.93	137.22	952.67	tonne	**2012.82**
12 mm nominal size	6.74	922.93	137.22	952.67	tonne	**2012.82**
16 mm nominal size	6.15	842.14	125.22	952.67	tonne	**1920.03**
20 mm nominal size	4.44	607.98	90.40	952.67	tonne	**1651.05**
25 mm nominal size	4.44	607.98	90.40	952.67	tonne	**1651.05**
32 mm nominal size	4.44	607.98	90.40	952.67	tonne	**1651.05**
40 mm nominal size	4.44	607.98	90.40	952.67	tonne	**1651.05**
Deformed high yield steel bars to BS 4449						
Bars; supplied in straight lengths						
6 mm nominal size	8.00	1095.47	162.87	961.59	tonne	**2219.93**
8 mm nominal size	6.74	922.93	137.22	961.58	tonne	**2021.73**
10 mm nominal size	6.74	922.93	137.22	961.59	tonne	**2021.74**
12 mm nominal size	6.74	922.93	137.22	961.59	tonne	**2021.74**
16 mm nominal size	6.15	842.14	125.22	961.59	tonne	**1928.95**
20 mm nominal size	4.44	607.98	90.40	961.59	tonne	**1659.97**
25 mm nominal size	4.44	607.98	90.40	961.59	tonne	**1659.97**
32 mm nominal size	4.44	607.98	90.40	961.59	tonne	**1659.97**
40 mm nominal size	4.44	607.98	90.40	961.59	tonne	**1659.97**
Bars; supplied in bent and cut lengths						
6 mm nominal size	8.00	1095.47	162.87	961.59	tonne	**2219.93**
8 mm nominal size	6.74	922.93	137.22	961.59	tonne	**2021.74**
10 mm nominal size	6.74	922.93	137.22	961.59	tonne	**2021.74**
12 mm nominal size	6.74	922.93	137.22	961.59	tonne	**2021.74**
16 mm nominal size	6.15	842.14	125.22	961.58	tonne	**1928.94**
20 mm nominal size	4.44	607.98	90.40	961.59	tonne	**1659.97**
25 mm nominal size	4.44	607.98	90.40	961.59	tonne	**1659.97**
32 mm nominal size	4.44	607.98	90.40	961.59	tonne	**1659.97**
40 mm nominal size	4.44	607.98	90.40	961.59	tonne	**1659.97**
Stainless steel bars; Alloy 1.4301						
Bars; supplied in straight lengths						
8 mm nominal size	6.74	922.93	137.22	5213.68	tonne	**6273.83**
10 mm nominal size	6.74	922.93	137.22	5213.68	tonne	**6273.83**
12 mm nominal size	6.74	922.93	137.22	5213.68	tonne	**6273.83**
16 mm nominal size	6.15	842.14	125.22	5213.68	tonne	**6181.04**
20 mm nominal size	4.44	607.98	90.40	5213.68	tonne	**5912.06**
25 mm nominal size	4.44	607.98	90.40	5213.68	tonne	**5912.06**
32 mm nominal size	4.44	607.98	90.40	5213.68	tonne	**5912.06**
Bars; supplied in bent and cut lengths						
8 mm nominal size	6.74	922.93	137.22	5213.68	tonne	**6273.83**
10 mm nominal size	6.74	922.93	137.22	5213.68	tonne	**6273.83**
12 mm nominal size	6.74	922.93	137.22	5213.68	tonne	**6273.83**
16 mm nominal size	6.15	842.14	125.22	5213.68	tonne	**6181.04**
20 mm nominal size	4.44	607.98	90.40	5213.68	tonne	**5912.06**
25 mm nominal size	4.44	607.98	90.40	5213.68	tonne	**5912.06**
32 mm nominal size	4.44	607.98	90.40	5213.68	tonne	**5912.06**

CLASS G: CONCRETE ANCILLARIES

Item	Gang hours	Labour £	Plant £	Material £	Unit	Total rate £
05 REINFORCEMENT – cont						
Additional allowances to bar reinforcement						
Add to the above bars						
12–15 m long; mild steel to BS 4449		–	–	81.30	tonne	**81.30**
12–15 m long; high yield steel to BS 4449		–	–	82.09	tonne	**82.09**
Over 15 m long, per 500 mm increment; mild steel to BS 4449		–	–	81.30	tonne	**81.30**
Over 15 m long, per 500 mm increment; high yield steel to BS 4449		–	–	82.09	tonne	**82.09**
Add for cutting, bending, tagging and baling reinforcement on site						
6 mm nominal size	5.90	250.16	99.16	3.24	tonne	**352.56**
8 mm nominal size	5.60	237.44	93.25	3.24	tonne	**333.93**
10 mm nominal size	4.20	178.08	69.63	3.24	tonne	**250.95**
12 mm nominal size	3.10	131.44	51.93	3.24	tonne	**186.60**
16 mm nominal size	2.50	106.00	41.32	3.24	tonne	**150.56**
20 mm nominal size	2.05	86.92	34.20	3.24	tonne	**124.36**
25 mm nominal size	2.05	86.92	34.20	3.24	tonne	**124.36**
32 mm nominal size	1.70	72.08	28.29	3.24	tonne	**103.61**
40 mm nominal size	1.70	72.08	28.29	3.24	tonne	**103.61**
Special joints						
Lenton type A couplers; threaded ends on reinforcing bars						
12 mm	0.09	12.32	–	11.65	nr	**23.97**
16 mm	0.09	12.32	–	13.77	nr	**26.09**
20 mm	0.09	12.32	–	20.41	nr	**32.73**
25 mm	0.09	12.32	–	28.15	nr	**40.48**
32 mm	0.09	12.32	–	37.96	nr	**50.28**
40 mm	0.09	12.32	–	52.74	nr	**65.07**
Lenton type B couplers; threaded ends on reinforcing bars						
12 mm	0.09	12.32	–	36.84	nr	**49.17**
16 mm	0.09	12.32	–	42.85	nr	**55.18**
20 mm	0.09	12.32	–	46.68	nr	**59.00**
25 mm	0.09	12.32	–	54.81	nr	**67.14**
32 mm	0.09	12.32	–	73.69	nr	**86.01**
40 mm	0.09	12.32	–	108.69	nr	**121.01**
Steel fabric to BS 4483						
nominal mass 0.77 kg/m^2; ref D49	0.02	3.29	0.41	1.42	m^2	**5.11**
nominal mass 1.54 kg/m^2; ref D98	0.02	3.29	0.41	2.84	m^2	**6.53**
nominal mass 1.54 kg/m^2; ref A98	0.03	4.38	0.62	5.55	m^2	**10.55**
nominal mass 2.22 kg/m^2; ref A142	0.03	4.24	0.62	5.12	m^2	**9.98**
nominal mass 2.61 kg/m^2; ref C283	0.03	4.11	0.62	5.98	m^2	**10.71**
nominal mass 3.02 kg/m^2; ref A193	0.04	5.48	0.81	6.93	m^2	**13.23**
nominal mass 3.05 kg/m^2; ref B196	0.04	5.48	0.81	12.16	m^2	**18.45**
nominal mass 3.41 kg/m^2; ref C385	0.04	5.48	0.81	7.86	m^2	**14.15**
nominal mass 3.73 kg/m^2; ref B283	0.04	5.48	0.81	9.44	m^2	**15.73**
nominal mass 3.95 kg/m^2; ref A252	0.04	5.48	0.81	7.68	m^2	**13.97**
nominal mass 4.34 kg/m^2; ref C503	0.05	6.85	1.03	9.96	m^2	**17.83**

CLASS G: CONCRETE ANCILLARIES

Item	Gang hours	Labour £	Plant £	Material £	Unit	**Total rate £**
nominal mass 4.35 kg/m²; ref B385	0.05	6.85	1.03	10.48	m²	**18.35**
nominal mass 5.55 kg/m²; ref C636	0.05	6.85	1.03	12.49	m²	**20.37**
nominal mass 5.93 kg/m²; ref B503	0.05	6.85	1.03	14.67	m²	**22.55**
nominal mass 6.16 kg/m²; ref A393	0.07	9.59	1.43	10.78	m²	**21.80**
nominal mass 6.72 kg/m²; ref C785	0.07	9.59	1.43	14.16	m²	**25.19**
nominal mass 8.14 kg/m²; ref B785	0.08	10.95	1.63	18.96	m²	**31.55**
nominal mass 10.90 kg/m²; ref B1131	0.09	12.32	1.82	24.96	m²	**39.10**
06 JOINTS						
Joint Filler						
Open surface plain; average width						
ne 0.5 m; scabbling concrete for subsequent pour	0.04	2.45	2.71	–	m²	**5.15**
0.5–1 m; scabbling concrete for subsequent pour	0.03	1.84	2.08	–	m²	**3.91**
Open surface with filler; average width						
ne 0.5 m; 12 mm Flexcell joint filler	0.04	2.45	2.71	12.09	m²	**17.24**
0.5–1 m; 12 mm Flexcell joint filler	0.04	2.45	2.71	12.09	m²	**17.24**
ne 0.5 m; 19 mm Flexcell joint filler	0.05	3.06	3.34	20.79	m²	**27.19**
0.5–1 m; 19 mm Flexcell joint filler	0.05	3.06	3.34	20.79	m²	**27.19**
Formed surface plain; average width (including formwork)						
ne 0.5 m	0.24	14.69	16.24	12.06	m²	**42.99**
0.5–1.0 m	0.24	14.69	16.24	12.06	m²	**42.99**
Formed surface with filler; average width						
ne 0.5 m; 10 mm Flexcell joint filler	0.40	24.49	20.21	24.15	m²	**68.84**
0.5–1 m; 10 mm Flexcell joint filler	0.41	25.10	27.70	24.15	m²	**76.95**
ne 0.5 m; 19 mm Flexcell joint filler	0.42	25.71	28.50	32.84	m²	**87.06**
0.5–1 m; 19 mm Flexcell joint filler	0.43	26.32	29.14	32.84	m²	**88.31**
ne 0.5 m; 25 mm Flexcell joint filler	0.45	27.55	30.41	37.25	m²	**95.20**
0.5–1 m; 25 mm Flexcell joint filler	0.47	28.77	31.85	37.25	m²	**97.87**
PVC						
Plastics or rubber waterstops						
160 mm centre bulb	0.04	2.45	–	4.88	m	**7.33**
junction piece	0.04	2.45	–	10.47	nr	**12.92**
210 mm centre bulb	0.05	3.06	–	6.98	m	**10.04**
junction piece	0.04	2.45	–	10.80	nr	**13.25**
260 mm centre bulb	0.05	3.06	–	8.37	m	**11.44**
junction piece	0.05	3.06	–	11.38	nr	**14.44**
170 mm flat dumbbell	0.04	2.45	–	6.28	m	**8.73**
junction piece	0.05	3.06	–	76.77	nr	**79.83**
210 mm flat dumbbell	0.04	2.45	–	8.37	m	**10.82**
junction piece	0.07	4.29	–	83.73	nr	**88.02**
250 mm flat dumbbell	0.05	3.06	–	10.47	m	**13.53**
junction piece	0.09	5.51	–	160.51	nr	**166.02**
Polysulphide sealant; gun grade						
Sealed rebates or grooves						
10 × 20 mm	0.05	3.06	–	0.03	m	**3.09**
20 × 20 mm	0.07	4.29	–	0.06	m	**4.35**
25 × 20 mm	0.08	4.90	–	0.08	m	**4.98**

CLASS G: CONCRETE ANCILLARIES

Item	Gang hours	Labour £	Plant £	Material £	Unit	**Total rate £**
06 JOINTS – cont						
Mild steel						
Dowels, plain or greased						
12 mm dia. × 500 mm long	0.04	2.45	–	0.91	nr	**3.36**
16 mm dia. × 750 mm long	0.05	2.75	–	2.80	nr	**5.55**
20 mm dia. × 750 mm long	0.05	2.75	–	4.35	nr	**7.11**
25 mm dia. × 750 mm long	0.05	2.75	–	4.54	nr	**7.29**
32 mm dia. × 750 mm long	0.05	2.75	–	5.91	nr	**8.66**
Dowels, sleeved or capped						
12 mm dia. × 500 mm long, debonding agent for 250 mm and capped with pvc dowel cap	0.05	2.75	–	1.04	nr	**3.79**
16 mm dia. × 750 mm long, debonding agent for 375 mm and capped with pvc dowel cap	0.05	3.24	–	2.92	nr	**6.17**
20 mm dia. × 750 mm long, debonding agent for 375 mm and capped with pvc dowel cap	0.05	3.24	–	4.48	nr	**7.72**
25 mm dia. × 750 mm long, debonding agent for 375 mm and capped with pvc dowel cap	0.06	3.67	–	4.66	nr	**8.34**
32 mm dia. × 750 mm long, debonding agent for 375 mm and capped with pvc dowel cap	0.07	3.98	–	6.04	nr	**10.02**
07 POST-TENSIONED PRESTRESSING						
The design of prestressing is based on standard patented systems, each of which has produced its own method of anchoring, joining and stressing the cables or wires. The companies marketing the systems will either supply all the materials and fittings required together with the sale or hire of suitable jacks and equipment for prestressing and grouting or they will undertake to complete the work on a subcontract basis. The rates given below are therefore indicative only of the probable labour and plant costs and do not include for any permanent materials. The advice of specialist contractors should be sought for more accurate rates based on the design for a particular contract. Pretensioned prestressing is normally used only in the manufacture of precast units utilizing special beds set up in the supplier's factory.						
Labour and plant cost in post-tensioning; material cost excluded						
form ducts to profile including supports and fixings; 50 mm internal dia.	0.04	0.85	3.98	–	m	**4.83**
extra for grout vents	0.50	9.80	–	–	nr	**9.80**
form ducts to profile including supports and fixings; 80 mm internal dia.	0.20	3.92	4.67	–	m	**8.59**
extra for grout vents	0.50	9.80	–	–	nr	**9.80**
form ducts to profile including supports and fixings; 100 mm internal dia.	0.30	5.88	4.91	–	m	**10.79**
extra for grout vents	0.50	9.80	–	–	nr	**9.80**
grout ducts including provision of equipment; 50 mm internal dia.	0.20	3.92	0.70	–	m	**4.62**

CLASS G: CONCRETE ANCILLARIES

Item	Gang hours	Labour £	Plant £	Material £	Unit	**Total rate £**
grout ducts including provision of equipment; 80 mm internal dia.	0.20	3.92	0.86	–	m	**4.78**
grout ducts including provision of equipment; 100 mm internal dia.	0.20	3.92	1.02	–	m	**4.94**
form tendons including spacers etc. and pull through ducts; 7 nr strands	0.60	11.76	4.72	–	m	**16.48**
form tendons including spacers etc. and pull through ducts; 12 nr strands	0.80	15.68	15.09	–	m	**30.78**
form tendons including spacers etc. and pull through ducts; 19 nr strands	1.20	23.52	22.64	–	m	**46.16**
dead end anchorage; 7 nr strands	1.20	23.52	22.64	–	nr	**46.16**
dead end anchorage; 12 nr strands	1.70	33.32	32.08	–	nr	**65.40**
dead end anchorage; 19 nr strands	2.00	39.20	33.02	–	nr	**72.22**
looped buried dead end anchorage; 7 nr strands	0.95	18.62	17.93	–	nr	**36.55**
looped buried dead end anchorage; 12 nr strands	1.20	23.52	22.64	–	nr	**46.16**
looped buried dead end anchorage; 19 nr strands	2.00	39.20	22.64	–	nr	**61.85**
end anchorage including reinforcement; 7 nr strands	3.00	58.81	56.61	–	nr	**115.41**
add to last for anchorage coupling	1.00	19.60	18.87	–	nr	**38.47**
end anchorage including reinforcement; 12 nr strands	3.30	64.69	62.27	–	nr	**126.95**
add to last for anchorage coupling	1.30	25.48	24.53	–	nr	**50.01**
end anchorage including reinforcement; 19 nr strands	4.50	88.21	81.14	–	nr	**169.34**
add to last for anchorage coupling	2.00	39.20	37.74	–	nr	**76.94**
stress and lock off including multimatic jack; 7 nr strands	3.00	183.64	56.61	–	nr	**240.25**
stress and lock off including multimatic jack; 12 nr strands	3.50	214.25	66.04	–	nr	**280.29**
stress and lock off including multimatic jack; 19 nr strands	4.00	244.86	75.47	–	nr	**320.33**
cut off and seal ends of tendons; 7 nr strands	0.50	9.80	9.43	–	nr	**19.24**
cut off and seal ends of tendons; 12 nr strands	0.90	17.64	16.98	–	nr	**34.62**
cut off and seal ends of tendons; 19 nr strands	1.00	19.60	18.87	–	nr	**38.47**
08 CONCRETE ACCESSORIES						
Finishing of top surfaces						
wood float; level	0.02	1.22	–	–	m^2	**1.22**
wood float; falls or cross-falls	0.03	1.84	–	–	m^2	**1.84**
steel trowel; level	0.03	1.84	–	–	m^2	**1.84**
steel trowel; falls or cross-falls	0.05	3.06	–	–	m^2	**3.06**
granolithic finish 20 mm thick laid monolithically	0.07	4.29	0.99	8.93	m^2	**14.21**
Finishing of formed surfaces						
aggregate exposure using retarder	0.05	3.06	–	1.44	m^2	**4.50**
bush hammering; kango hammer	0.28	17.14	6.65	–	m^2	**23.79**
rubbing down concrete surfaces after striking shutters	0.02	1.22	–	1.66	m^2	**2.88**

CLASS G: CONCRETE ANCILLARIES

Item	Gang hours	Labour £	Plant £	Material £	Unit	**Total rate £**
08 CONCRETE ACCESSORIES – cont						
Inserts totally within the concrete volume						
HDPE conduit 20 mm dia.	0.10	6.12	–	3.56	m	**9.69**
black enamelled steel conduit 20 mm dia.	0.10	6.12	–	3.11	m	**9.23**
galvanized steel conduit 20 mm dia.	0.10	6.12	–	4.06	m	**10.18**
Unistrut channel type P3270	0.20	12.24	–	13.55	m	**25.80**
Unistrut channel type P3370	0.20	12.24	–	10.64	m	**22.88**
Inserts projecting from surface(s) of the concrete						
expanding bolt; 10 mm dia. × 25 mm deep	0.05	3.06	–	2.15	nr	**5.21**
holding down bolt; 16 mm dia. × 250 mm deep	0.25	15.30	–	4.23	nr	**19.54**
holding down bolt; 16 mm dia. × 350 mm deep	0.25	15.30	–	4.56	nr	**19.86**
holding down bolt; 20 mm dia. × 250 mm deep	0.25	15.30	–	5.31	nr	**20.61**
holding down bolt; 20 mm dia. × 450 mm deep	0.25	15.30	–	5.54	nr	**20.84**
vitrified clay pipe to BS 65; 100 mm dia. × 1000 mm long	0.35	21.43	–	6.82	nr	**28.25**
cast iron pipe to BS 437; 100 mm dia. × 1000 mm long	0.25	15.30	–	35.72	nr	**51.02**
Grouting under plates; cement and sand (1:3)						
area ne 0.1 m^2	0.10	6.12	–	0.27	nr	**6.39**
area 0.1–0.5 m^2	0.45	27.55	–	1.31	nr	**28.86**
area 0.5–1.0 m^2	0.78	47.75	–	2.61	nr	**50.36**
Grouting under plates; non-shrink cementitious grout						
area ne 0.1 m^2	0.10	6.12	–	4.23	nr	**10.35**
area 0.1–0.5 m^2	0.45	27.55	–	40.63	nr	**68.18**
area 0.5–1.0 m^2	0.78	47.75	–	81.27	nr	**129.02**

CLASS H: PRECAST CONCRETE

Item	Gang hours	Labour £	Plant £	Material £	Unit	Total rate £
NOTES The cost of precast concrete items is very much dependent on the complexity of the moulds, the number of units to be cast from each mould and the size and weight of the unit to be handled. The unit rates below are for standard precast items that are often to be found on a civil engineering project. It would be misleading to quote for indicative costs for tailor-made precast concrete units and it is advisable to contact Specialist Manufacturers for guide prices.						
01 BEAMS						
Concrete mix C20						
Beams						
100 × 150 × 1050 mm long	0.50	73.75	–	8.31	nr	**82.07**
225 × 150 × 1200 mm long	0.50	73.75	2.39	32.11	nr	**108.26**
225 × 225 × 1800 mm long	0.60	88.51	3.00	44.93	nr	**136.43**
02 PRESTRESSED PRE-TENSIONED BEAMS						
Concrete mix C20						
Beams						
100 × 65 × 1050 mm long	0.30	44.25	–	4.99	nr	**49.24**
265 × 65 × 1800 mm long	0.50	73.75	2.31	20.79	nr	**96.86**
Bridge beams						
Inverted 'T' Beams, flange width 495 mm						
section T1; 8 m long, 380 mm deep; mass 1.88t		–	–	–	nr	**1402.80**
section T2; 9 m long, 420 mm deep; mass 2.29t		–	–	–	nr	**1729.03**
section T3; 11 m long, 535 mm deep; mass 3.02t		–	–	–	nr	**2038.95**
section T4; 12 m long, 575 mm deep; mass 3.54t		–	–	–	nr	**2120.53**
section T5; 13 m long, 615 mm deep; mass 4.08t		–	–	–	nr	**2365.19**
section T6; 13 m long, 655 mm deep; mass 4.33t		–	–	–	nr	**2365.20**
section T7; 14 m long, 695 mm deep; mass 4.95t		–	–	–	nr	**2528.31**
section T8; 15 m long, 735 mm deep; mass 5.60t		–	–	–	nr	**2854.54**
section T9; 16 m long, 775 mm deep; mass 6.28t		–	–	–	nr	**2936.09**
section T10; 18 m long, 815 mm deep; mass 7.43t		–	–	–	nr	**3458.04**
'M' beams, flange width 970 mm						
section M2; 17 m long, 720 mm deep; mass 12.95t		–	–	–	nr	**7829.58**
section M3; 18 m long, 800 mm deep; mass 15.11t		–	–	–	nr	**7258.66**
section M6; 22 m long, 1040 mm deep; mass 20.48t		–	–	–	nr	**11776.93**
section M8; 25 m long, 1200 mm deep; mass 23.68t		–	–	–	nr	**15169.58**

CLASS H: PRECAST CONCRETE

Item	Gang hours	Labour £	Plant £	Material £	Unit	Total rate £
02 PRESTRESSED PRE-TENSIONED BEAMS – cont						
Bridge beams – cont						
'U' beams, base width 970 mm						
section U3; 16 m long, 900 mm deep; mass 19.24t		–	–	–	nr	**12396.80**
section U5; 20 m long, 1000 mm deep; mass 25.64t		–	–	–	nr	**15577.56**
section U8; 24 m long, 1200 mm deep; mass 34.56t		–	–	–	nr	**17290.30**
section U12; 30 m long, 1600 mm deep; mass 52.74t		–	–	–	nr	**28830.73**
03 PRESTRESSED PRE-TENSIONED COLUMNS						
Concrete mix C32/40						
Columns; including shuttering						
dia.: not exceeding 300 mm	0.50	59.71	–	39.08	m	**98.80**
300 mm–450 mm	0.60	71.65	–	53.55	m	**125.20**
450 mm–600 mm	0.70	83.60	–	46.32	m	**129.91**
600 mm–750 mm	0.70	83.60	–	68.02	m	**151.61**
750 mm–900 mm	0.80	95.54	–	99.87	m	**195.41**
900 mm–1050 mm	1.15	133.12	–	125.89	m	**259.01**
1050 mm–1200 mm	1.30	152.44	–	134.57	m	**287.01**
Exceeding 1200 mm	2.00	238.84	–	212.18	m	**451.02**
05 SLABS						
Prestressed precast concrete flooring planks; Bison or similar; cement mortar grout between planks on bearings						
100 mm thick floor						
400 mm wide planks	0.21	17.03	23.60	44.33	m^2	**84.96**
1200 mm wide planks	0.12	9.50	13.16	55.63	m^2	**78.29**
150 mm thick floor						
400 mm wide planks	0.26	21.28	29.50	47.79	m^2	**98.58**
1200 mm wide planks	0.15	11.87	16.45	58.19	m^2	**86.52**
08 COPINGS, SILLS AND WEIR BLOCKS						
Concrete mix C30						
Coping; weathered and throated						
178 × 64 mm	0.25	15.30	5.93	5.48	m	**26.72**
305 × 76 mm	0.18	11.02	4.15	8.94	m	**24.10**

CLASS I: PIPEWORK – PIPES

Item	Gang hours	Labour £	Plant £	Material £	Unit	Total rate £
NOTES						
The rates assume the most efficient items of plant (excavator) and are optimum rates assuming continuous output with no delays caused by other operations or works.						
Ground conditions are assumed to be good easily worked soil with no abnormal conditions that would affect outputs and consistency of work.						
Multiplier Table for labour and plant for various site conditions for working:						
out of sequence × 2.75 (minimum)						
in hard clay × 1.75 to 2.00						
in running sand × 2.75 (minimum)						
in broken rock × 2.75 to 3.50						
below water table × 2.00 (minimum)						
Variance from CESMM4						
Fittings are included with the pipe concerned, for convenience of reference, rather than in Class J.						
RESOURCES – LABOUR						
Labour Resources Build-up Rate						
Drainage/pipework gang (small bore)						
1 ganger/chargehand (skill rate 4) – 50% of time		10.71	–	–		
1 skilled operative (skill rate 4)		20.20	–	–		
2 unskilled operatives (general)		39.20	–	–		
1 plant operator (skill rate 3)		29.17	–	–		
1 plant operator (skill rate 3) – 50% of time		14.59	–	–		
Total Gang Rate/Hour		113.86	–	–		
Drainage/pipework gang (small bore – not in trenches)						
1 ganger/chargehand (skill rate 4) – 50% of time		10.71	–	–		
1 skilled operative (skill rate 4)		20.20	–	–		
2 unskilled operatives (general)		39.20	–	–		
Total Gang Rate/Hour		70.11	–	–		
Drainage/pipework gang (large bore)						
Note: relates to pipes exceeding 700 mm dia.						
1 ganger/chargehand (skill rate 4) – 50% of time		10.71	–	–		
1 skilled operative (skill rate 4)		20.20	–	–		
2 unskilled operatives (general)		39.20	–	–		
1 plant operator (skill rate 3)		29.17	–	–		
Total Gang Rate/Hour		99.28	–	–		
Drainage/pipework gang (large bore – not in trenches)						
Note: relates to pipes exceeding 700 mm dia.						
1 ganger/chargehand (skill rate 4) – 50% of time		10.71	–	–		
1 skilled operative (skill rate 4)		20.20	–	–		
2 unskilled operatives (general)		39.20	–	–		
Total Gang Rate/Hour		70.11	–	–		

CLASS I: PIPEWORK – PIPES

Item	Gang hours	Labour £	Plant £	Material £	Unit	Total rate £
RESOURCES – MATERIAL						
Supply only ex works materials rates						
Clayware Pipes and Fittings						
Vitrified clay; socketed pipes to BS EN 295						
Straight pipes						
100 mm dia.		–	–	10.83	m	**10.83**
150 mm dia.		–	–	16.49	m	**16.49**
225 mm dia.		–	–	37.04	m	**37.04**
300 mm dia.		–	–	44.95	m	**44.95**
400 mm dia.		–	–	148.92	m	**148.92**
450 mm dia.		–	–	166.64	m	**166.64**
Vitrified clay; plain ended pipes to BS EN 295						
Straight pipes						
1.6 m lengths; 100 mm dia.		–	–	13.31	m	**13.31**
1.75 m lengths; 150 mm dia.		–	–	25.54	m	**25.54**
1.75 m lengths; 225 mm dia.		–	–	81.72	m	**81.72**
1.75 m lengths; 300 mm dia.		–	–	130.33	m	**130.33**
Couplings						
polypropylene coupling, 100 mm dia.		–	–	15.50	nr	**15.50**
polypropylene coupling, 150 mm dia.		–	–	29.25	nr	**29.25**
polypropylene coupling, 225 mm dia.		–	–	32.92	nr	**32.92**
polypropylene coupling, 300 mm dia.		–	–	67.58	nr	**67.58**
Tapers						
plain end; taper reducer, 225–150 mm		–	–	186.59	nr	**186.59**
plain end; taper reducer, 150–100 mm		–	–	81.35	nr	**81.35**
Field drains to BS 1196, butt joints, nominal bore; Hepworth						
300 mm lengths; 75 mm dia.		–	–	3.73	nr	**3.73**
300 mm lengths; 100 mm dia.		–	–	5.24	nr	**5.24**
300 mm lengths; 150 mm dia.		–	–	10.68	nr	**10.68**
300 mm lengths; 225 mm dia.		–	–	27.49	nr	**27.49**
Perforated vitrified clay sleeved pipe, BS EN295; Hepline						
1.6 m lengths;100 mm dia.		–	–	34.85	m	**34.85**
1.75 m lengths; 150 mm dia.		–	–	67.26	m	**67.26**
1.75 m lengths; 225 mm dia.		–	–	79.22	m	**79.22**
1.75 m lengths; 300 mm dia.		–	–	199.75	m	**199.75**
Concrete pipes						
Concrete pipes, BS5911 Class L						
300 mm dia.		–	–	30.31	m	**30.31**
375 mm dia.		–	–	37.52	m	**37.52**
450 mm dia.		–	–	44.27	m	**44.27**
525 mm dia.		–	–	58.13	m	**58.13**
750 mm dia.		–	–	143.80	m	**143.80**
900 mm dia.		–	–	192.91	m	**192.91**
1200 mm dia.		–	–	330.10	m	**330.10**
1500 mm dia.		–	–	584.64	m	**584.64**
1800 mm dia.		–	–	813.15	m	**813.15**
2100 mm dia.		–	–	979.62	m	**979.62**

CLASS I: PIPEWORK – PIPES

Item	Gang hours	Labour £	Plant £	Material £	Unit	**Total rate £**
Bends						
bend; 300 mm dia.; <45°		–	–	302.99	nr	**302.99**
bend; 375 mm dia.; <45°		–	–	375.07	nr	**375.07**
bend; 450 mm dia.; <45°		–	–	442.85	nr	**442.85**
bend; 525 mm dia.; <45°		–	–	581.18	nr	**581.18**
bend; 750 mm dia.; <45°		–	–	1438.18	nr	**1438.18**
bend; 900 mm dia.; 45–90°		–	–	1787.98	nr	**1787.98**
bend; 1200 mm dia.; <45°		–	–	3300.75	nr	**3300.75**
bend; 1500 mm dia.; <45°		–	–	5846.02	nr	**5846.02**
bend; 1800 mm dia.; <45°		–	–	8131.39	nr	**8131.39**
bend; 2100 mm dia.; <45°		–	–	9787.08	nr	**9787.08**
Cast iron pipes and fittings						
Cast iron drain BS EN 877; plain end						
70 mm dia.		–	–	29.16	m	**29.16**
100 mm dia.		–	–	35.72	m	**35.72**
150 mm dia.		–	–	14.94	m	**14.94**
250 mm dia.		–	–	134.77	m	**134.77**
Cast iron couplings complete with stainless steel nuts and bolts and synthetic rubber gaskets						
70 mm dia.		–	–	12.72	nr	**12.72**
100 mm dia.		–	–	16.70	nr	**16.70**
150 mm dia.		–	–	20.74	nr	**20.74**
250 mm dia.		–	–	107.24	nr	**107.24**
Bends; 45 degree						
70 mm dia.		–	–	22.69	nr	**22.69**
100 mm dia.		–	–	32.26	nr	**32.26**
150 mm dia.		–	–	48.26	nr	**48.26**
250 mm dia.		–	–	280.50	nr	**280.50**
Bends; short radius; 87.5°						
70 mm dia.		–	–	44.02	nr	**44.02**
100 mm dia.		–	–	64.29	nr	**64.29**
150 mm dia.		–	–	113.13	nr	**113.13**
250 mm dia.		–	–	282.06	nr	**282.06**
Single junction × 45°						
70 × 70 mm dia.		–	–	34.18	nr	**34.18**
100 × 100 mm dia.		–	–	51.12	nr	**51.12**
150 × 150 mm dia.		–	–	120.18	nr	**120.18**
250 × 250 mm dia.		–	–	738.10	nr	**738.10**
Taper reducer pipes						
70 mm dia.		–	–	32.15	nr	**32.15**
100 mm dia.		–	–	37.80	nr	**37.80**
150 mm dia.		–	–	72.56	nr	**72.56**
250 mm dia.		–	–	216.77	nr	**216.77**

CLASS I: PIPEWORK – PIPES

Item	Gang hours	Labour £	Plant £	Material £	Unit	**Total rate £**
RESOURCES – PLANT						
Plant Resources Build-up Rate						
Field drains						
0.4 m^3 hydraulic backacter		–	55.59	–		
2t dumper – 30% of time		–	3.24	–		
Stihl saw, 12", petrol – 30% of time		–	0.33	–		
small pump – 30% of time		–	1.67	–		
Total Rate/Hour		–	**60.84**	–		
Add to the above for trench supports appropriate to trench depth (see below).						
Field drains (not in trenches)						
2t dumper – 30% of time		–	3.24	–		
Stihl saw, 12", petrol – 30% of time		–	0.33	–		
Total Rate/Hour		–	**3.57**	–		
Drains/sewers (small bore)						
1.0 m^3 hydraulic backacter		–	84.68	–		
2t dumper – 30% of time		–	3.24	–		
2.80 m^3/min compressor, 2 tool – 30% of time		–	4.71	–		
disc saw – 30% of time		–	0.41	–		
extra 50ft / 15 m hose – 30% of time		–	0.14	–		
small pump – 30% of time		–	1.67	–		
sundry tools – 30% of time		–	0.48	–		
Total Rate/Hour		–	**95.34**	–		
Add to the above for trench supports appropriate to trench depth (see below)						
Drains/sewers (small bore – not in trenches)						
2t dumper – 30% of time		–	3.24	–		
2.80 m^3/min compressor, 2 tool – 30% of time		–	4.71	–		
disc saw – 30% of time		–	0.41	–		
extra 50ft / 15 m hose – 30% of time		–	0.14	–		
sundry tools – 30% of time		–	0.48	–		
Total Rate/Hour		–	**8.99**	–		
Drains/sewers (large bore)						
1.0 m^3 hydraulic backacter		–	84.68	–		
20t crawler crane – 50% of time		–	24.26	–		
2t dumper (30% of time)		–	3.24	–		
2.80 m^3/min compressor, 2 tool – 30% of time		–	4.71	–		
disc saw – 30% of time		–	0.41	–		
extra 50ft / 15 m hose – 30% of time		–	0.14	–		
small pump – 30% of time		–	1.67	–		
sundry tools – 30% of time		–	0.48	–		
Total Rate/Hour		–	**119.60**	–		
Add to the above for trench supports appropriate to trench depth (see below)						
Drains/sewers (large bore – not in trenches)						
2t dumper (30% of time)		–	3.24	–		
2.80 m^3/min compressor, 2 tool – 30% of time		–	4.71	–		
disc saw – 30% of time		–	0.41	–		
extra 50ft / 15 m hose – 30% of time		–	0.14	–		
sundry tools – 30% of time		–	0.48	–		
Total Rate/Hour		–	**8.99**	–		

CLASS I: PIPEWORK – PIPES

Item	Gang hours	Labour £	Plant £	Material £	Unit	**Total rate £**
Trench supports						
In addition to the above, the following allowances for close sheeted trench supports are included in the following unit rates, assuming that the ground conditions warrants it:						
ne 1.50 m deep		–	4.75	–	m	
1.50–2.00 m deep		–	6.62	–	m	
2.00–2.50 m deep		–	6.34	–	m	
2.50–3.00 m deep		–	7.80	–	m	
3.00–3.50 m deep		–	9.48	–	m	
3.50–4.00 m deep		–	10.44	–	m	
4.00–4.50 m deep		–	11.68	–	m	
4.50–5.00 m deep		–	20.70	–	m	
5.00–5.50 m deep		–	28.52	–	m	
01 CLAY PIPES						
Field drains to BS 1196, butt joints, nominal bore; excavation and supports, backfilling						
75 mm pipes; in trench, depth						
not in trenches	0.06	4.21	0.07	12.74	m	**17.02**
ne 1.50 m deep	0.09	10.25	5.48	12.74	m	**28.47**
1.50–2.00 m deep	0.13	14.80	7.95	12.74	m	**35.49**
2.00–2.50 m deep	0.18	20.50	11.00	12.74	m	**44.24**
2.50–3.00 m deep	0.23	26.19	14.06	12.74	m	**52.99**
100 mm pipes; in trench, depth						
not in trenches	0.06	4.21	0.07	17.89	m	**22.17**
ne 1.50 m deep	0.10	11.39	6.10	17.89	m	**35.37**
1.50–2.00 m deep	0.14	15.94	8.56	17.89	m	**42.39**
2.00–2.50 m deep	0.19	21.63	11.60	17.89	m	**51.12**
2.50–3.00 m deep	0.24	27.33	14.68	17.89	m	**59.89**
150 mm pipes; in trench, depth						
not in trenches	0.06	4.21	0.07	36.49	m	**40.77**
ne 1.50 m deep	0.11	12.53	6.69	36.49	m	**55.71**
1.50–2.00 m deep	0.15	17.08	9.16	36.49	m	**62.73**
2.00–2.50 m deep	0.20	22.77	12.22	36.49	m	**71.48**
2.50–3.00 m deep	0.25	28.47	15.29	36.49	m	**80.25**
225 mm pipes; in trench, depth						
not in trenches	0.06	4.21	0.07	93.91	m	**98.19**
ne 1.50 m deep	0.13	14.80	7.92	93.91	m	**116.63**
1.50–2.00 m deep	0.17	19.36	10.39	93.91	m	**123.66**
2.00–2.50 m deep	0.22	25.05	13.45	93.91	m	**132.41**
2.50–3.00 m deep	0.27	30.74	16.51	93.91	m	**141.16**

CLASS I: PIPEWORK – PIPES

Item	Gang hours	Labour £	Plant £	Material £	Unit	Total rate £
01 CLAY PIPES – cont						
Vitrified clay perforated field drains to BS EN295, sleeved joints; excavation and supports, backfilling						
100 mm pipes; in trench, depth						
not in trenches	0.25	17.53	0.07	22.32	m	**39.92**
ne 1.50 m deep	0.27	30.74	9.13	22.32	m	**62.20**
1.50–2.00 m deep	0.30	34.16	11.60	22.32	m	**68.09**
2.00–2.50 m deep	0.36	40.99	14.66	22.32	m	**77.97**
2.50–3.00 m deep	0.41	46.68	17.74	22.32	m	**86.75**
extra for bend	0.08	9.11	–	21.51	nr	**30.62**
extra for single junction	0.09	10.25	–	30.65	nr	**40.89**
150 mm pipes; in trench, depth						
not in trenches	0.47	32.95	0.07	40.56	m	**73.58**
ne 1.50 m deep	0.43	48.96	10.36	40.56	m	**99.87**
1.50–2.00 m deep	0.46	52.38	12.22	40.56	m	**105.15**
2.00–2.50 m deep	0.50	56.93	14.66	40.56	m	**112.14**
2.50–3.00 m deep	0.55	62.63	17.74	40.56	m	**120.92**
extra for bend	0.13	14.23	–	31.25	nr	**45.49**
extra for single junction	0.25	28.47	–	31.27	nr	**59.73**
225 mm pipes; in trench, depth						
not in trenches	0.55	38.56	0.07	46.40	m	**85.03**
ne 1.50 m deep	0.48	54.65	10.97	46.40	m	**112.02**
1.50–2.00 m deep	0.52	59.21	13.45	46.40	m	**119.06**
2.00–2.50 m deep	0.85	96.78	15.89	46.40	m	**159.07**
2.50–3.00 m deep	0.60	68.32	18.34	46.40	m	**133.05**
extra for bend	0.12	13.66	–	153.53	nr	**167.20**
extra for single junction	0.20	22.20	–	214.35	nr	**236.55**
300 mm pipes; in trench, depth						
not in trenches	1.20	84.13	0.36	83.57	m	**168.06**
ne 1.50 m deep	0.90	102.48	11.56	83.57	m	**197.61**
1.50–2.00 m deep	0.97	110.45	14.66	83.57	m	**208.68**
2.00–2.50 m deep	1.00	113.86	17.10	83.57	m	**214.54**
2.50–3.00 m deep	1.05	119.56	19.57	83.57	m	**222.70**
extra for bend	0.24	27.33	–	291.61	nr	**318.93**
extra for single junction	0.23	26.19	–	550.20	nr	**576.39**
Vitrified clay pipes to BS EN295, plain ends with push-fit polypropylene flexible couplings; excavation and supports, backfilling						
100 mm pipes; in trenches, depth						
not in trenches	0.15	10.52	0.53	16.75	m	**27.79**
ne 1.50 m deep	0.22	25.05	14.08	16.75	m	**55.88**
1.50–2.00 m deep	0.25	28.47	18.15	16.75	m	**63.37**
2.00–2.50 m deep	0.30	34.16	22.94	16.75	m	**73.85**
2.50–3.00 m deep	0.35	39.85	27.75	16.75	m	**84.35**
3.00–3.50 m deep	0.43	48.96	34.45	16.75	m	**100.16**
3.50–4.00 m deep	0.50	56.93	42.14	16.75	m	**115.82**
4.00–4.50 m deep	0.60	68.32	52.68	16.75	m	**137.75**
4.50–5.00 m deep	0.77	87.68	67.12	16.75	m	**171.54**
5.00–5.50 m deep	0.97	110.45	86.33	16.75	m	**213.53**

CLASS I: PIPEWORK – PIPES

Item	Gang hours	Labour £	Plant £	Material £	Unit	**Total rate £**
extra for bend	0.05	5.69	–	59.93	nr	**65.63**
extra for rest bend	0.06	6.83	–	79.19	nr	**86.02**
extra for single junction; equal	0.09	10.25	–	89.89	nr	**100.13**
extra for saddle; oblique	0.23	26.19	–	58.01	nr	**84.20**
150 mm pipes; in trenches, depth						
not in trenches	0.28	19.63	0.53	27.60	m	**47.76**
ne 1.50 m deep	0.32	36.44	16.23	27.60	m	**80.26**
1.50–2.00 m deep	0.20	22.77	19.12	27.60	m	**69.49**
2.00–2.50 m deep	0.31	35.30	22.94	27.60	m	**85.84**
2.50–3.00 m deep	0.35	39.85	27.75	27.60	m	**95.20**
3.00–3.50 m deep	0.39	44.41	37.31	27.60	m	**109.32**
3.50–4.00 m deep	0.45	51.24	43.11	27.60	m	**121.95**
4.00–4.50 m deep	0.58	66.04	55.54	27.60	m	**149.19**
4.50–5.00 m deep	0.75	85.40	71.88	27.60	m	**184.88**
5.00–5.50 m deep	0.96	109.31	92.06	27.60	m	**228.97**
extra for bend	0.15	17.08	–	93.32	nr	**110.40**
extra for rest bend	0.09	10.25	–	115.51	nr	**125.76**
extra for single junction; equal	0.11	12.53	–	149.29	nr	**161.81**
extra for taper reducer	0.07	7.97	–	130.36	nr	**138.33**
extra for saddle; oblique	0.29	33.02	–	90.56	nr	**123.58**
225 mm pipes; in trenches, depth						
not in trenches	0.83	58.19	0.72	49.26	m	**108.17**
ne 1.50 m deep	0.68	77.43	17.18	49.26	m	**143.87**
1.50–2.00 m deep	0.70	79.71	21.05	49.26	m	**150.02**
2.00–2.50 m deep	0.75	85.40	24.87	49.26	m	**159.53**
2.50–3.00 m deep	0.80	91.09	28.67	49.26	m	**169.03**
3.00–3.50 m deep	0.90	102.48	39.25	49.26	m	**190.99**
3.50–4.00 m deep	0.96	109.31	45.00	49.26	m	**203.57**
4.00–4.50 m deep	1.15	130.94	62.28	49.26	m	**242.49**
4.50–5.00 m deep	1.30	148.02	76.68	49.26	m	**273.97**
5.00–5.50 m deep	1.50	170.80	97.83	49.26	m	**317.89**
extra for bend	0.10	11.39	–	273.72	nr	**285.11**
extra for rest bend	0.11	12.53	–	286.98	nr	**299.51**
extra for single junction; equal	0.16	18.22	–	337.92	nr	**356.14**
extra for taper reducer	0.12	13.66	–	256.54	nr	**270.20**
extra for saddle; oblique	0.36	40.99	–	257.66	nr	**298.65**
300 mm pipes; in trenches, depth						
not in trenches	0.90	63.10	0.91	101.93	m	**165.94**
ne 1.50 m deep	0.72	81.98	18.11	101.93	m	**202.02**
1.50–2.00 m deep	0.77	87.68	22.94	101.93	m	**212.54**
2.00–2.50 m deep	0.82	93.37	26.76	101.93	m	**222.06**
2.50–3.00 m deep	0.85	96.78	30.61	101.93	m	**229.32**
3.00–3.50 m deep	0.95	108.17	40.21	101.93	m	**250.31**
3.50–4.00 m deep	1.05	119.56	49.80	101.93	m	**271.29**
4.00–4.50 m deep	1.23	140.05	67.08	101.93	m	**309.06**
4.50–5.00 m deep	1.45	165.10	86.29	101.93	m	**353.32**
5.00–5.50 m deep	1.65	187.88	107.40	101.93	m	**397.21**
extra for bend	0.15	17.08	–	385.99	nr	**403.06**
extra for rest bend	0.16	18.22	–	717.70	nr	**735.91**
extra for single junction; equal	0.19	21.63	–	828.27	nr	**849.91**
extra for saddle; oblique	0.44	50.10	–	458.38	nr	**508.48**

CLASS I: PIPEWORK – PIPES

Item	Gang hours	Labour £	Plant £	Material £	Unit	Total rate £
01 CLAY PIPES – cont						
Vitrified clay pipes to BS EN295, spigot and socket joints with sealing ring; excavation and supports, backfilling						
100 mm pipes; in trenches, depth						
not in trenches	0.06	4.21	0.60	11.10	m	**15.90**
ne 1.50 m deep	0.15	17.08	14.51	11.10	m	**42.69**
1.50–2.00 m deep	0.19	21.63	18.15	11.10	m	**50.88**
2.00–2.50 m deep	0.24	27.33	22.94	11.10	m	**61.37**
2.50–3.00 m deep	0.29	33.02	27.75	11.10	m	**71.87**
3.00–3.50 m deep	0.36	40.99	34.45	11.10	m	**86.54**
3.50–4.00 m deep	0.44	50.10	42.14	11.10	m	**103.34**
4.00–4.50 m deep	0.55	62.63	52.68	11.10	m	**126.40**
4.50–5.00 m deep	0.70	79.71	67.12	11.10	m	**157.92**
5.00–5.50 m deep	0.90	102.48	86.33	11.10	m	**199.90**
extra for bend	0.05	5.69	–	40.88	nr	**46.58**
extra for rest bend	0.12	13.66	–	34.84	nr	**48.50**
extra for single junction; equal	0.09	10.25	–	37.85	nr	**48.10**
extra for saddle; oblique	0.23	26.19	–	41.42	nr	**67.61**
150 mm pipes; in trenches, depth						
not in trenches	0.06	4.21	0.53	16.91	m	**21.64**
ne 1.50 m deep	0.17	19.36	16.23	16.91	m	**52.49**
1.50–2.00 m deep	0.20	22.77	19.12	16.91	m	**58.80**
2.00–2.50 m deep	0.24	27.33	22.94	16.91	m	**67.17**
2.50–3.00 m deep	0.29	33.02	27.75	16.91	m	**77.67**
3.00–3.50 m deep	0.39	44.41	37.31	16.91	m	**98.62**
3.50–4.00 m deep	0.45	51.24	43.11	16.91	m	**111.25**
4.00–4.50 m deep	0.58	66.04	55.54	16.91	m	**138.49**
4.50–5.00 m deep	0.75	85.40	71.88	16.91	m	**174.18**
5.00–5.50 m deep	0.96	109.31	92.06	16.91	m	**218.27**
extra for bend	0.08	9.11	–	67.03	nr	**76.14**
extra for rest bend	0.09	10.25	–	50.21	nr	**60.46**
extra for single junction; equal	0.11	12.53	–	62.66	nr	**75.18**
extra for double junction; equal	0.13	14.80	–	94.52	nr	**109.32**
extra for taper reducer	0.07	7.97	–	85.69	nr	**93.66**
extra for saddle; oblique	0.29	33.02	–	61.63	nr	**94.65**
225 mm pipes; in trenches, depth						
not in trenches	0.08	5.61	0.72	37.96	m	**44.29**
ne 1.50 m deep	0.18	20.50	17.19	37.96	m	**75.65**
1.50–2.00 m deep	0.22	25.05	21.05	37.96	m	**84.06**
2.00–2.50 m deep	0.26	29.60	24.87	37.96	m	**92.44**
2.50–3.00 m deep	0.30	34.16	28.67	37.96	m	**100.79**
3.00–3.50 m deep	0.41	46.68	39.25	37.96	m	**123.89**
3.50–4.00 m deep	0.47	53.52	44.97	37.96	m	**136.45**
4.00–4.50 m deep	0.65	74.01	62.28	37.96	m	**174.25**
4.50–5.00 m deep	0.80	91.09	76.68	37.96	m	**205.74**
5.00–5.50 m deep	1.02	116.14	97.83	37.96	m	**251.94**
extra for bend	0.10	11.39	–	205.13	nr	**216.51**

CLASS I: PIPEWORK – PIPES

Item	Gang hours	Labour £	Plant £	Material £	Unit	**Total rate £**
extra for rest bend	0.11	12.53	–	217.84	nr	**230.37**
extra for single junction; equal	0.83	94.51	–	180.39	nr	**274.90**
extra for double junction; equal	0.18	20.50	–	288.58	nr	**309.08**
extra for taper reducer	0.12	13.66	–	190.57	nr	**204.23**
extra for saddle; oblique	0.36	40.99	–	229.25	nr	**270.25**
300 mm pipes; in trenches, depth						
not in trenches	0.15	10.52	0.91	41.88	m	**53.31**
ne 1.50 m deep	0.23	26.19	18.11	41.88	m	**86.18**
1.50–2.00 m deep	0.27	30.74	22.94	41.88	m	**95.56**
2.00–2.50 m deep	0.32	36.44	26.76	41.88	m	**105.08**
2.50–3.00 m deep	0.35	39.85	30.61	41.88	m	**112.34**
3.00–3.50 m deep	0.42	47.82	40.21	41.88	m	**129.92**
3.50–4.00 m deep	0.52	59.21	49.80	41.88	m	**150.89**
4.00–4.50 m deep	0.70	79.71	67.08	41.88	m	**188.67**
4.50–5.00 m deep	0.90	102.48	86.29	41.88	m	**230.65**
5.00–5.50 m deep	1.12	127.53	107.40	41.88	m	**276.81**
extra for bend	0.15	17.08	–	378.32	nr	**395.40**
extra for rest bend	0.16	18.22	–	575.77	nr	**593.99**
extra for single junction; equal	0.19	21.63	–	575.77	nr	**597.40**
extra for double junction; equal	0.21	23.91	–	662.41	nr	**686.32**
extra for taper reducer	0.15	17.08	–	376.01	nr	**393.09**
extra for saddle; oblique	0.44	50.10	–	398.90	nr	**449.00**
400 mm pipes; in trenches, depth						
not in trenches	0.10	7.01	0.91	152.64	m	**160.56**
ne 1.50 m deep	0.23	26.19	21.93	152.64	m	**200.76**
1.50–2.00 m deep	0.29	33.02	27.73	152.64	m	**213.39**
2.00–2.50 m deep	0.32	36.44	30.59	152.64	m	**219.66**
2.50–3.00 m deep	0.38	43.27	36.37	152.64	m	**232.28**
3.00–3.50 m deep	0.46	52.38	44.04	152.64	m	**249.06**
3.50–4.00 m deep	0.58	66.04	55.53	152.64	m	**274.21**
4.00–4.50 m deep	0.75	85.40	71.84	152.64	m	**309.88**
4.50–5.00 m deep	0.95	108.17	91.05	152.64	m	**351.86**
5.00–5.50 m deep	1.20	136.64	115.07	152.64	m	**404.35**
extra for bend; 90 degree	0.24	27.33	–	1038.43	nr	**1065.76**
extra for bend; 45 degree	0.24	27.33	–	1039.45	nr	**1066.78**
extra for bend; 22.5 degree	0.24	27.33	–	1039.48	nr	**1066.80**
450 mm pipes; in trenches, depth						
not in trenches	0.23	16.12	2.06	170.80	m	**188.99**
ne 1.50 m deep	0.23	26.19	21.93	170.80	m	**218.92**
1.50–2.00 m deep	0.30	34.16	28.65	170.80	m	**233.62**
2.00–2.50 m deep	0.32	36.44	30.59	170.80	m	**237.83**
2.50–3.00 m deep	0.38	43.27	36.37	170.80	m	**250.44**
3.00–3.50 m deep	0.47	53.52	44.97	170.80	m	**269.29**
3.50–4.00 m deep	0.60	68.32	57.46	170.80	m	**296.58**
4.00–4.50 m deep	0.77	87.68	73.78	170.80	m	**332.26**
4.50–5.00 m deep	0.97	110.45	92.99	170.80	m	**374.24**
5.00–5.50 m deep	1.20	136.64	115.07	170.80	m	**422.51**
extra for bend; 90 degree	0.29	33.02	–	1368.79	nr	**1401.81**
extra for bend; 45 degree	0.29	33.02	–	1368.79	nr	**1401.81**
extra for bend; 22.5 degree	0.29	33.02	–	1368.79	nr	**1401.81**

CLASS I: PIPEWORK – PIPES

Item	Gang hours	Labour £	Plant £	Material £	Unit	**Total rate £**
02 CONCRETE PIPES						
Concrete porous pipes to BS 5911; excavation and supports, backfilling						
150 mm pipes; in trench, depth						
not in trenches	0.06	4.21	0.49	8.36	m	**13.05**
ne 1.50 m deep	0.17	19.36	10.36	8.36	m	**38.08**
1.50–2.00 m deep	0.20	22.77	12.22	8.36	m	**43.35**
2.00–2.50 m deep	0.24	27.33	14.66	8.36	m	**50.35**
2.50–3.00 m deep	0.29	33.02	17.74	8.36	m	**59.12**
3.00–3.50 m deep	0.39	44.41	23.86	8.36	m	**76.63**
3.50–4.00 m deep	0.45	51.24	27.58	8.36	m	**87.18**
4.00–4.50 m deep	0.58	66.04	35.55	8.36	m	**109.95**
4.50–5.00 m deep	0.75	85.40	46.01	8.36	m	**139.77**
5.00–5.50 m deep	0.96	109.31	58.94	8.36	m	**176.61**
225 mm pipes; in trench, depth						
not in trenches	0.08	5.61	0.29	9.14	m	**15.03**
ne 1.50 m deep	0.18	20.50	10.97	9.14	m	**40.60**
1.50–2.00 m deep	0.22	25.05	13.45	9.14	m	**47.64**
2.00–2.50 m deep	0.26	29.60	15.89	9.14	m	**54.64**
2.50–3.00 m deep	0.30	34.16	18.34	9.14	m	**61.63**
3.00–3.50 m deep	0.41	46.68	25.10	9.14	m	**80.92**
3.50–4.00 m deep	0.47	53.52	28.79	9.14	m	**91.45**
4.00–4.50 m deep	0.65	74.01	39.85	9.14	m	**123.00**
4.50–5.00 m deep	0.80	91.09	49.09	9.14	m	**149.32**
5.00–5.50 m deep	1.02	116.14	62.63	9.14	m	**187.92**
300 mm pipes; in trench, depth						
not in trenches	0.10	7.01	0.36	9.50	m	**16.87**
ne 1.50 m deep	0.19	21.63	11.56	9.50	m	**42.70**
1.50–2.00 m deep	0.24	27.33	14.66	9.50	m	**51.49**
2.00–2.50 m deep	0.28	31.88	17.10	9.50	m	**58.48**
2.50–3.00 m deep	0.32	36.44	19.57	9.50	m	**65.50**
3.00–3.50 m deep	0.42	47.82	25.71	9.50	m	**83.03**
3.50–4.00 m deep	0.52	59.21	31.86	9.50	m	**100.57**
4.00–4.50 m deep	0.70	79.71	42.92	9.50	m	**132.13**
4.50–5.00 m deep	0.90	102.48	55.23	9.50	m	**167.21**
5.00–5.50 m deep	1.12	127.53	68.77	9.50	m	**205.79**
Concrete pipes with rebated flexible joints to BS 5911 Class 120; excavation and supports, backfilling						
300 mm pipes; in trenches, depth						
not in trenches	0.12	8.41	0.53	30.31	m	**39.25**
ne 1.50 m deep	0.22	25.05	21.01	30.31	m	**76.36**
1.50–2.00 m deep	0.26	29.60	24.87	30.31	m	**84.78**
2.00–2.50 m deep	0.30	34.16	28.66	30.31	m	**93.12**
2.50–3.00 m deep	0.34	38.71	32.54	30.31	m	**101.56**
3.00–3.50 m deep	0.44	50.10	42.11	30.31	m	**122.51**
3.50–4.00 m deep	0.56	63.76	53.63	30.31	m	**147.70**

CLASS I: PIPEWORK – PIPES

Item	Gang hours	Labour £	Plant £	Material £	Unit	**Total rate £**
4.00–4.50 m deep	0.73	83.12	69.95	30.31	m	**183.37**
4.50–5.00 m deep	0.92	104.76	88.19	30.31	m	**223.25**
5.00–5.50 m deep	1.19	135.50	114.10	30.31	m	**279.91**
5.50–6.00 m deep	1.42	161.69	136.22	30.31	m	**328.22**
extra for bend	0.08	9.11	6.52	302.99	nr	**318.62**
375 mm pipes; in trenches, depth						
not in trenches	0.15	10.52	1.34	37.52	m	**49.37**
ne 1.50 m deep	0.24	27.33	22.89	37.52	m	**87.74**
1.50–2.00 m deep	0.29	33.02	27.73	37.52	m	**98.27**
2.00–2.50 m deep	0.33	37.58	31.55	37.52	m	**106.64**
2.50–3.00 m deep	0.38	43.27	36.37	37.52	m	**117.15**
3.00–3.50 m deep	0.46	52.38	44.04	37.52	m	**133.94**
3.50–4.00 m deep	0.58	66.04	55.53	37.52	m	**159.08**
4.00–4.50 m deep	0.75	85.40	71.84	37.52	m	**194.76**
4.50–5.00 m deep	0.95	108.17	91.05	37.52	m	**236.74**
5.00–5.50 m deep	1.20	136.64	115.07	37.52	m	**289.23**
5.50–6.00 m deep	1.45	165.10	139.13	37.52	m	**341.75**
extra for bend	0.10	11.39	9.10	375.07	nr	**395.56**
450 mm pipes; in trenches, depth						
not in trenches	0.17	11.92	1.53	44.27	m	**57.73**
ne 1.50 m deep	0.25	28.47	23.86	44.27	m	**96.60**
1.50–2.00 m deep	0.31	35.30	29.62	44.27	m	**109.19**
2.00–2.50 m deep	0.35	39.85	33.48	44.27	m	**117.61**
2.50–3.00 m deep	0.40	45.55	38.26	44.27	m	**128.08**
3.00–3.50 m deep	0.48	54.65	45.98	44.27	m	**144.91**
3.50–4.00 m deep	0.63	71.73	60.33	44.27	m	**176.33**
4.00–4.50 m deep	0.80	91.09	76.64	44.27	m	**212.01**
4.50–5.00 m deep	1.00	113.86	95.85	44.27	m	**253.99**
5.00–5.50 m deep	1.25	142.33	119.88	44.27	m	**306.48**
5.50–6.00 m deep	1.51	171.94	144.87	44.27	m	**361.08**
extra for bend	0.13	14.80	13.04	442.85	nr	**470.69**
525 mm pipes; in trenches, depth						
not in trenches	0.20	14.02	1.80	58.13	m	**73.95**
ne 1.50 m deep	0.27	30.74	25.74	58.13	m	**114.61**
1.50–2.00 m deep	0.33	37.58	31.55	58.13	m	**127.26**
2.00–2.50 m deep	0.37	42.13	35.38	58.13	m	**135.63**
2.50–3.00 m deep	0.42	47.82	40.19	58.13	m	**146.14**
3.00–3.50 m deep	0.49	55.79	46.90	58.13	m	**160.82**
3.50–4.00 m deep	0.65	74.01	62.26	58.13	m	**194.40**
4.00–4.50 m deep	0.83	94.51	79.51	58.13	m	**232.14**
4.50–5.00 m deep	1.03	117.28	98.72	58.13	m	**274.13**
5.00–5.50 m deep	1.28	145.75	122.75	58.13	m	**326.62**
5.50–6.00 m deep	1.55	176.49	148.71	58.13	m	**383.32**
extra for bend	0.16	18.22	16.94	581.18	nr	**616.34**
750 mm pipes; in trenches, depth						
not in trenches	0.22	15.42	1.99	143.80	m	**161.22**
ne 1.50 m deep	0.30	29.78	35.87	143.80	m	**209.46**
1.50–2.00 m deep	0.36	35.74	43.15	143.80	m	**222.69**
2.00–2.50 m deep	0.41	40.70	49.15	143.80	m	**233.66**
2.50–3.00 m deep	0.47	46.66	56.35	143.80	m	**246.81**
3.00–3.50 m deep	0.58	57.58	69.56	143.80	m	**270.94**
3.50–4.00 m deep	0.80	79.42	96.03	143.80	m	**319.25**

CLASS I: PIPEWORK – PIPES

Item	Gang hours	Labour £	Plant £	Material £	Unit	Total rate £
02 CONCRETE PIPES – cont						
Concrete pipes with rebated flexible joints to BS 5911 Class 120 – cont						
750 mm pipes – cont						
4.00–4.50 m deep	1.05	104.24	126.08	143.80	m	**374.13**
4.50–5.00 m deep	1.30	129.06	156.17	143.80	m	**429.04**
5.00–5.50 m deep	1.55	153.88	186.23	143.80	m	**483.92**
5.50–6.00 m deep	1.82	180.69	218.80	143.80	m	**543.29**
extra for bends	0.24	23.83	54.29	1438.18	nr	**1516.30**
900 mm pipes; in trenches, depth						
not in trenches	0.25	17.53	2.25	192.91	m	**212.69**
ne 1.50 m deep	0.33	32.76	39.50	192.91	m	**265.17**
1.50–2.00 m deep	0.40	39.71	47.94	192.91	m	**280.56**
2.00–2.50 m deep	0.46	45.67	55.15	192.91	m	**293.73**
2.50–3.00 m deep	0.53	52.62	63.56	192.91	m	**309.09**
3.00–3.50 m deep	0.70	69.50	83.99	192.91	m	**346.40**
3.50–4.00 m deep	0.92	91.34	110.43	192.91	m	**394.68**
4.00–4.50 m deep	1.20	119.13	144.08	192.91	m	**456.13**
4.50–5.00 m deep	1.50	148.92	180.20	192.91	m	**522.03**
5.00–5.50 m deep	1.80	178.70	216.28	192.91	m	**587.90**
5.50–6.00 m deep	2.10	208.49	252.46	192.91	m	**653.86**
extra for bends	0.39	38.72	88.76	1787.98	nr	**1915.46**
1200 mm pipes; in trenches, depth						
not in trenches	0.25	17.53	2.25	330.10	m	**349.88**
ne 1.50 m deep	0.46	45.67	55.06	330.10	m	**430.82**
1.50–2.00 m deep	0.53	52.62	63.53	330.10	m	**446.24**
2.00–2.50 m deep	0.60	59.57	71.91	330.10	m	**461.57**
2.50–3.00 m deep	0.70	69.50	83.96	330.10	m	**483.55**
3.00–3.50 m deep	0.85	84.39	101.98	330.10	m	**516.46**
3.50–4.00 m deep	1.12	111.19	134.44	330.10	m	**575.73**
4.00–4.50 m deep	1.45	143.95	174.10	330.10	m	**648.15**
4.50–5.00 m deep	1.75	173.74	210.19	330.10	m	**714.03**
5.00–5.50 m deep	2.05	203.52	246.33	330.10	m	**779.95**
5.50–6.00 m deep	2.36	234.30	283.69	330.10	m	**848.09**
extra for bends	0.51	50.63	116.32	3300.75	nr	**3467.70**
1500 mm pipes; in trenches, depth						
not in trenches	0.35	24.54	3.14	584.64	m	**612.32**
ne 1.50 m deep	0.60	59.57	71.79	584.64	m	**716.00**
1.50–2.00 m deep	0.70	69.50	83.91	584.64	m	**738.05**
2.00–2.50 m deep	0.81	80.42	97.09	584.64	m	**762.14**
2.50–3.00 m deep	0.92	91.34	110.32	584.64	m	**786.29**
3.00–3.50 m deep	1.05	104.24	125.97	584.64	m	**814.85**
3.50–4.00 m deep	1.27	126.08	152.43	584.64	m	**863.16**
4.00–4.50 m deep	1.70	168.77	204.13	584.64	m	**957.54**
4.50–5.00 m deep	2.05	203.52	246.25	584.64	m	**1034.41**
5.00–5.50 m deep	2.40	238.27	288.38	584.64	m	**1111.29**
5.50–6.00 m deep	2.75	273.02	327.57	589.43	m	**1190.03**
extra for bends	0.63	62.55	143.69	5846.02	nr	**6052.25**

CLASS I: PIPEWORK – PIPES

Item	Gang hours	Labour £	Plant £	Material £	Unit	**Total rate £**
1800 mm pipes; in trenches, depth						
not in trenches	0.40	28.04	3.59	813.15	m	**844.79**
ne 1.50 m deep	0.77	76.44	92.14	813.15	m	**981.73**
1.50–2.00 m deep	0.91	90.34	109.05	813.15	m	**1012.54**
2.00–2.50 m deep	1.03	102.26	123.43	813.15	m	**1038.84**
2.50–3.00 m deep	1.12	111.19	134.30	813.15	m	**1058.64**
3.00–3.50 m deep	1.20	119.13	143.95	813.15	m	**1076.24**
3.50–4.00 m deep	1.52	150.90	182.45	813.15	m	**1146.50**
4.00–4.50 m deep	2.00	198.56	240.13	813.15	m	**1251.84**
4.50–5.00 m deep	2.40	238.27	288.28	813.15	m	**1339.70**
5.00–5.50 m deep	2.80	277.98	336.44	813.15	m	**1427.57**
5.50–6.00 m deep	3.15	312.73	378.65	813.15	m	**1504.53**
extra for bends	0.77	76.44	175.64	8131.39	nr	**8383.48**
2100 mm pipes; in trenches, depth						
not in trenches	1.60	112.17	4.05	882.54	m	**998.76**
ne 1.50 m deep	1.84	182.67	117.24	882.54	m	**1182.45**
1.50–2.00 m deep	2.00	198.56	135.44	882.54	m	**1216.54**
2.00–2.50 m deep	2.08	206.50	147.25	882.54	m	**1236.29**
2.50–3.00 m deep	2.15	213.45	155.90	882.54	m	**1251.89**
3.00–3.50 m deep	2.35	233.31	179.96	882.54	m	**1295.81**
3.50–4.00 m deep	2.70	268.05	218.48	882.54	m	**1369.08**
4.00–4.50 m deep	3.20	317.69	282.14	882.54	m	**1482.38**
4.50–5.00 m deep	3.65	362.37	336.33	882.54	m	**1581.24**
5.00–5.50 m deep	4.05	402.08	384.50	882.54	m	**1669.13**
5.50–6.00 m deep	4.40	436.83	426.73	882.54	m	**1746.10**
extra for bends	0.89	88.36	203.14	9787.08	nr	**10078.58**
03 IRON PIPES						
Cast iron pipes to BS 437 plain ended pipe with Timesaver mechanical coupling joints; excavation and supports, backfilling						
75 mm pipes; in trenches, depth						
not in trenches	0.12	8.41	1.08	33.40	m	**42.89**
ne 1.50 m deep	0.19	21.63	17.84	33.40	m	**72.87**
1.50–2.00 m deep	0.21	23.91	20.08	33.40	m	**77.39**
2.00–2.50 m deep	0.28	31.88	26.76	33.40	m	**92.04**
2.50–3.00 m deep	0.34	38.71	32.54	33.40	m	**104.65**
3.00–3.50 m deep	0.43	48.96	41.14	33.40	m	**123.50**
3.50–4.00 m deep	0.55	62.63	52.66	33.40	m	**148.69**
4.00–4.50 m deep	0.71	80.84	68.01	33.40	m	**182.25**
4.50–5.00 m deep	0.89	101.34	85.32	33.40	m	**220.06**
5.00–5.50 m deep	1.15	130.94	110.27	33.40	m	**274.61**
extra for bend; 87.5 degree	0.31	35.30	2.63	69.45	nr	**107.37**
extra for bend; 45 degree	0.31	35.30	2.63	48.12	nr	**86.05**
extra for single junction; equal	0.48	54.65	5.22	72.32	nr	**132.20**
extra for taper reducer	0.27	30.74	2.63	56.56	nr	**89.94**

CLASS I: PIPEWORK – PIPES

Item	Gang hours	Labour £	Plant £	Material £	Unit	Total rate £
03 IRON PIPES – cont						
Cast iron pipes to BS 437 plain ended pipe with Timesaver mechanical coupling joints – cont						
100 mm pipes; in trenches, depth						
not in trenches	0.13	9.11	0.46	41.29	m	**50.86**
ne 1.50 m deep	0.21	23.91	20.04	41.29	m	**85.24**
1.50–2.00 m deep	0.23	26.19	21.97	41.29	m	**89.45**
2.00–2.50 m deep	0.30	34.16	28.65	41.29	m	**104.10**
2.50–3.00 m deep	0.37	42.13	35.40	41.29	m	**118.82**
3.00–3.50 m deep	0.48	54.65	45.94	41.29	m	**141.88**
3.50–4.00 m deep	0.60	68.32	57.46	41.29	m	**167.07**
4.00–4.50 m deep	0.75	85.40	71.84	41.29	m	**198.52**
4.50–5.00 m deep	0.95	108.17	91.05	41.29	m	**240.51**
5.00–5.50 m deep	1.22	138.91	117.01	41.29	m	**297.21**
extra for bend; 87.5 degree	0.38	43.27	3.97	97.68	nr	**144.92**
extra for bend; 45 degree	0.38	43.27	3.97	60.27	nr	**107.51**
extra for bend; long radius	0.38	43.27	3.97	101.71	nr	**148.95**
extra for single junction; equal	0.59	67.18	6.54	158.29	nr	**232.01**
extra for double junction; equal	0.80	91.09	10.44	131.76	nr	**233.30**
extra for taper reducer	0.40	45.55	6.54	67.21	nr	**119.30**
150 mm pipes; in trenches, depth						
not in trenches	0.18	12.62	0.46	21.85	m	**34.93**
ne 1.50 m deep	0.26	29.60	22.89	21.85	m	**74.35**
1.50–2.00 m deep	0.31	35.30	26.76	21.85	m	**83.92**
2.00–2.50 m deep	0.37	42.13	32.56	21.85	m	**96.54**
2.50–3.00 m deep	0.43	48.96	38.26	21.85	m	**109.07**
3.00–3.50 m deep	0.57	64.90	51.70	21.85	m	**138.45**
3.50–4.00 m deep	0.61	69.46	58.43	21.85	m	**149.74**
4.00–4.50 m deep	0.79	89.95	75.67	21.85	m	**187.48**
4.50–5.00 m deep	1.05	119.56	100.66	21.85	m	**242.07**
5.00–5.50 m deep	1.32	150.30	126.58	21.85	m	**298.74**
extra for bend; 87.5 degree	0.56	63.76	5.21	154.60	nr	**223.57**
extra for bend; 45 degree	0.75	85.40	5.21	89.73	nr	**180.34**
extra for bend; long radius	0.67	76.29	5.21	237.35	nr	**318.85**
extra for single junction; equal	1.03	117.28	10.42	182.40	nr	**310.09**
extra for taper reducer	0.56	63.76	10.42	123.22	nr	**197.40**
225 mm pipes; in trenches, depth						
not in trenches	0.15	10.52	0.46	170.51	m	**181.48**
ne 1.50 m deep	0.25	28.47	23.86	170.51	m	**222.83**
1.50–2.00 m deep	0.30	34.16	28.65	170.51	m	**233.32**
2.00–2.50 m deep	0.35	39.85	33.44	170.51	m	**243.81**
2.50–3.00 m deep	0.41	46.68	39.23	170.51	m	**256.42**
3.00–3.50 m deep	0.54	61.49	51.70	170.51	m	**283.69**
3.50–4.00 m deep	0.61	69.46	58.43	170.51	m	**298.40**
4.00–4.50 m deep	0.77	87.68	73.78	170.51	m	**331.96**
4.50–5.00 m deep	1.02	116.14	97.79	170.51	m	**384.44**
5.00–5.50 m deep	1.30	148.02	124.68	170.51	m	**443.22**
extra for bend; 87.5 degree	0.77	87.68	5.21	496.53	nr	**589.42**
extra for bend; 45 degree	0.77	87.68	5.21	494.97	nr	**587.85**
extra for single junction; equal	1.21	137.78	11.73	1059.81	nr	**1209.31**
extra for taper reducer	0.77	87.68	11.73	344.75	nr	**444.15**

CLASS I: PIPEWORK – PIPES

Item	Gang hours	Labour £	Plant £	Material £	Unit	**Total rate £**
Ductile iron pipes to BS 4772, Tyton joints; excavation and supports, backfilling						
100 mm pipes; in trenches, depth						
not in trenches	0.10	7.01	0.46	84.21	m	**91.68**
ne 1.50 m deep	0.19	21.63	17.19	84.21	m	**123.04**
1.50–2.00 m deep	0.20	22.77	19.12	84.21	m	**126.10**
2.00–2.50 m deep	0.26	29.60	24.87	84.21	m	**138.69**
2.50–3.00 m deep	0.32	36.44	30.61	84.21	m	**151.25**
3.00–3.50 m deep	0.42	47.82	40.21	84.21	m	**172.25**
3.50–4.00 m deep	0.53	60.35	50.77	84.21	m	**195.33**
4.00–4.50 m deep	0.66	75.15	63.25	84.21	m	**222.61**
4.50–5.00 m deep	0.84	95.65	80.52	84.21	m	**260.37**
5.00–5.50 m deep	1.08	122.97	103.57	84.21	m	**310.75**
extra for bend; 90 degree	0.38	43.27	26.76	156.50	nr	**226.53**
extra for single junction; equal	0.60	68.32	7.79	336.68	nr	**412.79**
150 mm pipes; in trenches, depth						
not in trenches	0.11	7.71	0.46	105.57	m	**113.74**
ne 1.50 m deep	0.21	23.91	20.04	105.57	m	**149.53**
1.50–2.00 m deep	0.25	28.47	23.91	105.57	m	**157.94**
2.00–2.50 m deep	0.30	34.16	28.65	105.57	m	**168.39**
2.50–3.00 m deep	0.37	42.13	35.40	105.57	m	**183.10**
3.00–3.50 m deep	0.49	55.79	46.90	105.57	m	**208.27**
3.50–4.00 m deep	0.55	62.63	52.66	105.57	m	**220.86**
4.00–4.50 m deep	0.72	81.98	68.98	105.57	m	**256.53**
4.50–5.00 m deep	0.94	107.03	90.08	105.57	m	**302.69**
5.00–5.50 m deep	1.19	135.50	114.10	105.57	m	**355.17**
extra for bend; 90 degree	0.57	64.90	5.21	240.88	nr	**310.99**
extra for single junction; equal	0.88	100.20	10.42	512.92	nr	**623.53**
250 mm pipes; in trenches, depth						
not in trenches	0.16	11.22	0.46	138.48	m	**150.15**
ne 1.50 m deep	0.24	27.33	22.89	138.48	m	**188.70**
1.50–2.00 m deep	0.30	34.16	28.65	138.48	m	**201.29**
2.00–2.50 m deep	0.35	39.85	33.48	138.48	m	**211.81**
2.50–3.00 m deep	0.42	47.82	40.19	138.48	m	**226.49**
3.00–3.50 m deep	0.56	63.76	53.59	138.48	m	**255.83**
3.50–4.00 m deep	0.64	72.87	61.29	138.48	m	**272.64**
4.00–4.50 m deep	0.81	92.23	77.61	138.48	m	**308.32**
4.50–5.00 m deep	1.02	116.14	97.79	138.48	m	**352.41**
5.00–5.50 m deep	1.27	144.61	121.78	138.48	m	**404.86**
extra for bend; 90 degree	0.96	109.31	7.79	708.25	nr	**825.35**
extra for single junction; equal	1.32	150.30	13.04	1478.39	nr	**1641.73**
400 mm pipes; in trenches, depth						
not in trenches	0.24	16.83	0.98	179.30	m	**197.10**
ne 1.50 m deep	0.33	37.58	31.49	179.30	m	**248.36**
1.50–2.00 m deep	0.43	48.96	41.09	179.30	m	**269.35**
2.00–2.50 m deep	0.48	54.65	45.88	179.30	m	**279.83**
2.50–3.00 m deep	0.57	64.90	54.53	179.30	m	**298.73**
3.00–3.50 m deep	0.71	80.84	67.94	179.30	m	**328.07**
3.50–4.00 m deep	0.91	103.62	87.14	179.30	m	**370.05**
4.00–4.50 m deep	1.16	132.08	111.13	179.30	m	**422.51**
4.50–5.00 m deep	1.47	167.38	140.89	179.30	m	**487.57**
5.00–5.50 m deep	1.77	201.54	169.74	179.30	m	**550.58**

CLASS I: PIPEWORK – PIPES

Item	Gang hours	Labour £	Plant £	Material £	Unit	**Total rate £**
03 IRON PIPES – cont						
Ductile iron pipes to BS 4772, Tyton joints – cont						
600 mm pipes; in trenches, depth						
not in trenches	0.34	23.84	1.34	228.38	m	**253.56**
ne 1.50 m deep	0.47	53.52	44.82	228.38	m	**326.72**
1.50–2.00 m deep	0.55	62.63	52.56	228.38	m	**343.57**
2.00–2.50 m deep	0.66	75.15	63.11	228.38	m	**366.64**
2.50–3.00 m deep	0.78	88.81	74.63	228.38	m	**391.82**
3.00–3.50 m deep	0.89	101.34	85.18	228.38	m	**414.90**
3.50–4.00 m deep	1.09	124.11	104.40	228.38	m	**456.90**
4.00–4.50 m deep	1.37	155.99	131.26	228.38	m	**515.64**
4.50–5.00 m deep	1.70	193.57	162.97	228.38	m	**584.92**
5.00–5.50 m deep	2.03	231.14	194.66	228.38	m	**654.18**
04 STEEL PIPES						
Carbon steel pipes to BS EN 10126; welded joints; (for protection and lining refer to manufacturer); excavation and supports, backfilling						
100 mm pipes; in trenches, depth						
not in trenches	0.07	4.91	0.62	43.70	m	**49.23**
ne 1.50 m deep	0.15	17.08	14.30	43.70	m	**75.08**
1.50–2.00 m deep	0.17	19.36	16.26	43.70	m	**79.32**
2.00–2.50 m deep	0.22	25.05	21.05	43.70	m	**89.80**
2.50–3.00 m deep	0.27	30.74	25.81	43.70	m	**100.26**
3.00–3.50 m deep	0.35	39.85	33.48	43.70	m	**117.04**
3.50–4.00 m deep	0.44	50.10	42.14	43.70	m	**135.94**
4.00–4.50 m deep	0.55	62.63	52.68	43.70	m	**159.01**
4.50–5.00 m deep	0.70	79.71	87.46	43.70	m	**210.87**
5.00–5.50 m deep	0.87	99.06	83.42	43.70	m	**226.18**
extra for bend; 45°	0.07	7.97	3.90	56.26	nr	**68.13**
extra for bend; 90°	0.07	7.97	3.90	58.64	nr	**70.51**
extra for single junction; equal	0.11	12.53	6.52	97.51	nr	**116.56**
150 mm pipes; in trenches, depth						
not in trenches	0.07	4.91	0.62	56.72	m	**62.25**
ne 1.50 m deep	0.17	19.36	16.23	56.72	m	**92.30**
1.50–2.00 m deep	0.20	22.77	19.12	56.72	m	**98.61**
2.00–2.50 m deep	0.24	27.33	22.94	56.72	m	**106.99**
2.50–3.00 m deep	0.29	33.02	27.75	56.72	m	**117.49**
3.00–3.50 m deep	0.39	44.41	37.31	56.72	m	**138.44**
3.50–4.00 m deep	0.44	50.10	42.14	56.72	m	**148.96**
4.00–4.50 m deep	0.57	64.90	54.62	56.72	m	**176.24**
4.50–5.00 m deep	0.72	81.98	89.94	56.72	m	**228.64**
5.00–5.50 m deep	0.90	102.48	86.33	56.72	m	**245.53**
extra for bend; 45°	0.18	20.50	5.21	92.55	nr	**118.25**
extra for bend; 90°	0.09	10.25	5.21	104.36	nr	**119.82**
extra for single junction; equal	0.16	18.22	10.42	167.76	nr	**196.39**

CLASS I: PIPEWORK – PIPES

Item	Gang hours	Labour £	Plant £	Material £	Unit	**Total rate £**
200 mm pipes; in trenches, depth						
not in trenches	0.09	6.31	13.09	88.33	m	**107.73**
ne 1.50 m deep	0.18	20.50	17.19	88.33	m	**126.01**
1.50–2.00 m deep	0.21	23.91	20.08	88.33	m	**132.32**
2.00–2.50 m deep	0.25	28.47	23.93	88.33	m	**140.73**
2.50–3.00 m deep	0.30	34.16	28.67	88.33	m	**151.16**
3.00–3.50 m deep	0.40	45.55	38.28	88.33	m	**172.15**
3.50–4.00 m deep	0.46	52.38	44.08	88.33	m	**184.78**
4.00–4.50 m deep	0.59	67.18	56.51	88.33	m	**212.02**
4.50–5.00 m deep	0.74	84.26	70.95	88.33	m	**243.54**
5.00–5.50 m deep	0.92	104.76	88.22	88.33	m	**281.31**
extra for bend; 45°	0.12	13.66	5.21	146.53	nr	**165.40**
extra for bend; 90°	0.12	13.66	5.21	152.76	nr	**171.64**
extra for single junction; equal	0.21	23.91	10.36	256.96	nr	**291.22**
250 mm pipes; in trenches, depth						
not in trenches	0.10	7.01	0.91	125.48	m	**133.40**
ne 1.50 m deep	0.18	20.50	23.46	125.48	m	**169.43**
1.50–2.00 m deep	0.22	25.05	27.97	125.48	m	**178.49**
2.00–2.50 m deep	0.26	29.60	31.67	125.48	m	**186.75**
2.50–3.00 m deep	0.31	35.30	39.35	125.48	m	**200.12**
3.00–3.50 m deep	0.41	46.68	52.07	125.48	m	**224.23**
3.50–4.00 m deep	0.47	53.52	58.18	125.48	m	**237.17**
4.00–4.50 m deep	0.60	68.32	72.39	125.48	m	**266.18**
4.50–5.00 m deep	0.75	85.40	93.67	125.48	m	**304.55**
5.00–5.50 m deep	0.94	107.03	117.40	125.48	m	**349.91**
extra for bend; 45°	0.13	14.80	5.21	265.42	nr	**285.43**
extra for bend; 90°	0.13	14.80	5.21	289.03	nr	**309.04**
extra for single junction; equal	0.23	26.19	10.42	385.23	nr	**421.84**
300 mm pipes; in trenches, depth						
not in trenches	0.11	7.71	0.98	154.59	m	**163.28**
ne 1.50 m deep	0.20	22.77	19.08	154.59	m	**196.44**
1.50–2.00 m deep	0.25	28.47	23.91	154.59	m	**206.96**
2.00–2.50 m deep	0.28	31.88	26.76	154.59	m	**213.23**
2.50–3.00 m deep	0.34	38.71	32.54	154.59	m	**225.84**
3.00–3.50 m deep	0.44	50.10	42.11	154.59	m	**246.79**
3.50–4.00 m deep	0.53	60.35	50.77	154.59	m	**265.70**
4.00–4.50 m deep	0.68	77.43	65.15	154.59	m	**297.16**
4.50–5.00 m deep	0.86	97.92	82.45	154.59	m	**334.96**
5.00–5.50 m deep	1.06	120.70	101.67	154.59	m	**376.95**
extra for bend; 45°	0.14	15.94	7.79	309.77	nr	**333.50**
extra for bend; 90°	0.38	43.27	7.79	366.20	nr	**417.26**
extra for single junction; equal	0.25	28.47	16.94	519.04	nr	**564.44**

CLASS I: PIPEWORK – PIPES

Item	Gang hours	Labour £	Plant £	Material £	Unit	**Total rate £**
05 POLYVINYL CHLORIDE PIPES						
Unplasticized pvc perforated pipes; ring seal sockets; excavation and supports, backfilling; 6 m pipe lengths unless stated otherwise						
82 mm pipes; in trench, depth						
not in trenches; 3.00 m pipe lengths	0.06	4.21	0.21	13.38	m	**17.80**
ne 1.50 m deep; 3.00 m pipe lengths	0.10	11.39	6.10	13.38	m	**30.86**
1.50–2.00 m deep	0.13	14.80	7.95	13.38	m	**36.13**
2.00–2.50 m deep	0.16	18.22	9.77	13.38	m	**41.37**
2.50–3.00 m deep	0.19	21.63	11.61	13.38	m	**46.63**
3.00–3.50 m deep	0.22	25.05	13.47	13.38	m	**51.90**
3.50–4.00 m deep	0.25	28.47	15.32	13.38	m	**57.17**
4.00–4.50 m deep	0.28	31.88	17.17	13.38	m	**62.43**
4.50–5.00 m deep	0.32	36.44	19.63	13.38	m	**69.45**
5.00–5.50 m deep	0.35	39.85	21.48	13.38	m	**74.72**
110 mm pipes; in trench, depth						
not in trenches	0.06	4.21	0.11	18.48	m	**22.79**
ne 1.50 m deep	0.10	11.39	6.10	18.48	m	**35.96**
1.50–2.00 m deep	0.13	14.80	7.95	18.48	m	**41.23**
2.00–2.50 m deep	0.16	18.22	9.77	18.48	m	**46.47**
2.50–3.00 m deep	0.19	21.63	11.61	18.48	m	**51.73**
3.00–3.50 m deep	0.22	25.05	13.47	18.48	m	**57.00**
3.50–4.00 m deep	0.25	28.47	15.32	18.48	m	**62.27**
4.00–4.50 m deep	0.28	31.88	17.17	18.48	m	**67.53**
4.50–5.00 m deep	0.32	36.44	19.63	18.48	m	**74.55**
5.00–5.50 m deep	0.36	40.99	22.10	18.48	m	**81.57**
160 mm pipes; in trench, depth						
not in trenches	0.06	4.21	0.21	47.34	m	**51.76**
ne 1.50 m deep	0.11	12.53	6.69	47.34	m	**66.56**
1.50–2.00 m deep	0.14	15.94	8.56	47.34	m	**71.84**
2.00–2.50 m deep	0.18	20.50	11.00	47.34	m	**78.84**
2.50–3.00 m deep	0.22	25.05	13.46	47.34	m	**85.85**
3.00–3.50 m deep	0.23	26.19	14.07	47.34	m	**87.60**
3.50–4.00 m deep	0.26	29.60	15.94	47.34	m	**92.89**
4.00–4.50 m deep	0.30	34.16	18.38	47.34	m	**99.88**
4.50–5.00 m deep	0.35	39.85	21.47	47.34	m	**108.66**
5.00–5.50 m deep	0.40	45.55	24.56	47.34	m	**117.44**
Unplasticized pvc pipes; ring seal sockets; excavation and supports, backfilling; 6 m pipe lengths unless stated otherwise						
82 mm pipes; in trenches, depth						
not in trenches; 3.00 m pipe lengths	0.06	4.21	0.53	15.60	m	**20.33**
ne 1.50 m deep; 3.00 m pipe lengths	0.10	11.39	9.40	15.60	m	**36.39**
1.50–2.00 m deep	0.13	14.80	12.44	15.60	m	**42.84**
2.00–2.50 m deep	0.16	18.22	15.29	15.60	m	**49.11**
2.50–3.00 m deep	0.19	21.63	18.16	15.60	m	**55.40**
3.00–3.50 m deep	0.22	25.05	21.07	15.60	m	**61.72**
3.50–4.00 m deep	0.25	28.47	23.95	15.60	m	**68.02**

CLASS I: PIPEWORK – PIPES

Item	Gang hours	Labour £	Plant £	Material £	Unit	**Total rate £**
4.00–4.50 m deep	0.28	31.88	26.82	15.60	m	**74.31**
4.50–5.00 m deep	0.32	36.44	30.67	15.60	m	**82.71**
5.00–5.50 m deep	0.35	39.85	33.55	15.60	m	**89.01**
extra for bend; short radius (socket/spigot)	0.05	5.69	–	39.71	nr	**45.41**
extra for branches; equal (socket/spigot)	0.07	7.97	–	50.22	nr	**58.19**
110 mm pipes; in trenches, depth						
not in trenches	0.06	4.21	0.53	20.59	m	**25.32**
ne 1.50 m deep	0.10	11.39	9.56	20.59	m	**41.54**
1.50–2.00 m deep	0.13	14.80	12.44	20.59	m	**47.83**
2.00–2.50 m deep	0.16	18.22	15.29	20.59	m	**54.10**
2.50–3.00 m deep	0.19	21.63	18.16	20.59	m	**60.39**
3.00–3.50 m deep	0.22	25.05	21.07	20.59	m	**66.72**
3.50–4.00 m deep	0.25	28.47	23.95	20.59	m	**73.01**
4.00–4.50 m deep	0.28	31.88	26.82	20.59	m	**79.30**
4.50–5.00 m deep	0.32	36.44	30.67	20.59	m	**87.70**
5.00–5.50 m deep	0.36	40.99	34.52	20.59	m	**96.11**
extra for bend; short redius (socket/spigot)	0.05	5.69	–	59.63	nr	**65.32**
extra for bend; adjustable (socket/spigot)	0.05	5.69	–	61.92	nr	**67.61**
extra for reducer	0.05	5.69	–	22.25	nr	**27.95**
extra for branches; equal (socket/spigot)	0.07	7.97	–	76.37	nr	**84.34**
160 mm pipes; in trenches, depth						
not in trenches	0.06	4.21	0.53	58.04	m	**62.78**
ne 1.50 m deep	0.11	12.53	10.48	58.04	m	**81.05**
1.50–2.00 m deep	0.14	15.94	13.40	58.04	m	**87.39**
2.00–2.50 m deep	0.18	20.50	17.25	58.04	m	**95.79**
2.50–3.00 m deep	0.22	25.05	21.06	58.04	m	**104.16**
3.00–3.50 m deep	0.23	26.19	22.00	58.04	m	**106.23**
3.50–4.00 m deep	0.26	29.60	24.92	58.04	m	**112.57**
4.00–4.50 m deep	0.30	34.16	28.72	58.04	m	**120.92**
4.50–5.00 m deep	0.35	39.85	33.54	58.04	m	**131.43**
5.00–5.50 m deep	0.40	45.55	38.36	58.04	m	**141.95**
extra for bend; short radius (socket/spigot)	0.05	5.69	–	233.38	nr	**239.07**
extra for branches; equal (socket/spigot)	0.07	7.97	–	262.04	nr	**270.01**
225 mm pipes; in trenches, depth						
not in trenches	0.07	4.91	0.62	110.78	m	**116.31**
ne 1.50 m deep	0.12	13.66	11.45	110.78	m	**135.89**
1.50–2.00 m deep	0.15	17.08	14.33	110.78	m	**142.19**
2.00–2.50 m deep	0.20	22.77	19.12	110.78	m	**152.67**
2.50–3.00 m deep	0.23	26.19	21.99	110.78	m	**158.96**
3.00–3.50 m deep	0.24	27.33	22.97	110.78	m	**161.08**
3.50–4.00 m deep	0.27	30.74	25.85	110.78	m	**167.37**
4.00–4.50 m deep	0.32	36.44	30.66	110.78	m	**177.88**
4.50–5.00 m deep	0.36	40.99	34.51	110.78	m	**186.28**
5.00–5.50 m deep	0.45	51.24	43.16	110.78	m	**205.19**
extra for bend; short radius 45° (double socket)	2.77	315.40	–	165.15	nr	**480.55**
extra for branches; equal (all socket)	0.09	10.25	–	542.43	nr	**552.68**

CLASS I: PIPEWORK – PIPES

Item	Gang hours	Labour £	Plant £	Material £	Unit	Total rate £
05 POLYVINYL CHLORIDE PIPES – cont						
Unplasticized pvc pipes – cont						
300 mm pipes; in trenches, depth						
not in trenches	0.08	5.61	0.72	172.10	m	**178.43**
ne 1.50 m deep	0.13	14.80	12.41	172.10	m	**199.32**
1.50–2.00 m deep	0.16	18.22	15.29	172.10	m	**205.62**
2.00–2.50 m deep	0.21	23.91	20.08	172.10	m	**216.10**
2.50–3.00 m deep	0.23	26.19	21.99	172.10	m	**220.28**
3.00–3.50 m deep	0.25	28.47	23.94	172.10	m	**224.51**
3.50–4.00 m deep	0.28	31.88	26.82	172.10	m	**230.80**
4.00–4.50 m deep	0.34	38.71	32.59	172.10	m	**243.41**
4.50–5.00 m deep	0.38	43.27	36.45	172.10	m	**251.82**
5.00–5.50 m deep	0.47	53.52	45.06	172.10	m	**270.68**
extra for bend; short radius 45° (double socket)	1.00	113.86	–	1407.22	nr	**1521.09**
extra for branches; unequal (all socket)	0.09	10.25	–	899.31	nr	**909.56**
Unplasticized pvc pipes; polypropylene couplings; excavation and supports, backfilling; 6 m pipe lengths unless stated otherwise						
110 mm pipes; in trenches, depth						
not in trenches	0.06	4.21	0.53	13.82	m	**18.56**
ne 1.50 m deep	0.10	11.39	9.56	13.82	m	**34.77**
1.50–2.00 m deep	0.13	14.80	12.44	13.82	m	**41.06**
2.00–2.50 m deep	0.16	18.22	15.29	13.82	m	**47.34**
2.50–3.00 m deep	0.19	21.63	18.16	13.82	m	**53.62**
3.00–3.50 m deep	0.22	25.05	21.07	13.82	m	**59.95**
3.50–4.00 m deep	0.25	28.47	23.95	13.82	m	**66.24**
4.00–4.50 m deep	0.28	31.88	26.82	13.82	m	**72.53**
4.50–5.00 m deep	0.32	36.44	30.67	13.82	m	**80.93**
5.00–5.50 m deep	0.36	40.99	34.52	13.82	m	**89.34**
160 mm pipes; in trenches, depth						
not in trenches	0.06	4.21	0.56	45.79	m	**50.55**
ne 1.50 m deep	0.11	12.53	10.48	45.79	m	**68.80**
1.50–2.00 m deep	0.14	15.94	13.40	45.79	m	**75.13**
2.00–2.50 m deep	0.18	20.50	17.25	45.79	m	**83.53**
2.50–3.00 m deep	0.22	25.05	21.06	45.79	m	**91.90**
3.00–3.50 m deep	0.23	26.19	22.00	45.79	m	**93.98**
3.50–4.00 m deep	0.26	29.60	24.92	45.79	m	**100.31**
4.00–4.50 m deep	0.30	34.16	28.72	45.79	m	**108.67**
4.50–5.00 m deep	0.35	39.85	33.54	45.79	m	**119.18**
5.00–5.50 m deep	0.40	45.55	38.36	45.79	m	**129.69**
225 mm pipes; in trenches, depth						
not in trenches	0.07	4.91	0.62	87.29	m	**92.82**
ne 1.50 m deep	0.12	13.66	11.45	87.29	m	**112.40**
1.50–2.00 m deep	0.15	17.08	14.33	87.29	m	**118.70**
2.00–2.50 m deep	0.20	22.77	19.12	87.29	m	**129.18**
2.50–3.00 m deep	0.23	26.19	21.99	87.29	m	**135.47**
3.00–3.50 m deep	0.24	27.33	22.97	87.29	m	**137.58**
3.50–4.00 m deep	0.27	30.74	25.85	87.29	m	**143.88**

CLASS I: PIPEWORK – PIPES

Item	Gang hours	Labour £	Plant £	Material £	Unit	**Total rate £**
4.00–4.50 m deep	0.32	36.44	30.66	87.29	m	**154.38**
4.50–5.00 m deep	0.36	40.99	34.51	87.29	m	**162.79**
5.00–5.50 m deep	0.45	51.24	43.16	87.29	m	**181.69**
300 mm pipes; in trenches, depth						
not in trenches	0.08	5.61	0.72	172.73	m	**179.06**
ne 1.50 m deep	0.13	14.80	12.41	172.73	m	**199.95**
1.50–2.00 m deep	0.16	18.22	15.29	172.73	m	**206.25**
2.00–2.50 m deep	0.21	23.91	20.08	172.73	m	**216.73**
2.50–3.00 m deep	0.23	26.19	21.99	172.73	m	**220.91**
3.00–3.50 m deep	0.25	28.47	23.94	172.73	m	**225.14**
3.50–4.00 m deep	0.28	31.88	26.82	172.73	m	**231.43**
4.00–4.50 m deep	0.34	38.71	32.59	172.73	m	**244.04**
4.50–5.00 m deep	0.38	43.27	36.45	172.73	m	**252.45**
5.00–5.50 m deep	0.47	53.52	45.06	172.73	m	**271.31**
Ultrarib unplasticized pvc pipes; ring seal joints; excavation and supports, backfilling						
150 mm pipes; in trenches, depth						
not in trenches	0.13	9.11	1.18	23.25	m	**33.54**
ne 1.50 m deep	0.16	18.22	15.26	24.19	m	**57.67**
1.50–2.00 m deep	0.19	21.63	18.15	23.88	m	**63.66**
2.00–2.50 m deep	0.22	25.05	21.07	23.56	m	**69.69**
2.50–3.00 m deep	0.24	27.33	22.96	23.25	m	**73.53**
3.00–3.50 m deep	0.24	27.33	22.97	23.69	m	**73.98**
3.50–4.00 m deep	0.27	30.74	25.85	23.25	m	**79.84**
4.00–4.50 m deep	0.32	36.44	30.66	23.25	m	**90.34**
4.50–5.00 m deep	0.36	40.99	34.51	23.69	m	**99.19**
5.00–5.50 m deep	0.45	51.24	43.16	23.69	m	**118.09**
extra for 45 deg bends; short radius (socket/spigot)	0.05	5.69	–	56.37	nr	**62.07**
extra for branches; equal (socket/spigot)	0.09	10.25	–	136.30	nr	**146.55**
225 mm pipes; in trenches, depth						
not in trenches	0.08	5.61	0.72	72.51	m	**78.84**
ne 1.50 m deep	0.36	40.99	12.41	48.94	m	**102.34**
1.50–2.00 m deep	0.38	43.27	15.29	48.94	m	**107.50**
2.00–2.50 m deep	0.44	50.10	20.08	48.94	m	**119.12**
2.50–3.00 m deep	0.46	52.38	21.99	48.94	m	**123.31**
3.00–3.50 m deep	0.47	53.52	23.94	48.94	m	**126.39**
3.50–4.00 m deep	0.51	58.07	26.82	48.94	m	**133.83**
4.00–4.50 m deep	0.57	64.90	32.59	48.94	m	**146.44**
4.50–5.00 m deep	0.61	69.46	36.45	48.94	m	**154.84**
5.00–5.50 m deep	0.70	79.71	45.06	48.94	m	**173.70**
extra for 45 deg bends; short radius (socket/spigot)	0.05	5.69	–	237.10	nr	**242.79**
extra for branches; equal (socket/spigot)	0.09	10.25	–	472.80	nr	**483.05**
300 mm pipes; in trenches, depth						
not in trenches	0.75	52.58	0.72	58.84	m	**112.14**
ne 1.50 m deep	0.57	64.90	12.41	58.84	m	**136.16**
1.50–2.00 m deep	0.60	68.32	15.29	58.84	m	**142.46**
2.00–2.50 m deep	0.65	74.01	20.08	58.84	m	**152.94**
2.50–3.00 m deep	0.67	76.29	21.99	58.84	m	**157.12**
3.00–3.50 m deep	0.60	68.32	23.94	58.84	m	**151.10**
3.50–4.00 m deep	0.72	81.98	26.82	58.84	m	**167.64**

CLASS I: PIPEWORK – PIPES

Item	Gang hours	Labour £	Plant £	Material £	Unit	**Total rate £**
05 POLYVINYL CHLORIDE PIPES – cont						
Ultrarib unplasticized pvc pipes – cont						
300 mm pipes – cont						
4.00–4.50 m deep	0.78	88.81	32.59	58.84	m	**180.25**
4.50–5.00 m deep	0.82	93.37	36.45	58.84	m	**188.66**
5.00–5.50 m deep	0.91	103.62	45.06	58.84	m	**207.52**
extra for 45 deg bends; short radius (socket/spigot)	0.35	39.85	–	430.58	nr	**470.44**
extra for branches; equal (socket/spigot)	0.09	10.25	–	1092.46	nr	**1102.71**

CLASS J: PIPEWORK – FITTINGS AND VALVES

Item	Gang hours	Labour £	Plant £	Material £	Unit	**Total rate £**
RESOURCES – LABOUR						
Labour Resources Build-up Rate						
Note						
Fittings on pipes shown with the appropriate pipe in Class I						
Fittings and valves gang						
1 ganger/chargehand (skill rate 4) – 50% of time		10.71	–	–		
1 skilled operative (skill rate 4)		20.20	–	–		
2 unskilled operatives (general)		39.20	–	–		
1 plant operator (skill rate 3)		29.17	–	–		
Total Gang Rate/Hour		**99.28**	–	–		
RESOURCES – PLANT						
Plant Resources Build-up Rate						
Fittings and valves						
1.0 m³ hydraulic backacter		–	84.68	–		
disc saw – 30% of time		–	0.41	–		
2.80 m³/min compressor, 2 tool – 30% of time		–	4.71	–		
2t dumper – 30% of time		–	3.24	–		
compressor tools, extra 50ft / 15 m hose – 30% of time		–	0.14	–		
small pump – 30% of time		–	1.67	–		
sundry tools – 30% of time		–	0.48	–		
Total Rate/Hour		–	**95.34**	–		
VALVES AND PENSTOCKS						
Valves						
Non-return valves; cast iron; single door; tidal flap						
250 mm	1.20	119.13	19.07	357.00	nr	**495.20**
350 mm	1.20	119.13	19.07	681.00	nr	**819.20**
450 mm	1.20	119.13	19.07	992.00	nr	**1130.20**
600 mm	1.60	158.85	28.58	1537.00	nr	**1724.43**
800 mm	1.80	178.70	33.36	2122.15	nr	**2334.21**
Penstocks; cast iron; wall mounted; hand operated						
250 mm	1.66	164.80	33.36	800.00	nr	**998.16**
350 mm	2.30	228.34	38.13	1125.94	nr	**1392.42**
450 mm	2.90	287.91	42.91	1343.20	nr	**1674.02**
600 × 600 mm	5.40	536.11	76.27	2172.84	nr	**2785.22**
1000 × 1000 mm	9.00	893.51	143.03	3753.10	nr	**4789.63**

CLASS K: PIPEWORK – MANHOLES AND PIPEWORK ANCILLARIES

Item	Gang hours	Labour £	Plant £	Material £	Unit	Total rate £
NOTES						
The rates assume the most efficient items of plant (excavator) and are optimum rates assuming continuous output with no delays caused by other operations or works.						
Ground conditions are assumed to be good easily worked soil with no abnormal conditions that would affect outputs and consistency of work.						
Multiplier Table for labour and plant for various site conditions for working:						
out of sequence × 2.75 (minimum)						
in hard clay × 1.75 to 2.00						
in running sand × 2.75 (minimum)						
in broken rock × 2.75 to 3.50						
below water table × 2.00 (minimum)						
RESOURCES – LABOUR						
Labour Resources Build-up Rate						
Gullies gang						
1 chargehand pipelayer (skill rate 4) – 50% of time		10.71	–	–		
1 skilled operative (skill rate 4)		20.20	–	–		
2 unskilled operatives (general)		39.20	–	–		
1 plant operator (skill rate 3)		29.17	–	–		
Total Gang Rate/Hour		**99.28**	–	–		
French/rubble drains, ditches and trenches gang; ducts and metal culverts gang						
1 chargehand pipelayer (skill rate 4) – 50% of time		10.71	–	–		
1 skilled operative (skill rate 4)		20.20	–	–		
2 unskilled operatives (general)		39.20	–	–		
1 plant operator (skill rate 3)		29.17	–	–		
Total Gang Rate/Hour		**99.28**	–	–		
RESOURCES – MATERIAL						
Supply only ex works materials rates						
Vitrified clay road gullies etc.						
Clay road gully; 450 mm dia. × 900 mm deep		–	–	469.78	nr	**469.78**
C.I. gully grating and frame; 434 × 434 mm to clay		–	–	283.25	nr	**283.25**
Clay yard gully; 225 mm dia.; including grating		–	–	429.80	nr	**429.80**
Clay grease interceptor; 600 × 450 mm		–	–	2324.73	nr	**2324.73**
Culvert; Corrugated steel						
1.6 mm thick; 1000 mm dia.		–	–	145.21	m	**145.21**
2.0 mm thick; 1600 mm dia.		–	–	231.61	m	**231.61**
2.0 mm thick; 2000 mm dia.		–	–	500.58	m	**500.58**
2.0 mm thick; 2200 mm dia.		–	–	564.79	m	**564.79**

CLASS K: PIPEWORK – MANHOLES AND PIPEWORK ANCILLARIES

Item	Gang hours	Labour £	Plant £	Material £	Unit	**Total rate £**
RESOURCES – PLANT						
Plant Resources Build-up Rate						
Gullies						
0.4 m^3 hydraulic excavator		–	55.59	–		
2t dumper (30% of time)		–	3.24	–		
2.80 m^3/min compressor, 2 tool (30% of time)		–	4.71	–		
compaction plate/roller (30% of time)		–	1.20	–		
2.40 m^3/min road breaker (30% of time)		–	1.95	–		
54 mm poker vibrator (30% of time)		–	0.56	–		
extra 15ft / 50 m hose (30% of time)		–	0.14	–		
disc saw (30% of time)		–	0.33	–		
small pump (30% of time)		–	1.67	–		
Total Rate/Hour		–	**69.40**	–		
French/rubble drains, ditches and trenches; ducts and metal culverts						
0.4 m^3 hydraulic backacter		–	55.59	–		
2t dumper (30% of time)		–	3.24	–		
disc saw (30% of time)		–	0.33	–		
compaction plate/roller (30% of time)		–	2.50	–		
2.80 m^3/min compressor, 2 tool (30% of time)		–	4.71	–		
small pump (30% of time)		–	1.67	–		
Total Rate/Hour		–	**68.05**	–		
01 MANHOLES						
Note: The rates assume the most efficient items of plant (excavator) and are optimum rates, assuming continuous output with no delays caused by other operations or works. Ground conditions are assumed to be good soil with no abnormal conditions that would affect outputs and consistency of work.						
Multiplier Factor Table for Labour and Plant to reflect various site conditions requiring additional works						
out of sequence with other works × 2.75 minimum						
in hard clay × 1.75–2.00						
in running sand × 2.75 minimum						
in broken rock × 2.75–3.50						
below water table × 2.00 minimum						
Brick construction						
Design criteria used in models:						
class A engineering bricks						
215 thick walls generally; 328 thick to chambers exceeding 2.5 m deep						
225 plain concrete C20/20 base slab						

CLASS K: PIPEWORK – MANHOLES AND PIPEWORK ANCILLARIES

Item	Gang hours	Labour £	Plant £	Material £	Unit	Total rate £
01 MANHOLES – cont						
Brick construction – cont						
Design criteria used in models: – cont						
300 reinforced concrete C20/20 reducing slab						
125 reinforced concrete C20/20 top slab						
maximum height of working chamber 2.0 m above benching						
750 × 750 access shaft						
plain concrete C15/20 benching, 150 clay main channel longitudinally and two 100 branch channels						
step irons at 300 mm centres, doubled if depth to invert exceeds 3000 mm						
heavy duty manhole cover and frame						
750 × 700 chamber 500 depth to invert						
excavation, support, backfilling and disposal	0.28	21.50	14.90	–	nr	**36.41**
concrete base	0.40	7.98	2.90	27.67	nr	**38.55**
brickwork chamber	3.81	72.72	10.03	80.85	nr	**163.60**
concrete cover slab	2.16	48.15	35.72	127.25	nr	**211.12**
concrete benching, main and branch channels	7.01	140.06	27.38	221.92	nr	**389.36**
step irons	0.21	4.01	0.56	39.04	nr	**43.62**
access cover and frame	1.44	28.80	3.59	275.50	nr	**307.89**
TOTAL	15.31	323.22	95.09	772.23	nr	**1190.53**
750 × 700 chamber 1000 depth to invert						
excavation, support, backfilling and disposal	0.55	42.44	29.59	–	nr	**72.03**
concrete base	0.40	7.98	2.90	27.67	nr	**38.55**
brickwork chamber	7.61	145.44	20.07	161.69	nr	**327.20**
concrete cover slab	2.16	48.15	35.72	127.25	nr	**211.12**
concrete benching, main and branch channels	7.01	140.06	27.38	221.92	nr	**389.36**
step irons	0.32	6.02	0.84	58.56	nr	**65.43**
access cover and frame	1.44	28.80	3.59	275.50	nr	**307.89**
TOTAL	19.49	418.88	120.09	872.60	nr	**1411.56**
750 × 700 chamber 1500 depth to invert						
excavation, support, backfilling and disposal	0.83	63.27	44.20	–	nr	**107.47**
concrete base	0.40	7.98	2.90	27.67	nr	**38.55**
brickwork chamber	12.18	232.70	32.09	242.54	nr	**507.33**
concrete cover slab	2.16	48.15	35.72	127.25	nr	**211.12**
concrete benching, main and branch channels	7.01	140.06	27.38	221.92	nr	**389.36**
step irons	0.42	8.02	1.12	78.09	nr	**87.23**
access cover and frame	1.44	28.80	3.59	275.50	nr	**307.89**
TOTAL	24.44	528.98	147.00	972.96	nr	**1648.94**
900 × 700 chamber 500 depth to invert						
excavation, support, backfilling and disposal	0.33	25.22	17.02	–	nr	**42.24**
concrete base	0.47	9.30	3.38	32.28	nr	**44.97**
brickwork chamber	4.20	80.24	11.07	89.21	nr	**180.52**
concrete cover slab	2.16	48.15	35.72	127.25	nr	**211.12**
concrete benching, main and branch channels	7.01	140.06	27.38	221.92	nr	**389.36**
step irons	0.11	2.01	0.28	19.52	nr	**21.81**
access cover and frame	1.44	28.80	3.59	275.50	nr	**307.89**
TOTAL	15.71	333.78	98.44	765.68	nr	**1197.90**

CLASS K: PIPEWORK – MANHOLES AND PIPEWORK ANCILLARIES

Item	Gang hours	Labour £	Plant £	Material £	Unit	**Total rate £**
900 × 700 chamber 1000 depth to invert						
excavation, support, backfilling and disposal	0.66	50.44	34.04	–	nr	**84.47**
concrete base	0.47	9.30	3.38	32.28	nr	**44.97**
brickwork chamber	8.40	160.48	22.14	178.42	nr	**361.05**
concrete cover slab	2.16	48.15	35.72	127.25	nr	**211.12**
concrete benching, main and branch channels	7.01	140.06	27.38	221.92	nr	**389.36**
step irons	0.32	6.02	0.84	58.56	nr	**65.43**
access cover and frame	1.44	28.80	3.59	275.50	nr	**307.89**
TOTAL	20.45	443.25	127.10	893.93	nr	**1464.28**
900 × 700 chamber 1500 depth to invert						
excavation, support, backfilling and disposal	0.98	75.09	50.83	–	nr	**125.92**
concrete base	0.47	9.30	3.38	32.28	nr	**44.97**
brickwork chamber	13.44	256.77	35.41	267.63	nr	**559.81**
concrete cover slab	2.16	48.15	35.72	127.25	nr	**211.12**
concrete benching, main and branch channels	7.01	140.06	27.38	221.92	nr	**389.36**
step irons	0.42	8.02	1.12	78.09	nr	**87.23**
access cover and frame	1.44	28.80	3.59	275.50	nr	**307.89**
TOTAL	25.92	566.20	157.44	1002.67	nr	**1726.30**
1050 × 700 chamber 1500 depth to invert						
excavation, support, backfilling and disposal	1.91	145.34	95.95	–	nr	**241.29**
concrete base	0.57	11.30	4.11	39.20	nr	**54.61**
brickwork chamber	14.70	280.85	38.73	292.72	nr	**612.29**
concrete cover slab	2.52	56.17	41.64	135.98	nr	**233.79**
concrete benching, main and branch channels	7.61	151.96	30.64	219.37	nr	**401.98**
step irons	0.84	16.05	2.25	156.17	nr	**174.47**
access cover and frame	1.44	28.80	3.59	275.50	nr	**307.89**
TOTAL	29.59	690.47	216.91	1118.94	nr	**2026.32**
1050 × 700 chamber 2500 depth to invert						
excavation, support, backfilling and disposal	1.99	153.20	117.73	–	nr	**270.93**
concrete base	0.57	11.30	4.11	39.20	nr	**54.61**
brickwork chamber and access shaft (700 × 700)	35.46	677.54	93.45	676.99	nr	**1447.98**
concrete cover slab	2.52	56.17	41.64	135.98	nr	**233.79**
concrete reducing slab	3.30	72.51	46.30	164.36	nr	**283.18**
concrete benching, main and branch channels	7.61	151.96	30.64	219.37	nr	**401.98**
step irons	0.84	16.05	2.25	156.17	nr	**174.47**
access cover and frame	1.44	28.80	3.59	275.50	nr	**307.89**
TOTAL	53.73	1167.54	339.71	1667.57	nr	**3174.82**
1050 × 700 chamber 3500 depth to invert						
excavation, support, backfilling and disposal	2.88	224.10	231.54	–	nr	**455.64**
concrete base	0.57	11.30	4.11	39.20	nr	**54.61**
brickwork chamber and access shaft (700 × 700)	45.57	870.62	120.07	872.14	nr	**1862.83**
concrete reducing slab	3.30	72.51	46.30	164.36	nr	**283.18**
concrete cover slab	2.52	56.17	41.64	135.98	nr	**233.79**
concrete benching, main and branch channels	9.41	187.67	42.87	231.83	nr	**462.38**
step irons	1.16	22.07	3.09	214.74	nr	**239.89**
access cover and frame	1.44	28.80	3.59	275.50	nr	**307.89**
TOTAL	66.84	1473.25	493.21	1933.74	nr	**3900.20**

CLASS K: PIPEWORK – MANHOLES AND PIPEWORK ANCILLARIES

Item	Gang hours	Labour £	Plant £	Material £	Unit	**Total rate £**
01 MANHOLES – cont						
Brick construction – cont						
1350 × 700 chamber 2500 depth to invert STOP						
excavation, support, backfilling and disposal	2.42	183.95	117.75	–	nr	**301.70**
concrete base	0.70	13.96	5.07	48.42	nr	**67.45**
brickwork chamber and access shaft (700 × 700)	41.54	793.69	109.47	793.05	nr	**1696.21**
concrete reducing slab	3.30	72.51	46.30	164.36	nr	**283.18**
concrete cover slab	2.80	62.42	46.30	169.15	nr	**277.86**
concrete benching, main and branch channels	11.14	221.90	54.63	243.27	nr	**519.80**
step irons	0.84	16.05	2.25	156.17	nr	**174.47**
access cover and frame	1.44	28.80	3.59	275.50	nr	**307.89**
TOTAL	64.18	1393.27	385.36	1849.92	nr	**3628.55**
1350 × 700 chamber 3500 depth to invert						
excavation, support, backfilling and disposal	3.65	283.66	279.79	–	nr	**563.44**
concrete base	0.70	13.96	5.07	48.42	nr	**67.45**
brickwork chamber and access shaft (700 × 700)	54.89	1048.66	144.64	1021.65	nr	**2214.95**
concrete reducing slab	3.30	72.51	46.30	164.36	nr	**283.18**
concrete cover slab	2.80	62.42	46.30	169.15	nr	**277.86**
concrete benching, main and branch channels	11.14	221.90	54.63	243.27	nr	**519.80**
step irons	1.16	22.07	3.09	214.74	nr	**239.89**
access cover and frame	1.44	28.80	3.59	275.50	nr	**307.89**
TOTAL	79.08	1753.96	583.42	2137.09	nr	**4474.47**
1350 × 700 chamber 4500 depth to invert						
excavation, support, backfilling and disposal	4.69	364.50	359.66	–	nr	**724.16**
concrete base	0.70	13.96	5.07	48.42	nr	**67.45**
brickwork chamber and access shaft (700 × 700)	67.37	1287.18	177.55	1250.25	nr	**2714.98**
concrete reducing slab	3.30	72.51	46.30	164.36	nr	**283.18**
concrete cover slab	2.80	62.42	46.30	169.15	nr	**277.86**
concrete benching, main and branch channels	11.14	221.90	54.63	243.27	nr	**519.80**
step irons	1.47	28.09	3.94	273.30	nr	**305.32**
access cover and frame	1.44	28.80	3.59	275.50	nr	**307.89**
TOTAL	92.92	2079.35	697.04	2424.25	nr	**5200.64**
Precast concrete construction						
Design criteria used in models:						
circular shafts						
150 plain concrete surround						
225 plain concrete C20/20 base slab						
precast top slab						
maximum height of working chamber 2.0 m above benching						
plain concrete C15/20 benching, 150 clay main channel longitudinally and two 100 branch channels						
step irons at 300 mm centres, doubled if depth to invert exceeds 3000 mm						
heavy duty manhole cover and frame						
in manholes over 6 m deep, landings at maximum intervals						

CLASS K: PIPEWORK – MANHOLES AND PIPEWORK ANCILLARIES

Item	Gang hours	Labour £	Plant £	Material £	Unit	**Total rate £**
675 dia. × 500 depth to invert						
excavation, support, backfilling and disposal	0.19	14.92	10.58	–	nr	**25.51**
concrete base	0.27	5.32	1.93	18.45	nr	**25.70**
main chamber rings	1.60	35.67	26.45	32.69	nr	**94.80**
cover slab	1.98	43.76	30.43	74.73	nr	**148.92**
concrete surround	0.37	7.30	2.98	18.41	nr	**28.69**
concrete benching, main and branch channels	2.38	47.44	10.59	112.04	nr	**170.06**
step irons	0.11	2.01	0.28	19.52	nr	**21.81**
access cover and frame	1.44	28.80	3.59	275.50	nr	**307.89**
TOTAL	8.34	185.21	86.83	551.34	nr	**823.38**
675 dia. × 750 depth to invert						
excavation, support, backfilling and disposal	0.28	21.82	15.65	–	nr	**37.47**
concrete base	0.27	5.32	1.93	18.45	nr	**25.70**
main chamber rings	2.40	53.50	39.67	49.03	nr	**142.20**
cover slab	1.98	43.76	30.43	74.73	nr	**148.92**
concrete surround	0.55	10.95	4.47	27.62	nr	**43.04**
concrete benching, main and branch channels	2.38	47.44	10.59	112.04	nr	**170.06**
step irons	0.21	4.01	0.56	39.04	nr	**43.62**
access cover and frame	1.44	28.80	3.59	275.50	nr	**307.89**
TOTAL	9.51	215.59	106.90	596.41	nr	**918.90**
675 dia. × 1000 depth to invert						
excavation, support, backfilling and disposal	0.38	29.28	20.94	–	nr	**50.23**
concrete base	0.27	5.32	1.93	18.45	nr	**25.70**
main chamber rings	3.20	71.33	52.89	65.37	nr	**189.60**
cover slab	1.98	43.76	30.43	74.73	nr	**148.92**
concrete surround	0.74	14.60	5.96	36.83	nr	**57.39**
concrete benching, main and branch channels	2.38	47.44	10.59	112.04	nr	**170.06**
step irons	0.32	6.02	0.84	58.56	nr	**65.43**
access cover and frame	1.44	28.80	3.59	275.50	nr	**307.89**
TOTAL	10.70	246.54	127.18	641.48	nr	**1015.20**
675 dia. × 1250 depth to invert						
excavation, support, backfilling and disposal	0.48	36.74	26.24	–	nr	**62.98**
concrete base	0.27	5.32	1.93	18.45	nr	**25.70**
main chamber rings	4.00	89.16	66.11	81.72	nr	**236.99**
cover slab	1.98	43.76	30.43	74.73	nr	**148.92**
concrete surround	0.92	18.25	7.45	46.03	nr	**71.73**
concrete benching, main and branch channels	2.38	47.44	10.59	112.04	nr	**170.06**
step irons	0.42	8.02	1.12	78.09	nr	**87.23**
access cover and frame	1.44	28.80	3.59	275.50	nr	**307.89**
TOTAL	11.89	277.49	147.47	686.55	nr	**1111.51**
900 dia. × 750 depth to invert						
excavation, support, backfilling and disposal	0.49	37.20	24.03	–	nr	**61.22**
concrete base	0.47	9.30	3.38	32.28	nr	**44.97**
main chamber rings	2.85	63.53	47.10	112.76	nr	**223.39**
cover slab	3.80	84.71	62.80	150.34	nr	**297.85**
concrete surround	0.74	14.60	5.96	36.83	nr	**57.39**
concrete benching, main and branch channels	3.38	67.28	17.92	120.80	nr	**205.99**
step irons	0.21	4.01	0.56	39.04	nr	**43.62**
access cover and frame	1.44	28.80	3.59	275.50	nr	**307.89**
TOTAL	13.37	309.43	165.33	767.55	nr	**1242.31**

CLASS K: PIPEWORK – MANHOLES AND PIPEWORK ANCILLARIES

Item	Gang hours	Labour £	Plant £	Material £	Unit	Total rate £
01 MANHOLES – cont						
Precast concrete construction – cont						
900 dia. × 1000 depth to invert						
excavation, support, backfilling and disposal	0.65	49.52	32.00	–	nr	**81.53**
concrete base	0.47	9.30	3.38	32.28	nr	**44.97**
main chamber rings	3.80	84.71	62.80	150.34	nr	**297.85**
cover slab	3.80	84.71	62.80	150.34	nr	**297.85**
concrete surround	0.97	19.17	7.82	48.33	nr	**75.32**
concrete benching, main and branch channels	3.38	67.28	17.92	120.80	nr	**205.99**
step irons	0.32	6.02	0.84	58.56	nr	**65.43**
access cover and frame	1.44	28.80	3.59	275.50	nr	**307.89**
TOTAL	14.82	349.50	191.15	836.16	nr	**1376.82**
900 dia. × 1500 depth to invert						
excavation, support, backfilling and disposal	0.99	74.96	48.27	–	nr	**123.23**
concrete base	0.47	9.30	3.38	32.28	nr	**44.97**
main chamber rings	5.70	127.06	94.19	225.52	nr	**446.77**
cover slab	2.16	48.15	35.72	127.25	nr	**211.12**
concrete surround	1.47	29.20	11.92	73.65	nr	**114.77**
concrete benching, main and branch channels	3.38	67.28	17.92	120.80	nr	**205.99**
step irons	0.42	8.02	1.12	78.09	nr	**87.23**
access cover and frame	1.44	28.80	3.59	275.50	nr	**307.89**
TOTAL	16.03	392.77	216.12	933.08	nr	**1541.98**
1200 dia. × 1500 depth to invert						
excavation, support, backfilling and disposal	1.70	128.56	75.52	–	nr	**204.08**
concrete base	0.84	16.62	6.04	57.65	nr	**80.30**
main chamber rings	7.02	156.48	116.04	225.29	nr	**497.81**
cover slab	2.80	62.42	46.30	169.15	nr	**277.86**
concrete surround	1.96	38.79	15.83	97.82	nr	**152.43**
concrete benching, main and branch channels	6.13	121.84	36.09	135.03	nr	**292.96**
step irons	0.42	8.02	1.12	78.09	nr	**87.23**
access cover and frame	1.44	28.80	3.59	275.50	nr	**307.89**
TOTAL	22.31	561.52	300.54	1038.51	nr	**1900.57**
1200 dia. × 2000 depth to invert						
excavation, support, backfilling and disposal	2.35	179.18	120.19	–	nr	**299.37**
concrete base	0.84	16.62	6.04	57.65	nr	**80.30**
main chamber rings	9.36	208.65	154.73	300.38	nr	**663.75**
cover slab	2.80	62.42	46.30	169.15	nr	**277.86**
concrete surround	2.60	51.56	21.04	130.04	nr	**202.64**
concrete benching, main and branch channels	6.13	121.84	36.09	135.03	nr	**292.96**
step irons	0.63	12.04	1.69	117.13	nr	**130.85**
access cover and frame	1.44	28.80	3.59	275.50	nr	**307.89**
TOTAL	26.14	681.09	389.67	1184.87	nr	**2255.63**

CLASS K: PIPEWORK – MANHOLES AND PIPEWORK ANCILLARIES

Item	Gang hours	Labour £	Plant £	Material £	Unit	**Total rate £**
1200 dia. × 2500 depth to invert						
excavation, support, backfilling and disposal	2.94	224.10	150.32	–	nr	**374.42**
concrete base	0.84	16.62	6.04	57.65	nr	**80.30**
main chamber rings	11.70	260.81	193.41	375.48	nr	**829.69**
cover slab	2.80	62.42	46.30	169.15	nr	**277.86**
concrete surround	3.24	64.34	26.25	162.26	nr	**252.85**
concrete benching, main and branch channels	6.13	121.84	36.09	135.03	nr	**292.96**
step irons	0.84	16.05	2.25	156.17	nr	**174.47**
access cover and frame	1.44	28.80	3.59	275.50	nr	**307.89**
TOTAL	29.93	794.96	464.25	1331.23	nr	**2590.44**
1200 dia. × 3000 depth to invert						
excavation, support, backfilling and disposal	3.53	269.12	180.44	–	nr	**449.56**
concrete base	0.84	16.62	6.04	57.65	nr	**80.30**
main chamber rings	14.04	312.97	232.09	450.57	nr	**995.63**
cover slab	2.80	62.42	46.30	169.15	nr	**277.86**
concrete surround	3.91	77.57	31.65	195.63	nr	**304.86**
concrete benching, main and branch channels	6.13	121.84	36.09	135.03	nr	**292.96**
step irons	0.95	18.05	2.53	175.69	nr	**196.28**
access cover and frame	1.44	28.80	3.59	275.50	nr	**307.89**
TOTAL	33.63	907.38	538.73	1459.22	nr	**2905.33**
1800 dia. × 2000 depth to invert						
excavation, support, backfilling and disposal	4.97	371.89	195.68	–	nr	**567.57**
concrete base	1.91	37.88	13.77	131.43	nr	**183.09**
main chamber rings	12.00	267.49	198.38	791.00	nr	**1256.88**
cover slab	3.60	80.25	59.52	456.04	nr	**595.81**
concrete surround	3.91	77.57	31.65	195.63	nr	**304.86**
concrete benching, main and branch channels	7.38	146.69	42.88	153.35	nr	**342.93**
step irons	0.63	12.04	1.69	117.13	nr	**130.85**
access cover and frame	1.44	28.80	3.59	275.50	nr	**307.89**
TOTAL	35.84	1022.62	547.17	2120.09	nr	**3689.88**
1800 dia. × 2500 depth to invert						
excavation, support, backfilling and disposal	6.42	485.15	284.67	–	nr	**769.82**
concrete base	1.91	37.88	13.77	131.43	nr	**183.09**
main chamber rings	15.00	334.37	247.98	988.75	nr	**1571.10**
cover slab	3.60	80.25	59.52	456.04	nr	**595.81**
concrete surround	4.88	96.74	39.47	243.97	nr	**380.18**
concrete benching, main and branch channels	7.38	146.69	42.88	153.35	nr	**342.93**
step irons	0.84	16.05	2.25	156.17	nr	**174.47**
access cover and frame	1.44	28.80	3.59	275.50	nr	**307.89**
TOTAL	41.46	1225.93	694.14	2405.22	nr	**4325.28**
1800 dia. × 3000 depth to invert						
excavation, support, backfilling and disposal	7.71	582.68	343.52	–	nr	**926.19**
concrete base	1.91	37.88	13.77	131.43	nr	**183.09**
main chamber rings	18.00	401.24	297.58	1186.51	nr	**1885.32**
cover slab	3.60	80.25	59.52	456.04	nr	**595.81**
concrete surround	5.84	115.90	47.29	292.30	nr	**455.49**
concrete benching, main and branch channels	7.38	146.69	42.88	153.35	nr	**342.93**
step irons	0.95	18.05	2.53	175.69	nr	**196.28**
access cover and frame	1.44	28.80	3.59	275.50	nr	**307.89**
TOTAL	46.82	1411.50	810.68	2670.82	nr	**4893.01**

CLASS K: PIPEWORK – MANHOLES AND PIPEWORK ANCILLARIES

Item	Gang hours	Labour £	Plant £	Material £	Unit	**Total rate £**
01 MANHOLES – cont						
Precast concrete construction – cont						
1800 dia. × 3500 depth to invert						
excavation, support, backfilling and disposal	9.26	706.81	513.76	–	nr	**1220.57**
concrete base	1.91	37.88	13.77	131.43	nr	**183.09**
main chamber rings	21.00	468.12	347.17	1384.26	nr	**2199.54**
cover slab	3.60	80.25	59.52	456.04	nr	**595.81**
concrete surround	6.83	135.52	55.30	341.78	nr	**532.60**
concrete benching, main and branch channels	7.38	146.69	42.88	153.35	nr	**342.93**
step irons	1.16	22.07	3.09	214.74	nr	**239.89**
access cover and frame	1.44	28.80	3.59	275.50	nr	**307.89**
TOTAL	52.58	1626.14	1039.10	2957.10	nr	**5622.34**
1800 dia. × 4000 depth to invert						
excavation, support, backfilling and disposal	10.58	807.31	586.99	–	nr	**1394.29**
concrete base	1.91	37.88	13.77	131.43	nr	**183.09**
main chamber rings	24.00	534.99	396.77	1582.01	nr	**2513.76**
cover slab	3.60	80.25	59.52	456.04	nr	**595.81**
concrete surround	7.80	154.69	63.12	390.12	nr	**607.92**
concrete benching, main and branch channels	7.38	146.69	42.88	153.35	nr	**342.93**
step irons	1.37	26.08	3.65	253.78	nr	**283.51**
access cover and frame	1.44	28.80	3.59	275.50	nr	**307.89**
TOTAL	58.07	1816.69	1170.30	3242.23	nr	**6229.21**
2400 dia. × 3000 depth to invert						
excavation, support, backfilling and disposal	4.69	465.59	514.41	–	nr	**979.99**
concrete base	3.42	67.79	24.64	235.20	nr	**327.63**
main chamber rings	21.60	596.44	1114.10	3392.23	nr	**5102.77**
cover slab	4.40	121.50	226.97	1293.29	nr	**1641.75**
concrete surround	7.80	154.69	63.12	390.12	nr	**607.92**
concrete benching, main and branch channels	7.88	156.79	42.88	153.35	nr	**353.03**
step irons	0.95	18.05	2.53	175.69	nr	**196.28**
access cover and frame	1.44	28.80	3.59	275.50	nr	**307.89**
TOTAL	52.17	1609.65	1992.24	5915.37	nr	**9517.26**
2400 dia. × 4500 depth to invert						
excavation, support, backfilling and disposal	20.82	1581.04	1030.46	–	nr	**2611.51**
concrete base	3.42	67.79	24.64	235.20	nr	**327.63**
main chamber rings	32.40	894.66	1671.15	5088.35	nr	**7654.15**
cover slab	4.40	121.50	226.97	1293.29	nr	**1641.75**
concrete surround	11.71	232.26	94.77	585.75	nr	**912.78**
concrete benching, main and branch channels	7.88	156.79	42.88	153.35	nr	**353.03**
step irons	1.47	28.09	3.94	273.30	nr	**305.32**
access cover and frame	1.44	28.80	3.59	275.50	nr	**307.89**
TOTAL	83.54	3110.93	3098.40	7904.73	nr	**14114.06**

CLASS K: PIPEWORK – MANHOLES AND PIPEWORK ANCILLARIES

Item	Gang hours	Labour £	Plant £	Material £	Unit	**Total rate £**
2700 dia. × 3000 depth to invert						
excavation, support, backfilling and disposal	16.98	1275.73	672.35	–	nr	**1948.08**
concrete base	4.32	85.74	31.17	297.45	nr	**414.36**
main chamber rings	24.00	662.71	1237.89	3660.54	nr	**5561.14**
cover slab	4.80	132.54	247.58	2601.54	nr	**2981.66**
concrete surround	8.79	174.31	71.12	439.60	nr	**685.03**
concrete benching, main and branch channels	8.08	160.83	42.88	153.35	nr	**357.07**
step irons	0.95	18.05	2.53	175.69	nr	**196.28**
access cover and frame	1.44	28.80	3.59	275.50	nr	**307.89**
TOTAL	69.35	2538.72	2309.10	7603.68	nr	**12451.51**
2700 dia. × 4500 depth to invert						
excavation, support, backfilling and disposal	26.21	1986.65	1243.54	–	nr	**3230.19**
concrete base	4.30	85.28	31.01	296.30	nr	**412.59**
main chamber rings	36.00	994.06	1856.83	5490.81	nr	**8341.71**
cover slab	4.80	132.54	247.58	2601.54	nr	**2981.66**
concrete surround	13.18	261.47	106.68	659.40	nr	**1027.55**
concrete benching, main and branch channels	8.08	160.83	42.88	153.35	nr	**357.07**
step irons	1.47	28.09	3.94	273.30	nr	**305.32**
access cover and frame	1.44	28.80	3.59	275.50	nr	**307.89**
TOTAL	95.48	3677.71	3536.05	9750.21	nr	**16963.98**
3000 dia. × 3000 depth to invert						
excavation, support, backfilling and disposal	20.88	1566.67	806.28	–	nr	**2372.95**
concrete base	5.33	105.68	38.42	366.63	nr	**510.72**
main chamber rings	25.80	712.41	1330.69	4692.04	nr	**6735.15**
cover slab	0.60	18.05	49.96	2482.94	nr	**2550.96**
concrete surround	9.75	193.48	78.94	487.93	nr	**760.35**
concrete benching, main and branch channels	8.88	176.69	42.88	153.35	nr	**372.93**
step irons	0.95	18.05	2.53	175.69	nr	**196.28**
access cover and frame	1.44	28.80	3.59	275.50	nr	**307.89**
TOTAL	73.63	2819.83	2353.30	8634.09	nr	**13807.21**
3000 dia. × 4500 depth to invert						
excavation, support, backfilling and disposal	32.21	2438.13	1475.19	–	nr	**3913.32**
concrete base	5.33	105.68	38.42	366.63	nr	**510.72**
main chamber rings	38.70	1068.62	1996.04	7038.06	nr	**10102.72**
cover slab	0.60	18.05	49.96	2482.94	nr	**2550.96**
concrete surround	14.63	290.21	118.41	731.90	nr	**1140.53**
concrete benching, main and branch channels	8.88	176.69	42.88	153.35	nr	**372.93**
step irons	1.47	28.09	3.94	273.30	nr	**305.32**
access cover and frame	1.44	28.80	3.59	275.50	nr	**307.89**
TOTAL	103.26	4154.26	3728.43	11321.68	nr	**19204.38**
3000 dia. × 6000 depth to invert						
excavation, support, backfilling and disposal	42.95	3250.64	1966.88	–	nr	**5217.52**
concrete base	5.33	105.68	38.42	366.63	nr	**510.72**
main chamber rings	51.60	1424.83	2661.39	9384.08	nr	**13470.29**
cover slab	0.60	18.05	49.96	2482.94	nr	**2550.96**
concrete surround	19.50	386.95	157.88	975.87	nr	**1520.70**
concrete benching, main and branch channels	8.88	176.69	42.88	153.35	nr	**372.93**
step irons	2.00	38.12	5.34	370.91	nr	**414.36**
access cover and frame	1.44	28.80	3.59	275.50	nr	**307.89**
TOTAL	132.29	5429.75	4926.34	14009.28	nr	**24365.36**

CLASS K: PIPEWORK – MANHOLES AND PIPEWORK ANCILLARIES

Item	Gang hours	Labour £	Plant £	Material £	Unit	Total rate £
01 MANHOLES – cont						
Precast concrete circular manhole; CPM Perfect Manhole; complete with preformed benching and outlets to base; elastomeric seal to joints to rings; single steps as required. Excavations and backfilling not included						
1200 mm dia. with up to four outlets 100 mm or 150 mm dia.; effective internal depth:						
1250 mm deep	2.25	139.41	41.79	992.20	nr	**1173.40**
1500 mm deep	2.25	129.32	41.79	1133.38	nr	**1304.48**
1800 mm deep	2.50	145.09	41.79	1192.67	nr	**1379.55**
2000 mm deep	2.50	145.09	41.79	1372.78	nr	**1559.66**
2500 mm deep	2.75	157.71	52.23	1573.25	nr	**1783.19**
1500 mm dia. with up to four outlets up to 450 mm dia.; effective internal depth:						
2000 mm deep	2.75	157.71	52.23	2884.69	nr	**3094.63**
2500 mm deep	3.05	174.74	58.50	3218.91	nr	**3452.14**
3000 mm deep	3.10	177.89	58.50	3285.66	nr	**3522.06**
3500 mm deep	3.40	194.29	66.86	3435.60	nr	**3696.75**
4000 mm deep	3.55	203.12	68.95	3894.79	nr	**4166.86**
4500 mm deep	3.55	203.12	68.95	4229.00	nr	**4501.07**
1500 mm dia. with up to four outlets up to 600 mm dia.; effective internal depth:						
4000 mm deep	3.95	227.73	73.13	3783.62	nr	**4084.47**
4500 mm deep	4.50	258.64	83.57	4020.68	nr	**4362.89**
03 GULLIES						
Vitrified clay; set in concrete grade C20, 150 mm thick; additional excavation and disposal						
Road gully						
450 mm dia. × 900 mm deep, 100 mm or 150 mm outlet; cast iron road gully grating and frame group 4 434 × 434 mm, on Class B engineering brick seating	0.50	49.64	1.78	840.17	nr	**891.59**
Yard gully (mud); trapped with rodding eye; galvanized bucket; stopper						
225 mm dia., 100 mm dia. outlet, cast iron hinged grate and frame	0.30	29.78	1.07	455.16	nr	**486.01**
Grease interceptors; internal access and bucket						
600 × 450 mm, metal tray and lid, square hopper with horizontal inlet	0.35	34.75	1.25	2353.14	nr	**2389.14**

CLASS K: PIPEWORK – MANHOLES AND PIPEWORK ANCILLARIES

Item	Gang hours	Labour £	Plant £	Material £	Unit	Total rate £
Precast concrete; set in concrete grade C20, 150 mm thick; additional excavation and disposal						
Road gully; trapped with rodding eye; galvanized bucket; stopper						
450 mm dia. × 750 mm deep; cast iron road gully grating and frame group 4, 434 × 434 mm, on Class B engineering brick seating	0.50	49.64	1.78	695.29	nr	**746.72**
450 mm dia. × 900 mm deep; cast iron road gully grating and frame group 4, 434 × 434 mm, on Class B engineering brick seating	0.54	53.61	1.93	720.45	nr	**775.99**
450 mm dia. × 1050 mm deep, cast iron road gully grating and frame group 4, 434 × 434 mm, on Class B engineering brick seating	0.58	57.58	2.07	737.08	nr	**796.74**
04 FRENCH DRAINS, RUBBLE DRAINS, DITCHES						
The rates assume the most efficient items of plant (excavator) and are optimum rates assuming continuous output with no delays caused by other operations or works.						
Ground conditions are assumed to be good easily worked soil with no abnormal conditions that would affect outputs and consistency of work.						
Multiplier Table for labour and plant for various site conditions for working:						
out of sequence × 2.75 (minimum)						
in hard clay × 1.75 to 2.00						
in running sand × 2.75 (minimum)						
in broken rock × 2.75 to 3.50						
below water table × 2.00 (minimum)						
Excavation						
Excavation of trenches for unpiped rubble drains (excluding trench support); cross-sectional area						
0.25–0.50 m^2	0.11	10.92	6.01	–	m	**16.93**
0.50–0.75 m^2	0.13	12.91	7.20	–	m	**20.11**
0.75–1.00 m^2	0.16	15.39	8.41	–	m	**23.80**
1.00–1.50 m^2	0.19	18.37	10.21	–	m	**28.58**
1.50–2.00 m^2	0.22	21.84	12.01	–	m	**33.85**
Excavation of rectangular section ditches; unlined; cross-sectional area						
0.25–0.50 m^2	0.11	10.92	7.24	–	m	**18.16**
0.50–0.75 m^2	0.14	13.90	8.03	–	m	**21.93**
0.75–1.00 m^2	0.17	16.88	9.80	–	m	**26.68**
1.00–1.50 m^2	0.21	20.85	12.15	–	m	**33.00**
1.50–2.00 m^2	0.26	25.81	15.41	–	m	**41.22**

CLASS K: PIPEWORK – MANHOLES AND PIPEWORK ANCILLARIES

Item	Gang hours	Labour £	Plant £	Material £	Unit	Total rate £
04 FRENCH DRAINS, RUBBLE DRAINS, DITCHES – cont						
Excavation – cont						
Excavation of rectangular ditches; lined with precast concrete slabs; cross-sectional area						
0.25–0.50 m^2	0.16	15.88	9.97	18.34	m	**44.20**
0.50–0.75 m^2	0.27	26.81	16.38	26.78	m	**69.97**
0.75–1.00 m^2	0.38	37.73	23.80	38.03	m	**99.55**
1.00–1.50 m^2	0.42	41.70	27.07	53.66	m	**122.43**
1.50–2.00 m^2	0.48	47.65	29.94	75.77	m	**153.36**
Excavation of vee section ditches; unlined; cross-sectional area						
0.25–0.50 m^2	0.11	10.92	7.22	–	m	**18.14**
0.50–0.75 m^2	0.13	12.91	8.56	–	m	**21.47**
0.75–1.00 m^2	0.15	14.89	9.94	–	m	**24.84**
1.00–1.50 m^2	0.19	18.86	12.67	–	m	**31.53**
1.50–2.00 m^2	0.24	23.83	15.78	–	m	**39.61**
Filling						
Filling French and rubble drains with graded material						
graded material; 20 mm stone aggregate	0.31	30.78	20.98	38.90	m^3	**90.66**
broken brick/concrete rubble	0.31	30.78	18.47	42.04	m^3	**91.29**
HydroBlox surface and sub-surface drains; recycled plastic; water conveyance plank format as a French/land drain alternative and or retrofit solution for existing drainage installations						
Sheet material to trenches providing 4560 litre/min/m^2 water permeability to trenches by trenching machine backfilled with excavated material						
Single unit 1.00 m × 220 mm deep × 45 mm thick × providing 1003 l/min permeability; backfilled with excavated material		–	–	–	m	**50.55**
Two units 1.00 m × 220 mm deep × 90 mm thick × providing 2006 l/min permeability; backfilled with excavated material		–	–	–	m	**94.52**
Four units 1.00 m × 220 mm deep × 180 mm thick × providing 4012 l/min permeability; backfilled with excavated material		–	–	–	m	**181.91**

CLASS K: PIPEWORK – MANHOLES AND PIPEWORK ANCILLARIES

Item	Gang hours	Labour £	Plant £	Material £	Unit	**Total rate £**
05 DUCTS AND METAL CULVERTS						
Galvanized steel culverts; bitumen coated						
Sectional corrugated metal culverts, nominal internal dia. 0.5–1 m; 1000 mm nominal internal dia., 1.6 mm thick						
not in trenches	0.15	14.89	1.24	154.65	m	**170.78**
in trenches, depth not exceeding 1.5 m	0.31	30.78	21.09	154.65	m	**206.51**
in trenches, depth 1.5–2 m	0.43	42.69	29.25	154.65	m	**226.59**
in trenches, depth 2–2.5 m	0.51	50.63	34.70	154.65	m	**239.97**
in trenches, depth 2.5–3 m	0.60	59.57	40.83	154.65	m	**255.05**
Sectional corrugated metal culverts, nominal internal dia. exceeding 1.5 m; 1600 mm nominal internal dia., 1.6 mm thick						
not in trenches	0.21	20.85	1.75	248.19	m	**270.78**
in trenches, depth 1.5–2 m	0.44	43.68	29.94	248.19	m	**321.81**
in trenches, depth 2–2.5 m	0.56	55.60	38.11	248.19	m	**341.89**
in trenches, depth 2.5–3 m	0.68	67.51	46.28	248.19	m	**361.97**
in trenches, depth 3–3.5 m	0.82	81.41	55.82	248.19	m	**385.42**
Sectional corrugated metal culverts, nominal internal dia. exceeding 1.5 m; 2000 mm nominal internal dia., 1.6 mm thick						
not in trenches	0.26	25.81	2.17	525.38	m	**553.36**
in trenches, depth 2–2.5 m	0.46	45.67	31.32	525.38	m	**602.37**
in trenches, depth 2.5–3 m	0.60	59.57	40.83	525.38	m	**625.78**
in trenches, depth 3–3.5 m	0.75	74.46	51.03	525.38	m	**650.87**
in trenches, depth 3.5–4 m	0.93	92.33	63.30	525.38	m	**681.01**
Sectional corrugated metal culverts, nominal internal dia. exceeding 1.5 m; 2200 mm nominal internal dia., 1.6 mm thick						
not in trenches	0.33	32.76	2.74	591.93	m	**627.43**
in trenches, depth 2.5–3 m	0.64	63.54	43.55	591.93	m	**699.02**
in trenches, depth 3–3.5 m	0.77	76.44	52.41	591.93	m	**720.78**
in trenches, depth 3.5–4 m	1.02	101.26	69.43	591.93	m	**762.63**
in trenches, depth exceeding 4 m	1.32	131.05	89.83	591.93	m	**812.81**
08 OTHER PIPEWORK ANCILLARIES						
Notes: Refer to Section G (Concrete and concrete ancillaries) for costs relevant to the construction of Headwall Structure.						
Build Ends in						
Connections to existing manholes and other chambers, pipe bore						
150 mm dia.	0.60	68.32	5.98	11.69	nr	**85.98**
225 mm dia.	0.95	108.17	9.45	21.75	nr	**139.37**
300 mm dia.	1.25	142.33	12.46	33.01	nr	**187.79**
375 mm dia.	1.45	165.10	14.45	43.49	nr	**223.04**
450 mm dia.	1.75	199.26	116.80	54.71	nr	**370.78**

CLASS L: PIPEWORK SUPPORTS AND PROTECTION, ANCILLARIES TO LAYING AND EXCAVATION

Item	Gang hours	Labour £	Plant £	Material £	Unit	**Total rate £**
NOTES						
The rates assume the most efficient items of plant (excavator) and are optimum rates assuming continuous output with no delays caused by other operations or works.						
Ground conditions are assumed to be good easily worked soil with no abnormal conditions that would affect outputs and consistency of work.						
Multiplier Table for labour and plant for various site conditions for working:						
out of sequence × 2.75 (minimum)						
in hard clay × 1.75 to 2.00						
in running sand × 2.75 (minimum)						
in broken rock × 2.75 to 3.50						
below water table × 2.00 (minimum)						
RESOURCES – LABOUR						
Labour Resources Build-up Rate						
Supports and protection gang						
2 unskilled operatives (general)		39.20	–	–		
1 plant operator (skill rate 3)		29.17	–	–		
Total Gang Rate/Hour		**68.37**	–	–		
RESOURCES – PLANT						
Plant Resources Build-up Rate						
Supports and protection						
0.40 m^3 hydraulic backacter		–	55.59	–		
Bomag BW 65S		–	9.07	–		
Total Rate/Hour		–	**64.66**	–		
01 EXTRAS TO EXCAVATION AND BACKFILLING						
Drainage sundries						
Extra over any item of drainage for excavation in						
rock	0.65	44.44	58.39	–	m^3	**102.84**
mass concrete	0.84	57.43	75.93	–	m^3	**133.37**
reinforced concrete	1.18	80.68	106.47	–	m^3	**187.16**
Excavation of soft spots, backfilling						
concrete grade C15P	0.30	20.51	26.95	114.10	m^3	**161.56**

CLASS L: PIPEWORK SUPPORTS AND PROTECTION, ANCILLARIES TO LAYING AND EXCAVATION

Item	Gang hours	Labour £	Plant £	Material £	Unit	**Total rate £**
02 SPECIAL PIPE LAYING METHODS						
There are many factors, apart from design consideration, which influence the cost of pipe jacking, so that it is only possible to give guide prices for a sample of the work involved. For more reliable estimates it is advisable to seek the advice of a Specialist Contractor.						
The main cost considerations are:						
* the nature of the ground						
* length of drive						
* location						
* presence of water						
* depth below surface						
Pipe Jacking						
Provision of all plant, equipment and labour establishing						
thrust pit; 6 m × 4 m × 8 m deep		–	–	–	item	**35535.00**
reception pit; 4 m × 4 m × 8 m deep		–	–	–	item	**26059.00**
mobilize and set up pipe jacking equipment		–	–	–	item	**52014.50**
Pipe jacking, excluding the cost of non-drainage materials; concrete pipes BS 5911 Part 1 Class 120 with rebated joints, steel reinforcing band; length not exceeding 50 m; in sand and gravel						
900 mm nominal bore		–	–	–	m	**3218.40**
1200 mm nominal bore		–	–	–	m	**3726.00**
1500 mm nominal bore		–	–	–	m	**2415.00**
1800 mm nominal bore		–	–	–	m	**2570.25**
03 BEDS						
Imported sand						
100 mm deep bed for pipes nominal bore						
100 mm	0.02	1.37	1.11	–	m	**6.19**
150 mm	0.02	1.37	1.11	–	m	**6.48**
225 mm	0.03	2.05	1.67	–	m	**8.38**
300 mm	0.04	2.74	2.22	–	m	**9.84**
150 mm deep bed for pipes nominal bore						
150 mm	0.06	4.10	3.34	–	m	**13.48**
225 mm	0.07	4.79	3.89	–	m	**15.66**
300 mm	0.09	6.15	5.00	–	m	**18.55**
400 mm	0.12	8.21	6.67	–	m	**23.14**
450 mm	0.14	9.57	7.78	–	m	**27.03**
600 mm	0.17	11.62	9.45	–	m	**33.04**
750 mm	0.19	12.99	10.56	–	m	**36.92**
900 mm	0.21	14.36	11.67	–	m	**41.69**
1200 mm	0.25	17.09	13.90	–	m	**49.41**

CLASS L: PIPEWORK SUPPORTS AND PROTECTION, ANCILLARIES TO LAYING AND EXCAVATION

Item	Gang hours	Labour £	Plant £	Material £	Unit	Total rate £
03 BEDS – cont						
Imported granular material						
100 mm deep bed for pipes nominal bore						
100 mm	0.02	1.37	1.11	2.32	m	**4.80**
150 mm	0.03	2.05	1.67	2.52	m	**6.23**
225 mm	0.04	2.74	2.22	2.88	m	**7.84**
300 mm	0.05	3.42	2.78	3.08	m	**9.28**
150 mm deep bed for pipes nominal bore						
150 mm	0.06	4.10	3.34	3.76	m	**11.20**
225 mm	0.08	5.47	4.45	4.33	m	**14.25**
300 mm	0.10	6.84	5.56	4.63	m	**17.02**
400 mm	0.13	8.89	7.23	5.21	m	**21.33**
450 mm	0.15	10.26	8.34	6.01	m	**24.61**
600 mm	0.18	12.31	10.01	7.49	m	**29.81**
750 mm	0.20	13.68	11.12	8.35	m	**33.15**
900 mm	0.22	15.04	12.23	9.80	m	**37.08**
1200 mm	0.26	17.78	14.45	11.51	m	**43.74**
Mass concrete						
100 mm deep bed for pipes nominal bore						
100 mm	0.07	4.79	3.89	8.73	m	**17.41**
150 mm	0.08	5.47	4.45	9.47	m	**19.39**
225 mm	0.09	6.15	5.00	10.94	m	**22.10**
300 mm	0.11	7.52	6.12	11.67	m	**25.30**
150 mm deep bed for pipes nominal bore						
100 mm	0.10	6.84	5.56	13.12	m	**25.52**
150 mm	0.12	8.21	6.67	14.21	m	**29.08**
225 mm	0.14	9.57	7.78	16.40	m	**33.76**
300 mm	0.16	10.94	8.89	17.50	m	**37.33**
400 mm	0.19	12.99	10.56	19.64	m	**43.19**
450 mm	0.21	14.36	11.67	22.96	m	**49.00**
600 mm	0.24	16.41	13.34	28.39	m	**58.14**
750 mm	0.26	17.78	14.45	31.67	m	**63.90**
900 mm	0.28	19.14	15.57	37.15	m	**71.86**
1200 mm	0.32	21.88	17.79	43.68	m	**83.35**
04 HAUNCHES						
The following items allow for dressing the haunching material half-way up the pipe barrel for the full width of the bed and then dressing in triangular fashion to the crown of the pipe. The items exclude the drain bed.						
Mass concrete						
Haunches for pipes nominal bore						
150 mm	0.24	16.41	13.34	8.02	m	**37.77**
225 mm	0.29	19.83	16.12	12.58	m	**48.53**
300 mm	0.36	24.61	20.01	15.85	m	**60.48**

CLASS L: PIPEWORK SUPPORTS AND PROTECTION, ANCILLARIES TO LAYING AND EXCAVATION

Item	Gang hours	Labour £	Plant £	Material £	Unit	**Total rate £**
400 mm	0.43	29.40	23.90	20.96	m	**74.27**
450 mm	0.50	34.19	27.80	28.39	m	**90.38**
600 mm	0.56	38.29	31.13	43.96	m	**113.38**
750 mm	0.62	42.39	34.47	54.36	m	**131.22**
900 mm	0.69	47.18	38.36	74.39	m	**159.92**
1200 mm	0.75	51.28	41.69	97.37	m	**190.35**
05 SURROUNDS						
The following items provide for dressing around the pipe above the bed. Sand and granular material is quantified on the basis of the full width of the bed to the stated distance above the crown, concrete as an ellipse from the top corners of the bed to a point at the stated distance above the crown. The items exclude the drain bed.						
Imported sand						
100 mm thick bed for pipes nominal bore						
100 mm	0.04	2.74	2.22	–	m	**11.67**
150 mm	0.05	3.42	2.78	–	m	**15.83**
225 mm	0.06	4.10	3.34	–	m	**22.86**
300 mm	0.08	5.47	4.45	–	m	**31.41**
150 mm thick bed for pipes nominal bore						
100 mm	0.10	6.84	5.56	–	m	**20.55**
150 mm	0.12	8.21	6.67	–	m	**26.05**
225 mm	0.14	9.57	7.78	–	m	**34.60**
300 mm	0.18	12.31	10.01	–	m	**45.65**
400 mm	0.24	16.41	13.34	–	m	**63.82**
450 mm	0.28	19.14	15.57	–	m	**78.11**
600 mm	0.34	23.25	18.90	–	m	**112.26**
750 mm	0.38	25.98	21.13	–	m	**144.86**
900 mm	0.42	28.72	23.35	–	m	**188.17**
1200 mm	0.50	34.19	27.80	–	m	**278.91**
Imported granular material						
100 mm thick bed for pipes nominal bore						
100 mm	0.12	8.21	3.89	4.19	m	**16.28**
150 mm	0.17	11.62	5.56	5.98	m	**23.17**
225 mm	0.19	12.99	9.54	9.59	m	**32.12**
300 mm	0.23	15.73	11.59	13.39	m	**40.70**
150 mm thick bed for pipes nominal bore						
100 mm	0.10	6.84	5.56	5.07	m	**17.47**
150 mm	0.12	8.21	6.67	6.97	m	**21.84**
225 mm	0.14	9.57	7.78	10.72	m	**28.08**
300 mm	0.18	12.31	10.01	14.59	m	**36.90**
400 mm	0.24	16.41	13.34	21.27	m	**51.02**
450 mm	0.28	19.14	15.57	27.14	m	**61.85**
600 mm	0.34	23.25	18.90	43.63	m	**85.78**
750 mm	0.38	25.98	21.13	61.03	m	**108.14**
900 mm	0.42	28.72	23.35	84.94	m	**137.01**
1200 mm	0.50	34.19	27.80	135.43	m	**197.41**

CLASS L: PIPEWORK SUPPORTS AND PROTECTION, ANCILLARIES TO LAYING AND EXCAVATION

Item	Gang hours	Labour £	Plant £	Material £	Unit	Total rate £
05 SURROUNDS – cont						
Mass concrete						
100 mm thick bed for pipes nominal bore						
100 mm	0.14	9.57	7.78	15.79	m	**33.15**
150 mm	0.16	10.94	8.89	22.67	m	**42.51**
225 mm	0.18	12.31	10.01	36.30	m	**58.62**
300 mm	0.22	15.04	12.23	50.65	m	**77.92**
150 mm thick bed for pipes nominal bore						
150 mm	0.23	15.73	12.79	26.36	m	**54.87**
225 mm	0.26	17.78	14.45	40.57	m	**72.80**
300 mm	0.30	20.51	16.68	55.21	m	**92.40**
400 mm	0.36	24.61	20.01	80.53	m	**125.16**
450 mm	0.40	27.35	22.24	102.64	m	**152.23**
600 mm	0.45	30.77	25.02	165.00	m	**220.79**
750 mm	0.50	34.19	27.80	230.86	m	**292.85**
900 mm	0.55	37.61	30.58	321.11	m	**389.29**
1200 mm	0.61	41.71	33.91	511.79	m	**587.41**
07 CONCRETE STOOLS AND THRUST BLOCKS						
Mass concrete						
Concrete stools or thrust blocks (net volume of concrete excluding volume occupied by pipes)						
0.1 m^3	0.18	12.31	10.01	13.08	nr	**35.39**
0.1–0.2 m^3	0.32	21.88	17.79	26.17	nr	**65.84**
0.2–0.5 m^3	0.62	42.39	34.47	65.55	nr	**142.41**
0.5–1.0 m^3	0.91	62.22	50.59	131.20	nr	**244.01**
1.0–2.0 m^3	1.29	88.20	71.71	262.65	nr	**422.57**
2.0–4.0 m^3	3.15	215.38	175.12	525.03	nr	**915.53**

CLASS M: STRUCTURAL METALWORK

Item	Gang hours	Labour £	Plant £	Material £	Unit	**Total rate £**
NOTES						
The following are guide prices for various structural members commonly found in a Civil Engineering contract. The list is by no means exhaustive and costs are very much dependent on the particular design and will vary greatly according to specific requirements.						
For more detailed prices, reference should be made to Specialist Contractors.						
03 FABRICATION OF MEMBERS FOR FRAMES						
Members for Frames						
Columns						
universal beams; straight on plan		–	–	–	t	**1607.45**
circular hollow sections; straight on plan		–	–	–	t	**1959.30**
rectangular hollow sections; straight on plan		–	–	–	t	**1861.92**
Beams						
universal beams; straight on plan		–	–	–	t	**1495.80**
universal beams; curved on plan		–	–	–	t	**1875.72**
channels; straight on plan		–	–	–	t	**1818.03**
channels; curved on plan		–	–	–	t	**3009.00**
castellated beams; straight on plan		–	–	–	t	**1943.00**
Portal frames						
straight on plan		–	–	–	t	**2256.85**
Trestles, towers and built-up columns						
straight on plan		–	–	–	t	**2495.09**
Trusses and built-up girders						
straight on plan		–	–	–	t	**2507.62**
curved on plan		–	–	–	t	**3134.53**
Bracings						
angles; straight on plan		–	–	–	t	**2139.00**
circular hollow sections; straight on plan		–	–	–	t	**2446.20**
Purlins and cladding rails						
straight on plan		–	–	–	t	**1479.00**
Cold rolled purlins and rails						
straight on plan		–	–	–	t	**3024.29**
Anchorages and holding down bolt assemblies						
base plate and bolt assemblies complete		–	–	–	t	**3071.83**

CLASS M: STRUCTURAL METALWORK

Item	Gang hours	Labour £	Plant £	Material £	Unit	**Total rate £**
06 ERECTION OF FABRICATED MEMBERS FOR FRAMES						
Erection of Frames						
Trial erection		–	–	–	t	**706.37**
Permanent erection		–	–	–	t	**706.37**
08 OFF SITE SURFACE TREATMENT						
Surface Treatment to Frames						
Note: The following preparation and painting systems have been calculated on the basis of 20 m^2 per tonne						
Blast cleaning		–	–	–	m^2	**3.68**
Galvanizing		–	–	–	m^2	**18.37**
Painting						
one coat zinc chromate primer		–	–	–	m^2	**6.11**
one coat two pack epoxy zinc phosphate primer (75 microns dry film thickness)		–	–	–	m^2	**6.11**
two coats epoxy micaceous iron oxide (100 microns dry film thickness per coat)		–	–	–	m^2	**18.37**

CLASS N: MISCELLANEOUS METALWORK

Item	Gang hours	Labour £	Plant £	Material £	Unit	**Total rate £**
NOTES						
The following are guide prices for various structural members commonly found in a Civil Engineering contract. The list is by no means exhaustive and costs are very much dependent on the particular design and will vary greatly according to specific requirements.						
For more detailed prices, reference should be made to Specialist Contractors.						
Cladding						
CESMM4 N.2.1 requires cladding to be measured in square metres, the item so produced being inclusive of all associated flashings at wall corners and bases, eaves, gables, ridges and around openings. As the relative quantities of these flashings will depend very much on the complexity of the building shape, the guide prices shown below for these items are shown separately to help with the accuracy of the estimate.						
Bridge bearings						
Bridge bearings are manufactured and installed to individual specifications. The following guide prices are for different sizes of simple bridge bearings. If requirements are known, then advice ought to be obtained from specialist manufacturers such as CCL.						
If there is a requirement for testing bridge bearings prior to their being installed then the tests should be enumerated separately.						
Specialist advice should be sought once details are known.						
RESOURCES – LABOUR						
Labour Resources Build-up Rate						
Roofing – cladding gang						
1 ganger/chargehand (skill rate 3) – 50% of time		11.84	–	–		
2 skilled operative (skill rate 3)		44.91	–	–		
1 unskilled operative (general) – 50% of time		9.80	–	–		
Total Gang Rate/Hour		**66.54**	–	–		
Bridge bearing gang						
1 skilled operative (skill rate 4)		20.20	–	–		
2 unskilled operatives (general)		39.20	–	–		
Total Gang Rate/Hour		**59.40**	–	–		

CLASS N: MISCELLANEOUS METALWORK

Item	Gang hours	Labour £	Plant £	Material £	Unit	Total rate £
RESOURCES – PLANT						
Plant Resources Build-up Rate						
Cladding to roofs						
15 m telescopic access platform – 50% of time		–	22.70	–		
Total Gang Rate/Hour		–	**22.70**	–		
Cladding to walls						
15 m telescopic access platform		–	39.66	–		
Total Gang Rate/Hour		–	**39.66**	–		
01 MILD STEEL						
Mild steel						
Stairways and landings		–	–	–	t	**7973.20**
Walkways and platforms		–	–	–	t	**8518.80**
Ladders						
cat ladder; 64 × 13 mm bar strings; 19 mm rungs at 250 mm centres; 450 mm wide with safety hoops		–	–	–	m	**146.32**
Miscellaneous framing						
angle section; 200 × 200 × 16 mm (equal)		–	–	–	m	**353.97**
angle section; 150 × 150 × 10 mm (equal)		–	–	–	m	**163.63**
angle section; 100 × 100 × 12 mm (equal)		–	–	–	m	**293.86**
angle section; 200 × 150 × 15 mm (unequal)		–	–	–	m	**233.74**
angle section; 150 × 75 × 10 mm (unequal)		–	–	–	m	**123.83**
universal beams; 914 × 419 mm		–	–	–	m	**2738.30**
universal beams; 533 × 210 mm		–	–	–	m	**935.02**
universal joists; 127 × 76 mm		–	–	–	m	**100.18**
channel section; 381 × 102 mm		–	–	–	m	**427.43**
channel section; 254 × 76 mm		–	–	–	m	**220.40**
channel section; 152 × 76 mm		–	–	–	m	**140.24**
tubular section; 100 × 100 × 10 mm		–	–	–	m	**200.38**
tubular section; 200 × 200 × 15 mm		–	–	–	m	**667.90**
tubular section; 76.1 × 5.0 mm		–	–	–	m	**69.45**
tubular section; 139.7 × 6.3 mm		–	–	–	m	**193.69**
Mild steel; galvanized						
Handrails						
76 mm dia. tubular handrail, 48 mm dia. standards at 750 mm centres, 48 mm dia. middle rail, 1070 mm high overall		–	–	–	m	**172.36**
Plate flooring						
8 mm (on plain) Durbar pattern floor plates, maximum weight each panel 50 kg		–	–	–	m^2	**451.86**

CLASS N: MISCELLANEOUS METALWORK

Item	Gang hours	Labour £	Plant £	Material £	Unit	Total rate £
Mild steel; internally and externally acid dipped, rinse and hot dip galvanized, epoxy internal paint						
Uncovered tanks						
1600 litre capacity open top water tank		–	–	–	nr	**1554.96**
18180 litre capacity open top water tank		–	–	–	nr	**14855.20**
Covered tanks						
1600 litre capacity open top water tank with loose fitting lid;		–	–	–	nr	**2285.32**
18180 litre capacity open top water tank with loose fitting lid		–	–	–	nr	**23014.40**
Corrugated steel plates to BS 1449 Pt 1, Gr H4, sealed and bolted; BS729 hot dip galvanized, epoxy internal and external paint						
Uncovered tanks						
713 m^3 capacity bolted cylindrical open top tank		–	–	–	nr	**57166.48**
02 PROPRIETARY WORK						
Galvanized steel troughed sheeting; 0.70 mm metal thickness, 75 mm deep corrugations; colour coating each side; fixing with plastic capped self-tapping screws to steel purlins or rails						
Cladding						
upper surfaces inclined at an angle ne 30° to the horizontal	0.15	9.98	0.52	54.34	m^2	**64.84**
Extra for:						
galvanized steel inner lining sheet, 0.40 mm thick, Plastisol colour coating	0.06	3.99	0.41	26.52	m^2	**30.93**
galvanized steel inner lining sheet, 0.40 mm thick, Plastisol colour coating; insulation, 80 mm thick	0.08	5.32	0.28	92.80	m^2	**98.40**
surfaces inclined at an angle exceeding 60° to the horizontal	0.16	10.65	1.10	46.40	m^2	**58.14**
Extra for:						
galvanized steel inner lining sheet, 0.40 mm thick, Plastisol colour coating	0.11	7.32	0.76	26.52	m^2	**34.60**
galvanized steel inner lining sheet, 0.40 mm thick, Plastisol colour coating; insulation, 80 mm thick	0.13	8.65	0.89	92.80	m^2	**102.35**
Galvanized steel flashings; 0.90 mm metal thickness; bent to profile; fixing with plastic capped self-tapping screws to steel purlins or rails; mastic sealant						
Flashings to cladding						
250 mm girth	0.12	7.99	0.83	29.85	m	**38.66**
500 mm girth	0.18	11.98	1.24	43.77	m	**56.99**
750 mm girth	0.22	14.64	1.51	48.41	m	**64.56**

CLASS N: MISCELLANEOUS METALWORK

Item	Gang hours	Labour £	Plant £	Material £	Unit	Total rate £
02 PROPRIETARY WORK – cont						
Aluminium profiled sheeting; 0.90 mm metal thickness, 75 mm deep corrugations; colour coating each side; fixing with plastic capped self-tapping screws to steel purlins or rails						
Cladding						
upper surfaces inclined at an angle exceeding 60° to the horizontal	0.20	13.31	1.38	53.03	m^2	**67.71**
Extra for:						
aluminium inner lining sheet, 0.70 mm thick, Plastisol colour coating	0.13	8.65	1.00	33.15	m^2	**42.80**
aluminium inner lining sheet, 0.70 mm thick, Plastisol colour coating; insulation, 80 mm thick	0.15	9.98	1.03	99.44	m^2	**110.45**
Aluminium profiled sheeting; 1.00 mm metal thickness, 75 mm deep corrugations; colour coating each side; fixing with plastic capped self-tapping screws to steel purlins or rails						
Cladding						
upper surfaces inclined at an angle ne 30° to the horizontal	0.17	11.31	0.58	66.28	m^2	**78.18**
Extra for:						
aluminium inner lining sheet, 0.70 mm thick, Plastisol colour coating	0.09	5.99	0.31	33.15	m^2	**39.45**
aluminium inner lining sheet, 0.70 mm thick, Plastisol colour coating; insulation, 80 mm thick	0.11	7.32	0.38	99.44	m^2	**107.14**
Aluminium flashings; 0.90 mm metal thickness; bent to profile; fixing with plastic capped self-tapping screws to steel purlins or rails; mastic sealant						
Flashings to cladding						
250 mm girth	0.12	7.99	0.83	29.99	m	**38.80**
500 mm girth	0.18	11.98	1.24	43.78	m	**56.99**
750 mm girth	0.22	14.64	1.51	49.08	m	**65.23**
Flooring; Eurogrid; galvanized mild steel						
Open grid flooring						
type 41/100; 3 × 25 mm bearer bar; 6 mm dia. transverse bar		–	–	–	m^2	**69.80**
type 41/100; 5 × 25 mm bearer bar; 6 mm dia. transverse bar		–	–	–	m^2	**85.68**
type 41/100; 3 × 30 mm bearer bar; 6 mm dia. transverse bar		–	–	–	m^2	**79.73**

CLASS N: MISCELLANEOUS METALWORK

Item	Gang hours	Labour £	Plant £	Material £	Unit	**Total rate £**
Duct covers; Stelduct; galvanized mild steel						
Duct covers; pedestrian duty						
225 mm clear opening		–	–	–	m	**147.56**
450 mm clear opening		–	–	–	m	**147.93**
750 mm clear opening		–	–	–	m	**176.70**
Duct covers; medium duty						
225 mm clear opening		–	–	–	m	**737.80**
450 mm clear opening		–	–	–	m	**739.66**
750 mm clear opening		–	–	–	m	**706.80**
Duct covers; heavy duty						
225 mm clear opening		–	–	–	m	**295.12**
450 mm clear opening		–	–	–	m	**369.83**
750 mm clear opening		–	–	–	m	**441.75**
Bridge bearings						
Supply plain rubber bearings (3 m and 5 m lengths)						
150 × 20 mm	0.35	20.79	–	55.99	m	**76.78**
150 × 25 mm	0.35	20.79	–	67.21	m	**88.00**
Supply and place in position laminated elastomeric rubber bearing						
250 × 150 × 19 mm	0.25	14.85	–	41.61	nr	**56.46**
300 × 200 × 19 mm	0.25	14.85	–	45.61	nr	**60.46**
300 × 200 × 30 mm	0.27	16.04	–	57.60	nr	**73.63**
300 × 200 × 41 mm	0.27	16.04	–	86.40	nr	**102.44**
300 × 250 × 41 mm	0.30	17.82	–	108.78	nr	**126.60**
300 × 250 × 63 mm	0.30	17.82	–	152.00	nr	**169.82**
400 × 250 × 19 mm	0.32	19.01	–	80.00	nr	**99.01**
400 × 250 × 52 mm	0.32	19.01	–	176.41	nr	**195.42**
400 × 300 × 19 mm	0.32	19.01	–	92.79	nr	**111.80**
600 × 450 × 24 mm	0.35	20.79	–	191.98	nr	**212.77**
Adhesive fixings to laminated elastomeric rubber bearings						
2 mm thick epoxy adhesive	1.00	59.40	–	64.21	m^2	**123.61**
15 mm thick epoxy mortar	1.50	89.10	–	429.58	m^2	**518.68**
15 mm thick epoxy pourable grout	2.00	118.80	–	408.90	m^2	**527.70**
Supply and install mechanical guides for laminated elastomeric rubber bearings						
500kN SLS design load; FP50 fixed pin Type 1	2.00	118.80	–	1235.45	nr	**1354.25**
500kN SLS design load; FP50 fixed pin Type 2	2.00	118.80	–	1292.72	nr	**1411.52**
750kN SLS design load; FP75 fixed pin Type 1	2.10	124.74	–	1466.34	nr	**1591.08**
750kN SLS design load; FP75 fixed pin Type 2	2.10	124.74	–	1652.02	nr	**1776.76**
300kN SLS design load; UG300 Uniguide Type 1	2.00	118.80	–	1503.18	nr	**1621.98**
300kN SLS design load; UG300 Uniguide Type 2	2.00	118.80	–	1726.41	nr	**1845.21**
Supply and install fixed pot bearings						
355 × 355 mm; PF200	2.00	118.80	–	1558.68	nr	**1677.48**
425 × 425 mm; PF300	2.10	124.74	–	1724.77	nr	**1849.51**
Supply and install free sliding pot bearings						
445 × 345 mm; PS200	2.10	124.74	–	1806.54	nr	**1931.28**
520 × 415 mm; PS300	2.20	130.68	–	2303.52	nr	**2434.20**

CLASS N: MISCELLANEOUS METALWORK

Item	Gang hours	Labour £	Plant £	Material £	Unit	Total rate £
02 PROPRIETARY WORK – cont						
Bridge bearings – cont						
Supply and install guided sliding pot bearings						
455 × 375 mm; PG200	2.20	130.68	–	2399.62	nr	**2530.30**
545 × 435 mm; PG300	2.30	136.62	–	2857.53	nr	**2994.15**
Testing; laminated elastomeric bearings						
compression test		–	–	83.96	nr	**83.96**
shear test		–	–	109.53	nr	**109.53**
bond test (Exclusive of cost of bearings as this is a destructive test)		–	–	416.24	nr	**416.24**

CLASS O: TIMBER

Item	Gang hours	Labour £	Plant £	Material £	Unit	**Total rate £**
RESOURCES – LABOUR						
Labour Resources Build-up Rate						
Timber gang						
1 foreman carpenter/joiner (craftsman)		26.14	–	–		
1 carpenter/joiner (craftsman)		22.80	–	–		
1 unskilled operative (general)		19.60	–	–		
1 plant operator (skill rate 3) – 50% of time		14.59	–	–		
Total Gang Rate/Hour		**83.12**	–	–		
Timber fixings gang						
1 foreman carpenter/joiner (craftsman)		26.14	–	–		
1 unskilled operative (general)		19.60	–	–		
Total Gang Rate/Hour		**45.74**	–	–		
RESOURCES – MATERIAL						
The timber material prices shown below are averages, actual prices being very much affected by availability of suitably sized forest timbers capable of conversion to the sizes shown. Apart from the practicality of being able to obtain the larger sizes in one timber, normal practice and drive for economy would lead to their being built up using smaller timbers.						
RESOURCES – PLANT						
Labour Resources Build-up Rate						
Timber						
tractor/trailer		–	57.68	–		
10t crawler crane (25% of time)		–	17.01	–		
5.6t rough terrain forklift (25% of time)		–	9.64	–		
7.5 KVA diesel generator		–	11.97	–		
two K637 rotary hammers		–	1.43	–		
two electric screwdrivers		–	3.39	–		
Total Gang Rate/Hour		–	**101.10**	–		
01 HARDWOOD COMPONENTS						
Greenheart						
100 × 75 mm						
length not exceeding 1.5 m	0.15	12.47	15.20	9.01	m	**36.68**
length 1.5–3 m	0.13	10.81	13.21	9.01	m	**33.03**
length 3–5 m	0.12	9.97	12.14	9.01	m	**31.12**
length 5–8 m	0.11	9.14	11.19	9.47	m	**29.80**
150 × 75 mm						
length not exceeding 1.5 m	0.17	14.13	17.47	11.19	m	**42.79**
length 1.5–3 m	0.15	12.47	15.20	10.92	m	**38.59**
length 3–5 m	0.14	11.64	14.16	10.92	m	**36.71**
length 5–8 m	0.12	9.97	15.33	11.47	m	**36.78**

CLASS O: TIMBER

Item	Gang hours	Labour £	Plant £	Material £	Unit	Total rate £
01 HARDWOOD COMPONENTS – cont						
Greenheart – cont						
200 × 100 mm						
length not exceeding 1.5 m	0.24	19.95	24.26	22.92	m	**67.13**
length 1.5–3 m	0.21	17.46	21.51	22.92	m	**61.89**
length 3–5 m	0.20	16.21	19.75	22.92	m	**58.88**
length 5–8 m	0.18	14.96	18.20	22.92	m	**56.08**
length 8–12 m	0.16	13.30	16.18	27.97	m	**57.44**
200 × 200 mm						
length not exceeding 1.5 m	0.42	34.91	42.46	37.44	m	**114.81**
length 1.5–3 m	0.40	33.25	40.44	39.68	m	**113.37**
length 3–5 m	0.38	31.59	38.42	39.68	m	**109.68**
length 5–8 m	0.34	28.26	34.37	39.68	m	**102.31**
length 8–12 m	0.30	24.94	30.86	42.77	m	**98.57**
225 × 100 mm						
length not exceeding 1.5 m	0.27	22.44	27.58	22.92	m	**72.94**
length 1.5–3 m	0.24	19.95	24.26	22.92	m	**67.13**
length 3–5 m	0.22	18.29	22.78	22.92	m	**63.98**
length 5–8 m	0.20	16.62	25.55	22.92	m	**65.09**
length 8–12 m	0.18	14.96	18.73	27.97	m	**61.66**
300 × 100 mm						
length not exceeding 1.5 m	0.36	29.92	36.40	28.76	m	**95.09**
length 1.5–3 m	0.33	27.43	33.65	28.77	m	**89.85**
length 3–5 m	0.30	24.94	30.86	28.77	m	**84.57**
length 5–8 m	0.27	22.44	27.58	26.61	m	**76.64**
length 8–12 m	0.24	19.95	24.26	40.66	m	**84.87**
300 × 200 mm						
length not exceeding 1.5 m	0.50	41.56	50.56	59.97	m	**152.08**
length 1.5–3 m	0.45	37.41	45.78	59.97	m	**143.15**
length 3–5 m	0.40	33.25	40.44	59.97	m	**133.66**
length 5–8 m	0.35	29.09	35.63	63.64	m	**128.37**
length 8–12 m	0.30	24.94	30.33	63.64	m	**118.91**
300 × 300 mm						
length not exceeding 1.5 m	0.52	43.22	52.58	89.93	m	**185.73**
length 1.5–3 m	0.48	39.90	48.53	89.93	m	**178.36**
length 3–5 m	0.44	36.57	44.49	89.93	m	**171.00**
length 5–8 m	0.40	33.25	40.44	95.65	m	**169.34**
length 8–12 m	0.36	29.92	36.40	95.65	m	**161.97**
450 × 450 mm						
length not exceeding 1.5 m	0.98	81.46	99.62	198.11	m	**379.18**
length 1.5–3 m	0.90	74.81	91.53	198.11	m	**364.44**
length 3–5 m	0.83	68.99	84.20	214.16	m	**367.35**
length 5–8 m	0.75	62.34	76.11	203.06	m	**341.51**
length 8–12 m	0.68	56.52	68.75	212.98	m	**338.25**

CLASS O: TIMBER

Item	Gang hours	Labour £	Plant £	Material £	Unit	**Total rate £**
02 SOFTWOOD COMPONENTS						
Softwood; stress graded SC3/4						
100 × 75 mm						
up to 3.00 m long	0.10	8.31	8.86	4.88	m	**22.06**
3.00–5.00 m long	0.10	8.31	8.86	6.94	m	**24.11**
150 × 75 mm						
up to 3.00 m long	0.13	10.81	11.50	9.66	m	**31.97**
3.00–5.00 m long	0.11	9.14	9.75	12.25	m	**31.14**
200 × 100 mm						
up to 3.00 m long	0.16	13.30	14.17	18.02	m	**45.48**
3.00–5.00 m long	0.14	11.64	12.41	26.95	m	**50.99**
200 × 200 mm						
up to 3.00 m long	0.25	20.78	22.13	54.14	m	**97.06**
3.00–5.00 m long	0.23	19.12	20.37	67.80	m	**107.29**
5.00–8.00 m long	0.20	16.62	17.71	59.69	m	**94.02**
300 × 200 mm						
up to 3.00 m long	0.27	22.44	23.91	91.28	m	**137.63**
3.00–5.00 m long	0.25	20.78	22.13	114.83	m	**157.74**
5.00–8.00 m long	0.23	19.12	20.37	92.40	m	**131.88**
300 × 300 mm						
up to 3.00 m long	0.30	24.94	26.55	147.78	m	**199.27**
3.00–5.00 m long	0.27	22.44	23.91	189.80	m	**236.16**
5.00–8.00 m long	0.25	20.78	22.13	148.73	m	**191.64**
450 × 450 mm						
up to 3.00 m long	0.35	29.09	30.99	422.65	m	**482.74**
3.00–5.00 m long	0.31	25.77	27.46	501.61	m	**554.84**
5.00–8.00 m long	0.28	23.27	24.79	417.49	m	**465.55**
600 × 600 mm						
3.00–5.00 m long	0.39	32.42	29.92	757.54	m	**819.87**
5.00–8.00 m long	0.37	30.76	32.75	762.21	m	**825.71**
ADD to the above prices for vacuum/pressure impregnating to minimum 5.30 kg/m^3 salt retention		–	–	49.25	m^3	**49.25**
03 HARDWOOD DECKING						
Greenheart; wrought finish						
Thickness 25–50 mm						
150 × 50 mm	0.58	48.21	51.34	71.50	m^2	**171.05**
Thickness 50–75 mm						
200 × 75 mm	0.75	62.34	66.41	91.19	m^2	**219.94**
Thickness 75–100 mm						
250 × 100 mm	0.95	78.97	82.10	122.59	m^2	**283.66**

CLASS O: TIMBER

Item	Gang hours	Labour £	Plant £	Material £	Unit	Total rate £
04 SOFTWOOD DECKING						
Douglas Fir						
Thickness 25–50 mm						
150 × 50 mm	0.39	32.42	34.54	36.79	m^2	**103.75**
Thickness 50–75 mm						
200 × 75 mm	0.50	41.56	44.29	48.25	m^2	**134.10**
Thickness 75–100 mm						
250 × 100 mm	0.65	54.03	73.51	53.28	m^2	**180.82**
05 FITTINGS AND FASTENINGS						
Metalwork						
Spikes; mild steel material rosehead						
14 × 14 × 275 mm long	0.13	5.95	–	1.44	nr	**7.39**
Metric mild steel bolts, nuts and washers						
M6 × 25 mm long	0.05	2.29	–	0.22	nr	**2.51**
M6 × 50 mm long	0.05	2.29	–	0.24	nr	**2.53**
M6 × 75 mm long	0.05	2.29	–	0.29	nr	**2.58**
M6 × 100 mm long	0.06	2.74	–	0.31	nr	**3.05**
M6 × 120 mm long	0.06	2.74	–	0.46	nr	**3.21**
M6 × 150 mm long	0.06	2.74	–	0.54	nr	**3.28**
M8 × 25 mm long	0.05	2.29	–	0.25	nr	**2.54**
M8 × 50 mm long	0.05	2.29	–	0.29	nr	**2.58**
M8 × 75 mm long	0.06	2.74	–	0.34	nr	**3.09**
M8 × 100 mm long	0.06	2.74	–	0.39	nr	**3.14**
M8 × 120 mm long	0.07	3.20	–	0.60	nr	**3.81**
M8 × 150 mm long	0.07	3.20	–	0.71	nr	**3.91**
M10 × 25 mm long	0.05	2.29	–	0.46	nr	**2.75**
M10 × 50 mm long	0.06	2.74	–	0.50	nr	**3.25**
M10 × 75 mm long	0.06	2.74	–	0.57	nr	**3.31**
M10 × 100 mm long	0.07	3.20	–	0.66	nr	**3.86**
M10 × 120 mm long	0.07	3.20	–	0.68	nr	**3.88**
M10 × 150 mm long	0.07	3.20	–	1.11	nr	**4.31**
M10 × 200 mm long	0.08	3.66	–	1.84	nr	**5.49**
M12 × 25 mm long	0.06	2.74	–	0.58	nr	**3.33**
M12 × 50 mm long	0.06	2.74	–	0.63	nr	**3.38**
M12 × 75 mm long	0.07	3.20	–	0.72	nr	**3.92**
M12 × 100 mm long	0.07	3.20	–	0.81	nr	**4.01**
M12 × 120 mm long	0.08	3.66	–	0.88	nr	**4.54**
M12 × 150 mm long	0.08	3.66	–	1.15	nr	**4.81**
M12 × 200 mm long	0.08	3.66	–	1.60	nr	**5.25**
M12 × 240 mm long	0.09	4.12	–	2.39	nr	**6.51**
M12 × 300 mm long	0.10	4.57	–	2.59	nr	**7.16**
M16 × 50 mm long	0.07	3.20	–	1.06	nr	**4.26**
M16 × 75 mm long	0.07	3.20	–	1.17	nr	**4.38**
M16 × 100 mm long	0.08	3.66	–	1.32	nr	**4.98**
M16 × 120 mm long	0.08	3.66	–	1.53	nr	**5.19**
M16 × 150 mm long	0.09	4.12	–	1.70	nr	**5.82**
M16 × 200 mm long	0.09	4.12	–	2.34	nr	**6.46**
M16 × 240 mm long	0.10	4.57	–	3.35	nr	**7.93**

CLASS O: TIMBER

Item	Gang hours	Labour £	Plant £	Material £	Unit	**Total rate £**
M16 × 300 mm long	0.10	4.57	–	3.64	nr	**8.22**
M20 × 50 mm long	0.07	3.20	–	1.17	nr	**4.37**
M20 × 75 mm long	0.08	3.66	–	1.84	nr	**5.50**
M20 × 100 mm long	0.08	3.66	–	2.08	nr	**5.74**
M20 × 120 mm long	0.09	4.12	–	2.54	nr	**6.66**
M20 × 150 mm long	0.09	4.12	–	2.70	nr	**6.82**
M20 × 200 mm long	0.10	4.57	–	3.53	nr	**8.10**
M20 × 240 mm long	0.10	4.57	–	4.69	nr	**9.26**
M20 × 300 mm long	0.11	5.03	–	5.19	nr	**10.22**
M24 × 50 mm long	0.08	3.66	–	3.69	nr	**7.35**
M24 × 75 mm long	0.08	3.66	–	3.95	nr	**7.61**
M24 × 100 mm long	0.09	4.12	–	4.08	nr	**8.20**
M24 × 120 mm long	0.09	4.12	–	4.37	nr	**8.49**
M24 × 150 mm long	0.10	4.57	–	4.89	nr	**9.46**
M24 × 200 mm long	0.11	5.03	–	5.66	nr	**10.69**
M24 × 240 mm long	0.11	5.03	–	7.21	nr	**12.24**
M24 × 300 mm long	0.12	5.49	–	7.98	nr	**13.47**
M30 × 100 mm long	0.09	4.12	–	9.75	nr	**13.87**
M30 × 120 mm long	0.10	4.57	–	10.14	nr	**14.71**
M30 × 150 mm long	0.11	5.03	–	10.73	nr	**15.76**
M30 × 200 mm long	0.11	5.03	–	11.71	nr	**16.74**
Carriage bolts, nuts and washer						
M6 × 25 mm long	0.05	2.29	–	1.14	nr	**3.43**
M6 × 50 mm long	0.05	2.29	–	0.62	nr	**2.90**
M6 × 75 mm long	0.05	2.29	–	0.66	nr	**2.94**
M6 × 100 mm long	0.06	2.74	–	0.81	nr	**3.55**
M6 × 150 mm long	0.06	2.74	–	1.02	nr	**3.76**
M8 × 25 mm long	0.05	2.29	–	0.72	nr	**3.01**
M8 × 50 mm long	0.05	2.29	–	0.76	nr	**3.05**
M8 × 75 mm long	0.06	2.74	–	0.79	nr	**3.53**
M8 × 100 mm long	0.06	2.74	–	0.87	nr	**3.62**
M8 × 150 mm long	0.07	3.20	–	0.90	nr	**4.10**
M8 × 200 mm long	0.07	3.20	–	1.42	nr	**4.63**
M10 × 25 mm long	0.05	2.29	–	0.92	nr	**3.20**
M10 × 50 mm long	0.06	2.74	–	0.88	nr	**3.62**
M10 × 75 mm long	0.06	2.74	–	0.90	nr	**3.65**
M10 × 100 mm long	0.07	3.20	–	0.93	nr	**4.14**
M10 × 150 mm long	0.07	3.20	–	1.15	nr	**4.35**
M10 × 200 mm long	0.08	3.66	–	1.64	nr	**5.30**
M10 × 240 mm long	0.08	3.66	–	3.13	nr	**6.79**
M10 × 300 mm long	0.09	4.12	–	3.40	nr	**7.51**
M12 × 25 mm long	0.06	2.74	–	1.34	nr	**4.09**
M12 × 50 mm long	0.06	2.74	–	1.70	nr	**4.45**
M12 × 75 mm long	0.07	3.20	–	1.85	nr	**5.05**
M12 × 100 mm long	0.07	3.20	–	1.92	nr	**5.13**
M12 × 150 mm long	0.08	3.66	–	2.35	nr	**6.00**
M12 × 200 mm long	0.08	3.66	–	2.93	nr	**6.59**
M12 × 240 mm long	0.09	4.12	–	4.29	nr	**8.41**
M12 × 300 mm long	0.10	4.57	–	4.57	nr	**9.14**

CLASS O: TIMBER

Item	Gang hours	Labour £	Plant £	Material £	Unit	**Total rate £**
05 FITTINGS AND FASTENINGS – cont						
Galvanized steel						
Straps						
30 × 2.5 × 600 mm girth	0.13	5.95	–	2.38	nr	**8.32**
30 × 2.5 × 800 mm girth	0.13	5.95	–	3.13	nr	**9.07**
30 × 2.5 × 1000 mm girth	0.13	5.95	–	3.90	nr	**9.84**
30 × 2.5 × 1200 mm girth	0.15	6.86	–	4.69	nr	**11.55**
30 × 2.5 × 1400 mm girth	0.13	5.95	–	5.47	nr	**11.42**
30 × 2.5 × 1600 mm girth	0.15	6.86	–	6.25	nr	**13.11**
30 × 2.5 × 1800 mm girth	0.15	6.86	–	7.03	nr	**13.90**
30 × 5 × 600 mm girth	0.13	5.95	–	4.64	nr	**10.58**
30 × 5 × 800 mm girth	0.13	5.95	–	6.20	nr	**12.14**
30 × 5 × 1000 mm girth	0.13	5.95	–	7.86	nr	**13.81**
30 × 5 × 1200 mm girth	0.15	6.86	–	9.36	nr	**16.22**
30 × 5 × 1400 mm girth	0.13	5.95	–	10.96	nr	**16.91**
30 × 5 × 1600 mm girth	0.15	6.86	–	12.45	nr	**19.31**
30 × 5 × 1800 mm girth	0.15	6.86	–	13.99	nr	**20.85**
Timber connectors; round toothed plate, single sided for 10 mm or 12 mm bolts						
38 mm dia.	0.01	0.23	–	2.13	nr	**2.36**
50 mm dia.	0.01	0.23	–	2.20	nr	**2.43**
63 mm dia.	0.01	0.37	–	2.39	nr	**2.75**
75 mm dia.	0.01	0.37	–	2.63	nr	**2.99**
Timber connectors; round toothed plate, double sided for 10 mm or 12 mm bolts						
38 mm dia.	0.01	0.23	–	2.16	nr	**2.39**
50 mm dia.	0.01	0.23	–	2.20	nr	**2.43**
63 mm dia.	0.01	0.37	–	2.39	nr	**2.75**
75 mm dia.	0.01	0.37	–	2.63	nr	**2.99**
Split ring connectors						
50 mm dia.	0.01	0.27	–	3.22	nr	**3.50**
63 mm dia.	0.01	0.27	–	3.33	nr	**3.60**
101 mm dia.	0.01	0.27	–	8.71	nr	**8.99**
Shear plate connectors						
67 mm dia.	0.01	0.27	–	5.38	nr	**5.65**
101 mm dia.	0.01	0.27	–	22.20	nr	**22.47**
Flitch plates						
200 × 75 × 10 mm	0.07	3.20	–	8.07	nr	**11.27**
300 × 100 × 10 mm	0.09	4.12	–	15.95	nr	**20.06**
450 × 150 × 12 mm	0.15	6.86	–	20.77	nr	**27.63**
Stainless steel						
Straps						
30 × 2.5 × 600 mm girth	0.13	5.95	–	5.07	nr	**11.02**
30 × 2.5 × 800 mm girth	0.13	5.95	–	6.95	nr	**12.89**
30 × 2.5 × 1000 mm girth	0.13	5.95	–	3.90	nr	**9.84**
30 × 2.5 × 1200 mm girth	0.15	6.86	–	4.69	nr	**11.55**
30 × 2.5 × 1400 mm girth	0.13	5.95	–	5.47	nr	**11.42**
30 × 2.5 × 1600 mm girth	0.15	6.86	–	6.25	nr	**13.11**
30 × 2.5 × 1800 mm girth	0.15	6.86	–	7.03	nr	**13.90**

CLASS O: TIMBER

Item	Gang hours	Labour £	Plant £	Material £	Unit	**Total rate £**
30 × 5 × 600 mm girth	0.13	5.95	–	8.15	nr	**14.10**
30 × 5 × 800 mm girth	0.13	5.95	–	10.77	nr	**16.72**
30 × 5 × 1000 mm girth	0.13	5.95	–	7.86	nr	**13.81**
30 × 5 × 1200 mm girth	0.15	6.86	–	9.36	nr	**16.22**
30 × 5 × 1400 mm girth	0.13	5.95	–	10.96	nr	**16.91**
30 × 5 × 1600 mm girth	0.15	6.86	–	12.45	nr	**19.31**
30 × 5 × 1800 mm girth	0.15	6.86	–	13.99	nr	**20.85**
Coach screws						
5.0 mm dia. × 75 mm long	0.04	1.83	–	–	nr	**1.83**
7.0 mm dia. × 105 mm long	0.05	2.29	–	–	nr	**2.29**
7.0 mm dia. × 140 mm long	0.06	2.74	–	–	nr	**2.75**
10.0 mm dia. × 95 mm long	0.06	2.74	–	0.01	nr	**2.75**
10.0 mm dia. × 165 mm long	0.07	3.20	–	0.01	nr	**3.21**
Timber connectors; round toothed plate, single sided for 10 mm or 12 mm bolts						
38 mm dia.	0.01	0.23	–	3.22	nr	**3.45**
50 mm dia.	0.01	0.23	–	3.35	nr	**3.58**
63 mm dia.	0.01	0.37	–	3.84	nr	**4.21**
75 mm dia.	0.01	0.37	–	4.27	nr	**4.63**
Timber connectors; round toothed plate, double sided for 10 mm or 12 mm bolts						
38 mm dia.	0.01	0.23	–	3.24	nr	**3.46**
50 mm dia.	0.01	0.23	–	3.35	nr	**3.58**
63 mm dia.	0.01	0.37	–	3.84	nr	**4.21**
75 mm dia.	0.01	0.37	–	4.27	nr	**4.63**
Split ring connectors						
63 mm dia.	0.06	2.74	–	14.83	nr	**17.57**
101 mm dia.	0.06	2.74	–	29.03	nr	**31.77**
Shear plate connectors						
67 mm dia.	0.06	2.74	–	0.58	nr	**3.32**
101 mm dia.	0.06	2.74	–	1.23	nr	**3.97**

CLASS P: PILES

Item	Gang hours	Labour £	Plant £	Material £	Unit	**Total rate £**
GENERALLY						
There are a number of different types of piling which are available for use in differing situations. Selection of the most suitable type of piling for a particular site will depend on a number of factors including the physical conditions likely to be encountered during driving, the loads to be carried, the design of superstructure, etc. The most commonly used systems are included in this section						
It is essential that a thorough and adequate site investigation is carried out to ascertain details of the ground strata and bearing capacities to enable a proper assessment to be made of the most suitable and economical type of piling to be adopted.						
There are so many factors, apart from design considerations, which influence the cost of piling that it is not possible to give more than an approximate indication of costs. To obtain reliable costs for a particular contract advice should be sought from a company specializing in the particular type of piling proposed. Some Specialist Contractors will also provide a design service if required.						
01 BORED CAST IN PLACE CONCRETE PILES						
Generally						
The items 'number of piles' are calculated based on the following:						
allowance for provision of all plant, equipment and labour including transporting to and from site and establishing and dismantling at £6,000 in total.						
moving the rig to and setting up at each pile position; preparing to commence driving; £40.00, £60.00 and 75.00 per 300 mm, 450 mm and 600 mm dia. piles using the tripod mounted percussion rig.						
Standing time is quoted at £110.00 per hour for tripod rig, £208.00 per hour for mobile rig and £145.00 per hour for continuous flight auger.						
Disposal of material arising from pile bores						
The disposal of excavated material is shown separately, partly as this task is generally carried out by the main contractor rather than the piling specialist, but also to allow for simple adjustment should contaminated ground be envisaged.						

CLASS P: PILES

Item	Gang hours	Labour £	Plant £	Material £	Unit	**Total rate £**
Disposal of material arising from pile bores; collection from around piling operations						
storage on site; 100 m maximum distance; using 22.5 t ADT	0.05	1.77	5.05	–	m^3	**6.82**
removal; 5 km distance; using 20 t tipper	0.12	4.25	14.22	–	m^3	**18.47**
removal; 15 km distance; using 20 t tipper	0.22	7.80	26.06	–	m^3	**33.86**
Add to the above rates where tipping charges apply (excluding Landfill Tax):						
non-hazardous waste		–	–	–	m^3	**57.07**
hazardous waste		–	–	–	m^3	**171.16**
special waste		–	–	–	m^3	**180.69**
contaminated liquid		–	–	–	m^3	**228.21**
contaminated sludge		–	–	–	m^3	**285.31**
Add to the above rates where Landfill Tax applies:						
inactive or inert waste material		–	–	–	m^3	**4.73**
other taxable waste material		–	–	–	m^3	**147.90**
Concrete 35 N/mm², 20 mm aggregate; installed by lorry/crawler-mounted rotary rig						
The following unit costs cover the construction of small dia. bored piles using lorry or crawler mounted rotary boring rigs. This type of plant is more mobile and faster in operation than the tripod rigs and is ideal for large contracts in cohesive ground. Construction of piles under bentonite suspension can be carried out to obviate the use of liners. Standard dia.s of 450–900 mm dia. can be constructed to depths of 30 m.						
The costs are based on installing 100 piles on a clear site with reasonable access.						
Dia.: 300 mm						
number of piles (see above)		–	–	–	nr	**346.30**
concreted length		–	–	–	m	**15.76**
depth bored to 10 m maximum depth		–	–	–	m	**48.83**
depth bored to 15 m maximum depth		–	–	–	m	**48.83**
depth bored to 20 m maximum depth		–	–	–	m	**48.83**
Dia.: 450 mm						
number of piles (see above)		–	–	–	nr	**346.30**
concreted length		–	–	–	m	**19.53**
depth bored to 10 m maximum depth		–	–	–	m	**61.08**
depth bored to 15 m maximum depth		–	–	–	m	**61.09**
depth bored to 20 m maximum depth		–	–	–	m	**61.09**
Dia.: 600 mm						
number of piles (see above)		–	–	–	nr	**346.30**
concreted length		–	–	–	m	**50.01**
depth bored to 10 m maximum depth		–	–	–	m	**108.41**
depth bored to 15 m maximum depth		–	–	–	m	**108.95**
depth bored to 20 m maximum depth		–	–	–	m	**108.95**

CLASS P: PILES

Item	Gang hours	Labour £	Plant £	Material £	Unit	**Total rate £**
01 BORED CAST IN PLACE CONCRETE PILES – cont						
Concrete 35 N/mm², 20 mm aggregate; installed by continuous flight auger						
The following unit costs cover the construction of piles by screwing a continuous flight auger into the ground to a design depth (Determined prior to commencement of piling operations and upon which the rates are based and subsequently varied to actual depths). Concrete is then pumped through the hollow stem of the auger to the bottom and the pile formed as the auger is withdrawn. Spoil is removed by the auger as it is withdrawn. This is a fast method of construction without causing disturbance or vibration to adjacent ground. No casing is required even in unsuitable soils. Reinforcement can be placed after grouting is complete.						
The costs are based on installing 100 piles on a clear site with reasonable access.						
Dia.: 300 mm						
number of piles		–	–	–	nr	**409.55**
concreted length		–	–	–	m	**20.49**
depth bored to 10 m maximum depth		–	–	–	nr	**24.41**
depth bored to 15 m maximum depth		–	–	–	nr	**19.53**
depth bored to 20 m maximum depth		–	–	–	nr	**19.53**
Dia.: 450 mm						
number of piles		–	–	–	nr	**409.55**
concreted length		–	–	–	m	**39.06**
depth bored to 10 m maximum depth		–	–	–	nr	**24.43**
depth bored to 15 m maximum depth		–	–	–	nr	**24.44**
depth bored to 20 m maximum depth		–	–	–	nr	**30.54**
Dia.: 600 mm						
number of piles		–	–	–	nr	**409.55**
concreted length		–	–	–	m	**60.01**
depth bored to 10 m maximum depth		–	–	–	nr	**32.52**
depth bored to 15 m maximum depth		–	–	–	nr	**30.51**
depth bored to 20 m maximum depth		–	–	–	nr	**27.24**

CLASS P: PILES

Item	Gang hours	Labour £	Plant £	Material £	Unit	**Total rate £**
02 DRIVEN CAST IN PLACE CONCRETE PILES						
Generally						
The items 'number of piles' are calculated based on the following:						
allowance for provision of all plant, equipment and labour including transporting to and from site and establishing and dismantling at £6,250 in total for piles using the Temporary Steel Casing Method and £10,500 in total for piles using the Segmental Casing Method.						
moving the rig to and setting up at each pile position; preparing to commence driving; £110.00 per pile.						
For the Temporary Steel Casing Method, obstructions (where within the capabilities of the normal plant) are quoted at £160.00 per hour and standing time at £152.50 per hour.						
For the Segmental Steel Casing Method, obstructions (where within the capabilities of the normal plant) are quoted at £275.00 per hour and standing time at £252.00 per hour.						
Temporary steel casing method; concrete 35 N/mm²; reinforced for 750 kN						
The following unit costs cover the construction of piles by driving a heavy steel tube into the ground either by using an internal hammer acting on a gravel or concrete plug, as is more usual, or by using an external hammer on a driving helmet at the top of the tube. After driving to the required depth an enlarged base is formed by hammering out successive charges of concrete down the tube. The tube is then filled with concrete which is compacted as the tube is vibrated and withdrawn. Piles of 350 to 500 mm dia. can be constructed with rakes up to 1 in 4 to carry working loads up to 120 t per pile. The costs are based on installing 100 piles on a clear site with reasonable access.						
Dia. 430 mm; drive shell and form pile						
number of piles		–	–	–	nr	**288.43**
concreted length		–	–	–	m	**35.34**
depth driven; bottom-driven method		–	–	–	m	**8.87**
depth driven; top-driven method		–	–	–	m	**5.10**

CLASS P: PILES

Item	Gang hours	Labour £	Plant £	Material £	Unit	Total rate £
02 DRIVEN CAST IN PLACE CONCRETE PILES – cont						
Segmental casing method; concrete 35 N/mm²; nominal reinforcement						
The following unit costs cover the construction of piles by driving into hard material using a serrated thick walled tube. It is oscillated and pressed into the hard material using a hydraulic attachment to the piling rig. The hard material is broken up using chiselling methods and is then removed by mechanical grab.						
The costs are based on installing 100 piles on a clear site with reasonable access.						
Dia. 620 mm						
number of piles		–	–	–	nr	**177.34**
concreted length		–	–	–	m	**205.00**
depth bored or driven to 15 m maximum depth		–	–	–	m	**12.94**
Dia. 1180 mm						
number of piles		–	–	–	nr	**177.34**
concreted length		–	–	–	m	**224.71**
depth bored or driven to 15 m maximum depth		–	–	–	m	**20.44**
Dia. 1500 mm						
number of piles		–	–	–	nr	**177.34**
concreted length		–	–	–	m	**302.78**
depth bored or driven to 15 m maximum depth		–	–	–	m	**32.37**
03 PREFORMED CONCRETE PILES						
The following unit costs cover the installation of driven precast concrete piles by using a hammer acting on shoe fitted or cast into the precast concrete pile unit.						
Single pile lengths are normally a maximum of 13 m long, at which point, a mechanical interlocking joint is required to extend the pile. These joints are most economically and practically formed at works.						
Lengths, sizes of sections, reinforcement details and concrete mixes vary for differing contractors, whose specialist advice should be sought for specific designs.						
The following unit costs are based on installing 100 piles on a clear site with reasonable access.						
The items 'number of piles' are calculated based on the following:						
allowance for provision of all plant, equipment and labour including transporting to and from site and establishing and dismantling at £3,150 in total for piles up to 275 × 275 mm and £3,850 in total for piles 350 × 350 mm and over.						

CLASS P: PILES

Item	Gang hours	Labour £	Plant £	Material £	Unit	**Total rate £**
moving the rig to and setting up at each pile position; preparing to commence driving; piles up to 275 × 275 mm £35.00 each; piles 350 × 350 mm and over, £55.00 each.						
an allowance for the cost of the driving head and shoe; £35.00 for 235 × 235 mm piles, £45.00 for 275 × 275 mm and £55.00 for 350 × 350 mm.						
cost of providing the pile of the stated length						
Typical allowances for standing time are £138.50 per hour for 235 × 235 mm piles, £157.50 for 275 × 275 mm and £189.50 for 350 × 350 mm.						
Concrete 50 N/mm^2; reinforced for 600 kN						
The costs are based on installing 100 piles on a clear site with reasonable access.						
Cross-sectional area: 0.05–0.1 m^2; 235 × 235 mm						
number of piles of 10 m length		–	–	–	nr	**454.46**
number of piles of 15 m length		–	–	–	nr	**602.61**
number of piles of 20 m length		–	–	–	nr	**750.76**
number of piles of 25 m length		–	–	–	nr	**898.90**
add for mechanical interlocking joint		–	–	–	nr	**83.62**
depth driven		–	–	–	m	**3.82**
Cross-sectional area: 0.05–0.1 m^2; 275 × 275 mm						
number of piles of 10 m length		–	–	–	nr	**505.91**
number of piles of 15 m length		–	–	–	nr	**679.78**
number of piles of 20 m length		–	–	–	nr	**853.65**
number of piles of 25 m length		–	–	–	nr	**1027.52**
add for mechanical interlocking joint		–	–	–	nr	**87.23**
depth driven		–	–	–	m	**4.03**
Cross-sectional area: 0.1–0.15 m^2; 350 × 350 mm						
number of piles of 10 m length		–	–	–	nr	**780.15**
number of piles of 15 m length		–	–	–	nr	**1054.21**
number of piles of 20 m length		–	–	–	nr	**1328.26**
number of piles of 25 m length		–	–	–	nr	**1602.32**
number of piles of 30 m length		–	–	–	nr	**1876.37**
add for mechanical interlocking joint		–	–	–	nr	**167.19**
depth driven		–	–	–	m	**5.72**

CLASS P: PILES

Item	Gang hours	Labour £	Plant £	Material £	Unit	Total rate £
05 TIMBER PILES						
The items 'number of piles' are calculated based on the following:						
allowance for provision of all plant, equipment and labour including transporting to and from site and establishing and dismantling at £6,000 in total.						
moving the rig to and setting up at each pile position and preparing to drive at £75.00 per pile.						
allowance for the cost of the driving head and shoe at £40.00 per 225 × 225 mm pile, £50.00 for 300 × 300 mm, £60.00 for 350 × 350 mm and £70.00 for 450 × 450 mm.						
cost of providing the pile of the stated length						
A typical allowance for standing time is £245.50 per hour.						
Douglas Fir; hewn to mean pile size						
The costs are based on installing 100 piles on a clear site with reasonable access.						
Cross-sectional area: 0.05–0.1 m^2; 225 × 225 mm						
number of piles of 10 m length		–	–	–	nr	**731.96**
number of piles of 15 m length		–	–	–	nr	**985.36**
number of piles of 20 m length		–	–	–	nr	**1238.75**
depth driven		–	–	–	m	**5.05**
Cross-sectional area: 0.05–0.1 m^2; 300 × 300 mm						
number of piles of 10 m length		–	–	–	nr	**1172.42**
number of piles of 15 m length		–	–	–	nr	**1531.78**
number of piles of 20 m length		–	–	–	nr	**1891.13**
depth driven		–	–	–	m	**8.73**
Cross-sectional area: 0.1–0.15 m^2; 350 × 350 mm						
number of piles of 10 m length		–	–	–	nr	**1626.43**
number of piles of 15 m length		–	–	–	nr	**2253.30**
number of piles of 20 m length		–	–	–	nr	**2342.00**
depth driven		–	–	–	m	**5.81**
Cross-sectional area: 0.15–0.25 m^2; 450 × 450 mm						
number of piles of 10 m length		–	–	–	nr	**1914.56**
number of piles of 15 m length		–	–	–	nr	**3109.89**
number of piles of 20 m length		–	–	–	nr	**3953.98**
number of piles of 25 m length		–	–	–	nr	**4902.91**
depth driven		–	–	–	m	**7.09**
Greenheart; hewn to mean pile size						
The costs are based on installing 100 piles on a clear site with reasonable access.						
Cross-sectional area: 0.05–0.1 m^2; 225 × 225 mm						
number of piles of 10 m length		–	–	–	nr	**1228.77**
number of piles of 15 m length		–	–	–	nr	**1461.48**
number of piles of 20 m length		–	–	–	nr	**1694.19**
depth driven		–	–	–	m	**4.21**

CLASS P: PILES

Item	Gang hours	Labour £	Plant £	Material £	Unit	**Total rate £**
Cross-sectional area: 0.05–0.1 m^2; 300 × 300 mm						
number of piles of 10 m length		–	–	–	nr	**1574.49**
number of piles of 15 m length		–	–	–	nr	**1973.42**
number of piles of 20 m length		–	–	–	nr	**2587.63**
depth driven		–	–	–	m	**7.27**
Cross-sectional area: 0.1–0.15 m^2; 350 × 350 mm						
number of piles of 10 m length		–	–	–	nr	**2190.06**
number of piles of 15 m length		–	–	–	nr	**2291.48**
number of piles of 20 m length		–	–	–	nr	**2500.55**
depth driven		–	–	–	m	**5.81**
Cross-sectional area: 0.15–0.25 m^2; 450 × 450 mm						
number of piles of 10 m length		–	–	–	nr	**2605.30**
number of piles of 15 m length		–	–	–	nr	**3503.01**
number of piles of 20 m length		–	–	–	nr	**4400.73**
number of piles of 25 m length		–	–	–	nr	**5298.44**
depth driven		–	–	–	m	**7.09**
06 ISOLATED STEEL PILES						
Steel bearing piles are commonly carried out by a Specialist Contractor and whose advice should be sought to arrive at accurate costing. However the following items can be used to assess a budget cost for such work.						
The items 'number of piles' are calculated based on the following:						
allowance for provision of all plant, equipment and labour including transporting to and from site and establishing and dismantling at £6,600 in total up to a maximum 100 miles radius from base and £16,050 up to a maximum 250 miles radius from base.						
moving the rig to and setting up at each pile position; preparing to commence driving; £193.15 per pile.						
cost of providing the pile of the stated length						
A typical allowance for standing time is £283.90 per hour.						
Steel EN 10025 grade S275; within 100 miles of steel plant						
The costs are based upon installing 100 nr on a clear site with reasonable access.						
Mass 45 kg/m; 203 × 203 mm						
number of piles: length 10 m		–	–	712.53	nr	**7917.92**
number of piles: length 15 m		–	–	1139.81	nr	**8345.20**
number of piles: length 20 m		–	–	1519.74	nr	**8725.14**
depth driven; vertical		–	–	–	m	**5.19**
depth driven; raking		–	–	–	m	**8.55**

CLASS P: PILES

Item	Gang hours	Labour £	Plant £	Material £	Unit	Total rate £
06 ISOLATED STEEL PILES – cont						
Steel EN 10025 grade S275 – cont						
Mass 54 kg/m; 203 × 203 mm						
number of piles; length 10 m		–	–	855.04	nr	**8060.43**
number of piles; length 15 m		–	–	1367.77	nr	**8573.16**
number of piles; length 20 m		–	–	1823.69	nr	**7842.75**
depth driven; vertical		–	–	–	m	**4.33**
depth driven; raking		–	–	–	m	**7.13**
Mass 63 kg/m; 254 × 254 mm						
number of piles; length 10 m		–	–	997.54	nr	**7016.60**
number of piles; length 15 m		–	–	1595.73	nr	**7614.79**
number of piles; length 20 m		–	–	2127.64	nr	**8146.70**
depth driven; vertical		–	–	–	m	**4.84**
depth driven; raking		–	–	–	m	**7.41**
Mass 71 kg/m; 254 × 254 mm						
number of piles; length 10 m		–	–	1124.21	nr	**7143.27**
number of piles; length 15 m		–	–	1798.36	nr	**7817.42**
number of piles; length 20 m		–	–	2397.82	nr	**8416.88**
depth driven; vertical		–	–	–	m	**4.84**
depth driven; raking		–	–	–	m	**7.41**
Mass 85 kg/m; 254 × 254 mm						
number of piles; length 10 m		–	–	1345.89	nr	**7364.95**
number of piles; length 15 m		–	–	2152.97	nr	**8172.03**
number of piles; length 20 m		–	–	2870.63	nr	**8889.69**
depth driven; vertical		–	–	–	m	**4.84**
depth driven; raking		–	–	–	m	**7.41**
Mass 79 kg/m; 305 × 305 mm						
number of piles; length 10 m		–	–	1250.88	nr	**8456.28**
number of piles; length 15 m		–	–	2000.99	nr	**9206.39**
number of piles; length 20 m		–	–	2667.99	nr	**9873.39**
depth driven; vertical		–	–	–	m	**5.81**
depth driven; raking		–	–	–	m	**8.90**
Mass 95 kg/m; 305 × 305 mm						
number of piles; length 10 m		–	–	1504.23	nr	**8709.62**
number of piles; length 15 m		–	–	2406.26	nr	**9611.65**
number of piles; length 20 m		–	–	3208.35	nr	**10413.74**
depth driven; vertical		–	–	–	m	**6.25**
depth driven; raking		–	–	–	m	**9.36**
Mass 110 kg/m; 305 × 305 mm						
number of piles; length 10 m		–	–	1741.74	nr	**7760.80**
number of piles; length 15 m		–	–	2786.20	nr	**8805.26**
number of piles; length 20 m		–	–	3714.93	nr	**9733.99**
depth driven; vertical		–	–	–	m	**5.21**
depth driven; raking		–	–	–	m	**7.80**
Mass 126 kg/m; 305 × 305 mm						
number of piles; length 10 m		–	–	1995.08	nr	**8014.14**
number of piles; length 15 m		–	–	3191.46	nr	**9210.52**
number of piles; length 20 m		–	–	4255.28	nr	**10274.34**
driving piles; vertical		–	–	–	m	**5.53**
driving piles; raking		–	–	–	m	**8.52**

CLASS P: PILES

Item	Gang hours	Labour £	Plant £	Material £	Unit	**Total rate £**
Mass 149 kg/m; 305 × 305 mm						
number of piles; length 10 m		–	–	2359.26	nr	**8378.33**
number of piles; length 15 m		–	–	3774.03	nr	**9793.09**
number of piles; length 20 m		–	–	5032.04	nr	**11051.10**
driving piles; vertical		–	–	–	m	**5.53**
driving piles; raking		–	–	–	m	**8.52**
Mass 186 kg/m; 305 × 305 mm						
number of piles; length 10 m		–	–	2945.12	nr	**8964.18**
number of piles; length 15 m		–	–	4711.20	nr	**10730.27**
number of piles; length 20 m		–	–	6281.60	nr	**12300.67**
driving piles; vertical		–	–	–	m	**6.47**
drive piles; raking		–	–	–	m	**9.18**
Mass 223 kg/m; 305 × 305 mm						
number of piles; length 10 m		–	–	3530.98	nr	**9550.04**
number of piles; length 15 m		–	–	5648.38	nr	**11667.44**
number of piles; length 20 m		–	–	7531.17	nr	**13550.23**
driving piles; vertical		–	–	–	m	**6.86**
driving piles; raking		–	–	–	m	**9.48**
Mass 109 kg/m; 356 × 368 mm						
number of piles; length 10 m		–	–	1751.98	nr	**7771.05**
number of piles; length 15 m		–	–	2799.99	nr	**8819.05**
number of piles; length 20 m		–	–	3733.32	nr	**9752.38**
depth driven; vertical		–	–	–	m	**5.21**
depth driven; raking		–	–	–	m	**7.80**
Mass 133 kg/m; 356 × 368 mm						
number of piles; length 10 m		–	–	2137.74	nr	**8156.80**
number of piles; length 15 m		–	–	3416.50	nr	**9435.56**
number of piles; length 20 m		–	–	4555.33	nr	**10574.39**
driving piles; vertical		–	–	–	m	**5.53**
driving piles; raking		–	–	–	m	**8.52**
Mass 152 kg/m; 356 × 368 mm						
number of piles; length 10 m		–	–	2443.13	nr	**8462.20**
number of piles; length 15 m		–	–	3904.57	nr	**9923.63**
number of piles; length 20 m		–	–	5206.09	nr	**11225.15**
driving piles; vertical		–	–	–	m	**6.22**
driving piles; raking		–	–	–	m	**8.66**
Mass 174 kg/m; 356 × 368 mm						
number of piles; length 10 m		–	–	2796.75	nr	**8815.81**
number of piles; length 15 m		–	–	4469.70	nr	**10488.77**
number of piles; length 20 m		–	–	5959.61	nr	**11978.67**
driving piles; vertical		–	–	–	m	**6.22**
driving piles; raking		–	–	–	m	**8.66**

CLASS P: PILES

Item	Gang hours	Labour £	Plant £	Material £	Unit	Total rate £
07 INTERLOCKING STEEL PILES						
Sheet steel piling is commonly carried out by a Specialist Contractor, whose advice should be sought to arrive at accurate costings. However, the following items can be used to assess a budget for such work.						
The following unit costs are based on driving/extracting 1,500 m^2 of sheet piling on a clear site with reasonable access.						
Note: area of driven piles will vary from area supplied dependent upon pitch line of piling and provision for such allowance has been made in PC for supply.						
The materials cost below includes the manufacturers tariffs for a 200 mile delivery radius from works, delivery in 5–10t loads and with an allowance of 10% to cover waste/projecting piles etc.						
Arcelor Mittal Z section steel piles; EN 10248 grade S270GP steel						
The following unit costs are based on driving/extracting 1,500 m^2 of sheet piling on a clear site with reasonable access.						
Provision of all plant, equipment and labour including transport to and from the site and establishing and dismantling for						
driving of sheet piling		–	–	–	item	**9740.42**
extraction of sheet piling		–	–	–	tem	**2622.42**
Standing time		–	–	–	hr	**523.31**
Section modulus 800–1200 cm^3/m; section reference AZ 12; mass 98.7 kg/m^2, sectional modulus 1200 cm^3/m; EN 10248 grade S270GP steel						
length of welded corner piles		–	–	–	m	**141.91**
length of welded junction piles		–	–	–	m	**223.17**
driven area		–	–	–	m^2	**67.83**
area of piles of length not exceeding 14 m		–	–	–	m^2	**153.54**
length 14–24 m		–	–	–	m^2	**144.49**
area of piles of length exceeding 24 m		–	–	–	m^2	**153.97**
Section modulus 1200–2000 cm^3/m; section reference AZ 17; mass 108.6 kg/m^2; sectional modulus 1665 cm^3/m; EN 10248 grade S270GP steel						
length of welded corner piles		–	–	–	m	**151.37**
length of welded junction piles		–	–	–	m	**223.17**
driven area		–	–	–	m^2	**67.56**
area of piles of length not exceeding 14 m		–	–	–	m^2	**179.13**
length 14–24 m		–	–	–	m^2	**133.37**
area of piles of length exceeding 24 m		–	–	–	m^2	**142.12**

CLASS P: PILES

Item	Gang hours	Labour £	Plant £	Material £	Unit	**Total rate £**
Section modulus 2000–3000 cm^3/m; section reference AZ 26; mass 155.2 kg/m^2; sectional modulus 2600 cm^3/m; EN 10248 grade S270GP steel						
driven area		–	–	–	m^2	**54.73**
area of piles of length 6–18 m		–	–	–	m^2	**98.26**
area of piles of length 18–24 m		–	–	–	m^2	**99.60**
Section modulus 3000–4000 cm^3/m; section reference AZ 36; mass 194.0 kg/m^2; sectional modulus 3600 cm^3/m; EN 10248 grade S270GP steel						
driven area		–	–	–	m^2	**55.63**
area of piles of length 6–18 m		–	–	–	m^2	**196.51**
area of piles of length 18–24 m		–	–	–	m^2	**199.20**
Straight section modulus ne 500 cm^3/m; section reference AS 500–12 mass 149 kg/m^2; sectional modulus 51 cm^3/m; EN 10248 grade S270GP steel						
driven area		–	–	–	m^2	**41.57**
area of piles of length 6–18 m		–	–	–	m^2	**216.17**
area of piles of length 18–24 m		–	–	–	m^2	**229.08**
One coat black tar vinyl (PC1) protective treatment applied all surfaces at shop to minimum dry film thickness up to 150 microns to steel piles						
section reference AZ 12; pile area		–	–	–	m^2	**24.04**
section reference AZ 17; pile area		–	–	–	m^2	**15.71**
section reference AZ 26; pile area		–	–	–	m^2	**17.11**
section reference AZ 36; pile area		–	–	–	m^2	**18.23**
section reference AS 500–12; pile area		–	–	–	m^2	**29.79**
One coat black high build isocyanate cured epoxy pitch (PC2) protective treatment applied all surfaces at shop to minimum dry film thickness up to 450 microns to steel piles						
section reference AZ 12; pile area		–	–	–	m^2	**24.04**
section reference AZ 17; pile area		–	–	–	m^2	**15.71**
section reference AZ 26; pile area		–	–	–	m^2	**17.11**
section reference AZ 36; pile area		–	–	–	m^2	**18.23**
section reference AS 500–12; pile area		–	–	–	m^2	**29.79**
Arcelor Mittal U section steel piles; EN 10248 grade S270GP steel						
The following unit costs are based on driving/ extracting 1,500 m^2 of sheet piling on a clear site with reasonable access.						
Provision of plant, equipment and labour including transport to and from the site and establishing and dismantling						
driving of sheet piling		–	–	–	item	**9740.42**
extraction of sheet piling		–	–	–	item	**2622.42**
Standing time		–	–	–	hr	**471.12**
Section modulus 500–800 cm^3/m; section reference PU 6; mass 76.0 kg/m^2; sectional modulus 600 cm^3/m						
driven area		–	–	–	m^2	**67.29**
area of piles of length 6–18 m		–	–	–	m^2	**98.26**
area of piles of length 18–24 m		–	–	–	m^2	**99.60**

CLASS P: PILES

Item	Gang hours	Labour £	Plant £	Material £	Unit	Total rate £
07 INTERLOCKING STEEL PILES – cont						
Arcelor Mittal U section steel piles – cont						
Section modulus 800–1200 cm^3/m; section reference PU 8; mass 90.9 kg/m^2; sectional modulus 830 cm^3/m						
driven area	–	–	–	m^2		**60.54**
area of piles of length 6–18 m	–	–	–	m^2		**98.26**
area of piles of length 18–24 m	–	–	–	m^2		**99.60**
Section modulus 1200–2000 cm^3/m; section reference PU 12; mass 110.1 kg/m^2; sectional modulus 1200 cm^3/m						
driven area	–	–	–	m^2		**51.83**
area of piles of length 6–18 m	–	–	–	m^2		**98.26**
area of piles of length 18–24 m	–	–	–	m^2		**99.60**
Section modulus 1200–2000 cm^3/m; section reference PU 18; mass 128.2 kg/m^2; sectional modulus 1800 cm^3/m						
driven area	–	–	–	m^2		**47.12**
area of piles of length 6–18 m	–	–	–	m^2		**98.26**
area of piles of length 18–24 m	–	–	–	m^2		**99.60**
Section modulus 2000–3000 cm^3/m; section reference PU 22; mass 143.6 kg/m^2; sectional modulus 2200 cm^3/m						
driven area	–	–	–	m^2		**42.40**
area of piles of length 6–18 m	–	–	–	m^2		**98.26**
area of piles of length 18–24 m	–	–	–	m^2		**99.60**
Section modulus 3000–4000 cm^3/m; section reference PU 32; mass 190.2 kg/m^2; sectional modulus 3200 cm^3/m						
driven area	–	–	–	m^2		**36.91**
area of piles of length 6–18 m	–	–	–	m^2		**98.26**
area of piles of length 18–24 m	–	–	–	m^2		**99.60**
One coat black tar vinyl (PC1) protective treatment applied all surfaces at shop to minimum dry film thickness up to 150 microns to steel piles						
section reference PU 6; pile area	–	–	–	m^2		**21.05**
section reference PU 8; pile area	–	–	–	m^2		**14.21**
section reference PU 12; pile area	–	–	–	m^2		**14.21**
section reference PU 18; pile area	–	–	–	m^2		**24.06**
section reference PU 22; pile area	–	–	–	m^2		**24.95**
section reference PU 32; pile area	–	–	–	m^2		**25.31**
One coat black high build isocyanate cured epoxy pitch (PC2) protective treatment applied all surfaces at shop to minimum dry film thickness up to 450 microns to steel piles						
section reference PU 6; pile area	–	–	–	m^2		**21.05**
section reference PU 8; pile area	–	–	–	m^2		**14.21**
section reference PU 12; pile area	–	–	–	m^2		**14.21**
section reference PU 18; pile area	–	–	–	m^2		**24.06**
section reference PU 22; pile area	–	–	–	m^2		**24.95**
section reference PU 32; pile area	–	–	–	m^2		**25.31**

CLASS Q: PILING ANCILLARIES

Item	Gang hours	Labour £	Plant £	Material £	Unit	**Total rate £**
01 CAST IN PLACE CONCRETE PILES						
Bored; lorry/crawler mounted rotary rig						
Backfilling empty bore with selected excavated material						
dia. 500 mm		–	–	–	m	**5.15**
Permanent casings; each length not exceeding 13 m						
dia. 500 mm		–	–	–	m	**119.55**
Permanent casings; each length exceeding 13 m						
dia. 500 mm		–	–	–	m	**127.55**
Enlarged bases						
dia. 1500 mm; to 500 mm dia. pile		–	–	–	nr	**398.56**
Cutting off surplus lengths						
dia. 500 mm		–	–	–	m	**39.23**
Preparing heads						
500 mm dia.		–	–	–	nr	**57.09**
Bored; continuous flight auger						
Backfilling empty bore with selected excavated material						
450 mm dia. piles		–	–	–	m	**4.06**
600 mm dia. piles		–	–	–	m	**5.54**
750 mm dia. piles		–	–	–	m	**6.05**
Permanent casings; each length not exceeding 13 m						
450 mm dia. piles		–	–	–	m	**105.45**
600 mm dia. piles		–	–	–	m	**143.47**
750 mm dia. piles		–	–	–	m	**183.36**
Permanent casings; each length exceeding 13 m						
450 mm dia. piles		–	–	–	m	**103.60**
600 mm dia. piles		–	–	–	m	**152.22**
750 mm dia. piles		–	–	–	m	**183.35**
Enlarged bases						
dia. 1400 mm; to 450 mm dia. piles		–	–	–	nr	**366.66**
dia. 1800 mm; to 600 mm dia. piles		–	–	–	nr	**446.38**
dia. 2100 mm; to 750 mm dia. piles		–	–	–	nr	**502.18**
Cutting off surplus lengths						
450 mm dia. piles		–	–	–	m	**34.25**
600 mm dia. piles		–	–	–	m	**48.53**
750 mm dia. piles		–	–	–	m	**57.80**
Preparing heads						
450 mm dia. piles		–	–	–	nr	**36.40**
600 mm dia. piles		–	–	–	nr	**64.20**
750 mm dia. piles		–	–	–	nr	**87.06**
Collection from around pile heads of spoil accruing from piling operations and depositing in spoil heaps (For final disposal see Class E – Excavation Ancillaries)		–	–	–	m^3	**4.26**

CLASS Q: PILING ANCILLARIES

Item	Gang hours	Labour £	Plant £	Material £	Unit	**Total rate £**
01 CAST IN PLACE CONCRETE PILES – cont						
Supply and fit reinforcement materials include for bars, tying wire, spacers, couplers and steel supports for bottom layer reinforcement (stools, chairs and risers).						
Plain round mild steel bars to BS 4449						
Bars; supplied in straight lengths						
6 mm nominal size	8.00	1095.47	162.87	844.66	tonne	**2103.00**
8 mm nominal size	6.74	922.93	137.22	844.66	tonne	**1904.81**
10 mm nominal size	6.74	922.93	137.22	844.63	tonne	**1904.78**
12 mm nominal size	6.74	922.93	137.22	844.63	tonne	**1904.78**
16 mm nominal size	6.15	842.14	125.22	844.63	tonne	**1811.99**
20 mm nominal size	4.44	607.98	90.40	844.63	tonne	**1543.01**
25 mm nominal size	4.44	607.98	90.40	844.63	tonne	**1543.01**
32 mm nominal size	4.44	607.98	90.40	844.63	tonne	**1543.01**
40 mm nominal size	4.44	607.98	90.40	844.63	tonne	**1543.01**
Helical bars, nominal size						
6 mm	8.00	1095.47	162.87	894.44	t	**2152.78**
8 mm	6.74	922.93	137.22	894.44	t	**1954.59**
10 mm	6.74	922.93	137.22	893.90	t	**1954.05**
12 mm	6.74	922.93	137.22	894.44	t	**1954.59**
Reinforcement; high tensile steel						
Straight bars, nominal size						
6 mm	8.40	1150.24	162.87	952.07	t	**2265.19**
8 mm	7.08	969.08	137.22	952.06	t	**2058.36**
10 mm	7.08	969.08	137.22	952.07	t	**2058.37**
12 mm	7.08	969.08	137.22	952.07	t	**2058.37**
16 mm	4.66	638.38	90.40	952.07	t	**1680.85**
25 mm	4.66	638.38	90.40	952.07	t	**1680.85**
32 mm	4.66	638.38	90.40	952.07	t	**1680.85**
50 mm	4.66	638.38	90.40	952.07	t	**1680.85**
Helical bars, nominal size						
6 mm	8.40	1150.24	162.87	1067.45	t	**2380.57**
8 mm	7.08	969.08	137.22	1067.45	t	**2173.75**
10 mm	7.08	969.08	137.22	1067.45	t	**2173.75**
12 mm	7.08	969.08	137.22	1067.47	t	**2173.76**
Couplers; Lenton type A; threaded ends on reinforcing bars						
12 mm	0.10	13.69	–	11.65	nr	**25.34**
16 mm	0.10	13.69	–	13.77	nr	**27.46**
20 mm	0.10	13.69	–	20.41	nr	**34.10**
25 mm	0.10	13.69	–	28.15	nr	**41.85**
32 mm	0.10	13.69	–	37.96	nr	**51.65**
40 mm	0.10	13.69	–	52.74	nr	**66.43**

CLASS Q: PILING ANCILLARIES

Item	Gang hours	Labour £	Plant £	Material £	Unit	**Total rate £**
Couplers; Lenton type B; threaded ends on reinforcing bars						
12 mm	0.10	13.69	–	36.84	nr	**50.54**
16 mm	0.10	13.69	–	42.85	nr	**56.55**
20 mm	0.10	13.69	–	46.68	nr	**60.37**
25 mm	0.10	13.69	–	54.81	nr	**68.51**
32 mm	0.10	13.69	–	73.69	nr	**87.38**
40 mm	0.09	12.32	–	108.69	nr	**121.01**
03 PREFORMED CONCRETE PILES						
General						
Preparing heads						
235 × 235 mm piles		–	–	–	nr	**40.58**
275 × 275 mm piles		–	–	–	nr	**55.47**
350 × 350 mm piles		–	–	–	nr	**85.89**
04 TIMBER PILES						
Douglas Fir						
Cutting off surplus lengths						
cross-sectional area: 0.025–0.05 m^2		–	–	–	nr	**3.56**
cross-sectional area: 0.05–0.1 m^2		–	–	–	nr	**6.13**
cross-sectional area: 0.1–0.15 m^2		–	–	–	nr	**7.93**
cross-sectional area: 0.15–0.25 m^2		–	–	–	nr	**15.49**
Preparing heads						
cross-sectional area: 0.025–0.05 m^2		–	–	–	nr	**3.56**
cross-sectional area: 0.05–0.1 m^2		–	–	–	nr	**6.13**
cross-sectional area: 0.1–0.15 m^2		–	–	–	nr	**7.93**
cross-sectional area: 0.15–0.25 m^2		–	–	–	nr	**15.49**
Greenheart						
Cutting off surplus lengths						
cross-sectional area: 0.025–0.05 m^2		–	–	–	nr	**12.91**
cross-sectional area: 0.05–0.1 m^2		–	–	–	nr	**12.91**
cross-sectional area: 0.1–0.15 m^2		–	–	–	nr	**16.16**
cross-sectional area: 0.15–0.25 m^2		–	–	–	nr	**29.07**
Preparing heads						
cross-sectional area: 0.025–0.05 m^2		–	–	–	nr	**12.91**
cross-sectional area: 0.05–0.1 m^2		–	–	–	nr	**12.91**
cross-sectional area: 0.1–0.15 m^2		–	–	–	nr	**16.16**
cross-sectional area: 0.15–0.25 m^2		–	–	–	nr	**29.07**

CLASS Q: PILING ANCILLARIES

Item	Gang hours	Labour £	Plant £	Material £	Unit	Total rate £
05 ISOLATED STEEL PILES						
Steel bearing piles						
Steel bearing piles are commonly carried out by a Specialist Contractor and whose advice should be sought to arrive at accurate costing. However the following items can be used to assess a budget cost for such work.						
The item for number of pile extensions includes for the cost of setting up the rig at the pile position together with welding the extension to the top of the steel bearing pile. The items for length of pile extension cover the material only, the driving cost being included in Class P.						
Number of pile extensions						
at each position		–	–	–	nr	**87.40**
Length of pile extensions, each length not exceeding 3 m; steel EN 10025 grade S275						
mass 45 kg/m		–	–	75.99	m	**75.99**
mass 54 kg/m		–	–	91.18	m	**91.18**
mass 63 kg/m		–	–	106.38	m	**106.38**
mass 71 kg/m		–	–	119.89	m	**119.89**
mass 79 kg/m		–	–	133.40	m	**133.40**
mass 85 kg/m		–	–	143.53	m	**143.53**
mass 95 kg/m		–	–	160.42	m	**160.42**
mass 109 kg/m		–	–	186.67	m	**186.67**
mass 110 kg/m		–	–	185.75	m	**185.75**
mass 126 kg/m		–	–	212.76	m	**212.76**
mass 133 kg/m		–	–	227.77	m	**227.77**
mass 149 kg/m		–	–	251.60	m	**251.60**
mass 152 kg/m		–	–	260.30	m	**260.30**
mass 174 kg/m		–	–	297.98	m	**297.98**
mass 186 kg/m		–	–	314.08	m	**314.08**
mass 223 kg/m		–	–	376.56	m	**376.56**
Length of pile extensions, each length exceeding 3 m; steel EN 10025 grade S275						
mass 45 kg/m		–	–	75.99	m	**75.99**
mass 54 kg/m		–	–	91.18	m	**91.18**
mass 63 kg/m		–	–	106.38	m	**106.38**
mass 71 kg/m		–	–	119.89	m	**119.89**
mass 79 kg/m		–	–	133.40	m	**133.40**
mass 85 kg/m		–	–	143.53	m	**143.53**
mass 95 kg/m		–	–	160.42	m	**160.42**
mass 109 kg/m		–	–	186.67	m	**186.67**
mass 110 kg/m		–	–	185.75	m	**185.75**
mass 126 kg/m		–	–	212.76	m	**212.76**
mass 133 kg/m		–	–	227.77	m	**227.77**
mass 149 kg/m		–	–	251.60	m	**251.60**
mass 152 kg/m		–	–	260.30	m	**260.30**
mass 174 kg/m		–	–	297.98	m	**297.98**
mass 186 kg/m		–	–	314.08	m	**314.08**
mass 223 kg/m		–	–	376.56	m	**376.56**

CLASS Q: PILING ANCILLARIES

Item	Gang hours	Labour £	Plant £	Material £	Unit	**Total rate £**
Number of pile extensions						
section size 203 × 203 × any kg/m		–	–	–	nr	**298.48**
section size 254 × 254 × any kg/m		–	–	–	nr	**353.47**
section size 305 × 305 × any kg/m		–	–	–	nr	**408.90**
section size 356 × 368 × any kg/m		–	–	–	nr	**450.47**
Cutting off surplus lengths						
mass 30–60 kg/m		–	–	–	nr	**103.95**
mass 60–120 kg/m		–	–	–	nr	**145.53**
mass 120–250 kg/m		–	–	–	nr	**200.99**
Burning off tops of piles to level						
mass 30–60 kg/m		–	–	–	nr	**103.95**
mass 60–120 kg/m		–	–	–	nr	**145.53**
mass 120–250 kg/m		–	–	–	nr	**200.99**
06 INTERLOCKING STEEL PILES						
Arcelor Mittal Z section steel piles; EN 10248 grade S270GP steel						
Steel bearing piles are commonly carried out by a Specialist Contractor and whose advice should be sought to arrive at accurate costing. However the following items can be used to assess a budget cost for such work.						
The item for number of pile extensions includes for welding the extension to the top of the steel bearing pile. The items for length of pile extension cover the material only, the driving cost being included in Class P.						
Number of pile extensions						
Cutting off surplus lengths						
section modulus 500–800 cm^3/m		–	–	–	m	**32.58**
section modulus 800–1200 cm^3/m		–	–	–	m	**33.23**
section modulus 1200–2000 cm^3/m		–	–	–	m	**32.58**
section modulus 2000–3000 cm^3/m		–	–	–	m	**32.58**
section modulus 3000–4000 cm^3/m		–	–	–	m	**32.58**
Extract piling and stacking on site						
section modulus 500–800 cm^3/m		–	–	–	m^2	**51.29**
section modulus 800–1200 cm^3/m		–	–	–	m^2	**25.65**
section modulus 1200–2000 cm^3/m		–	–	–	m^2	**25.65**
section modulus 2000–3000 cm^3/m		–	–	–	m^2	**25.65**
section modulus 3000–4000 cm^3/m		–	–	–	m^2	**25.65**

CLASS Q: PILING ANCILLARIES

Item	Gang hours	Labour £	Plant £	Material £	Unit	Total rate £
06 INTERLOCKING STEEL PILES – cont						
Arcelor Mittal U section steel piles; EN 10248 grade S270GP steel						
Steel bearing piles are commonly carried out by a Specialist Contractor and whose advice should be sought to arrive at accurate costing. However the following items can be used to assess a budget cost for such work.						
The item for number of pile extensions includes for welding the extension to the top of the steel bearing pile. The items for length of pile extension cover the material only, the driving cost being included in Class P.						
Number of pile extensions						
Cutting off surplus lengths						
section modulus 500–800 cm³/m		–	–	–	m	**32.58**
section modulus 800–1200 cm³/m		–	–	–	m	**33.23**
section modulus 1200–2000 cm³/m		–	–	–	m	**32.58**
section modulus 2000–3000 cm³/m		–	–	–	m	**32.58**
section modulus 3000–4000 cm³/m		–	–	–	m	**32.58**
Extract piling and stack on site						
section modulus 500–800 cm³/m		–	–	–	m²	**25.65**
section modulus 800–1200 cm³/m		–	–	–	m²	**25.65**
section modulus 800–1200 cm³/m PU12		–	–	–	m²	**25.65**
section modulus 800–1200 cm³/m PU18		–	–	–	m²	**25.65**
section modulus 2000–3000 cm³/m		–	–	–	m²	**25.65**
section modulus 3000–4000 cm³/m		–	–	–	m²	**25.65**
07 OBSTRUCTIONS						
General						
Obstructions		–	–	–	hr	**135.59**
08 PILE TESTS						
Cast in place						
Pile tests; 500 mm dia. working pile; maximum test load of 600kN using non-working tension piles as reaction tripod						
first pile		–	–	–	nr	**7055.30**
subsequent pile		–	–	–	nr	**5354.64**
Take and test undisturbed soil samples; tripod		–	–	–	nr	**242.59**
Make, cure and test concrete cubes; tripod		–	–	–	nr	**17.85**
Pile tests; working pile; maximum test load of 1½ times working load; first pile						
450 mm / 650kN		–	–	–	nr	**3372.96**
600 mm / 1400kN		–	–	–	nr	**7073.41**
750 mm / 2200kN		–	–	–	nr	**12754.70**

CLASS Q: PILING ANCILLARIES

Item	Gang hours	Labour £	Plant £	Material £	Unit	**Total rate £**
Pile tests; working pile; maximum test load of 1½ times working load; second and subsequent piles						
450 mm / 650kN		–	–	–	nr	**3143.73**
600 mm / 1400kN		–	–	–	nr	**6811.42**
750 mm / 2200kN		–	–	–	nr	**12273.51**
Pile tests; working pile; electronic integrity testing; each pile (minimum 40 piles per visit)		–	–	–	nr	**1552.82**
Make, cure and test concrete cubes		–	–	–	nr	**29.68**
Preformed						
Pile tests; working pile; maximum test load of 1.5 times working load		–	–	–	nr	**5681.91**
Pile tests; working pile; dynamic testing with piling hammer		–	–	–	nr	**378.80**
Steel bearing piles						
Steel bearing piles are commonly carried out by a Specialist Contractor and whose advice should be sought to arrive at accurate costing. However the following items can be used to assess a budget cost for such work.						
The following unit costs are based upon driving 100 nr steel bearing piles 15–24 m long on a clear site with reasonable access. Supply is based on delivery 75 miles from works, in loads over 20t.						
Establishment of pile testing equipment on site preliminary to any piling operation		–	–	–	item	**1980.20**
Carry out pile test on bearing piles irrespective of section using pile testing equipment on site up to 108 t load		–	–	–	nr	**6468.65**
Driven-temporary casing						
Pile tests; 430 mm dia. working pile; maximum test load of 1125kN using non-working tension piles as reaction; first piles						
bottom driven		–	–	–	nr	**5219.38**
top driven		–	–	–	nr	**4219.13**
Pile tests; 430 mm dia. working pile; maximum test load of 1125kN using non-working tension piles as reaction; subsequent piles						
bottom driven		–	–	–	nr	**2181.17**
top driven		–	–	–	nr	**2181.17**
Pile tests; working pile; electronic integrity testing; each pile (minimum 40 piles per visit)		–	–	–	nr	**20.43**
Make cure and test concrete cubes		–	–	–	nr	**14.99**
Driven – segmental casing						
Pile tests; 500 mm dia. working pile; maximum test load of 600kN using non-working tension piles as reaction						
first pile		–	–	–	nr	**7608.82**
subsequent piles		–	–	–	nr	**5784.48**

CLASS R: ROADS AND PAVINGS

Item	Gang hours	Labour £	Plant £	Material £	Unit	Total rate £
GENERAL						
Notes – Labour and Plant						
All outputs are based on clear runs without undue delay to two pavers with 75% utilization.						
The outputs can be adjusted as follows to take account of space or time influences on the utilization.						
Factors for varying utilization of Labour and Plant:						
1 paver @ 75 % utilization = × 2.00						
1 paver @ 100 % utilization = × 1.50						
2 paver @ 100 % utilization = × 0.75						
RESOURCES – LABOUR						
Labour Resources Build-up Rate						
Subbase laying gang						
1 ganger/chargehand (skill rate 4)		21.42	–	–		
1 skilled operative (skill rate 4)		20.20	–	–		
2 unskilled operatives (general)		39.20	–	–		
1 plant operator (skill rate 2)		35.46	–	–		
1 plant operator (skill rate 3)		29.17	–	–		
Total Gang Rate/Hour		**145.44**	–	–		
Flexible paving gang						
1 ganger/chargehand (skill rate 4)		21.42	–	–		
2 skilled operatives (skill rate 4)		40.39	–	–		
4 unskilled operatives (general)		78.41	–	–		
4 plant operators (skill rate 3)		116.68	–	–		
Total Gang Rate/Hour		**256.90**	–	–		
Concrete paving gang						
1 ganger/chargehand (skill rate 4)		21.42	–	–		
2 skilled operatives (skill rate 4)		40.39	–	–		
4 unskilled operatives (general)		78.41	–	–		
1 plant operator (skill rate 2)		35.46	–	–		
1 plant operator (skill rate 3)		29.17	–	–		
Total Gang Rate/Hour		**204.84**	–	–		
Road surface spraying gang						
1 plant operator (skill rate 3)		29.17	–	–		
Total Gang Rate/Hour		**29.17**	–	–		
Road chippings gang						
1 ganger/chargehand (skill rate 4) – 50% of time		10.71	–	–		
1 skilled operative (skill rate 4)		20.20	–	–		
2 unskilled operatives (general)		39.20	–	–		
3 plant operators (skill rate 3)		87.51	–	–		
Total Gang Rate/Hour		**157.62**	–	–		
Cutting slabs gang						
1 unskilled operative (generally)		19.60	–	–		
Total Gang Rate/Hour		**19.60**	–	–		

CLASS R: ROADS AND PAVINGS

Item	Gang hours	Labour £	Plant £	Material £	Unit	**Total rate £**
Concrete filled joints gang						
1 ganger/chargehand (skill rate 4) – 50% of time		10.71	–	–		
1 skilled operatives (skill rate 4)		20.20	–	–		
2 unskilled operatives (general)		39.20	–	–		
Total Gang Rate/Hour		**70.11**	–	–		
Milling gang						
1 ganger/chargehand (skill rate 4)		21.42	–	–		
2 skilled operatives (skill rate 4)		40.39	–	–		
4 unskilled operatives (general)		78.41	–	–		
1 plant operators (skill rate 3)		29.17	–	–		
1 plant operator (skill rate 2)		35.46	–	–		
Total Gang Rate/Hour		**204.84**	–	–		
Rake and compact planed material gang						
1 ganger/chargehand (skill rate 4)		21.42	–	–		
1 skilled operatives (skill rate 4)		20.20	–	–		
3 unskilled operatives (general)		58.81	–	–		
1 plant operator (skill rate 3)		29.17	–	–		
1 plant operator (skill rate 4)		26.30	–	–		
Total Gang Rate/Hour		**155.89**	–	–		
Kerb laying gang						
3 skilled operatives (skill rate 4)		60.59	–	–		
1 unskilled operative (general)		19.60	–	–		
1 plant operator (skill rate 3) – 25% of time		7.29	–	–		
Total Gang Rate/Hour		**87.48**	–	–		
Path subbase, bitmac and gravel laying gang						
1 ganger/chargehand (skill rate 4)		21.42	–	–		
2 unskilled operatives (general)		39.20	–	–		
1 plant operator (skill rate 3)		29.17	–	–		
Total Gang Rate/Hour		**89.79**	–	–		
Paviors and flagging gang						
1 skilled operative (skill rate 4)		20.20	–	–		
1 unskilled operative (general)		19.60	–	–		
Total Gang Rate/Hour		**39.80**	–	–		
Traffic signs gang						
1 ganger/chargehand (skill rate 3)		23.67	–	–		
1 skilled operative (skill rate 3)		22.45	–	–		
2 unskilled operatives (general)		39.20	–	–		
1 plant operator (skill rate 3) – 25% of time		7.29	–	–		
Total Gang Rate/Hour		**92.62**	–	–		
RESOURCES – PLANT						
Plant Resources Build-up Rate						
Subbase laying						
93 KW motor grader		–	55.26	–	hr	
0.80 m^3 tractor loader		–	55.43	–	hr	
6t towed roller		–	39.30	–	hr	
Total Rate/Hour		–	**149.99**	–	hr	

CLASS R: ROADS AND PAVINGS

Item	Gang hours	Labour £	Plant £	Material £	Unit	**Total rate £**
RESOURCES – PLANT – cont						
Plant Resources Build-up Rate – cont						
Flexible paving						
2 asphalt pavers, 35 kW, 4.0 m		–	129.29	–	hr	
2 deadweight rollers, 3 point, 10 t		–	126.76	–	hr	
tractor with front bucket and integral 2 tool compressor		–	325.42	–	hr	
compressor tools: scabbler		–	2.23	–	hr	
tar sprayer, 100 litre		–	25.51	–	hr	
self propelled chip spreader		–	60.31	–	hr	
channel (heat) iron		–	8.28	–	hr	
Total Rate/Hour		–	**677.81**	–	hr	
Concrete paving						
wheeled loader, 2.60 m^2		–	90.57	–	hr	
concrete paver, 6.0 m		–	299.42	–	hr	
compaction slipform finisher		–	58.48	–	hr	
Total Rate/Hour		–	**448.46**	–	hr	
Road surface spraying						
tar sprayer, 100 litre		–	25.51	–	hr	
Total Rate/Hour		–	**25.51**	–	hr	
Road chippings						
deadweight rollers, 3 point, 10 t		–	63.38	–	hr	
tar sprayer, 100 litre		–	25.51	–	hr	
self propelled chip spreader		–	60.31	–	hr	
channel (heat) iron		–	8.28	–	hr	
Total Rate/Hour		–	**63.38**	–	hr	
Cutting slabs						
compressor, 65 cfm		–	11.79	–	hr	
300 mm disc cutter		–	1.36	–	hr	
Total Rate/Hour		–	**13.15**	–	hr	
Milling						
cold planer, 2.10 m		–	320.00	–	hr	
wheeled loader, 2.60 m^2		–	90.57	–	hr	
Total Rate/Hour		–	**410.57**	–	hr	
Heat planing						
heat planer, 4.5 m		–	138.54	–	hr	
wheeled loader, 2.60 m^2		–	90.57	–	hr	
Total Rate/ Hour		–	229.11	–	hr	
Rake and compact planed material						
deadweight roller, 3 point, 10t		–	63.38	–	hr	
tractor with front bucket and integral 2 tool compressor		–	64.28	–	hr	
channel (heat) iron		–	4.14	–	hr	
Total Rate/Hour		–	**131.80**	–	hr	
Kerb laying						
backhoe JCB 3CX (25% of time)		–	13.90	–	hr	
300 mm Stihl saw		–	1.09	–	hr	
road forms		–	2.97	–	hr	
Total Rate/Hour		–	**14.99**	–	hr	

CLASS R: ROADS AND PAVINGS

Item	Gang hours	Labour £	Plant £	Material £	Unit	**Total rate £**
Path subbase, bitmac and gravel laying						
backhoe JCB 3CX		–	55.58	–	hr	
2 t dumper		–	10.81	–	hr	
pedestrian roller Bomag BW 90S		–	12.63	–	hr	
Total Rate/Hour		–	**79.03**	–	hr	
Paviors and flagging						
2 t dumper (33% of time)		–	3.60	–	hr	
Total Rate/Hour		–	**3.60**	–	hr	
Traffic signs						
JCB 3CX backhoe – 50% of time		–	27.79	–	hr	
125 cfm compressor – 50% of time		–	9.03	–	hr	
compressor tools: hand held hammer drill – 50% of time		–	0.84	–	hr	
compressor tools: clay spade – 50% of time		–	1.32	–	hr	
compressor tools: extra 15 m hose – 50% of time		–	0.23	–	hr	
8 t lorry with hiab lift – 50% of time		–	23.76	–	hr	
Total Rate/Hour		–	**62.96**	–	hr	
01 UNBOUND SUBBASE						
Type 1 Unbound Mixtures						
Subbase; spread and graded						
75 mm thick	0.04	5.82	6.00	30.23	m³	**42.04**
110 mm thick	0.04	5.82	6.00	34.95	m³	**46.77**
150 mm thick	0.04	5.82	6.00	37.98	m³	**49.79**
200 mm thick	0.04	5.82	6.00	40.91	m³	**52.73**
Type 2 Unbound Mixtures						
Subbase, spread and graded						
75 mm thick	0.04	5.09	5.25	37.56	m³	**47.90**
110 mm thick	0.04	5.82	6.00	37.56	m³	**49.37**
150 mm thick	0.05	6.55	6.75	37.56	m³	**50.85**
200 mm thick	0.05	7.27	7.50	37.56	m³	**52.33**
Type 3 (open graded) Unbound Mixtures						
Subbase; spread and graded						
75 mm thick	0.20	29.09	6.75	92.44	m³	**128.28**
110 mm thick	0.40	58.18	7.50	92.44	m³	**158.12**
150 mm thick	0.40	58.18	7.54	108.22	m³	**173.94**
200 mm thick	0.40	58.18	7.50	129.42	m³	**195.09**
Category B (closed graded) Unbound Mixtures						
Subbase; spread and graded						
75 mm thick	0.20	29.09	6.00	42.23	m³	**77.32**
110 mm thick	0.20	29.09	6.00	49.47	m³	**84.56**
150 mm thick	0.40	58.18	6.75	49.47	m³	**114.40**
200 mm thick	0.50	72.72	7.50	49.47	m³	**129.69**
Type 4 (asphalt arisings) Unbound Mixtures						
Subbase; spread and graded						
75 mm thick	0.15	21.82	6.75	92.44	m³	**121.01**
110 mm thick	0.35	50.91	7.50	73.95	m³	**132.36**
150 mm thick	0.35	50.91	7.50	83.20	m³	**141.60**
200 mm thick	0.35	50.91	7.50	92.44	m³	**150.84**

CLASS R: ROADS AND PAVINGS

Item	Gang hours	Labour £	Plant £	Material £	Unit	**Total rate £**
01 UNBOUND SUBBASE – cont						
Geotextiles refer to Class E, Filling Ancillaries						
150 mm deep						
200 mm deep						
02 CEMENT AND OTHER HYDRAULICALLY BOUND PAVEMENTS						
Cement Bound Granular Mixture 5 (CBGM 5) C3/4						
150 mm thick	0.10	14.54	7.95	127.20	m^3	**149.69**
200 mm thick	0.13	18.18	8.25	127.20	m^3	**153.63**
250 mm thick	0.15	21.82	8.70	127.20	m^3	**157.72**
Cement Bound Granular Mixture 5 (CBGM 5) C8/10						
Subbase, laid, compacted						
150 mm thick	0.13	18.18	8.25	135.47	m^3	**161.90**
200 mm thick	0.13	18.18	8.25	139.84	m^3	**166.27**
250 mm thick	0.13	18.18	8.25	144.21	m^3	**170.64**
Cement Bound Granular Mixture 1 (CBGM 1) C8/10						
Base, laid, compacted						
150 mm thick	0.13	18.18	8.25	140.12	m^3	**166.55**
200 mm thick	0.13	18.18	8.25	144.64	m^3	**171.07**
250 mm thick	0.13	18.18	8.25	149.16	m^3	**175.59**
Cement Bound Granular Mixture (CBGM 1) C12/16						
Base, laid, compacted						
150 mm thick	0.13	18.18	8.25	147.41	m^3	**173.83**
180 mm thick	0.13	18.18	8.25	152.16	m^3	**178.59**
Cement Bound Granular Mixture (CBGM 1) C15/20						
Base, laid, compacted						
150 mm thick	0.13	18.18	8.25	152.68	m^3	**179.10**
Bituminous bound mixtures, asphalt concrete						
Base Course AC 32 Dense Base 40/60 / 100/150 Des/Rec to PD 6691						
100 mm thick	0.02	5.14	13.56	18.89	m^2	**37.59**
150 mm thick	0.03	7.71	20.34	28.34	m^2	**56.39**
200 mm thick	0.03	7.71	20.34	37.79	m^2	**65.83**
Binder course AC 20 Dense Base 40/60 / 100/150 Des/Rec to PD 6691						
60 mm thick	0.02	3.85	10.17	12.30	m^2	**26.32**
100 mm thick	0.02	5.14	13.56	20.50	m^2	**39.19**
Surface course AC 10 Close Surf 100/150 to PD 6691						
30 mm thick	0.01	2.57	6.78	7.41	m^2	**16.77**
40 mm thick	0.02	3.85	10.17	9.89	m^2	**23.91**

CLASS R: ROADS AND PAVINGS

Item	Gang hours	Labour £	Plant £	Material £	Unit	**Total rate £**
03 BITUMINOUS BOUND PAVEMENTS						
Surface course						
SMA 10/14 Surf 40/60 to PD 6691						
20 mm thick	0.03	7.71	4.38	4.90	m^2	**16.98**
40 mm thick	0.03	7.71	6.41	9.80	m^2	**23.92**
50 mm thick	0.03	7.71	6.39	12.25	m^2	**26.35**
Thin Surface Course System						
30 mm thick	0.03	6.42	6.80	8.54	m^2	**21.76**
40 mm thick	0.02	5.14	8.18	11.38	m^2	**24.70**
50 mm thick	0.02	3.85	10.18	14.23	m^2	**28.27**
HRA 35/14 F Surf PMB Des TO PD 6691 WT. Class 2 with Pre Coated Chippings PSV 65						
50 mm thick	0.04	8.99	10.18	19.92	m^2	**39.09**
Binder course						
HRA 35/14 F Bin PMB Des TO PD 6691 WT. Class 2						
45 mm thick	0.02	3.85	10.18	14.08	m^2	**28.12**
Base and binder course						
AC 14/20 EME2 Base/Bin						
100 mm thick	0.02	5.14	13.56	24.85	m^2	**43.54**
150 mm thick	0.02	5.14	14.12	37.28	m^2	**56.54**
Hot rolled asphalt surface course						
HRA 35/14 F Surf 40/60 des to PD 6691 with pre-coated chippings						
50 mm thick	0.01	2.57	10.17	19.92	m^2	**32.65**
Slurry surfacing						
Slurry surfacing to BS EN 12273						
3 mm thick	0.02	0.44	0.38	1.98	m^2	**2.80**
4 mm thick	0.02	0.44	0.38	3.10	m^2	**3.92**
Surface dressing						
Surface dressing to BS EN 12271						
6 mm single surface dressing	0.01	1.58	1.57	2.50	m^2	**5.65**
10 mm & 6 mm racked-in surface dressing	0.01	1.58	1.57	2.70	m^2	**5.85**
10 mm & 6 mm double surface dressing	0.01	1.58	1.57	3.05	m^2	**6.20**
Bond coat						
Bond coat to BS 594987						
large areas; over 20 m^2 @0.2 kg/m^2	0.02	0.44	0.38	0.58	m^2	**1.40**
small areas; under 20 m^2 @0.2 kg/m^2	0.02	0.44	0.38	0.73	m^2	**1.55**
large areas; under 20 m^2 @0.35 kg/m^2	0.02	0.44	0.38	1.02	m^2	**1.84**
small areas; under 20 m^2 @0.35 kg/m^2	0.02	0.44	0.38	1.27	m^2	**2.09**

CLASS R: ROADS AND PAVINGS

Item	Gang hours	Labour £	Plant £	Material £	Unit	**Total rate £**
03 BITUMINOUS BOUND PAVEMENTS – cont						
Excavation, Trimming and Reinstatement of Existing Surfaces						
Trimming edges only of existing slabs, floors or similar surfaces (wet or dry); 6 mm cutting width						
50 mm deep	0.02	0.39	0.26	50.07	m	**50.72**
100 mm deep	0.03	0.59	0.41	122.18	m	**123.18**
Cutting existing slabs, floors or similar surfaces (wet or dry); 8 mm cutting width						
50 mm deep	0.03	0.49	0.32	50.07	m	**50.88**
100 mm deep	0.06	1.18	0.79	122.18	m	**124.15**
150 mm deep	0.08	1.57	1.05	164.60	m	**167.22**
Milling pavement (assumes disposal on site or reuse as fill but excludes transport if required)						
75 mm deep	0.03	5.53	114.76	–	m^2	**120.29**
100 mm deep	0.04	7.37	153.02	–	m^2	**160.39**
50 mm deep; scarifying surface	0.02	4.51	93.51	–	m^2	**98.02**
75 mm deep; scarifying surface	0.04	7.58	157.27	–	m^2	**164.85**
25 mm deep; heat planing for reuse	0.03	6.56	54.62	–	m^2	**61.17**
50 mm deep; heat planing for reuse	0.06	11.47	95.58	–	m^2	**107.06**
Raking over scarified or heat planed material; compacting with 10 t roller						
ne 50 mm deep	0.01	1.56	3.97	–	m^2	**5.53**
05 CONCRETE PAVEMENTS						
The following unit costs are for jointed reinforced concrete slabs, laid in reasonable areas (over 200 m^2) by paver train/slipformer.						
Designed mix; cement to BS EN 197–1; grade C30, 20 mm aggregate						
Dense base and binder course asphalt concrete (design mixture)						
180 mm deep	0.02	3.07	7.99	17.87	m^2	**28.93**
220 mm deep	0.02	3.69	9.59	21.84	m^2	**35.12**
260 mm deep	0.02	4.51	11.72	25.81	m^2	**42.04**
300 mm deep	0.03	5.12	13.32	29.78	m^2	**48.22**
Joints in concrete pavement						
Transverse Joints						
Depth; not exceeding 30 mm	0.07	7.17	0.74	1.99	m	**9.89**
60 mm–100 mm	0.02	3.69	9.59	21.84	m	**35.12**
100 mm–150 mm	0.02	4.51	11.72	25.81	m	**42.04**
200 mm–250 mm	0.03	5.12	13.32	29.78	m	**48.22**
Fabric						
Steel fabric reinforcement to BS 4483						
Ref A142 nominal mass 2.22 kg	0.03	6.15	–	5.12	m^2	**11.26**
Ref A252 nominal mass 3.95 kg	0.04	8.19	–	7.68	m^2	**15.88**
Ref B385 nominal mass 4.53 kg	0.04	8.19	–	10.48	m^2	**18.67**
Ref C636 nominal mass 5.55 kg	0.05	10.24	–	12.49	m^2	**22.73**
Ref B503 nominal mass 5.93 kg	0.05	10.24	–	14.67	m^2	**24.92**

CLASS R: ROADS AND PAVINGS

Item	Gang hours	Labour £	Plant £	Material £	Unit	**Total rate £**
Mild Steel bar reinforcement BS 4449						
Bars; supplied in bent and cut lengths						
6 mm nominal size	8.00	1638.75	–	952.67	t	**2591.42**
8 mm nominal size	6.74	1380.65	–	952.67	t	**2333.31**
10 mm nominal size	6.74	1380.65	–	952.67	t	**2333.31**
12 mm nominal size	6.74	1380.65	–	952.67	t	**2333.31**
16 mm nominal size	6.15	1259.79	–	952.67	t	**2212.46**
High yield steel bar reinforcement BS 4449 or 4461						
Bars; supplied in bent and cut lengths						
6 mm nominal size	8.00	1638.75	–	961.59	t	**2600.34**
8 mm nominal size	6.74	1380.65	–	961.59	t	**2342.24**
10 mm nominal size	6.74	1380.65	–	961.59	t	**2342.24**
12 mm nominal size	6.74	1380.65	–	961.59	t	**2342.24**
16 mm nominal size	6.15	1259.79	–	961.58	t	**2221.37**
Sheeting to prevent moisture loss						
Polyethylene sheeting; lapped joints; horizontal below concrete pavements						
250 micron	0.01	2.05	–	0.83	m^2	**2.87**
500 micron	0.01	2.05	–	2.10	m^2	**4.15**
06 JOINTS IN CONCRETE PAVEMENTS						
Joints						
Longitudinal joints						
180 mm deep	0.01	2.46	6.39	22.10	m	**30.95**
220 mm deep	0.01	2.46	6.39	25.97	m	**34.82**
260 mm deep	0.01	2.46	6.39	31.81	m	**40.66**
300 mm deep	0.01	2.46	6.39	37.02	m	**45.87**
Expansion joints						
180 mm deep	0.01	2.46	6.39	43.49	m	**52.34**
220 mm deep	0.01	2.46	6.39	50.64	m	**59.49**
260 mm deep	0.01	2.46	6.39	57.78	m	**66.63**
300 mm deep	0.01	2.46	6.39	59.21	m	**68.06**
Contraction joints						
180 mm deep	0.01	2.46	6.39	25.36	m	**34.21**
220 mm deep	0.01	2.46	6.39	26.25	m	**35.10**
260 mm deep	0.01	2.46	6.39	28.49	m	**37.34**
300 mm deep	0.01	2.46	6.39	33.96	m	**42.81**
Construction joints						
180 mm deep	0.01	2.46	6.39	16.76	m	**25.61**
220 mm deep	0.01	2.46	6.39	20.15	m	**29.00**
260 mm deep	0.01	2.46	6.39	19.58	m	**28.43**
300 mm deep	0.01	2.46	6.39	21.08	m	**29.93**
Open joints with filler						
ne 0.5 m; 10 mm flexcell joint filler	0.11	7.71	–	4.35	m	**12.06**
0.5–1.00 m; 10 mm flexcell joint filler	0.11	7.71	–	6.11	m	**13.82**
Joint sealants						
10 × 20 mm cold polysulphide sealant	0.14	9.82	–	4.82	m	**14.63**
20 × 20 mm cold polysulphide sealant	0.18	12.62	–	8.26	m	**20.88**

CLASS R: ROADS AND PAVINGS

Item	Gang hours	Labour £	Plant £	Material £	Unit	Total rate £
07 KERBS, CHANNELS AND EDGINGS						
Foundations to kerbs etc.						
Measurement Note: the following are shown separate from their associated kerb etc. to simplify the presentation of cost alternatives.						
Mass concrete						
200 × 100 mm	0.01	0.87	0.17	2.32	m	**3.36**
300 × 150 mm	0.02	1.31	0.24	5.31	m	**6.86**
450 × 150 mm	0.02	1.75	0.36	7.85	m	**9.96**
100 × 100 mm haunching, one side	0.01	0.44	0.09	0.57	m	**1.09**
Precast Concrete kerbs to BS 7263						
Kerbs; bullnosed; splayed or half battered; straight or curved over 12 m radius						
125 × 150 mm	0.06	5.25	1.07	12.74	m	**19.07**
125 × 255 mm	0.07	6.12	1.25	14.25	m	**21.63**
150 × 305 mm	0.07	6.12	1.28	21.32	m	**28.72**
Kerbs; bullnosed; splayed or half battered; curved ne 12 m radius						
125 × 150 mm	0.07	5.69	1.16	12.74	m	**19.59**
125 × 255 mm	0.08	6.56	1.34	14.25	m	**22.15**
150 × 305 mm	0.08	6.56	1.37	21.32	m	**29.25**
Quadrants						
305 × 305 × 150 mm	0.08	7.00	1.46	19.02	nr	**27.48**
455 × 455 × 255 mm	0.10	8.75	1.83	28.21	nr	**38.79**
Drop kerbs						
125 × 255 mm	0.07	6.12	1.28	31.71	m	**39.11**
150 × 305 mm	0.07	6.12	1.28	52.71	m	**60.12**
Channel; straight or curved over 12 m radius						
255 × 125 mm	0.07	6.12	1.28	8.34	m	**15.75**
Channel; curved radius ne 12 m						
255 × 125 mm	0.07	6.12	1.28	8.34	m	**15.75**
Tensitions						
50 × 150 mm	0.04	3.50	0.73	4.22	m	**8.45**
Edging; curved ne 12 m radius						
50 × 150 mm	0.05	3.94	0.82	4.22	m	**8.97**
Precast concrete drainage channels; Charcon Safeticurb; channels jointed with plastic rings and bedded; jointed and pointed in cement mortar						
Channel unit; straight; Type DBA/3						
250 × 254 mm; medium duty	0.08	6.56	1.37	67.48	m	**75.41**
305 × 305 mm; heavy duty	0.10	8.31	1.74	142.39	m	**152.43**
Precast concrete Ellis Trief safety kerb; bedded jointed and pointed in cement mortar						
Kerb; straight or curved over 12 m radius						
415 × 380 mm	0.23	19.68	3.93	95.55	m	**119.16**
Kerb; curved ne 12 m radius						
415 × 380 mm	0.25	21.87	4.37	110.35	m	**136.60**

CLASS R: ROADS AND PAVINGS

Item	Gang hours	Labour £	Plant £	Material £	Unit	**Total rate £**
Precast concrete combined kerb and drainage block Beany block system; bedded jointed and pointed in cement mortar						
Kerb; top block, shallow base unit, standard cover plate and frame						
straight or curved over 12 m radius	0.15	13.12	2.63	149.59	m	**165.34**
curved ne 12 m radius	0.20	17.50	3.48	228.41	m	**249.39**
Kerb; top block, standard base unit, standard cover plate and frame						
straight or curved over 12 m radius	0.15	13.12	2.65	149.59	m	**165.37**
curved ne 12 m radius	0.20	17.50	3.48	228.41	m	**249.39**
Kerb; top block, deep base unit, standard cover plate and frame						
straight or curved over 12 m radius	0.15	13.12	2.60	333.90	m	**349.62**
curved ne 12 m radius	0.20	17.50	3.43	426.90	m	**447.82**
base block depth tapers	0.10	9.08	1.77	36.03	m	**46.88**
Extruded asphalt kerbs to BS 5931; extruded and slip formed						
Kerb; straight or curved over 12 m radius						
75 mm kerb height		–	–	4.69	m	**4.69**
100 mm kerb height		–	–	10.08	m	**10.08**
125 mm kerb height		–	–	13.43	m	**13.43**
Channel; straight or curved over 12 m radius						
300 mm channel width		–	–	16.81	m	**16.81**
250 mm channel width		–	–	13.43	m	**13.43**
Kerb; curved to radius ne 12 m						
75 mm kerb height		–	–	4.47	m	**4.47**
100 mm kerb height		–	–	9.58	m	**9.58**
125 mm kerb height		–	–	12.77	m	**12.77**
Channel; curved to radius ne 12 m						
300 mm channel width		–	–	16.72	m	**16.72**
250 mm channel width		–	–	12.77	m	**12.77**
Extruded concrete; slip formed						
Kerb; straight or curved over 12 m radius						
100 mm kerb height		–	–	10.08	m	**10.08**
125 mm kerb height		–	–	13.43	m	**13.43**
Kerb; curved to radius ne 12 m						
100 mm kerb height		–	–	14.75	m	**14.75**
125 mm kerb height		–	–	17.84	m	**17.84**

CLASS R: ROADS AND PAVINGS

Item	Gang hours	Labour £	Plant £	Material £	Unit	**Total rate £**
07 KERBS, CHANNELS AND EDGINGS – cont						
Light Duty Pavements (Generally Pedestrian)						
Subbases						
Measurement Note: the following are shown separate from their associated paving to simplify the presentation of cost alternatives.						
To paved area; sloping not exceeding 10° to the horizontal						
100 mm thick Type 1 Unbound Mixtures	0.01	0.81	0.71	3.32	–	**4.84**
150 mm thick Type 1 Unbound Mixtures	0.01	1.08	0.95	4.98	–	**7.00**
100 mm thick Concrete GEN3 or ST4 to BS 8500	0.02	1.89	1.66	11.55	–	**15.10**
150 mm thick Concrete GEN3 or ST4 to BS 8500	0.03	2.87	2.53	17.33	–	**22.73**
Binder Course AC 20 Dense Bin 100/150 Rec to PD6691; Surface Course AC 6 Dense Surf 100/150 to PD6691.						
Paved area comprising binder course 50 mm thick, surface course 20 mm thick						
sloping not exceeding 10° to the horizontal	0.09	7.63	6.72	9.18	m^2	**23.53**
sloping not exceeding 10° to the horizontal; red additives	0.09	7.63	6.72	37.73	m^2	**52.08**
sloping not exceeding 10° to the horizontal; green additives	0.09	7.63	6.72	13.85	m^2	**28.19**
sloping exceeding 10° to the horizontal	0.10	8.54	7.52	9.18	m^2	**25.23**
sloping exceeding 10° to the horizontal; red additives	0.10	8.54	7.52	37.73	m^2	**53.78**
sloping exceeding 10° to the horizontal; green additives	0.10	8.54	7.52	13.85	m^2	**29.90**
Granular base surfacing; Central Reserve Treatments Limestone, graded 10 mm down; laid and compacted						
Paved area 100 mm thick; surface sprayed twice with two coats of cold bituminous emulsion; blinded with 6 mm quartzite fine gravel						
sloping not exceeding 10° to the horizontal	0.02	1.80	1.58	3.72	m^2	**7.09**
Ennstone Johnston Golden gravel; graded 13 mm to fines; rolled wet						
Paved area 50 mm thick; single layer						
sloping not exceeding 10° to the horizontal	0.03	2.69	2.37	10.87	m^2	**15.94**

CLASS R: ROADS AND PAVINGS

Item	Gang hours	Labour £	Plant £	Material £	Unit	**Total rate £**
Precast concrete flags; BS 7263; grey; bedding in cement mortar						
Paved area; sloping not exceeding 10° to the horizontal						
900 × 600 × 63 mm	0.21	8.36	0.76	12.11	m^2	**21.23**
900 × 600 × 50 mm	0.20	7.96	0.72	12.06	m^2	**20.74**
600 × 600 × 63 mm	0.25	9.95	0.90	10.63	m^2	**21.48**
600 × 600 × 50 mm	0.24	9.55	0.87	10.57	m^2	**20.98**
600 × 450 × 50 mm	0.28	11.14	1.01	10.23	m^2	**22.38**
Extra for coloured, 50 mm thick		–	–	3.19	m^2	**3.19**
Precast concrete rectangular paving blocks; BS 6717; grey; bedding on 50 mm thick dry sharp sand; filling joints; excluding subbase						
Paved area; sloping not exceeding 10° to the horizontal						
200 × 100 × 80 mm thick	0.30	11.94	1.08	27.25	m^2	**40.27**
200 × 100 × 80 mm thick; coloured blocks	0.30	11.94	1.08	33.96	m^2	**46.98**
Brick paviors; bedding on 20 mm thick mortar; excluding subbase						
Paved area; sloping not exceeding 10° to the horizontal						
200 × 100 × 65 mm	0.30	11.94	1.08	27.25	m^2	**40.27**
Granite setts; bedding on 25 mm cement mortar; excluding subbase						
Paved area; sloping not exceeding 10° to the horizontal						
to random pattern	0.90	35.82	3.24	63.19	m^2	**102.26**
to specific pattern	1.20	47.76	4.33	63.19	m^2	**115.28**
Cobble paving; 50–75 mm; bedding on 25 mm cement mortar; filling joints; excluding subbase						
Paved area; sloping not exceeding 10° to the horizontal						
50–75 mm dia. cobbles	1.00	39.80	3.57	19.80	m^2	**63.17**
08 ANCILLARIES						
Traffic signal installation is carried out exclusively by specialist contractors, although certain items are dealt with by the main contractor or a subcontractor.						
Signal pedestals						
Installation of signal pedestals, loop detector unit pedestals, controller unit boxes and cable connection pillars						
signal pedestal		–	–	–	nr	**61.28**
loop detector unit pedestral		–	–	–	nr	**27.23**

CLASS R: ROADS AND PAVINGS

Item	Gang hours	Labour £	Plant £	Material £	Unit	Total rate £
08 ANCILLARIES – cont						
Signal pedestals – cont						
Excavate trench for traffic signal cable, depth ne 1.50 m; supports, backfilling						
450 mm wide		–	–	–	m	**20.42**
Extra for excavating in hard material		–	–	–	m^3	**40.87**
Saw cutting grooves in pavement for detector loops and feeder cables; seal with hot bitumen sealant after installation						
25 mm deep		–	–	–	m	**29.97**
Traffic signs						
In this section prices will vary depending upon the diagram configurations. The following are average costs of signs and bollards. Diagram numbers refer to the Traffic Signs Regulations and General Directions 2002 and the figure numbers refer to the Traffic Sign Manual.						
Examples of Prime Costs for Class 1 (High Intensity) traffic and road signs (ex works).						
600 × 450 mm		–	–	46.56	nr	**46.56**
600 mm dia.		–	–	180.94	nr	**180.94**
600 mm triangular		–	–	150.75	nr	**150.75**
500 × 500 mm		–	–	46.49	nr	**46.49**
450 × 450 mm		–	–	44.18	nr	**44.18**
450 × 300 mm		–	–	41.94	nr	**41.94**
1200 × 400 mm (CHEVRONS)		–	–	69.98	nr	**69.98**
Examples of Prime Costs for Class 21 (Engineering Grade) traffic and road signs (ex works).						
600 × 450 mm		–	–	46.56	nr	**46.56**
600 mm dia.		–	–	180.94	nr	**180.94**
600 mm triangular		–	–	150.75	nr	**150.75**
500 × 500 mm		–	–	46.49	nr	**46.49**
450 × 450 mm		–	–	44.18	nr	**44.18**
450 × 300 mm		–	–	41.94	nr	**41.94**
1200 × 400 mm (CHEVRONS)		–	–	74.52	nr	**74.52**
Standard reflectorised traffic signs						
Note: Unit costs do not include concrete foundations						
Standard one post signs; 600 × 450 mm type C1 signs						
fixed back to back to another sign (measured separately) with aluminium clips to existing post (measured separately)	0.04	3.70	5.61	48.89	nr	**58.20**
extra for fixing singly with aluminium clips	0.01	0.93	0.52	1.30	nr	**2.74**
extra for fixing singly with stainless steel clips	0.01	0.93	1.89	5.01	nr	**7.83**

CLASS R: ROADS AND PAVINGS

Item	Gang hours	Labour £	Plant £	Material £	Unit	**Total rate £**
fixed back to back to another sign (measured separately) with stainless steel clips to one new 76 mm dia. plastic coated steel posts 1.75 m long	0.27	25.01	37.86	114.04	nr	**176.91**
extra for fixing singly to one face only	0.01	0.93	0.52	–	nr	**1.45**
extra for 76 mm dia. 1.75 m long aluminium post	0.02	1.85	1.09	36.23	nr	**39.17**
extra for 76 mm dia. 3.5 m long plastic coated steel post	0.02	1.85	1.09	56.05	nr	**58.99**
extra for 76 mm dia. 3.5 m long aluminium post	0.02	1.85	1.09	68.42	nr	**71.36**
extra for excavation for post, in hard material	1.10	101.89	58.51	–	nr	**160.40**
extra for single external illumination unit with fitted photo cell (excluding trenching and cabling); unit cost per face illuminated	0.33	30.57	17.54	107.47	nr	**155.57**
Standard two post signs; 1200 × 400 mm;						
fixed back to back to another sign (measured separately) with stainless steel clips to two new 76 mm dia. plastic coated steel posts	0.51	47.24	71.50	205.31	nr	**324.05**
extra for fixing singly to one face only	0.02	1.85	1.09	–	nr	**2.94**
extra for two 76 mm dia. 1.75 m long aluminium posts	0.04	3.70	2.13	72.47	nr	**78.30**
extra for two 76 mm dia. 1.75 m long plastic coated steel posts	0.04	3.70	2.13	112.10	nr	**117.94**
extra for two 76 mm dia. 3.5 m long aluminium posts	0.04	3.70	2.13	136.84	nr	**142.67**
extra for excavation for post, in hard material	1.10	101.89	58.51	–	nr	**160.40**
extra for single external illumination unit with fitted photo cell (including trenching and cabling); unite per face illuminated	0.58	53.72	30.82	166.11	nr	**250.65**
Standard internally illuminated traffic signs						
Bollard with integral mould-in translucent graphics (excluding trenching and cabling)						
fixing to concrete base	0.48	44.46	67.29	529.36	nr	**641.11**
Special traffic signs						
Note: Unit costs do not include concrete foundations or trenching and cabling						
Externally illuminated reflectorised traffic signs manufactured to order						
special signs, surface area 1.50 m^2 on two 100 mm dia. steel posts		–	–	–	nr	**1038.40**
special signs, surface area 4.00 m^2 on three 100 mm dia. steel posts		–	–	–	nr	**1557.60**
Internally illuminated traffic signs manufactured to order						
special signs, surface area 0.25 m^2 on one new 76 mm dia. post		–	–	–	nr	**260.05**
special signs, surface area 0.75 m^2 on one new 100 mm dia. steel posts		–	–	–	nr	**319.91**
special signs, surface area 4.00 m^2 on four new 120 mm dia. steel posts		–	–	–	nr	**1636.79**

CLASS R: ROADS AND PAVINGS

Item	Gang hours	Labour £	Plant £	Material £	Unit	**Total rate £**
08 ANCILLARIES – cont						
Signs on gantries						
Externally illuminated reflectorised signs						
1.50 m^2	1.79	165.33	354.99	1333.80	nr	**1854.12**
2.50 m^2	2.15	199.14	427.58	1366.74	nr	**1993.46**
3.00 m^2	3.07	284.35	610.54	1497.10	nr	**2391.99**
Internally illuminated sign with translucent optical reflective sheeting and remote light source						
0.75 m^2	1.56	144.49	310.24	1486.08	nr	**1940.81**
1.00 m^2	1.70	157.46	338.08	1968.84	nr	**2464.38**
1.50 m^2	2.41	223.22	479.28	3033.72	nr	**3736.23**
Existing signs						
Take from store and re-erect						
3.0 m high road sign	0.28	25.93	55.68	53.01	nr	**134.63**
road sign on two posts	0.50	46.31	99.44	106.01	nr	**251.76**
Surface markings; reflectorised white						
Letters and shapes						
triangles; 1.6 m high		–	–	–	nr	**18.25**
triangles; 2.0 m high		–	–	–	nr	**22.59**
triangles; 3.75 m high		–	–	–	nr	**29.19**
circles with enclosing arrows; 1.6 m dia.		–	–	–	nr	**87.58**
arrows; 4.0 m long; straight		–	–	–	nr	**36.50**
arrows; 4.0 m long; turning		–	–	–	nr	**36.50**
arrows; 6.0 m long; straight		–	–	–	nr	**43.79**
arrows; 6.0 m long; turning		–	–	–	nr	**43.79**
arrows; 6.0 m long; curved		–	–	–	nr	**43.79**
arrows; 6.0 m long; double headed		–	–	–	nr	**61.30**
arrows; 8.0 m long; double headed		–	–	–	nr	**87.58**
arrows; 16.0 m long; double headed		–	–	–	nr	**131.37**
arrows; 32.0 m long; double headed		–	–	–	nr	**175.15**
letters or numerals; 1.6 m high		–	–	–	nr	**11.29**
letters or numerals; 2.0 m high		–	–	–	nr	**16.80**
letters or numerals; 3.75 m high		–	–	–	nr	**29.19**
Continuous lines						
150 mm wide		–	–	–	m	**2.17**
200 mm wide		–	–	–	m	**2.54**
Intermittent lines						
60 mm wide; 0.60 m line and 0.60 m gap		–	–	–	m	**1.14**
100 mm wide; 1.0 m line and 5.0 m gap		–	–	–	m	**1.17**
100 mm wide; 2.0 m line and 7.0 m gap		–	–	–	m	**1.26**
100 mm wide; 4.0 m line and 2.0 m gap		–	–	–	m	**1.29**
100 mm wide; 6.0 m line and 3.0 m gap		–	–	–	m	**1.31**
150 mm wide; 1.0 m line and 5.0 m gap		–	–	–	m	**1.65**
150 mm wide; 6.0 m line and 3.0 m gap		–	–	–	m	**1.71**
150 mm wide; 0.60 m line and 0.30 m gap		–	–	–	m	**2.10**
200 mm wide; 0.60 m line and 0.30 m gap		–	–	–	m	**2.10**
200 mm wide; 1.0 m line and 1.0 m gap		–	–	–	m	**2.10**

CLASS R: ROADS AND PAVINGS

Item	Gang hours	Labour £	Plant £	Material £	Unit	**Total rate £**
Surface markings; reflectorised yellow						
Continuous lines						
100 mm wide		–	–	–	m	**2.40**
150 mm wide		–	–	–	m	**2.69**
Intermittent lines						
kerb marking; 0.25 m long		–	–	–	nr	**2.53**
Surface markings; Vinyl strip markings						
Note: Unit costs based upon new road with clean surface closed to traffic.						
Vinyl strip markings (pedestrian crossings and similar locations)						
200 mm wide line		–	–	–	m	**10.86**
600 × 300 mm single stud tile		–	–	–	nr	**18.95**
Removal of existing reflectorised thermoplastic markings						
100 mm wide line		–	–	–	m	**1.91**
150 mm wide line		–	–	–	m	**2.82**
200 mm wide line		–	–	–	m	**3.80**
arrow or letter ne 6.0 m long		–	–	–	nr	**29.33**
arrow or letter 6.0–16.00 m long		–	–	–	nr	**114.92**
Reflecting road studs						
General						
100 × 100 mm square bi-directional reflecting road studs with amber corner cube reflectors		–	–	–	nr	**1.42**
140 × 254 mm rectangular one way reflecting road studs with red catseye reflectors		–	–	–	nr	**3.23**
140 × 254 mm rectangular one way reflecting road studs with green catseye reflectors		–	–	–	nr	**3.34**
140 × 254 mm rectangular bi-directional reflecting road studs with white catseye reflectors		–	–	–	nr	**3.34**
140 × 254 mm rectangular bi-directional reflecting road studs with amber catseye reflectors		–	–	–	nr	**3.34**
140 × 254 mm rectangular bi-directional reflecting road stud without catseye reflectors		–	–	–	nr	**2.08**
Removal of road studs						
Remove from store and re-install 100 × 100 mm square bi-directional reflecting road stud with corner cube reflectors		–	–	–	nr	**0.69**
Remove from store and re-install 140 × 254 mm rectangular one way reflecting road stud with catseye reflectors		–	–	–	nr	**1.64**

CLASS S: RAIL TRACK

Item	Gang hours	Labour £	Plant £	Material £	Unit	**Total rate £**
NOTES The following unit costs are for guidance only. For more reliable estimates it is advisable to seek the advice of a Specialist Contractor. These rates are for the supply and laying of track in connection with the Permanent Way.						
Permanent Way The following rates would not reflect work carried out on the existing public track infrastructure (Permanent Way), which tends to be more costly not merely due to enclosure complexity and control standards, but also due to a number of logistical factors, such as: * access to the works for personnel, plant and machinery would be via approved access points to the rail followed by travel along the rail to the work area; this calls for the use of Engineering Trains as well as reducing the effective shift time of the works gang * effect of track possession periods will dictate when the work can be carried out and could well force night-time or weekend working and perhaps severely reducing the effective shift hours where coupled to long travel to work distances and the need to clear away before the resumption of traffic. * the labour gang will be composed of more highly paid personnel, reflecting the additional training received; in addition there may well be additional gang members acting as look-outs; this could add 30% to the gang rates shown below * plant will tend to cost more, especially if the circumstances of the work call for rail/road plant; this could add 20 % to the gang rates shown						
Possession costs Any cost related to planned safety arrangements to allow railway possession for the construction works and controls or prevent the normal movement of rail traffic on the network between defined location and pre defined periods including any speed restrictions. During this period the contractor is given possession and rail traffic stops or speed is restricted. Possessions of the Operational Safety Zone may well be fragmented rather than a single continuous period, dependent upon windows in the pattern of rail traffic Costs for working alongside an operational rail system are high, the need for safety demanding a high degree of supervision, look-outs, the induction of labour gangs and may involve temporary works such as safety barriers.						

CLASS S: RAIL TRACK

Item	Gang hours	Labour £	Plant £	Material £	Unit	**Total rate £**
01 TRACK FOUNDATIONS						
Ballast						
Imported crushed granite						
Bottom ballast		–	–	74.89	m^3	**74.89**
Top ballast		–	–	74.89	m^3	**74.89**
Imported granular material						
Blankets; 150 mm thick		–	–	62.42	m^3	**62.42**
Imported sand						
Blinding; 100 mm thick		–	–	37.45	m^3	**37.45**
Waterproof Membranes						
Polythene sheeting						
Waterproof membrane; 1200 gauge		–	–	4.65	m^2	**4.65**
Ballast Cleaning						
Machine ballast cleaning, on straight or curved track; formation depth 450 mm	0.30	22.33	641.91	–	m^2	**664.24**
Tamping						
Machine tamping; straight track	0.50	37.21	207.64	–	m	**244.85**
Pneumatic Ballast Injection						
Pneumatic ballast injection: Stone blowing	0.50	37.21	20.38	–	m^2	**57.59**
02 TAKING UP						
Taking up; dismantling into individual components; sorting; storing on site where directed						
Bullhead or flat bottom rails						
plain track; fishplated; timber sleepers	0.06	5.59	–	–	m	**5.59**
plain track; fishplated; concrete sleepers	0.08	7.83	–	–	m	**7.83**
plain track; welded; timber sleepers	0.06	5.59	–	–	m	**5.59**
plain track; welded; concrete sleepers	0.08	7.83	–	–	m	**7.83**
turnouts; fishplated; concrete sleepers	6.36	628.50	–	–	nr	**628.50**
diamond crossings; fishplated; timber sleepers	10.34	1021.30	–	–	nr	**1021.30**
Dock and crane rails						
plain track; welded; base plates	0.10	10.06	–	–	m	**10.06**
turnouts; welded; base plates	6.36	628.50	–	–	nr	**628.50**
diamonds; welded; base plates	10.34	1021.30	–	–	nr	**1021.30**
Check and guard rails						
plain track; fishplated	0.06	5.59	–	–	m	**5.59**
Conductor rails						
plain track; fishplated	0.06	5.87	–	–	m	**5.87**
Sundries						
buffer stops	1.33	130.94	–	–	nr	**130.94**
retarders	1.86	183.31	–	–	nr	**183.31**
wheel stops	0.56	54.99	–	–	nr	**54.99**
lubricators	0.80	78.56	–	–	nr	**78.56**
swith heaters	1.27	125.70	–	–	nr	**125.70**
switch levers	0.80	78.56	–	–	nr	**78.56**

CLASS S: RAIL TRACK

Item	Gang hours	Labour £	Plant £	Material £	Unit	**Total rate £**
03 LIFTING, PACKING AND SLEWING						
Maximum distance of slew 300 mm; maximum lift 100 mm; no extra ballast allowed						
Bullhead rail track; fishplated; timber sleepers	4.00	395.28	–	–	m	**395.28**
Bullhead rail track; fishplated; concrete sleepers	4.00	395.28	–	–	m	**395.28**
Bullhead rail track with turnout; timber sleepers	2.50	247.05	–	–	nr	**247.05**
Flat bottom rail track; welded; timber sleepers	4.00	395.28	–	–	m	**395.28**
Flat bottom rail track; welded; concrete sleepers	4.00	395.28	–	–	m	**395.28**
Flat bottom rail track with turnout; concrete sleepers	3.33	329.37	–	–	nr	**329.37**
Spot replacement of sleepers; Timber sleeper	1.00	98.82	5.49	23.56	nr	**127.87**
Spot replacement of sleepers; Concrete sleeper	0.50	49.41	5.11	70.00	nr	**124.52**
Existing rail turning and refixing; plain with conductor rail & turnouts	2.50	247.05	191.78	–	nr	**438.83**
Rerailing	0.40	39.53	–	–	m	**39.53**
Stressing rail	0.83	82.35	–	243.18	m	**325.53**
Buffer stops	2.50	247.05	64.40	–	nr	**311.45**
04 SUPPLYING (STANDARD GAUGE TRACK)						
Bullhead rails; BS 11; delivered in standard 18.288 m lengths						
BS95R section, 47 kg/m; for jointed track		–	–	1181.17	tonne	**1181.17**
Flat bottom rails; BS 11; delivered in standard 18.288 m lengths						
BS113 'A' section, 56 kg/m; for jointed track		–	–	1181.17	tonne	**1181.17**
BS113 'A' section, 40 kg/m; for welded track		–	–	1181.17	tonne	**1181.17**
Extra for curved rails to form super elevation; radius over 600 m		–	–	–	%	**23.00**
Check and guard rails; BS 11; delivered in standard 18.288 m lengths; flange planed to allow 50 mm free wheel clearance						
BS113 'A' section, 56 kg/m; for bolting		–	–	1410.01	tonne	**1410.01**
Conductor rails; BS 11; delivered in standard 18.288 m lengths						
BS113 'A' section, 56 kg/m; for bolting		–	–	1410.01	tonne	**1410.01**
Twist rails; BS 11; delivered in standard 18.288 m lengths						
BS113 'A' section, 56 kg/m; for bolting		–	–	1410.01	tonne	**1410.01**
Sleepers; bullhead track timber sleepers						
2600 × 250 × 130 mm		–	–	38.40	nr	**38.40**
Sleepers; bullhead track concrete sleepers						
2600 × 250 × 130 mm		–	–	68.12	nr	**68.12**
Sleepers; bullhead track steel sleepers						
2525 × 264 × 204 mm		–	–	91.54	nr	**91.54**
Sleepers; Flat bottom track timber sleepers						
2600 × 250 × 130 mm		–	–	38.40	nr	**38.40**
Sleepers; Flat bottom track concrete sleepers						
2600 × 250 × 130 mm		–	–	68.12	nr	**68.12**

CLASS S: RAIL TRACK

Item	Gang hours	Labour £	Plant £	Material £	Unit	**Total rate £**
Sleepers; Flat bottom track steel sleepers						
2525 × 264 × 204 mm		–	–	91.54	nr	**91.54**
Fittings						
cast iron chairs complete with chair screws, plastic ferrules, spring steels and keys; BR type S1		–	–	76.78	nr	**76.78**
cast iron chairs complete with chair screws, plastic ferrules, spring steels and keys; BR type CC		–	–	107.76	nr	**107.76**
cast iron chairs complete with resilient pad, chair screws, ferrules, rail clips and nylon insulators; BR type PAN 6		–	–	58.68	nr	**58.68**
cast iron chairs complete with resilient pad, chair screws, ferrules, rail clips and nylon insulators; BR type VN		–	–	66.43	nr	**66.43**
cast iron chairs complete with resilient pad, chair screws, ferrules, rail clips and nylon insulators; BR type C		–	–	50.20	nr	**50.20**
Pandrol rail clips and nylon insulator	0.83	52.81	–	4.15	nr	**56.96**
plain fishplates; for BS95R section rail, skirted pattern; sets of two; complete with fishbolts, nuts and washers		–	–	147.64	nr	**147.64**
plain fishplates; for BS95R section rail, joggled pattern; sets of two; complete with fishbolts, nuts and washers		–	–	169.80	nr	**169.80**
plain fishplates; for BS113 'A' section rail, shallow section; sets of two; complete with fishbolts, nuts and washers		–	–	147.64	nr	**147.64**
insulated fishplates; BS95R section rail, steel billet pattern; sets of two; complete with high tensile steel bolts, nuts and washers		–	–	85.63	nr	**85.63**
cast iron spacer block between running and guard rails; for BS95R section rail; M25 × 220 mm bolt, nut and washers		–	–	66.44	nr	**66.44**
cast iron spacer block between running and guard rails; for BS113 'A' section rail; M25 × 220 mm bolt, nut and washers		–	–	66.44	nr	**66.44**
Prefabricated Turnouts; complete with closures, check rails, fittings, etc.						
Type B8; BS 95R bullhead rail with timber bearers		–	–	31005.50	nr	**31005.50**
Type C10; BS 95R bullhead rail with concrete bearers		–	–	2436.16	nr	**2436.16**
Type C10; BS 95R bullhead rail with steel bearers		–	–	2436.16	nr	**2436.16**
Type Bx8; BS 113 'A' section flat bottom rail with timber bearers		–	–	20375.95	nr	**20375.95**
Type Cv9.25; BS 113 'A' section flat bottom rail with concrete bearers		–	–	32328.76	nr	**32328.76**
Type Cv9.25; BS 113 'A' section flat bottom rail with sleeper bearers		–	–	32328.76	nr	**32328.76**

CLASS S: RAIL TRACK

Item	Gang hours	Labour £	Plant £	Material £	Unit	**Total rate £**
04 SUPPLYING (STANDARD GAUGE TRACK) – cont						
Prefabricated Diamond crossings; complete with closures, check rails, fittings, timber sleepers						
RT standard design, angle 1 in 4; BS95R bullhead rail		–	–	135833.59	nr	**135833.59**
RT standard design, angle 1 in 4; BS113 'A' section flat bottom rail		–	–	147645.19	nr	**147645.19**
Sundries						
buffer stops; single raker, steel rail and timber; 2 tonnes approximate weight		–	–	16219.10	nr	**16219.10**
buffer stops; double raker, steel rail and timber; 2.5 tonnes approximate weight		–	–	18127.23	nr	**18127.23**
wheel stops; steel; 100 kg approximate weight		–	–	763.25	nr	**763.25**
lubricators; single rail		–	–	1462.91	nr	**1462.91**
lubricators; double rail		–	–	2862.21	nr	**2862.21**
switch heaters; single rail	2.00	197.64	–	92.36	nr	**290.00**
switch heater; double rail	3.00	296.46	–	106.03	nr	**402.49**
switch levers; upright pattern		–	–	973.15	nr	**973.15**
switch levers; flush type		–	–	973.15	nr	**973.15**
conductor rail guard boards	1.00	98.82	–	–	m	**98.82**
07 TRACK (SUPPLY & INSTALLATION)						
Bullhead rails; BS11 Delivered in standard 18.288 m lengths; Jointed with fishplates; Softwood sleepers; BS95R section, 47 kg/m; for jointed track						
plain track	0.23	37.06	378.92	545.96	m	**961.93**
form curve in plain track, radius ne 300 m	0.25	41.18	378.92	545.92	m	**966.01**
form curve in plain track, radius over 300 m	0.40	65.88	378.92	545.92	m	**990.72**
extra for curved rails to form super elevation; Radius over 600 mm		–	–	–	%	**23.00**
Bullhead rails; BS11 Delivered in standard 18.288 m lengths; Jointed with fishplates; Concrete sleepers; BS95R section, 47 kg/m; for jointed track						
plain track	0.20	32.94	378.92	601.28	m	**1013.14**
form curve in plain track, radius ne 300 m	0.25	41.18	378.92	601.28	m	**1021.37**
form curve in plain track, radius over 300 m	0.28	45.29	378.92	601.28	m	**1025.49**
extra for curved rails to form super elevation; Radius over 600 mm		–	–	–	%	**23.00**
Bullhead rails; BS11 Delivered in standard 18.288 m lengths; Welded joints; Softwood sleepers; BS95R section, 47 kg/m; for jointed track						
plain track	0.20	32.94	378.92	529.83	m	**941.68**
form curve in plain track, radius ne 300 m	0.20	32.94	378.92	529.83	m	**941.68**
form curve in plain track, radius over 300 m	0.20	32.94	378.92	529.83	m	**941.68**
extra for curved rails to form super elevation; Radius over 600 mm		–	–	–	%	**23.00**

CLASS S: RAIL TRACK

Item	Gang hours	Labour £	Plant £	Material £	Unit	**Total rate £**
Bullhead rails; BS11 Delivered in standard 18.288 m lengths; Welded joints; Concrete sleepers; BS95R section, 47 kg/m; for jointed track						
plain track	0.26	43.32	378.92	585.19	m	**1007.42**
form curve in plain track, radius ne 300 m	0.28	45.29	378.92	585.19	m	**1009.40**
form curve in plain track, radius over 300 m	0.30	49.41	378.92	585.30	m	**1013.63**
extra for curved rails to form super elevation; Radius over 600 mm		–	–	–	%	**23.00**
Flat bottom rails; BS11 Delivered in standard 18.288 m lengths; Jointed with fishplates; Softwood sleepers; BS113 'A' section, 56 kg/m; for jointed track						
plain track	0.20	32.94	378.92	563.92	m	**975.78**
form curve in plain track, radius ne 300 m	0.23	37.06	378.92	578.58	m	**994.56**
form curve in plain track, radius over 300 m	0.23	37.06	378.92	587.57	m	**1003.54**
extra for curved rails to form super elevation; Radius over 600 mm		–	–	–	%	**23.00**
Flat bottom rails; BS11 Delivered in standard 18.288 m lengths; Jointed with fishplates; Concrete sleepers; BS113 'A' section, 56 kg/m; for jointed track						
plain track	0.20	32.94	378.92	619.28	m	**1031.14**
form curve in plain track, radius ne 300 m	0.23	37.06	378.92	633.95	m	**1049.92**
form curve in plain track, radius over 300 m	0.23	37.06	378.92	642.93	m	**1058.90**
extra for curved rails to form super elevation; Radius over 600 mm		–	–	–	%	**23.00**
Flat bottom rails; BS11 Delivered in standard 18.288 m lengths; Welded joints; Softwood sleepers; BS113 'A' section, 56 kg/m; for jointed track						
plain track	0.18	28.82	378.92	547.83	m	**955.57**
form curve in plain track, radius ne 300 m	0.20	32.94	378.92	562.49	m	**974.35**
form curve in plain track, radius over 300 m	0.20	32.94	378.92	571.47	m	**983.33**
extra for curved rails to form super elevation; Radius over 600 mm		–	–	–	%	**23.00**
Flat bottom rails; BS11 Delivered in standard 18.288 m lengths; Welded joints; Concrete sleepers; BS113 'A' section, 56 kg/m; for jointed track						
plain track	0.20	32.94	378.92	603.19	m	**1015.05**
form curve in plain track, radius ne 300 m	0.23	37.06	378.92	617.85	m	**1033.83**
form curve in plain track, radius over 300 m	0.23	37.06	378.92	626.83	m	**1042.81**
extra for curved rails to form super elevation; Radius over 600 mm		–	–	–	%	**23.00**
Flat bottom rails; BS11 Delivered in standard 18.288 m lengths; Jointed with fishplates; Softwood sleepers; BS113 'A' section, 40 kg/m; for jointed track						
plain track	0.23	37.06	316.30	526.41	m	**879.76**
form curve in plain track, radius ne 300 m	0.23	37.06	316.30	536.92	m	**890.27**
form curve in plain track, radius over 300 m	0.23	37.06	316.30	543.09	m	**896.45**
extra for curved rails to form super elevation; Radius over 600 mm		–	–	–	%	**23.00**

CLASS S: RAIL TRACK

Item	Gang hours	Labour £	Plant £	Material £	Unit	**Total rate £**
07 TRACK (SUPPLY & INSTALLATION) – cont						
Flat bottom rails; BS11 Delivered in standard 18.288 m lengths; Jointed with fishplates; Concrete sleepers; BS113 'A' section, 40 kg/m; for jointed track						
plain track	0.23	37.06	316.30	581.77	m	**935.12**
form curve in plain track, radius ne 300 m	0.23	37.14	316.30	592.28	m	**945.72**
form curve in plain track, radius over 300 m	0.23	37.06	316.30	598.46	m	**951.81**
extra for curved rails to form super elevation; Radius over 600 mm		–	–	–	%	**23.00**
Flat bottom rails; BS11 Delivered in standard 18.288 m lengths; Welded joints; Softwood sleepers; BS113 'A' section, 40 kg/m; for jointed track						
plain track	0.23	37.06	316.30	510.31	m	**863.67**
form curve in plain track, radius ne 300 m	0.23	37.06	316.30	520.83	m	**874.18**
form curve in plain track, radius over 300 m	0.25	41.18	316.30	527.00	m	**884.47**
extra for curved rails to form super elevation; Radius over 600 mm		–	–	–	%	**23.00**
Flat bottom rails; BS11 Delivered in standard 18.288 m lengths; Welded joints; Concrete sleepers; BS113 'A' section, 40 kg/m; for jointed track						
plain track	0.23	37.06	316.30	565.68	m	**919.03**
form curve in plain track, radius ne 300 m	0.23	37.06	316.30	576.19	m	**929.54**
form curve in plain track, radius over 300 m	0.23	37.06	316.30	582.36	m	**935.72**
extra for curved rails to form super elevation; Radius over 600 mm		–	–	–	%	**23.00**
Prefabricated turnouts; Complete with closures, check rails, fittings, etc.						
type B8; BS 95R bullhead rail with timber bearers	36.00	5929.20	240.62	31005.50	nr	**37175.32**
type C10; BS 95R bullhead rail with concrete bearers	40.00	6588.00	240.62	24827.93	nr	**31656.55**
type C10; BS 95R bullhead rail with steel bearers	40.00	6588.00	240.62	27221.90	nr	**34050.52**
type Bx8; BS 113 'A' section flat bottom rail with timber bearers	36.00	5929.20	240.62	20375.95	nr	**26545.77**
type Cv9; BS 113 'A' section flat bottom rail with concrete bearers	40.00	6588.00	240.62	32328.76	nr	**39157.37**
type Cv9; BS 113 'A' section flat bottom rail with steel bearers	40.00	6588.00	240.62	35348.84	nr	**42177.46**
Prefabricated diamond crossings; Complete with closures, check rails, fittings						
RT standard design, angle 1 in 4; BS95R bullhead rail; jointed with fishplates; timber sleepers	95.00	15646.50	240.62	135833.59	nr	**151720.71**
RT standard design, angle 1 in 4; BS95R bullhead rail; welded joints; timber sleepers	90.00	14823.00	240.62	135833.59	nr	**150897.21**
RT standard design, angle 1 in 4; BS95R bullhead rail; jointed with fishplates; concrete sleepers	120.00	19764.00	240.62	135833.59	nr	**155838.21**

CLASS S: RAIL TRACK

Item	Gang hours	Labour £	Plant £	Material £	Unit	**Total rate £**
RT standard design, angle 1 in 4; BS95R bullhead rail; welded joints; concrete sleepers	120.00	19764.00	240.62	135833.59	nr	**155838.21**
RT standard design, angle 1 in 4; BS113 'A' section flat bottom rail; jointed with fishplates; timber sleepers	95.00	15646.50	240.62	147644.40	nr	**163531.52**
RT standard design, angle 1 in 4; BS113 'A' section flat bottom rail; welded joints; timber sleepers	90.00	14823.00	240.62	147644.40	nr	**162708.02**
RT standard design, angle 1 in 4; BS113 'A' section flat bottom rail; jointed with fishplates; concrete sleepers	120.00	19764.00	240.62	147644.40	nr	**167649.02**
RT standard design, angle 1 in 4; BS113 'A' section flat bottom rail; welded joints; concrete sleepers	120.00	19764.00	240.62	147644.40	nr	**167649.02**
Buffer stops						
single raker, steel rail and timber; 2 tonnes approximate weight	10.00	1647.00	–	16219.10	nr	**17866.10**
double raker, steel rail and timber; 2.5 tonnes approximate weight	10.00	1647.00	–	18127.23	nr	**19774.23**
Wheel stops						
wheel stops; steel; 100 kg approximate weight	10.00	1647.00	–	763.25	nr	**2410.25**
Lubricators						
lubricators; single rail	2.50	411.75	–	1462.91	nr	**1874.66**
lubricators; double rail	2.50	411.75	–	2862.21	nr	**3273.96**
Check and guard rails; BS 11; delivered in standard 18.288 m lengths; flange planed to allow 50 mm free wheel clearance						
BS113 'A' section, 56 kg/m; for bolting	0.30	49.41	178.00	86.59	m	**314.00**
BS113 'A' section, 47 kg/m; for bolting	0.25	41.18	178.00	82.36	m	**301.53**
BS113 'A' section, 40 kg/m; for bolting	0.25	41.18	178.00	72.49	m	**291.66**

CLASS T: TUNNELS

Item	Gang hours	Labour £	Plant £	Material £	Unit	Total rate £
NOTES						
There are so many factors, apart from design considerations, which influence the cost of tunnelling that it is only possible to give guide prices for a sample of the work involved. For more reliable estimates it is advisable to seek the advice of a Specialist Contractor. We have included Viaducts in this section for ease of reference.						
The main cost considerations are						
* the nature of the ground						
* size of tunnel						
* length of drive						
* depth below surface						
* anticipated overbreak						
* support of face and roof of tunnel (rock bolting etc.)						
* necessity for pre-grouting						
* ventilation						
* presence of water						
* use of compressed air working						
The following rates for mass concrete work cast in primary and secondary linings to tunnels and access shafts are based on a 5.0 m depth of shaft and 15.0 m head of tunnel.						
Apply the following factors for differing depths and lengths:						
HEAD LENGTH: 15 m 30 m 60 m 90 m						
Shaft depth 5 m +0% +10% +20% +32½%						
Shaft depth 10 m +5% +12½% +27½% +35%						
Shaft depth 15 m +10% +17½% +32½% +40%						
Shaft depth 20 m +15% +20% +37½% +42½%						
01 EXCAVATION						
NB: Rates dependent on ground conditions. These prices are for guidance only and adjustments for different factors should be considered when pricing for a particular project.						
Excavation						
Excavating tunnels in rock						
1.5 m dia.		–	–	–	m^3	**75.48**
3.0 m dia.		–	–	–	m^3	**62.16**
Excavating tunnels in soft material						
1.5 m dia.		–	–	–	m^3	**75.48**
3.0 m dia.		–	–	–	m^3	**62.16**
Excavating shafts in rock						
3.0 m dia.		–	–	–	m^3	**294.15**
4.5 m dia.		–	–	–	m^3	**249.75**

CLASS T: TUNNELS

Item	Gang hours	Labour £	Plant £	Material £	Unit	**Total rate £**
Excavating shafts in soft material						
3.0 m dia.		–	–	–	m^3	**176.03**
4.5 m dia.		–	–	–	m^3	**151.50**
Excavating other cavities in rock						
1.5 m dia.		–	–	–	m^3	**75.48**
3.0 m dia.		–	–	–	m^3	**62.16**
Excavating other cavities in soft material						
1.5 m dia.		–	–	–	m^3	**75.48**
3.0 m dia.		–	–	–	m^3	**62.16**
Excavated surfaces in rock		–	–	–	m^2	**29.50**
Excavated surfaces in soft material		–	–	–	m^2	**28.65**
Bore Tunnels						
Excavation in any stated materials; internal dia. >8 m		–	–	–	m^3	**74.37**
Muckaway of inert material to spoil heaps withing 250 m		–	–	–	tonne	**29.04**
Cross Passages						
Excavation; Tunnels in any other stated materials; internal dia. 4–5 m and temporary support; cross passage.		–	–	–	m^3	**1186.59**
Muckaway of excavated inert material; disposal by road to landfill (Landfill excluded Tax) within 15 km radius		–	–	–	m^3	**49.63**
02 IN SITU LINING TO TUNNELS						
Note: The following rates for mass concrete work cast in primary and secondary linings to tunnels and access shafts are based on a 5.0 m depth of shaft and 15.0 m head of tunnel.						
See above for additions for differing shaft depths and tunnel lengths.						
NB: Rates dependent on ground conditions						
Mass concrete; grade C30, 20 mm aggregate						
Cast primary lining to tunnels						
1.5 m dia.		–	–	–	m^3	**498.01**
3.0 m dia.		–	–	–	m^3	**515.17**
Secondary lining to tunnels						
1.5 m dia.		–	–	–	m^3	**579.58**
3.0 m dia.		–	–	–	m^3	**459.37**
Formwork; rough finish						
Tunnel lining						
1.5 m dia.		–	–	–	m^2	**84.54**
3.0 m dia.		–	–	–	m^2	**84.54**

CLASS T: TUNNELS

Item	Gang hours	Labour £	Plant £	Material £	Unit	Total rate £
02 IN SITU LINING TO TUNNELS – cont						
Sprayed concrete primary						
In situ lining to tunnels; sprayed concrete primary; external dia. 4-5 m; 8 cm thick		–	–	–	m^2	**132.66**
Sprayed concrete primary; 15 cm thick		–	–	–	m^2	**188.34**
Cast concrete secondary						
In situ lining to tunnels; cast concrete secondary		–	–	–	m^3	**2066.95**
Provision of concrete – designated concrete						
Provision of concrete – Designated concrete; Strength RC30/37; maximum agg. size 20 mm		–	–	–	m^3	**133.85**
Placing of concrete – reinforced; bases, footings, pile caps and ground slabs						
Placing of reinforced concrete; bases, footings, pile caps & ground slabs; – thickness > 500 mm: cross passage floor slab		–	–	–	m^3	**1742.40**
Spray concrete lining tunnels						
In situ lining to tunnels; sprayed concrete primary lining; external dia. 5–6 m ; thickness ne 300 mm thick		–	–	–	m^3	**408.25**
In situ lining to tunnels; cast concrete secondary lining; external dia. 5–6 m		–	–	–	m^3	**208.15**
In situ lining to tunnels; formwork to smooth finish; external dia. 5–6 m		–	–	–	m^3	**140.07**
03 IN SITU LINING TO SHAFTS						
Note: The following rates for mass concrete work cast in primary and secondary linings to tunnels and access shafts are based on a 5.0 m depth of shaft and 15.0 m head of tunnel. NB rates vary dependent on ground conditions.						
Constructed in in situ concrete						
Provision of complete standard shaft in in situ concrete; including piling, walls, floors, roof, drainage, finishes etc.		–	–	–	nr	**42262500.00**
Mass concrete; grade C30, 20 mm aggregate						
Secondary linings to shafts						
3.0 m int dia.		–	–	–	m^3	**459.37**
4.5 m int dia.		–	–	–	m^3	**447.43**
Cast primary lining to shafts						
3.0 m int dia.		–	–	–	m^3	**489.41**
4.5 m int dia.		–	–	–	m^3	**467.93**

CLASS T: TUNNELS

Item	Gang hours	Labour £	Plant £	Material £	Unit	**Total rate £**
Formwork; rough finish						
Shaft lining						
3.0 m int dia.		–	–	–	m^2	**127.68**
4.5 m int dia.		–	–	–	m^2	**86.32**
Waterproofing system for cross passage						
Sheet linings membrane system; waterproofing system for shaft; junction with tunnel		–	–	–	m^2	**311.54**
Sheet linings membrane prop system; waterproofing system for shaft; junction with tunnel		–	–	–	m^2	**86.94**
Depressurization allowances						
Depressurization measures allowances		–	–	–	nr	**633937.50**
04 IN SITU LINING TO OTHER CAVITIES						
Note: The following rates for mass concrete work cast in primary and secondary linings to tunnels and access shafts are based on a 5.0 m depth of shaft and 15.0 m head of tunnel						
See above for additions for differing shaft depths and tunnel lengths.						
NB: Rates dependent on ground conditions						
Mass concrete; grade C30, 20 mm aggregate						
Cast primary lining to other cavities						
1.5 m int dia.		–	–	–	m^3	**50.33**
3.0 m int dia.		–	–	–	m^3	**429.32**
Secondary linings to other cavities						
1.5 m int dia.		–	–	–	m^3	**579.58**
3.0 m int dia.		–	–	–	m^3	**459.37**
Formwork; rough finish						
Other cavities lining						
1.5 m int dia.		–	–	–	m^2	**66.70**
3.0 m int dia.		–	–	–	m^2	**84.54**

CLASS T: TUNNELS

Item	Gang hours	Labour £	Plant £	Material £	Unit	Total rate £
05 PREFORMED SEGMENTAL LININGS TO TUNNELS						
This is a very specialist area and users are advised to contact specialist contractors directly for associated costs. Please note that these prices are for guidance only and adjustment for different factors should be considered when pricing for a particular project.						
Precast concrete bolted rings						
preformed segment lining to tunels; precast concrete boltd rings; – external dia.. > 8 m; 2.0 m steel fiber reinforced rings	–	–	–		nr	**8694.00**
preformed segment lining to tunels; precast concrete boltd rings; – external dia.. > 8 m; 2.0 m standard reinforced rings 120 kg /m^3	–	–	–		nr	**13639.00**
preformed segment lining to tunels; 2.0 m standard reinforced rings 80 kg /m^3	–	–	–		nr	**11350.50**
Lining ancillaries						
preformed segment lining to tunels; ancillaries ; grout behind lining	–	–	–		m^3	**471.50**
06 PREFORMED SEGMENTAL LININGS TO SHAFTS						
This is a very specialist area and users are advised to contact specialist contractors directly for associated costs						
07 PREFORMED SEGMENTAL LININGS TO OTHER CAVITIES						
Precast concrete bolted rings; flanged; including packing; guide price/ring based upon standard bolted concrete segmental rings; ring width 610 mm						
NB: Rates dependent on ground conditions						
Linings						
Linings to tunnels						
1.5 m int dia.; Ring 1.52 m ID × 1.77 m OD; 6 segments, maximum piece weigth 139 kg	–	–	–		nr	**973.58**
3.0 m int dia.; Ring 3.05 m ID × 3.35 m OD; 7 segments , maximum piece weight 247 kg	–	–	–		nr	**1743.31**
Linings to shafts						
3.0 m int dia.; Ring 3.05 m ID × 3.35 m OD; 7 segments, maximum piece weight 247 kg	–	–	–		nr	**1379.59**

CLASS T: TUNNELS

Item	Gang hours	Labour £	Plant £	Material £	Unit	Total rate £
Lining ancillaries; bitumen impregnated fibreboard						
Parallel circumferential packing						
1.5 m int dia.		–	–	–	nr	**8.34**
3.0 m int dia.		–	–	–	nr	**16.66**
Lining ancillaries; PC4 AF caulking compound						
Caulking						
1.5 m int dia.		–	–	–	m	**15.13**
3.0 m int dia.		–	–	–	m	**15.14**
08 SUPPORT AND STABILIZATION						
NB: Rates dependent on ground conditions						
Rock bolts						
mechanical		–	–	–	m	**38.28**
mechanical grouted		–	–	–	m	**59.74**
pre-grouted impacted		–	–	–	m	**56.86**
chemical end anchor		–	–	–	m	**57.48**
chemical grouted		–	–	–	m	**40.90**
chemically filled		–	–	–	m	**64.31**
Internal support						
steel arches; supply		–	–	–	t	**1681.56**
steel arches; erection		–	–	–	t	**693.04**
timber supports; supply		–	–	–	m^3	**489.70**
timber supports; erection		–	–	–	m^3	**441.05**
lagging		–	–	–	m^2	**35.13**
sprayed concrete		–	–	–	m^2	**40.23**
mesh or link		–	–	–	m^2	**12.98**
Pressure grouting						
sets of drilling and grouting plant		–	–	–	nr	**1842.75**
face packers		–	–	–	nr	**81.64**
deep packers		–	–	–	nr	**140.02**
drilling and flushing to 40 mm dia.		–	–	–	m	**39.93**
re-drilling and flushing		–	–	–	m	**24.55**
injection of grout materials; chemical grout		–	–	–	t	**980.07**
Forward probing		–	–	–	m	**33.43**
09 VIADUCTS						
Viaduct						
Segmental units						
Segmental units; – mass 1–2 tonne; precast concrete retaining wall average height 1.50 m		–	–	–	nr	**3453.45**
Fabrication and erection						
Handrails		–	–	–	m	**155.77**
Bored cast in place concrete piles; 1800 mm dia.						
bored cast in place concrete piles for piers; dia. 1800 mm; concreted length; on-shore works; no permanent casing		–	–	–	m	**2753.10**
bored cast in place concret piles for piers; dia. 1800 mm; concreted length; off-shore works; with permanent casing		–	–	–	m	**2742.23**

CLASS T: TUNNELS

Item	Gang hours	Labour £	Plant £	Material £	Unit	Total rate £
09 VIADUCTS – cont						
Viaduct – cont						
Permanent casings each length						
cast in place concrete piles; permanent casings each length > 13 m; dia. 1500 mm & 1800 mm		–	–	–	m	**1370.80**
cast in place conc piles; cutting off surplus lengths; dia. 1500 mm		–	–	–	m	**521.64**
cast in place conc piles; cutting off surplus lengths; dia. 1800 mm		–	–	–	m	**668.15**
In situ reinforced concrete piers on piles; off shore						
standard vertical pier		–	–	–	nr	**525262.50**
V shaped pier		–	–	–	nr	**2058787.50**
fixed buttress 3 legs pier		–	–	–	nr	**966000.00**
In situ reinforced concrete piers on piles; on shore						
standard vertical pier		–	–	–	nr	**507150.00**
fixed vertical pier		–	–	–	nr	**1034344.50**
expansion buttress pier		–	–	–	nr	**698002.85**
fixed buttress 3 legs pier		–	–	–	nr	**962377.50**
Segmental units; bridge parapet						
segmental units; mass 500 kg – 1tonne; uhpc parapet; csa 0.31 m^2; average 0.80 tonne per m		–	–	–	nr	**779.70**
segmental units; mass 1–2 tonne ; parapet; average csa 0.45 m^2; average 1.10 tonne per m		–	–	–	nr	**377.20**

CLASS U: BRICKWORK, BLOCKWORK AND MASONRY

Item	Gang hours	Labour £	Plant £	Material £	Unit	**Total rate £**
NOTES						
Apply the following multipliers to both labour and plant for rubble walls:						
height 2 to 5 m × 1.21						
height 5 to 10 m × 1.37						
wall to radius small × 1.75						
wall to radius large × 1.50						
wall to rake or batter × 1.15						
wall in piers or stanchion × 1.50						
wall in buttresses × 1.15						
RESOURCES – LABOUR						
Labour Resources Build-up Rate						
Brickwork, blockwork and masonry gang						
1 foreman bricklayer (craftsman)		26.80	–	–		
4 bricklayers (craftsman)		91.20	–	–		
1 unskilled operative (general)		19.60	–	–		
Total Gang Rate/Hour		**26.80**	–	–		
RESOURCES – MATERIAL						
Supply only materials rates						
Clay bricks						
Common bricks; PC £ per thousand	–	–	461.07	1000	**461.07**	
Facing bricks; PC £ per thousand	–	–	760.00	1000	**760.00**	
Class A engineering bricks; PC £ per thousand	–	–	621.90	1000	**621.90**	
Class B engineering bricks; PC £ per thousand	–	–	580.00	1000	**580.00**	
Concrete blocks to BS EN 771–3						
Lightweight						
solid; 3.5 N/mm^2; 100 mm thick	–	–	14.32	m^2	**14.32**	
solid; 3.5 N/mm^2; 140 mm thick	–	–	17.36	m^2	**17.36**	
solid 3.5 N/mm^2; 215 mm thick	–	–	25.47	m^2	**25.47**	
Dense						
solid 7 N/mm^2; 100 mm thick	–	–	8.49	m^2	**8.49**	
solid 7 N/mm^2; 140 mm thick	–	–	13.66	m^2	**13.66**	
solid 7 N/mm^2; 215 mm thick	–	–	17.12	m^2	**17.12**	
RESOURCES – PLANT						
Plant Resources Build-up Rate						
Brickwork, blockwork and masonry						
2t dumper (50% of time)		–	5.41	–		
mixer (50% of time)		–	2.06	–		
small power tools		–	0.97	–		
scaffold, etc.		–	1.48	–		
Total Rate/Hour		–	**9.92**	–		

CLASS U: BRICKWORK, BLOCKWORK AND MASONRY

Item	Gang hours	Labour £	Plant £	Material £	Unit	Total rate £
01 COMMON BRICKWORK						
Common bricks in cement mortar designation (ii)						
Thickness 103 mm						
vertical straight walls	0.23	31.22	2.26	31.57	m^2	**65.06**
vertical curved walls	0.30	40.53	2.90	31.57	m^2	**75.00**
battered straight walls	0.33	45.05	3.26	31.57	m^2	**79.88**
battered curved walls	0.37	50.67	3.67	31.57	m^2	**85.90**
vertical facing to concrete	0.25	33.82	2.45	37.84	m^2	**74.12**
battered facing to concrete	0.37	50.67	3.67	37.86	m^2	**92.20**
casing to metal sections	0.32	43.13	3.12	37.86	m^2	**84.12**
Thickness 215 mm						
vertical straight walls	0.44	60.52	4.38	63.77	m^2	**128.68**
vertical curved walls	0.57	78.05	5.65	63.77	m^2	**147.47**
battered straight walls	0.63	86.27	6.25	63.77	m^2	**156.29**
battered curved walls	0.71	97.22	6.99	63.77	m^2	**167.98**
vertical facing to concrete	0.48	65.73	4.74	70.04	m^2	**140.51**
battered facing to concrete	0.71	97.22	6.99	70.04	m^2	**174.26**
casing to metal sections	0.61	83.53	6.00	70.04	m^2	**159.58**
Thickness 328 mm						
vertical straight walls	0.64	87.64	6.38	96.42	m^2	**190.44**
vertical curved walls	0.82	112.29	8.16	96.42	m^2	**216.87**
battered straight walls	0.91	124.61	8.99	96.42	m^2	**230.02**
battered curved walls	1.01	138.30	10.01	96.42	m^2	**244.74**
vertical facing to concrete	0.69	94.48	6.88	102.69	m^2	**204.06**
battered facing to concrete	1.01	138.30	10.01	102.69	m^2	**251.01**
casing to metal sections	0.87	119.13	8.63	102.69	m^2	**230.45**
Thickness 440 mm						
vertical straight walls	0.84	115.02	8.27	128.89	m^2	**252.19**
vertical curved walls	1.06	145.15	10.49	128.89	m^2	**284.53**
battered straight walls	1.16	158.84	11.52	128.89	m^2	**299.26**
battered curved walls	1.29	176.64	12.77	128.89	m^2	**318.31**
vertical facing to concrete	0.90	123.24	8.90	135.10	m^2	**267.24**
battered facing to concrete	1.29	176.64	12.77	135.17	m^2	**324.58**
casing to metal sections	1.12	153.37	11.09	135.17	m^2	**299.62**
Thickness 890 mm						
vertical straight walls	1.50	205.40	14.88	258.98	m^2	**479.26**
vertical curved walls	1.85	253.33	18.37	258.98	m^2	**530.68**
battered straight walls	2.00	273.87	19.92	258.98	m^2	**552.77**
battered curved walls	2.20	301.25	21.77	258.98	m^2	**582.00**
vertical facing to concrete	1.60	219.09	15.89	265.25	m^2	**500.23**
battered facing to concrete	2.20	301.25	21.77	265.25	m^2	**588.27**
casing to metal sections	1.94	265.65	19.27	265.25	m^2	**550.17**
Thickness exceeding 1 m						
vertical straight walls	1.64	224.57	16.22	289.91	m^2	**530.70**
vertical curved walls	2.00	273.87	19.91	289.91	m^2	**583.69**
battered straight walls	2.17	297.15	21.54	289.91	m^2	**608.60**
battered curved walls	2.37	324.53	23.47	289.91	m^2	**637.91**
vertical facing to concrete	1.74	238.26	17.29	296.19	m^2	**551.74**
battered facing to concrete	2.37	324.53	23.47	296.19	m^2	**644.18**
casing to metal sections	2.10	287.56	20.86	296.19	m^2	**604.60**

CLASS U: BRICKWORK, BLOCKWORK AND MASONRY

Item	Gang hours	Labour £	Plant £	Material £	Unit	**Total rate £**
Columns and piers						
215 × 215 mm	0.13	17.80	1.29	14.08	m	**33.17**
440 × 215 mm	0.24	32.86	2.38	28.43	m	**63.67**
665 × 328 mm	0.44	60.25	4.36	64.31	m	**128.92**
890 × 890 mm	1.10	150.63	10.91	230.24	m	**391.78**
Surface features						
copings; standard header-on-edge; 215 mm wide × 103 mm high	0.10	13.69	1.04	7.08	m	**21.82**
sills; standard header-on-edge; 215 mm wide × 103 mm high	0.13	17.80	1.24	6.90	m	**25.95**
rebates	0.30	41.08	2.97	–	m	**44.05**
chases	0.35	47.93	3.47	–	m	**51.40**
band courses;flush; 215 mm wide	0.05	6.85	0.45	–	m	**7.30**
band courses;projection 103 mm; 215 mm wide	0.05	6.85	0.49	–	m	**7.34**
corbels;maximum projection 103 mm; 215 mm wide	0.15	20.54	1.49	6.81	m	**28.84**
pilasters; 328 mm wide × 103 mm projection	–	–	0.70	14.08	m	**14.78**
pilasters; 440 mm wide × 215 mm projection	0.12	16.43	1.19	31.75	m	**49.37**
plinths; projection 103 mm × 900 mm wide	0.19	26.02	1.89	28.88	m	**56.78**
fair facing	0.06	8.22	0.56	–	m^2	**8.78**
Ancillaries						
bonds to existing work; to brickwork	1.50	205.40	14.88	32.65	m^2	**252.93**
built-in pipes and ducts, cross-sectional area not exceeding 0.05 m^2; excluding supply; brickwork 103 mm thick	0.06	8.22	0.60	1.09	nr	**9.90**
built-in pipes and ducts, cross-sectional area not exceeding 0.05 m^2; excluding supply; brickwork 215 mm thick	0.12	16.43	1.14	1.60	nr	**19.17**
built-in pipes and ducts, cross-sectional area 0.05–0.25 m^2; excluding supply; brickwork 103 mm thick	0.15	20.54	1.49	1.63	nr	**23.66**
built-in pipes and ducts, cross-sectional area 0.05–0.25 m^2; excluding supply; brickwork 215 mm thick	0.29	39.71	2.87	2.41	nr	**44.99**
built-in pipes and ducts, cross-sectional area 0. 25–0.5 m^2; excluding supply; brickwork 103 mm thick	0.18	24.65	1.74	1.90	nr	**28.28**
built-in pipes and ducts, cross-sectional area 0. 25–0.5 m^2; excluding supply; brickwork 215 mm thick	0.34	46.56	3.38	2.77	nr	**52.70**

CLASS U: BRICKWORK, BLOCKWORK AND MASONRY

Item	Gang hours	Labour £	Plant £	Material £	Unit	**Total rate £**
02 FACING BRICKWORK						
Facing bricks; in plasticized cement mortar designation (iii)						
Thickness 103 mm						
vertical straight walls	0.34	46.56	3.38	52.75	m^2	**102.68**
vertical curved walls	0.45	61.62	4.46	52.75	m^2	**118.83**
battered straight walls	0.45	61.62	4.46	52.75	m^2	**118.83**
battered curved walls	0.56	76.68	5.58	52.75	m^2	**135.01**
vertical facing to concrete	0.37	50.67	3.48	59.02	m^2	**113.17**
battered facing to concrete	0.56	76.68	5.58	59.02	m^2	**141.28**
casing to metal sections	0.49	67.10	4.86	59.02	m^2	**130.97**
Thickness 215 mm						
vertical straight walls	0.57	78.05	8.09	106.12	m^2	**192.27**
vertical curved walls	0.84	115.02	8.33	106.12	m^2	**229.48**
battered straight walls	0.84	115.02	8.33	106.12	m^2	**229.48**
battered curved walls	1.02	139.67	10.15	106.12	m^2	**255.94**
vertical facing to concrete	0.66	90.38	6.55	112.39	m^2	**209.32**
battered facing to concrete	1.02	139.67	10.15	112.39	m^2	**262.21**
casing to metal sections	0.82	112.29	8.14	112.39	m^2	**232.82**
Thickness 328 mm						
vertical straight walls	0.83	113.65	8.20	159.95	m^2	**281.80**
vertical curved walls	1.12	153.37	11.05	159.95	m^2	**324.37**
battered straight walls	1.12	153.37	11.05	159.95	m^2	**324.37**
battered curved walls	1.36	186.23	13.49	159.95	m^2	**359.66**
vertical facing to concrete	0.88	120.50	8.69	166.22	m^2	**295.41**
battered facing to concrete	1.36	186.23	13.49	166.22	m^2	**365.94**
casing to metal sections	1.18	161.58	11.70	166.22	m^2	**339.50**
Thickness 440 mm						
vertical straight walls	1.08	147.89	10.71	213.59	m^2	**372.19**
vertical curved walls	1.39	190.34	13.79	213.59	m^2	**417.72**
battered straight walls	1.39	190.34	13.79	213.59	m^2	**417.72**
vertical facing to concrete	1.09	149.26	10.83	219.87	m^2	**379.96**
battered facing to concrete	1.39	190.34	13.79	219.87	m^2	**423.99**
casing to metal sections	1.39	190.34	13.79	219.87	m^2	**423.99**
Columns and piers						
215 × 215 mm	0.17	23.28	1.66	23.63	m	**48.56**
440 × 215 mm	0.29	39.71	2.87	47.49	m	**90.07**
665 × 328 mm	0.58	79.42	5.72	107.19	m	**192.33**
Surface features						
copings; standard header-on-edge; standard bricks; 215 mm wide × 103 mm high	0.13	17.80	1.33	11.06	m	**30.20**
flat arches; standard stretcher-on-end; 215 mm wide × 103 mm high	0.21	28.76	2.08	10.79	m	**41.63**
flat arches; standard stretcher-on-end; bullnosed special bricks; 103 mm × 215 mm high	0.22	30.13	2.19	6.50	m	**38.81**

CLASS U: BRICKWORK, BLOCKWORK AND MASONRY

Item	Gang hours	Labour £	Plant £	Material £	Unit	Total rate £
segmental arches; single ring;standard bricks; 103 mm wide × 215 mm high	0.37	50.67	3.67	10.79	m	**65.13**
segmental arches; two ring;standard bricks; 103 mm wide × 440 mm high	0.49	67.10	4.86	21.78	m	**93.73**
segmental arches; cut voussoirs; 103 mm wide × 215 mm high	0.39	53.40	3.87	26.35	m	**83.62**
rebates	0.33	45.19	3.27	–	m	**48.46**
chases	0.37	50.67	3.67	–	m	**54.33**
cornices; maximum projection 103 mm; 215 mm wide	0.37	50.67	3.67	11.25	m	**65.58**
band courses; projection 113 mm; 215 mm wide	0.06	8.22	0.55	–	m	**8.76**
corbels; maximum projection 113 mm; 215 mm wide	0.37	50.67	3.67	11.25	m	**65.58**
pilasters; 328 mm wide × 113 mm projection	0.05	6.85	0.49	23.61	m	**30.95**
pilasters; 440 mm wide × 215 mm projection	0.06	8.22	0.55	53.19	m	**61.95**
plinths; projection 113 mm × 900 mm wide	0.24	32.86	2.38	47.94	m	**83.18**
fair facing; pointing as work proceeds	0.06	8.22	0.60	–	m^2	**8.81**
Ancillaries						
bonds to existing work; to brickwork	1.50	205.40	14.88	14.38	m	**234.66**
built-in pipes and ducts, cross-sectional area not exceeding 0.05 m^2; excluding supply; brickwork half brick thick	0.10	13.69	0.94	1.18	nr	**15.81**
built-in pipes and ducts, cross-sectional area not exceeding 0.05 m^2; excluding supply; brickwork one brick thick	0.15	20.54	1.10	2.08	nr	**23.73**
built-in pipes and ducts, cross-sectional area 0.05–0.25 m^2; excluding supply; brickwork half brick thick	0.19	26.02	1.89	1.86	nr	**29.76**
built-in pipes and ducts, cross-sectional area 0.05–0.25 m^2; excluding supply; brickwork one brick thick	0.33	45.19	3.27	2.98	nr	**51.44**
built-in pipes and ducts, cross-sectional area 0.25–0.5 m^2; excluding supply; brickwork half brick thick	0.23	31.49	2.28	2.13	nr	**35.90**
built-in pipes and ducts, cross-sectional area 0.25–0.5 m^2; excluding supply; brickwork one brick thick	0.40	54.77	3.91	3.43	nr	**62.12**
03 ENGINEERING BRICKWORK						
Class A engineering bricks, solid; in cement mortar designation (ii)						
Thickness 103 mm						
vertical straight walls	0.27	36.97	2.67	41.58	m^2	**81.22**
vertical curved walls	0.37	50.67	3.67	41.58	m^2	**95.91**
battered straight walls	0.37	50.67	3.67	41.58	m^2	**95.91**
battered curved walls	0.41	56.14	4.09	41.58	m^2	**101.81**
vertical facing to concrete	0.32	43.82	3.12	47.85	m^2	**94.79**
battered facing to concrete	0.46	62.99	4.59	47.85	m^2	**115.43**
casing to metal sections	0.41	56.14	4.09	47.85	m^2	**108.08**

CLASS U: BRICKWORK, BLOCKWORK AND MASONRY

Item	Gang hours	Labour £	Plant £	Material £	Unit	Total rate £
03 ENGINEERING BRICKWORK – cont						
Class A engineering bricks, solid – cont						
Thickness 215 mm						
vertical straight walls	0.52	71.21	5.16	83.79	m^2	**160.15**
vertical curved walls	0.71	97.22	6.99	83.79	m^2	**188.00**
battered straight walls	0.71	97.22	6.99	83.79	m^2	**188.00**
battered curved walls	0.78	106.81	7.74	83.79	m^2	**198.33**
vertical facing to concrete	0.61	83.53	6.00	90.06	m^2	**179.59**
battered facing to concrete	0.87	119.13	8.64	90.06	m^2	**217.83**
casing to metal sections	0.78	106.81	7.74	90.06	m^2	**204.61**
Thickness 328 mm						
vertical straight walls	0.75	102.70	7.47	126.45	m^2	**236.61**
vertical curved walls	1.01	138.30	10.01	126.45	m^2	**274.76**
battered straight walls	1.01	138.30	10.01	126.45	m^2	**274.76**
battered curved walls	1.11	152.00	11.02	126.45	m^2	**289.46**
vertical facing to concrete	0.87	119.13	8.64	132.72	m^2	**260.49**
battered facing to concrete	1.24	169.80	12.25	132.72	m^2	**314.76**
casing to metal sections	1.11	152.00	11.02	132.72	m^2	**295.73**
Thickness 440 mm						
vertical straight walls	0.97	132.83	9.63	168.92	m^2	**311.38**
vertical curved walls	1.29	176.64	12.77	168.92	m^2	**358.34**
battered straight walls	1.29	176.64	12.77	168.92	m^2	**358.34**
battered curved walls	1.41	193.08	13.99	168.92	m^2	**375.99**
vertical facing to concrete	1.12	153.37	11.09	175.20	m^2	**339.65**
battered facing to concrete	1.56	213.62	15.46	175.20	m^2	**404.27**
casing to metal sections	1.41	193.08	13.99	175.20	m^2	**382.26**
Thickness 890 mm						
vertical straight walls	1.72	235.53	17.04	338.83	m^2	**591.40**
vertical curved walls	2.20	301.25	21.77	338.83	m^2	**661.85**
battered straight walls	2.20	301.25	21.77	338.83	m^2	**661.85**
battered curved walls	2.37	324.53	23.50	338.83	m^2	**686.87**
vertical facing to concrete	1.94	265.65	19.27	345.10	m^2	**630.02**
battered facing to concrete	2.58	353.29	25.56	345.10	m^2	**723.95**
casing to metal sections	2.37	324.53	23.50	345.10	m^2	**693.14**
Thickness exceeding 1 m						
vertical straight walls	1.87	256.07	18.50	379.59	m^3	**654.15**
vertical curved walls	2.37	324.53	23.47	379.59	m^3	**727.58**
battered straight walls	2.37	324.53	23.47	379.59	m^3	**727.58**
battered curved walls	2.55	349.18	25.27	379.59	m^3	**754.04**
vertical facing to concrete	2.10	287.56	20.86	385.77	m^3	**694.19**
battered facing to concrete	2.76	377.94	27.37	385.77	m^3	**791.08**
casing to metal sections	2.55	349.18	25.27	385.77	m^3	**760.22**
Columns and piers						
215 × 215 mm	0.16	21.91	1.58	18.59	m	**42.07**
440 × 215 mm	0.28	38.34	2.76	37.44	m	**78.54**
665 × 328 mm	0.56	76.68	5.52	84.57	m	**166.77**
890 × 890 mm	1.78	243.74	17.63	302.30	m	**563.67**

CLASS U: BRICKWORK, BLOCKWORK AND MASONRY

Item	Gang hours	Labour £	Plant £	Material £	Unit	**Total rate £**
Surface features						
copings; standard header-on-edge; 215 mm wide × 103 mm high	0.11	15.06	1.04	9.34	m	**25.44**
sills; standard header-on-edge; 215 mm wide × 103 mm high	0.13	17.80	1.24	9.16	m	**28.20**
rebates	0.33	45.19	3.27	–	m	**48.46**
chases	0.37	50.67	3.67	–	m	**54.33**
band courses; flush; 215 mm wide	0.05	6.85	0.45	–	m	**7.30**
band courses; projection 103 mm; 215 mm wide	0.05	6.85	0.49	–	m	**7.34**
corbels; maximum projection 103 mm; 215 mm wide	0.15	20.54	1.49	9.06	m	**31.09**
pilasters; 328 mm wide × 103 mm projection	0.07	9.59	0.70	18.59	m	**28.87**
pilasters; 440 mm wide × 215 mm projection	0.12	16.43	1.19	41.88	m	**59.50**
plinths; projection 103 mm × 900 mm wide	0.19	26.02	1.89	37.89	m	**65.79**
fair facing	0.06	8.22	0.60	–	m^2	**8.81**
Ancillaries						
bonds to existing brickwork; to brickwork	0.29	39.71	2.87	19.67	m^2	**62.25**
built-in pipes and ducts, cross-sectional area not exceeding 0.05 m^2; excluding supply; brickwork half brick thick	0.10	13.69	0.94	1.20	nr	**15.84**
built-in pipes and ducts, cross-sectional area not exceeding 0.05 m^2; excluding supply; brickwork one brick thick	0.15	20.54	1.49	1.67	nr	**23.70**
built-in pipes and ducts, cross-sectional area 0.05–0.25 m^2; excluding supply; brickwork half brick thick	0.19	26.02	1.89	1.81	nr	**29.71**
built-in pipes and ducts, cross-sectional area 0.05–0.25 m^2; excluding supply; brickwork one brick thick	0.33	45.19	3.27	2.68	nr	**51.14**
built-in pipes and ducts, cross-sectional area 0.025–0.5 m^2; excluding supply; brickwork half brick thick	0.23	31.49	2.28	2.11	nr	**35.88**
built-in pipes and ducts, cross-sectional area 0.025–0.5 m^2; excluding supply; brickwork one brick thick	0.40	54.77	3.91	3.14	nr	**61.83**
Class B engineering bricks, perforated; in cement mortar designation (ii)						
Thickness 103 mm						
vertical straight walls	0.27	36.97	2.67	27.57	m^2	**67.21**
vertical curved walls	0.37	50.67	3.67	27.57	m^2	**81.90**
battered straight walls	0.37	50.67	3.67	27.57	m^2	**81.90**
vertical facing to concrete	0.32	43.82	3.12	33.84	m^2	**80.78**
casings to metal sections	0.41	56.14	4.09	33.84	m^2	**94.07**
Thickness 215 mm						
vertical straight walls	0.52	71.21	5.16	55.76	m^2	**132.13**
vertical curved walls	0.71	97.22	6.99	55.76	m^2	**159.98**
battered straight walls	0.71	97.22	6.99	55.76	m^2	**159.98**
vertical facing to concrete	0.61	83.53	6.00	62.04	m^2	**151.57**
casings to metal sections	0.78	106.81	7.74	62.04	m^2	**176.58**

CLASS U: BRICKWORK, BLOCKWORK AND MASONRY

Item	Gang hours	Labour £	Plant £	Material £	Unit	Total rate £
03 ENGINEERING BRICKWORK – cont						
Class B engineering bricks, perforated – cont						
Thickness 328 mm						
vertical straight walls	0.76	104.07	7.50	84.41	m^2	**195.98**
vertical curved walls	1.01	138.30	10.01	84.41	m^2	**232.73**
battered straight walls	1.01	138.30	10.01	84.41	m^2	**232.73**
vertical facing to concrete	0.87	119.13	8.64	90.69	m^2	**218.46**
casings to metal sections	1.11	152.00	11.02	90.69	m^2	**253.70**
Thickness 440 mm						
vertical straight walls	0.97	132.83	9.63	112.88	m^2	**255.33**
vertical curved walls	1.29	176.64	12.77	112.88	m^2	**302.30**
battered straight walls	1.29	176.64	12.77	112.88	m^2	**302.30**
vertical facing to concrete	1.12	153.37	11.09	119.15	m^2	**283.61**
casings to metal sections	1.41	193.08	13.99	119.15	m^2	**326.22**
Thickness 890 mm						
vertical straight walls	1.72	235.53	17.04	226.75	m^2	**479.31**
vertical curved walls	2.20	301.25	21.77	226.75	m^2	**549.77**
battered straight walls	2.20	301.25	21.77	226.75	m^2	**549.77**
battered curved walls	2.37	324.53	23.50	226.75	m^2	**574.78**
vertical facing to concrete	1.94	265.65	19.27	233.02	m^2	**517.94**
battered facing to concrete	2.58	353.29	25.56	233.02	m^2	**611.86**
casing to metal sections	2.37	324.53	23.50	233.02	m^2	**581.05**
Thickness exceeding 1 m						
vertical straight walls	1.87	256.07	18.50	254.28	m^3	**528.85**
vertical curved walls	2.37	324.53	23.47	254.28	m^3	**602.28**
battered straight walls	2.37	324.53	23.47	254.28	m^3	**602.28**
battered curved walls	2.55	349.18	25.27	254.28	m^3	**628.73**
vertical facing to concrete	2.10	287.56	20.86	260.46	m^3	**568.88**
battered facing to concrete	2.76	377.94	27.37	260.46	m^3	**665.77**
casing to metal sections	2.55	349.18	25.27	260.46	m^3	**634.91**
Columns and piers						
215 × 215 mm	0.16	21.91	1.58	12.28	m	**35.77**
440 × 215 mm	0.28	38.34	2.76	24.83	m	**65.93**
665 × 328 mm	0.56	76.68	5.52	56.20	m	**138.40**
890 × 890 mm	1.78	243.74	17.63	201.42	m	**462.79**
Surface features						
copings; standard header-on-edge; 215 mm wide × 103 mm high	0.11	15.06	1.04	6.18	m	**22.29**
sills; standard header-on-edge; 215 mm wide × 103 mm high	0.13	17.80	1.24	6.00	m	**25.05**
rebates	0.33	45.19	3.27	–	m	**48.46**
chases	0.37	50.67	3.67	–	m	**54.33**
band courses; flush; 215 mm wide	0.05	6.85	0.45	–	m	**7.30**
band courses; projection 103 mm; 215 mm wide	0.05	6.85	0.49	–	m	**7.34**
corbels; maximum projection 103 mm; 215 mm wide	0.15	20.54	1.49	5.91	m	**27.94**
pilasters; 328 mm wide × 103 mm projection	0.07	9.59	0.70	12.28	m	**22.56**
pilasters; 440 mm wide × 215 mm projection	0.12	16.43	1.19	27.70	m	**45.32**
plinths; projection 103 mm × 900 mm wide	0.19	26.02	1.89	25.28	m	**53.18**
fair facing	0.06	8.22	0.60	–	m^2	**8.81**

CLASS U: BRICKWORK, BLOCKWORK AND MASONRY

Item	Gang hours	Labour £	Plant £	Material £	Unit	**Total rate £**
Ancillaries						
bonds to existing brickwork; to brickwork	0.29	39.71	2.87	19.43	m^2	**62.01**
built-in pipes and ducts, cross-sectional area not exceeding 0.05 m^2; excluding supply; brickwork half brick thick	0.10	13.69	0.94	1.20	nr	**15.84**
built-in pipes and ducts, cross-sectional area not exceeding 0.05 m^2; excluding supply; brickwork one brick thick	0.15	20.54	1.49	1.67	nr	**23.70**
built-in pipes and ducts, cross-sectional area 0.05–0.25 m^2; excluding supply; brickwork half brick thick	0.19	26.02	1.89	1.81	nr	**29.71**
built-in pipes and ducts, cross-sectional area 0.05–0.25 m^2; excluding supply; brickwork one brick thick	0.33	45.19	3.27	2.68	nr	**51.14**
built-in pipes and ducts, cross-sectional area 0. 25–0.5 m^2; excluding supply; brickwork half brick thick	0.23	31.49	2.28	2.11	nr	**35.88**
built-in pipes and ducts, cross-sectional area 0. 25–0.5 m^2; excluding supply; brickwork one brick thick	0.40	54.77	3.91	3.11	nr	**61.79**
04 LIGHTWEIGHT BLOCKWORK						
Lightweight concrete blocks; 3.5 N/mm²; in cement-lime mortar designation (iii)						
Thickness 100 mm;						
vertical straight walls	0.17	23.28	1.73	15.25	m^2	**40.25**
vertical curved walls	0.23	31.49	2.29	15.25	m^2	**49.03**
vertical facework to concrete	0.18	24.65	1.77	18.95	m^2	**45.37**
casing to metal sections	0.21	28.76	2.08	18.95	m^2	**49.79**
Thickness 140 mm;						
vertical straight walls	0.23	31.49	2.23	18.52	m^2	**52.24**
vertical curved walls	0.30	41.08	2.96	18.52	m^2	**62.56**
vertical facework to concrete	0.23	31.49	2.30	21.76	m^2	**55.55**
casing to metal sections	0.27	36.97	2.68	21.76	m^2	**61.41**
Thickness 215 mm;						
vertical straight walls	0.28	38.34	2.72	27.29	m^2	**68.35**
vertical curved walls	0.37	50.67	3.63	27.29	m^2	**81.59**
vertical facework to concrete	0.28	38.34	2.80	30.38	m^2	**71.53**
casing to metal sections	0.31	42.45	3.08	30.38	m^2	**75.91**
Columns and piers						
440 × 100 mm	0.08	10.95	0.79	6.72	m	**18.47**
890 × 140 mm	0.22	30.13	2.19	16.62	m	**48.93**
Surface features						
fair facing	0.06	8.22	0.60	–	m^2	**8.81**

CLASS U: BRICKWORK, BLOCKWORK AND MASONRY

Item	Gang hours	Labour £	Plant £	Material £	Unit	**Total rate £**
04 LIGHTWEIGHT BLOCKWORK – cont						
Lightweight concrete blocks – cont						
Ancillaries – cont						
Ancillaries						
built-in pipes and ducts, cross-sectional area not exceeding 0.05 m^2; excluding supply; blockwork 100 mm thick	0.04	5.48	0.40	0.39	nr	**6.26**
built-in pipes and ducts, cross-sectional area not exceeding 0.05 m^2; excluding supply; blockwork 140 mm thick	0.09	12.32	0.89	1.22	nr	**14.43**
built-in pipes and ducts, cross-sectional area not exceeding 0.05 m^2; excluding supply; blockwork 215 mm thick	0.13	17.80	1.25	1.57	nr	**20.62**
built-in pipes and ducts, cross-sectional area 0.05–0.25 m^2; excluding supply; blockwork 100 mm thick	0.13	17.80	1.25	1.81	nr	**20.86**
built-in pipes and ducts, cross-sectional area 0.05–0.25 m^2; excluding supply; blockwork 140 mm thick	0.18	24.65	1.75	2.61	nr	**29.00**
built-in pipes and ducts, cross-sectional area 0.05–0.25 m^2; excluding supply; blockwork 215 mm thick	0.25	34.23	2.48	3.32	nr	**40.03**
built-in pipes and ducts, cross-sectional area 0. 25–0.5 m^2; excluding supply; blockwork 100 mm thick	0.15	20.54	1.49	2.29	nr	**24.32**
built-in pipes and ducts, cross-sectional area 0. 25–0.5 m^2; excluding supply; blockwork 140 mm thick	0.21	28.76	2.08	2.61	nr	**33.44**
built-in pipes and ducts, cross-sectional area 0. 25–0.5 m^2; excluding supply; blockwork 215 mm thick	0.30	41.08	2.93	3.93	nr	**47.93**
05 DENSE CONCRETE BLOCKWORK						
Dense concrete blocks; 7 N/mm²; in cement mortar designation (iii)						
Walls, built vertical and straight						
100 mm thick	0.19	26.02	1.89	9.24	m^2	**37.15**
140 mm thick	0.25	34.23	2.46	14.69	m^2	**51.38**
215 mm thick	0.34	46.56	3.38	18.73	m^2	**68.66**
Walls, built vertical and curved						
100 mm thick	0.25	34.23	2.52	9.28	m^2	**46.03**
140 mm thick	0.33	45.19	3.27	14.69	m^2	**63.15**
215 mm thick	0.45	61.62	4.49	18.77	m^2	**84.87**
Walls, built vertical in facework to concrete						
100 mm thick	0.20	27.39	1.96	11.49	m^2	**40.84**
140 mm thick	0.26	35.60	2.53	17.28	m^2	**55.42**
215 mm thick	0.35	47.93	3.47	20.80	m^2	**72.20**

CLASS U: BRICKWORK, BLOCKWORK AND MASONRY

Item	Gang hours	Labour £	Plant £	Material £	Unit	**Total rate £**
Walls, as casings to metal sections, built vertical and straight						
100 mm thick	0.23	31.49	2.28	11.49	m^2	**45.27**
140 mm thick	0.30	41.08	2.93	17.28	m^2	**61.29**
Columns and piers						
440 × 100 mm	0.08	10.95	0.79	3.72	m	**15.46**
890 × 140 mm thick	0.22	30.13	2.19	13.21	m	**45.52**
Ancillaries						
built-in pipes and ducts, cross-sectional area not exceeding 0.05 m^2; excluding supply; blockwork 100 mm thick	0.05	6.85	0.45	0.35	nr	**7.65**
built-in pipes and ducts, cross-sectional area not exceeding 0.05 m^2; excluding supply; blockwork 140 mm thick	0.10	13.69	1.00	1.00	nr	**15.69**
built-in pipes and ducts, cross-sectional area not exceeding 0.05 m^2; excluding supply; blockwork 215 mm thick	0.14	19.17	1.34	1.18	nr	**21.69**
built-in pipes and ducts, cross-sectional area 0.05–0.25 m^2; excluding supply; blockwork 100 mm thick	0.14	19.17	1.34	1.25	nr	**21.76**
built-in pipes and ducts, cross-sectional area 0.05–0.25 m^2; excluding supply; blockwork 140 mm thick	0.19	26.02	1.89	2.13	nr	**30.03**
built-in pipes and ducts, cross-sectional area 0.05–0.25 m^2; excluding supply; blockwork 215 mm thick	0.27	36.97	2.68	2.39	nr	**42.04**
built-in pipes and ducts, cross-sectional area 0.25–0.5 m^2; excluding supply; blockwork 100 mm thick	0.16	21.91	1.61	1.55	nr	**25.07**
built-in pipes and ducts, cross-sectional area 0.25–0.5 m^2; excluding supply; blockwork 140 mm thick	0.23	31.49	2.28	2.17	nr	**35.95**
built-in pipes and ducts, cross-sectional area 0.25–0.5 m^2; excluding supply; blockwork 215 mm thick	0.32	43.82	3.17	2.85	nr	**49.84**
06 ARTIFICIAL STONE BLOCKWORK						
Reconstituted stone masonry blocks; Bradstone 100 bed weathered Cotswold or North Cerney rough hewn rockfaced blocks; in coloured cement-lime mortar designation (iii)						
Thickness 100 mm vertical facing						
vertical straight walls	0.30	41.08	2.97	62.26	m^2	**106.31**
vertical curved walls	0.39	53.40	3.87	62.26	m^2	**119.54**
vertical facework to concrete	0.31	42.45	3.09	62.30	m^2	**107.84**
casing to metal sections	0.40	54.77	3.97	62.30	m^2	**121.04**

CLASS U: BRICKWORK, BLOCKWORK AND MASONRY

Item	Gang hours	Labour £	Plant £	Material £	Unit	Total rate £
06 ARTIFICIAL STONE BLOCKWORK – cont						
Reconstituted stone masonry blocks – cont						
Ancillaries						
built-in pipes and ducts, cross-sectional area not exceeding 0.05 m^2; excluding supply	0.07	9.59	0.70	0.64	nr	**10.93**
built-in pipes and ducts, cross-sectional area 0.05–0.25 m^2; excluding supply	0.22	30.13	2.19	6.51	nr	**38.82**
built-in pipes and ducts, cross-sectional area 0.25–0.5 m^2; excluding supply	0.26	35.60	2.58	8.17	nr	**46.35**
Reconstituted stone masonry blocks; Bradstone Architectural dressing in weathered Cotswold or North Cerney shades; in coloured cement-lime mortar designation (iii)						
Surface features; Pier Caps						
305 × 305 mm, weathered and throated	0.09	12.32	0.89	24.92	nr	**38.14**
381 × 381 mm, weathered and throated	0.11	15.06	1.09	35.32	nr	**51.48**
457 × 457 mm, weathered and throated	0.13	17.80	1.29	48.37	nr	**67.46**
533 × 533 mm, weathered and throated	0.15	20.54	1.49	67.31	nr	**89.34**
Surface features; Copings						
152 × 76 mm, twice weathered and throated	0.08	10.95	0.79	10.87	m	**22.62**
152 × 76 mm, curved on plan, twice weathered and throated	0.11	15.06	1.05	37.18	m	**53.30**
305 × 76 mm, twice weathered and throated	0.10	13.69	3.97	18.32	m	**35.98**
305 × 76 mm, curved on plan, twice weathered and throated	0.13	17.80	1.32	99.36	m	**118.49**
Surface features; Pilasters						
440 × 100 mm	0.14	19.17	1.39	27.41	m	**47.97**
Surface features; Corbels						
479 × 100 × 215 mm, splayed	0.49	67.10	4.86	167.61	nr	**239.56**
665 × 100 × 215 mm, splayed	0.55	75.31	5.46	120.90	nr	**201.67**
Surface features; Lintels						
100 × 140 mm	0.11	15.06	1.09	51.92	m	**68.07**
100 × 215 mm	0.16	21.91	1.59	67.67	m	**91.17**
07 ASHLAR MASONRY						
Portland Whitbed limestone; in cement-lime mortar designation (iv); pointed one side cement-lime mortar designation (iii) incorporating stone dust						
Thickness 50 mm						
vertical facing to concrete	0.85	116.39	8.43	124.72	m^2	**249.54**
Thickness 75 mm						
vertical straight walls	0.95	130.09	9.42	220.87	m^2	**360.38**
vertical curved walls	1.45	198.55	14.38	391.88	m^2	**604.81**

CLASS U: BRICKWORK, BLOCKWORK AND MASONRY

Item	Gang hours	Labour £	Plant £	Material £	Unit	**Total rate £**
Surface features						
copings; weathered and twice throated; 250 × 150 mm	0.45	61.62	4.46	165.01	m	**231.09**
copings; weathered and twice throated; 250 × 150 mm; curved on plan	0.45	61.62	4.46	197.98	m	**264.06**
copings; weathered and twice throated; 400 × 150 mm	0.49	67.10	4.86	242.56	m	**314.52**
copings; weathered and twice throated; 400 × 150 mm; curved on plan	0.49	67.10	4.86	291.03	m	**362.98**
string courses; shaped and dressed; 75 mm projection × 150 mm high	0.45	61.62	4.46	132.43	m	**198.51**
corbel; shaped and dressed; 500 × 450 × 300 mm	0.55	75.31	5.46	218.46	nr	**299.23**
keystone; shaped and dressed; 750 × 900 × 300 mm (extreme)	1.30	178.01	12.90	691.63	nr	**882.54**
08 RUBBLE MASONRY						
Rubble masonry; random stones; in cement-lime mortar designation (iii)						
Walls, built vertical and straight; not exceeding 2 m high						
300 mm thick	1.25	171.17	12.39	329.72	m^2	**513.28**
450 mm thick	1.80	246.48	17.85	569.46	m^2	**833.79**
600 mm thick	2.40	328.64	23.80	659.38	m^2	**1011.82**
Walls, built vertical, curved on plan; not exceeding 2 m high						
300 mm thick	1.40	191.71	4.48	337.41	m^2	**533.59**
450 mm thick	2.00	273.87	19.83	494.58	m^2	**788.28**
600 mm thick	2.65	362.87	26.28	659.38	m^2	**1048.53**
Walls, built with battered face; not exceeding 2 m high						
300 mm thick	1.40	191.71	13.88	329.72	m^2	**535.31**
450 mm thick	2.00	273.87	19.83	494.58	m^2	**788.28**
600 mm thick	2.65	362.87	26.28	659.38	m^2	**1048.53**
Rubble masonry; squared stones; in cement-lime mortar designation (iii)						
Walls, built vertical and straight; not exceeding 2 m high						
300 mm thick	1.25	171.17	12.39	595.56	m^2	**779.12**
450 mm thick	1.80	246.48	17.85	893.32	m^2	**1157.65**
600 mm thick	2.40	328.64	23.80	1191.12	m^2	**1543.56**
Walls, built vertical, curved on plan; not exceeding 2 m high						
300 mm thick	1.40	191.71	13.88	595.56	m^2	**801.15**
450 mm thick	2.00	273.87	19.83	893.32	m^2	**1187.02**
600 mm thick	2.40	328.64	23.80	1254.09	m^2	**1606.53**
Walls, built with battered face; not exceeding 2 m high						
300 mm thick	1.40	191.71	13.88	595.56	m^2	**801.15**
450 mm thick	2.00	273.87	19.83	893.32	m^2	**1187.02**
600 mm thick	2.40	328.64	23.80	1254.09	m^2	**1606.53**

CLASS U: BRICKWORK, BLOCKWORK AND MASONRY

Item	Gang hours	Labour £	Plant £	Material £	Unit	Total rate £
08 RUBBLE MASONRY – cont						
Dry stone walling; random stones						
Average thickness 300 mm						
battered straight walls	1.15	157.47	6.22	304.38	m^2	**468.07**
Average thickness 450 mm						
battered straight walls	1.65	225.94	8.92	456.61	m^2	**691.47**
Average thickness 600 mm						
battered straight walls	2.15	294.41	11.63	608.72	m^2	**914.75**
Surface features						
copings; formed of rough stones 275 × 200 mm (average) high	0.45	61.62	4.46	78.50	m	**144.58**
copings; formed of rough stones 500 × 200 mm (average) high	0.55	75.31	5.46	142.75	m	**223.52**
09 ANCILLARIES COMMON TO ALL DIVISIONS						
Expamet joint reinforcement						
Ancillaries						
joint reinforcement; 65 mm wide	0.01	1.37	0.06	1.89	m	**3.32**
joint reinforcement; 115 mm wide	0.01	1.37	0.08	3.16	m	**4.61**
joint reinforcement; 175 mm wide	0.01	1.37	0.11	4.95	m	**6.43**
joint reinforcement; 225 mm wide	0.01	1.37	0.14	6.37	m	**7.88**
Hyload pitch polymer damp-proof course; lapped joints; in cement mortar						
Ancillaries						
103 mm wide; horizontal	0.01	1.37	0.11	8.38	m	**9.85**
103 mm wide; vertical	0.02	2.74	0.17	8.38	m	**11.28**
215 mm wide; horizontal	0.03	4.11	0.26	18.93	m	**23.30**
215 mm wide; vertical	0.04	5.48	0.40	18.93	m	**24.81**
328 mm wide; horizontal	0.04	5.48	0.38	26.67	m	**32.52**
328 mm wide; vertical	0.06	8.22	0.56	26.67	m	**35.45**
Preformed closed cell joint filler; pointing with polysulphide sealant						
Ancillaries						
movement joints; 12 mm filler 90 wide; 12 × 12 mm sealant one side	0.06	8.22	0.55	0.94	m	**9.70**
movement joints; 12 mm filler 200 wide; 12 × 12 mm sealant one side	0.07	9.59	0.64	1.41	m	**11.64**
Dritherm cavity insulation						
Infills						
50 mm thick	0.04	5.48	0.40	6.15	m^2	**12.02**
75 mm thick	0.05	6.85	0.45	8.44	m^2	**15.73**

CLASS U: BRICKWORK, BLOCKWORK AND MASONRY

Item	Gang hours	Labour £	Plant £	Material £	Unit	**Total rate £**
Concrete						
Infills						
50 mm thick	0.06	8.22	0.56	6.00	m^2	**14.78**
Galvanized steel wall ties						
Fixings and ties						
vertical twist strip type; 900 mm horizontal and 450 mm vertical staggered spacings	0.02	2.74	0.15	1.54	m^2	**4.43**
Stainless steel wall ties						
Fixings and ties						
vertical twist strip type; 900 mm horizontal and 450 mm vertical staggered spacings	0.02	2.74	0.15	3.39	m^2	**6.28**

CLASS V: PAINTING

Item	Gang hours	Labour £	Plant £	Material £	Unit	**Total rate £**
RESOURCES – LABOUR						
Labour Resources Build-up Rate						
Painting gang						
1 ganger (skill rate 3)		23.67	–	–		
3 painters (skill rate 3)		67.36	–	–		
1 unskilled operative (general)		19.60	–	–		
Total Gang Rate/Hour		**110.64**	–	–		
RESOURCES – PLANT						
Plant Resources Build-up Rate						
Painting						
1.5 KVA diesel generator		–	6.71	–		
transformers/cables; junction box		–	0.76	–		
125 mm electric grinder		–	0.48	–		
transit van (50% of time)		–	11.96	–		
ladders		–	1.48	–		
Total Rate/Hour		–	**21.38**	–		
01 IRON BASED PRIMER PAINT						
One coat iron oxide primer						
Metal						
upper surfaces inclined at an angle ne 30° to the horizontal	0.03	2.82	0.53	0.52	m^2	**3.87**
upper surfaces inclined at an angle 30–60° to the horizontal	0.03	2.82	0.53	0.52	m^2	**3.87**
surfaces inclined at an angle exceeding 60° to the horizontal	0.03	2.82	0.53	0.52	m^2	**3.87**
soffit surfaces and lower surfaces inclined at an angle ne 60° to the horizontal	0.03	3.43	0.65	0.52	m^2	**4.60**
surfaces of width ne 300 mm	0.01	1.55	0.23	0.16	m	**1.93**
surfaces of width 300 mm–1 m	0.03	2.77	0.53	0.35	m	**3.64**
pipework	0.03	3.35	0.64	0.52	m^2	**4.51**
02 ETCH PRIMER PRIMER PAINT						
One coat calcium plumbate primer						
Metal						
upper surfaces inclined at an angle ne 30° to the horizontal	0.03	2.77	0.53	0.62	m^2	**3.91**
upper surfaces inclined at an angle 30–60° to the horizontal	0.03	2.77	0.53	0.62	m^2	**3.91**
surfaces inclined at an angle exceeding 60° to the horizontal	0.03	2.77	0.53	0.62	m^2	**3.91**
soffit surfaces and lower surfaces inclined at an angle ne 60° to the horizontal	0.03	3.32	0.65	0.62	m^2	**4.59**
surfaces of width ne 300 mm	0.01	1.11	0.23	0.19	m	**1.52**
surfaces of width 300 mm–1 m	0.03	2.77	0.53	0.40	m	**3.70**
pipework	0.03	3.32	0.64	0.62	m^2	**4.58**

CLASS V: PAINTING

Item	Gang hours	Labour £	Plant £	Material £	Unit	**Total rate £**
One coat acrylic wood primer						
Timber						
upper surfaces inclined at an angle ne 30° to the horizontal	0.02	2.21	0.43	0.42	m^2	**3.06**
upper surfaces inclined at an angle 30–60° to the horizontal	0.02	2.21	0.43	0.42	m^2	**3.06**
surfaces inclined at an angle exceeding 60° to the horizontal	0.02	2.21	0.41	0.42	m^2	**3.04**
soffit surfaces and lower surfaces inclined at an angle ne 60° to the horizontal	0.03	2.77	0.53	0.42	m^2	**3.71**
surfaces of width ne 300 mm	0.01	1.11	0.21	0.12	m	**1.44**
surfaces of width 300 mm–1000 mm	0.03	2.77	0.53	0.27	m	**3.57**
04 GLOSS PAINT						
One coat calcium plumbate primer; one undercoat and one finishing coat of gloss paint						
Metal						
upper surfaces inclined at an angle ne 30° to the horizontal	0.07	7.74	1.51	1.53	m^2	**10.79**
upper surfaces inclined at an angle 30–60° to the horizontal	0.08	8.30	1.61	1.53	m^2	**11.44**
surfaces inclined at an angle exceeding 60° to the horizontal	0.08	8.85	1.71	1.53	m^2	**12.10**
soffit surfaces and lower surfaces inclined at an angle ne 60° to the horizontal	0.09	9.40	1.81	1.53	m^2	**12.75**
surfaces of width ne 300 mm	0.03	2.77	0.53	0.46	m	**3.75**
surfaces of width 300 mm–1000 mm	0.05	5.53	1.08	1.00	m	**7.61**
Metal sections	0.09	9.96	1.87	1.53	m^2	**13.36**
Pipework	0.09	9.96	1.87	1.53	m^2	**13.36**
One coat acrylic wood primer; one undercoat and one finishing coat of gloss paint						
Timber						
upper surfaces inclined at an angle ne 30° to the horizontal	0.07	7.19	1.26	1.33	m^2	**9.78**
upper surfaces inclined at an angle 30–60° to the horizontal	0.07	7.74	1.51	1.33	m^2	**10.58**
surfaces inclined at an angle exceeding 60° to the horizontal	0.08	8.30	1.61	1.34	m^2	**11.24**
soffit surfaces and lower surfaces inclined at an angle ne 60° to the horizontal	0.08	8.85	1.71	1.33	m^2	**11.89**
surfaces of width ne 300 mm	0.03	2.77	0.34	0.40	m	**3.50**
surfaces of width 300 mm–1000 mm	0.05	5.53	1.08	0.86	m	**7.48**

CLASS V: PAINTING

Item	Gang hours	Labour £	Plant £	Material £	Unit	Total rate £
04 GLOSS PAINT – cont						
One coat calcium plumbate primer; two undercoats and one finishing coat of gloss paint						
Metal						
upper surfaces inclined at an angle ne 30° to the horizontal	0.06	6.64	1.28	1.82	m^2	**9.74**
upper surfaces inclined at an angle 30–60° to the horizontal	0.07	7.19	1.38	1.82	m^2	**10.40**
surfaces inclined at an angle exceeding 60° to the horizontal	0.07	7.74	1.51	1.82	m^2	**11.07**
soffit surfaces and lower surfaces inclined at an angle ne 60° to the horizontal	0.08	8.30	1.61	1.77	m^2	**11.68**
surfaces of width ne 300 mm	0.03	3.32	0.65	0.55	m	**4.52**
surfaces of width 300 mm–1000 mm	0.06	6.64	1.28	1.18	m	**9.10**
One coat acrylic wood primer; two undercoats and one finishing coat of gloss paint						
Timber						
upper surfaces inclined at an angle ne 30° to the horizontal	0.07	7.19	1.38	1.62	m^2	**10.19**
upper surfaces inclined at an angle 30–60° to the horizontal	0.07	7.74	1.51	1.62	m^2	**10.87**
surfaces inclined at an angle exceeding 60° to the horizontal	0.08	8.30	1.61	1.62	m^2	**11.53**
soffit surfaces and lower surfaces inclined at an angle ne 60° to the horizontal	0.08	8.85	1.61	1.62	m^2	**12.08**
surfaces of width ne 300 mm	0.03	3.32	1.51	0.48	m	**5.31**
surfaces of width 300 mm–1000 mm	0.06	6.64	1.28	1.05	m	**8.97**
One coat alkali resisting primer; two undercoats and one finishing coat of gloss paint						
Smooth concrete						
upper surfaces inclined at an angle ne 30° to the horizontal	0.04	3.87	0.75	1.61	m^2	**6.24**
upper surfaces inclined at an angle 30–60° to the horizontal	0.04	4.43	0.86	1.61	m^2	**6.89**
surfaces inclined at an angle exceeding 60° to the horizontal	0.05	4.98	0.96	1.62	m^2	**7.55**
soffit surfaces and lower surfaces inclined at an angle ne 60° to the horizontal	0.05	5.53	1.08	1.61	m^2	**8.22**
surfaces of width ne 300 mm	0.02	1.66	0.33	0.48	m	**2.47**
surfaces of width 300 mm–1000 mm	0.03	3.32	0.65	1.05	m	**5.02**
Brickwork and rough concrete						
upper surfaces inclined at an angle ne 30° to the horizontal	0.04	4.43	0.86	1.63	m^2	**6.91**
upper surfaces inclined at an angle 30–60° to the horizontal	0.05	4.98	0.96	1.64	m^2	**7.57**
surfaces inclined at an angle exceeding 60° to the horizontal	0.05	5.53	1.08	1.67	m^2	**8.29**

CLASS V: PAINTING

Item	Gang hours	Labour £	Plant £	Material £	Unit	**Total rate £**
soffit surfaces and lower surfaces inclined at an angle ne 60° to the horizontal	0.06	6.08	1.18	1.67	m^2	**8.94**
surfaces of width ne 300 mm	0.02	2.21	0.43	0.50	m	**3.14**
surfaces of width 300 mm–1000 mm	0.04	3.87	0.75	1.08	m	**5.71**
Blockwork						
upper surfaces inclined at an angle ne 30° to the horizontal	0.06	6.08	1.18	1.67	m^2	**8.94**
soffit surfaces and lower surfaces inclined at an angle ne 60° to the horizontal	0.06	6.64	1.28	1.67	m^2	**9.59**
surfaces of width ne 300 mm	0.03	2.77	0.53	0.50	m	**3.80**
surfaces of width 300 mm–1000 mm	0.04	4.43	0.86	1.09	m	**6.37**
Two coats anti-condensation paint						
Metal						
upper surfaces inclined at an angle ne 30° to the horizontal	0.08	8.30	1.61	1.93	m^2	**11.84**
upper surfaces inclined at an angle 30–60° to the horizontal	0.08	8.30	1.61	1.93	m^2	**11.84**
surfaces inclined at an angle exceeding 60° to the horizontal	0.08	8.30	1.61	1.93	m^2	**11.84**
soffit surfaces and lower surfaces inclined at an angle ne 60° to the horizontal	0.08	8.85	1.71	1.93	m^2	**12.49**
surfaces of width ne 300 mm	0.03	3.32	0.65	0.58	m	**4.55**
surfaces of width 300 mm–1000 mm	0.05	5.53	1.08	1.25	m	**7.87**
Metal sections	0.09	9.96	1.91	1.93	m^2	**13.80**
05 EMULSION PAINT						
One thinned coat, two coats vinyl emulsion paint						
Smooth concrete						
upper surfaces inclined at an angle ne 30° to the horizontal	0.07	7.19	1.38	1.08	m^2	**9.65**
upper surfaces inclined at an angle 30–60° to the horizontal	0.07	7.74	1.51	1.08	m^2	**10.33**
surfaces inclined at an angle exceeding 60° to the horizontal	0.08	8.30	1.61	1.08	m^2	**10.99**
soffit surfaces and lower surfaces inclined at an angle ne 60° to the horizontal	0.08	8.85	1.71	1.08	m^2	**11.64**
surfaces of width ne 300 mm	0.03	3.32	0.65	0.32	m	**4.30**
surfaces of width 300 mm–1000 mm	0.07	7.74	1.51	0.70	m	**9.95**
Brickwork and rough concrete						
upper surfaces inclined at an angle ne 30° to the horizontal	0.08	8.30	1.61	1.11	m^2	**11.02**
upper surfaces inclined at an angle 30–60° to the horizontal	0.08	8.85	1.71	1.11	m^2	**11.67**
surfaces inclined at an angle exceeding 60° to the horizontal	0.09	9.40	1.81	1.11	m^2	**12.33**
soffit surfaces and lower surfaces inclined at an angle ne 60° to the horizontal	0.09	9.96	1.91	1.11	m^2	**12.98**
surfaces of width ne 300 mm	0.03	3.32	0.65	0.33	m	**4.31**
surfaces of width 300 mm–1000 mm	0.08	8.85	1.71	0.76	m	**11.33**

CLASS V: PAINTING

Item	Gang hours	Labour £	Plant £	Material £	Unit	**Total rate £**
05 EMULSION PAINT – cont						
One thinned coat, two coats vinyl emulsion paint – cont						
Blockwork						
surfaces inclined at an angle exceeding 60° to the horizontal	0.09	9.96	1.91	1.14	m^2	**13.01**
soffit surfaces and lower surfaces inclined at an angle ne 60° to the horizontal	0.10	11.06	2.14	1.14	m^2	**14.34**
surfaces of width ne 300 mm	0.04	4.43	0.86	0.34	m	**5.62**
surfaces of width 300 mm–1000 mm	0.08	8.85	1.71	0.74	m	**11.30**
06 CEMENT PAINT						
Two coats masonry paint						
Smooth concrete						
upper surfaces inclined at an angle ne 30° to the horizontal	0.07	7.19	1.38	0.87	m^2	**9.44**
upper surfaces inclined at an angle 30–60° to the horizontal	0.07	7.74	1.51	0.87	m^2	**10.12**
surfaces inclined at an angle exceeding 60° to the horizontal	0.08	8.30	1.61	0.87	m^2	**10.77**
soffit surfaces and lower surfaces inclined at an angle ne 60° to the horizontal	0.08	8.85	1.71	0.87	m^2	**11.43**
surfaces of width ne 300 mm	0.03	3.32	0.65	0.26	m	**4.23**
surfaces of width 300 mm–1000 mm	0.07	7.74	1.51	0.56	m	**9.82**
Brickwork and rough concrete						
upper surfaces inclined at an angle ne 30° to the horizontal	0.08	8.30	1.61	1.00	m^2	**10.90**
upper surfaces inclined at an angle 30–60° to the horizontal	0.08	8.85	1.71	1.00	m^2	**11.56**
surfaces inclined at an angle exceeding 60° to the horizontal	0.09	9.40	1.81	1.00	m^2	**12.21**
soffit surfaces and lower surfaces inclined at an angle ne 60° to the horizontal	0.09	9.96	1.91	1.00	m^2	**12.87**
surfaces of width ne 300 mm	0.03	3.32	0.65	0.30	m	**4.27**
surfaces of width 300 mm–1000 mm	0.08	8.85	1.71	0.65	m	**11.21**
Blockwork						
surfaces inclined at an angle exceeding 60° to the horizontal	0.09	9.96	1.91	1.30	m^2	**13.17**
soffit surfaces and lower surfaces inclined at an angle ne 60° to the horizontal	0.10	11.06	2.14	1.30	m^2	**14.50**
surfaces of width ne 300 mm	0.04	4.43	0.86	0.39	m	**5.67**
surfaces of width 300 mm–1000 mm	0.08	8.85	1.71	0.84	m	**11.40**

CLASS V: PAINTING

Item	Gang hours	Labour £	Plant £	Material £	Unit	**Total rate £**
One thinned coat, two coats concrete floor paint						
Smooth concrete						
upper surfaces inclined at an angle ne 30° to the horizontal	0.07	7.19	1.38	3.10	m^2	**11.67**
upper surfaces inclined at an angle 30–60° to the horizontal	0.07	7.74	1.51	3.10	m^2	**12.35**
surfaces inclined at an angle exceeding 60° to the horizontal	0.08	8.30	1.61	3.10	m^2	**13.01**
soffit surfaces and lower surfaces inclined at an angle ne 60° to the horizontal	0.08	8.85	1.71	2.66	m^2	**13.22**
surfaces of width ne 300 mm	0.03	3.32	0.65	0.96	m	**4.93**
surfaces of width 300 mm–1000 mm	0.07	7.74	1.51	2.01	m	**11.27**
Additional coats						
width exceeding 1000 mm	0.03	3.32	0.65	1.24	m^2	**5.21**
surfaces of width ne 300 mm	0.01	1.11	0.23	0.37	m	**1.70**
surfaces of width 300 mm–1000 mm	0.02	1.66	0.53	0.81	m	**2.99**
07 EPOXY OR POLYURETHANE PAINT						
Blast clean to BS 7079; one coat zinc chromate etc. primer, two coats zinc phosphate CR/alkyd undercoat off site; one coat MIO CR undercoat and one coat CR finish on site						
Metal sections		–	–	–	m^2	**29.47**
Blast clean to BS 7079; one coat zinc rich 2 pack primer, one coat MIO high build epoxy 2 pack paint off site; one coat polyurethane 2 pack undercoat and one coat polyurethane 2 pack finish on site						
Metal sections		–	–	–	m^2	**19.89**
Two coats clear polyurethane varnish						
Timber						
upper surfaces inclined at an angle ne 30° to the horizontal	0.04	4.43	0.86	0.66	m^2	**5.94**
upper surfaces inclined at an angle 30–60° to the horizontal	0.05	4.98	0.96	0.66	m^2	**6.59**
surfaces inclined at an angle exceeding 60° to the horizontal	0.05	5.53	1.08	0.66	m^2	**7.27**
soffit surfaces and lower surfaces inclined at an angle ne 60° to the horizontal	0.06	6.64	1.28	0.66	m^2	**8.58**
surfaces of width ne 300 mm	0.03	2.77	0.53	0.20	m	**3.49**
surfaces of width 300 mm–1000 mm	0.05	5.53	1.08	0.43	m	**7.04**

CLASS V: PAINTING

Item	Gang hours	Labour £	Plant £	Material £	Unit	**Total rate £**
07 EPOXY OR POLYURETHANE PAINT – cont						
Two coats colour stained polyurethane varnish						
Timber						
upper surfaces inclined at an angle ne 30° to the horizontal	0.04	4.43	0.86	1.03	m^2	**6.31**
upper surfaces inclined at an angle 30–60° to the horizontal	0.05	4.98	0.96	1.03	m^2	**6.97**
surfaces inclined at an angle exceeding 60° to the horizontal	0.05	5.53	1.08	1.03	m^2	**7.64**
soffit surfaces and lower surfaces inclined at an angle ne 60° to the horizontal	0.06	6.64	1.28	1.03	m^2	**8.95**
surfaces of width ne 300 mm	0.03	2.77	0.53	0.31	m	**3.60**
surfaces of width 300 mm–1000 mm	0.05	5.53	1.08	0.67	m	**7.28**
Three coats colour stained polyurethane varnish						
Timber						
upper surfaces inclined at an angle ne 30° to the horizontal	0.06	6.08	1.18	1.55	m^2	**8.81**
upper surfaces inclined at an angle 30–60° to the horizontal	0.07	7.19	1.38	1.55	m^2	**10.12**
surfaces inclined at an angle exceeding 60° to the horizontal	0.07	7.74	1.51	1.55	m^2	**10.80**
soffit surfaces and lower surfaces inclined at an angle ne 60° to the horizontal	0.09	9.40	1.81	1.55	m^2	**12.76**
surfaces of width ne 300 mm	0.04	3.87	0.75	0.46	m	**5.09**
surfaces of width 300 mm–1000 mm	0.07	7.74	1.51	1.00	m	**10.26**
One coat hardwood stain basecoat and two coats hardwood woodstain						
Timber						
upper surfaces inclined at an angle ne 30° to the horizontal	0.06	6.64	1.28	1.67	m^2	**9.59**
upper surfaces inclined at an angle 30–60° to the horizontal	0.07	7.19	1.38	1.67	m^2	**10.25**
surfaces inclined at an angle exceeding 60° to the horizontal	0.07	7.74	1.51	1.67	m^2	**10.92**
soffit surfaces and lower surfaces inclined at an angle ne 60° to the horizontal	0.08	8.85	1.71	1.78	m^2	**12.34**
surfaces of width ne 300 mm	0.04	3.87	0.75	0.50	m	**5.13**
surfaces of width 300 mm–1000 mm	0.07	7.74	1.51	1.09	m	**10.34**

CLASS V: PAINTING

Item	Gang hours	Labour £	Plant £	Material £	Unit	**Total rate £**
08 BITUMINOUS OR PRESERVATIVE PAINT						
Two coats golden brown wood preservative						
Timber						
upper surfaces inclined at an angle ne 30° to the horizontal	0.05	4.98	0.96	1.33	m^2	**7.27**
upper surfaces inclined at an angle 30–60° to the horizontal	0.05	5.53	1.08	1.33	m^2	**7.94**
surfaces inclined at an angle exceeding 60° to the horizontal	0.06	6.08	1.18	1.33	m^2	**8.60**
soffit surfaces and lower surfaces inclined at an angle ne 60° to the horizontal	0.07	7.19	1.38	1.33	m^2	**9.91**
surfaces of width ne 300 mm	0.03	3.32	0.65	0.40	m	**4.37**
surfaces of width 300 mm–1000 mm	0.06	6.08	1.18	0.86	m	**8.13**
Two coats bituminous paint						
Metal						
metal sections	0.08	8.85	1.71	0.77	m^2	**11.33**
pipework	0.09	9.96	2.04	0.77	m^2	**12.76**

CLASS W: WATERPROOFING

Item	Gang hours	Labour £	Plant £	Material £	Unit	Total rate £
RESOURCES – LABOUR						
Labour Resources Build-up Rate						
Roofing – cladding gang						
1 ganger/chargehand (skill rate 3) – 50% of time		11.84	–	–		
2 skilled operative (skill rate 3)		44.91	–	–		
1 unskilled operative (general) – 50% of time		9.80	–	–		
Total Gang Rate/Hour		**–**	–	–		
Damp-proofing gang						
1 ganger (skill rate 4)		21.42	–	–		
1 skilled operative (skill rate 4)		20.20	–	–		
1 unskilled labour (general)		19.60	–	–		
Total Gang Rate/Hour		**61.21**	–	–		
Roofing – asphalt gang						
1 ganger/chargehand (skill rate 4) – 50% of time		10.71	–	–		
1 skilled operative (skill rate 4)		20.20	–	–		
1 unskilled operative (general)		19.60	–	–		
Total Gang Rate/Hour		**50.51**	–	–		
Tanking – asphalt gang						
1 ganger/chargehand (skill rate 4) – 50% of time		10.71	–	–		
1 skilled operative (skill rate 4)		20.20	–	–		
1 unskilled operative (general)		19.60	–	–		
Total Gang Rate/Hour		**50.51**	–	–		
Tanking – waterproof sheeting gang						
1 skilled operative (skill rate 4)		20.20	–	–		
Total Gang Rate/Hour		**20.20**	–	–		
Tanking – rendering gang						
1 ganger/chargehand (skill rate 4)		21.42	–	–		
1 skilled operative (skill rate 4)		20.20	–	–		
1 unskilled operative (generally)		19.60	–	–		
Total Gang Rate/Hour		**61.21**	–	–		
Protective layers – flexible sheeting, sand and pea gravel coverings gang						
1 unskilled operative (generally)		19.60	–	–		
Total Gang Rate/Hour		**19.60**	–	–		
Protective layers – screed gang						
1 ganger/chargehand (skill rate 4)		21.42	–	–		
1 skilled operative (skill rate 4)		20.20	–	–		
1 unskilled operative (general)		19.60	–	–		
Total Gang Rate/Hour		**61.21**	–	–		
Sprayed or brushed waterproofing gang						
1 ganger/chargehand (skill rate 4) – 30% of time		6.42	–	–		
1 skilled operative (skill rate 4)		20.20	–	–		
Total Gang Rate/Hour		**26.62**	–	–		

CLASS W: WATERPROOFING

Item	Gang hours	Labour £	Plant £	Material £	Unit	**Total rate £**
RESOURCES – PLANT						
Plant Resources Build-up Rate						
Damp-proofing						
2t dumper (50% of time)		–	5.41	–		
Total Rate/Hour		–	**5.41**	–		
Tanking – asphalt						
tar boiler (50% of time)		–	22.29	–		
2t dumper (50% of time)		–	5.41	–		
Total Rate/Hour		–	**27.69**	–		
Tanking – waterproof sheeting						
2t dumper (50% of time)		–	5.41	–		
Total Rate/Hour		–	**5.41**	–		
Tanking – rendering						
mixer		–	62.73	–		
2t dumper (50% time)		–	5.41	–		
Total Rate/Hour		–	**68.14**	–		
Protective layers – flexible sheeting, sand and pea gravel coverings						
2t dumper (50% of time)		–	5.41	–		
Total Rate/Hour		–	**5.41**	–		
Protective layers – screed						
mixer		–	62.73	–		
2t dumper (50% time)		–	5.41	–		
Total Rate/Hour		–	**68.14**	–		
Sprayed or brushed waterproofing						
2t dumper (50% of time)		–	5.41	–		
Total Rate/Hour		–	**5.41**	–		
01 DAMP-PROOFING						
Waterproof sheeting						
0.3 mm polythene sheet						
ne 300 mm wide	0.01	0.18	0.05	0.46	m	**0.69**
300 mm–1 m wide	0.01	0.20	0.05	0.91	m	**1.17**
on horizontal or included surfaces	0.01	0.24	0.06	1.39	m²	**1.70**
on vertical surfaces	0.01	0.20	0.05	1.39	m²	**1.65**
02 TANKING						
Asphalt						
13 mm Mastic asphalt to BS 6925, Type T 1097; two coats; on concrete surface						
upper surfaces inclined at an angle ne 30° to the horizontal	0.45	22.73	13.47	13.15	m²	**49.35**
upper surfaces inclined at an angle 30–60° to the horizontal	0.60	30.30	17.96	13.15	m²	**61.41**
upper surfaces inclined at an angle exceeding 60° to the horizontal	1.00	50.51	29.94	13.15	m²	**93.59**

CLASS W: WATERPROOFING

Item	Gang hours	Labour £	Plant £	Material £	Unit	**Total rate £**
02 TANKING – cont						
Asphalt – cont						
13 mm Mastic asphalt to BS 6925, Type T 1097 – cont						
curved surfaces	1.20	60.61	35.92	13.15	m^2	**109.68**
domed surfaces	1.50	75.76	44.90	13.15	m^2	**133.81**
ne 300 mm wide	0.20	10.10	5.99	3.96	m	**20.05**
300 mm–1 m wide	0.45	22.73	13.47	13.15	m	**49.35**
20 mm Mastic asphalt to BS 6925, Type T 1097; two coats; on concrete surface						
upper surfaces inclined at an angle ne 30° to the horizontal	0.50	25.25	14.97	20.23	m^2	**60.45**
upper surfaces inclined at an angle 30–60° to the horizontal	0.70	35.35	20.96	20.23	m^2	**76.54**
upper surfaces inclined at an angle exceeding 60° to the horizontal	1.20	60.61	35.92	20.23	m^2	**116.76**
curved surfaces	1.40	70.71	41.91	20.23	m^2	**132.84**
domed surfaces	1.75	88.39	52.39	20.23	m^2	**161.00**
ne 300 mm wide	0.23	11.62	6.89	6.07	m	**24.57**
300 mm–1 m wide	0.50	25.25	14.97	20.23	m	**60.45**
13 mm Mastic asphalt to BS 6925, Type T 1097; two coats; on brickwork surface; raking joints to form key						
upper surfaces inclined at an angle 30–60° to the horizontal	0.90	45.46	18.14	20.31	m^2	**83.91**
upper surfaces inclined at an angle exceeding 60° to the horizontal	1.30	65.66	38.92	20.31	m^2	**124.88**
curved surfaces	1.50	75.76	44.90	20.31	m^2	**140.97**
domed surfaces	1.80	90.91	53.88	20.31	m^2	**165.11**
ne 300 mm wide	0.30	15.15	5.99	6.11	m	**27.25**
300 mm–1 m wide	0.75	37.88	22.69	20.31	m	**80.88**
Waterproof sheeting						
Bituthene 4000; lapped joints						
upper surfaces inclined at an angle ne 30° to the horizontal	0.05	3.06	0.27	13.38	m^2	**16.71**
upper surfaces inclined at an angle 30–60° to the horizontal	0.05	3.06	0.27	13.38	m^2	**16.71**
ne 300 mm wide	0.03	1.53	0.14	3.52	m	**5.19**
300 mm–1 m wide	0.04	2.45	0.22	7.52	m	**10.18**
Bituthene 4000; lapped joints; primer coat						
upper surfaces inclined at an angle exceeding 60° to the horizontal	0.06	3.67	0.32	14.59	m^2	**18.59**
Rendering in waterproof cement mortar						
19 mm render in waterproof cement mortar (1:3); two coat work						
upper surfaces inclined at an angle exceeding 60° to the horizontal	0.11	6.73	7.50	9.51	m^2	**23.74**
ne 300 mm wide	0.07	4.29	4.77	3.53	m	**12.59**
300 mm–1 m wide	0.11	6.73	7.50	6.32	m	**20.55**

CLASS W: WATERPROOFING

Item	Gang hours	Labour £	Plant £	Material £	Unit	**Total rate £**
32 mm render in waterproof cement mortar (1:3); one coat work						
upper surfaces inclined at an angle ne 30° to the horizontal	0.11	6.73	7.50	14.04	m^2	**28.27**
ne 300 mm wide	0.07	4.29	4.77	4.08	m	**13.14**
300 mm–1 m wide	0.11	6.73	7.50	7.12	m	**21.35**
03 ROOFING						
Asphalt roofing						
Work has been presented in more detail than CESMM4 to allow the user greater freedom to access an appropriate rate to suit the complexity of their work						
13 mm Mastic asphalt to BS 6925 Type R 988; two coats; on concrete surface						
ne 300 mm wide	0.20	10.10	5.99	3.67	m	**19.76**
20 mm Mastic asphalt to BS 6925 Type R 988; two coats; on concrete surface						
upper surfaces inclined at an angle ne 30° to the horizontal	0.50	25.25	14.97	19.57	m^2	**59.80**
upper surfaces inclined at an angle 30–60° to the horizontal	0.70	35.35	20.96	19.57	m^2	**75.88**
surfaces inclined at an angle exceeding 60° to the horizontal	1.20	60.61	35.92	19.57	m^2	**116.10**
curved surfaces	1.50	75.76	44.90	19.57	m^2	**140.24**
domed surfaces	1.75	88.39	52.39	19.57	m^2	**160.35**
ne 300 mm wide	0.23	11.62	6.89	5.87	m	**24.37**
300 mm–1 m wide	0.50	25.25	14.97	19.57	m	**59.80**
Extra for:						
10 mm thick limestone chippings bedded in hot bitumen	0.05	2.53	0.90	2.73	m^2	**6.16**
dressing with solar reflective paint	0.05	2.53	0.90	7.60	m^2	**11.02**
300 × 300 × 8 mm GRP tiles bedded in hot bitumen	0.30	15.15	5.40	155.82	m^2	**176.38**
04 PROTECTIVE LAYERS						
Flexible sheeting						
3 mm Servi-pak protection board to Bituthene						
upper surfaces inclined at an angle ne 30° to the horizontal	0.20	3.92	1.08	23.52	m^2	**28.52**
ne 300 mm wide	0.10	1.96	0.54	10.04	m	**12.54**
300 mm–1 m wide	0.20	3.92	1.08	23.52	m	**28.52**
3 mm Servi-pak protection board to Bituthene; fixing with adhesive dabs						
upper surfaces inclined at an angle exceeding 60° to the horizontal	0.25	4.90	1.35	24.91	m^2	**31.16**
ne 300 mm wide	0.12	2.35	0.65	10.57	m	**13.57**
300 mm–1 m wide	0.25	4.90	1.35	24.91	m	**31.16**

CLASS W: WATERPROOFING

Item	Gang hours	Labour £	Plant £	Material £	Unit	**Total rate £**
04 PROTECTIVE LAYERS – cont						
Flexible sheeting – cont						
6 mm Servi-pak protection board to Bituthene						
upper surfaces inclined at an angle ne 30° to the horizontal	0.20	3.92	1.08	44.90	m^2	**49.90**
ne 300 mm wide	0.10	1.96	0.54	18.48	m	**20.98**
300 mm 1 m wide	0.20	3.92	1.08	44.90	m	**49.90**
6 mm Servi-pak protection board to Bituthene; fixing with adhesive dabs						
upper surfaces inclined at an angle exceeding 60° to the horizontal	0.25	4.90	1.35	47.10	m^2	**53.35**
ne 300 mm wide	0.12	2.35	0.65	19.83	m	**22.83**
300 mm 1 m wide	0.25	4.90	1.35	47.10	m	**53.35**
Sand covering						
25 mm thick						
upper surfaces inclined at an angle ne 30° to the horizontal	0.02	0.39	0.05	4.34	m^2	**4.78**
Pea gravel covering						
50 mm thick						
upper surfaces inclined at an angle ne 30° to the horizontal	0.02	0.39	0.05	13.99	m^2	**14.43**
Sand and cement screed						
50 mm screed in cement mortar (1:4); one coat work						
upper surfaces inclined at an angle ne 30° to the horizontal	0.13	7.96	8.68	21.79	m^2	**38.43**
05 SPRAYED OR BRUSHED WATERPROOFING						
Two coats RIW liquid asphaltic composition						
on horizontal or vertical surfaces	0.06	1.60	0.32	90.48	m^2	**92.40**
Two coats Aquaseal						
on horizontal or vertical surfaces	0.06	1.60	0.32	1.33	m^2	**3.25**
One coat Ventrot primer; one coat Ventrot hot applied damp-proof membrane						
on horizontal or vertical surfaces	0.05	1.33	0.27	7.39	m^2	**9.00**
06 SHEET LININGS MEMBRANE						
Butyl rubber two coats RIW liquid asphaltic composition						
upper surfaces inclined at an angle not exceeding 30, to horizontal	0.05	1.33	0.32	90.48	m^2	**92.14**
upper surfaces inclined at 30–60, to horizontal	0.06	1.60	0.32	90.48	m^2	**92.40**
upper surfaces inclined at an angle exceeding 60 to horizontal	0.06	1.60	0.32	90.48	m^2	**92.40**
curved surface	0.06	1.60	0.32	90.48	m^2	**92.40**
domed surface	0.06	1.60	0.32	90.48	m^2	**92.40**
surfaces of width 300 m–1000 mm	0.06	1.60	0.32	90.48	m^2	**92.40**

CLASS W: WATERPROOFING

Item	Gang hours	Labour £	Plant £	Material £	Unit	**Total rate £**
EPDM rubber						
upper surfaces inclined at an angle not exceeding 30 deg to horizontal	0.06	1.60	0.32	90.48	m^2	**92.40**
upper surfaces inclined at 30–60 deg to horizontal	0.06	1.60	0.32	90.48	m^2	**92.40**
upper surfaces inclined at an angle exceeding 60 deg to horizontal	0.06	1.60	0.32	90.48	m^2	**92.40**
curved surface	0.06	1.60	0.32	90.48	m^2	**92.40**
domed surface	0.06	1.60	0.32	90.48	m^2	**92.40**
surfaces of width 300 m–1000 mm	0.06	1.60	0.32	90.48	m^2	**92.40**
PVC						
upper surfaces inclined at an angle not exceeding 30, to horizontal	0.06	1.60	0.32	90.48	m^2	**92.40**
upper surfaces inclined at 30–60 deg to horizontal	0.06	1.60	0.32	90.48	m^2	**92.40**
upper surfaces inclined at an angle exceeding 60 deg to horizontal	0.06	1.60	0.32	90.48	m^2	**92.40**
curved surface	0.06	1.60	0.32	90.48	m^2	**92.40**
domed surface	0.06	1.60	0.32	90.48	m^2	**92.40**
Polypropylene						
upper surfaces inclined at an angle not exceeding 30, to horizontal	0.06	1.60	0.32	90.48	m^2	**92.40**
upper surfaces inclined at an angle exceeding 60 to horizontal	0.06	1.60	0.32	90.48	m^2	**92.40**
curved surface	0.06	1.60	0.32	90.48	m^2	**92.40**
domed surface	0.06	1.60	0.32	90.48	m^2	**92.40**
surfaces of width 300 m–1000 mm	0.06	1.60	0.32	90.48	m^2	**92.40**
upper surfaces inclined at an angle not exceeding 30, horizontal	0.06	1.60	0.32	90.48	m^2	**92.40**
Polyethylene						
upper surfaces inclined at an angle not exceeding 30, to horizontal	0.06	1.60	0.32	90.48	m^2	**92.40**
upper surfaces inclined at an angle exceeding 60 to horizontal	0.06	1.60	0.32	90.48	m^2	**92.40**
curved surface	0.06	1.60	0.32	90.48	m^2	**92.40**
domed surface	0.06	1.60	0.32	90.48	m^2	**92.40**
surfaces of width 300 m–1000 mm	0.06	1.60	0.32	90.48	m^2	**92.40**
upper surfaces inclined at an angle not exceeding 30, horizontal	0.06	1.60	0.32	90.48	m^2	**92.40**
Proprietary System						
upper surfaces inclined at an angle not exceeding 30, to horizontal	0.06	1.60	0.32	90.48	m^2	**92.40**
upper surfaces inclined at an angle exceeding 60 to horizontal	0.06	1.60	0.32	90.48	m^2	**92.40**
curved surface	0.06	1.60	0.32	90.48	m^2	**92.40**
domed surface	0.06	1.60	0.32	90.48	m^2	**92.40**
surfaces of width 300 m–1000 mm	0.06	1.60	0.32	90.48	m^2	**92.40**
upper surfaces inclined at an angle not exceeding 30, horizontal	0.06	1.60	0.32	90.48	m^2	**92.40**

CLASS X: MISCELLANEOUS WORK

Item	Gang hours	Labour £	Plant £	Material £	Unit	Total rate £
RESOURCES – LABOUR						
Labour Resources Build-up Rate						
Fencing/barrier gang						
1 ganger/chargehand (skill rate 4) – 50% of time		10.71	–	–		
1 skilled operative (skill rate 4) – 50% of time		10.10	–	–		
1 unskilled operative (general)		19.60	–	–		
1 plant operator (skill rate 4)		26.30	–	–		
Total Gang Rate/Hour		**66.71**	–	–		
Safety fencing gang						
1 ganger/chargehand (skill rate 4)		21.42	–	–		
1 skilled operative (skill rate 4)		20.20	–	–		
2 unskilled operatives (general)		39.20	–	–		
1 plant operator (skill rate 4)		26.30	–	–		
Total Gang Rate/Hour		**107.11**	–	–		
Guttering gang						
2 skilled operatives (skill rate 4)		41.64	–	–		
Total Gang Rate/Hour		**41.64**	–	–		
Rock filled gabions gang						
1 ganger/chargehand (skill rate 4)		21.42	–	–		
4 unskilled operatives (general)		78.41	–	–		
1 plant operator (skill rate 3) – 50% of time		14.59	–	–		
Total Gang Rate/Hour		**114.41**	–	–		
RESOURCES – PLANT						
Plant Resources Build-up Rate						
Fencing/barriers						
agricultural type tractor; fencing auger		–	31.75	–		
gas oil for ditto		–	16.19	–		
drop sided trailer; two axles		–	0.48	–		
power tools (fencing)		–	1.48	–		
Total Rate/Hour		–	**49.89**	–		
Guttering						
ladders		–	1.48	–		
Total Rate/Hour		–	**1.48**	–		
Rock filled gabions						
16 tonne crawler backacter (50% of time)		–	34.06	–		
Total Rate/Hour		–	**34.06**	–		
01 FENCES						
Timber fencing						
Timber post and wire						
1.20 m high; DfT type 3; timber posts, driven; cleft chestnut paling	0.07	4.67	3.49	11.73	m	**19.89**
0.90 m high; DfT type 4; galvanized rectangular wire mesh	0.13	8.67	6.48	11.73	m	**26.88**
1.275 m high; DfT type 1; galvanized wire, 2 barbed, 4 plain	0.06	4.00	3.00	6.51	m	**13.51**
1.275 m high; DfT type 2; galvanized wire, 2 barbed, 4 plain	0.06	4.00	3.00	9.77	m	**16.77**

CLASS X: MISCELLANEOUS WORK

Item	Gang hours	Labour £	Plant £	Material £	Unit	Total rate £
Concrete post and wire						
1.20 m high; DfT type 3; timber posts, driven; cleft chestnut paling	0.07	4.67	3.49	11.73	m	**19.89**
0.90 m high; DfT type 4; galvanized rectangular wire mesh	0.13	8.67	6.48	11.73	m	**26.88**
1.275 m high; DfT type 1; galvanized wire, 2 barbed, 4 plain	0.06	4.00	3.00	6.51	m	**13.51**
1.275 m high; DfT type 2; galvanized wire, 2 barbed, 4 plain	0.06	4.00	3.00	9.77	m	**16.77**
Metal post and wire						
1.20 m high; DfT type 3; timber posts, driven; cleft chestnut paling	0.07	4.67	3.49	11.73	m	**19.89**
0.90 m high; DfT type 4; galvanized rectangular wire mesh	0.13	8.67	6.48	11.73	m	**26.88**
1.275 m high; DfT type 1; galvanized wire, 2 barbed, 4 plain	0.06	4.00	3.00	6.51	m	**13.51**
1.275 m high; DfT type 2; galvanized wire, 2 barbed, 4 plain	0.06	4.00	3.00	9.77	m	**16.77**
Timber post and wire						
1.20 m high; DfT type 3; timber posts, driven; cleft chestnut paling	0.07	4.67	3.49	11.73	m	**19.89**
0.90 m high; DfT type 4; galvanized rectangular wire mesh	0.13	8.67	6.48	11.73	m	**26.88**
1.275 m high; DfT type 1; galvanized wire, 2 barbed, 4 plain	0.06	4.00	3.00	6.51	m	**13.51**
1.275 m high; DfT type 2; galvanized wire, 2 barbed, 4 plain	0.06	4.00	3.00	9.77	m	**16.77**
Timber closeboarded; concrete posts						
Timber closeboarded						
1.80 m high; 125 × 125 mm posts	0.30	20.01	14.97	54.89	m	**89.87**
Wire rope safety fencing to BS 5750; based on 600 m lengths						
Metal crash barriers						
600 mm high; 4 wire ropes; long line posts at 2.40 m general spacings, driven	0.57	61.06	7.98	35.32	m	**104.36**
600 mm high; 4 wire ropes; short line posts at 2.40 m general spacings, 400 × 400 × 600 mm concrete footing	0.70	74.98	13.47	36.85	m	**125.30**
600 mm high; 4 wire ropes; short line posts at 2.40 m general spacings, 400 × 400 × 600 mm concrete footing, socketed	0.74	79.26	15.96	36.79	m	**132.02**
600 mm high; 4 wire ropes; short line posts at 2.40 m general spacings, bolted to structure	0.62	66.41	9.98	32.86	m	**109.25**
Pedestrian guard rails						
Metal guard rails						
1000 mm high; tubular galvanized mild steel to BS 3049, mesh infill (105 swg, 50 × 50 mm mesh); steel posts with concrete footing	0.16	17.14	7.98	96.55	m	**121.67**

CLASS X: MISCELLANEOUS WORK

Item	Gang hours	Labour £	Plant £	Material £	Unit	**Total rate £**
01 FENCES – cont						
Beam safety fencing; based on 600 m lengths						
Metal crash barriers						
600 mm high; untensioned corrugated beam, single sided; long posts at 3.20 m general spacings, driven	0.10	10.71	4.99	104.43	m	**120.13**
600 mm high; untensioned corrugated beam, double sided; long posts at 3.20 m general spacings, driven	0.26	27.85	12.97	189.18	m	**230.00**
600 mm high; untensioned open box beam, single sided; long posts at 3.20 m general spacings, driven	0.10	10.71	4.99	96.89	m	**112.59**
600 mm high; untensioned open box beam, double sided; long posts at 3.20 m general spacings, driven	0.26	27.85	12.97	174.09	m	**214.91**
600 mm high; untensioned open box beam, double height; long posts at 3.20 m general spacings, driven	0.30	32.13	14.97	202.18	m	**249.28**
600 mm high; tensioned corrugated beam, single sided; long posts at 3.20 m general spacings, driven	0.13	13.92	6.48	92.32	m	**112.73**
600 mm high; tensioned corrugated beam, double sided; long posts at 3.20 m general spacings, driven	0.37	39.63	18.46	164.96	m	**223.05**
02 GATES AND STILES						
Timber gates						
single; 3.00 m wide × 1.27 m high	1.84	122.74	90.93	404.60	nr	**618.27**
single; 3.60 m wide × 1.27 m high	1.90	126.74	93.89	426.53	nr	**647.17**
single; 4.10 m wide × 1.27 m high	2.00	133.41	98.84	440.79	nr	**673.04**
single; 4.71 m wide × 1.27 m high	2.00	133.41	98.84	451.75	nr	**684.00**
Metal gates						
single; steel tubular; 3.60 m wide × 1.175 m high	0.30	20.01	14.83	312.60	nr	**347.44**
single; steel tubular; 4.50 m wide × 1.175 m high	0.30	20.01	14.83	307.96	nr	**342.79**
single; steel tubular; half mesh; 3.60 m wide × 1.175 m high	0.30	20.01	14.83	312.60	nr	**347.44**
single; steel tubular; half mesh; 4.50 m wide × 1.175 m high	0.30	20.01	14.83	307.96	nr	**342.79**
single; steel tubular; extra wide; 4.88 m wide × 1.175 m high	0.30	20.01	14.83	334.68	nr	**369.52**
double; steel tubular; 5.02 m wide × 1.175 m high	0.60	40.02	29.65	363.74	nr	**433.41**
Stiles						
1.00 m wide × 1.45 m high; DfT Type 1	1.50	100.06	74.13	152.51	nr	**326.70**
1.00 m wide × 1.45 m high; DfT Type 2	1.40	93.39	69.19	131.58	nr	**294.15**

CLASS X: MISCELLANEOUS WORK

Item	Gang hours	Labour £	Plant £	Material £	Unit	**Total rate £**
03 DRAINAGE TO STRUCTURES ABOVE GROUND						
Note						
Outputs are based on heights up to 3 m above ground and exclude time spent on erecting access equipment, but include marking, cutting, drilling to wood, brick or concrete and all fixings. Testing of finished work is not included.						
Output multipliers for labour and plant for heights over 3 m:						
3–6 m × 1.25						
6–9 m × 1.50						
9–12 m × 1.75						
12–15 m × 2.00						
Cast iron gutters and fittings; BS 460 2002						
100 × 75 mm gutters; support brackets	0.25	10.10	0.37	72.85	m	**83.32**
stop end	0.03	1.09	0.04	57.10	nr	**58.23**
running outlet	0.04	1.41	0.05	90.89	nr	**92.36**
angle	0.04	1.41	0.05	90.89	nr	**92.36**
125 × 75 mm gutters; support brackets	0.30	12.12	0.44	90.25	m	**102.81**
stop end	0.03	1.29	0.05	66.72	nr	**68.06**
running outlet	0.04	1.62	0.06	111.61	nr	**113.28**
angle	0.04	1.62	0.06	110.31	nr	**111.98**
Cast iron rainwater pipes and fittings; BS 460 2002						
65 mm dia.; support brackets	0.26	10.50	0.38	50.61	m	**61.50**
bend	0.28	11.31	0.41	26.35	nr	**38.07**
offset, 75 mm projection	0.30	12.12	0.44	40.35	nr	**52.91**
offset, 225 mm projection	0.30	12.12	0.44	46.97	nr	**59.53**
offset, 455 mm projection	0.30	12.12	0.44	128.54	nr	**141.10**
shoe	0.18	7.27	0.27	43.09	nr	**50.62**
75 mm dia.; support brackets	0.28	11.31	0.41	50.82	m	**62.55**
bend	0.28	11.31	0.41	32.01	nr	**43.73**
offset, 75 mm projection	0.30	12.12	0.44	40.35	nr	**52.91**
offset, 225 mm projection	0.30	12.12	0.44	46.97	nr	**59.53**
offset, 455 mm projection	0.30	12.12	0.44	128.54	nr	**141.10**
shoe	0.18	7.27	0.27	43.09	nr	**50.62**
PVC-U gutters and fittings; Marley						
116 × 75 mm gutters; support brackets	0.18	7.27	0.27	18.67	m	**26.21**
stop end	0.05	2.02	0.07	6.14	nr	**8.23**
running outlet	0.05	2.02	0.07	21.90	nr	**23.99**
angle	0.14	5.65	0.21	26.70	nr	**32.56**
150 mm half round gutters; support brackets	0.18	7.27	0.27	22.01	m	**29.55**
stop end	0.05	2.02	0.07	9.71	nr	**11.81**
running outlet	0.05	2.02	0.07	26.95	nr	**29.04**
angle	0.14	5.65	0.21	24.63	nr	**30.49**

CLASS X: MISCELLANEOUS WORK

Item	Gang hours	Labour £	Plant £	Material £	Unit	**Total rate £**
03 DRAINAGE TO STRUCTURES ABOVE GROUND – cont						
PVC-U external rainwater pipes and fittings; Marley						
68 mm dia.; support brackets	0.15	6.06	0.22	11.23	m	**17.52**
bend	0.16	6.46	0.24	11.16	nr	**17.86**
offset bend	0.18	7.27	0.27	6.83	nr	**14.37**
shoe	0.10	4.04	0.15	10.68	nr	**14.87**
110 mm dia.; support brackets	0.15	6.06	0.22	25.61	m	**31.89**
bend	0.16	6.46	0.24	30.38	nr	**37.08**
offset bend	0.18	7.27	0.27	29.23	nr	**36.77**
shoe	0.10	4.04	0.15	24.54	nr	**28.73**
04 ROCK FILLED GABIONS						
Gabions						
PVC coated galvanized wire mesh box gabions, wire laced; graded broken stone filling						
1.0 × 1.0 m module sizes	0.65	74.37	22.14	88.25	m^3	**184.75**
1.0 × 0.5 m module sizes	0.80	91.53	27.25	113.14	m^3	**231.91**
Heavily galvanized woven wire mesh box gabions, wire laced; graded broken stone filling						
1.0 × 1.0 m module sizes	0.65	74.37	22.14	77.92	m^3	**174.43**
1.0 × 0.5 m module sizes	0.80	91.53	27.25	98.70	m^3	**217.48**
Reno mattresses						
PVC coated woven wire mesh mattresses, wire tied; graded broken stone filling						
230 mm deep	0.15	17.16	5.11	13.66	m^2	**50.16**
Heavily galvanized woven wire mesh, wire tied; graded broken stone filling						
300 mm deep	0.15	17.16	5.11	17.82	m^2	**57.55**

CLASS Y: SEWER AND WATER MAIN RENOVATION AND ANCILLARY WORKS

Item	Gang hours	Labour £	Plant £	Material £	Unit	Total rate £
RESOURCES – LABOUR						
Labour Resources Build-up Rate						
Drain repair gang						
1 chargehand pipelayer (skill rate 4) – 50% of time		10.71	–	–		
1 skilled operative (skill rate 4)		20.20	–	–		
2 unskilled operatives (generally)		39.20	–	–		
1 plant operator (skill rate 3)		29.17	–	–		
Total Gang Rate/Hour		**99.28**	–	–		
RESOURCES – PLANT						
Plant Resources Build-up Rate						
Drainage						
0.40 m^3 hydraulic excavator		–	55.59	–		
trench sheets, shores, props etc.		–	34.86	–		
2t dumper (30% of time)		–	3.24	–		
compaction/plate roller (30% of time)		–	0.96	–		
7.30 m^3/min compressor		–	11.68	–		
small pump		–	1.67	–		
Total Rate/Hour		–	**108.01**	–		
01 PREPARATION OF EXISTING SEWERS						
Cleaning						
eggshape sewer 1300 mm high		–	–	–	m	**23.38**
Removing intrusions						
brickwork		–	–	–	m^3	**117.04**
concrete		–	–	–	m^3	**201.84**
reinforced concrete		–	–	–	m^3	**257.74**
Plugging laterals with concrete plug						
bore not exceeding 300 mm		–	–	–	nr	**113.10**
bore 450 mm		–	–	–	nr	**174.12**
Plugging laterals with brickwork plug						
bore 750 mm		–	–	–	nr	**524.04**
Local internal repairs to brickwork						
area not exceeding 0.1 m^2		–	–	–	nr	**23.20**
area 0.1–0.25 m^2		–	–	–	nr	**60.69**
area 1 m^2		–	–	–	nr	**147.25**
area 10 m^2		–	–	–	nr	**680.37**
Grouting ends of redundant drains and sewers						
100 mm dia.	0.03	2.98	3.21	6.90	nr	**13.10**
300 mm dia.	0.13	12.41	13.46	27.62	nr	**53.49**
450 mm dia.	0.26	25.81	28.14	60.28	nr	**114.23**
600 mm dia.	0.50	49.64	53.97	121.98	nr	**225.60**
1200 mm dia.	1.70	168.77	183.67	345.24	nr	**697.68**

CLASS Y: SEWER AND WATER MAIN RENOVATION AND ANCILLARY WORKS

Item	Gang hours	Labour £	Plant £	Material £	Unit	Total rate £
02 STABILIZATION OF EXISTING SEWERS						
Pointing, cement mortar (1:3)						
faces of brickwork		–	–	–	m^2	**49.62**
03 RENOVATION OF EXISTING SEWERS						
Sliplining						
GRP one piece unit; eggshape sewer 1300 mm high		–	–	–	m	**519.77**
Renovation of existing sewers						
sliplining		–	–	–	hr	**126.43**
04 LATERALS TO RENOVATED SEWERS						
Jointing						
bore not exceeding 150 mm		–	–	–	nr	**80.97**
bore 150–300 mm		–	–	–	nr	**143.73**
bore 450 mm		–	–	–	nr	**201.14**
06 NEW MANHOLES						
Note: The rates assume the most efficient items of plant (excavator) and are optimum rates, assuming continuous output with no delays caused by other operations or works. Ground conditions are assumed to be good soil with no abnormal conditions that would affect outputs and consistency of work.						
Multiplier Factor Table for Labour and Plant to reflect various site conditions requiring additional works						
out of sequence with other works × 2.75 minimum						
in hard clay × 1.75–2.00						
in running sand × 2.75 minimum						
in broken rock × 2.75–3.50						
below water table × 2.00 minimum						
Brick construction						
Design criteria used in models:						
class A engineering bricks						
215 thick walls generally; 328 thick to chambers exceeding 2.5 m deep						
225 plain concrete C20/20 base slab						
300 reinforced concrete C20/20 reducing slab						
125 reinforced concrete C20/20 top slab						
maximum height of working chamber 2.0 m above benching						
750 × 750 access shaft						
plain concrete C15/20 benching, 150 clay main channel longitudinally and two 100 branch channels						
step irons at 300 mm centres, doubled if depth to invert exceeds 3000 mm						
heavy duty manhole cover and frame						

CLASS Y: SEWER AND WATER MAIN RENOVATION AND ANCILLARY WORKS

Item	Gang hours	Labour £	Plant £	Material £	Unit	**Total rate £**
750 × 700 chamber 500 depth to invert						
excavation, support, backfilling and disposal	0.28	21.50	14.90	–	nr	**36.41**
concrete base	0.40	7.98	2.90	27.67	nr	**38.55**
brickwork chamber	3.81	72.72	10.03	80.85	nr	**163.60**
concrete cover slab	2.16	48.15	35.72	127.25	nr	**211.12**
concrete benching, main and branch channels	7.01	140.06	27.38	221.92	nr	**389.36**
step irons	0.21	4.01	0.56	39.04	nr	**43.62**
access cover and frame	1.44	28.80	3.59	275.50	nr	**307.89**
TOTAL	15.31	323.22	95.09	772.23	nr	**1190.53**
750 × 700 chamber 1000 depth to invert						
excavation, support, backfilling and disposal	0.55	42.44	29.59	–	nr	**72.03**
concrete base	0.40	7.98	2.90	27.67	nr	**38.55**
brickwork chamber	7.61	145.44	20.07	161.69	nr	**327.20**
concrete cover slab	2.16	48.15	35.72	127.25	nr	**211.12**
concrete benching, main and branch channels	7.01	140.06	27.38	221.92	nr	**389.36**
step irons	0.32	6.02	0.84	58.56	nr	**65.43**
access cover and frame	1.44	28.80	3.59	275.50	nr	**307.89**
TOTAL	19.49	418.88	120.09	872.60	nr	**1411.56**
750 × 700 chamber 1500 depth to invert						
excavation, support, backfilling and disposal	0.83	63.27	44.20	–	nr	**107.47**
concrete base	0.40	7.98	2.90	27.67	nr	**38.55**
brickwork chamber	12.18	232.70	32.09	242.54	nr	**507.33**
concrete cover slab	2.16	48.15	35.72	127.25	nr	**211.12**
concrete benching, main and branch channels	7.01	140.06	27.38	221.92	nr	**389.36**
step irons	0.42	8.02	1.12	78.09	nr	**87.23**
access cover and frame	1.44	28.80	3.59	275.50	nr	**307.89**
TOTAL	24.44	528.98	147.00	972.96	nr	**1648.94**
900 × 700 chamber 500 depth to invert						
excavation, support, backfilling and disposal	0.33	25.22	17.02	–	nr	**42.24**
concrete base	0.47	9.30	3.38	32.28	nr	**44.97**
brickwork chamber	4.20	80.24	11.07	89.21	nr	**180.52**
concrete cover slab	2.16	48.15	35.72	127.25	nr	**211.12**
concrete benching, main and branch channels	7.01	140.06	27.38	221.92	nr	**389.36**
step irons	0.11	2.01	0.28	19.52	nr	**21.81**
access cover and frame	1.44	28.80	3.59	275.50	nr	**307.89**
TOTAL	15.71	333.78	98.44	765.68	nr	**1197.90**
900 × 700 chamber 1000 depth to invert						
excavation, support, backfilling and disposal	0.66	50.44	34.04	–	nr	**84.47**
concrete base	0.47	9.30	3.38	32.28	nr	**44.97**
brickwork chamber	8.40	160.48	22.14	178.42	nr	**361.05**
concrete cover slab	2.16	48.15	35.72	127.25	nr	**211.12**
concrete benching, main and branch channels	7.01	140.06	27.38	221.92	nr	**389.36**
step irons	0.32	6.02	0.84	58.56	nr	**65.43**
access cover and frame	1.44	28.80	3.59	275.50	nr	**307.89**
TOTAL	20.45	443.25	127.10	893.93	nr	**1464.28**

CLASS Y: SEWER AND WATER MAIN RENOVATION AND ANCILLARY WORKS

Item	Gang hours	Labour £	Plant £	Material £	Unit	**Total rate £**
06 NEW MANHOLES – cont						
Brick construction – cont						
900 × 700 chamber 1500 depth to invert						
excavation, support, backfilling and disposal	0.98	75.09	50.83	–	nr	**125.92**
concrete base	0.47	9.30	3.38	32.28	nr	**44.97**
brickwork chamber	13.44	256.77	35.41	267.63	nr	**559.81**
concrete cover slab	2.16	48.15	35.72	127.25	nr	**211.12**
concrete benching, main and branch channels	7.01	140.06	27.38	221.92	nr	**389.36**
step irons	0.42	8.02	1.12	78.09	nr	**87.23**
access cover and frame	1.44	28.80	3.59	275.50	nr	**307.89**
TOTAL	25.92	566.20	157.44	1002.67	nr	**1726.30**
1050 × 700 chamber 1500 depth to invert						
excavation, support, backfilling and disposal	1.91	145.34	95.95	–	nr	**241.29**
concrete base	0.57	11.30	4.11	39.20	nr	**54.61**
brickwork chamber	14.70	280.85	38.72	292.72	nr	**612.29**
concrete cover slab	2.52	56.17	41.64	135.98	nr	**233.79**
concrete benching, main and branch channels	7.61	151.96	30.64	219.37	nr	**401.98**
step irons	0.84	16.05	2.25	156.17	nr	**174.47**
access cover and frame	1.44	28.80	3.59	275.50	nr	**307.89**
TOTAL	29.59	690.47	216.91	1118.94	nr	**2026.32**
1050 × 700 chamber 2500 depth to invert						
excavation, support, backfilling and disposal	1.99	153.20	117.73	–	nr	**270.93**
concrete base	0.57	11.30	4.11	39.20	nr	**54.61**
brickwork chamber and access shaft (700 × 700)	35.46	677.54	93.45	676.99	nr	**1447.98**
concrete cover slab	2.52	56.17	41.64	135.98	nr	**233.79**
concrete reducing slab	3.30	72.51	46.30	164.36	nr	**283.18**
concrete benching, main and branch channels	7.61	151.96	30.64	219.37	nr	**401.98**
step irons	0.84	16.05	2.25	156.17	nr	**174.47**
access cover and frame	1.44	28.80	3.59	275.50	nr	**307.89**
TOTAL	53.73	1167.54	339.71	1667.57	nr	**3174.82**
1050 × 700 chamber 3500 depth to invert						
excavation, support, backfilling and disposal	2.88	224.10	231.54	–	nr	**455.64**
concrete base	0.57	11.30	4.11	39.20	nr	**54.61**
brickwork chamber and access shaft (700 × 700)	45.57	870.62	120.06	872.14	nr	**1862.83**
concrete reducing slab	3.30	72.51	46.30	164.36	nr	**283.18**
concrete cover slab	2.52	56.17	41.64	135.98	nr	**233.79**
concrete benching, main and branch channels	9.41	187.67	42.87	231.83	nr	**462.38**
step irons	1.16	22.07	3.09	214.74	nr	**239.89**
access cover and frame	1.44	28.80	3.59	275.50	nr	**307.89**
TOTAL	66.84	1473.25	493.21	1933.74	nr	**3900.20**
1350 × 700 chamber 2500 depth to invertSTOP						
excavation, support, backfilling and disposal	2.42	183.95	117.75	–	nr	**301.70**
concrete base	0.70	13.96	5.07	48.42	nr	**67.45**
brickwork chamber and access shaft (700 × 700)	41.54	793.69	109.47	793.05	nr	**1696.21**
concrete reducing slab	3.30	72.51	46.30	164.36	nr	**283.18**
concrete cover slab	2.80	62.42	46.30	169.15	nr	**277.86**
concrete benching, main and branch channels	11.14	221.90	54.63	243.27	nr	**519.80**
step irons	0.84	16.05	2.25	156.17	nr	**174.47**
access cover and frame	1.44	28.80	3.59	275.50	nr	**307.89**
TOTAL	64.18	1393.27	385.36	1849.92	nr	**3628.55**

CLASS Y: SEWER AND WATER MAIN RENOVATION AND ANCILLARY WORKS

Item	Gang hours	Labour £	Plant £	Material £	Unit	**Total rate £**
1350 × 700 chamber 3500 depth to invert						
excavation, support, backfilling and disposal	3.65	283.66	279.79	–	nr	**563.44**
concrete base	0.70	13.96	5.07	48.42	nr	**67.45**
brickwork chamber and access shaft (700 × 700)	54.89	1048.66	144.64	1021.65	nr	**2214.95**
concrete reducing slab	3.30	72.51	46.30	164.36	nr	**283.18**
concrete cover slab	2.80	62.42	46.30	169.15	nr	**277.86**
concrete benching, main and branch channels	11.14	221.90	54.63	243.27	nr	**519.80**
step irons	1.16	22.07	3.09	214.74	nr	**239.89**
access cover and frame	1.44	28.80	3.59	275.50	nr	**307.89**
TOTAL	79.08	1753.96	583.42	2137.09	nr	**4474.46**
1350 × 700 chamber 4500 depth to invert						
excavation, support, backfilling and disposal	4.69	364.50	359.66	–	nr	**724.16**
concrete base	0.70	13.96	5.07	48.42	nr	**67.45**
brickwork chamber and access shaft (700 × 700)	67.37	1287.18	177.55	1250.25	nr	**2714.98**
concrete reducing slab	3.30	72.51	46.30	164.36	nr	**283.18**
concrete cover slab	2.80	62.42	46.30	169.15	nr	**277.86**
concrete benching, main and branch channels	11.14	221.90	54.63	243.27	nr	**519.80**
step irons	1.47	28.09	3.93	273.30	nr	**305.32**
access cover and frame	1.44	28.80	3.59	275.50	nr	**307.89**
TOTAL	92.92	2079.35	697.04	2424.25	nr	**5200.64**
Precast concrete construction						
Design criteria used in models:						
circular shafts						
150 plain concrete surround						
225 plain concrete C20/20 base slab						
precast top slab						
maximum height of working chamber 2.0 m above benching						
plain concrete C15/20 benching, 150 clay main channel longitudinally and two 100 branch channels						
step irons at 300 mm centres, doubled if depth to invert exceeds 3000 mm						
heavy duty manhole cover and frame						
in manholes over 6 m deep, landings at maximum intervals						
675 dia. × 500 depth to invert						
excavation, support, backfilling and disposal	0.19	14.92	10.58	–	nr	**25.51**
concrete base	0.27	5.32	1.93	18.45	nr	**25.70**
main chamber rings	1.60	35.67	26.45	32.69	nr	**94.80**
cover slab	1.98	43.76	30.43	74.73	nr	**148.92**
concrete surround	0.37	7.30	2.98	18.41	nr	**28.69**
concrete benching, main and branch channels	2.38	47.44	10.59	112.04	nr	**170.06**
step irons	0.11	2.01	0.28	19.52	nr	**21.81**
access cover and frame	1.44	28.80	3.59	275.50	nr	**307.89**
TOTAL	8.34	185.21	86.83	551.34	nr	**823.38**

CLASS Y: SEWER AND WATER MAIN RENOVATION AND ANCILLARY WORKS

Item	Gang hours	Labour £	Plant £	Material £	Unit	Total rate £
06 NEW MANHOLES – cont						
Precast concrete construction – cont						
675 dia. × 750 depth to invert						
excavation, support, backfilling and disposal	0.28	21.82	15.65	–	nr	**37.47**
concrete base	0.27	5.32	1.93	18.45	nr	**25.70**
main chamber rings	2.40	53.50	39.67	49.03	nr	**142.20**
cover slab	1.98	43.76	30.43	74.73	nr	**148.92**
concrete surround	0.55	10.95	4.47	27.62	nr	**43.04**
concrete benching, main and branch channels	2.38	47.44	10.59	112.04	nr	**170.06**
step irons	0.21	4.01	0.56	39.04	nr	**43.62**
access cover and frame	1.44	28.80	3.59	275.50	nr	**307.89**
TOTAL	9.51	215.59	106.90	596.41	nr	**918.90**
675 dia. × 1000 depth to invert						
excavation, support, backfilling and disposal	0.38	29.28	20.94	–	nr	**50.22**
concrete base	0.27	5.32	1.93	18.45	nr	**25.70**
main chamber rings	3.20	71.33	52.89	65.37	nr	**189.60**
cover slab	1.98	43.76	30.43	74.73	nr	**148.92**
concrete surround	0.74	14.60	5.96	36.83	nr	**57.39**
concrete benching, main and branch channels	2.38	47.44	10.59	112.04	nr	**170.06**
step irons	0.32	6.02	0.84	58.56	nr	**65.43**
access cover and frame	1.44	28.80	3.59	275.50	nr	**307.89**
TOTAL	10.70	246.54	127.18	641.48	nr	**1015.20**
675 dia. × 1250 depth to invert						
excavation, support, backfilling and disposal	0.48	36.74	26.24	–	nr	**62.98**
concrete base	0.27	5.32	1.93	18.45	nr	**25.70**
main chamber rings	4.00	89.16	66.11	81.72	nr	**236.99**
cover slab	1.98	43.76	30.43	74.73	nr	**148.92**
concrete surround	0.92	18.25	7.45	46.03	nr	**71.73**
concrete benching, main and branch channels	2.38	47.44	10.59	112.04	nr	**170.06**
step irons	0.42	8.02	1.12	78.09	nr	**87.23**
access cover and frame	1.44	28.80	3.59	275.50	nr	**307.89**
TOTAL	11.89	277.49	147.47	686.55	nr	**1111.51**
900 dia. × 750 depth to invert						
excavation, support, backfilling and disposal	0.49	37.20	24.03	–	nr	**61.22**
concrete base	0.47	9.30	3.38	32.28	nr	**44.97**
main chamber rings	2.85	63.53	47.10	112.76	nr	**223.38**
cover slab	3.80	84.71	62.80	150.34	nr	**297.85**
concrete surround	0.74	14.60	5.96	36.83	nr	**57.39**
concrete benching, main and branch channels	3.38	67.28	17.92	120.80	nr	**205.99**
step irons	0.21	4.01	0.56	39.04	nr	**43.62**
access cover and frame	1.44	28.80	3.59	275.50	nr	**307.89**
TOTAL	13.37	309.43	165.33	767.55	nr	**1242.31**

CLASS Y: SEWER AND WATER MAIN RENOVATION AND ANCILLARY WORKS

Item	Gang hours	Labour £	Plant £	Material £	Unit	**Total rate £**
900 dia. × 1000 depth to invert						
excavation, support, backfilling and disposal	0.65	49.52	32.00	–	nr	**81.53**
concrete base	0.47	9.30	3.38	32.28	nr	**44.97**
main chamber rings	3.80	84.71	62.80	150.34	nr	**297.85**
cover slab	3.80	84.71	62.80	150.34	nr	**297.85**
concrete surround	0.97	19.17	7.82	48.33	nr	**75.32**
concrete benching, main and branch channels	3.38	67.28	17.92	120.80	nr	**205.99**
step irons	0.32	6.02	0.84	58.56	nr	**65.43**
access cover and frame	1.44	28.80	3.59	275.50	nr	**307.89**
TOTAL	14.82	349.50	191.15	836.16	nr	**1376.82**
900 dia. × 1500 depth to invert						
excavation, support, backfilling and disposal	0.99	74.96	48.27	–	nr	**123.23**
concrete base	0.47	9.30	3.38	32.28	nr	**44.97**
main chamber rings	5.70	127.06	94.19	225.52	nr	**446.77**
cover slab	2.16	48.15	35.72	127.25	nr	**211.12**
concrete surround	1.47	29.20	11.92	73.65	nr	**114.77**
concrete benching, main and branch channels	3.38	67.28	17.92	120.80	nr	**205.99**
step irons	0.42	8.02	1.12	78.09	nr	**87.23**
access cover and frame	1.44	28.80	3.59	275.50	nr	**307.89**
TOTAL	16.03	392.77	216.12	933.08	nr	**1541.98**
1200 dia. × 1500 depth to invert						
excavation, support, backfilling and disposal	1.70	128.56	75.52	–	nr	**204.08**
concrete base	0.84	16.62	6.04	57.65	nr	**80.30**
main chamber rings	7.02	156.48	116.04	225.29	nr	**497.81**
cover slab	2.80	62.42	46.30	169.15	nr	**277.86**
concrete surround	1.96	38.79	15.83	97.82	nr	**152.43**
concrete benching, main and branch channels	6.13	121.84	36.09	135.03	nr	**292.96**
step irons	0.42	8.02	1.12	78.09	nr	**87.23**
access cover and frame	1.44	28.80	3.59	275.50	nr	**307.89**
TOTAL	22.31	561.52	300.54	1038.51	nr	**1900.57**
1200 dia. × 2000 depth to invert						
excavation, support, backfilling and disposal	2.35	179.18	120.19	–	nr	**299.37**
concrete base	0.84	16.62	6.04	57.65	nr	**80.30**
main chamber rings	9.36	208.65	154.73	300.38	nr	**663.75**
cover slab	2.80	62.42	46.30	169.15	nr	**277.86**
concrete surround	2.60	51.56	21.04	130.04	nr	**202.64**
concrete benching, main and branch channels	6.13	121.84	36.09	135.03	nr	**292.96**
step irons	0.63	12.04	1.69	117.13	nr	**130.85**
access cover and frame	1.44	28.80	3.59	275.50	nr	**307.89**
TOTAL	26.14	681.09	389.67	1184.87	nr	**2255.63**
1200 dia. × 2500 depth to invert						
excavation, support, backfilling and disposal	2.94	224.10	150.32	–	nr	**374.42**
concrete base	0.84	16.62	6.04	57.65	nr	**80.30**
main chamber rings	11.70	260.81	193.41	375.47	nr	**829.69**
cover slab	2.80	62.42	46.30	169.15	nr	**277.86**
concrete surround	3.24	64.34	26.25	162.26	nr	**252.85**
concrete benching, main and branch channels	6.13	121.84	36.09	135.03	nr	**292.96**
step irons	0.84	16.05	2.25	156.17	nr	**174.47**
access cover and frame	1.44	28.80	3.59	275.50	nr	**307.89**
TOTAL	29.93	794.96	464.25	1331.23	nr	**2590.44**

CLASS Y: SEWER AND WATER MAIN RENOVATION AND ANCILLARY WORKS

Item	Gang hours	Labour £	Plant £	Material £	Unit	Total rate £
06 NEW MANHOLES – cont						
Precast concrete construction – cont						
1200 dia. × 3000 depth to invert						
excavation, support, backfilling and disposal	3.53	269.12	180.44	–	nr	**449.56**
concrete base	0.84	16.62	6.04	57.65	nr	**80.30**
main chamber rings	14.04	312.97	232.09	450.57	nr	**995.63**
cover slab	2.80	62.42	46.30	169.15	nr	**277.86**
concrete surround	3.91	77.57	31.65	195.63	nr	**304.86**
concrete benching, main and branch channels	6.13	121.84	36.09	135.03	nr	**292.96**
step irons	0.95	18.05	2.53	175.69	nr	**196.28**
access cover and frame	1.44	28.80	3.59	275.50	nr	**307.89**
TOTAL	33.63	907.38	538.73	1459.22	nr	**2905.33**
1800 dia. × 2000 depth to invert						
excavation, support, backfilling and disposal	4.97	371.89	195.68	–	nr	**567.57**
concrete base	1.91	37.88	13.77	131.43	nr	**183.09**
main chamber rings	12.00	267.49	198.38	791.00	nr	**1256.88**
cover slab	3.60	80.25	59.52	456.04	nr	**595.81**
concrete surround	3.91	77.57	31.65	195.63	nr	**304.86**
concrete benching, main and branch channels	7.38	146.69	42.88	153.35	nr	**342.93**
step irons	0.63	12.04	1.69	117.13	nr	**130.85**
access cover and frame	1.44	28.80	3.59	275.50	nr	**307.89**
TOTAL	35.84	1022.62	547.17	2120.09	nr	**3689.88**
1800 dia. × 2500 depth to invert						
excavation, support, backfilling and disposal	6.42	485.15	284.67	–	nr	**769.82**
concrete base	1.91	37.88	13.77	131.43	nr	**183.09**
main chamber rings	15.00	334.37	247.98	988.75	nr	**1571.10**
cover slab	3.60	80.25	59.52	456.04	nr	**595.81**
concrete surround	4.88	96.74	39.47	243.97	nr	**380.18**
concrete benching, main and branch channels	7.38	146.69	42.88	153.35	nr	**342.93**
step irons	0.84	16.05	2.25	156.17	nr	**174.47**
access cover and frame	1.44	28.80	3.59	275.50	nr	**307.89**
TOTAL	41.46	1225.93	694.14	2405.22	nr	**4325.28**
1800 dia. × 3000 depth to invert						
excavation, support, backfilling and disposal	7.71	582.68	343.52	–	nr	**926.19**
concrete base	1.91	37.88	13.77	131.43	nr	**183.09**
main chamber rings	18.00	401.24	297.58	1186.51	nr	**1885.32**
cover slab	3.60	80.25	59.52	456.04	nr	**595.81**
concrete surround	5.84	115.90	47.29	292.30	nr	**455.49**
concrete benching, main and branch channels	7.38	146.69	42.88	153.35	nr	**342.93**
step irons	0.95	18.05	2.53	175.69	nr	**196.28**
access cover and frame	1.44	28.80	3.59	275.50	nr	**307.89**
TOTAL	46.82	1411.50	810.68	2670.82	nr	**4893.01**

CLASS Y: SEWER AND WATER MAIN RENOVATION AND ANCILLARY WORKS

Item	Gang hours	Labour £	Plant £	Material £	Unit	**Total rate £**
1800 dia. × 3500 depth to invert						
excavation, support, backfilling and disposal	9.26	706.81	513.76	–	nr	**1220.57**
concrete base	1.91	37.88	13.77	131.43	nr	**183.09**
main chamber rings	21.00	468.12	347.17	1384.26	nr	**2199.54**
cover slab	3.60	80.25	59.52	456.04	nr	**595.81**
concrete surround	6.83	135.52	55.30	341.78	nr	**532.60**
concrete benching, main and branch channels	7.38	146.69	42.88	153.35	nr	**342.93**
step irons	1.16	22.07	3.09	214.74	nr	**239.89**
access cover and frame	1.44	28.80	3.59	275.50	nr	**307.89**
TOTAL	52.58	1626.14	1039.09	2957.10	nr	**5622.33**
1800 dia. × 4000 depth to invert						
excavation, support, backfilling and disposal	10.58	807.31	586.99	–	nr	**1394.29**
concrete base	1.91	37.88	13.77	131.43	nr	**183.09**
main chamber rings	24.00	534.99	396.77	1582.01	nr	**2513.76**
cover slab	3.60	80.25	59.52	456.04	nr	**595.81**
concrete surround	7.80	154.69	63.12	390.12	nr	**607.92**
concrete benching, main and branch channels	7.38	146.69	42.88	153.35	nr	**342.93**
step irons	1.37	26.08	3.65	253.78	nr	**283.51**
access cover and frame	1.44	28.80	3.59	275.50	nr	**307.89**
TOTAL	58.07	1816.69	1170.30	3242.23	nr	**6229.21**
2400 dia. × 3000 depth to invert						
excavation, support, backfilling and disposal	4.69	465.59	514.41	–	nr	**979.99**
concrete base	3.42	67.79	24.64	235.20	nr	**327.63**
main chamber rings	21.60	596.44	1114.10	3392.23	nr	**5102.77**
cover slab	4.40	121.50	226.97	1293.29	nr	**1641.75**
concrete surround	7.80	154.69	63.12	390.12	nr	**607.92**
concrete benching, main and branch channels	7.88	156.79	42.88	153.35	nr	**353.03**
step irons	0.95	18.05	2.53	175.69	nr	**196.28**
access cover and frame	1.44	28.80	3.59	275.50	nr	**307.89**
TOTAL	52.17	1609.65	1992.24	5915.37	nr	**9517.26**
2400 dia. × 4500 depth to invert						
excavation, support, backfilling and disposal	20.82	1581.04	1030.46	–	nr	**2611.50**
concrete base	3.42	67.79	24.64	235.20	nr	**327.63**
main chamber rings	32.40	894.66	1671.15	5088.35	nr	**7654.15**
cover slab	4.40	121.50	226.97	1293.29	nr	**1641.75**
concrete surround	11.71	232.26	94.77	585.75	nr	**912.78**
concrete benching, main and branch channels	7.88	156.79	42.88	153.35	nr	**353.03**
step irons	1.47	28.09	3.93	273.30	nr	**305.32**
access cover and frame	1.44	28.80	3.59	275.50	nr	**307.89**
TOTAL	83.54	3110.93	3098.40	7904.73	nr	**14114.06**
2700 dia. × 3000 depth to invert						
excavation, support, backfilling and disposal	16.98	1275.73	672.34	–	nr	**1948.08**
concrete base	4.32	85.74	31.17	297.45	nr	**414.36**
main chamber rings	24.00	662.71	1237.89	3660.54	nr	**5561.14**
cover slab	4.80	132.54	247.58	2601.54	nr	**2981.66**
concrete surround	8.79	174.31	71.12	439.60	nr	**685.03**
concrete benching, main and branch channels	8.08	160.83	42.88	153.35	nr	**357.07**
step irons	0.95	18.05	2.53	175.69	nr	**196.28**
access cover and frame	1.44	28.80	3.59	275.50	nr	**307.89**
TOTAL	69.35	2538.72	2309.10	7603.68	nr	**12451.50**

CLASS Y: SEWER AND WATER MAIN RENOVATION AND ANCILLARY WORKS

Item	Gang hours	Labour £	Plant £	Material £	Unit	**Total rate £**
06 NEW MANHOLES – cont						
Precast concrete construction – cont						
2700 dia. × 4500 depth to invert						
excavation, support, backfilling and disposal	26.21	1986.65	1243.54	–	nr	**3230.19**
concrete base	4.30	85.28	31.01	296.30	nr	**412.59**
main chamber rings	36.00	994.06	1856.83	5490.81	nr	**8341.71**
cover slab	4.80	132.54	247.58	2601.54	nr	**2981.66**
concrete surround	13.18	261.47	106.68	659.40	nr	**1027.55**
concrete benching, main and branch channels	8.08	160.83	42.88	153.35	nr	**357.07**
step irons	1.47	28.09	3.93	273.30	nr	**305.32**
access cover and frame	1.44	28.80	3.59	275.50	nr	**307.89**
TOTAL	95.48	3677.71	3536.05	9750.21	nr	**16963.97**
3000 dia. × 3000 depth to invert						
excavation, support, backfilling and disposal	20.88	1566.67	806.28	–	nr	**2372.94**
concrete base	5.33	105.68	38.42	366.63	nr	**510.72**
main chamber rings	25.80	712.41	1330.69	4692.04	nr	**6735.15**
cover slab	0.60	18.05	49.96	2482.94	nr	**2550.96**
concrete surround	9.75	193.48	78.94	487.93	nr	**760.35**
concrete benching, main and branch channels	8.88	176.69	42.88	153.35	nr	**372.93**
step irons	0.95	18.05	2.53	175.69	nr	**196.28**
access cover and frame	1.44	28.80	3.59	275.50	nr	**307.89**
TOTAL	73.63	2819.83	2353.29	8634.09	nr	**13807.21**
3000 dia. × 4500 depth to invert						
excavation, support, backfilling and disposal	32.21	2438.13	1475.19	–	nr	**3913.32**
concrete base	5.33	105.68	38.42	366.63	nr	**510.72**
main chamber rings	38.70	1068.62	1996.04	7038.06	nr	**10102.72**
cover slab	0.60	18.05	49.96	2482.94	nr	**2550.96**
concrete surround	14.63	290.21	118.41	731.90	nr	**1140.53**
concrete benching, main and branch channels	8.88	176.69	42.88	153.35	nr	**372.93**
step irons	1.47	28.09	3.93	273.30	nr	**305.32**
access cover and frame	1.44	28.80	3.59	275.50	nr	**307.89**
TOTAL	103.26	4154.26	3728.43	11321.68	nr	**19204.37**
3000 dia. × 6000 depth to invert						
excavation, support, backfilling and disposal	42.95	3250.64	1966.87	–	nr	**5217.51**
concrete base	5.33	105.68	38.42	366.63	nr	**510.72**
main chamber rings	51.60	1424.83	2661.39	9384.08	nr	**13470.29**
cover slab	0.60	18.05	49.96	2482.94	nr	**2550.96**
concrete surround	19.50	386.95	157.89	975.87	nr	**1520.70**
concrete benching, main and branch channels	8.88	176.69	42.88	153.35	nr	**372.93**
step irons	2.00	38.12	5.34	370.91	nr	**414.36**
access cover and frame	1.44	28.80	3.59	275.50	nr	**307.89**
TOTAL	132.29	5429.75	4926.34	14009.28	nr	**24365.36**

CLASS Y: SEWER AND WATER MAIN RENOVATION AND ANCILLARY WORKS

Item	Gang hours	Labour £	Plant £	Material £	Unit	**Total rate £**
Precast concrete circular manhole; CPM Perfect Manhole; complete with preformed benching and outlets to base; elastomeric seal to joints to rings; single steps as required. Excavations and backfilling not included.						
1200 mm dia. with up to four outlets 100 mm or 150 mm dia.; effective internal depth:						
1250 mm deep	2.25	139.41	41.79	992.20	nr	**1173.40**
1500 mm deep	2.25	129.32	41.79	1133.38	nr	**1304.48**
1800 mm deep	2.50	145.09	41.79	1192.67	nr	**1379.55**
2000 mm deep	2.50	145.09	41.79	1372.78	nr	**1559.66**
2500 mm deep	2.75	157.71	52.23	1573.25	nr	**1783.19**
1500 mm dia. with up to four outlets up to 450 mm dia.; effective internal depth:						
2000 mm deep	2.75	157.71	52.23	2884.69	nr	**3094.63**
2500 mm deep	3.05	174.74	58.50	3218.91	nr	**3452.14**
3000 mm deep	3.10	177.89	58.50	3285.66	nr	**3522.06**
3500 mm deep	3.40	194.29	66.86	3435.60	nr	**3696.75**
4000 mm deep	3.55	203.12	68.95	3894.79	nr	**4166.86**
4500 mm deep	3.55	203.12	68.95	4229.00	nr	**4501.07**
1500 mm dia. with up to four outlets up to 600 mm dia.; effective internal depth:						
4000 mm deep	3.95	227.73	73.13	3783.62	nr	**4084.47**
4500 mm deep	4.50	258.64	83.57	4020.68	nr	**4362.89**
07 EXISTING MANHOLES						
Abandonment						
sealing redundant road gullies with grade C15 concrete	0.02	2.28	2.51	21.87	nr	**26.66**
sealing redundant chambers with grade C15 concrete						
ne 1.0 m deep to invert	0.09	8.94	9.75	88.04	nr	**106.72**
1.0–2.0 m deep to invert	0.21	20.85	22.71	142.12	nr	**185.68**
2.0–3.0 m deep to invert	0.55	54.60	59.38	236.49	nr	**350.47**
Alteration						
100 × 100 mm water stop tap boxes on 100 × 100 mm brick chambers						
raising the level by 150 mm or less	0.06	5.96	6.43	27.89	nr	**40.27**
lowering the level by 150 mm or less	0.04	3.97	4.32	16.38	nr	**24.67**
420 × 420 mm cover and frame on 420 × 420 mm in situ concrete chamber						
raising the level by 150 mm or less	0.10	9.93	10.86	50.16	nr	**70.94**
lowering the level by 150 mm or less	0.06	5.96	6.43	33.80	nr	**46.18**
Raising the level of 700 × 700 mm cover and frame on 700 × 500 mm in situ concrete chamber						
by 150 mm or less	0.17	16.88	18.39	72.16	nr	**107.42**

CLASS Y: SEWER AND WATER MAIN RENOVATION AND ANCILLARY WORKS

Item	Gang hours	Labour £	Plant £	Material £	Unit	**Total rate £**
07 EXISTING MANHOLES – cont						
Alteration – cont						
600 × 600 mm grade ‘A’ heavy duty manhole cover and frame on 600 × 600 mm brick chamber						
raising the level by 150 mm or less	0.17	16.88	18.39	72.16	nr	**107.42**
raising the level by 150–300 mm	0.21	20.85	22.71	89.42	nr	**132.98**
lowering the level by 150 mm or less	0.10	9.93	10.86	46.26	nr	**67.05**
08 INTERRUPTIONS						
Preparation of existing sewers						
cleaning		–	–	–	hr	**537.82**
Stabilization of existing sewers						
pointing		–	–	–	hr	**108.67**

PART 7

Unit Costs – Highway Works

INTRODUCTORY NOTES

The Unit Costs in this part represent the net cost to the Contractor of executing the work on site; they are not the prices which would be entered in a tender Bill of Quantities.

It must be emphasized that the unit rates are averages calculated on unit outputs for typical site conditions. *Costs can vary considerably from contract to contract depending on individual Contractors, site conditions, working methods and other factors.* Reference should be made to Part 1 for a general discussion on Civil Engineering Estimating.

Guidance prices are included for work normally executed by specialists, with a brief description where necessary of the assumptions upon which the costs have been based. Should the actual circumstances differ, it would be prudent to obtain check prices from the specialists concerned, on the basis of actual/likely quantity of the work, nature of site conditions, geographical location, time constraints, etc.

The method of measurement adopted in this section is the **Method of Measurement for Highway Works**, subject to variances where this has been felt to be of advantage to produce more helpful price guidance.

We have structured this Unit Cost section to cover as many aspects of Civil and Highway works as possible.

The Gang hours column shows the output per measured unit in actual time, not the total labour hours; thus for an item involving a gang of 5 men each for 0.3 hours, the total labour hours would naturally be 1.5, whereas the Gang hours shown would be 0.3.

This section is structured in such a manner as to provide the User with adequate background information on how the rates have been calculated, so as to allow them to be readily adjusted to suit other conditions to the example presented:

Alternative gang structures as well as the effect of varying bonus levels, travelling costs etc.

Other types of plant or else different running costs from the medium usage presumed

Other types of materials or else different discount/waste allowances from the levels presumed

GUIDANCE NOTES

Generally

Adjustments should be made to the rates shown for time, location, local conditions, site constraints and any other factors likely to affect the costs of a specific scheme.

Method of Measurement

Although this part of the book is primarily based on MMHW, the specific rules have been varied in cases where it has been felt that an alternative presentation would be of value to the book's main purpose of providing guidance on prices. This is especially so with a number of specialist contractors but also in the cases of work where a more detailed presentation will enable the user to allow for ancillary items.

Materials Cost

Materials costs within the rates have generally been calculated using the list prices contained in Part 4: Resources, adjusted to allow for delivery charges (if any) and a reasonable level of discount obtainable by the contractor, this will vary very much depending on the contractor's standing, the potential size of the order and the supplier's eagerness and will vary also between raw traded goods such as timber which will attract a low discount of perhaps 3%, if at all, and manufactured goods where the room for bargaining is much greater and can reach levels of 30% to 40%. High demand for a product at the time of pricing can dramatically reduce the potential discount, as can the world economy in the case of imported goods such as timber and copper. Allowance has also been made for wastage on site (generally 2½% to 5%) dependent upon the risk, the actual level should take account of the nature of the material and its method of storage and distribution about the site.

Labour Cost

The composition of the labour and type of plant is generally stated at the beginning of each section, more detailed information on the calculation of the labour rates is given in Part 4: Resources. In addition is a summary of labour grades and responsibilities extracted from the Working Rule Agreement adjusted for the premium paid in the infrastructure projects. Within Parts 6 and 7 each section is prefaced by a detailed build-up of the labour gang assumed for each type of work. This should allow the user to see the cost impact of a different gang as well as different levels of bonus payments, allowances for skills and conditions, travelling allowances etc. The user should be aware that the output constants are based on the gangs shown and would also need to be changed.

Plant Cost

A rate build-up of suitable plant is generally stated at the beginning of each section, with more detailed information on alternative machines and their average fuel costs being given in Part 4: Resources, Within Parts 6 and 7 each section is prefaced by a detailed build-up of plant assumed for each type of work This should allow the user to see the cost impact of using alternative plant as well as different levels of usage. The user should be aware that the output constants are based on the plant shown and would also need to be changed.

Outputs

The user is directed to Part 10: Outputs, which contains a selection of output constants and in particular a chart of haulage times for various capacities of tippers.

LEVEL 1 DIVISION	LEVEL 2 CONSTRUCTION HEADING	LEVEL 3 MMHW SERIES HEADINGS
(i) Preliminaries	Preliminaries	Series 100
(ii) Road Works	Road Works General	Series 200 Series 300 Series 400 Series 600
	Main Carriageway	Series 500 Series 700 Series 1100
	Interchanges	Series 500 Series 700 Series 1100
	Side Roads	Series 500 Series 700 Series 1100
	Signs, Motorway Communications and Lighting	Series 1200 Series 1300 Series 1400 Series 1500
	Landscape and Ecology	Series 3000

LEVEL 1 DIVISION		LEVEL 2 CONSTRUCTION HEADING	LEVEL 3 MMHW SERIES HEADINGS
(iii) Structures	Structure in form of Bridge or Viaduct; Name or Reference	Special Preliminaries	Series 2700
		Piling	Series 1600
		Substructure – End Supports	Series 500 Series 600 Series 1100 Series 1700 Series 1800 Series 1900 Series 2300 Series 2400
		Substructure – Intermediate Supports Substructure – Main Span Substructure – Approach Spans	As for End Supports
		Superstructure – Main Span Superstructure – Approach Spans Superstructure – Arch Ribs	Series 500 Series 1700 Series 1800 Series 1900 Series 2100 Series 2300 Series 2400
		Finishing	Series 400 Series 600 Series 700 Series 1100 Series 2000 Series 2400 Series 5000

LEVEL 1 DIVISION		LEVEL 2 CONSTRUCTION HEADING	LEVEL 3 MMHW SERIES HEADINGS
	Retaining Wall, Culvert, Subway, Gantry, Large Headwall, Gabion Wall, Diaphragm Wall, Pocket Type Reinforced Brickwork, Retaining Wall and the like; Name or Reference	Special Preliminaries	
		Main Construction	Series 500 Series 600 Series 1100 Series 1600 Series 1700 Series 1800 Series 1900 Series 2300 Series 2400
		Finishings	Series 400 Series 600 Series 700 Series 1100 Series 2000 Series 2400 Series 5000
(iv) Structures where a choice of designs is offered	Structure Designed by the Overseeing Organization; Name or Reference	To comply with the principles set down above for Structures	
	Structure Designed by the Contractor; Name or Reference		
(v) Structures Designed by the Contractor	Structure; Name or Reference		
(vi) Service Areas		Road works Structures	To comply with the principles set down above for Road Works and Structures
(vii) Maintenance Compounds		Road works Structures	
(viii) Accommodation Works		Interest; Name or Reference	
(ix) Works for Statutory or Other Bodies		Body; Name or Reference	To comply with the principles set down above for Road Works and Structures
(x) Day Works		Day Works	
(xi) PC & Provisional Sum		PC & Provisional Sum	

SERIES 100: PRELIMINARIES

Item	Gang hours	Labour £	Plant £	Material £	Unit	Total rate £
NOTES						
General						
Refer also to the example calculations of Preliminaries and Oncosts and profit-outlined in Part 2						
TEMPORARY ACCOMMODATION						
Erection of principal offices for the Overseeing Organization						
prefabricated unit; connect to services		–	637.70	–	nr	**637.70**
Erection of offices and messes for the Contractor						
prefabricated unit; connect to services		–	637.70	–	nr	**637.70**
Erection of stores and workshops for the Contractor						
prefabricated unit; connect to services		–	637.70	–	nr	**637.70**
Servicing of principal offices for the Overseeing Organization						
jack leg hutment; 3.7 m × 2.6 m		–	82.28	–	week	**82.28**
jack leg hutment; 7.35 m × 3.1 m		–	158.24	–	week	**158.24**
jack leg hutment; 14.7 m × 3.7 m		–	363.73	–	week	**363.73**
jack leg toilet unit; 4.9 m × 2.6 m		–	46.30	–	week	**46.30**
Servicing of portable offices for the Overseeing Organization						
wheeled cabin; 3.7 m × 2.3 m		–	122.34	–	week	**122.34**
wheeled cabin; 6.7 m × 2.3 m		–	158.24	–	week	**158.24**
Servicing of offices and messes for the Overseeing Organization						
jack leg hutment; 3.7 m × 2.6 m		–	82.28	–	week	**82.28**
jack leg hutment; 7.35 m × 3.1 m		–	158.24	–	week	**158.24**
jack leg hutment; 14.7 m × 3.7 m		–	363.73	–	week	**363.73**
wheeled cabin; 3.7 m × 2.3 m		–	122.34	–	week	**122.34**
wheeled cabin; 6.7 m × 2.3 m		–	158.24	–	week	**158.24**
jack leg toilet unit; 4.9 m × 2.6 m		–	46.30	–	week	**46.30**
canteen unit; 9.8 m × 2.6 m		–	146.67	–	week	**146.67**
Servicing of stores and workshops for the Contractor						
jack leg hutment; 3.7 m × 2.6 m		–	82.28	–	week	**82.28**
jack leg hutment; 7.35 m × 3.1 m		–	158.24	–	week	**158.24**
wheeled cabin; 3.7 m × 2.3 m		–	122.34	–	week	**122.34**
wheeled cabin; 6.7 m × 2.3 m		–	158.24	–	week	**158.24**
pollution decontamination unit; 6.7 m × 2.3 m		–	223.19	–	week	**223.19**
Dismantling of principal offices for the Overseeing Organization						
prefabricated unit; disconnect from services; removing		–	637.70	–	nr	**637.70**
Dismantling of offices and messes for the Contractor						
prefabricated unit; disconnect from services; removing		–	637.70	–	nr	**637.70**

SERIES 100: PRELIMINARIES

Item	Gang hours	Labour £	Plant £	Material £	Unit	Total rate £
TEMPORARY ACCOMMODATION – cont						
Dismantling of stores and workshops for the Contractor						
prefabricated unit; disconnect from services; removing		–	637.70	–	nr	**637.70**
VEHICLES FOR THE OVERSEEING ORGANIZATION						
Typical vehicle hire rates						
Land Rover or similar, short wheelbase		–	1125.71	–	week	**1125.71**
Land Rover or similar, long wheelbase		–	990.07	–	week	**990.07**
OPERATIVES FOR THE ENGINEER						
OPERATIVES FOR THE OVERSEEING ORGANIZATION						
Chainman for the Overseeing Organization	40.00	767.66	–	–	week	**767.66**
Driver for the Overseeing Organization	40.00	976.32	–	–	week	**976.32**
Laboratory Assistant for the Overseeing Organization	40.00	1052.35	–	–	week	**1052.35**

SERIES 200: SITE CLEARANCE

Item	Gang hours	Labour £	Plant £	Material £	Unit	Total rate £
NOTES						
Notes						
The prices in this section are to include for the removal of superficial obstructions down to existing ground level.						
Demolition of individual or groups of buildings or structures						
MMHW states that individual structures should be itemized. The following rates are given as £ per m^3 to simplify the pricing of different sized structures.						
RESOURCES – LABOUR						
Clearance gang						
1 ganger/chargehand (skill rate 4)	2.00	21.42	–	–	-	**21.42**
1 skilled operative (skill rate 4)	1.00	20.20	–	–	-	**20.20**
2 unskilled operatives (general)	2.00	39.20	–	–	-	**39.20**
1 plant operator (skill rate 3)	1.00	29.17	–	–	-	**29.17**
Total Gang Rate/Hour	1.00	109.99	–	–	-	**109.99**
RESOURCES – PLANT						
Clearance						
0.8 m^3 tractor loader – 50% of time		–	27.71	–	-	**27.71**
8 t lorry with hiab – 25% of time		–	9.31	–	-	**9.31**
4 t dumper – 50% of time		–	10.51	–	-	**10.51**
20 t mobile crane – 25% of time		–	48.95	–	-	**48.95**
compressor, 11.3 m^3/min (450 cfm), 4 tool		–	31.70	–	-	**31.70**
compressor tools: two brick hammers/picks – 50% of time		–	2.23	–	-	**2.23**
compressor tools: chipping hammer		–	0.99	–	-	**0.99**
compressor tools: medium rock drill 30		–	1.19	–	-	**1.19**
compressor tools: road breaker		–	3.25	–	-	**3.25**
compressor tools: two 15 m lengths hose		–	0.46	–	-	**0.46**
Total Rate/Hour		–	136.30	–	-	**136.30**
SITE CLEARANCE						
General site clearance						
open field site	9.91	1089.97	1343.32	–	ha	**2433.30**
medium density wooded	20.61	2266.84	2793.32	–	ha	**5060.16**
heavy density wooded	32.09	3529.49	4349.23	–	ha	**7878.71**
urban areas (town centre)	30.60	3365.61	4147.76	–	ha	**7513.37**
live dual carriageway	30.60	3365.61	4147.76	–	ha	**7513.37**
Demolition of building or structure						
building; brick construction with timber floor and roof		–	–	–	m^3	**9.49**
building; brick construction with concrete floor and roof		–	–	–	m^3	**16.74**
building; masonry construction with timber floor and roof		–	–	–	m^3	**12.33**

SERIES 200: SITE CLEARANCE

Item	Gang hours	Labour £	Plant £	Material £	Unit	Total rate £
SITE CLEARANCE – cont						
building; reinforced concrete frame construction with brick infill		–	–	–	m^3	**16.39**
building; steel frame construction with brick infill		–	–	–	m^3	**8.91**
building; steel frame construction with cladding		–	–	–	m^3	**8.51**
building; timber		–	–	–	m^3	**7.56**
reinforced concrete bridge deck or superstructure		–	–	–	m^3	**77.36**
reinforced concrete bridge abutment or bank seat		–	–	–	m^3	**70.31**
reinforced concrete retaining wall		–	–	–	m^3	**210.93**
brick or masonry retaining wall		–	–	–	m^3	**118.81**
brick or masonry boundary wall		–	–	–	m^3	**98.41**
dry stone boundary wall		–	–	–	m^3	**105.47**
TAKE UP OR TAKE DOWN AND SET ASIDE FOR RE-USE OR REMOVE TO STORE OR TIP OFF SITE						
Take up or down and set aside for reuse						
precast concrete kerbs and channels	0.02	1.87	4.49	–	m	**6.36**
precast concrete edgings	0.01	1.41	3.53	–	m	**4.94**
precast concrete drainage and kerb blocks	0.02	2.34	5.51	–	m	**7.85**
precast concrete drainage channel systems	0.02	2.34	5.51	–	m	**7.85**
tensioned single sided corrugated beam safety fence	0.14	15.40	19.18	–	m	**34.58**
timber post and 4 rail fence	0.08	8.80	10.90	–	m	**19.70**
bench seat	0.13	14.30	15.14	–	nr	**29.43**
cattle trough	0.16	17.16	18.09	–	nr	**35.25**
permanent bollard	0.13	14.30	15.14	–	nr	**29.43**
pedestrian crossing lights; pair	0.26	28.60	35.53	–	nr	**64.13**
lighting column including bracket arm and lantern; 5 m high	0.58	63.79	78.96	–	nr	**142.75**
lighting column including bracket arm and lantern; 10 m high	0.61	67.09	83.09	–	nr	**150.19**
traffic sign	0.26	28.60	35.53	–	nr	**64.13**
timber gate	0.13	14.30	15.14	–	nr	**29.43**
stile	0.13	14.30	15.14	–	nr	**29.43**
road stud	0.03	2.86	2.96	–	nr	**5.82**
chamber cover and frame	0.03	2.86	2.96	–	nr	**5.82**
gully grating and frame	0.03	2.86	2.96	–	nr	**5.82**
feeder pillars	0.03	2.86	2.96	–	nr	**5.82**
Take up or down and remove to store off site						
precast concrete kerbs and channels	0.02	1.87	7.58	–	m	**9.45**
precast concrete edgings	0.01	1.41	5.85	–	m	**7.25**
precast concrete drainage and kerb blocks	0.02	2.34	9.37	–	m	**11.71**
precast concrete drainage channel systems	0.02	2.34	9.37	–	m	**11.71**
tensioned single sided corrugated beam safety fence	0.14	15.40	21.72	–	m	**37.12**
timber post and 4 rail fence	0.08	8.80	12.36	–	m	**21.16**
bench seat	0.13	14.30	24.59	–	nr	**38.89**

SERIES 200: SITE CLEARANCE

Item	Gang hours	Labour £	Plant £	Material £	Unit	**Total rate £**
cattle trough	0.16	17.16	29.43	–	nr	**46.59**
permanent bollard	0.13	14.30	24.59	–	nr	**38.89**
pedestrian crossing lights; pair	0.26	28.60	40.26	–	nr	**68.86**
lighting column including bracket arm and lantern; 5 m high	0.58	63.79	89.50	–	nr	**153.30**
lighting column including bracket arm and lantern; 10 m high	0.61	67.09	94.18	–	nr	**161.27**
traffic sign	0.26	28.60	40.26	–	nr	**68.85**
timber gate	0.13	14.30	24.59	–	nr	**38.89**
stile	0.13	14.30	24.59	–	nr	**38.89**
road stud	0.03	2.86	4.85	–	nr	**7.71**
chamber cover and frame	0.03	2.86	4.85	–	nr	**7.71**
gully grating and frame	0.03	2.86	4.85	–	nr	**7.71**
feeder pillars	0.03	2.86	4.85	–	nr	**7.71**
Take up or down and remove to tip off site						
tensioned single sided corrugated beam safety fence	0.17	18.70	23.10	–	m	**41.80**
timber post and 4 rail fence	0.09	9.90	12.20	–	m	**22.10**
low pressure gas mains up to 150 mm dia.	0.04	4.40	5.45	–	m	**9.85**
low pressure water mains up to 75 mm dia.	0.03	3.30	4.14	–	m	**7.44**
power cable laid singly	0.03	3.30	4.14	–	m	**7.44**
lighting column including bracket arm and lantern; 5 m high	0.82	90.19	111.74	–	nr	**201.93**
lighting column including bracket arm and lantern; 10 m high	0.85	93.49	115.69	–	nr	**209.18**
traffic sign including posts	0.38	41.80	51.21	–	nr	**93.01**
Removal of existing reflectorised thermoplastic road markings						
100 mm wide line	–	–	–	–	m	**1.91**
150 mm wide line	–	–	–	–	m	**2.82**
200 mm wide line	–	–	–	–	m	**3.80**
arrow or letter ne 6.0 m long	–	–	–	–	nr	**31.27**
arrow or letter 6.0–16.0 m long	–	–	–	–	nr	**145.26**

SERIES 300: FENCING

Item	Gang hours	Labour £	Plant £	Material £	Unit	Total rate £
NOTES						
This section is restricted to those fences and barriers which are most commonly found on Highway Works. Hedges have been included, despite not being specifically catered for in the MMHW.						
The re-erection cost for fencing taken from store assumes that major components are in good condition; the prices below allow a sum of 20% of the value of new materials to cover minor repairs, new fixings and touching up any coatings.						
RESOURCES – LABOUR						
Fencing/barrier gang						
1 ganger/chargehand (skill rate 4) – 50% of time	1.00	10.71	–	–	-	**10.71**
1 skilled operative (skill rate 4) – 50% of time	0.50	10.10	–	–	-	**10.10**
1 unskilled operative (general)	1.00	19.60	–	–	-	**19.60**
1 plant operator (skill rate 4)	1.00	26.30	–	–	-	**26.30**
Total Gang Rate/Hour	1.00	66.71	–	–	-	**66.71**
Horticultural works gang						
1 skilled operative (skill rate 4)	1.00	20.20	–	–	-	**20.20**
1 unskilled operative (general)	1.00	19.60	–	–	-	**19.60**
Total Gang Rate/Hour	1.00	39.80	–	–	-	**39.80**
RESOURCES – PLANT						
Fencing/Barriers						
agricultural type tractor; fencing auger		–	31.75	–	-	**31.75**
gas oil for ditto		–	16.19	–	-	**16.19**
drop sided trailer, two axles		–	0.48	–	-	**0.48**
power tools etc. (fencing)		–	1.50	–	-	**1.50**
Total Rate/Hour		–	49.91	–	-	**49.91**
RESOURCES – MATERIALS						
All rates for materials are based on the most economically available materials with a minimum waste allowance of 2.5% and supplier's discount of 15%.						
ENVIRONMENTAL BARRIERS						
Hedges						
Set out, nick out and excavate trench minimum 400 mm deep and break up subsoil to minimum depth 300 mm	0.15	5.97	–	–	m	**5.97**
Supply and plant hedging plants; backfilling with excavated topsoil						
single row plants at 200 mm centres	0.25	9.95	–	8.47	m	**18.42**
single row plants at 300 mm centres	0.17	6.77	–	5.64	m	**12.41**
single row plants at 400 mm centres	0.13	4.97	–	4.24	m	**9.21**
single row plants at 500 mm centres	0.10	3.98	–	3.39	m	**7.37**
single row plants at 600 mm centres	0.08	3.18	–	2.81	m	**6.00**
double row plants at 200 mm centres	0.50	19.90	–	16.94	m	**36.84**
double row plants at 300 mm centres	0.34	13.53	–	11.28	m	**24.81**

SERIES 300: FENCING

Item	Gang hours	Labour £	Plant £	Material £	Unit	**Total rate £**
double row plants at 400 mm centres	0.25	9.95	–	8.47	m	**18.42**
double row plants at 500 mm centres	0.20	7.96	–	6.78	m	**14.74**
double row plants at 600 mm centres	0.16	6.37	–	5.62	m	**11.99**
extra for incorporating manure at 1 m^3 / 30 m^3	0.60	11.76	–	0.33	m^3	**12.09**
Noise barriers						
Noise barriers consist of the erection of reflective or absorbent acoustical screening to reduce nuisance from noise. Due to the divergence in performance requirements and specification for various locations it is not practical to state all inclusive unit costs. Therefore advice should be sought from Specialist Contractors in order to obtain accurate costings. However listed below are examples of sample specification together with approximate costings in order to obtain budget prices.						
NB: The following unit costs are based upon a 2.0 m high barrier						
Noise reflective barriers						
Barrier with architectural precast concrete panels and integral posts		–	–	–	m^2	**241.73**
Barrier with acoustical timber planks post support system		–	–	–	m^2	**219.36**
Sound absorptive barriers						
Barrier with architectural precast wood fibre concrete panels and integral posts		–	–	–	m^2	**285.14**
Barrier with perforated steel and mineral wool blankets in self-supporting system		–	–	–	m^2	**312.02**
FENCING, GATES AND STILES						
Temporary fencing						
Type 1; 1.275 m high, timber posts and two strands of galvanized barbed wire and four strands of galvanized plain wire	0.06	4.00	3.00	6.51	m	**13.51**
Type 2; 1.275 m high, timber posts and two strands of galvanized barbed wire and four strands of galvanized plain wire	0.06	4.00	3.00	9.77	m	**16.77**
Type 3; 1.2 m high, timber posts and cleft chestnut paling	0.07	4.67	3.49	11.73	m	**19.89**
Type 4; 0.9 m high, timber posts and galvanized rectangular wire mesh	0.13	8.67	6.49	12.90	m	**28.06**
Timber rail fencing						
1.4 m high, timber posts and four rails	0.13	8.67	6.49	17.19	m	**32.35**
Plastic coated heavy pattern chain link fencing						
1.40 m high with 125 × 125 mm concrete posts	0.05	3.34	2.49	10.83	m	**16.66**
1.80 m high with 125 × 125 mm concrete posts	0.06	4.00	3.00	20.75	m	**27.74**

SERIES 300: FENCING

Item	Gang hours	Labour £	Plant £	Material £	Unit	**Total rate £**
FENCING, GATES AND STILES – cont						
Plastic coated strained wire fencing						
1.35 m high, nine strand with 40 × 40 × 3 mm plastic coated RHS posts	0.16	10.67	7.99	20.38	m	**39.04**
1.80 m high, eleven strand with 50 × 50 × 3 mm plastic coated RHS posts	0.20	13.34	9.99	27.40	m	**50.73**
2.10 m high, fifteen strand with 50 × 50 × 3 mm plastic coated RHS posts	0.22	14.68	10.98	31.63	m	**57.28**
Woven wire fencing						
1.23 m high, galvanized wire with 75 × 150 mm timber posts	0.06	4.00	3.00	9.13	m	**16.12**
Closeboarded fencing						
1.80 m high with 125 × 125 mm concrete posts	0.30	20.01	14.97	55.69	m	**90.68**
Concrete foundation						
to main posts	0.09	6.00	0.18	6.64	nr	**12.83**
to straining posts	0.09	6.00	0.18	6.64	nr	**12.83**
to struts	0.09	6.00	0.18	6.64	nr	**12.83**
to intermediate posts	0.09	6.00	0.18	6.64	nr	**12.83**
Steel tubular frame single field gates						
1.175 m high 3.60 m wide	0.30	20.01	14.97	312.60	nr	**347.59**
1.175 m high 4.50 m wide	0.30	20.01	14.97	307.96	nr	**342.94**
Steel tubular frame half mesh single field gates						
1.175 m high 3.60 m wide	0.30	20.01	14.97	312.60	nr	**347.59**
1.175 m high 4.50 m wide	0.30	20.01	14.97	307.96	nr	**342.94**
Steel tubular frame extra wide single field gates						
1.175 m high 4.88 m wide	0.30	20.01	14.97	334.68	nr	**369.67**
Steel tubular frame double field gates						
1.175 m high 5.02 m wide	0.60	40.02	29.95	363.74	nr	**433.71**
Timber single field gates						
1.27 m high 3.00 m wide	1.84	122.74	91.84	404.60	nr	**619.19**
1.27 m high 3.60 m wide	1.90	126.74	94.84	426.53	nr	**648.11**
1.27 m high 4.10 m wide	2.00	133.41	99.83	440.79	nr	**674.03**
1.27 m high 4.71 m wide	2.00	133.41	99.83	451.75	nr	**684.99**
Timber Type 1 wicket gates						
1.27 m high 1.20 m wide	1.20	80.05	59.90	185.31	nr	**325.25**
Timber Type 2 wicket gates						
1.27 m high 1.02 m wide	1.20	80.05	59.90	269.74	nr	**409.68**
Timber kissing gates						
1.27 m high 1.77 m wide	2.00	133.41	99.83	313.60	nr	**546.84**
Timber stiles Type 1						
1.45 m high 1.00 m wide	1.50	100.06	74.87	128.29	nr	**303.22**
Timber stiles Type 2						
1.45 m high 1.00 m wide	1.40	93.39	69.88	91.23	nr	**254.49**
Extra for						
sheep netting on post and wire	0.04	2.33	0.07	1.14	m	**3.54**
pig netting on post and wire	0.04	2.33	0.07	1.14	m	**3.54**

SERIES 300: FENCING

Item	Gang hours	Labour £	Plant £	Material £	Unit	**Total rate £**
REMOVE FROM STORE AND RE-ERECT FENCING, GATES AND STILES						
Timber rail fencing						
1.4 m high, timber posts and four rails	0.13	8.67	6.49	3.54	m	**18.70**
Plastic coated heavy pattern chain link fencing						
1.40 m high with 125 × 125 mm concrete posts	0.05	3.34	2.49	2.24	m	**8.07**
1.80 m high with 125 × 125 mm concrete posts	0.06	4.00	3.00	2.88	m	**9.88**
Plastic coated strained wire fencing						
1.35 m high, nine strand with 40 × 40 × 3 mm plastic coated RHS posts	0.16	10.67	7.99	4.05	m	**22.71**
1.80 m high, eleven strand with 50 × 50 × 3 mm plastic coated RHS posts	0.20	13.34	9.99	5.35	m	**28.67**
2.10 m high, fifteen strand with 50 × 50 × 3 mm plastic coated RHS posts	0.22	14.68	10.98	5.73	m	**31.39**
Woven wire fencing						
1.23 m high, galvanized wire with 75 × 150 mm timber posts	0.06	4.00	3.00	1.77	m	**8.76**
Closeboarded fencing						
1.80 m high with 125 × 125 mm concrete posts	0.30	20.01	14.97	9.65	m	**44.64**
Steel tubular frame single field gates						
1.175 m high 3.60 m wide	0.30	20.01	14.97	15.93	nr	**50.92**
1.175 m high 4.50 m wide	0.30	20.01	14.97	20.45	nr	**55.44**
Steel tubular frame half mesh single field gates						
1.175 m high 3.60 m wide	0.30	20.01	14.97	20.45	nr	**55.44**
1.175 m high 4.50 m wide	0.30	20.01	14.97	24.85	nr	**59.83**
Steel tubular frame extra wide single field gates						
1.175 m high 4.88 m wide	0.30	20.01	14.97	22.75	nr	**57.74**
Steel tubular frame double field gates						
1.175 m high 5.02 m wide	0.60	40.02	29.95	34.08	nr	**104.06**
Timber single field gates						
1.27 m high 3.00 m wide	1.84	122.74	91.84	33.57	nr	**248.16**
1.27 m high 3.60 m wide	1.90	126.74	94.84	43.89	nr	**265.47**
1.27 m high 4.10 m wide	2.00	133.41	99.83	52.92	nr	**286.16**
1.27 m high 4.71 m wide	2.00	133.41	99.83	54.55	nr	**287.79**
Timber Type 1 wicket gates						
1.27 m high 1.20 m wide	1.20	80.05	59.90	13.10	nr	**153.05**
Timber Type 2 wicket gates						
1.27 m high 1.02 m wide	1.20	80.05	59.90	13.84	nr	**153.78**
Timber kissing gates						
1.27 m high 1.77 m wide	2.00	133.41	99.83	31.26	nr	**264.50**
Timber stiles Type 1						
1.45 m high 1.00 m wide	1.50	100.06	74.87	31.26	nr	**206.19**
Timber stiles Type 2						
1.45 m high 1.00 m wide	1.40	93.39	69.88	23.45	nr	**186.72**
Excavation in hard material						
Extra over excavation for excavation in hard material in fencing works	0.50	33.35	–	–	m^3	**33.35**

SERIES 400: ROAD RESTRAINT SYSTEMS (VEHICLE AND PEDESTRIAN)

Item	Gang hours	Labour £	Plant £	Material £	Unit	Total rate £
NOTES						
The re-erection cost for safety fencing taken from store assumes that major components are in good condition; the prices below allow a sum of 20% of the value of new materials to cover minor repairs, new fixings and touching up any coatings.						
The heights of the following parapets are in accordance with the Standard Designs and DfT requirements. The rates include for all anchorages and fixings and in the case of steel, galvanizing at works and painting four coat paint system on site together with etching the galvanized surface, as necessary.						
RESOURCES – LABOUR						
Safety barrier gang						
1 ganger/chargehand (skill rate 4)	2.00	21.42	–	–	–	**21.42**
1 skilled operative (skill rate 4)	1.00	20.20	–	–	–	**20.20**
2 unskilled operatives (general)	2.00	39.20	–	–	–	**39.20**
1 plant operator (skill rate 4)	1.00	26.30	–	–	–	**26.30**
Total Gang Rate/Hour	1.00	107.11	–	–	–	**107.11**
RESOURCES – PLANT						
Safety barriers						
agricultural type tractor; fencing auger		–	31.75	–	–	**31.75**
gas oil for ditto		–	16.19	–	–	**16.19**
drop sided trailer, two axles		–	0.48	–	–	**0.48**
small tools (part time)		–	1.50	–	–	**1.50**
Total Rate/Hour		–	49.91	–	–	**49.91**
Parapets						
agricultural type tractor; front bucket – 50% of time		–	8.91	–	–	**8.91**
gas oil for ditto		–	9.01	–	–	**9.01**
2.80 m^3/min (100 cfm) compressor; two tool		–	4.91	–	–	**4.91**
gas oil for ditto		–	10.79	–	–	**10.79**
compressor tools: heavy rock drill 33, 84 cfm		–	2.79	–	–	**2.79**
compressor tools: rotary drill, 10 cfm		–	2.38	–	–	**2.38**
8 t lorry with 1 t hiab – 50% of time		–	11.88	–	–	**11.88**
gas oil for ditto		–	6.75	–	–	**6.75**
Total Rate/Hour		–	57.42	–	–	**57.42**
BEAM SAFETY BARRIERS						
Prices generally are for beams ‘straight or curved exceeding 120 m radius’, for work to a tighter radius						
50–120 m radius – Add 15%						
not exceeding 50 m radius – Add 40%						
Untensioned beams						
single sided corrugated beam	0.10	10.71	3.49	25.33	m	**39.53**
double sided corrugated beam	0.27	28.92	10.98	50.65	m	**90.55**
single sided open box beam	0.20	21.42	3.49	42.90	m	**67.82**
single sided double rail open box beam	0.47	50.34	10.98	85.81	m	**147.13**
double height open box beam	0.51	54.63	11.98	85.81	m	**152.42**

SERIES 400: ROAD RESTRAINT SYSTEMS (VEHICLE AND PEDESTRIAN)

Item	Gang hours	Labour £	Plant £	Material £	Unit	**Total rate £**
Tensioned beams						
single sided corrugated beam	0.12	13.07	4.99	27.93	m	**45.99**
double sided corrugated beam	0.17	18.21	5.99	56.44	m	**80.64**
Long driven post						
for single sided tensioned corrugated beam	0.20	21.42	2.74	49.12	nr	**73.28**
for double sided tensioned corrugated beam	0.20	21.42	2.74	49.12	nr	**73.28**
for single sided open box beam	0.20	21.42	2.74	49.12	nr	**73.28**
for double sided open box beam	0.20	21.42	2.74	49.12	nr	**73.28**
for double height open box beam	0.20	21.42	2.74	49.12	nr	**73.28**
Short post for setting in concrete or socket						
for single sided tensioned corrugated beam	0.12	12.85	3.00	19.51	nr	**35.36**
for double sided tensioned corrugated beam	0.13	13.92	3.00	19.51	nr	**36.43**
for single sided open box beam	0.16	17.14	3.00	19.51	nr	**39.64**
for double sided open box beam	0.21	22.49	3.00	19.51	nr	**45.00**
Mounting bracket fixed to structure						
for single sided open box beam	0.60	64.27	7.99	50.59	nr	**122.84**
Terminal section						
for untensioned single sided corrugated beam	2.60	278.50	35.44	502.76	nr	**816.70**
for untensioned double sided corrugated beam	0.50	53.56	43.21	1005.53	nr	**1102.30**
for untensioned single sided open box beam	1.01	108.19	50.42	925.22	nr	**1083.82**
for untensioned double sided open box beam	1.78	190.66	88.85	1514.99	nr	**1794.50**
for tensioned single sided corrugated beam	2.10	224.94	47.91	384.04	nr	**656.89**
for tensioned double sided corrugated beam	3.60	385.61	84.85	578.18	nr	**1048.64**
Full height anchorage						
for single sided tensioned corrugated beam	5.20	556.99	197.16	948.85	nr	**1703.01**
for double sided tensioned corrugated beam	7.20	771.22	217.13	948.85	nr	**1937.20**
for single sided open box beam	4.33	463.81	189.67	553.85	nr	**1207.33**
for double sided open box beam	7.90	846.20	209.64	553.85	nr	**1609.70**
Expansion joint anchorage						
for single sided open box beam	10.62	1137.55	225.61	1370.15	nr	**2733.32**
for double sided open box beam	13.35	1429.98	257.05	1399.07	nr	**3086.10**
Type 048 connection to bridge parapet						
for single sided open box beam	0.70	74.98	34.94	245.03	nr	**354.94**
Connection piece for single sided open box beam to single sided corrugated beam	1.70	182.09	38.93	141.52	nr	**362.55**
Standard concrete foundation						
for post for corrugated beam	0.23	24.64	11.28	10.46	nr	**46.37**
for post for open box beam	0.23	24.64	11.48	12.72	nr	**48.83**
Concrete foundation Type 1 spanning filter drain						
for post for corrugated beam	0.25	26.78	12.48	17.27	nr	**56.53**
for post for open box beam	0.25	26.78	12.48	14.80	nr	**54.06**
Standard socketed foundation for post for open box beam	0.33	35.35	14.97	12.84	nr	**63.17**

SERIES 400: ROAD RESTRAINT SYSTEMS (VEHICLE AND PEDESTRIAN)

Item	Gang hours	Labour £	Plant £	Material £	Unit	Total rate £
CONCRETE SAFETY BARRIERS						
Permanent vertical concrete safety barrier; TRL Design – DfT Approved						
Intermediate units Type V01 & V02; 3 m long						
straight or curved exceeding 50 m radius	0.16	17.14	7.99	617.79	nr	**642.91**
curved not exceeding 50 m radius	0.20	21.42	9.99	616.87	nr	**648.27**
Make up units Type V05 & V06; 1 m long						
straight or curved exceeding 50 m radius	0.30	32.13	14.97	782.38	nr	**829.49**
Termination units Type V03 & V04; 3 m long	0.50	53.56	24.96	764.63	nr	**843.15**
Transition to single sided open box beam unit Type V08 & V09; 1.5 m long	0.37	39.63	18.47	904.90	nr	**963.00**
Transition to rectangular hollow section beam unit Type V10, V11 & V12; 1.5 m long	0.37	39.63	18.47	794.13	nr	**852.23**
Transition to double sided open box beam; 1.5 m long unit Type V07	0.37	39.63	18.47	879.61	nr	**937.71**
Anchor plate sets (normally two plates per first and last three units in any run)	0.15	16.07	14.49	60.40	nr	**90.95**
Temporary concrete safety barrier; TRL Design – DfT Approved						
Intermediate units Type V28; 3 m long						
straight or curved exceeding 50 m radius	0.16	17.14	7.99	564.02	nr	**589.14**
curved not exceeding 50 m radius	0.20	21.42	9.99	564.02	nr	**595.43**
Termination units Type V29; 3 m long	0.50	53.56	24.96	637.55	nr	**716.07**
WIRE ROPE SAFETY FENCES						
Brifen wire rope safety fencing DfT approved; based on 600 m lengths; 4 rope system; posts at 2.40 m general spacings.						
Wire rope	0.14	15.00	1.50	5.44	m	**21.93**
Long driven line posts	0.05	5.36	2.49	32.57	nr	**40.42**
Long driven deflection posts	0.05	5.36	2.49	39.46	nr	**47.31**
Long driven height restraining posts	0.05	5.36	2.49	28.80	nr	**36.65**
Short line post for setting in concrete or socket	0.06	6.43	3.00	26.67	nr	**36.10**
Short deflection post for setting in concrete or socket	0.06	6.43	3.00	26.65	nr	**36.08**
Short height restraining post for setting in concrete or socket	0.06	6.43	3.00	26.65	nr	**36.08**
Fixed height surface mounted post fixed to structure or foundation	0.09	9.64	4.49	35.06	nr	**49.19**
Standard intermediate anchorage	2.00	214.23	99.83	391.85	nr	**705.90**
Standard end anchorage	2.00	214.23	99.83	141.44	nr	**455.49**
In situ standard concrete foundation for post	0.23	24.64	11.48	9.56	nr	**45.68**
In situ standard socketed foundation for post	0.33	35.35	16.47	9.43	nr	**61.25**
Concrete foundation Type 1 spanning filter drain for post	0.37	39.63	18.47	12.85	nr	**70.95**

SERIES 400: ROAD RESTRAINT SYSTEMS (VEHICLE AND PEDESTRIAN)

Item	Gang hours	Labour £	Plant £	Material £	Unit	**Total rate £**
VEHICLE PARAPETS						
The heights of the following parapets are in accordance with the Standard Designs and DfT requirements. The rates include for all anchorages and fixings and in the case of steel, galvanizing at works and painting four coat paint system on site, together with etc.hing the galvanized surface as necessary						
Steel Parapets						
Metal parapet Group P1; 1.0 m high; comprising steel yielding posts and steel horizontal rails						
straight or curved exceeding 50 m radius		–	–	224.94	m	**224.94**
curved not exceeding 50 m radius		–	–	231.97	m	**231.97**
Metal parapet Group P2 (48 Kph); 1.0 m high; comprising steel yielding posts and steel horizontal rails with vertical infill bars						
straight or curved exceeding 50 m radius		–	–	295.23	m	**295.23**
curved not exceeding 50 m radius		–	–	316.32	m	**316.32**
Metal parapet Group P2 (80 Kph); 1.0 m high; comprising steel yielding posts and steel horizontal rails						
straight or curved exceeding 50 m radius		–	–	323.35	m	**323.35**
curved not exceeding 50 m radius		–	–	330.38	m	**330.38**
Metal parapet Group P2 (113 Kph); 1.0 m high; comprising steel yielding posts and steel horizontal rails						
straight or curved exceeding 50 m radius		–	–	351.46	m	**351.46**
curved not exceeding 50 m radius		–	–	330.38	m	**330.38**
Metal parapet Group P4; 1.15 m high; comprising steel yielding posts and steel horizontal rails						
straight or curved exceeding 50 m radius		–	–	456.90	m	**456.90**
curved not exceeding 50 m radius		–	–	492.03	m	**492.03**
Metal parapet Group P5; 1.25 m high; comprising steel yielding posts and steel horizontal rails						
straight or curved exceeding 50 m radius		–	–	535.22	m	**535.22**
curved not exceeding 50 m radius		–	–	632.64	m	**632.64**
Metal parapet Group P5; 1.50 m high; comprising steel yielding posts and steel horizontal rails						
straight or curved exceeding 50 m radius		–	–	913.79	m	**913.79**
curved not exceeding 50 m radius		–	–	984.09	m	**984.09**
Metal parapet Group P6; 1.50 m high; comprising steel yielding posts and steel horizontal rails						
straight or curved exceeding 50 m radius		–	–	1054.38	m	**1054.38**
curved not exceeding 50 m radius		–	–	1124.68	m	**1124.68**

SERIES 400: ROAD RESTRAINT SYSTEMS (VEHICLE AND PEDESTRIAN)

Item	Gang hours	Labour £	Plant £	Material £	Unit	Total rate £
VEHICLE PARAPETS – cont						
Aluminium parapets						
Metal parapet Group P1; 1.0 m high; comprising aluminium yielding/frangible posts and aluminium horizontal rails						
straight or curved exceeding 50 m radius		–	–	220.55	m	**220.55**
curved not exceeding 50 m radius		–	–	231.44	m	**231.44**
Metal parapet Group P2 (80 Kph); 1.0 m high; comprising aluminium yielding/frangible posts and aluminium horizontal rails with mesh infill						
straight or curved exceeding 50 m radius		–	–	238.24	m	**238.24**
curved not exceeding 50 m radius		–	–	241.02	m	**241.02**
Metal parapet Group P2 (113 Kph); 1.0 m high; comprising aluminium yielding/frangible posts and aluminium horizontal rails with mesh infill						
straight or curved exceeding 50 m radius		–	–	235.60	m	**235.60**
curved not exceeding 50 m radius		–	–	252.36	m	**252.36**
Metal parapet Group P4; 1.15 m high; comprising aluminium yielding/frangible posts and aluminium horizontal rails						
straight or curved exceeding 50 m radius		–	–	231.44	m	**231.44**
curved not exceeding 50 m radius		–	–	240.29	m	**240.29**
Metal parapet Group P5; 1.25 m high; comprising aluminium yielding/frangible posts and aluminium horizontal rails with solid sheet infill, anti-access panels						
straight or curved exceeding 50 m radius		–	–	272.28	m	**272.28**
curved not exceeding 50 m radius		–	–	292.70	m	**292.70**
Metal parapet Group P5; 1.50 m high; comprising aluminium yielding/frangible posts and aluminium horizontal rails with solid sheet infill, anti-access panels						
straight or curved exceeding 50 m radius		–	–	313.14	m	**313.14**
curved not exceeding 50 m radius		–	–	319.94	m	**319.94**
Crash cushions						
Static crash cushion system to BS EN1317–3, Class 110 km/hr		–	–	–	nr	**46920.00**
REMOVE FROM STORE AND RE-ERECT BEAM SAFETY BARRIERS						
Prices generally are for beams straight or curved exceeding 120 m radius, for work to a tighter radius						
50–120 m radius – Add 15%						
not exceeding 50 m radius – Add 25%						
Untensioned beams						
single sided corrugated beam	0.07	7.50	3.49	5.61	m	**16.60**
double sided corrugated beam	0.22	23.57	10.98	11.44	m	**45.99**
single sided open box beam	0.07	7.50	3.49	11.20	m	**22.19**
single sided double rail open box beam	0.22	23.57	10.98	22.96	m	**57.51**
double height open box beam	0.24	25.71	11.98	23.22	m	**60.90**

SERIES 400: ROAD RESTRAINT SYSTEMS (VEHICLE AND PEDESTRIAN)

Item	Gang hours	Labour £	Plant £	Material £	Unit	**Total rate £**
Tensioned beams						
single sided corrugated beam	0.10	10.71	4.99	5.76	m	**21.46**
double sided corrugated beam	0.12	12.85	5.99	12.09	m	**30.94**
Long driven post						
for single sided tensioned corrugated beam	0.06	5.89	2.74	14.22	nr	**22.85**
for double sided tensioned corrugated beam	0.06	5.89	2.74	14.84	nr	**23.48**
for single sided open box beam	0.06	5.89	2.74	13.70	nr	**22.33**
for double sided open box beam	0.06	5.89	2.74	16.20	nr	**24.83**
for double height open box beam	0.06	5.89	2.74	39.69	nr	**48.33**
Short post for setting in concrete or socket						
for single sided tensioned corrugated beam	0.06	6.43	3.00	26.72	nr	**36.14**
for double sided tensioned corrugated beam	0.06	6.43	3.00	27.40	nr	**36.82**
for single sided open box beam	0.06	6.43	3.00	32.81	nr	**42.23**
for double sided open box beam	0.06	6.43	3.00	39.68	nr	**49.10**
Mounting bracket fixed to structure						
for single sided open box beam	0.16	17.14	7.99	10.47	nr	**35.59**
Terminal section						
for untensioned single sided corrugated beam	0.71	76.05	35.44	147.71	nr	**259.20**
for untensioned double sided corrugated beam	1.25	133.89	62.39	155.09	nr	**351.38**
for untensioned single sided open box beam	1.01	108.19	50.42	158.79	nr	**317.39**
for untensioned double sided open box beam	1.78	190.66	88.85	258.50	nr	**538.01**
for tensioned single sided corrugated beam	0.96	102.83	47.91	96.01	nr	**246.75**
for tensioned double sided corrugated beam	1.70	182.09	84.85	155.09	nr	**422.04**
Full height anchorage						
for single sided tensioned corrugated beam	3.95	423.10	197.16	191.99	nr	**812.25**
for double sided tensioned corrugated beam	4.35	465.95	217.13	251.07	nr	**934.14**
for single sided open box beam	3.80	407.03	189.67	118.89	nr	**715.59**
for double sided open box beam	4.20	449.88	209.64	183.58	nr	**843.10**
Expansion joint anchorage						
for single sided open box beam	4.52	484.16	225.61	393.50	nr	**1103.27**
for double sided open box beam	5.15	551.64	257.05	498.55	nr	**1307.24**
Type 048 connection to bridge parapet						
for single sided open box beam	0.70	74.98	34.94	10.96	nr	**120.88**
Connection piece for single sided open box beam to single sided corrugated beam	0.78	83.55	38.93	141.52	nr	**264.00**
Standard concrete foundation						
for post for corrugated beam	0.23	24.64	11.28	10.46	nr	**46.37**
for post for open box beam	0.23	24.64	11.48	12.72	nr	**48.83**
Concrete foundation Type 1 spanning filter drain						
for post for corrugated beam	0.25	26.78	12.48	17.27	nr	**56.53**
for post for open box beam	0.25	26.78	12.48	14.80	nr	**54.06**
Standard socketed foundation for post for open box beam	0.33	35.35	14.97	12.84	nr	**63.17**

SERIES 400: ROAD RESTRAINT SYSTEMS (VEHICLE AND PEDESTRIAN)

Item	Gang hours	Labour £	Plant £	Material £	Unit	Total rate £
PEDESTRIAN GUARD RAILS AND HANDRAILS						
New work						
Tubular galvanized mild steel pedestrian guard rails to BS 7818 with mesh infill (105 swg, 50 × 50 mm mesh); 1.0 m high						
mounted on posts with concrete footing	0.16	17.14	7.99	360.27	m	**385.39**
mounted on posts bolted to structure or ground beam	0.14	15.00	6.99	440.48	m	**462.46**
Solid section galvanized steel pedestrian guard rails with vertical rails (group P4 parapet); 1.0 m high						
mounted on posts with concrete footing	0.19	20.35	9.48	360.27	m	**390.10**
mounted on posts bolted to structure or ground beam	0.17	18.21	8.49	440.48	m	**467.18**
Tubular double ball galvanized steel handrail						
50 mm dia.; 1.20 m high posts	0.15	16.07	7.48	233.83	m	**257.39**
63 mm dia.; 1.20 m high posts	0.15	16.07	7.48	233.83	m	**257.39**
Extra for concrete footings for handrail support posts	0.05	5.36	–	6.09	m	**11.45**
Existing guard rails						
Take from store and re-erect						
pedestrian guard railing, 3.0 m long × 1.0 m high panels	0.15	16.07	7.48	6.99	nr	**30.54**

SERIES 500: DRAINAGE AND SERVICE DUCTS

Item	Gang hours	Labour £	Plant £	Material £	Unit	Total rate £
NOTES						
The re-erection cost for covers and grating complete with frames taken from store assumes that major components are in good condition; the prices below allow a sum of 10% of the value of new materials to cover minor repairs, new fixings and touching up any coatings.						
RESOURCES – LABOUR						
Drains/sewers/culverts gang (small bore)						
1 ganger/chargehand (skill rate 4) – 50% of time	1.00	10.71	–	–	-	**10.71**
1 skilled operative (skill rate 4)	1.00	20.20	–	–	-	**20.20**
2 unskilled operatives (general)	2.00	39.20	–	–	-	**39.20**
1 plant operator (skill rate 3)	1.00	29.17	–	–	-	**29.17**
Total Gang Rate/Hour	1.00	99.28	–	–	-	**99.28**
Drains/sewers/culverts gang (large bore)						
1 ganger/chargehand (skill rate 4) – 50% of time	1.00	10.71	–	–	-	**10.71**
1 skilled operative (skill rate 4)	1.00	20.20	–	–	-	**20.20**
2 unskilled operatives (general)	2.00	39.20	–	–	-	**39.20**
1 plant operator (skill rate 3)	1.00	29.17	–	–	-	**29.17**
1 plant operator (skill rate 3) – 30% of time	0.30	8.75	–	–	-	**8.75**
Total Gang Rate/Hour	1.00	108.03	–	–	-	**108.03**
Gullies gang						
1 ganger/chargehand (skill rate 4) – 50% of time	1.50	20.81	–	–	-	**20.81**
1 plant operator (skill rate 3)	1.00	29.17	–	–	-	**29.17**
2 unskilled operatives (general)	2.00	39.20	–	–	-	**39.20**
Total Gang Rate/Hour	1.00	99.28	–	–	-	**99.28**
RESOURCES – PLANT						
Drains/sewers/culverts gang (small bore)						
0.4 m^3 hydraulic excavator		–	55.59	–	-	**55.59**
2t dumper – 30% of time		–	3.24	–	-	**3.24**
360 mm compaction plate – 30% of time		–	1.85	–	-	**1.85**
2.80 m^3/min compressor, 2 tool – 30% of time		–	4.71	–	-	**4.71**
disc saw – 30% of time		–	0.41	–	-	**0.41**
extra 15ft/50 m hose – 30% of time		–	0.14	–	-	**0.14**
small pump – 30% of time		–	1.67	–	-	**1.67**
Total Rate/Hour		–	67.62	–	-	**67.62**
Note: in addition to the above are the following allowances for trench struts/props/sheeting, assuming the need for closeboarded earth support:		–	–	–	-	**–**
average 1.00 m deep		–	5.93	–	m	**5.93**
average 1.50 m deep		–	6.62	–	m	**6.62**
average 2.00 m deep		–	6.97	–	m	**6.97**
average 2.50 m deep		–	7.13	–	m	**7.13**
average 3.00 m deep		–	10.22	–	m	**10.22**
average 3.50 m deep		–	12.62	–	m	**12.62**
Note: excavation in hard materials as above but with the addition of breaker attachments to the excavator as follows:		–	–	–	-	**–**
generally: BRH91 (141 kg)		–	5.04	–	-	**5.04**
reinforced concrete; BRH125 (310 kg)		–	9.23	–	-	**9.23**

SERIES 500: DRAINAGE AND SERVICE DUCTS

Item	Gang hours	Labour £	Plant £	Material £	Unit	Total rate £
RESOURCES – PLANT – cont						
Drains/sewers/culverts gang (large bore – i.e. greater than 700 mm bore)						
0.4 m^3 hydraulic excavator		–	55.59	–	-	**55.59**
2t dumper – 30% of time		–	3.24	–	-	**3.24**
2.80 m^3/min compressor, 2 tool – 30% of time		–	4.71	–	-	**4.71**
compaction plate/roller – 30% of time		–	1.85	–	-	**1.85**
disc saw – 30% of time		–	0.33	–	-	**0.33**
small pump – 30% of time		–	1.67	–	-	**1.67**
10 t crawler crane – 30% of time		–	20.37	–	-	**20.37**
Total Rate/Hour		–	87.78	–	-	**87.78**
Gullies						
2t dumper – 30% of time		–	3.24	–	-	**3.24**
Stihl disc saw – 30% of time		–	0.33	–	-	**0.33**
Total Rate/Hour		–	3.57	–	-	**3.57**
RESOURCES – MATERIALS						
For the purposes of bedding widths for pipe bedding materials, trenches have been taken as exceeding 1.50 m in depth; trenches to lesser depths are generally 150 mm narrower than those given here so that the rates need to be reduced proportionately.						
DRAINS AND SERVICE DUCTS (EXCLUDING FILTER, NARROW FILTER DRAINS AND FIN DRAINS)						
Vitrified clay pipes to BS 65, plain ends with push-fit polypropylene flexible couplings						
150 mm dia. drain or sewer in trench, depth to invert						
average 1.00 m deep	0.26	25.81	22.22	34.76	m	**82.80**
average 1.50 m deep	0.28	27.80	24.64	34.76	m	**87.20**
average 2.00 m deep	0.31	30.78	28.28	34.76	m	**93.82**
average 2.50 m deep	0.35	34.75	33.23	34.76	m	**102.74**
average 3.00 m deep	0.44	43.68	43.13	34.76	m	**121.58**
average 3.50 m deep	0.51	50.63	53.27	34.76	m	**138.67**
Extra for						
Type N sand bed 650 × 100 mm	0.04	3.97	5.26	2.45	m	**11.69**
Type T sand surround 650 wide × 100 mm	0.08	7.94	10.53	5.88	m	**24.35**
Type F granular bed 650 × 100 mm	0.05	4.96	6.59	2.62	m	**14.17**
Type S granular surround 650 wide × 100 mm	0.16	15.88	21.05	6.25	m	**43.18**
Type A concrete bed 650 × 100 mm	0.11	10.92	13.68	9.47	m	**34.08**
Type B 100 mm concrete bed and haunch	0.24	23.83	29.87	8.09	m	**61.78**
Type Z concrete surround 650 wide × 100 mm	0.22	21.84	27.39	22.70	m	**71.93**
225 mm dia. drain or sewer in trench, depth to invert						
average 1.50 m deep	0.20	19.86	17.85	72.46	m	**110.16**
average 2.00 m deep	0.24	23.83	30.82	72.46	m	**127.10**
average 2.50 m deep	0.28	27.80	34.47	72.46	m	**134.73**
average 3.00 m deep	0.35	34.75	44.35	72.46	m	**151.56**
average 3.50 m deep	0.44	43.68	55.78	72.46	m	**171.92**
average 4.00 m deep	0.56	55.60	70.99	72.46	m	**199.04**

SERIES 500: DRAINAGE AND SERVICE DUCTS

Item	Gang hours	Labour £	Plant £	Material £	Unit	**Total rate £**
Extra for						
Type N sand bed 750 × 150 mm	0.12	11.91	14.93	4.24	m	**31.09**
Type T sand surround 750 wide × 150 mm	0.24	23.83	29.87	10.48	m	**64.18**
Type F granular bed 750 × 150 mm	0.13	12.91	16.18	4.50	m	**33.59**
Type S granular surround 750 wide × 150 mm	0.24	23.83	29.87	11.16	m	**64.85**
Type A concrete bed 750 × 150 mm	0.19	18.86	23.64	16.40	m	**58.91**
Type B 150 mm concrete bed and haunch	0.29	28.79	36.10	12.62	m	**77.51**
Type Z concrete surround 750 wide × 150 mm	0.36	35.74	44.80	40.66	m	**121.20**
300 mm dia. drain or sewer in trench, depth to invert						
average 1.50 m deep	0.21	20.85	27.64	108.68	m	**157.17**
average 2.00 m deep	0.26	25.81	33.41	108.68	m	**167.90**
average 2.50 m deep	0.30	29.78	36.91	108.68	m	**175.37**
average 3.00 m deep	0.37	36.73	46.97	108.68	m	**192.38**
average 4.00 m deep	0.61	60.56	75.93	108.68	m	**245.17**
average 5.00 m deep	1.01	100.27	126.21	108.68	m	**335.16**
Extra for						
Type N sand bed 800 × 150 mm	0.13	12.91	16.18	4.53	m	**33.62**
Type T sand surround 800 × 150 mm	0.26	25.81	29.15	14.58	m	**69.54**
Type F granular bed 800 × 150 mm	0.14	13.90	17.43	4.78	m	**36.11**
Type S granular surround 800 wide × 150 mm	0.26	25.81	32.37	15.20	m	**73.37**
Type A concrete bed 800 × 150 mm	0.20	19.86	24.89	17.50	m	**62.25**
Type B 150 mm concrete bed and haunch	0.36	35.74	44.80	15.94	m	**96.48**
Type Z concrete surround 800 wide × 150 mm	0.22	21.84	66.72	52.21	m	**140.77**
Vitrified clay pipes to BS 65, spigot and socket joints with sealing ring						
400 mm dia. drain or sewer in trench, depth to invert						
average 2.00 m deep	0.31	30.78	39.79	152.64	m	**223.21**
average 2.50 m deep	0.35	34.75	43.08	152.64	m	**230.46**
average 3.00 m deep	0.42	41.70	53.27	152.64	m	**247.61**
average 4.00 m deep	0.66	65.52	82.14	152.64	m	**300.31**
average 5.00 m deep	1.07	106.23	133.95	152.64	m	**392.82**
average 6.00 m deep	1.85	183.67	231.65	152.64	m	**567.96**
Extra for						
Type N sand bed 900 × 150 mm	0.14	13.90	17.43	5.09	m	**36.42**
Type T sand surround 900 wide × 150 mm	0.28	27.80	34.85	20.84	m	**83.48**
Type F granular bed 900 × 150 mm	0.15	14.89	18.67	5.44	m	**38.99**
Type S granular surround 900 × 150 mm	0.28	27.80	34.85	22.18	m	**84.83**
Type A concrete bed 900 × 150 mm	0.21	20.85	26.14	19.66	m	**66.65**
Type B, 150 mm concrete bed and haunch	0.43	42.69	53.51	21.00	m	**117.20**
Type Z concrete surround 900 wide × 150 mm	0.40	39.71	49.78	80.68	m	**170.17**
Concrete pipes with rebated flexible joints to BS 5911–1						
450 mm dia. piped culvert in trench, depth to invert						
average 2.00 m deep	0.33	32.76	42.39	100.65	m	**175.80**
average 3.00 m deep	0.44	43.68	55.60	100.65	m	**199.93**
average 4.00 m deep	0.72	71.48	89.27	100.65	m	**261.40**
average 6.00 m deep	1.65	163.81	199.27	100.65	m	**463.73**

SERIES 500: DRAINAGE AND SERVICE DUCTS

Item	Gang hours	Labour £	Plant £	Material £	Unit	Total rate £
DRAINS AND SERVICE DUCTS (EXCLUDING FILTER, NARROW FILTER DRAINS AND FIN DRAINS) – cont						
Concrete pipes with rebated flexible joints to BS 5911–1 – cont						
Extra for						
Type A concrete bed 1050 × 150 mm	0.24	23.83	29.87	22.94	m	**76.64**
Type Z concrete surround 1050 wide × 150 mm	0.45	44.68	56.01	102.80	m	**203.48**
750 mm dia. piped culvert in trench, depth to invert						
average 2.00 m deep	0.38	41.05	63.87	309.83	m	**414.75**
average 3.00 m deep	0.52	56.18	86.54	309.83	m	**452.55**
average 4.00 m deep	0.92	99.39	151.03	309.83	m	**560.25**
average 6.00 m deep	2.05	221.46	328.97	309.83	m	**860.26**
Extra for						
Type A concrete bed 1250 × 150 mm	0.26	28.09	32.37	31.76	m	**92.22**
Type Z concrete surround 1250 wide × 150 mm	0.50	54.02	62.22	231.01	m	**347.25**
900 mm dia. piped culvert in trench, depth to invert						
average 2.00 m deep	0.43	46.45	72.22	458.32	m	**577.00**
average 3.00 m deep	0.62	66.98	103.52	458.32	m	**628.82**
average 4.00 m deep	1.06	114.51	173.94	458.32	m	**746.77**
average 6.00 m deep	2.25	243.07	360.98	458.32	m	**1062.37**
Extra for						
Type A concrete bed 1500 × 150 mm	0.28	30.25	34.85	37.19	m	**102.29**
Type Z concrete surround 1500 wide × 150 mm	0.55	59.42	68.45	321.22	m	**449.08**
Corrugated steel pipes galvanized, hot dip bitumen coated (Armco type)						
1000 mm dia. piped culvert in trench, Type S granular surround, depth to invert						
average 2.00 m deep	0.99	106.95	165.99	167.22	m	**440.16**
1600 mm dia. piped culvert in trench, Type S granular surround, depth to invert						
average 2.00 m deep	2.13	230.10	357.22	276.99	m	**864.32**
2000 mm dia. piped culvert in trench, Type S granular surround, depth to invert						
average 3.00 m deep	2.97	320.85	498.02	556.38	m	**1375.25**
2200 mm dia. piped culvert in trench, Type S granular surround, depth to invert						
average 3.00 m deep	3.29	355.42	551.46	627.99	m	**1534.87**
Clay cable ducts; Hepduct						
100 mm dia. service duct in trench, Type S granular surround, depth to invert						
average 1.00 m deep	0.14	13.90	17.50	24.98	m	**56.38**
Two 100 mm dia. service ducts in trench, Type S granular surround, depth to invert						
average 1.00 m deep	0.24	23.83	30.31	49.63	m	**103.76**

SERIES 500: DRAINAGE AND SERVICE DUCTS

Item	Gang hours	Labour £	Plant £	Material £	Unit	**Total rate £**
Three 100 mm dia. service ducts in trench, Type S granular surround, depth to invert						
average 1.00 m deep	0.32	31.77	40.19	72.42	m	**144.39**
Four 100 mm dia. service ducts in trench, Type S granular surround, depth to invert						
average 1.00 m deep	0.40	39.71	50.40	98.81	m	**188.92**
Six 100 mm dia. service ducts in trench, Type S granular surround, depth to invert						
average 1.00 m deep	0.60	59.57	75.28	139.87	m	**274.72**
Extra for						
Type Z concrete surround on single duct	0.08	7.94	9.96	12.50	m	**30.40**
Type Z concrete surround on additional ways	0.08	7.94	9.96	12.50	m	**30.40**
150 mm dia. conduit, per way	0.01	0.99	1.25	26.84	m	**29.09**
225 mm dia. conduit, per way	0.01	0.99	1.25	73.81	m	**76.06**
FILTER DRAINS						
Vitrified clay perforated pipes to BS 65, sleeved joints						
150 mm dia. filter drain in trench with Type A bed and Type A fill filter material,						
average 1.00 m deep	0.55	54.60	31.57	56.76	m	**142.93**
average 2.00 m deep	0.60	59.57	36.18	68.16	m	**163.90**
average 3.00 m deep	0.65	64.53	44.56	79.56	m	**188.65**
150 mm pipes with Type A bed and Type B fill, depth						
average 1.00 m deep	0.55	54.60	31.57	68.76	m	**154.93**
average 2.00 m deep	0.59	58.57	36.18	91.34	m	**186.09**
average 3.00 m deep	0.64	63.54	44.56	113.82	m	**221.91**
225 mm pipes with Type A bed and Type A fill, depth						
average 1.00 m deep	0.58	57.58	31.68	68.08	m	**157.34**
average 2.00 m deep	0.62	61.55	37.10	81.95	m	**180.61**
average 3.00 m deep	0.67	66.52	43.93	94.92	m	**205.37**
225 mm pipes with Type A bed and Type B fill, depth						
average 1.00 m deep	0.58	57.58	31.68	80.56	m	**169.82**
average 2.00 m deep	0.63	62.55	37.10	107.49	m	**207.14**
average 3.00 m deep	0.68	67.51	43.93	133.34	m	**244.78**
300 mm pipes with Type A bed and Type A fill, depth						
average 1.00 m deep	0.92	91.34	32.78	120.54	m	**244.66**
average 2.00 m deep	0.96	95.31	38.66	136.49	m	**270.46**
average 3.00 m deep	1.00	99.28	44.86	151.76	m	**295.91**
300 mm pipes with Type A bed and Type B fill, depth						
average 1.00 m deep	0.93	92.33	32.78	131.31	m	**256.42**
average 2.00 m deep	0.96	95.31	38.66	161.56	m	**295.53**
average 3.00 m deep	1.01	100.27	44.86	191.55	m	**336.68**

SERIES 500: DRAINAGE AND SERVICE DUCTS

Item	Gang hours	Labour £	Plant £	Material £	Unit	**Total rate £**
FILTER DRAINS – cont						
Concrete porous pipe BS 5911–114, sleeved joints						
150 mm pipes with Type A bed and Type A fill, depth						
average 1.00 m deep	0.26	25.81	31.57	24.56	m	**81.94**
average 2.00 m deep	0.30	29.78	36.18	35.96	m	**101.92**
average 3.00 m deep	0.35	34.75	44.56	47.37	m	**126.67**
150 mm pipes with Type A bed and Type B fill, depth						
average 1.00 m deep	0.26	25.81	31.57	36.57	m	**93.95**
average 2.00 m deep	0.30	29.78	36.18	59.15	m	**125.11**
average 3.00 m deep	0.35	34.75	44.56	81.62	m	**160.93**
225 mm pipes with Type A bed and Type A fill, depth						
average 1.00 m deep	0.27	26.81	31.68	26.70	m	**85.18**
average 2.00 m deep	0.31	30.78	37.10	40.57	m	**108.45**
average 3.00 m deep	0.36	35.74	43.93	53.54	m	**133.21**
225 mm pipes with Type A bed and Type B fill, depth						
average 1.00 m deep	0.27	26.81	31.68	39.17	m	**97.66**
average 2.00 m deep	0.31	30.78	37.10	66.11	m	**133.98**
average 3.00 m deep	0.36	35.74	43.93	91.95	m	**171.62**
300 mm pipes with Type A bed and Type A fill, depth						
average 1.00 m deep	0.28	27.80	32.78	26.32	m	**86.90**
average 2.00 m deep	0.32	31.77	38.66	42.26	m	**112.70**
average 3.00 m deep	0.37	36.73	44.86	57.54	m	**139.13**
300 mm pipes with Type A bed and Type B fill, depth						
average 1.00 m deep	0.28	27.80	32.78	37.08	m	**97.66**
average 2.00 m deep	0.32	31.77	38.66	67.33	m	**137.76**
average 3.00 m deep	0.37	36.73	44.86	97.32	m	**178.91**
Filter material contiguous with filter drains, subbase material and lightweight aggregate infill						
Type A	0.07	6.95	7.07	25.86	m^3	**39.89**
Type B	0.07	6.95	7.64	51.53	m^3	**66.12**
Excavate and replace filter material contiguous with filter drain						
Type A	0.47	23.53	39.96	25.86	m^3	**89.35**
Type B	0.47	23.53	39.96	51.53	m^3	**115.02**
FIN DRAINS AND NARROW FILTER DRAINS						
Fin Drain DfT type 6 using 'Trammel' drainage fabrics and perforated clay drain; surrounding pipe with sand and granular fill						
100 mm clay perforated pipes, depth						
average 1.00 m deep	0.24	23.83	21.97	24.10	m	**69.89**
average 2.00 m deep	0.31	30.78	29.60	31.84	m	**92.22**

SERIES 500: DRAINAGE AND SERVICE DUCTS

Item	Gang hours	Labour £	Plant £	Material £	Unit	**Total rate £**
Fin Drain DfT type 7 using 'Trammel' drainage fabrics and slotted UPVC drain; surrounding pipe with sand and granular fill, backfilling with selected suitable material						
100 mm UPVC slotted pipes, depth						
average 1.00 m deep	0.17	16.88	21.97	39.43	m	**78.27**
average 2.00 m deep	0.23	22.83	29.60	47.17	m	**99.61**
Narrow Filter Drain DfT type 8 using 'Trammel' drainage fabrics and perforated UPVC drain; surrounding pipe with sand and granular fill, backfilling with granular material						
110 mm UPVC perforated pipes, depth						
average 1.00 m deep	0.17	16.88	21.97	25.16	m	**64.00**
average 2.00 m deep	0.22	21.84	28.48	30.63	m	**80.95**
Narrow Filter Drain DfT type 9 using 'Trammel' drainage fabrics and perforated UPVC drain; surrounding pipe with sand and granular fill, backfilling with granular material						
110 mm UPVC perforated pipes, depth						
average 1.00 m deep	0.17	16.88	21.97	27.90	m	**66.74**
average 2.00 m deep	0.23	22.83	29.60	35.73	m	**88.17**
CONNECTIONS						
Note: excavation presumed covered by new trench.						
Connection of pipe to existing drain, sewer or piped culvert						
150 mm	0.42	41.70	37.15	63.18	nr	**142.03**
225 mm	0.60	59.57	63.93	253.55	nr	**377.05**
300 mm	1.15	114.17	128.21	409.61	nr	**651.99**
Connection of pipes to existing chambers						
150 mm to one brick	1.20	119.13	154.68	4.77	nr	**278.59**
150 mm to precast	0.60	59.57	77.30	4.77	nr	**141.64**
300 mm to one and a half brick	2.40	238.27	309.05	7.79	nr	**555.11**

SERIES 500: DRAINAGE AND SERVICE DUCTS

Item	Gang hours	Labour £	Plant £	Material £	Unit	**Total rate £**
CHAMBERS AND GULLIES						
Notes						
The rates assume the most efficient items of plant (excavator) and are optimum rates, assuming continuous output with no delays caused by other operations or works. Ground conditions are assumed to be good soil with no abnormal conditions that would affect outputs and consistency of work.						
Multiplier Factor Table for Labour and Plant to reflect various site conditions requiring additional works						
out of sequence with other works × 2.75 minimum						
in hard clay × 1.75–2.00						
in running sand × 2.75 minimum						
in broken rock × 2.75–3.50						
below water table × 2.00 minimum						
Brick construction						
Design criteria used in models:						
class A engineering bricks						
215 thick walls generally; 328 thick to chambers exceeding 2.5 m deep						
225 plain concrete C20/20 base slab						
300 reinforced concrete C20/20 reducing slab						
125 reinforced concrete C20/20 top slab						
maximum height of working chamber 2.0 m above benching						
750 × 750 access shaft						
plain concrete C15/20 benching, 150 clay main channel longitudinally and two 100 branch channels						
step irons at 300 mm centres, doubled if depth to invert exceeds 3000 mm						
heavy duty manhole cover and frame						
750 × 700 chamber 500 depth to invert						
excavation, support, backfilling and disposal	0.28	21.50	14.90	–	nr	**36.41**
concrete base	0.40	7.98	2.90	27.67	nr	**38.55**
brickwork chamber	3.81	72.72	10.03	80.85	nr	**163.60**
concrete cover slab	2.16	48.15	35.72	127.25	nr	**211.12**
concrete benching, main and branch channels	6.01	120.22	20.31	212.77	nr	**353.30**
step irons	0.21	4.01	0.56	39.04	nr	**43.62**
access cover and frame	1.44	28.80	3.59	275.50	nr	**307.89**
TOTAL	14.31	303.38	88.02	763.07	nr	**1154.47**

SERIES 500: DRAINAGE AND SERVICE DUCTS

Item	Gang hours	Labour £	Plant £	Material £	Unit	**Total rate £**
750 × 700 chamber 1000 depth to invert						
excavation, support, backfilling and disposal	0.55	42.44	29.59	–	nr	**72.03**
concrete base	0.40	7.98	2.90	27.67	nr	**38.55**
brickwork chamber	7.61	145.44	20.07	161.69	nr	**327.20**
concrete cover slab	2.16	48.15	35.72	127.25	nr	**211.12**
concrete benching, main and branch channels	6.01	120.22	20.31	212.77	nr	**353.30**
step irons	0.32	6.02	0.84	58.56	nr	**65.43**
access cover and frame	0.79	15.84	1.98	151.52	nr	**169.34**
TOTAL	15.61	343.87	105.41	802.11	nr	**1251.39**
750 × 700 chamber 1500 depth to invert						
excavation, support, backfilling and disposal	0.83	63.27	44.20	–	nr	**107.47**
concrete base	0.40	7.98	2.90	27.67	nr	**38.55**
brickwork chamber	12.18	232.70	32.09	242.54	nr	**507.33**
concrete cover slab	2.16	48.15	35.72	127.25	nr	**211.12**
concrete benching, main and branch channels	6.81	136.09	25.96	220.09	nr	**382.14**
step irons	0.42	8.02	1.12	78.09	nr	**87.23**
access cover and frame	0.78	15.55	1.94	148.77	nr	**166.26**
TOTAL	21.33	440.27	107.85	932.03	nr	**1480.14**
900 × 700 chamber 500 depth to invert						
excavation, support, backfilling and disposal	0.33	25.22	17.02	–	nr	**42.24**
concrete base	0.47	9.30	3.38	32.28	nr	**44.97**
brickwork chamber	4.20	80.24	11.07	89.21	nr	**180.52**
concrete cover slab	2.16	48.15	35.72	127.25	nr	**211.12**
concrete benching, main and branch channels	6.81	136.09	25.96	220.09	nr	**382.14**
step irons	0.11	2.01	0.28	19.52	nr	**21.81**
access cover and frame	0.78	15.55	1.94	148.77	nr	**166.26**
TOTAL	11.83	258.77	83.77	695.19	nr	**1037.73**
900 × 700 chamber 1000 depth to invert						
excavation, support, backfilling and disposal	0.66	50.44	34.04	–	nr	**84.47**
concrete base	0.47	9.30	3.38	32.28	nr	**44.97**
brickwork chamber	8.40	160.48	22.14	178.42	nr	**361.05**
concrete cover slab	2.16	48.15	35.72	127.25	nr	**211.12**
concrete benching, main and branch channels	6.81	136.09	25.96	220.09	nr	**382.14**
step irons	0.32	6.02	0.84	58.56	nr	**65.43**
access cover and frame	0.78	15.55	1.94	148.77	nr	**166.26**
TOTAL	16.57	368.24	112.42	823.45	nr	**1304.11**
900 × 700 chamber 1500 depth to invert						
excavation, support, backfilling and disposal	0.98	75.09	50.83	–	nr	**125.92**
concrete base	0.47	9.30	3.38	32.28	nr	**44.97**
brickwork chamber	13.44	256.77	35.41	267.63	nr	**559.81**
concrete cover slab	2.16	48.15	35.72	127.25	nr	**211.12**
concrete benching, main and branch channels	6.81	136.09	25.96	220.09	nr	**382.14**
step irons	0.42	8.02	1.12	78.09	nr	**87.23**
access cover and frame	0.78	15.55	1.94	148.77	nr	**166.26**
TOTAL	21.84	487.51	142.26	932.18	nr	**1561.95**

SERIES 500: DRAINAGE AND SERVICE DUCTS

Item	Gang hours	Labour £	Plant £	Material £	Unit	Total rate £
CHAMBERS AND GULLIES – cont						
Brick construction – cont						
1050 × 700 chamber 1500 depth to invert						
excavation, support, backfilling and disposal	1.91	145.34	95.95	–	nr	**241.29**
concrete base	0.57	11.30	4.11	39.20	nr	**54.61**
brickwork chamber	14.70	280.85	38.73	292.72	nr	**612.29**
concrete cover slab	2.52	56.17	41.64	135.98	nr	**233.79**
concrete benching, main and branch channels	8.51	169.82	36.76	225.60	nr	**432.18**
step irons	0.84	16.05	2.25	156.17	nr	**174.47**
access cover and frame	0.78	15.55	1.94	148.77	nr	**166.26**
TOTAL	26.65	634.30	209.16	1060.40	nr	**1903.86**
1050 × 700 chamber 2500 depth to invert						
excavation, support, backfilling and disposal	1.99	153.20	117.73	–	nr	**270.93**
concrete base	0.57	11.30	4.11	39.20	nr	**54.61**
brickwork chamber and access shaft (700 × 700)	35.46	677.54	93.45	676.99	nr	**1447.98**
concrete cover slab	2.52	56.17	41.64	135.98	nr	**233.79**
concrete reducing slab	3.43	75.41	48.15	170.94	nr	**294.50**
concrete benching, main and branch channels	8.51	169.82	36.76	225.60	nr	**432.18**
step irons	0.84	16.05	2.25	156.17	nr	**174.47**
access cover and frame	0.78	15.55	1.94	148.77	nr	**166.26**
TOTAL	47.34	1036.05	285.27	1445.27	nr	**2766.59**
1050 × 700 chamber 3500 depth to invert						
excavation, support, backfilling and disposal	2.88	224.10	231.54	–	nr	**455.64**
concrete base	0.57	11.30	4.11	39.20	nr	**54.61**
brickwork chamber and access shaft (700 × 700)	45.57	870.62	120.07	872.14	nr	**1862.83**
concrete reducing slab	3.43	75.41	48.15	170.94	nr	**294.50**
concrete cover slab	2.52	56.17	41.64	135.98	nr	**233.79**
concrete benching, main and branch channels	8.44	168.39	36.27	225.10	nr	**429.76**
step irons	1.16	22.07	3.09	214.74	nr	**239.89**
access cover and frame	0.78	15.55	1.94	148.77	nr	**166.26**
TOTAL	58.65	1306.05	426.55	1698.98	nr	**3431.57**
1350 × 700 chamber 2500 depth to invert STOP						
excavation, support, backfilling and disposal	2.42	183.95	117.75	–	nr	**301.70**
concrete base	0.70	13.96	5.07	48.42	nr	**67.45**
brickwork chamber and access shaft (700 × 700)	41.54	793.69	109.47	793.05	nr	**1696.21**
concrete reducing slab	3.43	75.41	48.15	170.94	nr	**294.50**
concrete cover slab	2.80	62.42	46.30	169.15	nr	**277.86**
concrete benching, main and branch channels	9.62	191.74	44.29	232.86	nr	**468.89**
step irons	0.84	16.05	2.25	156.17	nr	**174.47**
access cover and frame	0.78	15.55	1.94	148.77	nr	**166.26**
TOTAL	55.42	1214.66	314.78	1611.34	nr	**3140.78**
1350 × 700 chamber 3500 depth to invert						
excavation, support, backfilling and disposal	3.65	283.66	279.79	–	nr	**563.44**
concrete base	0.70	13.96	5.07	48.42	nr	**67.45**
brickwork chamber and access shaft (700 × 700)	54.89	1048.66	144.64	1021.65	nr	**2214.95**
concrete reducing slab	3.43	75.41	48.15	170.94	nr	**294.50**
concrete cover slab	2.80	62.42	46.30	169.15	nr	**277.86**
concrete benching, main and branch channels	9.57	190.80	43.97	232.53	nr	**467.30**
step irons	1.16	22.07	3.09	214.74	nr	**239.89**
access cover and frame	0.78	15.55	1.94	148.77	nr	**166.26**
TOTAL	73.61	1647.87	559.14	2062.87	nr	**4269.87**

SERIES 500: DRAINAGE AND SERVICE DUCTS

Item	Gang hours	Labour £	Plant £	Material £	Unit	**Total rate £**
1350 × 700 chamber 4500 depth to invert						
excavation, support, backfilling and disposal	4.69	364.50	359.66	–	nr	**724.16**
concrete base	0.70	13.96	5.07	48.42	nr	**67.45**
brickwork chamber and access shaft (700 × 700)	67.37	1287.18	177.55	1250.25	nr	**2714.98**
concrete reducing slab	3.43	75.41	48.15	170.94	nr	**294.50**
concrete cover slab	2.80	62.42	46.30	169.15	nr	**277.86**
concrete benching, main and branch channels	9.57	190.80	43.97	232.53	nr	**467.30**
step irons	1.47	28.09	3.94	273.30	nr	**305.32**
access cover and frame	0.78	15.55	1.94	148.77	nr	**166.26**
TOTAL	87.45	1973.25	672.76	2350.04	nr	**4996.05**
Precast concrete construction						
Design criteria used in models:						
circular shafts						
150 plain concrete surround						
225 plain concrete C20/20 base slab						
precast top slab						
maximum height of working chamber 2.0 m above benching						
plain concrete C15/20 benching, 150 clay main channel longitudinally and two 100 branch channels						
step irons at 300 mm centres, doubled if depth to invert exceeds 3000 mm						
heavy duty manhole cover and frame						
in manholes over 6 m deep, landings at maximum intervals						
675 dia. × 500 depth to invert						
excavation, support, backfilling and disposal	0.19	14.92	10.58	–	nr	**25.51**
concrete base	0.27	5.32	1.93	18.45	nr	**25.70**
main chamber rings	1.60	35.67	26.45	32.69	nr	**94.80**
cover slab	1.98	43.76	30.43	74.73	nr	**148.92**
concrete surround	0.37	7.30	2.98	18.41	nr	**28.69**
concrete benching, main and branch channels	2.38	47.44	10.59	112.04	nr	**170.06**
step irons	0.11	2.01	0.28	19.52	nr	**21.81**
access cover and frame	0.78	15.55	1.94	148.77	nr	**166.26**
TOTAL	6.32	141.56	61.30	467.51	nr	**670.37**
675 dia. × 750 depth to invert						
excavation, support, backfilling and disposal	0.28	21.82	15.65	–	nr	**37.47**
concrete base	0.27	5.32	1.93	18.45	nr	**25.70**
main chamber rings	2.40	53.50	39.67	49.03	nr	**142.20**
cover slab	1.98	43.76	30.43	74.73	nr	**148.92**
concrete surround	0.55	10.95	4.47	27.62	nr	**43.04**
concrete benching, main and branch channels	2.38	47.44	10.59	112.04	nr	**170.06**
step irons	0.21	4.01	0.56	39.04	nr	**43.62**
access cover and frame	0.78	15.55	1.94	148.77	nr	**166.26**
TOTAL	9.51	215.59	106.90	596.41	nr	**918.90**

SERIES 500: DRAINAGE AND SERVICE DUCTS

Item	Gang hours	Labour £	Plant £	Material £	Unit	Total rate £
CHAMBERS AND GULLIES – cont						
Precast concrete construction – cont						
675 dia. × 1000 depth to invert						
excavation, support, backfilling and disposal	0.38	29.28	20.94	–	nr	**50.23**
concrete base	0.27	5.32	1.93	18.45	nr	**25.70**
main chamber rings	3.20	71.33	52.89	65.37	nr	**189.60**
cover slab	1.98	43.76	30.43	74.73	nr	**148.92**
concrete surround	0.74	14.60	5.96	36.83	nr	**57.39**
concrete benching, main and branch channels	2.38	47.44	10.59	112.04	nr	**170.06**
step irons	0.32	6.02	0.84	58.56	nr	**65.43**
access cover and frame	0.78	15.55	1.94	148.77	nr	**166.26**
TOTAL	10.70	246.54	127.18	641.48	nr	**1015.20**
675 dia. × 1250 depth to invert						
excavation, support, backfilling and disposal	0.48	36.74	26.24	–	nr	**62.98**
concrete base	0.27	5.32	1.93	18.45	nr	**25.70**
main chamber rings	4.00	89.16	66.11	81.72	nr	**236.99**
cover slab	1.98	43.76	30.43	74.73	nr	**148.92**
concrete surround	0.92	18.25	7.45	46.03	nr	**71.73**
concrete benching, main and branch channels	2.38	47.44	10.59	112.04	nr	**170.06**
step irons	0.42	8.02	1.12	78.09	nr	**87.23**
access cover and frame	0.78	15.55	1.94	148.77	nr	**166.26**
TOTAL	11.89	277.49	147.47	686.55	nr	**1111.51**
900 dia. × 750 depth to invert						
excavation, support, backfilling and disposal	0.49	37.20	24.03	–	nr	**61.22**
concrete base	0.47	9.30	3.38	32.28	nr	**44.97**
main chamber rings	2.85	63.53	47.10	112.76	nr	**223.39**
cover slab	3.80	84.71	62.80	150.34	nr	**297.85**
concrete surround	0.74	14.60	5.96	36.83	nr	**57.39**
concrete benching, main and branch channels	3.38	67.28	17.92	120.80	nr	**205.99**
step irons	0.21	4.01	0.56	39.04	nr	**43.62**
access cover and frame	0.78	15.55	1.94	148.77	nr	**166.26**
TOTAL	13.37	309.43	165.33	767.55	nr	**1242.31**
900 dia. × 1000 depth to invert						
excavation, support, backfilling and disposal	0.65	49.52	32.00	–	nr	**81.53**
concrete base	0.47	9.30	3.38	32.28	nr	**44.97**
main chamber rings	3.80	84.71	62.80	150.34	nr	**297.85**
cover slab	3.80	84.71	62.80	150.34	nr	**297.85**
concrete surround	0.97	19.17	7.82	48.33	nr	**75.32**
concrete benching, main and branch channels	3.38	67.28	17.92	120.80	nr	**205.99**
step irons	0.32	6.02	0.84	58.56	nr	**65.43**
access cover and frame	0.78	15.55	1.94	148.77	nr	**166.26**
TOTAL	14.82	349.50	191.15	836.16	nr	**1376.82**

SERIES 500: DRAINAGE AND SERVICE DUCTS

Item	Gang hours	Labour £	Plant £	Material £	Unit	**Total rate £**
900 dia. × 1500 depth to invert						
excavation, support, backfilling and disposal	0.99	74.96	48.27	–	nr	**123.23**
concrete base	0.47	9.30	3.38	32.28	nr	**44.97**
main chamber rings	5.70	127.06	94.19	225.52	nr	**446.77**
cover slab	2.16	48.15	35.72	127.25	nr	**211.12**
concrete surround	1.47	29.20	11.92	73.65	nr	**114.77**
concrete benching, main and branch channels	3.38	67.28	17.92	120.80	nr	**205.99**
step irons	0.42	8.02	1.12	78.09	nr	**87.23**
access cover and frame	0.78	15.55	1.94	148.77	nr	**166.26**
TOTAL	16.03	392.77	216.12	933.08	nr	**1541.98**
1200 dia. × 1500 depth to invert						
excavation, support, backfilling and disposal	1.70	128.56	75.52	–	nr	**204.08**
concrete base	0.84	16.62	6.04	57.65	nr	**80.30**
main chamber rings	7.02	156.48	116.04	225.29	nr	**497.81**
cover slab	2.80	62.42	46.30	169.15	nr	**277.86**
concrete surround	1.96	38.79	15.83	97.82	nr	**152.43**
concrete benching, main and branch channels	6.13	121.84	36.09	135.03	nr	**292.96**
step irons	0.42	8.02	1.12	78.09	nr	**87.23**
access cover and frame	0.78	15.55	1.94	148.77	nr	**166.26**
TOTAL	22.31	561.52	300.54	1038.51	nr	**1900.57**
1200 dia. × 2000 depth to invert						
excavation, support, backfilling and disposal	2.35	179.18	120.19	–	nr	**299.37**
concrete base	0.84	16.62	6.04	57.65	nr	**80.30**
main chamber rings	9.36	208.65	154.73	300.38	nr	**663.75**
cover slab	2.80	62.42	46.30	169.15	nr	**277.86**
concrete surround	2.60	51.56	21.04	130.04	nr	**202.64**
concrete benching, main and branch channels	6.13	121.84	36.09	135.03	nr	**292.96**
step irons	0.63	12.04	1.69	117.13	nr	**130.85**
access cover and frame	0.78	15.55	1.94	148.77	nr	**166.26**
TOTAL	26.14	681.09	389.67	1184.87	nr	**2255.63**
1200 dia. × 2500 depth to invert						
excavation, support, backfilling and disposal	2.94	224.10	150.32	–	nr	**374.42**
concrete base	0.84	16.62	6.04	57.65	nr	**80.30**
main chamber rings	11.70	260.81	193.41	375.48	nr	**829.69**
cover slab	2.80	62.42	46.30	169.15	nr	**277.86**
concrete surround	3.24	64.34	26.25	162.26	nr	**252.85**
concrete benching, main and branch channels	6.13	121.84	36.09	135.03	nr	**292.96**
step irons	0.84	16.05	2.25	156.17	nr	**174.47**
access cover and frame	0.78	15.55	1.94	148.77	nr	**166.26**
TOTAL	29.93	794.96	464.25	1331.23	nr	**2590.44**
1200 dia. × 3000 depth to invert						
excavation, support, backfilling and disposal	3.53	269.12	180.44	–	nr	**449.56**
concrete base	0.84	16.62	6.04	57.65	nr	**80.30**
main chamber rings	14.04	312.97	232.09	450.57	nr	**995.63**
cover slab	2.80	62.42	46.30	169.15	nr	**277.86**
concrete surround	3.91	77.57	31.65	195.63	nr	**304.86**
concrete benching, main and branch channels	6.13	121.84	36.09	135.03	nr	**292.96**
step irons	0.95	18.05	2.53	175.69	nr	**196.28**
access cover and frame	0.78	15.55	1.94	148.77	nr	**166.26**
TOTAL	33.63	907.38	538.73	1459.22	nr	**2905.33**

SERIES 500: DRAINAGE AND SERVICE DUCTS

Item	Gang hours	Labour £	Plant £	Material £	Unit	**Total rate £**
CHAMBERS AND GULLIES – cont						
Precast concrete construction – cont						
1800 dia. × 2000 depth to invert						
excavation, support, backfilling and disposal	4.97	371.89	195.68	–	nr	**567.57**
concrete base	1.91	37.88	13.77	131.43	nr	**183.09**
main chamber rings	12.00	267.49	198.38	791.00	nr	**1256.88**
cover slab	3.60	80.25	59.52	456.04	nr	**595.81**
concrete surround	3.91	77.57	31.65	195.63	nr	**304.86**
concrete benching, main and branch channels	7.38	146.69	42.88	153.35	nr	**342.93**
step irons	0.63	12.04	1.69	117.13	nr	**130.85**
access cover and frame	0.78	15.55	1.94	148.77	nr	**166.26**
TOTAL	35.84	1022.62	547.17	2120.09	nr	**3689.88**
1800 dia. × 2500 depth to invert						
excavation, support, backfilling and disposal	6.42	485.15	284.67	–	nr	**769.82**
concrete base	1.91	37.88	13.77	131.43	nr	**183.09**
main chamber rings	15.00	334.37	247.98	988.75	nr	**1571.10**
cover slab	3.60	80.25	59.52	456.04	nr	**595.81**
concrete surround	4.88	96.74	39.47	243.97	nr	**380.18**
concrete benching, main and branch channels	7.38	146.69	42.88	153.35	nr	**342.93**
step irons	0.84	16.05	2.25	156.17	nr	**174.47**
access cover and frame	0.78	15.55	1.94	148.77	nr	**166.26**
TOTAL	41.46	1225.93	694.14	2405.22	nr	**4325.28**
1800 dia. × 3000 depth to invert						
excavation, support, backfilling and disposal	7.70	582.36	341.62	–	nr	**923.98**
concrete base	1.91	37.88	13.77	131.43	nr	**183.09**
main chamber rings	18.00	401.24	297.58	1186.51	nr	**1885.32**
cover slab	3.60	80.25	59.52	456.04	nr	**595.81**
concrete surround	5.84	115.90	47.29	292.30	nr	**455.49**
concrete benching, main and branch channels	7.38	146.69	42.88	153.35	nr	**342.93**
step irons	0.95	18.05	2.53	175.69	nr	**196.28**
access cover and frame	0.78	15.55	1.94	148.77	nr	**166.26**
TOTAL	46.82	1411.18	808.78	2670.82	nr	**4890.79**
1800 dia. × 3500 depth to invert						
excavation, support, backfilling and disposal	9.26	706.81	513.76	–	nr	**1220.57**
concrete base	1.91	37.88	13.77	131.43	nr	**183.09**
main chamber rings	21.00	468.12	347.17	1384.26	nr	**2199.54**
cover slab	3.60	80.25	59.52	456.04	nr	**595.81**
concrete surround	6.83	135.52	55.30	341.78	nr	**532.60**
concrete benching, main and branch channels	7.38	146.69	42.88	153.35	nr	**342.93**
step irons	1.16	22.07	3.09	214.74	nr	**239.89**
access cover and frame	0.78	15.55	1.94	148.77	nr	**166.26**
TOTAL	52.58	1626.14	1039.10	2957.10	nr	**5622.34**

SERIES 500: DRAINAGE AND SERVICE DUCTS

Item	Gang hours	Labour £	Plant £	Material £	Unit	**Total rate £**
1800 dia. × 4000 depth to invert						
excavation, support, backfilling and disposal	10.58	807.31	586.99	–	nr	**1394.29**
concrete base	1.91	37.88	13.77	131.43	nr	**183.09**
main chamber rings	24.00	534.99	396.77	1582.01	nr	**2513.76**
cover slab	3.60	80.25	59.52	456.04	nr	**595.81**
concrete surround	7.80	154.69	63.12	390.12	nr	**607.92**
concrete benching, main and branch channels	7.38	146.69	42.88	153.35	nr	**342.93**
step irons	1.37	26.08	3.65	253.78	nr	**283.51**
access cover and frame	0.78	15.55	1.94	148.77	nr	**166.26**
TOTAL	58.07	1816.69	1170.30	3242.23	nr	**6229.21**
2400 dia. × 3000 depth to invert						
excavation, support, backfilling and disposal	4.69	465.59	514.41	–	nr	**979.99**
concrete base	3.42	67.79	24.64	235.20	nr	**327.63**
main chamber rings	21.60	596.44	1114.10	3392.23	nr	**5102.77**
cover slab	4.40	121.50	226.97	1293.29	nr	**1641.75**
concrete surround	7.80	154.69	63.12	390.12	nr	**607.92**
concrete benching, main and branch channels	7.88	156.79	42.88	153.35	nr	**353.03**
step irons	0.95	18.05	2.53	175.69	nr	**196.28**
access cover and frame	0.78	15.55	1.94	148.77	nr	**166.26**
TOTAL	52.17	1609.65	1992.24	5915.37	nr	**9517.26**
2400 dia. × 4500 depth to invert						
excavation, support, backfilling and disposal	20.82	1581.04	1030.46	–	nr	**2611.51**
concrete base	3.42	67.79	24.64	235.20	nr	**327.63**
main chamber rings	32.40	894.66	1671.15	5088.35	nr	**7654.15**
cover slab	4.40	121.50	226.97	1293.29	nr	**1641.75**
concrete surround	11.71	232.26	94.77	585.75	nr	**912.78**
concrete benching, main and branch channels	7.88	156.79	42.88	153.35	nr	**353.03**
step irons	1.47	28.09	3.94	273.30	nr	**305.32**
access cover and frame	0.78	15.55	1.94	148.77	nr	**166.26**
TOTAL	83.54	3110.93	3098.40	7904.73	nr	**14114.06**
2700 dia. × 3000 depth to invert						
excavation, support, backfilling and disposal	16.98	1275.73	672.35	–	nr	**1948.08**
concrete base	4.32	85.74	31.17	297.45	nr	**414.36**
main chamber rings	24.00	662.71	1237.89	3660.54	nr	**5561.14**
cover slab	4.80	132.54	247.58	2601.54	nr	**2981.66**
concrete surround	8.79	174.31	71.12	439.60	nr	**685.03**
concrete benching, main and branch channels	8.08	160.83	42.88	153.35	nr	**357.07**
step irons	0.95	18.05	2.53	175.69	nr	**196.28**
access cover and frame	0.78	15.55	1.94	148.77	nr	**166.26**
TOTAL	69.35	2538.72	2309.10	7603.68	nr	**12451.51**
2700 dia. × 4500 depth to invert						
excavation, support, backfilling and disposal	26.21	1986.65	1243.54	–	nr	**3230.19**
concrete base	4.30	85.28	31.01	296.30	nr	**412.59**
main chamber rings	36.00	994.06	1856.83	5490.81	nr	**8341.71**
cover slab	4.80	132.54	247.58	2601.54	nr	**2981.66**
concrete surround	13.18	261.47	106.68	659.40	nr	**1027.55**
concrete benching, main and branch channels	8.08	160.83	42.88	153.35	nr	**357.07**
step irons	1.47	28.09	3.94	273.30	nr	**305.32**
access cover and frame	0.78	15.55	1.94	148.77	nr	**166.26**
TOTAL	95.48	3677.71	3536.05	9750.21	nr	**16963.98**

SERIES 500: DRAINAGE AND SERVICE DUCTS

Item	Gang hours	Labour £	Plant £	Material £	Unit	Total rate £
CHAMBERS AND GULLIES – cont						
Precast concrete construction – cont						
3000 dia. × 3000 depth to invert						
excavation, support, backfilling and disposal	20.88	1566.67	806.28	–	nr	**2372.95**
concrete base	5.33	105.68	38.42	366.63	nr	**510.72**
main chamber rings	25.80	712.41	1330.69	4692.04	nr	**6735.15**
cover slab	0.60	18.05	49.96	2482.94	nr	**2550.96**
concrete surround	9.75	193.48	78.94	487.93	nr	**760.35**
concrete benching, main and branch channels	8.88	176.69	42.88	153.35	nr	**372.93**
step irons	0.95	18.05	2.53	175.69	nr	**196.28**
access cover and frame	0.78	15.55	1.94	148.77	nr	**166.26**
TOTAL	73.63	2819.83	2353.30	8634.09	nr	**13807.21**
3000 dia. × 4500 depth to invert						
excavation, support, backfilling and disposal	32.21	2438.13	1475.19	–	nr	**3913.32**
concrete base	5.33	105.68	38.42	366.63	nr	**510.72**
main chamber rings	38.70	1068.62	1996.04	7038.06	nr	**10102.72**
cover slab	0.60	18.05	49.96	2482.94	nr	**2550.96**
concrete surround	14.63	290.21	118.41	731.90	nr	**1140.53**
concrete benching, main and branch channels	8.88	176.69	42.88	153.35	nr	**372.93**
step irons	1.47	28.09	3.94	273.30	nr	**305.32**
access cover and frame	0.78	15.55	1.94	148.77	nr	**166.26**
TOTAL	103.26	4154.26	3728.43	11321.68	nr	**19204.38**
3000 dia. × 6000 depth to invert						
excavation, support, backfilling and disposal	42.95	3250.64	1966.88	–	nr	**5217.52**
concrete base	5.33	105.68	38.42	366.63	nr	**510.72**
main chamber rings	51.60	1424.83	2661.39	9384.08	nr	**13470.29**
cover slab	0.60	18.05	49.96	2482.94	nr	**2550.96**
concrete surround	19.50	386.95	157.88	975.87	nr	**1520.70**
concrete benching, main and branch channels	8.88	176.69	42.88	153.35	nr	**372.93**
step irons	2.00	38.12	5.34	370.91	nr	**414.36**
access cover and frame	0.78	15.55	1.94	148.77	nr	**166.26**
TOTAL	132.29	5429.75	4926.34	14009.28	nr	**24365.37**
Precast concrete circular manhole; CPM Perfect Manhole; complete with preformed benching and outlets to base; elastomeric seal to joints to rings; single steps as required. Excavations and backfilling not included.						
1200 mm dia. with up to four outlets 100 mm or 150 mm dia.; effective internal depth:						
1250 mm deep	2.25	139.41	41.79	992.20	nr	**1173.40**
1500 mm deep	2.25	129.32	41.79	1133.38	nr	**1304.48**
1800 mm deep	2.50	145.09	41.79	1192.67	nr	**1379.55**
2000 mm deep	2.50	145.09	41.79	1372.78	nr	**1559.66**
2500 mm deep	2.75	157.71	52.23	1573.25	nr	**1783.19**

SERIES 500: DRAINAGE AND SERVICE DUCTS

Item	Gang hours	Labour £	Plant £	Material £	Unit	**Total rate £**
1500 mm dia. with up to four outlets up to 450 mm dia.; effective internal depth:						
2000 mm deep	2.75	157.71	52.23	2884.69	nr	**3094.63**
2500 mm deep	3.05	174.74	58.50	3218.91	nr	**3452.14**
3000 mm deep	3.10	177.89	58.50	3285.66	nr	**3522.06**
3500 mm deep	3.40	194.29	66.86	3435.60	nr	**3696.75**
4000 mm deep	3.55	203.12	68.95	3894.79	nr	**4166.86**
4500 mm deep	3.55	203.12	68.95	4229.00	nr	**4501.07**
1500 mm dia. with up to four outlets up to 600 mm dia.; effective internal depth:						
4000 mm deep	3.95	227.73	73.13	3783.62	nr	**4084.47**
4500 mm deep	4.50	258.64	83.57	4020.68	nr	**4362.89**
Gullies						
Vitrified clay; set in concrete grade C20, 150 mm thick; additional excavation and disposal						
Road gully 450 mm dia. × 900 mm deep, 100 mm or 150 mm outlet; cast iron road gulley grating and frame group 4, 434 × 434 mm, on Class B engineering brick seating	0.50	49.64	1.78	846.08	nr	**897.51**
Yard gully (mud); trapped with rodding eye; galvanized bucket; stopper, 225 mm dia., 100 mm dia. outlet, cast iron hinged grate and frame	0.30	29.78	1.11	673.02	nr	**703.92**
Grease interceptors; internal access and bucket 600 × 450 mm, metal tray and lid, square hopper with horizontal inlet	0.35	34.75	1.25	2424.77	nr	**2460.77**
Precast concrete; set in concrete grade C20P, 150 mm thick; additional excavation and disposal						
Road gully; trapped with rodding eye; galvanized bucket; stopper 450 mm dia. × 900 mm deep, cast iron road gully grating and frame group 4, 434 × 434 mm, on Class B engineering brick seating	0.54	53.61	1.93	749.46	nr	**805.00**
SOFT SPOTS AND OTHER VOIDS						
Excavation of soft spots and other voids in bottom of trenches, chambers and gullies	0.07	6.95	8.76	–	m^3	**15.71**
Filling of soft spots and other voids in bottom of trenches, chambers and gullies with imported selected sand	0.30	29.78	10.65	26.61	m^3	**67.04**
Filling of soft spots and other voids in bottom of trenches, chambers and gullies with imported natural gravel	0.20	19.86	10.65	43.67	m^3	**74.17**
Filling of soft spots and other voids in bottom of trenches, chambers and gullies with concrete Grade C15, 40 mm aggregate	0.09	8.94	10.54	104.21	m^3	**123.68**
Filling of soft spots and other voids in bottom of trenches, chambers and gullies with concrete Grade C20, 20 mm aggregate	0.09	8.94	10.54	115.51	m^3	**134.98**

SERIES 500: DRAINAGE AND SERVICE DUCTS

Item	Gang hours	Labour £	Plant £	Material £	Unit	Total rate £
SUPPORTS LEFT IN EXCAVATION						
Timber closeboarded supports left in						
trench		–	9.86	–	m²	**9.86**
pits		–	9.86	–	m²	**9.86**
Steel trench sheeting supports left in						
trench		–	46.04	–	m²	**46.04**
pits		–	46.04	–	m²	**46.04**
RENEWAL, RAISING OR LOWERING OF COVERS AND GRATINGS ON EXISTING CHAMBERS						
Raising the level of 100 × 100 mm water stop tap boxes on 100 × 100 mm brick chambers						
by 150 mm or less	0.06	5.96	0.73	26.59	nr	**33.27**
Lowering the level of 100 × 100 mm water stop tap boxes on 100 × 100 mm brick chambers						
by 150 mm or less	0.04	3.97	0.50	26.59	nr	**31.06**
Raising the level of 420 × 420 mm cover and frame on 420 × 420 mm in situ concrete chamber						
by 150 mm or less	0.10	9.93	1.28	26.59	nr	**37.79**
Lowering the level of 420 × 420 mm British Telecom cover and frame on 420 × 420 mm in situ concrete chamber						
by 150 mm or less	0.08	7.45	3.60	26.59	nr	**37.63**
Raising the level of 700 × 500 mm cover and frame on 700 × 500 mm in situ concrete chamber						
by 150 mm or less	0.17	16.88	2.14	48.84	nr	**67.86**
Raising the level of 600 × 600 mm grade A heavy duty manhole cover and frame on 600 × 600 mm brick chamber						
by 150 mm or less	0.17	16.88	2.14	48.84	nr	**67.86**
by 150–300 mm	0.21	20.85	2.64	71.32	nr	**94.81**
Lowering the level of 600 × 600 mm grade A heavy duty manhole cover and frame on 600 × 600 mm brick chamber						
by 150 mm or less	0.10	9.93	1.28	48.84	nr	**60.04**
REMOVE FROM STORE AND REINSTALL COVERS						
600 × 600 mm; Group 5; super heavy duty E600 cast iron	0.25	24.82	0.97	270.93	nr	**296.73**
600 × 600 mm; Group 4; heavy duty triangular D400 cast iron	0.25	24.82	0.97	283.21	nr	**309.00**
600 × 600 × 75 mm; Group 2; light duty single seal B125 cast iron	0.25	24.82	0.97	298.73	nr	**324.53**
600 × 600 × 100 mm; Group 2; medium duty single seal B125 cast iron	0.25	24.82	0.97	308.00	nr	**333.79**

SERIES 500: DRAINAGE AND SERVICE DUCTS

Item	Gang hours	Labour £	Plant £	Material £	Unit	**Total rate £**
GROUTING UP OF EXISTING DRAINS AND SERVICE DUCTS						
Concrete grade C15						
Sealing redundant road gullies	0.02	1.99	0.20	17.84	m^3	**20.03**
Filling redundant chambers with						
ne 1.0 m deep to invert	0.09	8.94	8.50	71.58	nr	**89.01**
1.0–2.0 m deep to invert	0.21	20.85	19.81	115.68	nr	**156.34**
2.0–3.0 m deep to invert	0.55	54.60	51.87	192.76	nr	**299.23**
Grouting up of existing drains and service ducts						
100 mm dia.	0.03	2.98	2.87	5.91	m	**11.76**
300 mm dia.	0.13	12.91	12.49	22.35	m	**47.75**
450 mm dia.	0.26	25.81	24.98	44.41	m	**95.20**
600 mm dia.	0.50	49.64	47.98	89.65	m	**187.27**
1200 mm dia.	1.70	168.77	163.22	250.15	m	**582.14**
EXCAVATION IN HARD MATERIAL						
Extra over excavation for excavation in hard material in drainage:						
existing pavement, brickwork, concrete, masonry and the like	0.15	14.89	19.12	–	m^3	**34.01**
rock	0.35	34.75	44.63	–	m^3	**79.38**
reinforced concrete	0.60	59.57	79.07	–	m^3	**138.63**
Reinstatement of pavement construction; extra over excavation for breaking up and subsequently reinstating 150 mm flexible surfacing and 280 mm subbase						
100 mm dia. sewer, drain or service duct	0.09	8.94	11.90	17.64	m	**38.48**
150 mm dia. sewer, drain or service duct	0.10	9.93	13.23	18.82	m	**41.98**
225 mm dia. sewer, drain or service duct	0.10	9.93	13.23	20.59	m	**43.75**
300 mm dia. sewer, drain or service duct	0.10	9.93	13.23	21.17	m	**44.33**
375 mm dia. sewer, drain or service duct	0.10	9.93	13.23	21.76	m	**44.92**
450 mm dia. sewer, drain or service duct	0.12	11.91	14.51	22.94	m	**49.37**
2 way 100 mm dia. service ducts	0.10	9.93	13.23	18.24	m	**41.39**
3 way 100 mm dia. service ducts	0.10	9.93	13.23	18.82	m	**41.98**
4 way 100 mm dia. service ducts	0.10	9.93	13.23	20.59	m	**43.75**
MANHOLES						
Materials						
In order to arrive at an all in rate for enumerated manholes the various constituent elements have been broken down and unit rates given against each item.						
Earthworks						
Excavation by machine						
ne 2.0 m deep	0.22	21.84	13.72	–	m^3	**35.56**
2.0–3.0 m deep	0.24	23.83	14.97	–	m^3	**38.79**
3.0–4.0 m deep	0.26	25.81	16.23	–	m^3	**42.04**

SERIES 500: DRAINAGE AND SERVICE DUCTS

Item	Gang hours	Labour £	Plant £	Material £	Unit	Total rate £
MANHOLES – cont						
Earthworks – cont						
Backfilling						
ne 2.0 m deep	0.11	10.92	6.85	–	m^3	**17.77**
2.0–3.0 m deep	0.11	10.92	6.85	–	m^3	**17.77**
3.0–4.0 m deep	0.11	10.92	6.85	–	m^3	**17.77**
Earthwork support						
ne 2.0 m deep	0.01	0.70	4.38	–	m^2	**5.08**
2.0–3.0 m deep	0.01	1.09	6.55	–	m^2	**7.65**
3.0–4.0 m deep	0.01	1.39	11.69	–	m^2	**13.08**
Disposal of excavated material						
off site	0.55	34.70	8.29	–	m^3	**42.99**
to spoil heap average 50 m distant	0.19	5.00	2.05	–	m^3	**7.05**
Mass concrete grade C15						
Blinding; thickness						
ne 150 mm deep	1.05	20.83	8.53	115.08	m^3	**144.44**
Surrounds to manholes; thickness						
ne 150 mm thick	2.30	45.63	18.62	115.08	m^3	**179.33**
150–500 mm thick	1.70	33.73	13.76	115.08	m^3	**162.57**
Benching to bottom of manhole						
675 mm dia.	1.00	19.84	6.79	9.55	nr	**36.19**
900 mm dia.	2.00	39.68	14.13	18.31	nr	**72.12**
1050 mm dia.	3.60	71.42	24.46	24.91	nr	**120.80**
1200 mm dia.	4.75	94.24	32.30	32.54	nr	**159.08**
1500 mm dia.	5.75	114.08	39.09	50.86	nr	**204.03**
1800 mm dia.	6.00	119.10	39.09	50.86	nr	**209.05**
2400 mm dia.	6.50	129.20	39.09	50.86	nr	**219.15**
2700 mm dia.	6.70	133.23	39.09	50.86	nr	**223.19**
3000 mm dia.	7.50	149.09	39.09	50.86	nr	**239.05**
rectangular	3.50	69.44	23.78	28.77	m^2	**121.99**
Reinforced concrete grade C20						
Bases; thickness						
150–300 mm deep	2.30	45.63	15.63	115.51	m^3	**176.76**
300–500 mm deep	2.50	49.60	16.99	115.51	m^3	**182.09**
Suspended slabs; thickness						
ne 150 mm deep	3.45	68.45	23.46	115.51	m^3	**207.42**
150–300 mm deep	2.70	53.57	18.35	115.51	m^3	**187.42**
Formwork; GRP forms						
Plain curved, vertical						
to base and benching	0.60	11.90	4.08	12.70	m^2	**28.68**
to surrounds	0.40	7.94	2.72	9.35	hr	**20.00**
Bar reinforcement						
See Series 1700						

SERIES 500: DRAINAGE AND SERVICE DUCTS

Item	Gang hours	Labour £	Plant £	Material £	Unit	**Total rate £**
Precast concrete circular manhole rings; BS 5911 Part 1						
Shaft rings						
675 mm; plain/reinforced	3.20	71.33	52.89	65.37	m	**189.60**
900 mm; plain/reinforced	3.80	84.71	62.80	150.34	m	**297.85**
1050 mm; plain/reinforced	4.20	93.62	83.40	125.18	m	**302.20**
1200 mm; plain/reinforced	4.68	104.32	77.36	150.19	m	**331.88**
1500 mm; reinforced	5.28	117.70	87.24	257.54	m	**462.48**
1800 mm; reinforced	6.00	133.75	99.19	395.50	m	**628.44**
2100 mm; reinforced	6.60	182.25	331.30	763.17	m	**1276.72**
2400 mm; reinforced	7.20	198.81	371.37	1130.74	m	**1700.92**
2700 mm; reinforced	8.00	220.90	412.63	1220.18	m	**1853.71**
3000 mm; reinforced	8.60	237.47	443.56	1564.01	m	**2245.05**
Reducing slabs + landing slabs; 900 mm dia. access						
1050 mm; plain/reinforced	2.52	56.17	41.64	141.60	nr	**239.42**
1200 mm; plain/reinforced	3.30	72.51	46.30	164.36	nr	**283.18**
1500 mm; reinforced	3.16	70.44	52.22	506.00	nr	**628.65**
1800 mm; reinforced	3.60	80.25	59.52	579.91	nr	**719.69**
2100 mm; reinforced	4.00	110.45	206.31	1211.83	nr	**1528.59**
2400 mm; reinforced	4.40	121.50	226.97	1661.51	nr	**2009.98**
2700 mm; reinforced	4.80	132.54	247.58	2601.54	nr	**2981.66**
3000 mm; reinforced	5.20	143.59	268.23	3278.37	nr	**3690.18**
Cover slabs						
675 mm plain/reinforced	1.98	43.76	30.43	74.73	nr	**148.92**
900 mm; plain/reinforced	2.16	48.15	35.72	127.25	nr	**211.12**
1050 mm; plain/reinforced	2.52	56.17	41.64	135.98	nr	**233.79**
1200 mm; plain/reinforced	2.80	62.42	46.30	169.15	nr	**277.86**
1500 mm; reinforced	3.16	70.44	52.22	296.24	nr	**418.90**
1800 mm; reinforced	3.60	80.25	59.52	456.04	nr	**595.81**
2100 mm; reinforced	4.00	110.45	206.31	944.62	nr	**1261.38**
2400 mm; reinforced	4.40	121.50	226.97	1293.29	nr	**1641.75**
2700 mm; reinforced	4.80	132.54	247.58	2017.95	nr	**2398.07**
3000 mm; reinforced	0.60	18.05	49.96	2482.94	nr	**2550.96**
Engineering bricks, Class B; PC £790.00/1000; in cement mortar (1:3)						
Walls, built vertical and straight						
one brick thick; depth ne 1.0 m	2.63	50.15	6.92	55.76	m^2	**112.83**
one brick thick; depth 1.0–2.0 m	2.80	53.49	7.38	55.76	m^2	**116.63**
one brick thick; depth 2.0–3.0 m	3.05	58.18	8.03	55.76	m^2	**121.96**
one brick thick; depth 3.0–4.0 m	3.29	62.86	8.67	55.76	m^2	**127.28**
one brick thick; depth 4.0–5.0 m	3.68	70.21	9.68	55.76	m^2	**135.65**
one and a half bricks thick; depth ne 1.0 m	3.61	68.87	9.50	82.77	m^2	**161.15**
one and a half bricks thick; depth 1.0–2.0 m	4.03	76.90	10.61	82.77	m^2	**170.28**
one and a half bricks thick; depth 2.0–3.0 m	4.41	84.25	11.62	82.77	m^2	**178.65**
one and a half bricks thick; depth 3.0–4.0 m	4.80	91.61	12.63	82.77	m^2	**187.02**
one and a half bricks thick; depth 4.0–5.0 m	5.18	98.96	13.65	82.77	m^2	**195.38**

SERIES 500: DRAINAGE AND SERVICE DUCTS

Item	Gang hours	Labour £	Plant £	Material £	Unit	Total rate £
MANHOLES – cont						
Step irons; BS 1247; malleable; galvanized						
Built into joints; at depths						
ne 1.0 m	0.11	2.01	0.28	19.52	nr	**21.81**
1.0–2.0 m	0.11	2.01	0.28	19.52	nr	**21.81**
2.0–3.0 m	0.14	2.67	0.37	19.52	nr	**22.56**
3.0–4.0 m	0.18	3.34	0.46	19.52	nr	**23.33**
4.0–5.0 m	0.21	4.01	0.78	19.52	nr	**24.31**
Cast into concrete; at depths						
ne 1.0 m	0.09	1.80	0.28	19.52	nr	**21.60**
1.0–2.0 m	0.09	1.80	0.28	19.52	nr	**21.60**
2.0–3.0 m	0.12	2.40	0.37	19.52	nr	**22.29**
3.0–4.0 m	0.15	3.00	0.46	19.52	nr	**22.98**
4.0–5.0 m	0.18	3.60	0.78	19.52	nr	**23.90**
Vitrified clayware channels; bedding in cement mortar in bottom of manhole						
Half section straight						
100 mm dia.	0.54	10.80	1.48	30.05	nr	**42.33**
150 mm dia.	0.69	13.80	1.90	60.70	nr	**76.40**
225 mm dia.	0.96	19.20	2.63	195.99	nr	**217.82**
300 mm dia.	1.29	25.80	3.54	342.83	nr	**372.17**
Channel bend						
100 mm dia.	0.42	8.40	1.15	36.22	nr	**45.77**
150 mm dia.	0.57	11.40	1.57	45.42	nr	**58.39**
225 mm dia.	4.68	94.35	2.31	231.22	nr	**327.88**
300 mm dia.	7.75	156.28	3.29	348.52	nr	**508.10**
Channel branch						
100 mm dia.	0.20	3.98	1.32	48.07	nr	**53.36**
150 mm dia.	1.66	33.37	2.14	66.88	nr	**102.39**
225 mm dia.	2.01	40.20	5.52	239.24	nr	**284.95**
300 mm dia.	3.21	64.19	8.81	464.86	nr	**537.87**
Half section straight taper						
150 mm dia.	0.66	13.20	1.81	22.53	nr	**37.54**
225 mm dia.	0.99	19.80	2.72	175.68	nr	**198.19**
300 mm dia.	1.44	28.80	3.95	470.33	nr	**503.08**
Half section branch channel bend						
100 mm dia.	0.42	8.40	1.15	47.63	nr	**57.18**
150 mm dia.	0.57	11.40	1.57	71.12	nr	**84.08**
225 mm dia.	0.84	16.80	2.31	231.87	nr	**250.97**
300 mm dia.	1.20	24.00	3.29	348.56	nr	**375.85**
Three quarter section branch channel bend						
100 mm dia.	0.42	8.40	1.15	52.56	nr	**62.11**
150 mm dia.	0.57	11.40	1.57	88.17	nr	**101.14**
225 mm dia.	0.84	16.80	2.31	276.36	nr	**295.47**
300 mm dia.	1.20	24.00	3.29	415.09	nr	**442.38**

SERIES 500: DRAINAGE AND SERVICE DUCTS

Item	Gang hours	Labour £	Plant £	Material £	Unit	**Total rate £**
Spun concrete channels; bedding in cement mortar in bottom of manhole						
Half section straight						
300 mm dia.	1.14	22.80	3.51	13.16	nr	**39.46**
450 mm dia.	2.25	45.00	6.92	22.34	nr	**74.25**
600 mm dia.	3.15	62.99	7.88	40.63	nr	**111.51**
900 mm dia.	5.91	118.19	18.17	78.24	nr	**214.60**
1200 mm dia.	9.42	188.38	28.95	137.19	nr	**354.53**
Labours						
Build in pipes; to brickwork						
300 mm dia.	0.28	5.35	0.72	1.13	m^2	**7.19**
450 mm dia.	0.39	7.36	0.99	1.58	m^2	**9.92**
600 mm dia.	0.53	10.03	1.35	2.12	m^2	**13.50**
900 mm dia.	0.70	13.37	1.80	3.74	m^2	**18.91**
1200 mm dia.	0.77	14.71	1.98	7.52	m^2	**24.21**
Build in pipes; to precast concrete						
300 mm dia.	1.35	27.00	4.04	1.58	nr	**32.62**
450 mm dia.	2.01	40.20	6.02	2.25	nr	**48.47**
600 mm dia.	2.61	52.19	7.82	3.15	nr	**63.17**
900 mm dia.	3.84	76.79	11.50	6.80	nr	**95.09**
1200 mm dia.	4.83	96.59	14.46	11.93	nr	**122.99**
Access covers and frames; coated; BS EN 124; bed frame in cement mortar; set cover in grease and sand						
Group 5; super heavy duty; solid top						
600 × 600 mm; E600	1.80	36.00	4.49	554.75	nr	**595.24**
Group 4; heavy duty; solid top						
600 × 600 mm; double trianglar, hinged; D400	1.44	28.80	3.59	275.50	nr	**307.89**
Group 2; medium duty						
600 × 600 mm; single seal; B125	1.17	23.40	2.92	326.30	nr	**352.62**
Group 2; light duty						
600 × 600 mm; single seal; B125	0.87	17.40	2.17	246.84	nr	**266.41**
Manhole ancillaries						
Manhole interceptors						
100 mm dia.	2.00	39.80	1.62	328.08	nr	**369.50**
150 mm dia.	2.00	39.80	1.80	363.28	nr	**404.88**
225 mm dia.	2.00	39.80	2.70	1220.01	nr	**1262.51**
300 mm dia.	10.00	201.37	3.59	1749.31	nr	**1954.27**
Rodding eye points (excluding pipework)						
100 mm dia.	2.00	39.80	5.05	343.04	nr	**387.88**
150 mm dia.	2.00	39.80	3.71	411.48	nr	**454.99**

SERIES 600: EARTHWORKS

Item	Gang hours	Labour £	Plant £	Material £	Unit	Total rate £
NOTES The cost of earth moving and other associated works is dependent on matching the overall quantities and production rates called for by the programme of works with the most appropriate plant and assessing the most suitable version of that plant that will: deal with the site conditions (e.g. type of ground, type of excavation, length of haul, prevailing weather, etc.); comply with the specification requirements (e.g. compaction, separation of materials, surface tolerances, etc.); complete the work economically (e.g. provide surface tolerances which will avoid undue excessive thickness of expensive imported materials).						
Excavation rates Unless stated otherwise the units for excavation in material other than topsoil, rock or artificial hard materials are based on excavation in firm gravel soils. Factors for alternative types of soil multiply the rates by: Scrapers Stiff clay × 1.5 Chalk × 2.5 Soft rock × 3.5 Broken rock × 3.7 Tractor dozers and loaders Stiff clay × 2.0 Chalk × 3.0 Soft rock × 2.5 Broken rock × 2.5 Backacter (minimum bucket size 0.5 m^3) Stiff clay × 1.7 Chalk × 2.0 Soft rock × 2.0 Broken rock × 1.7						
Disposal rates The other important consideration in excavation of material is bulkage of material after it is dug and loaded onto transport. All pricing and estimating for disposal is based on the volume of solid material excavated and rates for disposal should be adjusted by the following factors for bulkage:						

SERIES 600: EARTHWORKS

Item	Gang hours	Labour £	Plant £	Material £	Unit	Total rate £
Sand bulkage × 1.10						
Gravel bulkage × 1.20						
Compacted soil bulkage × 1.30						
Compacted subbase, acceptable fill etc. bulkage × 1.30						
Stiff clay bulkage × 1.20						
Fill rates						
The price for filling presume the material being supplied in loose state.						
RESOURCES – LABOUR						
Scraper gang						
1 plant operator (skill rate 2)	1.00	35.46	–	–	-	**35.46**
Total Gang Rate/Hour	1.00	35.46	–	–	-	**35.46**
Scraper and ripper bulldozer gang (hard material)						
2 plant operator (skill rate 2)	2.00	70.91	–	–	-	**70.91**
Total Gang Rate/Hour	1.00	70.91	–	–	-	**70.91**
General excavation gang						
1 plant operator (skill rate 3)	1.00	29.17	–	–	-	**29.17**
1 plant operator (skill rate 3) – 25% of time	0.25	7.29	–	–	-	**7.29**
1 banksman (skill rate 4)	1.00	20.20	–	–	-	**20.20**
Total Gang Rate/Hour	1.00	56.66	–	–	-	**56.66**
Shore defences (armour stones) gang						
1 plant operator (skill rate 2)	1.00	35.46	–	–	-	**35.46**
1 banksman (skill rate 4)	1.00	20.20	–	–	-	**20.20**
Total Gang Rate/Hour	1.00	55.65	–	–	-	**55.65**
Filling gang						
1 plant operator (skill rate 4)	1.00	26.30	–	–	-	**26.30**
2 unskilled operatives (general)	2.00	39.20	–	–	-	**39.20**
Total Gang Rate/Hour	1.00	65.50	–	–	-	**65.50**
Treatment of filled surfaces gang						
1 plant operator (skill rate 2)	1.00	35.46	–	–	-	**35.46**
Total Gang Rate/Hour	1.00	35.46	–	–	-	**35.46**
Geotextiles (light sheets) gang						
1 ganger/chargehand (skill rate 4) – 50% of time	1.00	10.71	–	–	-	**10.71**
2 unskilled operatives (general)	2.00	39.20	–	–	-	**39.20**
Total Gang Rate/Hour	1.00	49.91	–	–	-	**49.91**
Geotextiles (medium sheets) gang						
1 ganger/chargehand (skill rate 4) – 20% of time	0.40	4.28	–	–	-	**4.28**
3 unskilled operatives (general)	3.00	58.81	–	–	-	**58.81**
Total Gang Rate/Hour	1.00	63.09	–	–	-	**63.09**
Geotextiles (heavy sheets) gang						
1 ganger/chargehand (skill rate 4) – 50% of time	1.00	10.71	–	–	-	**10.71**
2 unskilled operatives (general)	2.00	39.20	–	–	-	**39.20**
1 plant operator (skill rate 4)	1.00	26.30	–	–	-	**26.30**
Total Gang Rate/Hour	1.00	76.21	–	–	-	**76.21**

SERIES 600: EARTHWORKS

Item	Gang hours	Labour £	Plant £	Material £	Unit	Total rate £
RESOURCES – LABOUR – cont						
Horticultural works gang						
1 skilled operative (skill rate 4)	1.00	20.20	–	–	-	**20.20**
1 unskilled operative (general)	1.00	19.60	–	–	-	**19.60**
Total Gang Rate/Hour	1.00	39.80	–	–	-	**39.80**
RESOURCES – PLANT						
Scraper excavation						
motor scraper, 16.80 m^3, elevating		–	219.41	–	-	**219.41**
Total Rate/Hour		–	219.41	–	-	**219.41**
Scraper and ripper bulldozer excavation						
motor scraper, 16.80 m^3, elevating		–	219.41	–	-	**219.41**
D8 tractor dozer		–	199.98	–	-	**199.98**
dozer attachment: triple shank ripper		–	8.65	–	-	**8.65**
Total Rate/Hour		–	428.05	–	-	**428.05**
General excavation						
hydraulic crawler backacter, 0.40 m^3		–	55.59	–	-	**55.59**
backacter attachments: percussion breaker (25% of time)		–	2.33	–	-	**2.33**
tractor loader, 1.50 m^3 (25% of time)		–	23.43	–	-	**23.43**
loader attachments: ripper (25% of time)		–	2.23	–	-	**2.23**
Total Rate/Hour		–	83.57	–	-	**83.57**
Shore defences (armour stones)						
hydraulic crawler backacter, 1.20 m^3		–	110.29	–	-	**110.29**
backacter attachments: rock bucket		–	3.75	–	-	**3.75**
backacter attachments: clamshell grab		–	5.12	–	-	**5.12**
Total Rate/Hour		–	119.15	–	-	**119.15**
Geotextiles (heavy sheets)						
tractor loader, 1.5 m^3		–	93.72	–	-	**93.72**
Total Gang Rate/Hour		–	93.72	–	-	**93.72**
Filling						
tractor loader, 1.5 m^3		–	93.72	–	-	**93.72**
Total Rate/Hour		–	93.72	–	-	**93.72**
Treatment of filled surfaces						
tractor loader, 1.50 m^3		–	93.72	–	-	**93.72**
3 wheel deadweight roller, 10 tonne		–	63.38	–	-	**63.38**
Total Rate/Hour		–	157.10	–	-	**157.10**
EXCAVATION						
Typical motorway cutting generally using motorized scrapers and/or dozers on an average haul of 2000 m (one way)						
Excavation of acceptable material Class 5 A	0.01	0.28	1.76	–	m^3	**2.04**
Excavation of acceptable material excluding Class 5 A in cutting and other excavation	0.01	0.44	4.19	–	m^3	**4.64**

SERIES 600: EARTHWORKS

Item	Gang hours	Labour £	Plant £	Material £	Unit	**Total rate £**
General excavation using backacters						
Excavation of acceptable material excluding Class 5 A in new watercourses	0.06	3.40	5.50	–	m³	**8.90**
Excavation of unacceptable material Class U1 / U2 in new watercourses	0.07	3.68	5.79	–	m³	**9.48**
Excavation of unacceptable material Class U1/U2 in clearing abandoned watercourses	0.06	3.12	4.92	–	m³	**8.04**
General excavation using backacters and tractor loaders						
Excavation of acceptable material Class 5 A	0.04	2.27	3.49	–	m³	**5.76**
Excavation of acceptable material excluding Class 5 A in structural foundations						
ne 3.0 m deep	0.08	4.25	6.67	–	m³	**10.92**
ne 6.0 m deep	0.20	11.33	17.51	–	m³	**28.84**
Excavation of acceptable material excluding Class 5 A in foundations for corrugated steel buried structures and the like						
ne 3.0 m deep	0.08	4.25	6.67	–	m³	**10.92**
ne 6.0 m deep	0.20	11.33	17.51	–	m³	**28.84**
Excavation of unacceptable material Class U1 / U2 in structural foundations						
ne 3.0 m deep	0.09	4.82	7.59	–	m³	**12.41**
ne 6.0 m deep	0.21	11.90	18.52	–	m³	**30.42**
Excavation of unacceptable material Class U1 / U2 in foundations for corrugated steel buried structures and the like						
ne 3.0 m deep	0.09	4.82	7.59	–	m³	**12.41**
ne 6.0 m deep	0.21	11.90	18.52	–	m³	**30.42**
General excavation using backacters and tractor loader with ripper						
Excavation of acceptable material Class 5 A	0.03	1.70	2.61	–	m³	**4.31**
Excavation of acceptable material excluding Class 5 A in cutting and other excavation	0.04	2.27	3.49	–	m³	**5.76**
Excavate unacceptable material Class U1 / U2 in cutting and other excavation	0.04	2.27	3.49	–	m³	**5.76**
EXCAVATION IN HARD MATERIAL						
Typical motorway cutting generally using motorized scrapers and/or dozers on an average haul of 2000 m (one way)						
Excavate in hard material; using scraper and ripper bulldozer						
mass concrete/medium hard rock	0.09	6.38	38.52	–	m³	**44.91**
reinforced concrete/hard rock	0.15	10.64	64.21	–	m³	**74.84**
tarmacadam	0.16	11.35	6.85	–	m³	**18.19**

SERIES 600: EARTHWORKS

Item	Gang hours	Labour £	Plant £	Material £	Unit	Total rate £
EXCAVATION IN HARD MATERIAL – cont						
General excavation using backacters						
Extra over excavation in new watercourses for excavation in hard material						
rock	0.74	41.93	64.06	–	m^3	**105.99**
pavements, brickwork, concrete and masonry	0.69	39.10	59.67	–	m^3	**98.77**
reinforced concrete	1.12	63.46	96.82	–	m^3	**160.28**
General excavation using backacters and tractor loaders						
Extra over excavation in structural foundations for excavation in hard material						
rock	0.74	41.93	64.06	–	m^3	**105.99**
pavements, brickwork, concrete and masonry	0.69	39.10	59.67	–	m^3	**98.77**
reinforced concrete	1.12	63.46	96.82	–	m^3	**160.28**
Extra over excavation for excavation in foundations for corrugated steel buried structures and the like for excavation in hard material						
rock	0.74	41.93	64.06	–	m^3	**105.99**
pavements, brickwork, concrete and masonry	0.69	39.10	59.67	–	m^3	**98.77**
reinforced concrete	1.12	63.46	96.82	–	m^3	**160.28**
General excavation using backacters and tractor loader with ripper						
Extra over excavation for excavation in cutting and other excavation for excavation in hard material						
rock	0.71	40.23	61.38	–	m^3	**101.61**
pavements, brickwork, concrete and masonry	0.66	37.40	57.08	–	m^3	**94.47**
reinforced concrete	1.04	58.93	89.93	–	m^3	**148.85**
DEPOSITION OF FILL						
Deposition of acceptable material Class 1C/6B in						
embankments and other areas of fill	0.01	0.72	1.03	–	m^3	**1.75**
strengthened embankments	0.01	0.72	1.03	–	m^3	**1.75**
reinforced earth structures	0.01	0.72	1.03	–	m^3	**1.75**
anchored earth structures	0.01	0.72	1.03	–	m^3	**1.75**
landscaped areas	0.01	0.72	1.03	–	m^3	**1.75**
environmental bunds	0.01	0.72	1.03	–	m^3	**1.75**
fill to structures	0.01	0.72	1.03	–	m^3	**1.75**
fill above structural concrete foundations	0.01	0.72	1.03	–	m^3	**1.75**
DISPOSAL OF MATERIAL						
Disposal of acceptable material excluding Class 5A						
using 10 tonnes capacity tipping lorry for on-site or off-site use; haul distance to tip not exceeding 1 km	0.06	1.75	4.59	–	m^3	**6.34**
ADD per further km haul	0.03	0.88	2.29	–	m^3/km	**3.17**
using 10–15 tonnes capacity tipping lorry for on-site or off-site use; haul distance to tip not exceeding 1 km	0.05	1.46	3.82	–	m^3	**5.28**

SERIES 600: EARTHWORKS

Item	Gang hours	Labour £	Plant £	Material £	Unit	**Total rate £**
ADD per further km haul	0.03	0.73	1.91	–	m^3/km	**2.64**
using 15–25 tonnes capacity tipping lorry for on-site or off-site use; haul distance to tip not exceeding 1 km	0.05	1.31	3.44	–	m^3	**4.75**
ADD per further km haul	0.02	0.64	1.68	–	m^3/km	**2.32**
Disposal of acceptable material Class 5 A (excluding resale value of soil)						
using 10 tonnes capacity tipping lorry for on-site or off-site use; haul distance to tip not exceeding 1 km	0.06	1.75	4.59	–	m^3	**6.34**
ADD per further km haul	0.03	0.88	2.29	–	m^3/km	**3.17**
using 10–15 tonnes capacity tipping lorry for on-site or off-site use; haul distance to tip not exceeding 1 km	0.05	1.46	3.82	–	m^3	**5.28**
ADD per further km haul	0.03	0.73	1.91	–	m^3/km	**2.64**
using 15–25 tonnes capacity tipping lorry for on-site or off-site use; haul distance to tip not exceeding 1 km	0.05	1.31	3.44	–	m^3	**4.75**
ADD per further km haul	0.02	0.64	1.68	–	m^3/km	**2.32**
Disposal of unacceptable material Class U1						
using 10 tonnes capacity tipping lorry for on-site or off-site use; haul distance to tip not exceeding 1 km	0.06	1.75	4.59	–	m^3	**6.34**
ADD per further km haul	0.03	0.88	2.29	–	m^3/km	**3.17**
using 10–15 tonnes capacity tipping lorry for on-site or off-site use; haul distance to tip not exceeding 1 km	0.05	1.46	3.82	–	m^3	**5.28**
ADD per further km haul	0.03	0.73	1.91	–	m^3/km	**2.64**
using 15–25 tonnes capacity tipping lorry for on-site or off-site use; haul distance to tip not exceeding 1 km	0.05	1.31	3.44	–	m^3	**4.75**
ADD per further km haul	0.02	0.64	1.68	–	m^3/km	**2.32**
Disposal of unacceptable material Class U2						
using 10 tonnes capacity tipping lorry for on-site or off-site use; haul distance to tip not exceeding 1 km	0.06	1.75	4.59	–	m^3	**6.34**
ADD per further km haul	0.03	0.88	2.29	–	m^3/km	**3.17**
using 10–15 tonnes capacity tipping lorry for on-site or off-site use; haul distance to tip not exceeding 1 km	0.05	1.46	3.82	–	m^3	**5.28**
ADD per further km haul	0.03	0.73	1.91	–	m^3/km	**2.64**
using 15–25 tonnes capacity tipping lorry for on-site or off-site use; haul distance to tip not exceeding 1 km	0.05	1.31	3.44	–	m^3	**4.75**
ADD per further km haul	0.02	0.64	1.68	–	m^3/km	**2.32**
Add to the above rates where tipping charges apply:						
non-hazardous waste		–	–	–	m^3	**38.05**
hazardous waste		–	–	–	m^3	**114.11**
special waste		–	–	–	m^3	**120.46**
contaminated liquid		–	–	–	m^3	**152.14**
contaminated sludge		–	–	–	m^3	**190.21**
Add to the above rates where Landfill Tax applies:						
inactive or inert material		–	–	–	tonne	**3.15**
other taxable waste material		–	–	–	tonne	**98.60**

SERIES 600: EARTHWORKS

Item	Gang hours	Labour £	Plant £	Material £	Unit	**Total rate £**
IMPORTED FILL						
Imported graded granular fill, natural gravels DfT Class 1 A/B/C [1.9 t/m^3]						
Imported acceptable material in						
embankments and other areas of fill	0.02	1.18	1.69	28.65	m^3	**31.52**
extra for Aggregate Tax		–	–	3.80	m^3	**3.80**
Imported graded granular fill, crushed gravels or rock DfT Class 1 A/B/C [1.9 t/m^3]; using tractor loader						
Imported acceptable material in						
embankments and other areas of fill	0.02	1.18	1.69	28.65	m^3	**31.52**
extra for Aggregate Tax		–	–	3.80	m^3	**3.80**
Cohesive material DfT Class 2 A/B/C/D [1.8 t/m^3]; using tractor loader						
Imported acceptable material in						
embankments and other areas of fill	0.02	1.18	1.69	29.77	m^3	**32.64**
landscaped areas	0.02	1.18	1.69	29.77	m^3	**32.64**
environmental bunds	0.02	1.38	1.97	29.77	m^3	**33.11**
fill to structures	0.02	1.57	2.25	29.77	m^3	**33.59**
fill above structural concrete foundations	0.02	1.24	1.78	29.77	m^3	**32.80**
Reclaimed pulverised fuel ash DfT Class 2E [2.8 t/m^3]; using tractor loader						
Imported acceptable material in						
embankments and other areas of fill	0.02	1.18	1.69	32.13	m^3	**34.99**
Reclaimed quarry waste DfT Class 2E [1.8 t/m^3]; using tractor loader						
Imported acceptable material in						
embankments and other areas of fill	0.02	1.18	1.69	20.80	m^3	**23.67**
extra for Aggregate Tax (fill other than from acceptable process)		–	–	3.60	m^3	**3.60**
Imported topsoil DfT Class 5B [1.44 t/m^3]; using tractor loader						
Imported topsoil Class 5B	0.02	1.18	1.69	27.89	m^3	**30.75**
Imported selected well graded granular fill DfT Class 6 A (1.90 t/m^3); using tractor loader						
Imported acceptable material						
embankments and other areas of fill	–	–	1.97	37.05	m^3	**39.01**
landscape areas	0.02	1.38	1.97	37.05	m^3	**40.39**
environmental bunds	0.02	1.18	1.69	37.05	m^3	**39.91**
fill to structures	0.02	1.57	2.25	37.05	m^3	**40.87**
fill above structural concrete foundations	0.02	1.24	1.78	37.05	m^3	**40.07**
extra for Aggregate Tax		–	–	3.80	m^3	**3.80**

SERIES 600: EARTHWORKS

Item	Gang hours	Labour £	Plant £	Material £	Unit	Total rate £
Imported selected granular fill, DfT Class 6F (1.9 t/m³); using tractor loader						
Imported acceptable material in						
embankments and other areas of fill	0.10	6.22	8.90	36.79	m^3	**51.92**
extra for Aggregate Tax	–	–	–	3.80	m^3	**3.80**
Imported selected well graded fill DfT Class 6I (1.90 t/m³); using tractor loader						
Imported acceptable material						
reinforced earth structures	0.02	1.44	2.06	40.50	m^3	**44.01**
extra for Aggregate Tax		–	–	3.80	m^3	**3.80**
Imported rock fill (1.90 t/m³); using tractor loader						
Imported acceptable material						
in embankments and other areas of fill	0.02	1.31	1.87	48.94	m^3	**52.12**
extra for Aggregate Tax		–	–	3.80	m^3	**3.80**
Imported well graded granular material (1.90 t/m³); (bedding/free draining materials under shore protection) using tractor loader						
Imported acceptable material						
in embankments and other areas of fill	0.02	1.31	1.87	37.73	m^3	**40.92**
Imported rock fill (as a core embankment) (1.90 t/m³); using tractor loader						
Imported acceptable material						
in embankments and other areas of fill	0.03	2.16	3.09	48.94	m^3	**54.19**
extra for Aggregate Tax		–	–	3.80	m^3	**3.80**
Imported Armour Stones (1.90 t/m³) (shore protection of individual rocks up to 0.5 t each); using backacter						
Imported acceptable material						
in embankments and other areas of fill	0.06	3.12	6.67	55.93	m^3	**65.72**
extra for Aggregate Tax		–	–	3.80	m^3	**3.80**
Imported Armour Stones (1.90 t/m³) (shore protection of individual rocks up to 1.0 t each); using backacter						
Imported acceptable material						
in embankments and other areas of fill	0.03	1.78	3.81	62.95	m^3	**68.54**
extra for Aggregate Tax		–	–	3.80	m^3	**3.80**
Imported Armour Stones (1.90 t/m³) (shore protection of individual rocks up to 3.0 t each); using backacter						
Imported acceptable material						
in embankments and other areas of fill	0.03	1.45	3.08	69.24	m^3	**73.77**
extra for Aggregate Tax		–	–	3.80	m^3	**3.80**

SERIES 600: EARTHWORKS

Item	Gang hours	Labour £	Plant £	Material £	Unit	Total rate £
COMPACTION OF FILL						
Compaction of granular fill material						
in embankments and other areas of fill	0.01	0.18	0.79	–	m^3	**0.96**
adjacent to structures	0.01	0.28	1.26	–	m^3	**1.54**
above structural concrete foundations	0.01	0.43	1.89	–	m^3	**2.31**
Compaction of fill material						
in subbase or capping layers under verges, central reserves or side slopes	0.02	0.71	3.14	–	m^3	**3.85**
adjacent to structures	0.04	1.24	5.50	–	m^3	**6.74**
above structural concrete foundations	0.04	1.42	6.28	–	m^3	**7.70**
Compaction of graded fill material						
in embankments and other areas of fill	0.01	0.21	0.94	–	m^3	**1.16**
adjacent to structures	0.01	0.35	1.57	–	m^3	**1.93**
above structural concrete foundations	0.02	0.53	2.36	–	m^3	**2.89**
Compaction of rock fill materials						
in embankments and other areas of fill	0.01	0.32	1.41	–	m^3	**1.73**
adjacent to structures	0.01	0.50	2.20	–	m^3	**2.70**
above structural concrete foundations	0.02	0.60	2.67	–	m^3	**3.27**
Compaction of clay fill material						
in embankments and other areas of fill	0.03	1.06	4.71	–	m^3	**5.78**
adjacent to structures	0.05	1.70	7.54	–	m^3	**9.24**
above structural concrete foundations	0.05	1.70	7.54	–	m^3	**9.24**
GEOTEXTILES						
Notes						
The geotextile products mentioned below are not specifically confined to the individual uses stated but are examples of one of many scenarios to which they may be applied. Conversely, the scenarios are not limited to the geotextile used as an example.						
The heavier grades of sheeting will need to be manipulated into place by machine and cutting would be by hacksaw rather than knife.						
Care should be taken in assessing the wastage of the more expensive sheeting.						
The prices include for preparing surfaces, overlaps and turnups, jointing and sealing, fixing material in place if required and reasonable waste between 5% and 10%.						
Stabilization applications for reinforcement of granular subbases and capping layers placed over weak and variable soils.						
For use in weak soils with moderate traffic intensities e.g. light access roads: Tensar SS20 Polypropylene Geogrid						
horizontal	0.04	2.10	–	2.19	m^2	**4.29**
inclined at an angle 10–45° to the horizontal	0.05	2.65	–	2.19	m^2	**4.84**

SERIES 600: EARTHWORKS

Item	Gang hours	Labour £	Plant £	Material £	Unit	**Total rate £**
For use in weak soils with high traffic intensities and/or high axle loadings: Tensar SS30 Polypropylene Geogrid						
horizontal	0.05	3.03	–	2.24	m^2	**5.27**
inclined at an angle 10–45° to the horizontal	0.06	3.53	–	2.24	m^2	**5.77**
For construction over very weak soils e.g. alluvium, marsh or peat, or firmer soil subject to exceptionally high axle loadings: Tensar SS40 Polypropylene Geogrid						
horizontal	0.05	3.66	4.22	3.93	m^2	**11.81**
inclined at an angle 10–45° to the horizontal	0.06	4.27	5.25	3.93	m^2	**13.45**
For trafficked areas where fill comprises of aggregate exceeding 100 mm: Tensar SSLA20 Polypropylene Geogrid						
horizontal	0.04	2.10	–	4.06	m^2	**6.16**
inclined at an angle 10–45° to the horizontal	0.05	2.65	–	4.06	m^2	**6.71**
For stabilization and separation of granular fill from soft sub grade to prevent intermixing: Terram 1000						
horizontal	0.05	3.22	–	0.68	m^2	**3.90**
inclined at an angle 10–45° to the horizontal	0.06	4.04	–	0.68	m^2	**4.72**
For stabilization and separation of granular fill from soft sub grade to prevent intermixing: Terram 2000						
horizontal	0.04	3.20	3.94	1.57	m^2	**8.70**
inclined at an angle 10–45° to the horizontal	0.05	4.04	4.97	1.57	m^2	**10.57**
Reinforcement applications for asphalt pavements						
For roads, hardstandings and airfield pavements: Tensar AR-G grid bonded to a geotextile						
horizontal	0.05	2.84	–	3.66	m^2	**6.50**
inclined at an angle 10–45° to the horizontal	0.06	3.53	–	3.66	m^2	**7.19**
Slope reinforcement and embankment support; for use where soils can only withstand limited shear stresses, therefore steep slopes require external support						
Paragrid 30/155; 330 g/m^2						
horizontal	0.04	2.10	–	2.28	m^2	**4.38**
inclined at an angle 10–45° to the horizontal	0.05	2.65	–	2.28	m^2	**4.93**
Paragrid 100/255; 330 g/m^2						
horizontal	0.04	2.10	–	1.40	m^2	**3.50**
inclined at an angle 10–45° to the horizontal	0.05	2.65	–	1.40	m^2	**4.05**
Paralink 200; 1120 g/m^2						
horizontal	0.05	4.12	5.06	4.07	m^2	**13.25**
inclined at an angle 10–45° to the horizontal	0.07	5.18	6.37	4.07	m^2	**15.63**
Paralink 600; 2040 g/m^2						
horizontal	0.06	4.80	5.90	8.96	m^2	**19.66**
inclined at an angle 10–45° to the horizontal	0.08	6.02	7.40	8.96	m^2	**22.38**
TerramGrid 3/3 W						
horizontal	0.06	4.34	5.34	3.27	m^2	**12.96**
inclined at an angle 10–45° to the horizontal	0.07	5.41	6.65	3.27	m^2	**15.33**

SERIES 600: EARTHWORKS

Item	Gang hours	Labour £	Plant £	Material £	Unit	Total rate £
GEOTEXTILES – cont						
Scour and erosion protection						
For use where erosion protection is required to the surface of a slope once its geotechnical stability has been achieved, and to allow grass establishment: Tensar 'Mat' Polyethylene mesh, fixed with Tensar pegs						
horizontal	0.04	2.59	–	4.83	m^2	**7.42**
inclined at an angle 10–45° from the horizontal	0.05	3.22	–	5.18	m^2	**8.39**
For use where hydraulic action exists, such as coastline protection from pressures exerted by waves, currents and tides: Typar SF56						
horizontal	0.05	3.89	4.78	0.60	m^2	**9.27**
inclined at an angle 10–45° from the horizontal	0.06	4.88	6.00	0.60	m^2	**11.48**
For protection against puncturing to reservoir liner: Typar SF56						
horizontal	0.05	3.89	4.78	0.60	m^2	**9.27**
inclined at an angle 10–45° from the horizontal	0.06	4.88	6.00	0.60	m^2	**11.48**
Temporary parking areas						
For reinforcement of grassed areas subject to wear from excessive pedestrian and light motor vehicle traffic: Netlon CE131 high density polyethylene geogrid, including fixing pegs						
sheeting	0.04	2.20	–	5.76	m^2	**7.95**
Landscaping applications						
For prevention of weed growth in planted areas by incorporating a geotextile over topsoil: Typar SF20, including pegs						
horizontal	0.08	4.73	–	0.70	m^2	**5.43**
inclined at an angle 10–45° to the horizontal	0.09	5.93	–	0.70	m^2	**6.63**
For root growth control/prevention of lateral spread of roots and mixing of road base and humus: Typar SF20						
horizontal	0.08	4.73	–	0.35	m^2	**5.09**
inclined at an angle 10–45° to the horizontal	0.09	5.93	–	0.35	m^2	**6.28**
SOFT SPOTS AND OTHER VOIDS						
Excavation of soft spots and other voids using motorized scrapers and/or dozers						
Excavate below cuttings or under embankments	0.05	1.77	10.97	–	m^3	**12.74**
Excavate in side slopes	0.06	2.13	13.16	–	m^3	**15.29**
Excavation of soft spots and other voids using backacters and tractor loader						
Excavate below structural foundations and foundations for corrugated steel buried structures	0.09	2.62	3.49	–	m^3	**6.11**

SERIES 600: EARTHWORKS

Item	Gang hours	Labour £	Plant £	Material £	Unit	**Total rate £**
Imported graded granular fill; deposition using tractor loader and towed roller						
Filling of soft spots and other voids below cuttings or under embankments	0.04	1.00	2.83	28.96	m^3	**32.79**
Filling of soft spots and other voids in side slopes	0.04	1.22	3.46	28.96	m^3	**33.64**
Filling of soft spots and other voids below structural foundations and foundations for corrugated steel buried structures	0.04	1.00	2.83	28.96	m^3	**32.79**
Imported rock fill; 1.9 t/m³; using tractor loader						
Deposition into soft areas (rock punching)	0.03	1.64	2.34	48.94	m^3	**52.92**
DISUSED SEWERS, DRAINS, CABLES, DUCTS, PIPELINES AND THE LIKE						
Removal of disused sewer or drain						
100 mm internal dia.; with less than 1 m of cover to formation level	0.16	3.95	7.13	–	m	**11.08**
150 mm internal dia.; with 1 to 2 m of cover to formation level	0.20	4.84	8.74	–	m	**13.57**
Backfilling with acceptable material of disused sewer or drain						
100 mm internal dia.; with less than 1 m of cover to formation level	0.23	5.68	10.25	–	m^3	**15.93**
150 mm internal dia.; with 1 to 2 m of cover to formation level	0.23	5.68	10.25	–	m^3	**15.93**
Backfilling of disused basements, cellars and the like						
with acceptable material	0.26	6.42	11.59	–	m^3	**18.01**
Backfilling of disused gullies						
with concrete Grade C15	0.10	2.47	4.46	13.81	nr	**20.73**
SUPPORTS LEFT IN EXCAVATION						
Timber closeboarded supports left in excavation		–	9.86	–	m^2	**9.86**
Steel trench sheeting supports left in excavation		–	46.04	–	m^2	**46.04**
TOPSOILING AND STORAGE OF TOPSOIL						
Topsoiling 150 mm thick to surfaces						
at 10° or less to horizontal	0.04	1.24	7.68	–	m^2	**8.92**
more than 10° to horizontal	0.04	1.42	8.78	–	m^2	**10.19**
Topsoiling 350 mm thick to surfaces						
at 10° or less to horizontal	0.05	1.60	9.87	–	m^2	**11.47**
more than 10° to horizontal	0.05	1.77	10.97	–	m^2	**12.74**
Topsoiling 450 mm thick to surfaces						
at 10° or less to horizontal	0.05	1.77	10.97	–	m^2	**12.74**
more than 10° to horizontal	0.06	1.95	12.07	–	m^2	**14.02**
Topsoiling 600 mm thick to surfaces						
at 10° or less to horizontal	0.05	1.77	10.97	–	m^2	**12.74**
more than 10° to horizontal	0.06	1.95	12.07	–	m^2	**14.02**
Permanent storage of topsoil	0.06	2.13	13.16	–	m^3	**15.29**

SERIES 600: EARTHWORKS

Item	Gang hours	Labour £	Plant £	Material £	Unit	**Total rate £**
COMPLETION OF FORMATION AND SUB-FORMATION						
Completion of sub-formation						
on material other than Class 1C, 6B or rock in cuttings	0.01	0.39	1.63	–	m^2	**2.02**
Completion of formation						
on material other than Class 1C, 6B or rock in cuttings	0.01	0.39	1.03	–	m^2	**1.42**
LINING OF WATERCOURSES						
Lining new watercourse invert with						
precast concrete units 63 mm thick	0.40	15.92	0.76	25.56	m^2	**42.24**
Lining new watercourse side slopes with						
precast concrete units 63 mm thick	0.50	19.90	0.86	24.27	m^2	**45.02**
Lining enlarged watercourse invert with						
precast concrete units 63 mm thick	0.80	31.84	0.87	25.56	m^2	**58.27**
Lining enlarged watercourse side slopes with						
precast concrete units 63 mm thick	1.00	39.80	1.05	24.27	m^2	**65.11**
GROUND IMPROVEMENT – DYNAMIC COMPACTION						
Ground consolidation by dynamic compaction is a technique which involves the dropping of a steel or concrete pounder several times in each location in a grid pattern that covers the whole site. For a ground compaction of up to 10 m, a 15 t pounder with a free fall of 20 m would be typical. Several passes over the site are normally required to achieve full compaction. The process is recommended for naturally cohesive soils and is usually uneconomic for areas of less than 4000 m^2, for sites with granular or mixed granular cohesive soils and 6000 m^2 for a site with weak cohesive soils. The main considerations to be taken into account when using this method of consolidation are:						
* sufficient area to be viable						
* proximity and condition of adjacent property and services						
* need for blanket layer of granular material for a working surface and as backfill to offset induced settlement						
* water table level						

SERIES 600: EARTHWORKS

Item	Gang hours	Labour £	Plant £	Material £	Unit	**Total rate £**
The granular blanket layer performs the dual functions of working surface and backfill material. Generally 300 mm thickness is required.						
The final bearing capacity and settlement criteria that can be achieved depends on the nature of the material being compacted. Allowable bearing capacity may be increased by up to twice the pre-treated value for the same settlement. Control testing can be by crater volume measurements, site levelling between passes, penetration tests or plate loading tests.						
DYNAMIC COMPACTION						
Dynamic compaction in main compaction with a 15 t pounder		–	–	–	m²	**10.21**
Dynamic compaction plant standing time		–	–	–	hr	**37.38**
Free-draining granular material in granular blanket		–	–	–	tonne	**22.50**
Control testing including levelling, piezometers and penetrameter testing		–	–	–	m²	**8.71**
Kentledge load test		–	–	–	nr	**16317.00**
ESTABLISHMENT OF PLANT						
Establishment of dynamic compaction plant		–	–	–	sum	**15750.00**
VIBRATED STONE COLUMNS						
Refer to Civil Engineering Works, Class C: Geotechnical and Other Specialist Services, for an explanation and items for vibrated stone columns						
GABION WALLING AND MATTRESSES						
Gabion walling with plastic coated galvanized wire mesh, wire laced; filled with 50 mm Class 6G material						
2.0 × 1.0 × 1.0 m module sizes	5.40	100.58	22.14	48.62	m³	**171.34**
2.0 × 1.0 × 0.5 m module sizes	6.20	113.31	54.49	82.14	m³	**249.94**
Gabion walling with heavily galvanized woven wire mesh, wire laced; filled with 50 mm Class 6G material						
2.0 × 1.0 × 1.0 m module sizes	4.23	74.37	44.27	41.81	m³	**160.45**
2.0 × 1.0 × 0.5 m module sizes	6.20	113.31	54.49	65.73	m³	**233.54**
Mattress with plastic coated galvanized wire mesh; filled with 50 mm Class 6G material installed at 10° or less to the horizontal						
6.0 × 2.0 × 0.23 m module sizes	4.23	74.37	44.27	32.55	m³	**151.19**

SERIES 600: EARTHWORKS

Item	Gang hours	Labour £	Plant £	Material £	Unit	Total rate £
CRIB WALLING						
Notes						
There are a number of products specially designated for large scale earth control. Crib walling consists of a rigid unit built of rectangular interlocking timber or precast concrete members forming a skeleton of cells laid on top of each other and filled with earth or rock. Prices for these items depend on quantity, difficulty of access to the site and availability of suitable filling material; estimates should be obtained from the manufacturer when the site conditions have been determined.						
Crib walling of timber components laid with a battering face; stone infill						
ne 1.5 m high	5.28	98.78	59.04	292.29	m^2	**450.11**
ne 3.7 m high	7.11	133.01	74.48	300.44	m^2	**507.93**
ne 4.2 m high	8.13	152.09	83.08	308.59	m^2	**543.76**
ne 5.9 m high	10.71	200.36	104.85	324.86	m^2	**630.07**
ne 7.4 m high	17.16	321.02	159.27	341.14	m^2	**821.43**
Crib walling of precast concrete crib units, laid dry jointed with a battered face; stone infill						
1.0 m high; no dowels	6.75	126.28	69.92	300.76	m	**496.95**
1.5 m high; no dowels	9.36	175.10	96.95	451.14	m	**723.19**
2.5 m high; no dowels	22.11	413.62	364.57	910.34	m	**1688.53**
4.0 m high; no dowels	34.11	638.11	562.44	1456.54	m	**2657.09**
GROUND ANCHORAGES						
Ground anchorages consist of the installation of a cable or solid bar tendon fixed in the ground by grouting and tensioned to exceed the working load to be carried. Ground anchors may be of a permanent or temporary nature and can be used in conjunction with diaphragm walling or sheet piling to eliminate the use of strutting etc.						
The following costs are based on the installation of 50 nr ground anchors						
GROUND ANCHORAGE PLANT						
Establishment of ground anchorage plant		–	–	–	sum	**14059.50**
Ground anchorage plant standing time		–	–	–	hr	**299.03**
GROUND ANCHORAGES						
Ground anchorages; temporary or permanent						
15.0 m maximum depth; in rock, alluvial or clay; 0–50 t load		–	–	–	nr	**130.62**
15.0 m maximum depth; in rock or alluvial; 50–90 t load		–	–	–	nr	**156.74**
15.0 m maximum depth; in rock only; 90–150 t load		–	–	–	nr	**210.41**

SERIES 600: EARTHWORKS

Item	Gang hours	Labour £	Plant £	Material £	Unit	**Total rate £**
Temporary tendons						
in rock, alluvial or clay; 0–50 t load		–	–	–	nr	**97.96**
in rock or alluvial; 50–90 t load		–	–	–	nr	**152.40**
in rock only; 90–150 t load		–	–	–	nr	**206.08**
Permanent tendons						
in rock, alluvial or clay; 0–50 t load		–	–	–	nr	**152.40**
in rock or alluvial; 50–90 t load		–	–	–	nr	**210.42**
in rock only; 90–150 t load		–	–	–	nr	**264.14**
GROUND WATER LOWERING						
The following unit costs are for dewatering pervious ground only and are for sets of equipment comprising:						
*hire of 1 nr diesel driven pump (WP 150/60 or similar) complete with allowance of £50 for fuel		–	–	–	day	**135.88**
*hire of 50 m of 150 mm dia. header pipe		–	–	–	day	**24.70**
*purchase of 35 nr of disposable well points		–	–	–	sum	**11889.44**
*hire of 18 m of delivery pipe		–	–	–	day	**9.57**
*hire of 1 nr diesel driven standby pump		–	–	–	day	**76.58**
*hire of 1 nr jetting pump with hoses (for installation of wellpoints only)		–	–	–	day	**80.29**
*cost of attendant labour and plant (2 hrs per day) inclusive of small dumper and bowser		–	–	–	day	**139.25**
Costs are based on 24 hr operation in 12 hr shifts with attendant operators (specialist advice)						
Guide price for single set of equipment comprising pump, 150 mm dia. header pipe, 35 nr well points, delivery pipe and attendant labour and plant						
Bring to site equipment and remove upon completion		–	–	–	sum	**3205.09**
Installation: 3 day hire of jetting pump		–	–	–	sum	**240.88**
Operating costs						
purchase of well points		–	–	–	nr	**339.70**
hire of pump, header pipe, delivery pipe and standby pump complete with fuel etc. and attendant labour and plant		–	–	–	day	**385.98**
TRIAL PITS						
The following costs assume the use of mechanical plant and excavating and backfilling on the same day						
Trial pit						
ne 1.0 m deep	1.44	26.94	111.48	–	nr	**138.42**
1.0–2.0 m deep	2.62	48.83	218.33	–	nr	**267.16**
over 2.0 m deep	3.18	59.49	246.20	–	nr	**305.69**

SERIES 600: EARTHWORKS

Item	Gang hours	Labour £	Plant £	Material £	Unit	**Total rate £**
BREAKING UP AND PERFORATION OF REDUNDANT PAVEMENTS						
Using scraper and ripper bulldozer						
Breaking up of redundant concrete slab						
ne 100 mm deep	0.05	1.77	21.40	–	m^2	**23.18**
100 to 200 mm deep	0.10	3.55	42.80	–	m^2	**46.35**
Breaking up of redundant flexible pavement						
ne 100 mm deep	0.02	0.71	8.56	–	m^2	**9.27**
100 to 200 mm deep	0.03	1.06	12.84	–	m^2	**13.91**
Using backacters and tractor loader with ripper						
Breaking up of redundant reinforced concrete pavement						
ne 100 mm deep	0.28	7.09	10.05	–	m^2	**17.14**
100 to 200 mm deep	0.49	12.41	17.59	–	m^2	**30.00**
200 to 300 mm deep	0.70	17.73	25.18	–	m^2	**42.91**
Using scraper and ripper bulldozer						
Breaking up of redundant flexible pavement using scraper and ripper bulldozer						
ne 100 mm deep	0.02	0.71	8.56	–	m^2	**9.27**
100 to 200 mm deep	0.03	1.06	12.84	–	m^2	**13.91**
Using backacters and breakers						
Perforation of redundant reinforced concrete pavement						
ne 100 mm deep	0.02	0.58	1.15	–	m^2	**1.73**
100 to 200 mm deep	0.03	0.88	1.74	–	m^2	**2.62**
200 to 300 mm deep	0.05	1.46	2.89	–	m^2	**4.35**
Perforation of redundant flexible pavement						
ne 100 mm deep	0.01	0.29	0.59	–	m^2	**0.88**
100 to 200 mm deep	0.02	0.44	0.87	–	m^2	**1.31**
PERFORATION OF REDUNDANT SLABS, BASEMENTS AND THE LIKE						
Using backacters and breakers						
Perforation of redundant reinforced concrete slab						
ne 100 mm deep	0.02	0.58	1.15	–	m^2	**1.73**
100 to 200 mm deep	0.03	0.88	1.74	–	m^2	**2.62**
200 to 300 mm deep	0.05	1.46	2.89	–	m^2	**4.35**
Perforation of redundant reinforced concrete basement						
ne 100 mm deep	0.02	0.61	1.20	–	m^2	**1.82**
100 to 200 mm deep	0.03	0.92	1.83	–	m^2	**2.74**
200 to 300 mm deep	0.05	1.53	3.03	–	m^2	**4.56**

SERIES 600: EARTHWORKS

Item	Gang hours	Labour £	Plant £	Material £	Unit	Total rate £
REINFORCED AND ANCHORED EARTH STRUCTURES						
SPECIALIST ADVICE						
As each structure is different, it is virtually impossible to give accurate unit cost prices, as they will vary with the following parameters:						
* Type of structure						
* Where located (in water, dry condition)						
* Where geographically in the country						
* Type of fill						
* Duration of structure						
* Size of structure, etc.						
To arrive at the unit costs below assumptions have been made for a structure with the following characteristics:						
* Structure – retaining wall 6 m high × 150 m in length						
* Construction – as DfT Specification BE 3/78						
* Site conditions – good foundations						
* Fill – 5 m^3 per m^2 of wall face						
* Fill costs – DfT Specification average £9.50 / tonne						
Therefore specialist advice should be sought in order to give accurate budget costings for individual projects						
Retaining wall (per m^2 of face)						
concrete faced using ribbed strip	0.30	18.36	5.27	249.39	m^2	**273.03**
concrete faced using flat strip	0.29	17.75	5.28	278.74	m^2	**301.77**
concrete faced using polyester strip	0.73	44.38	11.00	271.40	m^2	**326.78**
concrete faced using geogrid reinforcement	0.60	36.73	12.06	272.13	m^2	**320.91**
preformed mesh using ribbed strip	0.33	20.38	8.31	243.50	m^2	**272.20**

SERIES 700: PAVEMENTS

Item	Gang hours	Labour £	Plant £	Material £	Unit	Total rate £
RESOURCES – LABOUR						
Subbase laying gang						
1 ganger/chargehand (skill rate 4)	2.00	21.42	–	–	-	**21.42**
1 skilled operative (skill rate 4)	1.00	20.20	–	–	-	**20.20**
2 unskilled operatives (general)	2.00	39.20	–	–	-	**39.20**
1 plant operator (skill rate 2)	1.00	35.46	–	–	-	**35.46**
1 plant operator (skill rate 3)	1.00	29.17	–	–	-	**29.17**
Total Gang Rate/Hour	1.00	145.44	–	–	-	**145.44**
Flexible paving gang						
1 ganger/chargehand (skill rate 4)	2.00	21.42	–	–	-	**21.42**
2 skilled operatives (skill rate 4)	2.00	40.39	–	–	-	**40.39**
4 unskilled operatives (general)	4.00	78.41	–	–	-	**78.41**
4 plant operators (skill rate 3)	4.00	116.68	–	–	-	**116.68**
Total Gang Rate/Hour	1.00	256.90	–	–	-	**256.90**
Concrete paving gang						
1 ganger/chargehand (skill rate 4)	2.00	21.42	–	–	-	**21.42**
2 skilled operatives (skill rate 4)	2.00	40.39	–	–	-	**40.39**
4 unskilled operatives (general)	4.00	78.41	–	–	-	**78.41**
1 plant operator (skill rate 2)	1.00	35.46	–	–	-	**35.46**
1 plant operator (skill rate 3)	1.00	29.17	–	–	-	**29.17**
Total Rate/Hour	1.00	204.84	–	–	-	**204.84**
Road surface spraying gang						
1 plant operator (skill rate 3)	1.00	29.17	–	–	-	**29.17**
Total Gang Cost/Hour	1.00	29.17	–	–	-	**29.17**
Road chippings gang						
1 ganger/chargehand (skill rate 4) – 50% of time	1.00	10.71	–	–	-	**10.71**
1 skilled operative (skill rate 4)	1.00	20.20	–	–	-	**20.20**
2 unskilled operatives (general)	2.00	39.20	–	–	-	**39.20**
3 plant operators (skill rate 3)	3.00	87.51	–	–	-	**87.51**
Total Gang Rate/Hour	1.00	157.62	–	–	-	**157.62**
Cutting slabs gang						
1 unskilled operative (general)	1.00	19.60	–	–	-	**19.60**
Total Gang Rate/Hour	1.00	19.60	–	–	-	**19.60**
Concrete filled joints gang						
1 ganger/chargehand – 50% of time	1.00	10.71	–	–	-	**10.71**
1 skilled operative (skill rate 4)	1.00	20.20	–	–	-	**20.20**
2 unskilled operatives (general)	2.00	39.20	–	–	-	**39.20**
Total Gang Rate/Hour	1.00	70.11	–	–	-	**70.11**
Milling gang						
1 ganger/chargehand (skill rate 4)	2.00	21.42	–	–	-	**21.42**
2 skilled operatives (skill rate 4)	2.00	40.39	–	–	-	**40.39**
4 unskilled operatives (general)	4.00	78.41	–	–	-	**78.41**
1 plant operator (skill rate 2)	1.00	35.46	–	–	-	**35.46**
1 plant operator (skill rate 3)	1.00	29.17	–	–	-	**29.17**
Total Gang Rate/Hour	1.00	204.84	–	–	-	**204.84**
Rake and compact planed material gang						
1 ganger/chargehand (skill rate 4)	2.00	21.42	–	–	-	**21.42**
1 skilled operative (skill rate 4)	1.00	20.20	–	–	-	**20.20**
3 unskilled operatives (general)	3.00	58.81	–	–	-	**58.81**
1 plant operator (skill rate 4)	1.00	26.30	–	–	-	**26.30**
1 plant operator (skill rate 3)	1.00	29.17	–	–	-	**29.17**
Total Gang Rate/Hour	1.00	155.89	–	–	-	**155.89**

SERIES 700: PAVEMENTS

Item	Gang hours	Labour £	Plant £	Material £	Unit	Total rate £
RESOURCES – PLANT						
Subbase laying						
93 kW motor grader		–	114.63	–	-	**114.63**
0.8 m^3 tractor loader		–	55.43	–	-	**55.43**
6 t towed vibratory roller		–	39.30	–	-	**39.30**
Total Rate/Hour		–	209.36	–	-	**209.36**
Flexible paving						
2 asphalt pavers, 35 kW, 4.0 m		–	129.29	–	-	**129.29**
2 deadweight rollers, 3 point, 10 t		–	126.76	–	-	**126.76**
tractor with front bucket and integral 2 tool compressor		–	325.42	–	-	**325.42**
compressor tools: rammer		–	3.85	–	-	**3.85**
compressor tools: poker vibrator		–	2.55	–	-	**2.55**
compressor tools: extra 15 m hose		–	0.46	–	-	**0.46**
tar sprayer, 100 litre		–	25.51	–	-	**25.51**
self propelled chip spreader		–	60.31	–	-	**60.31**
channel (heat) iron		–	4.14	–	-	**4.14**
Total Rate/Hour		–	678.30	–	-	**678.30**
Concrete paving						
wheeled loader, 2.60 m^3		–	111.50	–	-	**111.50**
concrete paver, 6.0 m		–	239.66	–	-	**239.66**
concrete slipform finisher		–	47.23	–	-	**47.23**
Total Gang Rate/Hour		–	398.38	–	-	**398.38**
Road surface spraying						
tar sprayer; 100 litre		–	25.51	–	-	**25.51**
Total Rate/Hour		–	25.51	–	-	**25.51**
Road chippings						
deadweight roller, 3 point, 10 t		–	63.38	–	-	**63.38**
tar sprayer, 100 litre		–	25.51	–	-	**25.51**
self propelled chip spreader		–	60.31	–	-	**60.31**
channel (heat) iron		–	4.14	–	-	**4.14**
Total Rate/Hour		–	153.35	–	-	**153.35**
Cutting slabs						
compressor, 65 cfm		–	11.79	–	-	**11.79**
12" disc cutter		–	1.36	–	-	**1.36**
Total Rate/Hour		–	13.15	–	-	**13.15**
Cold milling						
cold planer, 2.10 m		–	340.02	–	-	**340.02**
wheeled loader, 2.60 m^3		–	160.07	–	-	**160.07**
Total Rate/Hour		–	500.09	–	-	**500.09**
Heat planing						
heat planer, 4.5 m		–	151.43	–	-	**151.43**
wheeled loader, 2.60 m^3		–	133.08	–	-	**133.08**
Total Rate/Hour		–	284.51	–	-	**284.51**
Rake and compact planed material						
deadweight roller, 3 point, 10 t		–	63.38	–	-	**63.38**
tractor with front bucket and integral 2 tool compressor		–	46.29	–	-	**46.29**
channel (heat) iron		–	4.14	–	-	**4.14**
Total Rate/Hour		–	113.81	–	-	**113.81**

SERIES 700: PAVEMENTS

Item	Gang hours	Labour £	Plant £	Material £	Unit	**Total rate £**
SUBBASE						
The following unit costs are generally based on the National Highways Specification for Highway Works.						
Type 1 Unbound Mixtures						
Subbase; spread and graded						
150 mm thick	0.05	6.55	6.75	2.91	m^2	**16.20**
250 mm thick	0.04	5.82	6.00	4.84	m^2	**16.66**
200 mm thick	0.05	7.27	7.50	3.88	m^2	**18.65**
Cement Bound Granular Mixtures 5 (CBGM 5) C3/4						
Subbase, laid and compacted						
100 mm thick	0.04	5.82	6.00	7.95	m^2	**19.77**
Cement Bound Granular Mixtures 5 (CBGM 5) C8/10						
Subbase, laid and compacted						
100 mm thick	0.04	5.09	7.95	8.74	m^2	**21.78**
PAVEMENT (FLEXIBLE)						
Notes – Labour and Plant						
All outputs are based on clear runs without undue delay to two pavers with a 75% utilization.						
The outputs can be adjusted as follows to take account of space or time influences on the utilization.						
Factors for varying utilization of Labour and Plant:						
1 paver @ 75% utilization = × 2.00						
1 paver @ 100% utilization = × 1.50						
2 pavers @ 100% utilization = × 0.75						
Dense asphalt concrete						
AC 32 Dense Base 40/60 / 100/150 Des/Rec to PD 6691						
100 mm thick	0.02	5.14	13.57	18.89	m^2	**37.60**
150 mm thick	0.05	11.56	16.95	28.34	m^2	**56.85**
200 mm thick	0.07	16.70	20.36	37.79	m^2	**74.85**
AC 14/20 EME2 Base/Bin to PD 6691						
100 mm thick	0.05	11.56	13.57	24.85	m^2	**49.98**
150 mm thick	0.08	20.55	16.95	37.28	m^2	**74.78**
200 mm thick	0.11	28.26	20.36	49.70	m^2	**98.32**
AC 20 Dense Bin 40/60 / 100/150 Des/Rec to PD 6691						
60 mm thick	0.01	2.57	10.18	12.30	m^2	**25.05**
100 mm thick	0.03	6.42	13.57	20.50	m^2	**40.49**
AC 10 Close Surf 100/150 to PD 6691						
30 mm thick	0.01	2.57	6.80	7.41	m^2	**16.78**
40 mm thick	0.01	3.08	8.64	9.89	m^2	**21.61**

SERIES 700: PAVEMENTS

Item	Gang hours	Labour £	Plant £	Material £	Unit	Total rate £
Stone Mixed Asphalt						
SMA 10/14 Surf 40/60 to PD 6691						
20 mm thick	0.03	7.71	4.38	4.90	m^2	**16.98**
50 mm thick	0.03	7.71	6.39	12.25	m^2	**26.35**
2EF71491" group="1">40 mm thick	0.03	7.71	6.41	9.80	m^2	**23.92**
Thin Surface Course System						
TSCS 14, PSV 65, Category B						
30 mm thick	0.03	6.42	6.80	8.54	m^2	**21.76**
40 mm thick	0.02	5.14	8.18	11.38	m^2	**24.70**
50 mm thick	0.02	3.85	10.18	14.23	m^2	**28.27**
Hot Rolled Asphalt						
HRA 35/14 F Bin PMB Des to PD 6691 WT. Class 2						
45 mm thick	0.02	3.85	10.18	14.08	m^2	**28.12**
HRA 35/14 F Surf PMB Des to PD 6691 WT. Class 2 WITH Pre coated chippings PSV 65						
50 mm thick	0.04	8.99	10.18	19.92	m^2	**39.09**
PAVEMENT (CONCRETE)						
The following unit costs are for jointed reinforced concrete slabs, laid in reasonable areas (over 200 m^2) by paver train/slipformer						
Designed mix; cement to BS EN 197–1; grade C30, 20 mm aggregate						
Slab, runway, access roads or similar						
180 mm deep	0.02	3.07	23.27	9.93	m^2	**36.27**
220 mm deep	0.02	3.69	32.42	11.91	m^2	**48.02**
260 mm deep	0.02	4.51	30.14	11.91	m^2	**46.56**
300 mm deep	0.03	5.12	31.80	14.89	m^2	**51.82**
Fabric reinforcement						
Steel fabric reinforcement to BS4483						
Ref A142 nominal mass 2.22 kg	0.03	6.15	–	5.12	m^2	**11.26**
Ref A252 nominal mass 3.95 kg	0.04	8.19	–	7.68	m^2	**15.88**
Ref B385 nominal mass 4.53 kg	0.04	8.19	–	10.48	m^2	**18.67**
Ref C636 nominal mass 5.55 kg	0.05	10.24	–	12.49	m^2	**22.73**
Ref B503 nominal mass 5.93 kg	0.05	10.24	–	14.67	m^2	**24.92**
Mild steel bar reinforcement BS 4449						
Bars; supplied in bent and cut lengths						
6 mm nominal size	8.00	1638.75	–	952.69	tonne	**2591.44**
8 mm nominal size	6.74	1380.65	–	952.69	tonne	**2333.34**
10 mm nominal size	6.74	1380.65	–	952.69	tonne	**2333.34**
12 mm nominal size	6.74	1380.65	–	952.69	tonne	**2333.34**
16 mm nominal size	6.15	1259.79	–	951.74	tonne	**2211.53**

SERIES 700: PAVEMENTS

Item	Gang hours	Labour £	Plant £	Material £	Unit	Total rate £
PAVEMENT (CONCRETE) – cont						
High yield steel bar reinforcement BS 4449 or 4461						
Bars; supplied in bent and cut lengths						
6 mm nominal size	8.00	1638.75	–	961.62	tonne	**2600.36**
8 mm nominal size	6.74	1380.65	–	961.62	tonne	**2342.26**
10 mm nominal size	6.74	1380.65	–	961.62	tonne	**2342.26**
12 mm nominal size	6.74	1380.65	–	961.62	tonne	**2342.26**
16 mm nominal size	6.15	1259.79	–	961.60	tonne	**2221.39**
Sheeting to prevent moisture loss						
Polyethylene sheeting; lapped joints; horizontal below concrete pavements						
1000 gauge	0.01	2.05	–	1.07	m²	**3.12**
2000 gauge	0.01	2.05	–	1.97	m²	**4.02**
Joints in concrete slabs						
Longitudinal joints						
180 mm deep concrete	0.01	2.46	37.18	22.10	m	**61.74**
220 mm deep concrete	0.01	2.46	37.18	25.97	m	**65.61**
260 mm deep concrete	0.01	2.46	37.18	31.81	m	**71.45**
300 mm deep concrete	0.01	2.46	37.18	37.02	m	**76.66**
Expansion joints						
180 mm deep concrete	0.01	2.46	37.18	43.49	m	**83.13**
220 mm deep concrete	0.01	2.46	37.18	50.64	m	**90.28**
260 mm deep concrete	0.01	2.46	37.18	57.78	m	**97.42**
300 mm deep concrete	0.01	2.46	37.18	59.21	m	**98.86**
Contraction joints						
180 mm deep concrete	0.01	2.46	37.18	25.36	m	**65.01**
220 mm deep concrete	0.01	2.46	37.18	26.25	m	**65.89**
260 mm deep concrete	0.01	2.46	37.18	28.49	m	**68.13**
300 mm deep concrete	0.01	2.46	37.18	33.96	m	**73.60**
Construction joints						
180 mm deep concrete	0.01	2.46	37.18	16.76	m	**56.40**
220 mm deep concrete	0.01	2.46	37.18	20.15	m	**59.79**
260 mm deep concrete	0.01	2.46	37.18	19.58	m	**59.22**
300 mm deep concrete	0.01	2.46	37.18	21.08	m	**60.72**
Open joints with filler						
ne 0.5 m; 10 mm flexcell joint filler	0.11	7.71	–	4.42	m	**12.13**
0.5–1 m; 10 mm flexcell joint filler	0.11	7.71	–	6.25	m	**13.97**
Joint sealants						
10 × 20 mm hot bitumen sealant	0.14	9.82	–	4.82	m	**14.63**
20 × 20 mm cold polysulphide sealant	0.18	12.62	–	8.26	m	**20.88**
Trimming edges only of existing slabs, floors or similar surfaces (wet or dry); 6 mm cutting width						
50 mm deep	0.02	0.39	0.26	50.07	m	**50.72**
100 mm deep	0.03	0.59	0.41	122.18	m	**123.18**

SERIES 700: PAVEMENTS

Item	Gang hours	Labour £	Plant £	Material £	Unit	Total rate £
Cutting existing slabs, floors or similar surfaces (wet or dry); 8 mm cutting width						
50 mm deep	0.03	0.49	0.32	50.08	m	**50.89**
100 mm deep	0.06	1.18	0.79	130.14	m	**132.11**
150 mm deep	0.08	1.57	1.05	164.60	m	**167.22**
SURFACE TREATMENT						
Slurry surfacing to BS EN 12273						
3 mm deep	0.02	0.44	0.38	7.20	m^2	**8.02**
4 mm deep	0.02	0.44	0.38	7.20	m^2	**8.02**
Surface dressing to BS EN 12271						
30701468" group="1"> 6 mm single surface dressing	0.01	1.58	1.53	2.50	m^2	**5.61**
31701468" group="1">10 mm & 6 mm racked-in surface dressing	0.01	1.58	1.53	2.70	m^2	**5.81**
32701468" group="1">10 mm & 6 mm double surface dressing	0.01	1.58	1.53	3.05	m^2	**6.16**
High friction surfacing to BS 8870						
Anti Skid surfacing system						
High friction surfacing to BS 8870	–	–	28.90	m^2		**28.90**
TACK COAT						
Bond coat						
Bond coat to BS 594987						
large areas; over 20 m^2	0.02	0.44	0.38	0.58	m^2	**1.40**
large areas; over 20 m^2 @ 0.35 kg/m^2						
COLD MILLING (PLANING)						
Milling pavement (assumes disposal on site or reuse as fill but excludes transport if required)						
75 mm deep	0.03	5.53	11.65	–	m^2	**17.18**
100 mm deep	0.04	7.37	15.48	–	m^2	**22.86**
50 mm deep; scarifying surface	0.07	14.34	10.02	–	m^2	**24.36**
75 mm deep; scarifying surface	0.04	7.58	15.97	–	m^2	**23.54**
25 mm deep; heat planing for reuse	0.03	6.56	9.64	–	m^2	**16.19**
50 mm deep; heat planing for reuse	0.06	11.47	14.33	–	m^2	**25.80**
IN SITU RECYCLING						
Raking over scarified or heat planed material; compacting with 10 t roller						
50 mm deep	0.01	1.56	3.93	–	m^2	**5.49**

SERIES 1100: KERBS, FOOTWAYS AND PAVED AREAS

Item	Gang hours	Labour £	Plant £	Material £	Unit	Total rate £
NOTES						
Measurement Note: subbases are shown separate from their associated paving to simplify the presentation of cost alternatives.						
Measurement Note: bases are shown separate from their associated kerb etc. to simplify the presentation of cost alternatives.						
Kerb quadrants, droppers are shown separately						
The re-erection cost for kerbs, channels and edgings etc. taken from store assumes that major components are in good condition; the prices below allow a sum of 20% of the value of new materials to cover minor repairs together with an allowance for replacing a proportion of units.						
RESOURCES – LABOUR						
Kerb laying gang						
3 skilled operatives (skill rate 4)	3.00	60.59	–	–	-	**60.59**
1 unskilled operative (general)	1.00	19.60	–	–	-	**19.60**
1 plant operator (skill rate 3) – 25% of time	0.25	7.29	–	–	-	**7.29**
Total Gang Rate/Hour	1.00	87.48	–	–	-	**87.48**
Path subbase, bitmac and gravel laying gang						
1 skilled operative (skill rate 4)	1.00	20.20	–	–	-	**20.20**
2 unskilled operatives (general)	2.00	39.20	–	–	-	**39.20**
1 plant operator (skill rate 3)	1.00	29.17	–	–	-	**29.17**
Total Gang Rate/Hour	1.00	88.57	–	–	-	**88.57**
Paviors and flagging gang						
1 skilled operative (skill rate 4)	1.00	20.20	–	–	-	**20.20**
1 unskilled operative (general)	1.00	19.60	–	–	-	**19.60**
Total Gang Rate/Hour	1.00	39.80	–	–	-	**39.80**
RESOURCES – PLANT						
Kerb laying						
backhoe JCB 3CX (25% of time)		–	12.08	–	-	**12.08**
12" Stihl saw		–	1.09	–	-	**1.09**
road forms		–	2.97	–	-	**2.97**
Total Rate/Hour		–	16.14	–	-	**16.14**
Path subbase, bitmac and gravel laying						
backhoe JCB3CX		–	48.31	–	-	**48.31**
2 t dumper		–	10.81	–	-	**10.81**
pedestrian roller Bomag BW90S		–	12.75	–	-	**12.75**
Total Rate/Hour		–	70.92	–	-	**70.92**
Paviors and flagging						
2 t dumper (33% of time)		–	3.60	–	-	**3.60**
Total Rate/Hour		–	3.60	–	-	**3.60**

SERIES 1100: KERBS, FOOTWAYS AND PAVED AREAS

Item	Gang hours	Labour £	Plant £	Material £	Unit	**Total rate £**
KERBS, CHANNELS, EDGINGS, COMBINED DRAINAGE AND KERB BLOCKS AND LINEAR DRAINAGE						
Foundations to kerbs etc.						
Mass concrete						
200 × 100 mm	0.01	0.87	0.15	2.32	m	**3.34**
300 × 150 mm	0.02	1.31	0.24	5.31	m	**6.87**
450 × 150 mm	0.02	1.75	0.32	7.85	m	**9.93**
100 × 100 mm haunching, per side	0.01	0.44	0.08	0.57	m	**1.09**
Precast concrete units; BS 7263; bedded jointed and pointed in cement mortar						
Kerbs; bullnosed, splayed or half battered; laid straight or curved exceeding 12 m radius						
125 × 150 mm	0.06	5.25	0.97	12.68	m	**18.90**
125 × 255 mm	0.07	6.12	1.13	14.25	m	**21.51**
150 × 305 mm	0.07	6.12	1.13	21.32	m	**28.58**
Kerbs; bullnosed, splayed or half battered; laid to curves not exceeding 12 m radius						
125 × 150 mm	0.07	5.69	1.05	11.49	m	**18.23**
125 × 255 mm	0.08	6.56	1.21	12.92	m	**20.69**
150 × 305 mm	0.08	6.56	1.21	19.28	m	**27.05**
Quadrants (normally included in general rate for kerbs; shown separately for estimating purposes)						
305 × 305 × 150 mm	0.08	7.00	1.29	21.76	nr	**30.05**
455 × 455 × 255 mm	0.10	8.75	1.61	28.21	nr	**38.58**
Drop kerbs (normally included in general rate for kerbs; shown separately for estimating purposes)						
125 × 255 mm	0.07	6.12	1.13	29.09	nr	**36.35**
150 × 305 mm	0.07	6.12	1.13	48.31	nr	**55.57**
Channels; laid straight or curved exceeding 12 m radius						
125 × 255 mm	0.07	6.12	1.13	56.46	m	**63.71**
Channels; laid to curves not exceeding 12 m radius						
255 × 125 mm	0.07	6.12	1.13	60.42	m	**67.68**
Edgings; laid straight or curved exceeding 12 m radius						
150 × 50 mm	0.04	3.50	0.65	3.54	m	**7.68**
Edgings; laid to curves not exceeding 12 m radius						
150 × 50 mm	0.05	3.94	0.72	4.82	m	**9.48**
Precast concrete drainage channels; Charcon Safeticurb; channels jointed with plastic rings and bedded, jointed and pointed in cement mortar						
Channel unit; Type DBA/3; laid straight or curved exceeding 12 m radius						
250 × 250 mm; medium duty	0.08	6.56	1.21	67.48	m	**75.25**
305 × 305 mm; heavy duty	0.10	8.31	1.54	142.39	m	**152.23**

SERIES 1100: KERBS, FOOTWAYS AND PAVED AREAS

Item	Gang hours	Labour £	Plant £	Material £	Unit	Total rate £
KERBS, CHANNELS, EDGINGS, COMBINED DRAINAGE AND KERB BLOCKS AND LINEAR DRAINAGE – cont						
Precast concrete Ellis Trief safety kerb; bedded jointed and pointed in cement mortar						
Kerbs; laid straight or curved exceeding 12 m radius						
415 × 380 mm	0.23	19.68	3.63	104.98	m	**128.30**
Kerbs; laid to curves not exceeding 12 m radius						
415 × 380 mm	0.25	21.87	4.04	126.79	m	**152.69**
Precast concrete combined kerb and drainage block Beany Block System; bedded jointed and pointed in cement mortar						
Kerb; top block, shallow base unit, standard cover plate and frame						
laid straight or curved exceeding 12 m radius	0.15	13.12	2.42	149.59	m	**165.13**
laid to curves not exceeding 12 m radius	0.20	17.50	3.23	228.41	m	**249.14**
Kerb; top block, standard base unit, standard cover plate and frame						
laid straight or curved exceeding 12 m radius	0.15	13.12	2.42	149.59	m	**165.13**
laid to curves not exceeding 12 m radius	0.20	17.50	3.23	228.41	m	**249.14**
Kerb; top block, deep base unit, standard cover plate and frame						
Straight or curved over 12 m radius	0.15	13.12	2.42	333.90	m	**349.44**
laid to curves not exceeding 12 m radius	0.20	17.50	3.23	426.90	m	**447.62**
Base block depth tapers	0.10	8.75	1.61	36.03	m	**46.40**
Extruded asphalt edgings to pavings; slip formed BS 5931						
Kerb; laid straight or curved exceeding 12 m radius						
75 mm kerb height		–	–	10.03	m	**10.03**
100 mm kerb height		–	–	13.53	m	**13.53**
125 mm kerb height		–	–	16.88	m	**16.88**
Channel; laid straight or curved exceeding 12 m radius						
300 mm channel width		–	–	18.60	m	**18.60**
250 mm channel width		–	–	17.59	m	**17.59**
Kerb; laid to curves not exceeding 12 m radius						
75 mm kerb height		–	–	11.50	m	**11.50**
100 mm kerb height		–	–	9.58	m	**9.58**
125 mm kerb height		–	–	12.77	m	**12.77**
Channel; laid to curves not exceeding 12 m radius						
300 mm channel width		–	–	19.95	m	**19.95**
250 mm channel width		–	–	17.69	m	**17.69**
Extruded concrete; slip formed						
Kerb; laid straight or curved exceeding 12 m radius						
100 mm kerb height		–	–	11.65	m	**11.65**
125 mm kerb height		–	–	17.84	m	**17.84**

SERIES 1100: KERBS, FOOTWAYS AND PAVED AREAS

Item	Gang hours	Labour £	Plant £	Material £	Unit	**Total rate £**
Kerb; laid to curves not exceeding 12 m radius						
100 mm kerb height		–	–	14.75	m	**14.75**
125 mm kerb height		–	–	17.84	m	**17.84**
Additional concrete for kerbs, channels, edgings, combined drainage and kerb blocks and linear drainage channel systems						
Additional in situ concrete						
kerbs	0.50	43.74	7.41	115.51	m^3	**166.66**
channels	0.50	43.74	7.41	115.51	m^3	**166.66**
edgings	0.50	43.74	7.41	115.51	m^3	**166.66**
Remove from store and relay kerbs, channels, edgings, combined drainage and kerb blocks and linear drainage channel systems						
Remove from store and relay precast concrete units; bedded jointed and pointed in cement mortar						
Kerbs; laid straight or curved exceeding 12 m radius						
125 × 150 mm	0.06	5.25	0.97	3.27	m	**9.49**
125 × 255 mm	0.07	6.12	1.13	3.58	m	**10.84**
150 × 305 mm	0.07	6.12	1.13	5.00	m	**12.25**
Kerbs; laid to curves not exceeding 12 m radius						
125 × 150 mm	0.07	5.69	1.05	3.03	m	**9.77**
125 × 255 mm	0.08	6.56	1.21	3.32	m	**11.09**
150 × 305 mm	0.08	6.56	1.21	4.59	m	**12.36**
Quadrants (normally included in general rate for kerbs; shown separately for estimating purposes)						
305 × 305 × 150 mm	0.08	7.00	1.29	5.08	nr	**13.37**
455 × 455 × 255 mm	0.10	8.75	1.61	6.37	nr	**16.74**
Drop kerbs (normally included in general rate for kerbs; shown separately for estimating purposes)						
125 × 255 mm	0.07	6.12	1.13	6.55	nr	**13.81**
150 × 305 mm	0.07	6.12	1.13	10.39	nr	**17.65**
Channels; laid straight or curved exceeding 12 m radius						
125 × 255 mm	0.07	6.12	1.13	12.02	m	**19.28**
Channels; laid to curves not exceeding 12 m radius						
255 × 125 mm	0.07	6.12	1.13	12.82	m	**20.07**
Edgings; laid straight or curved exceeding 12 m radius						
150 × 50 mm	0.04	3.50	0.65	1.44	m	**5.58**
Edgings; laid to curves not exceeding 12 m radius						
150 × 50 mm	0.05	3.94	0.72	1.44	m	**6.10**

SERIES 1100: KERBS, FOOTWAYS AND PAVED AREAS

Item	Gang hours	Labour £	Plant £	Material £	Unit	Total rate £
KERBS, CHANNELS, EDGINGS, COMBINED DRAINAGE AND KERB BLOCKS AND LINEAR DRAINAGE – cont						
Remove from store and relay precast concrete drainage channels; Charcon 'Safeticurb'; channels jointed with plastic rings and bedded, jointed and pointed in cement mortar						
Channel unit; Type DBA/3; laid straight or curved exceeding 12 m radius						
250 × 254 mm; medium duty	0.08	6.56	1.21	14.23	m	**22.00**
305 × 305 mm; heavy duty	0.10	8.31	1.54	29.21	m	**39.06**
Remove from store and relay precast concrete 'Ellis Trief' safety kerb; bedded jointed and pointed in cement mortar						
Kerbs; laid straight or curved exceeding 12 m radius						
415 × 380 mm	0.23	19.68	3.63	21.73	m	**45.05**
Kerbs; laid to curves not exceeding 12 m radius						
415 × 380 mm	0.25	21.87	4.04	26.09	m	**52.00**
Remove from store and relay precast concrete combined kerb and drainage block 'Beany Block System'; bedded jointed and pointed in cement mortar						
Kerb; top block, shallow base unit, standard cover plate and frame						
laid straight or curved exceeding 12 m radius	0.15	13.12	2.42	30.65	m	**46.19**
laid to curves not exceeding 12 m radius	0.20	17.50	3.23	46.41	m	**67.14**
Kerb; top block, standard base unit, standard cover plate and frame						
laid straight or curved exceeding 12 m radius	0.15	13.12	2.42	30.65	m	**46.19**
laid to curves not exceeding 12 m radius	0.20	17.50	3.23	46.41	m	**67.14**
Kerb; top block, deep base unit, standard cover plate and frame						
Straight or curved over 12 m radius	0.15	13.12	2.42	67.51	m	**83.05**
laid to curves not exceeding 12 m radius	0.20	17.50	3.23	86.11	m	**106.84**
Base block depth tapers	0.10	8.75	1.61	7.94	m	**18.30**
FOOTWAYS AND PAVED AREAS						
Subbases						
100 mm thick Type 1 Unbound Mixture	0.01	0.80	0.64	3.32	m²	**4.75**
150 mm thick Type 1 Unbound Mixture	0.01	1.06	0.85	4.98	m²	**6.89**
100 mm thick concrete GEN3 or ST4 to BS 8500	0.05	4.43	1.49	10.40	m²	**16.32**
150 mm thick concrete GEN3 or ST4 to BS 8500	0.08	6.64	2.27	15.60	m²	**24.51**

SERIES 1100: KERBS, FOOTWAYS AND PAVED AREAS

Item	Gang hours	Labour £	Plant £	Material £	Unit	Total rate £
Asphalt surfacing. Binder course AC 20 Dense bin 100/150 Rec to PD 6691, surface course AC 6 Dense surf 100/150 to PD 6691; excluding subbase						
Paved area 70 mm thick; comprising binder course 50 mm thick surface course 20 mm thick						
To paved area; sloping not exceeding 10° to the horizontal						
Red coloured asphalt surfacing. Binder course AC 20 Dense bin 100/150 Rec to PD 6691, proprietary red coloured surface course; excluding subbase						
sloping at 10° or less to the horizontal	0.09	7.53	6.03	14.27	m²	**27.83**
sloping at more than 10° to the horizontal	0.10	8.41	6.74	14.27	m²	**29.42**
Paved area 70 mm thick; comprising binder course 50 mm thick proprietary red coloured surface course 20 mm thick.						
sloping at 10° or less to the horizontal	0.09	7.53	6.03	42.82	m²	**56.38**
sloping at more than 10° to the horizontal	0.10	8.41	6.74	42.82	m²	**57.97**
Green coloured asphalt surfacing. Binder course AC 20 Dense bin 100/150 Rec to PD 6691, proprietary green coloured surface course; excluding subbase						
Paved area 70 mm thick; comprising binder course 50 mm thick proprietary green coloured surface course 20 mm thick.						
sloping at 10° or less to the horizontal	0.09	7.53	6.03	18.94	m²	**32.49**
sloping at more than 10° to the horizontal	0.10	8.41	6.74	18.94	m²	**34.09**
Granular base surfacing; Central Reserve Treatments Limestone, graded 10 mm down laid and compacted; excluding subbase						
Paved area 100 mm thick; surface sprayed with two coats of cold bituminous emulsion; blinded with 6 mm quartzite fine gravel						
sloping not exceeding 10° to the horizontal	0.02	1.77	1.42	8.22	m²	**11.41**
Ennstone Johnston Golden gravel; graded 13 mm to fines; rolled wet						
Paved area 50 mm thick; single layer						
sloping not exceeding 10° to the horizontal	0.03	2.66	2.13	10.97	m²	**15.75**

SERIES 1100: KERBS, FOOTWAYS AND PAVED AREAS

Item	Gang hours	Labour £	Plant £	Material £	Unit	Total rate £
FOOTWAYS AND PAVED AREAS – cont						
Precast concrete slabs; BS 7263; grey; 5 point bedding and pointing joints in cement mortar; excluding subbase						
Paved area 50 mm thick; comprising 600 × 450 × 50 mm units						
sloping at 10° or less to the horizontal	0.28	11.14	1.01	10.23	m^2	**22.38**
Paved area 50 mm thick; comprising 600 × 600 × 50 mm units						
sloping at 10° or less to the horizontal	0.24	9.55	0.87	10.57	m^2	**20.98**
Paved area 50 mm thick; comprising 900 × 600 × 50 mm units						
sloping at 10° or less to the horizontal	0.20	7.96	0.71	12.06	m^2	**20.73**
Extra for coloured, 50 mm thick		–	–	3.43	m^2	**3.43**
Paved area 63 mm thick; comprising 600 × 600 × 63 mm units						
sloping at 10° or less to the horizontal	0.25	9.95	0.90	10.63	m^2	**21.47**
Paved area 63 mm thick; comprising 900 × 600 × 63 mm units						
sloping at 10° or less to the horizontal	0.21	8.36	0.76	12.11	m^2	**21.23**
Precast concrete rectangular paving blocks; BS 6717; grey; bedding on 50 mm thick dry sharp sand; filling joints; excluding subbase						
Paved area 80 mm thick; comprising 200 × 100 × 80 mm units						
sloping at 10° or less to the horizontal	0.30	11.94	1.08	18.64	m^2	**31.66**
Precast concrete rectangular paving blocks; BS 6717; coloured; bedding on 50 mm thick dry sharp sand; filling joints; excluding subbase						
Paved area 80 mm thick; comprising 200 × 100 × 80 mm units						
sloping at 10° or less to the horizontal	0.30	11.94	1.08	35.39	m^2	**48.41**
Brick paviors delivered to site; bedding on 20 mm thick mortar; excluding subbase						
Paved area 85 mm thick; comprising 215 × 103 × 65 mm units						
sloping at 10° or less to the horizontal	0.30	11.94	1.08	28.93	m^2	**41.95**
Granite setts (2.88 kg/mm thickness/m^2); bedding on 25 mm cement mortar; excluding subbase						
Paved area 100 mm thick; comprising 100 × 100 × 100 mm units; laid to random pattern						
sloping at 10° or less to the horizontal	0.90	35.82	3.24	75.94	m^2	**115.01**
Paved area 100 mm thick; comprising 100 × 100 × 100 mm units; laid to specific pattern						
sloping at 10° or less to the horizontal	1.20	47.76	4.33	75.94	m^2	**128.03**

SERIES 1100: KERBS, FOOTWAYS AND PAVED AREAS

Item	Gang hours	Labour £	Plant £	Material £	Unit	Total rate £
Cobble paving; 50–75 mm stones; bedding on 25 mm cement mortar; filling joints; excluding subbase						
Paved area; comprising 50–75 mm stones; laid to random pattern						
sloping at 10° or less to the horizontal	1.00	39.80	3.57	21.90	m^2	**65.26**
REMOVE FROM STORE AND RELAY PAVING FLAGS, SLABS AND BLOCKS						
Note						
An allowance of 20% of the cost of providing new pavings has been included in the rates below to allow for units which have to be replaced through unacceptable damage.						
Remove from store and relay precast concrete slabs; 5 point bedding and pointing joints in cement mortar; excluding subbase						
Paved area 50 mm thick; comprising 600 × 450 × 50 mm units						
sloping at 10° or less to the horizontal	0.28	11.14	1.01	2.46	m^2	**14.61**
Paved area 50 mm thick; comprising 600 × 600 × 50 mm units						
sloping at 10° or less to the horizontal	0.24	9.55	0.87	2.53	m^2	**12.95**
Paved area 50 mm thick; comprising 900 × 600 × 50 mm units						
sloping at 10° or less to the horizontal	0.20	7.96	0.71	2.83	m^2	**11.50**
Paved area 63 mm thick; comprising 600 × 600 × 63 mm units						
sloping at 10° or less to the horizontal	0.25	9.95	0.90	2.54	m^2	**13.39**
Paved area 63 mm thick; comprising 900 × 600 × 63 mm units						
sloping at 10° or less to the horizontal	0.21	8.36	0.76	2.84	m^2	**11.95**
Remove from store and relay precast concrete rectangular paving blocks; bedding on 50 mm thick dry sharp sand; filling joints; excluding subbase						
Paved area 80 mm thick; comprising 200 × 100 × 80 mm units						
sloping at 10° or less to the horizontal	0.30	11.94	1.08	4.87	m^2	**17.89**
Remove from store and relay brick paviors; bedding on 20 mm thick mortar; excluding subbase						
Paved area 85 mm thick; comprising 215 × 103 × 65 mm units						
sloping at 10° or less to the horizontal	0.30	11.94	1.08	7.00	m^2	**20.02**

SERIES 1200: TRAFFIC SIGNS AND ROAD MARKINGS

Item	Gang hours	Labour £	Plant £	Material £	Unit	**Total rate £**
NOTES						
The re-erection cost for traffic signs taken from store assumes that major components are in good condition; the prices below allow a sum of 20% of the value of new materials to cover minor repairs, new fixings and touching up any coatings.						
RESOURCES – LABOUR						
Traffic signs gang						
1 ganger/chargehand (skill rate 3)	2.00	23.67	–	–	-	**23.67**
1 skilled operative (skill rate 3)	1.00	22.45	–	–	-	**22.45**
2 unskilled operatives (general)	2.00	39.20	–	–	-	**39.20**
1 plant operator (skill rate 3) – 25% of time	0.25	7.29	–	–	-	**7.29**
Total Gang Rate/Hour	1.00	92.62	–	–	-	**92.62**
Bollards, furniture gang						
1 ganger/chargehand (skill rate 4)	2.00	21.42	–	–	-	**21.42**
1 skilled operative (skill rate 4)	1.00	20.20	–	–	-	**20.20**
2 unskilled operatives (general)	2.00	39.20	–	–	-	**39.20**
Total Gang Rate/Hour	1.00	80.82	–	–	-	**80.82**
RESOURCES – PLANT						
Traffic signs						
JCB 3CX backhoe – 50% of time		–	24.16	–	-	**24.16**
125 cfm compressor – 50% of time		–	9.03	–	-	**9.03**
compressor tools: hand held hammer drill – 50% of time		–	0.84	–	-	**0.84**
compressor tools: clay spade – 50% of time		–	1.32	–	-	**1.32**
compressor tools: extra 15 m hose – 50% of time		–	0.23	–	-	**0.23**
8 t lorry with hiab lift – 50% of time		–	18.63	–	-	**18.63**
Total/Hour		–	54.20	–	-	**54.20**
Bollards, furniture						
125 cfm compressor – 50% of time		–	9.03	–	-	**9.03**
compressor tools: hand held hammer drill – 50% of time		–	0.84	–	-	**0.84**
compressor tools: clay spade – 50% of time		–	1.32	–	-	**1.32**
compressor tools: extra 15 m hose – 50% of time		–	0.23	–	-	**0.23**
8 t lorry with hiab lift – 25% of time		–	9.31	–	-	**9.31**
Total Rate/Hour		–	20.73	–	-	**20.73**
TRAFFIC SIGNS						
In this section prices will vary depending upon the diagram configurations. The following are average costs of signs and bollards. Diagram numbers refer to the Traffic Signs Regulations and General Directions 2002 and the figure numbers refer to the Traffic Signs Manual.						
Examples of Prime Costs for Class 1 (High Intensity) traffic and road signs (ex works) for orders exceeding £1,000.						

SERIES 1200: TRAFFIC SIGNS AND ROAD MARKINGS

Item	Gang hours	Labour £	Plant £	Material £	Unit	**Total rate £**
Permanent traffic sign as non-lit unit on						
600 × 450 mm		–	–	46.56	nr	**46.56**
600 mm dia.		–	–	180.94	nr	**180.94**
600 mm triangular		–	–	150.75	nr	**150.75**
500 × 500 mm		–	–	46.49	nr	**46.49**
450 × 450 mm		–	–	44.18	nr	**44.18**
450 × 300 mm		–	–	41.94	nr	**41.94**
1200 × 400 mm (CHEVRONS)		–	–	74.52	nr	**74.52**
Examples of Prime Costs for Class 2 (Engineering Grade) traffic and road signs (ex works) for orders exceeding £1,000.						
600 × 450 mm		–	–	46.56	nr	**46.56**
600 mm dia.		–	–	180.94	nr	**180.94**
600 mm triangular		–	–	150.75	nr	**150.75**
500 × 500 mm		–	–	46.49	nr	**46.49**
450 × 450 mm		–	–	44.18	nr	**44.18**
450 × 300 mm		–	–	41.94	nr	**41.94**
1200 × 400 mm (CHEVRONS)		–	–	74.52	nr	**74.52**
Standard reflectorised traffic signs						
Note: Unit costs do not include concrete foundations (see Series 1700)						
Standard one post signs; 600 × 450 mm type C1 signs						
fixed back to back to another sign (measured separately) with aluminium clips to existing post (measured separately)	0.04	3.70	2.17	48.89	nr	**54.76**
Extra for fixing singly with aluminium clips	0.01	0.93	0.48	1.30	nr	**2.71**
Extra for fixing singly with stainless steel clips	0.01	0.93	1.85	5.01	nr	**7.79**
fixed back to back to another sign (measured separately) with stainless steel clips to one new 76 mm dia. plastic coated steel posts 1.75 m long	0.27	25.01	14.64	114.04	nr	**153.70**
Extra for fixing singly to one face only	0.01	0.93	0.48	–	nr	**1.41**
Extra for 76 mm dia. 1.75 m long aluminium post	0.02	1.85	1.01	36.23	nr	**39.10**
Extra for 76 mm dia. 3.5 m long plastic coated steel post	0.02	1.85	1.01	56.05	nr	**58.92**
Extra for 76 mm dia. 3.5 m long aluminium post	0.02	1.85	1.01	68.42	nr	**71.28**
Extra for excavation for post, in hard material	1.10	101.89	54.51	–	nr	**156.40**
Extra for single external illumination unit with fitted photo cell (excluding trenching and cabling – see Series 1400); unit cost per face illuminated	0.33	30.57	16.34	107.47	nr	**154.37**
Standard two post signs; 1200 × 400 mm, signs						
fixed back to back to another sign (measured separately) with stainless steel clips to two new 76 mm dia. plastic coated steel posts 1.75 m long	0.51	47.24	27.65	209.84	nr	**284.73**
Extra for fixing singly to one face only	0.02	1.85	1.01	–	nr	**2.86**
Extra for two 76 mm dia. 1.75 m long aluminium posts	0.04	3.70	1.98	72.47	nr	**78.15**

SERIES 1200: TRAFFIC SIGNS AND ROAD MARKINGS

Item	Gang hours	Labour £	Plant £	Material £	Unit	Total rate £
TRAFFIC SIGNS – cont						
Standard reflectorised traffic signs – cont						
Standard two post signs – cont						
Extra for two 76 mm dia. 3.5 m long plastic coated steel posts	0.04	3.70	1.98	112.10	nr	**117.79**
Extra for two 76 mm dia. 3.5 m long aluminium post	0.04	3.70	1.98	136.84	nr	**142.52**
Extra for excavation for post, in hard material	1.10	101.89	54.51	–	nr	**156.40**
Extra for single external illumination unit with fitted photo cell (excluding trenching and cabling – see Series 1400); unit cost per face illuminated	0.58	53.72	28.71	166.11	nr	**248.55**
Standard internally illuminated traffic signs						
Bollard with integral mould-in translucent graphics (excluding trenching and cabling)						
fixing to concrete base	1.00	92.62	26.01	529.36	nr	**648.00**
Special traffic signs						
Note: Unit costs do not include concrete foundations (see Series 1700) or trenching and cabling (see Series 1400)						
Externally illuminated reflectorised traffic signs manufactured to order						
special signs, surface area 1.50 m² on two 100 mm dia. steel posts		–	–	–	nr	**1038.40**
special signs, surface area 4.00 m² on three 100 mm dia. steel posts		–	–	–	nr	**1557.60**
Internally illuminated traffic signs manufactured to order						
special signs, surface area 0.25 m² on one new 76 mm dia. steel post		–	–	–	nr	**250.29**
special signs, surface area 0.75 m² on one new 100 mm dia. steel post		–	–	–	nr	**326.10**
special signs, surface area 4.00 m² on four new 120 mm dia. steel posts		–	–	–	nr	**1589.13**
Signs on gantries						
Externally illuminated reflectorised signs						
1.50 m²	1.79	165.33	207.99	1333.80	nr	**1707.13**
2.50 m²	2.15	199.14	250.52	1366.74	nr	**1816.40**
3.00 m²	3.07	284.35	357.73	1497.10	nr	**2139.18**
Internally illuminated sign with translucent optical reflective sheeting and remote light source						
0.75 m²	1.56	144.49	181.78	1486.08	nr	**1812.35**
1.00 m²	1.70	157.46	198.09	1968.84	nr	**2324.39**
1.50 m²	2.41	223.22	280.82	3033.72	nr	**3537.76**

SERIES 1200: TRAFFIC SIGNS AND ROAD MARKINGS

Item	Gang hours	Labour £	Plant £	Material £	Unit	Total rate £
Remove from store and re-erect traffic signs						
Take from store and re-erect						
3.0 m high road sign	1.00	92.62	32.63	53.01	nr	**178.26**
road sign on two posts	0.50	46.31	58.26	106.01	nr	**210.59**
ROAD MARKINGS						
Thermoplastic screed or spray						
Note: Unit costs based upon new road with clean surface closed to traffic)						
Continuous line in reflectorised white						
150 mm wide		–	–	–	m	**1.67**
200 mm wide		–	–	–	m	**2.23**
Continuous line in reflectorised yellow						
100 mm wide		–	–	–	m	**1.12**
150 mm wide		–	–	–	m	**1.67**
Intermittent line in reflectorised white						
60 mm wide with 0.60 m line and 0.60 m gap		–	–	–	m	**1.18**
100 mm wide with 1.0 m line and 5.0 m gap		–	–	–	m	**1.12**
100 mm wide with 2.0 m line and 7.0 m gap		–	–	–	m	**1.12**
100 mm wide with 4.0 m line and 2.0 m gap		–	–	–	m	**1.12**
100 mm wide with 6.0 m line and 3.0 m gap		–	–	–	m	**1.12**
150 mm wide with 1.0 m line and 5.0 m gap		–	–	–	m	**1.67**
150 mm wide with 6.0 m line and 3.0 m gap		–	–	–	m	**1.67**
150 mm wide with 0.60 m line and 0.30 m gap		–	–	–	m	**1.67**
200 mm wide with 0.60 m line and 0.30 m gap		–	–	–	m	**2.23**
200 mm wide with 1.0 m line and 1.0 m gap		–	–	–	m	**2.23**
Ancillary line in reflectorised white						
150 mm wide in hatched areas		–	–	–	m	**1.41**
200 mm wide in hatched areas		–	–	–	m	**2.13**
Ancillary line in yellow						
150 mm wide in hatched areas		–	–	–	m	**1.41**
Triangles in reflectorised white						
1.6 m high		–	–	–	nr	**14.78**
2.0 m high		–	–	–	nr	**17.75**
3.75 m high		–	–	–	nr	**22.92**
Circles with enclosing arrows in reflectorised white						
1.6 m dia.		–	–	–	nr	**91.86**
Arrows in reflectorised white						
4.0 m long straight or turning		–	–	–	nr	**36.04**
6.0 m long straight or turning		–	–	–	nr	**44.33**
6.0 m long curved		–	–	–	nr	**44.33**
6.0 m long double headed		–	–	–	nr	**66.52**
8.0 m long double headed		–	–	–	nr	**91.85**
16.0 m long double headed		–	–	–	nr	**133.04**
32.0 m long double headed		–	–	–	nr	**184.75**
Kerb markings in yellow						
250 mm long		–	–	–	nr	**2.50**

SERIES 1200: TRAFFIC SIGNS AND ROAD MARKINGS

Item	Gang hours	Labour £	Plant £	Material £	Unit	Total rate £
ROAD MARKINGS – cont						
Thermoplastic screed or spray – cont						
Letters or numerals in reflectorised white						
1.6 m high		–	–	–	nr	**14.78**
2.0 m high		–	–	–	nr	**17.75**
3.75 m high		–	–	–	nr	**22.92**
Vinyl strip markings						
Note: Unit costs based upon new road with clean surface closed to traffic						
Verynyl strip markings (pedestrian crossings and similar locations)						
200 mm wide line		–	–	–	m	**12.05**
600 × 300 mm single stud tile		–	–	–	nr	**19.96**
REFLECTING ROAD STUDS						
100 × 100 mm square bi-directional reflecting road stud with amber corner cube reflectors		–	–	–	nr	**11.10**
140 × 254 mm rectangular one way reflecting road stud with red catseye reflectors		–	–	–	nr	**22.16**
140 × 254 mm rectangular one way reflecting road stud with green catseye reflectors		–	–	–	nr	**234.34**
140 × 254 mm rectangular bi-directional reflecting road stud with white catseye reflectors		–	–	–	nr	**22.16**
140 × 254 mm rectangular bi-directional reflecting road stud with amber catseye reflectors		–	–	–	nr	**22.16**
140 × 254 mm rectangular bi-directional reflecting road stud without catseye reflectors		–	–	–	nr	**13.31**
Remove from store and re-install road studs						
100 × 100 mm square bi-directional reflecting road stud with corner cube reflectors		–	–	–	nr	**4.36**
140 × 254 mm rectangular one way reflecting road stud with catseye reflectors		–	–	–	nr	**10.34**
TRAFFIC SIGNAL INSTALLATIONS						
Traffic signal installation is carried out exclusively by specialist contractors, although certain items are dealt with by the main contractor or a subcontractor. The following detailed prices are given to assist in the calculation of the total installation cost.						
Installation of signal pedestals, loop detector unit pedestals, controller unit boxes and cable connection pillars						
signal pedestal		–	–	–	nr	**61.28**
loop detector unit pedestal		–	–	–	nr	**27.76**
controller unit box		–	–	–	nr	**62.47**

SERIES 1200: TRAFFIC SIGNS AND ROAD MARKINGS

Item	Gang hours	Labour £	Plant £	Material £	Unit	Total rate £
Excavate trench for traffic signal cable, depth ne 1.50 m; supports, backfilling						
450 mm wide		–	–	–	m	**20.42**
Extra for excavating in hard material		–	–	–	m^3	**40.87**
Saw cutting grooves in pavement for detector loops and feeder cables; seal with hot bitumen sealant after installation						
25 mm deep		–	–	–	m	**29.97**
MARKER POSTS						
Standard reflectorised traffic cylinder 1000 mm high 125 mm dia.; mounted in cats eye base (deliniator)		–	–	–	nr	**38.94**
Glass reinforced plastic marker posts						
types 1,2,3 or 4		–	–	–	nr	**24.53**
types 5,6,7 or 8		–	–	–	nr	**19.37**
Line posts for emergency crossing		–	–	–	nr	**12.27**
PERMANENT BOLLARDS						
Permanent bollard; non-illuminated; precast concrete						
150 mm minimum dia. 750 mm high	0.80	64.65	16.00	178.58	nr	**259.24**
300 mm minimum dia. 750 mm high	0.80	64.65	16.00	205.37	nr	**286.03**
Extra for exposed aggregate finish		–	–	15.40	nr	**15.40**
Permanent bollard; non-illuminated; galvanized steel						
removable and lockable pattern	0.80	64.65	25.67	153.02	nr	**243.34**
MISCELLANEOUS FURNITURE						
Galvanized steel lifting traffic barrier						
4.0 m wide	2.40	193.96	48.01	1030.15	nr	**1272.12**
Precast concrete seats						
bench seat 2.0 m long	0.75	60.61	15.02	1091.77	nr	**1167.40**
bench seat with concrete ends and timber slats 2.0 m long	0.75	60.61	15.02	836.81	nr	**912.44**
Timber seat fixed to concrete base						
bench seat 2.0 m long	0.45	36.37	6.66	857.16	nr	**900.19**
Metal seat						
bench seat 2.0 m long	0.75	60.61	15.02	789.45	nr	**865.08**

SERIES 1300: ROAD LIGHTING COLUMNS AND BRACKETS, CCTV MASTS AND CANTILEVER MASTS

Item	Gang hours	Labour £	Plant £	Material £	Unit	Total rate £
NOTES						
For convenience in pricing this section departs from Series 1300 requirements and shows lighting column costs broken down into main components of columns, brackets, and lamps and cabling.						
The outputs assume operations are continuous and are based on at least 10 complete units and do not include any allowance for on site remedial works after erection of the columns.						
Painting and protection of the columns apart from galvanizing is not included in the following prices or outputs.						
The re-erection cost for lighting columns taken from store assumes that major components are in good condition; the prices below allow a sum of 20% of the value of new materials to cover minor repairs, new fixings and touching up any coatings.						
RESOURCES – LABOUR						
Column erection gang						
1 ganger/chargehand (skill rate 4)	2.00	21.42	–	–	-	**21.42**
1 skilled operative (skill rate 4)	1.00	20.20	–	–	-	**20.20**
1 unskilled operative (general)	1.00	19.60	–	–	-	**19.60**
1 plant operator (craftsman) – 50% of time	0.50	25.23	–	–	-	**25.23**
Total Gang Rate/Hour	1.00	86.44	–	–	-	**86.44**
Bracket erection gang						
1 ganger/chargehand (skill rate 4)	2.00	21.42	–	–	-	**21.42**
1 skilled operative (skill rate 4)	1.00	20.20	–	–	-	**20.20**
1 plant operator (skill rate 4)	1.00	26.30	–	–	-	**26.30**
1 plant operator (craftsman)	1.00	50.46	–	–	-	**50.46**
Total Gang Rate/Hour	1.00	118.37	–	–	-	**118.37**
Lanterns gang						
1 skilled operative (skill rate 3)	1.00	22.45	–	–	-	**22.45**
1 skilled operative (skill rate 4)	1.00	20.20	–	–	-	**20.20**
1 plant operator (skill rate 4)	1.00	26.30	–	–	-	**26.30**
1 plant operator (craftsman)	1.00	50.46	–	–	-	**50.46**
Total Gang Rate/Hour	1.00	119.40	–	–	-	**119.40**
RESOURCES – PLANT						
Columns and bracket arms						
15 t mobile crane – 50% of time		–	56.74	–	-	**56.74**
125 cfm compressor – 50% of time		–	9.03	–	-	**9.03**
compressor tools: 2 single head scabbler – 50% of time		–	2.23	–	-	**2.23**
2 t dumper – 50% of time		–	5.41	–	-	**5.41**
Total Rate/Hour		–	73.40	–	-	**73.40**
Bracket arms						
15 t mobile crane		–	113.48	–	-	**113.48**
access platform, Simon hoist (50 ft)		–	33.91	–	-	**33.91**
Total Rate/Hour		–	147.38	–	-	**147.38**

SERIES 1300: ROAD LIGHTING COLUMNS AND BRACKETS, CCTV MASTS AND CANTILEVER MASTS

Item	Gang hours	Labour £	Plant £	Material £	Unit	Total rate £
Lanterns						
15 t mobile crane		–	113.48	–	-	**113.48**
access platform, Simon hoist (50 ft)		–	33.91	–	-	**33.91**
Total Rate/Hour		–	147.38	–	-	**147.38**
ROAD LIGHTING COLUMNS, BRACKETS, WALL MOUNTINGS, CCTV MASTS AND CANTILEVER MASTS						
Galvanized steel road lighting columns to BS EN 40 with flange plate base (including all control gear, switching, fuses and internal wiring)						
4.0 m nominal height	0.75	64.83	55.06	159.99	nr	**279.89**
6.0 m nominal height	0.80	69.15	78.05	438.22	nr	**585.42**
8.0 m nominal height	0.96	82.99	93.66	556.46	nr	**733.11**
10.0 m nominal height	1.28	110.65	124.88	695.56	nr	**931.09**
12.0 m nominal height	1.44	124.48	140.49	1071.19	nr	**1336.16**
15.0 m nominal height	1.76	152.14	171.71	1669.36	nr	**1993.21**
3.0 m cast iron column (pedestrian/landscape area)	0.75	64.83	73.18	1947.61	nr	**2085.63**
Precast concrete lighting columns to BS EN 40 with flange plate base (including all control gear, switching, fuses and internal wiring)						
5.0 m nominal height	0.75	64.83	73.18	319.72	nr	**457.74**
10.0 m nominal height	1.28	110.65	124.92	813.22	nr	**1048.79**
Galvanized steel bracket arm to BS EN 40; with 5° uplift						
0.5 m projection, single arm	0.16	18.94	54.87	324.52	nr	**398.33**
1.0 m projection, single arm	0.19	22.49	28.00	454.36	nr	**504.85**
1.5 m projection, single arm	0.21	18.15	30.59	519.25	nr	**567.99**
2.0 m projection, single arm	0.27	31.96	39.79	584.15	nr	**655.90**
1.0 m projection, double arm	0.29	34.33	42.74	584.15	nr	**661.21**
2.0 m projection, double arm	0.32	37.88	47.16	649.05	nr	**734.09**
Precast concrete bracket arm to BS EN 40; with 5° uplift						
1.0 m projection, single arm	0.24	28.41	35.37	191.75	nr	**255.53**
2.0 m projection, single arm	0.32	37.88	47.16	225.99	nr	**311.03**
1.0 m projection, double arm	0.35	41.43	51.58	273.91	nr	**366.93**
2.0 m projection, doube arm	0.37	43.80	54.53	342.40	nr	**440.73**
Lantern unit with photo-electric control set to switch on at 100 lux; lamps						
55 W SON (P226); to suit 4 m and 5 m columns	0.40	47.76	36.26	343.42	nr	**427.44**
70 W SON (P236); to suit 5 m and 6 m columns	0.40	47.76	36.26	364.04	nr	**448.06**
250 W SON (P426); to suit 8 m, 10 m and 12 m columns	0.50	59.70	45.32	535.76	nr	**640.78**
Sphere 70 W SON; to suit 3 m columns (P456)	0.50	59.70	45.32	638.79	nr	**743.81**
400 W SON High pressure sodium; to suit 12 m and 15 m columns	0.50	59.70	45.32	638.79	nr	**743.81**

SERIES 1300: ROAD LIGHTING COLUMNS AND BRACKETS, CCTV MASTS AND CANTILEVER MASTS

Item	Gang hours	Labour £	Plant £	Material £	Unit	Total rate £
REMOVE FROM STORE AND RE-ERECT ROAD LIGHTING COLUMNS AND BRACKETS AND MASTS						
Re-erection of galvanized steel road lighting columns with flange plate base; including all control gear, switching, fuses and internal wiring						
4.0 m nominal height	0.75	64.83	55.08	24.73	nr	**144.65**
6.0 m nominal height	0.80	69.15	58.77	32.48	nr	**160.41**
8.0 m nominal height	0.96	82.99	70.47	40.10	nr	**193.56**
10.0 m nominal height	1.28	110.65	94.00	47.72	nr	**252.37**
12.0 m nominal height	1.44	124.48	105.75	54.58	nr	**284.80**
15.0 m nominal height	1.76	152.14	129.24	61.31	nr	**342.68**
3.0 m cast iron column (pedestrian/landscape area)	0.75	64.83	55.11	20.89	nr	**140.83**
Re-erection of precast concrete lighting columns with flange plate base; including all control gear, switching, fuses and internal wiring						
5.0 m nominal height	0.75	64.83	55.08	28.83	nr	**148.75**
10.0 m nominal height	1.28	110.65	94.00	47.72	nr	**252.37**
Re-erection of galvanized steel bracket arms						
0.5 m projection, single arm	0.16	18.94	23.58	41.11	nr	**83.63**
1.0 m projection, single arm	0.19	22.49	28.00	41.11	nr	**91.61**
1.5 m projection, single arm	0.21	24.86	30.95	41.11	nr	**96.92**
2.0 m projection, single arm	0.27	31.96	39.79	41.11	nr	**112.87**
1.0 m projection, double arm	0.29	34.33	42.74	41.11	nr	**118.18**
2.0 m projection, double arm	0.32	37.88	47.16	41.11	nr	**126.15**
Re-erection of precast concrete bracket arms						
1.0 m projection, single arm	0.24	28.41	35.37	41.11	nr	**104.89**
2.0 m projection, single arm	0.32	37.88	47.16	41.11	nr	**126.15**
1.0 m projection, double arm	0.35	41.43	52.79	41.11	nr	**135.33**
2.0 m projection, double arm	0.37	43.80	51.58	41.11	nr	**136.49**
Re-installing lantern unit with photo-electric control set to switch on at 100 lux; lamps						
55 W SON (P226); to suit 4 m and 5 m columns	0.40	47.76	36.26	–	nr	**84.02**
70 W SON (P236); to suit 5 m and 6 m columns	0.22	26.27	36.26	–	nr	**62.53**
250 W SON (P426); to suit 8 m, 10 m and 12 m columns	0.50	59.70	45.32	–	nr	**105.03**
Sphere 70 W SON; to suit 3 m (P456)	0.50	59.70	45.32	–	nr	**105.03**
400 W SON High pressure sodium; to suit 12 m and 15 m columns	0.50	59.70	45.32	–	nr	**105.03**

SERIES 1400: ELECTRICAL WORK FOR ROAD LIGHTING AND TRAFFIC SIGNS

Item	Gang hours	Labour £	Plant £	Material £	Unit	Total rate £
RESOURCES – LABOUR						
Trenching gang						
1 ganger/chargehand (skill rate 4)	2.00	21.42	–	–	-	**21.42**
1 skilled operative (skill rate 4)	1.00	20.20	–	–	-	**20.20**
2 unskilled operatives (general)	2.00	39.20	–	–	-	**39.20**
1 plant operator (skill rate 3) – 75% of time	0.75	21.88	–	–	-	**21.88**
Total Gang Rate/Hour	1.00	102.69	–	–	-	**102.69**
Cable laying gang						
1 ganger/chargehand (skill rate 4)	2.00	21.42	–	–	-	**21.42**
1 skilled operative (skill rate 4)	1.00	20.20	–	–	-	**20.20**
2 skilled operatives (skill rate 3)	2.00	44.91	–	–	-	**44.91**
Total Gang Rate/Hour	1.00	86.52	–	–	-	**86.52**
RESOURCES – PLANT						
Service trenching						
JCB 3CX backhoe – 50% of time		–	24.16	–	-	**24.16**
125 cfm compressor – 50% of time		–	9.03	–	-	**9.03**
compressor tools: 2 single head scabbler – 50% of time		–	2.23	–	-	**2.23**
2 t dumper – 50% of time		–	5.41	–	-	**5.41**
trench excavator – 25% of time		–	32.05	–	-	**32.05**
Total Rate/Hour		–	72.46	–	-	**72.46**
Cable laying						
8 t IVECO chassis or similar – 50% of time		–	13.97	–	-	**13.97**
Total Rate/Hour		–	13.97	–	-	**13.97**
LOCATING BURIED ROAD LIGHTING AND TRAFFIC SIGNS CABLES						
Locating buried road lighting and traffic signs cable						
in carriageways, footways, bridge decks and paved areas	0.25	25.67	12.79	–	m	**38.46**
in verges and central reserves	0.20	20.54	10.23	–	m	**30.77**
in side slopes of cuttings or side slopes of embankments	0.15	15.40	7.68	–	m	**23.08**
TRENCH FOR CABLE OR DUCT						
Excavate trench for cable or duct						
300 to 450 mm wide; depth not exceeding 1.5 m	0.15	15.40	7.68	–	m	**23.08**
450 to 600 mm wide; depth not exceeding 1.5 m	0.20	20.54	10.23	–	m	**30.77**
Extra for excavating rock or reinforced concrete in trench	0.50	51.35	25.60	–	m^3	**76.94**
Extra for excavating brickwork or mass concrete in trench	0.40	41.08	20.47	–	m^3	**61.54**
Extra for backfilling with pea gravel	0.02	2.05	1.04	–	m^3	**64.28**
Extra for 450 × 100 mm sand cable bedding and covering	0.02	2.05	1.04	2.48	m	**5.57**
Extra for PVC marker tape	0.01	0.51	0.26	2.44	m	**3.22**
Extra for 150 × 300 clay cable tiles	0.05	5.13	2.56	5.52	m	**13.21**
Extra for 150 × 900 concrete cable tiles	0.03	3.08	1.55	9.21	m	**13.84**

SERIES 1400: ELECTRICAL WORK FOR ROAD LIGHTING AND TRAFFIC SIGNS

Item	Gang hours	Labour £	Plant £	Material £	Unit	Total rate £
CABLE AND DUCT						
600/1000 V 2 core, PVC/SWA/PVC cable with copper conductors						
Cable; in trench not exceeding 1.5 m deep						
2.5 mm^2	0.01	0.87	0.14	3.10	m	**4.10**
4 mm^2	0.02	1.73	0.28	4.15	m	**6.16**
6 mm^2	0.02	1.73	0.28	4.95	m	**6.96**
10 mm^2	0.02	1.73	0.28	6.80	m	**8.81**
16 mm^2	0.02	1.73	0.28	8.57	m	**10.58**
25 mm^2	0.03	2.60	0.42	10.83	m	**13.85**
600/1000 V 4 core, PVC/SWA/PVC cable with copper conductors						
Cable; in trench not exceeding 1.5 m deep						
16 mm^2	0.03	2.60	0.42	12.53	m	**15.54**
35 mm^2	0.13	11.25	1.82	15.19	m	**28.25**
70 mm^2	0.15	12.98	2.10	20.45	m	**35.53**
600/1000 V 2 core, PVC/SWA/PVC cable drawn into ducts, pipe bays or troughs						
Cable; in trench not exceeding 1.5 m deep						
2.5 mm^2	0.01	0.87	0.14	3.10	m	**4.10**
4 mm^2	0.02	1.73	0.28	4.15	m	**6.16**
6 mm^2	0.02	1.73	0.28	4.95	m	**6.96**
10 mm^2	0.02	1.73	0.28	6.43	m	**8.44**
16 mm^2	0.03	2.60	0.42	8.57	m	**11.58**
35 mm^2	0.12	10.38	1.68	10.81	m	**22.87**
70 mm^2	0.14	12.11	1.96	16.51	m	**30.57**
CABLE JOINTS AND CABLE TERMINATIONS						
Straight joint in 2 core PVC/SWA/PVC cable						
2.5 mm^2	0.30	25.96	4.19	63.33	nr	**93.48**
4 mm^2	0.30	25.96	4.19	63.33	nr	**93.48**
6 mm^2	0.32	27.69	4.47	63.33	nr	**95.49**
10 mm^2	0.42	36.34	5.87	64.66	nr	**106.87**
16 mm^2	0.50	43.26	6.98	67.81	nr	**118.05**
35 mm^2	0.80	69.22	11.17	86.25	nr	**166.64**
70 mm^2	1.15	99.50	16.06	95.46	nr	**211.02**
Tee joint in 2 core PVC/SWA/PVC cable						
2.5 mm^2	0.46	39.80	6.43	90.03	nr	**136.25**
4 mm^2	0.46	39.80	6.43	90.01	nr	**136.24**
6 mm^2	0.48	41.53	6.70	90.01	nr	**138.25**
10 mm^2	0.62	53.64	8.66	92.36	nr	**154.66**
16 mm^2	0.74	64.02	10.34	96.40	nr	**170.76**
25 mm^2	0.91	78.73	12.71	103.24	nr	**194.68**
Tee joint in 4 core PVC/SWA/PVC cable						
16 mm^2	1.10	95.17	15.36	106.03	nr	**216.56**
35 mm^2	1.30	112.47	18.16	118.71	nr	**249.35**
70 mm^2	1.60	138.43	22.35	170.50	nr	**331.28**

SERIES 1400: ELECTRICAL WORK FOR ROAD LIGHTING AND TRAFFIC SIGNS

Item	Gang hours	Labour £	Plant £	Material £	Unit	**Total rate £**
Looped terminations of 2 core PVC/SWA/PVC cable in lit sign units, traffic signals installation control unit, pedestrian crossing control unit, road lighting column, wall mounting, subway distribution box, gantry distribution box or feeder pillar.						
2.5 mm^2	0.15	12.98	2.10	8.82	nr	**23.89**
6 mm^2	0.15	12.98	2.10	11.96	nr	**27.03**
10 mm^2	0.16	13.84	2.23	15.54	nr	**31.62**
16 mm^2	0.25	21.63	3.49	15.87	nr	**40.99**
25 mm^2	0.30	25.96	4.19	22.14	nr	**52.28**
Terminations of 4 core PVC/SWA/PVC cable in lit sign units, traffic signals installation control unit, pedestrian crossing control unit, road lighting column, wall mounting, subway distribution box, gantry distribution box or feeder pillar.						
16 mm^2	0.35	30.28	4.89	17.14	nr	**52.31**
35 mm^2	0.55	47.59	7.68	31.27	nr	**86.53**
70 mm^2	0.68	58.83	9.50	59.35	nr	**127.68**
FEEDER PILLARS						
Galvanized steel feeder pillars						
411 × 610 mm	4.64	401.45	64.81	512.50	nr	**978.76**
611 × 810 mm	4.24	366.84	59.22	576.54	nr	**1002.61**
811 × 1110 mm	4.96	429.13	69.28	1153.12	nr	**1651.53**
1111 × 1203 mm	4.19	362.51	58.67	1537.49	nr	**1958.67**
EARTH ELECTRODES						
Earth electrodes providing minimal protection using earth rods, plates or stops and protective tape and joint						
to suit columns ne 12.0 m	0.40	34.61	–	353.77	nr	**388.37**
to suit columns ne 15.0 m	0.40	34.61	–	464.74	nr	**499.35**
to suit superstructure or buildings using copper lead conductor (per 23 m height)	1.00	86.52	–	783.86	nr	**870.38**
CHAMBERS						
Brick chamber with galvanized steel cover and frame; depth to uppermost surface of base slab						
ne 1.0 m deep		–	–	–	nr	**1005.99**

SERIES 1500: MOTORWAY COMMUNICATIONS

Item	Gang hours	Labour £	Plant £	Material £	Unit	Total rate £
RESOURCES – LABOUR						
Service trenching gang						
1 ganger/chargehand (skill rate 4)	2.00	21.42	–	–	-	**21.42**
1 skilled operative (skill rate 4)	1.00	20.20	–	–	-	**20.20**
2 unskilled operatives (general)	2.00	39.20	–	–	-	**39.20**
1 plant operator (skill rate 3) – 75% of time	0.75	21.88	–	–	-	**21.88**
Total Gang Rate/Hour	1.00	102.69	–	–	-	**102.69**
RESOURCES – PLANT						
Trenching						
JCB 3CX backhoe (50% of time)		–	24.16	–	-	**24.16**
125 cfm compressor (50% of time)		–	9.03	–	-	**9.03**
compressor tools: 2 single head scabbler (50% of time)		–	2.23	–	-	**2.23**
2 t dumper (50% of time)		–	5.41	–	-	**5.41**
trench excavator (25% of time)		–	25.30	–	-	**25.30**
Total Rate/Hour		–	66.12	–	-	**66.12**
LOCATING BURIED COMMUNICATIONS CABLING						
Locating buried road lighting and traffic signs cable						
in carriageways, footways, bridge decks and paved areas	0.25	25.67	12.79	–	m	**38.46**
in verges and central reserves	0.20	20.54	10.23	–	m	**30.77**
in side slopes of cuttings or side slopes of embankments	0.15	15.40	7.68	–	m	**23.08**
TRENCH FOR COMMUNICATIONS CABLE OR DUCT						
Trench for cable						
300 to 450 mm wide; depth not exceeding 1.5 m	0.15	15.40	7.68	–	m	**23.08**
450 to 600 mm wide; depth not exceeding 1.5 m	0.20	20.54	10.23	–	m	**30.77**
Extra for excavating rock or reinforced concrete in trench	0.50	51.35	25.60	–	m^3	**76.94**
Extra for excavating brickwork or mass concrete in trench	0.40	41.08	20.47	–	m^3	**61.54**
Extra for backfilling with pea gravel	0.02	2.05	1.04	–	m^3	**64.28**
Extra for 450 × 100 mm sand cable bedding and covering	0.02	2.05	1.04	2.10	m	**5.19**
Extra for PVC marker tape	0.01	0.51	0.26	2.44	m	**3.22**
Extra for 150 × 300 clay cable tiles	0.05	5.13	2.56	5.52	m	**13.21**
Extra for 150 × 900 concrete cable tiles	0.03	3.08	1.55	9.21	m	**13.84**
COMMUNICATIONS CABLING AND DUCT						
Communication cables laid in trench						
type A1 2 pair 0.9 mm^2 armoured multi-pair		–	–	–	m	**5.60**
type A2 20 pair 0.9 mm^2 armoured multi-pair		–	–	–	m	**10.06**
type A3 30 pair 0.9 mm^2 armoured multi-pair		–	–	–	m	**12.35**
Power cables laid in trench						
type A4 10.0 mm^2 armoured split concentric		–	–	–	m	**9.94**

SERIES 1500: MOTORWAY COMMUNICATIONS

Item	Gang hours	Labour £	Plant £	Material £	Unit	Total rate £
Detector feeder cables laid in trench						
type A6 50/0.25 m² single core detector feeder cable		–	–	–	m	**12.57**
type A7 50/0.25 m² single core detector feeder cable		–	–	–	m	**12.57**
COMMUNICATIONS CABLE JOINTS AND TERMINATIONS						
Cable terminations						
of type 1 cable		–	–	–	nr	**79.95**
of type 2 cable		–	–	–	nr	**357.21**
of type 3 cable		–	–	–	nr	**512.40**
of type 4 cable		–	–	–	nr	**85.67**
of type 6 cable		–	–	–	nr	**51.05**
of type 7 cable		–	–	–	nr	**51.05**
COMMUNICATIONS EQUIPMENT						
Cabinet bases						
foundation plinth		–	–	–	nr	**189.84**
Matrix signal post bases						
foundation plinth		–	–	–	nr	**197.75**
CCTV camera bases						
foundation plinth		–	–	–	nr	**201.14**
Wall mounted brackets						
at maximum 15.0 m height		–	–	–	nr	**73.45**
Fix only the following equipment						
communication equipment cabinet, 600 type series		–	–	–	nr	**188.21**
terminator type II		–	–	–	nr	**199.52**
emergency telephone post		–	–	–	nr	**99.94**
telephone housing		–	–	–	nr	**31.20**
matrix signal post		–	–	–	nr	**99.94**
Motorwarn/fogwarn		–	–	–	nr	**199.76**
distributor on gantry		–	–	–	nr	**254.34**
isolator switch for gantry		–	–	–	nr	**553.92**
heater unit mounted on gantry; Henleys' 65 W type 22501		–	–	–	nr	**37.95**
Terminal blocks						
Klippon BK6		–	–	–	nr	**9.48**
Klippon BK12		–	–	–	nr	**10.71**
Work to pavement for loop detection circuits						
cut or form grooves in pavement for detector loops and feeders		–	–	–	m	**7.32**
additional cost for sealing with hot bitumen sealant		–	–	–	m	**0.96**
CHAMBERS						
Brick chamber with galvanized steel cover and frame; depth to uppermost surface of base slab						
ne 1.0 m deep		–	–	–	nr	**1005.99**

SERIES 1600: PILING AND EMBEDDED RETAINING WALLS

Item	Gang hours	Labour £	Plant £	Material £	Unit	Total rate £
NOTES						
There are a number of different types of piling which are available for use in differing situations. Selection of the most suitable type of piling for a particular site will depend on a number of factors including the physical conditions likely to be encountered during driving, the loads to be carried, the design of superstructure, etc.						
The most commonly used systems are included in this section.						
It is essential that a thorough and adequate site investigation is carried out to ascertain details of the ground strata and bearing capacities to enable a proper assessment to be made of the most suitable and economical type of piling to be adopted.						
There are so many factors, apart from design considerations, which influence the cost of piling that it is not possible to give more than an approximate indication of costs. To obtain reliable costs for a particular contract advice should be sought from a company specializing in the particular type of piling proposed. Some Specialist Contractors will also provide a design service if required.						
PILING PLANT						
Driven precast concrete reinforced piles						
Establishment of piling plant for						
235 × 235 mm precast reinforced and prestressed concrete piles in main piling		–	–	–	item	**6396.30**
275 × 275 mm precast reinforced and prestressed concrete piles in main piling		–	–	–	item	**6396.30**
350 × 350 mm precast reinforced and prestressed concrete piles in main piling		–	–	–	item	**6980.31**
Moving piling plant for						
235 × 235 mm precast reinforced and prestressed concrete piles in main piling		–	–	–	nr	**71.13**
275 × 275 mm precast reinforced and prestressed concrete piles in main piling		–	–	–	nr	**71.13**
350 × 350 mm precast reinforced and prestressed concrete piles in main piling		–	–	–	nr	**78.26**
Bored in situ reinforced concrete piling (tripod rig)						
Establishment of piling plant for 500 mm dia. cast-in-place concrete piles (tripod rig) in main piling		–	–	–	item	**9736.92**
Moving piling plant for 500 mm dia. cast-in-place concrete piles (tripod rig) in main piling		–	–	–	nr	**249.72**

SERIES 1600: PILING AND EMBEDDED RETAINING WALLS

Item	Gang hours	Labour £	Plant £	Material £	Unit	Total rate £
Bored in situ reinforced concrete piling (mobile rig)						
Establishment of piling plant for 500 mm dia. cast-in-place concrete piles (mobile rig) in main piling		–	–	–	item	**16843.75**
Moving piling plant for 500 mm dia. cast-in-place concrete piles in (mobile rig) main piling		–	–	–	nr	**396.27**
Concrete injected piles (continuous flight augured)						
Establishment of piling plant for cast-in-place concrete piles (CFA) in main piling						
450 mm dia.; 650kN		–	–	–	item	**11003.03**
600 mm dia.; 1400kN		–	–	–	item	**11003.03**
750 mm dia.; 2200kN		–	–	–	item	**11003.03**
Moving piling plant for cast-in-place concrete piles (CFA) in main piling						
450 mm dia.; 650kN		–	–	–	nr	**53.97**
600 mm dia. 1400kN		–	–	–	nr	**53.97**
750 mm dia. 2200kN		–	–	–	nr	**53.97**
Driven cast in place piles; segmental casing method						
Establishment of piling plant for cast-in-place concrete piles in main piling		–	–	–	item	**16843.75**
Moving piling plant for cast-in-place concrete piles in main piling		–	–	–	nr	**396.27**
Establishment of piling plant for cast-in-place concrete piles in main piling						
bottom driven		–	–	–	item	**7360.00**
top driven		–	–	–	item	**7360.00**
Moving piling plant for 430 mm dia. cast-in-place concrete piles in main piling						
bottom driven		–	–	–	nr	**128.80**
top driven		–	–	–	nr	**87.40**
Steel bearing piles						
Establishment of piling plant for steel bearing piles in main piling						
maximum 100 miles radius from base		–	–	–	item	**5569.20**
maximum 250 miles radius from base		–	–	–	item	**12316.50**
Moving piling plant for steel bearing piles in main piling		–	–	–	nr	**85.20**
Z section sheet steel piles						
Provision of all plant, equipment and labour including transport to and from the site and establishing and dismantling for						
driving of sheet piling		–	–	–	item	**9740.42**
extraction of sheet piling		–	–	–	item	**2622.42**

SERIES 1600: PILING AND EMBEDDED RETAINING WALLS

Item	Gang hours	Labour £	Plant £	Material £	Unit	Total rate £
PILING PLANT – cont						
U section sheet steel piles						
Provision of plant, equipment and labour including transport to and from the site and establishing and dismantling for						
driving of sheet piling		–	–	–	item	**9740.42**
extraction of sheet piling		–	–	–	item	**2622.42**
Establishment of piling plant for steel tubular piles in main piling						
maximum 100 miles radius from base		–	–	–	item	**10500.00**
maximum 250 miles radius from base		–	–	–	item	**26250.00**
Moving piling plant for steel tubular piles in main piling		–	–	–	nr	**396.27**
PRECAST CONCRETE PILES						
Driven precast reinforced concrete piles						
The following unit costs cover the installation of driven precast concrete piles by using a hammer acting on a shoe fitted onto or cast into the pile unit. The costs are based installing 100 piles of nominal sizes stated, and a concrete strength of 50 N/mm² suitably reinforced for a working load not exceeding 600kN, with piles average 15 m long, on a clear site with reasonable access.						
Single pile lengths are normally a maximum of 13 m long, at which point, a mechanical interlocking joint is required to extend the pile. These joints are most economically and practically formed at works.						
Lengths, sizes of sections, reinforcement details and concrete mixes vary for differing contractors, whose specialist advice should be sought for specific designs.						
Precast concrete piles; concrete 50 N/mm²						
235 × 235 mm; 5–10 m in length; main piling		–	–	–	m	**44.44**
275 × 275 mm; 5–10 m in length; main piling		–	–	–	m	**41.73**
350 × 350 mm; 5–10 m in length; main piling		–	–	–	m	**71.25**
Mechanical interlocking joint						
235 × 235 mm		–	–	–	nr	**100.34**
275 × 275 mm		–	–	–	nr	**104.68**
350 × 350 mm		–	–	–	nr	**200.63**
Driving vertical precast piles						
235 × 235 mm; 5–10 m in length; in main piling		–	–	–	m	**5.73**
275 × 275 mm; 5–10 m in length; in main piling		–	–	–	m	**5.64**
350 × 350 mm; 5–10 m in length; in main piling		–	–	–	m	**8.00**

SERIES 1600: PILING AND EMBEDDED RETAINING WALLS

Item	Gang hours	Labour £	Plant £	Material £	Unit	**Total rate £**
Stripping vertical precast concrete pile heads						
235 × 235 mm piles in main piling		–	–	–	nr	**61.97**
275 × 275 mm piles in main piling		–	–	–	nr	**64.02**
350 × 350 mm piles in main piling		–	–	–	nr	**93.18**
Standing time						
275 × 275 mm		–	–	–	hr	**248.98**
350 × 350 mm		–	–	–	hr	**248.98**
CAST IN PLACE PILES						
Bored in situ reinforced concrete piling (tripod rig)						
The following unit costs cover the construction of small dia. bored piling using light and compact tripod rigs requiring no expensive site levelling or access ways. Piling can be constructed in very restricted headroom or on confined and difficult sites. Standard dias are between 400 and 600 mm with a normal maximum depth of 30 m.						
The costs are based on installing 100 piles of 500 mm nominal dia., a concrete strength of 20 N/mm² with nominal reinforcement, on a clear site with reasonable access. Disposal of excavated material is included separately.						
Vertical 500 mm dia. cast-in-place piles; 20 N/mm² concrete; nominal reinforcement; in main piling		–	–	–	m	**288.25**
Vertical 500 mm dia. empty bores in main piling		–	–	–	m	**57.09**
Add for boring through obstructions		–	–	–	hr	**135.56**
Standing time		–	–	–	hr	**74.76**
Bored in situ reinforced concrete piling (mobile rig)						
The following unit costs cover the construction of small dia. bored piles using lorry or crawler mounted rotary boring rigs. This type of plant is more mobile and faster in operation than the tripod rigs and is ideal for large contracts in cohesive ground. Construction of piles under bentonite suspension can be carried out to obviate the use of liners. Standard dias of 450 to 900 mm dia. can be constructed to depths of 30 m.						
The costs are based on installing 100 piles of 500 mm nominal dia., a concrete strength of 20 N/mm² with nominal reinforcement, on a clear site with reasonable access. Disposal of excavated material is included separately.						
Vertical 500 mm dia. cast-in-place piles; 20 N/mm² concrete; nominal reinforcement		–	–	–	m	**187.84**
Vertical 500 mm dia. empty bores		–	–	–	m	**57.09**
Add for boring through obstructions		–	–	–	hr	**135.56**
Standing time		–	–	–	hr	**74.76**

SERIES 1600: PILING AND EMBEDDED RETAINING WALLS

Item	Gang hours	Labour £	Plant £	Material £	Unit	Total rate £
CAST IN PLACE PILES – cont						
Concrete injected piles (continuous flight augured)						
The following unit costs cover the construction of piles by screwing a continuous flight auger into the ground to a design depth (determined prior to commencement of piling operations and upon which the rates are based and subsequently varied to actual depths). Concrete is then pumped through the hollow stem of the auger to the bottom and the pile formed as the auger is withdrawn. Spoil is removed by the auger as it is withdrawn. This is a fast method of construction without causing disturbance or vibration to adjacent ground. No casing is required even in unsuitable soils. Reinforcement can be placed after grouting is complete.						
The costs are based on installing 100 piles on a clear site with reasonable access. Disposal of excavated material is included separately.						
Vertical cast-in-place piles; 20 N/mm² concrete						
450 mm dia.; 650kN; 10–15 m in length; main piling		–	–	–	m	**29.38**
600 mm dia.; 1400kN; 10–15 m in length; main piling		–	–	–	m	**68.77**
750 mm dia.; 2200kN; 10–15 m in length; main piling		–	–	–	m	**74.96**
Vertical empty bores						
450 mm dia.		–	–	–	m	**35.53**
600 mm dia.		–	–	–	m	**35.96**
750 mm dia.		–	–	–	m	**41.93**
Standing time/Boring through obstructions time						
450 mm dia.		–	–	–	hr	**472.22**
600 mm dia.		–	–	–	hr	**472.22**
750 mm dia.		–	–	–	hr	**472.22**
DRIVEN CAST-IN-PLACE PILES						
Driven cast-in-place piles; segmental casing method						
The following unit costs cover the construction of piles by driving into hard material using a serrated thick wall tube. It is oscillated and pressed into the hard material using a hydraulic attachment to the piling rig. The hard material is broken up using chiselling methods and is then removed by mechanical grab.						

SERIES 1600: PILING AND EMBEDDED RETAINING WALLS

Item	Gang hours	Labour £	Plant £	Material £	Unit	Total rate £
Vertical cast-in-place piles; 20 N/mm² concrete						
620 mm dia.; 10–15 m in length; main piling	–	–	–		m	**263.05**
1180 mm dia.; 10–15 m in length; main piling	–	–	–		m	**286.26**
1500 mm dia.; 10–15 m in length; main piling	–	–	–		m	**394.58**
Standing time	–	–	–		hr	**74.76**
Add for driving through obstructions	–	–	–		hr	**429.41**
Driven in situ reinforced concrete piling						
The following unit costs cover the construction of piles by driving a tube into the ground either by using an internal hammer acting on a gravel or concrete plug or, as is more usual, by using an external hammer on a driving helmet at the top of the tube. After driving to the required depth an enlarged base is formed by hammering out successive charges of concrete down the tube. The tube is then filled with concrete which is compacted as the tube is vibrated and withdrawn. Piles of 350 to 500 mm dia. can be constructed with rakes up to 1 in 4 to carry working loads up to 120t per pile.						
The costs are based on installing 100 piles of 430 mm nominal dia., a concrete strength of 20 N/mm² suitably reinforced for a working load not exceeding 750kN, on a clear site with reasonable access.						
Establishment of piling plant for cast-in-place concrete piles in main piling						
bottom driven	–	–	–		item	**15750.00**
top driven	–	–	–		item	**15750.00**
Moving piling plant for 430 mm dia. cast-in-place concrete piles in main piling						
bottom driven	–	–	–		nr	**57.07**
top driven	–	–	–		nr	**57.07**
Vertical 430 mm dia. cast-in-place piles 20 N/mm² concrete; reinforcement for 750kN maximum load						
bottom driven	–	–	–		m	**30.89**
top driven	–	–	–		m	**31.16**
Standing time	–	–	–		hr	**74.76**
Add for driving through obstructions where within the capabilities of the normal plant	–	–	–		hr	**249.72**
Stripping vertical concrete pile heads						
430 mm dia. heads	–	–	–		nr	**107.05**
REINFORCEMENT FOR CAST-IN-PLACE PILES						
Mild steel bars BS 4449; Grade 250						
Bar reinforcement nominal size 16 mm and under ne 12 m in length	6.74	922.93	137.49	953.42	tonne	**2013.84**
Bar reinforcement nominal size 20 mm and over; not exceeding 12 m in length						
20 mm nominal size	4.44	607.98	90.57	952.67	tonne	**1651.22**
25 mm nominal size	4.44	607.98	90.57	952.67	tonne	**1651.22**
32 mm nominal size	4.44	607.98	90.57	952.67	tonne	**1651.22**
40 mm nominal size	4.44	607.98	90.57	952.67	tonne	**1651.22**

SERIES 1600: PILING AND EMBEDDED RETAINING WALLS

Item	Gang hours	Labour £	Plant £	Material £	Unit	Total rate £
REINFORCEMENT FOR CAST-IN-PLACE PILES – cont						
Mild steel bars BS 4449 – cont						
ADD to the above for bars						
12–13.5 m long		–	–	143.60	tonne	**143.60**
13.5–15 m long		–	–	143.60	tonne	**143.60**
over 15 m long, per 500 mm increment		–	–	14.34	500 mm	**14.34**
High yield steel bars BS 4449; deformed, Grade 500C						
Bar reinforcement nominal size 16 mm and under not exceeding 12 m in length	6.74	922.93	137.49	952.07	tonne	**2012.49**
Bar reinforcement nominal size 20 mm and over not exceeding 12 m in length						
20 mm nominal size	4.44	607.98	90.57	952.07	tonne	**1650.63**
25 mm nominal size	4.44	607.98	90.57	952.07	tonne	**1650.63**
32 mm nominal size	4.44	607.98	90.57	952.07	tonne	**1650.63**
40 mm nominal size	4.44	607.98	90.57	952.07	tonne	**1650.63**
ADD to the above for bars						
12–13.5 m long		–	–	143.60	tonne	**143.60**
13.5–15 m long		–	–	143.60	tonne	**143.60**
over 15 m long; per 500 mm increment		–	–	14.34	500 mm	**14.34**
Helical reinforcement nominal size 16 mm and under						
ne 12 m in length	6.74	922.93	137.49	991.48	tonne	**2051.90**
Helical reinforcement nominal size 20 mm and over						
20 mm nominal size	4.44	607.98	90.57	961.59	tonne	**1660.15**
25 mm nominal size	4.44	607.98	90.57	961.59	tonne	**1660.15**
32 mm nominal size	4.44	607.98	90.57	961.59	tonne	**1660.15**
40 mm nominal size	4.44	607.98	90.57	961.59	tonne	**1660.15**
Dowels						
16 mm dia. × 600 mm long	0.10	13.69	2.04	2.32	nr	**18.05**
20 mm dia. × 600 mm long	0.11	14.38	2.04	4.17	nr	**20.59**
25 mm dia. × 600 mm long	0.11	14.38	2.04	4.78	nr	**21.19**
32 mm dia. × 600 mm long	0.10	13.69	2.04	5.84	nr	**21.57**
STEEL BEARING PILES						
Steel bearing piles are commonly carried out by a Specialist Contractor and whose advice should be sought to arrive at accurate costing. However the following items can be used to assess a budget cost for such work.						
The following unit costs are based upon driving 100nr steel bearing piles on a clear site with reasonable access. Supply is based on delivery 75 miles from works, in loads over 20 t.						

SERIES 1600: PILING AND EMBEDDED RETAINING WALLS

Item	Gang hours	Labour £	Plant £	Material £	Unit	**Total rate £**
Steel bearing piles						
Standing time		–	–	–	hr	**747.59**
203 × 203 × 45 kg/m steel bearing piles; Grade S275						
not exceeding 5 m in length in main piling		–	–	72.50	m	**72.50**
5–10 m in length in main piling		–	–	71.25	m	**71.25**
10–15 m in length in main piling		–	–	71.25	m	**71.25**
15–20 m in length in main piling		–	–	75.99	m	**75.99**
203 × 203 × 54 kg/m steel bearing piles; Grade S275						
not exceeding 5 m in length in main piling		–	–	86.30	m	**86.30**
5–10 m in length in main piling		–	–	84.81	m	**84.81**
10–15 m in length in main piling		–	–	84.81	m	**84.81**
15–20 m in length in main piling		–	–	90.49	m	**90.49**
254 × 254 × 63 kg/m steel bearing piles; Grade S275						
ne 5 m in length		–	–	100.69	m	**100.69**
5–10 m in length		–	–	98.95	m	**98.95**
10–15 m in length		–	–	98.95	m	**98.95**
15–20 m in length		–	–	105.57	m	**105.57**
254 × 254 × 71 kg/m steel bearing piles; Grade S275						
ne 5 m in length		–	–	113.47	m	**113.47**
5–10 m in length		–	–	111.51	m	**111.51**
10–15 m in length		–	–	111.51	m	**111.51**
15–20 m in length		–	–	118.98	m	**118.98**
254 × 254 × 85 kg/m steel bearing piles; Grade S275						
ne 5 m in length		–	–	135.84	m	**135.84**
5–10 m in length		–	–	133.50	m	**133.50**
10–15 m in length		–	–	133.50	m	**133.50**
15–20 m in length		–	–	142.44	m	**142.44**
305 × 305 × 79 kg/m steel bearing piles; Grade S275						
ne 5 m in length		–	–	126.26	m	**126.26**
5–10 m in length		–	–	124.07	m	**124.07**
10–15 m in length		–	–	124.07	m	**124.07**
15–20 m in length		–	–	132.39	m	**132.39**
305 × 305 × 95 kg/m steel bearing piles; Grade S275						
ne 5 m in length		–	–	151.83	m	**151.83**
5–10 m in length		–	–	149.20	m	**149.20**
10–15 m in length		–	–	149.20	m	**149.20**
15–20 m in length		–	–	159.20	m	**159.20**
305 × 305 × 110 kg/m steel bearing piles; Grade S275						
ne 5 m in length		–	–	177.21	m	**177.21**
5–10 m in length		–	–	174.17	m	**174.17**
10–15 m in length		–	–	174.17	m	**174.17**
15–20 m in length		–	–	185.75	m	**185.75**
305 × 305 × 126 kg/m steel bearing piles; Grade S275						
ne 5 m in length		–	–	202.99	m	**202.99**
5–10 m in length		–	–	199.51	m	**199.51**
10–15 m in length		–	–	199.51	m	**199.51**
15–20 m in length		–	–	212.76	m	**212.76**
305 × 305 × 149 kg/m steel bearing piles; Grade S275						
ne 5 m in length		–	–	240.04	m	**240.04**
5–10 m in length		–	–	235.93	m	**235.93**
10–15 m in length		–	–	235.93	m	**235.93**
15–20 m in length		–	–	251.60	m	**251.60**

SERIES 1600: PILING AND EMBEDDED RETAINING WALLS

Item	Gang hours	Labour £	Plant £	Material £	Unit	Total rate £
STEEL BEARING PILES – cont						
305 × 305 × 186 kg/m steel bearing piles; Grade S275						
ne 5 m in length		–	–	299.65	m	**299.65**
5–10 m in length		–	–	294.51	m	**294.51**
10–15 m in length		–	–	294.51	m	**294.51**
15–20 m in length		–	–	314.08	m	**314.08**
305 × 305 × 233 kg/m steel bearing piles; Grade S275						
ne 5 m in length		–	–	359.25	m	**359.25**
5–10 m in length		–	–	353.10	m	**353.10**
10–15 m in length		–	–	353.10	m	**353.10**
15–20 m in length		–	–	376.56	m	**376.56**
356 × 368 × 109 kg/m steel bearing piles; Grade S275						
ne 5 m in length		–	–	178.21	m	**178.21**
5–10 m in length		–	–	175.20	m	**175.20**
10–15 m in length		–	–	175.20	m	**175.20**
15–20 m in length		–	–	186.67	m	**186.67**
356 × 368 × 133 kg/m steel bearing piles; Grade S275						
ne 5 m in length		–	–	217.45	m	**217.45**
5–10 m in length		–	–	213.77	m	**213.77**
10–15 m in length		–	–	213.77	m	**213.77**
15–20 m in length		–	–	227.77	m	**227.77**
356 × 368 × 152 kg/m steel bearing piles; Grade S275						
ne 5 m in length		–	–	248.51	m	**248.51**
5–10 m in length		–	–	244.31	m	**244.31**
10–15 m in length		–	–	244.31	m	**244.31**
15–20 m in length		–	–	260.30	m	**260.30**
356 × 368 × 174 kg/m steel bearing piles; Grade S275						
ne 5 m in length		–	–	284.48	m	**284.48**
5–10 m in length		–	–	279.67	m	**279.67**
10–15 m in length		–	–	279.67	m	**279.67**
15–20 m in length		–	–	297.98	m	**297.98**
Driving vertical steel bearing piles						
section weight not exceeding 70 kg/m		–	–	–	m	**4.16**
section weight 70–90 kg/m		–	–	–	m	**4.51**
section weight 90–110 kg/m		–	–	–	m	**4.83**
section weight 110–130 kg/m		–	–	–	m	**5.27**
section weight 130–150 kg/m		–	–	–	m	**5.27**
section weight 150–170 kg/m		–	–	–	m	**5.68**
Driving raking steel bearing piles						
section weight not exceeding 70 kg/m		–	–	–	m	**6.58**
section weight 70–90 kg/m		–	–	–	m	**6.94**
section weight 90–110 kg/m		–	–	–	m	**7.33**
section weight 110–130 kg/m		–	–	–	m	**7.33**
section weight 130–150 kg/m		–	–	–	m	**5.27**
section weight 150–170 kg/m		–	–	–	m	**5.68**
section weight 170–190 kg/m		–	–	–	m	**6.03**
section weight 190–210 kg/m		–	–	–	m	**6.38**
Allow 30% of the respective item above for the lengthened section only						

SERIES 1600: PILING AND EMBEDDED RETAINING WALLS

Item	Gang hours	Labour £	Plant £	Material £	Unit	Total rate £
Welding on lengthening pieces to vertical steel bearing piles						
203 × 203 × any kg/m		–	–	–	nr	**298.48**
254 × 254 × any kg/m		–	–	–	nr	**353.47**
305 × 305 × any kg/m		–	–	–	nr	**408.90**
356 × 368 × any kg/m		–	–	–	nr	**450.47**
Cutting or burning off surplus length of vertical steel bearing piles						
203 × 203 × any kg/m		–	–	–	nr	**103.95**
254 × 254 × any kg/m		–	–	–	nr	**145.53**
305 × 305 × any kg/m		–	–	–	nr	**156.85**
356 × 368 × any kg/m		–	–	–	nr	**200.99**
STEEL TUBULAR PILES						
Steel tubular piles are commonly carried out by a Specialist Contractor and whose advice should be sought to arrive at accurate costings. However the following items can be used to assess a budget cost for each work.						
The following unit costs are based upon driving 100nr steel tubular piles on a clear site with reasonable access.						
Standing time		–	–	–	hr	**523.31**
Steel Grade 275; delivered in 10–20 t loads; mass 60–120 kg/m						
section 508 mm × 8 mm × 98.6 kg/m		–	–	133.13	m	**133.13**
section 559 mm × 8 mm × 109 kg/m		–	–	147.17	m	**147.17**
Steel Grade 275; delivered in 10–20 t loads; mass 120–250 kg/m						
section 508 mm × 10 mm × 123 kg/m		–	–	166.08	m	**166.08**
section 508 mm × 12.5 mm × 153 kg/m		–	–	206.58	m	**206.58**
section 508 mm × 16 mm × 194 kg/m		–	–	261.94	m	**261.94**
section 508 mm × 20 mm × 241 kg/m		–	–	325.40	m	**325.40**
section 559 mm × 10 mm × 135 kg/m		–	–	182.28	m	**182.28**
section 559 mm × 12.5 mm × 168 kg/m		–	–	226.84	m	**226.84**
section 559 mm × 16 mm × 214 kg/m		–	–	288.95	m	**288.95**
section 610 mm × 8 mm × 119 kg/m		–	–	160.68	m	**160.68**
section 610 mm × 10 mm × 148 kg/m		–	–	199.83	m	**199.83**
section 610 mm × 12.5 mm × 184 kg/m		–	–	248.44	m	**248.44**
section 610 mm × 16 mm × 234 kg/m		–	–	315.95	m	**315.95**
section 660 mm × 8 mm × 129 kg/m		–	–	174.18	m	**174.18**
section 660 mm × 10 mm × 160 kg/m		–	–	216.04	m	**216.04**
section 660 mm × 12.5 mm × 200 kg/m		–	–	270.04	m	**270.04**
section 711 mm × 8 mm × 134 kg/m		–	–	180.93	m	**180.93**
section 711 mm × 10 mm × 173 kg/m		–	–	233.59	m	**233.59**
section 711 mm × 12 mm × 215 kg/m		–	–	290.30	m	**290.30**
section 762 mm × 8 mm × 149 kg/m		–	–	201.18	m	**201.18**
section 762 mm × 10 mm × 185 kg/m		–	–	249.79	m	**249.79**
section 762 mm × 12.5 mm × 231 kg/m		–	–	311.90	m	**311.90**

SERIES 1600: PILING AND EMBEDDED RETAINING WALLS

Item	Gang hours	Labour £	Plant £	Material £	Unit	Total rate £
STEEL TUBULAR PILES – cont						
Steel Grade 275; delivered in 10–20 t loads; mass 250–500 kg/m						
section 559 mm × 20 mm × 266 kg/m		–	–	359.16	m	**359.16**
section 610 mm × 20 mm × 291 kg/m		–	–	392.91	m	**392.91**
section 660 mm × 16 mm × 254 kg/m		–	–	342.96	m	**342.96**
section 660 mm × 20 mm × 316 kg/m		–	–	426.67	m	**426.67**
section 660 mm × 25 mm × 392 kg/m		–	–	529.29	m	**529.29**
section 711 mm × 16 mm × 274 kg/m		–	–	369.96	m	**369.96**
section 711 mm × 20 mm × 341 kg/m		–	–	460.42	m	**460.42**
section 711 mm × 25 mm × 423 kg/m		–	–	571.14	m	**571.14**
section 762 mm × 16 mm × 294 kg/m		–	–	396.96	m	**396.96**
section 762 mm × 20 mm × 366 kg/m		–	–	494.18	m	**494.18**
section 762 mm × 25 mm × 454 kg/m		–	–	613.00	m	**613.00**
Driving vertical steel tubular piles						
mass 60–120 kg/m		–	–	–	m	**9.99**
mass 120–150 kg/m		–	–	–	m	**10.26**
mass 150–160 kg/m		–	–	–	m	**10.32**
mass 160–190 kg/m		–	–	–	m	**10.51**
mass 190–220 kg/m		–	–	–	m	**10.73**
mass 220–250 kg/m		–	–	–	m	**10.88**
mass 250–280 kg/m		–	–	–	m	**11.01**
mass 280–310 kg/m		–	–	–	m	**11.18**
mass 310–340 kg/m		–	–	–	m	**11.28**
mass 340 −370 kg/m		–	–	–	m	**11.50**
mass 370–400 kg/m		–	–	–	m	**11.94**
mass 400–430 kg/m		–	–	–	m	**11.97**
mass 430–460 kg/m		–	–	–	m	**12.28**
Driving raking steel tubular piles						
mass 60–120 kg/m		–	–	–	m	**10.59**
mass 120–150 kg/m		–	–	–	m	**10.71**
mass 150–160 kg/m		–	–	–	m	**10.88**
mass 160–190 kg/m		–	–	–	m	**11.01**
mass 190–220 kg/m		–	–	–	m	**11.29**
mass 220–250 kg/m		–	–	–	m	**11.43**
mass 250–280 kg/m		–	–	–	m	**11.56**
mass 280–310 kg/m		–	–	–	m	**11.70**
mass 310–340 kg/m		–	–	–	m	**11.91**
mass 340 −370 kg/m		–	–	–	m	**12.13**
mass 370–400 kg/m		–	–	–	m	**12.40**
mass 400–430 kg/m		–	–	–	m	**12.61**
mass 430–460 kg/m		–	–	–	m	**12.96**
Driving lengthened vertical steel tubular piles						
mass 60–120 kg/m		–	–	–	m	**13.03**
mass 120–150 kg/m		–	–	–	m	**13.44**
mass 150–160 kg/m		–	–	–	m	**13.53**
mass 160–190 kg/m		–	–	–	m	**13.64**
mass 190–220 kg/m		–	–	–	m	**13.84**
mass 220–250 kg/m		–	–	–	m	**14.05**
mass 250–280 kg/m		–	–	–	m	**14.07**

SERIES 1600: PILING AND EMBEDDED RETAINING WALLS

Item	Gang hours	Labour £	Plant £	Material £	Unit	**Total rate £**
mass 280–310 kg/m		–	–	–	m	**14.51**
mass 310–340 kg/m		–	–	–	m	**14.65**
mass 340 –370 kg/m		–	–	–	m	**15.03**
mass 370–400 kg/m		–	–	–	m	**15.23**
mass 400–430 kg/m		–	–	–	m	**15.46**
mass 430–460 kg/m		–	–	–	m	**15.87**
Driving lengthened raking steel tubular piles						
mass 60–120 kg/m		–	–	–	m	**13.64**
mass 120–150 kg/m		–	–	–	m	**13.84**
mass 150–160 kg/m		–	–	–	m	**14.11**
mass 160–190 kg/m		–	–	–	m	**14.33**
mass 190–220 kg/m		–	–	–	m	**14.55**
mass 220–250 kg/m		–	–	–	m	**14.77**
mass 250–280 kg/m		–	–	–	m	**15.04**
mass 280–310 kg/m		–	–	–	m	**15.15**
mass 310–340 kg/m		–	–	–	m	**15.45**
mass 340 –370 kg/m		–	–	–	m	**15.65**
mass 370–400 kg/m		–	–	–	m	**15.94**
mass 400–430 kg/m		–	–	–	m	**16.28**
mass 430–460 kg/m		–	–	–	m	**16.62**
Welding on lengthening piece to steel tubular piles						
section dia. 508 × any thickness		–	–	–	nr	**221.70**
section dia. 559 × any thickness		–	–	–	nr	**235.54**
section dia. 610 × any thickness		–	–	–	nr	**249.42**
section dia. 660 × any thickness		–	–	–	nr	**260.49**
section dia. 711 × any thickness		–	–	–	nr	**274.36**
section dia. 762 × any thickness		–	–	–	nr	**287.51**
Cutting or burning off surplus length of steel tubular piles						
section dia. 508 × any thickness		–	–	–	nr	**14.55**
section dia. 559 × any thickness		–	–	–	nr	**14.69**
section dia. 610 × any thickness		–	–	–	nr	**14.83**
section dia. 660 × any thickness		–	–	–	nr	**14.90**
section dia. 711 × any thickness		–	–	–	nr	**14.96**
section dia. 762 × any thickness		–	–	–	nr	**14.17**
PROOF LOADING OF PILES						
Driven precast concrete piles						
Establishment of proof loading equipment; proof loading of vertical precast concrete piles with maintained load to 900 kN		–	–	–	item	**4322.75**
Establishment of proof loading equipment; proof loading of vertical precast concrete piles by dynamic testing with piling hammer		–	–	–	nr	**993.24**

SERIES 1600: PILING AND EMBEDDED RETAINING WALLS

Item	Gang hours	Labour £	Plant £	Material £	Unit	Total rate £
PROOF LOADING OF PILES – cont						
Bored in situ reinforced concrete piling (tripod rig)						
Establishment of proof loading equipment for bored cast-in-place piles		–	–	–	item	**1896.44**
Proof loading of vertical cast-in-place piles with maximum test load of 600kN on a working pile 500 mm dia. using tension piles as reaction		–	–	–	nr	**4981.68**
Bored in situ reinforced concrete piling (mobile rig)						
Establishment of proof loading equipment for bored cast-in-place piles		–	–	–	item	**1979.03**
Proof loading of vertical cast-in-place piles with maximum test load of 600kN on a working pile 500 mm dia. using tension piles as reaction		–	–	–	nr	**4947.56**
Concrete injected piles (continuous flight augured)						
Establishment of proof loading equipment for bored cast-in-place piles						
450 mm dia.; 650kN		–	–	–	item	**2357.91**
600 mm dia.; 1400kN		–	–	–	item	**2357.79**
750 mm dia.; 2200kN		–	–	–	item	**2357.79**
Proof loading of vertical cast-in-place piles to 1.5 times working load						
450 mm dia.; 650kN		–	–	–	nr	**3372.96**
600 mm dia.; 1400kN		–	–	–	nr	**7073.41**
750 mm dia.; 2200kN		–	–	–	nr	**12754.70**
Electronic integrity testing						
cost per pile (minimum 40 piles per visit)		–	–	–	nr	**1552.82**
Segmental casing method piles						
Establishment of proof loading equipment for driven cast-in-place piles		–	–	–	item	**2037.82**
Proof loading of vertical cast-in-place piles with maximum test load of 600kN on a working pile 500 mm dia. using non-working tension piles as reaction		–	–	–	nr	**5822.37**
Driven in situ reinforced concrete piling						
Establishment of proof loading equipment for driven cast-in-place piles						
bottom driven		–	–	–	item	**1427.03**
top driven		–	–	–	item	**1427.03**

SERIES 1600: PILING AND EMBEDDED RETAINING WALLS

Item	Gang hours	Labour £	Plant £	Material £	Unit	**Total rate £**
Proof loading of vertical cast-in-place piles with maximum test load of 1125 kN on a working pile 430 mm dia. using non-working tension piles as reaction						
bottom driven		–	–	–	nr	**4994.68**
top driven		–	–	–	nr	**4994.68**
Electronic integrity testing						
Cost per pile (minimum 40 piles per visit)		–	–	–	nr	**20.43**
Steel bearing piles						
Establishment of proof loading equipment for steel bearing piles in main piling		–	–	–	item	**1896.44**
Proof loading of vertical steel bearing piles with maximum test load of 108 t load on a working pile using non-working tension piles as reaction		–	–	–	nr	**6011.98**
Steel tubular piles						
Establishment of proof loading equipment for steel tubular piles		–	–	–	item	**9803.91**
Proof loading of steel tubular piles with maximum test load of 108 t load on a working pile using non-working tension piles as reaction		–	–	–	nr	**9150.31**
STEEL SHEET PILES						
Sheet steel piling is commonly carried out by a Specialist Contractor, whose advice should be sought to arrive at accurate costings. However, the following items can be used to assess a budget for such work.						
The following unit costs are based on driving/extracting 1,500 m² of sheet piling on a clear site with reasonable access.						
Note: area of driven piles will vary from area supplied dependent upon pitch line of piling and provision for such allowance has been made in PC for supply.						
The materials cost below includes the manufacturers tariffs for a 200 mile delivery radius from works, delivery in 5–10 t loads and with an allowance of 10% to cover waste/projecting piles etc.						
Arcelor Mittal Z section steel piles; EN 10248 grade S270GP steel						
The following unit costs are based on driving/extracting 1,500 m² of sheet piling on a clear site with reasonable access.						

SERIES 1600: PILING AND EMBEDDED RETAINING WALLS

Item	Gang hours	Labour £	Plant £	Material £	Unit	Total rate £
STEEL SHEET PILES – cont						
Arcelor Mittal Z section steel piles – cont						
Provision of all plant, equipment and labour including transport to and from the site and establishing and dismantling for						
driving of sheet piling		–	–	–	sum	**9740.42**
extraction of sheet piling		–	–	–	sum	**2622.42**
Standing time		–	–	–	hr	**523.31**
Section modulus 800–1200 cm³/m; section reference AZ 12; mass 98.7 kg/m², sectional modulus 1200 cm³/m; EN 10248 grade S270GP steel						
length of welded corner piles		–	–	–	m	**141.91**
length of welded junction piles		–	–	–	m	**223.17**
driven area		–	–	–	m²	**67.83**
area of piles of length not exceeding 14 m		–	–	–	m²	**153.54**
length 14–24 m		–	–	–	m²	**144.49**
area of piles of length exceeding 24 m		–	–	–	m²	**153.97**
Section modulus 1200–2000 cm³/m; section reference AZ 17; mass 108.6 kg/m²; sectional modulus 1665 cm³/m; EN 10248 grade S270GP steel						
length of welded corner piles		–	–	–	m	**151.37**
length of welded junction piles		–	–	–	m	**223.17**
driven area		–	–	–	m²	**67.56**
area of piles of length not exceeding 14 m		–	–	–	m²	**179.13**
length 14–24 m		–	–	–	m²	**133.37**
area of piles of length exceeding 24 m		–	–	–	m²	**142.12**
Section modulus 2000–3000 cm³/m; section reference AZ 26; mass 155.2 kg/m²; sectional modulus 2600 cm³/m; EN 10248 grade S270GP steel						
driven area		–	–	–	m²	**54.73**
area of piles of length 6–18 m		–	–	–	m²	**98.26**
area of piles of length 18–24 m		–	–	–	m²	**99.60**
Section modulus 3000–4000 cm³/m; section reference AZ 36; mass 194.0 kg/m²; sectional modulus 3600 cm³/m; EN 10248 grade S270GP steel						
driven area		–	–	–	m²	**55.63**
area of piles of length 6–18 m		–	–	–	m²	**196.51**
area of piles of length 18–24 m		–	–	–	m²	**199.20**
Straight section modulus ne 500 cm³/m; section reference AS 500–12 mass 149 kg/m²; sectional modulus 51 cm³/m; EN 10248 grade S270GP steel						
driven area		–	–	–	m²	**41.57**
area of piles of length 6–18 m		–	–	–	m²	**216.17**
area of piles of length 18–24 m		–	–	–	m²	**229.08**

SERIES 1600: PILING AND EMBEDDED RETAINING WALLS

Item	Gang hours	Labour £	Plant £	Material £	Unit	**Total rate £**
One coat black tar vinyl (PC1) protective treatment applied all surfaces at shop to minimum dry film thickness up to 150 microns to steel piles						
section reference AZ 12; pile area		–	–	–	m^2	**24.04**
section reference AZ 17; pile area		–	–	–	m^2	**15.71**
section reference AZ 26; pile area		–	–	–	m^2	**17.11**
section reference AZ 36; pile area		–	–	–	m^2	**18.23**
section reference AS 500–12; pile area		–	–	–	m^2	**29.79**
One coat black high build isocyanate cured epoxy pitch (PC2) protective treatment applied all surfaces at shop to minimum dry film thickness up to 450 microns to steel piles						
section reference AZ 12; pile area		–	–	–	m^2	**24.04**
section reference AZ 17; pile area		–	–	–	m^2	**15.71**
section reference AZ 26; pile area		–	–	–	m^2	**17.11**
section reference AZ 36; pile area		–	–	–	m^2	**18.23**
section reference AS 500–12; pile area		–	–	–	m^2	**29.79**
Arcelor Mittal U section steel piles; EN 10248 grade S270GP steel						
The following unit costs are based on driving/ extracting 1,500 m^2 of sheet piling on a clear site with reasonable access.						
Provision of plant, equipment and labour including transport to and from the site and establishing and dismantling						
driving of sheet piling		–	–	–	sum	**9740.42**
extraction of sheet piling		–	–	–	sum	**2622.42**
Standing time		–	–	–	hr	**471.12**
Section modulus 500–800 cm^3/m; section reference PU 6; mass 76.0 kg/m^2; sectional modulus 600 cm^3/m						
driven area		–	–	–	m^2	**67.29**
area of piles of length 6–18 m		–	–	–	m^2	**98.26**
area of piles of length 18–24 m		–	–	–	m^2	**99.60**
Section modulus 800–1200 cm^3/m; section reference PU 8; mass 90.9 kg/m^2; sectional modulus 830 cm^3/m						
driven area		–	–	–	m^2	**60.54**
area of piles of length 6–18 m		–	–	–	m^2	**98.26**
area of piles of length 18–24 m		–	–	–	m^2	**99.60**
Section modulus 1200–2000 cm^3/m; section reference PU 12; mass 110.1 kg/m^2; sectional modulus 1200 cm^3/m						
driven area		–	–	–	m^2	**51.83**
area of piles of length 6–18 m		–	–	–	m^2	**98.26**
area of piles of length 18–24 m		–	–	–	m^2	**99.60**
Section modulus 1200–2000 cm^3/m; section reference PU 18; mass 128.2 kg/m^2; sectional modulus 1800 cm^3/m						
driven area		–	–	–	m^2	**47.12**
area of piles of length 6–18 m		–	–	–	m^2	**98.26**
area of piles of length 18–24 m		–	–	–	m^2	**99.60**

SERIES 1600: PILING AND EMBEDDED RETAINING WALLS

Item	Gang hours	Labour £	Plant £	Material £	Unit	Total rate £
STEEL SHEET PILES – cont						
Arcelor Mittal U section steel piles – cont						
Section modulus 2000–3000 cm^3/m; section reference PU 22; mass 143.6 kg/m^2; sectional modulus 2200 cm^3/m						
driven area		–	–	–	m^2	**42.40**
area of piles of length 6–18 m		–	–	–	m^2	**98.26**
area of piles of length 18–24 m		–	–	–	m^2	**99.60**
Section modulus 3000–4000 cm^3/m; section reference PU 32; mass 190.2 kg/m^2; sectional modulus 3200 cm^3/m						
driven area		–	–	–	m^2	**36.91**
area of piles of length 6–18 m		–	–	–	m^2	**98.26**
area of piles of length 18–24 m		–	–	–	m^2	**99.60**
One coat black tar vinyl (PC1) protective treatment applied all surfaces at shop to minimum dry film thickness up to 150 microns to steel piles						
section reference PU 6; pile area		–	–	–	m^2	**21.05**
section reference PU 8; pile area		–	–	–	m^2	**14.21**
section reference PU 12; pile area		–	–	–	m^2	**14.21**
section reference PU 18; pile area		–	–	–	m^2	**24.06**
section reference PU 22; pile area		–	–	–	m^2	**24.95**
section reference PU 32; pile area		–	–	–	m^2	**25.31**
One coat black high build isocyanate cured epoxy pitch (PC2) protective treatment applied all surfaces at shop to minimum dry film thickness up to 450 microns to steel piles						
section reference PU 6; pile area		–	–	–	m^2	**21.05**
section reference PU 8; pile area		–	–	–	m^2	**14.21**
section reference PU 12; pile area		–	–	–	m^2	**14.21**
section reference PU 18; pile area		–	–	–	m^2	**24.06**
section reference PU 22; pile area		–	–	–	m^2	**24.95**
section reference PU 32; pile area		–	–	–	m^2	**25.31**

SERIES 1600: PILING AND EMBEDDED RETAINING WALLS

Item	Gang hours	Labour £	Plant £	Material £	Unit	**Total rate £**
EMBEDDED RETAINING WALL PLANT						
Diaphragm walls are the construction of vertical walls, cast in place in a trench excavation. They can be formed in reinforced concrete to provide structural elements for temporary or permanent retaining walls. Wall thicknesses of 500 to 1500 mm up to 40 m deep may be constructed. Special equipment such as the Hydrofraise can construct walls up to 100 m deep. Restricted urban sites will significantly increase the costs.						
The following costs are based on constructing a diaphragm wall with an excavated volume of 4000 m^3 using a grab. Typical progress would be up to 500 m per week.						
Establishment of standard diaphragm walling plant, including bentonite storage tanks.		–	–	–	item	**16843.75**
Standing time		–	–	–	hr	**1121.38**
Guide walls (twin)		–	–	–	m	**648.00**
Waterproofed joints		–	–	–	m	**13.80**
DIAPHRAGM WALLS						
Excavation for walls 1000 mm thick, disposal of soil and placing of concrete		–	–	–	m^3	**537.05**
Provide and place reinforcement cages		–	–	–	tonne	**915.40**
Excavate/chisel in hard materials/rock		–	–	–	hr	**1191.40**

SERIES 1700: STRUCTURAL CONCRETE

Item	Gang hours	Labour £	Plant £	Material £	Unit	Total rate £
NOTES						
Refer also to Civil Engineering – Concrete, Formwork, Reinforcement and Precast Concrete, although this section is fundamentally different in that the provision of concrete of different classes and its placement is combined in the unit costs.						
RESOURCES – LABOUR						
Concreting gang						
1 ganger/chargehand (skill rate 4)	2.00	21.42	–	–	-	**21.42**
2 skilled operatives (skill rate 4)	2.00	40.39	–	–	-	**40.39**
4 unskilled operatives (general)	4.00	78.41	–	–	-	**78.41**
1 plant operator (skill rate 3) – 25% of time	0.25	7.29	–	–	-	**7.29**
Total Gang Rate/Hour	1.00	147.51	–	–	-	**147.51**
Formwork gang						
1 foreman (craftsman)	2.00	26.14	–	–	-	**26.14**
2 joiners (craftsman)	2.00	45.60	–	–	-	**45.60**
1 unskilled operative (general)	1.00	19.60	–	–	-	**19.60**
1 plant operator (craftsman) – 25% of time	0.25	12.61	–	–	-	**12.61**
Total Gang Rate/Hour	1.12	102.61	–	–	-	**102.61**
Reinforcement gang						
1 foreman (craftsman)	2.00	26.14	–	–	-	**26.14**
4 steel fixers (craftsman)	4.00	91.20	–	–	-	**91.20**
1 unskilled operative (general)	1.00	19.60	–	–	-	**19.60**
1 plant operator (craftsman) – 25% of time	0.25	12.61	–	–	-	**12.61**
Total Gang Rate/Hour	1.08	148.08	–	–	-	**148.08**
RESOURCES – PLANT						
Concreting						
10 t crane – 25% of time		–	15.63	–	-	**15.63**
gas oil for ditto		–	1.35	–	-	**1.35**
0.76 m^3 concrete skip – 25% of time		–	1.08	–	-	**1.08**
11.3 m^3/min compressor, 4 tool		–	36.41	–	-	**36.41**
gas oil for ditto		–	26.99	–	-	**26.99**
4 poker vibrators P5475 mm or less in thickness		–	7.48	–	-	**7.48**
Total Rate/Hour		–	88.93	–	-	**88.93**
Formwork						
20 t crawler crane – 25% of time		–	18.23	–	-	**18.23**
gas oil for ditto		–	1.57	–	-	**1.57**
255 mm saw bench		–	2.23	–	-	**2.23**
gas oil for ditto		–	1.80	–	-	**1.80**
small power tools (formwork)		–	0.96	–	-	**0.96**
Total Rate/Hour		–	24.79	–	-	**24.79**
Reinforcement						
30 t crawler crane (25% of time)		–	11.14	–	-	**11.14**
gas oil for ditto		–	1.57	–	-	**1.57**
bar cropper		–	4.97	–	-	**4.97**
small power tools (reinforcement)		–	1.92	–	-	**1.92**
tirfors, kentledge etc.		–	0.98	–	-	**0.98**
Total Rate/Hour		–	20.62	–	-	**20.62**

SERIES 1700: STRUCTURAL CONCRETE

Item	Gang hours	Labour £	Plant £	Material £	Unit	Total rate £
IN SITU CONCRETE						
In situ concrete Grade C10						
Blinding						
75 mm or less in thickness	0.18	26.55	16.00	121.26	m^3	**163.81**
Blinding; in narrow widths up to 1.0 m wide or in bottoms of trenches up to 2.5 m wide; excluding formwork						
75 mm or less in thickness	0.20	29.50	17.79	121.26	m^3	**168.55**
In situ concrete Grade C15						
Blinding; excluding formwork						
75 mm or less in thickness	0.16	23.60	14.23	109.43	m^3	**147.26**
Blinding; in narrow widths up to 1.0 m wide or in bottoms of trenches up to 2.5 m wide; excluding formwork						
75 mm or less in thickness	0.18	26.55	16.00	109.43	m^3	**151.98**
In situ concrete Grade C20/20						
Bases, footings, pile caps and ground beams; thickness						
ne 150 mm	0.20	29.50	17.79	118.16	m^3	**165.46**
150–300 mm	0.17	25.08	15.20	118.16	m^3	**158.44**
300–500 mm	0.15	22.13	13.26	118.16	m^3	**153.55**
exceeding 500 mm	0.14	20.65	12.45	118.16	m^3	**151.27**
Walls; thickness						
ne 150 mm	0.21	30.98	18.74	118.16	m^3	**167.88**
150–300 mm	0.15	22.13	13.26	118.16	m^3	**153.55**
300–500 mm	0.13	19.18	11.63	118.16	m^3	**148.97**
exceeding 500 mm	0.12	17.70	10.68	118.16	m^3	**146.54**
Suspended slabs; thickness						
ne 150 mm	0.27	39.83	24.08	118.16	m^3	**182.07**
150–300 mm	0.21	30.98	18.74	118.16	m^3	**167.88**
300–500 mm	0.19	28.03	16.97	118.16	m^3	**163.16**
exceeding 500 mm	0.19	28.03	16.97	118.16	m^3	**163.16**
Columns, piers and beams; cross-sectional area						
ne 0.03 m^2	0.50	73.75	44.45	118.16	m^3	**236.37**
0.03–0.10 m^2	0.40	59.00	35.57	118.16	m^3	**212.73**
0.10–0.25 m^2	0.35	51.63	31.05	118.16	m^3	**200.84**
0.12–1.00 m^2	0.35	51.63	31.05	118.16	m^3	**200.84**
exceeding 1 m^2	0.28	41.30	24.91	118.16	m^3	**184.37**
ADD to the above prices for						
sulphate-resisting cement		–	–	13.27	m^3	**13.27**
air entrained concrete		–	–	7.48	m^3	**7.48**
water repellant concrete		–	–	7.14	m^3	**7.14**
In situ concrete Grade C30/20						
Bases, footings, pile caps and ground beams; thickness						
ne 150 mm	0.21	30.98	18.74	137.23	m^3	**186.95**
150–300 mm	0.18	26.55	16.00	137.23	m^3	**179.78**
300–500 mm	0.15	22.13	13.26	137.23	m^3	**172.62**
exceeding 500 mm	0.14	20.65	12.45	137.23	m^3	**170.33**

SERIES 1700: STRUCTURAL CONCRETE

Item	Gang hours	Labour £	Plant £	Material £	Unit	Total rate £
IN SITU CONCRETE – cont						
In situ concrete Grade C30/20 – cont						
Walls; thickness						
ne 150 mm	0.22	32.45	19.57	137.23	m^3	**189.25**
150–300 mm	0.16	23.60	17.11	137.23	m^3	**177.95**
300–500 mm	0.13	19.18	13.97	137.23	m^3	**170.38**
exceeding 500 mm	0.12	17.70	12.84	137.23	m^3	**167.77**
Suspended slabs; thickness						
ne 150 mm	0.28	41.30	29.95	137.23	m^3	**208.48**
150–300 mm	0.22	32.45	23.54	137.23	m^3	**193.22**
300–500 mm	0.19	28.03	20.40	137.23	m^3	**185.66**
exceeding 500 mm	0.18	26.55	19.24	137.23	m^3	**183.03**
Columns, piers and beams; cross-sectional area						
ne 0.03 m^2	0.53	78.18	56.76	137.23	m^3	**272.17**
0.03–0.10 m^2	0.42	61.95	44.93	137.23	m^3	**244.12**
0.10–0.25 m^2	0.36	53.10	38.51	137.23	m^3	**228.84**
0.25–1.00 m^2	0.35	51.63	37.37	137.23	m^3	**226.23**
exceeding 1 m^2	0.28	41.30	29.95	137.23	m^3	**208.48**
ADD to the above prices for						
sulphate-resisting cement		–	–	13.27	m^3	**13.27**
air entrained concrete		–	–	7.48	m^3	**7.48**
water repellant concrete		–	–	7.14	m^3	**7.14**
PRECAST CONCRETE						
The cost of precast concrete item is very much dependent on the complexity of the moulds, the number of units to be cast from each mould and the size and the weight of the unit to be handled. The unit rates below are for standard precast items that are often to be found on a Civil Engineering project. It would be misleading to quote for indicative costs for tailor-made precast concrete units and it is advisable to contact specialist manufacturers for guide prices.						
Pretensioned prestressed beams; concrete Grade C20						
Beams						
100 × 65 × 1050 mm long	0.50	10.10	1.98	16.63	nr	**28.71**
265 × 65 × 1800 mm long	0.50	10.10	4.62	41.59	nr	**56.31**
Inverted 'T' beams, flange width 495 mm						
section T1; 8 m long, 380 mm deep; mass 1.88t		–	–	–	nr	**1295.24**
section T2; 9 m long, 420 mm deep; mass 2.29t		–	–	–	nr	**1554.26**
section T3; 11 m long, 535 mm deep; mass 3.02t		–	–	–	nr	**1813.05**
section T4; 12 m long, 575 mm deep; mass 3.54t		–	–	–	nr	**1985.85**
section T5; 13 m long, 615 mm deep; mass 4.08t		–	–	–	nr	**2158.65**
section T6; 13 m long, 655 mm deep; mass 4.33t		–	–	–	nr	**2158.65**
section T7; 12 m long, 695 mm deep; mass 4.95t		–	–	–	nr	**2590.43**
section T8; 15 m long, 735 mm deep; mass 5.60t		–	–	–	nr	**2935.83**
section T9; 16 m long, 775 mm deep; mass 6.28t		–	–	–	nr	**3107.70**
section T10; 18 m long, 815 mm deep; mass 7.43t		–	–	–	nr	**3453.93**

SERIES 1700: STRUCTURAL CONCRETE

Item	Gang hours	Labour £	Plant £	Material £	Unit	Total rate £
'M' beams, flange width 970 mm						
section M2 ; 17 m long, 720 mm deep; mass 12.95t		–	–	–	nr	**6907.84**
section M3 ; 18 m long, 800 mm deep; mass 15.11t		–	–	–	nr	**7339.61**
section M6 ; 22 m long, 1040 mm deep; mass 20.48t		–	–	–	nr	**12434.15**
section M8 ; 25 m long, 1200 mm deep; mass 23.68t		–	–	–	nr	**14161.11**
'U' beams, base width 970 mm						
section U3 ; 16 m long, 900 mm deep; mass 19.24t		–	–	–	nr	**11684.14**
section U5 ; 20 m long, 1000 mm deep; mass 25.64t		–	–	–	nr	**15024.60**
section U8 ; 24 m long, 1200 mm deep; mass 34.56t		–	–	–	nr	**20723.58**
section U12 ; 30 m long, 1600 mm deep; mass 52.74t		–	–	–	nr	**28408.60**
Precast concrete culverts, cattle creeps and subway units; rebated joints						
Rectangular cross section						
500 mm high × 1000 mm wide	0.14	8.45	5.73	472.15	m	**486.33**
1000 mm high × 1500 mm wide	0.34	20.81	12.76	773.74	m	**807.31**
1500 mm high × 1500 mm wide	0.53	32.14	23.62	895.90	m	**951.66**
2000 mm high × 2750 mm wide	0.90	55.09	38.08	2602.30	m	**2695.47**
2750 mm high × 3000 mm wide	1.15	70.40	75.43	2907.70	m	**3053.53**
Extra for units curved on plan to less than 20 m radius		–	–	480.55	m	**480.55**
SURFACE FINISH OF CONCRETE – FORMWORK						
Materials						
Formwork materials include for shutter, bracing, ties, support, kentledge and all consumables.						
These unit costs are based upon those outputs and prices detailed in Civil Engineering – Concrete Formwork but are referenced to The Specification for Highway Works, clause 1708.						
The following unit rates do not include for formwork outside the payline and are based on an optimum of a minimum 8 uses with 10% per use towards the cost of repairs/replacement of components damaged during disassembly.						
ADJUST formwork material costs generally depending on the number of uses:						

Nr of Uses	% Adjustment	Waste
1	Add 90–170%	7%
2	Add 50–180%	7%
3	Add 15–30%	6%
6	Add 5–10%	5%
8	No change	5%
10	Deduct 5–7%	5%

SERIES 1700: STRUCTURAL CONCRETE

Item	Gang hours	Labour £	Plant £	Material £	Unit	Total rate £
SURFACE FINISH OF CONCRETE – FORMWORK – cont						
Definitions						
Class F1 formwork is rough finish						
Class F2 formwork is fair finish						
Class F3 formwork is extra smooth finish						
Formwork Class F1						
Horizontal more than 300 mm wide	0.52	47.49	12.89	9.74	m^2	**70.12**
Inclined more than 300 mm wide	0.55	50.23	13.64	13.36	m^2	**77.23**
Vertical more than 300 mm wide	0.61	55.71	15.12	13.34	m^2	**84.17**
300 mm wide or less at any inclination	0.72	65.76	17.85	13.34	m^2	**96.95**
Curved of both girth and width more than 300 mm at any inclination	0.95	86.77	23.55	15.14	m^2	**125.46**
Curved of girth or width of 300 mm or less at any inclination	0.72	65.76	17.85	15.14	m^2	**98.75**
Domed	1.20	109.60	29.75	18.45	m^2	**157.80**
Void former cross-section 100 × 100 mm	0.07	6.39	0.35	4.60	m	**11.35**
Void former cross-section 250 × 250 mm	0.12	10.96	0.60	8.06	m	**19.62**
Void former cross-section 500 × 500 mm	0.30	27.40	1.50	14.45	m	**43.35**
Formwork Class F2						
Horizontal more than 300 mm wide	0.54	49.32	13.39	26.44	m^2	**89.15**
Inclined more than 300 mm wide	0.57	52.06	14.13	34.36	m^2	**100.55**
Vertical more than 300 mm wide	0.63	57.54	15.62	34.36	m^2	**107.53**
300 mm wide or less at any inclination	0.74	67.59	18.34	34.36	m^2	**120.30**
Curved of both girth and width more than 300 mm at any inclination	0.98	89.51	24.29	38.33	m^2	**152.13**
Curved of girth or width of 300 mm or less at any inclination	0.75	68.50	18.60	38.33	m^2	**125.42**
Domed	1.40	127.87	34.71	45.60	m^2	**208.18**
Formwork Class F3						
Horizontal more than 300 mm wide	0.56	51.15	13.88	34.20	m^2	**99.23**
Inclined more than 300 mm wide	0.59	53.89	14.62	42.12	m^2	**110.63**
Vertical more than 300 mm wide	0.65	59.37	16.12	42.12	m^2	**117.61**
300 mm wide or less at any inclination	0.76	69.42	18.84	42.12	m^2	**130.38**
Curved of both girth and width more than 300 mm at any inclination	0.99	90.42	24.55	46.08	m^2	**161.05**
Curved of girth or width of 300 mm or less at any inclination	0.77	70.33	19.09	46.08	m^2	**135.50**
Domed	1.45	132.44	35.94	53.36	m^2	**221.74**
Formwork ancillaries						
Allowance for additional craneage and rub up where required	0.13	11.87	3.23	4.16	m^2	**19.26**

SERIES 1700: STRUCTURAL CONCRETE

Item	Gang hours	Labour £	Plant £	Material £	Unit	**Total rate £**
PATTERNED PROFILE FORMWORK						
Extra over formwork for patterned profile formliners, INSITEX or similar		–	–	–	m^2	**43.70**
STEEL REINFORCEMENT FOR STRUCTURES						
Stainless steel bars						
Bar reinforcement nominal size 16 mm and under not exceeding 12 m in length	6.74	922.93	137.49	5213.68	tonne	**6274.10**
Bar reinforcement nominal size 20 mm and over not exceeding 12 m in length						
20 mm nominal size	4.44	607.98	90.57	5213.68	tonne	**5912.24**
25 mm nominal size	4.44	607.98	90.57	5213.68	tonne	**5912.24**
32 mm nominal size	4.44	607.98	90.57	5213.68	tonne	**5912.24**
ADD to the above for bars						
12–13.5 m long		–	–	143.60	tonne	**143.60**
13.5–15 m long		–	–	143.60	tonne	**143.60**
over 15 m long; per 500 mm increment		–	–	14.34	tonne	**14.34**
High yield steel bars BS 4449; deformed, Grade 500C						
Bar reinforcement nominal size 16 mm and under not exceeding 12 m in length	6.74	922.93	137.49	952.07	tonne	**2012.49**
Bar reinforcement nominal size 20 mm and over not exceeding 12 m in length						
20 mm nominal size	4.44	607.98	90.57	961.59	tonne	**1660.15**
25 mm nominal size	4.44	607.98	90.57	961.59	tonne	**1660.15**
32 mm nominal size	4.44	607.98	90.57	961.59	tonne	**1660.15**
40 mm nominal size	4.44	607.98	90.57	961.59	tonne	**1660.15**
ADD to the above for bars						
12–13.5 m long		–	–	143.60	tonne	**143.60**
13.5–15 m long		–	–	143.60	tonne	**143.60**
over 15 m long; per 500 mm increment		–	–	14.34	tonne	**14.34**
Helical reinforcement nominal size 16 mm and under						
ne 12 m in length	6.74	922.93	137.49	991.48	tonne	**2051.90**
Helical reinforcement nominal size 20 mm and over						
20 mm nominal size	4.44	607.98	90.57	961.59	tonne	**1660.15**
25 mm nominal size	4.44	607.98	90.57	961.59	tonne	**1660.15**
32 mm nominal size	4.44	607.98	90.57	961.59	tonne	**1660.15**
40 mm nominal size	4.44	607.98	90.57	961.59	tonne	**1660.15**
Dowels						
16 mm dia. × 600 mm long	0.11	14.38	2.04	2.32	nr	**18.74**
20 mm dia. × 600 mm long	0.11	14.38	2.04	4.17	nr	**20.59**
25 mm dia. × 600 mm long	0.11	14.38	2.04	4.78	nr	**21.19**
32 mm dia. × 600 mm long	0.10	13.69	2.04	5.84	nr	**21.57**

SERIES 1700: STRUCTURAL CONCRETE

Item	Gang hours	Labour £	Plant £	Material £	Unit	Total rate £
STEEL REINFORCEMENT FOR STRUCTURES – cont						
Mild steel bars BS 4449; Grade 250						
Bar reinforcement nominal size 16 mm and under						
ne 12 m in length	6.74	922.93	137.58	844.63	tonne	**1905.14**
Bar reinforcement nominal size 20 mm and over; not exceeding 12 m in length						
20 mm nominal size	4.44	607.98	90.57	952.67	tonne	**1651.22**
25 mm nominal size	4.44	607.98	90.57	952.67	tonne	**1651.22**
32 mm nominal size	4.44	607.98	90.57	952.67	tonne	**1651.22**
40 mm nominal size	4.44	607.98	90.57	952.67	tonne	**1651.22**
ADD to the above for bars						
12–13.5 m long		–	–	143.60	tonne	**143.60**
13.5–15 m long		–	–	143.60	tonne	**143.60**
over 15 m long, per 500 mm increment		–	–	14.34	tonne	**14.34**
ADD for cutting, bending, tagging and baling reinforcement on site						
6 mm nominal size	4.87	206.49	99.35	3.24	tonne	**309.08**
8 mm nominal size	4.58	194.20	93.43	3.24	tonne	**290.86**
10 mm nominal size	3.42	145.01	69.76	3.24	tonne	**218.01**
12 mm nominal size	2.55	108.12	52.03	3.24	tonne	**163.38**
16 mm nominal size	2.03	86.07	41.40	3.24	tonne	**130.71**
20 mm nominal size	1.68	71.23	34.27	3.24	tonne	**108.74**
25 mm nominal size	1.68	71.23	34.27	3.24	tonne	**108.74**
32 mm nominal size	1.39	58.94	28.34	3.24	tonne	**90.52**
40 mm nominal size	1.39	58.94	28.34	3.24	tonne	**90.52**
Fabric reinforcement; high yield steel BS 4483						
Fabric reinforcement						
BS ref A98; nominal mass 1.54 kg/m^2	0.03	4.38	0.62	5.55	m^2	**10.55**
BS ref A142; nominal mass 2.22 kg/m^2	0.03	4.11	0.62	5.12	m^2	**9.85**
BS ref A193; nominal mass 3.02 kg/m^2	0.04	5.48	0.82	6.93	m^2	**13.23**
BS ref A252; nominal mass 3.95 kg/m^2	0.04	5.48	0.82	7.68	m^2	**13.98**
BS ref A393; nominal mass 6.16 kg/m^2	0.07	9.59	1.44	10.78	m^2	**21.80**
BS ref B196; nominal mass 3.05 kg/m^2	0.04	5.48	0.82	12.16	m^2	**18.46**
BS ref B283; nominal mass 3.73 kg/m^2	0.04	5.48	0.82	9.44	m^2	**15.73**
BS ref B385; nominal mass 4.53 kg/m^2	0.05	6.85	1.03	10.48	m^2	**18.35**
BS ref B503; nominal mass 5.93 kg/m^2	0.05	6.85	1.03	14.67	m^2	**22.55**
BS ref B785; nominal mass 8.14 kg/m^2	0.08	10.95	1.63	18.96	m^2	**31.55**
BS ref B1131; nominal mass 10.90 kg/m^2	0.09	12.32	1.83	24.96	m^2	**39.11**
BS ref C282; nominal mass 2.61 kg/m^2	0.03	4.11	0.62	5.98	m^2	**10.71**
BS ref C385; nominal mass 3.41 kg/m^2	0.04	5.48	0.82	7.86	m^2	**14.15**
BS ref C503; nominal mass 4.34 kg/m^2	0.05	6.85	1.03	9.96	m^2	**17.84**
BS ref C636; nominal mass 5.55 kg/m^2	0.05	6.85	1.03	12.49	m^2	**20.37**
BS ref C785; nominal mass 6.72 kg/m^2	0.07	9.59	1.44	14.16	m^2	**25.19**
BS ref D49; nominal mass 0.77 kg/m^2	0.02	2.88	0.41	1.42	m^2	**4.70**
BS ref D98; nominal mass 1.54 kg/m^2	0.02	2.88	0.41	2.84	m^2	**6.12**

SERIES 1800: STEELWORK FOR STRUCTURES

Item	Gang hours	Labour £	Plant £	Material £	Unit	Total rate £
FABRICATION OF STEELWORK						
Steelwork to BS EN10025; Grade S275						
Fabrication of main members						
rolled sections		–	–	–	tonne	**2581.68**
plated rolled sections		–	–	–	tonne	**6884.43**
plated girders		–	–	–	tonne	**6022.59**
box girders		–	–	–	tonne	**2581.66**
Fabrication of deck panels						
rolled sections		–	–	–	tonne	**4474.89**
plated rolled sections		–	–	–	tonne	**4474.89**
plated girders		–	–	–	tonne	**4474.89**
Fabrication of subsidiary steelwork						
rolled sections		–	–	–	tonne	**4733.04**
plated rolled sections		–	–	–	tonne	**4733.04**
plated girders		–	–	–	tonne	**4733.03**
ERECTION OF STEELWORK						
Trial erection at the place of fabrication		–	–	–	tonne	**302.00**
Permanent erection of steelwork; substructure		–	–	–	tonne	**267.00**
Permanent erection of steelwork; superstructure		–	–	–	tonne	**267.00**
MISCELLANEOUS METALWORK						
Mild steel						
Ladders						
Cat ladder; 64 × 13 mm bar strings; 19 mm rungs at 250 mm centres; 450 mm wide with safety hoops		–	–	–	m	**146.32**
Handrails						
Galvanized tubular metal; 76 mm dia. handrail, 48 mm dia. standards at 750 mm centres, 48 mm dia. rail; 1070 mm high overall		–	–	–	m	**199.03**
Metal access cover and frame						
Group 4, ductile iron, single seal 610 × 610 × 100 mm depth; D400		–	–	–	nr	**333.08**
Group 2, ductile iron, double seal single piece cover 600 × 450 mm; B125		–	–	–	nr	**275.98**

SERIES 1900: PROTECTION OF STEELWORK AGAINST CORROSION

Item	Gang hours	Labour £	Plant £	Material £	Unit	Total rate £
PROTECTIVE SYSTEM – cont						
Blast clean to BS 7079 (surface preparation) – cont						
RESOURCES – LABOUR						
Protective painting gang						
1 ganger/chargehand (skill rate 4)	2.00	21.42	–	–	-	**21.42**
2 skilled operatives (skill rate 4)	2.00	40.39	–	–	-	**40.39**
2 unskilled operatives (general)	2.00	39.20	–	–	-	**39.20**
Total Gang rate/Hour	1.00	101.01	–	–	-	**101.01**
RESOURCES – PLANT						
Protective painting						
power tools (protection of steelwork)		–	3.73	–	-	**3.73**
access scaffolding, trestles and ladders		–	5.71	–	-	**5.71**
5 t transit van (50% of time)		–	5.67	–	-	**5.67**
gas oil for ditto		–	6.30	–	-	**6.30**
Total Rate/Hour		–	21.40	–	-	**21.40**
RESOURCES – MATERIALS						
All coats applied off site except as noted						
The external environment has been taken as Inland 'B' Exposed.						
PROTECTIVE SYSTEM						
Galvanizing to BS EN ISO 1461; apply protective coatings comprising: 1 st coat: Mordant T wash; 2nd coat: Zinc rich epoxy primer; 3rd coat: Zinc phosphate, CR/Alkyd Undercoat; 4th coat; MIO CR Undercoat-on site externally; 5th coat: CR coloured finish on site externally						
To metal parapets and fencing, lighting columns, brackets						
by brush or airless spray to dry film thickness 200 microns	0.20	20.20	4.28	–	m^2	**51.49**
Blast clean to BS 7079 (surface preparation); apply protective coatings comprising: 1st coat: Zinc Chromate, Red Oxide Blast Primer; 2nd coat: Zinc Phosphate, Epoxy Ester Undercoat; 3rd coat: MIO Undercoat; 4th coat: MIO coloured finish-on site externally						
To subsidiary steelwork, interior finishes						
By brush or airless spray to dry film thickness 175 microns	0.15	15.15	3.20	–	m^2	**39.44**

SERIES 1900: PROTECTION OF STEELWORK AGAINST CORROSION

Item	Gang hours	Labour £	Plant £	Material £	Unit	**Total rate £**
Blast clean to BS 7079 (surface preparation); apply metal coating of aluminium spray at works; apply protective coatings comprising: 1 st coat: Zinc Chromate Etch Primer (2 pack); 2nd coat: Zinc Phosphate, CR/Alkyd Undercoat; 3rd coat: Zinc Phosphate, CR/alkyd Undercoat; 4th coat: MIO CR Undercoat; 5th coat: CR finish-on site externally						
To main steel members						
By brush or airless spray to dry film thickness 250 microns	0.18	18.18	3.85	–	m^2	**44.42**
Blast clean to BS 7079 (second quality surface preparation); remove all surface defects to BS EN10025;apply protective coatings comprising: 1 st coat: Zinc rich primer (2 pack); 2nd coat: Epoxy High Build M10 (2 pack); 3rd coat: Polyurethane Undercoat (2 pack)-on site internally; 4th coat: Finish coat polyurethane (2 pack)-on site externally						
To internal steel members						
By brush or airless spray	0.15	15.15	3.20	–	m^2	**49.79**
ALTERNATIVE SURFACE TREATMENTS						
Galvanizing (Hot dip) to BS EN ISO 1461, assuming average depth 20 m^2 per tonne of steel		–	–	–	m^2	**21.31**
Shot blasting (at works)		–	–	–	m^2	**4.55**
Grit blasting (at works)		–	–	–	m^2	**6.92**
Sand blasting (at works)		–	–	–	m^2	**9.88**
Shot blasting (on site)		–	–	–	m^2	**6.63**
Grit blasting (on site)		–	–	–	m^2	**9.32**
Sand blasting (on site)		–	–	–	m^2	**13.76**

SERIES 2000: WATERPROOFING FOR STRUCTURES

Item	Gang hours	Labour £	Plant £	Material £	Unit	Total rate £
NOTES						
This section is based around the installation of proprietary systems to new/recently completed works as part of major scheme, for minor works/ repairs outputs will be many times more. Outputs are also based on use of skilled labour, therefore efficiency is high and wastage low (5–7% only allowed) excepting laps where required.						
RESOURCES – LABOUR						
Asphalting gang						
1 ganger/chargehand (skill rate 4)	2.00	21.42	–	–	-	**21.42**
2 unskilled operative (general)	2.00	39.20	–	–	-	**39.20**
Total Gang Rate/Hour	1.00	60.62	–	–	-	**60.62**
Damp-proofing gang						
1 ganger/chargehand (skill rate 4)	2.00	21.42	–	–	-	**21.42**
1 skilled operative (skill rate 4)	1.00	20.20	–	–	-	**20.20**
1 unskilled operative (general)	1.00	19.60	–	–	-	**19.60**
Total Gang Rate/Hour	1.00	61.21	–	–	-	**61.21**
Sprayed/brushed waterproofing gang						
1 ganger/chargehand (skill rate 4) – 30% of time	0.60	6.42	–	–	-	**6.42**
1 skilled operative (skill rate 4)	1.00	20.20	–	–	-	**20.20**
Total Gang Rate/Hour	1.00	26.62	–	–	-	**26.62**
Protective layers – screed gang						
1 ganger/chargehand (skill rate 4)	2.00	21.42	–	–	-	**21.42**
1 skilled operative (skill rate 4)	1.00	20.20	–	–	-	**20.20**
1 unskilled operative (general)	1.00	19.60	–	–	-	**19.60**
Total Gang Rate/Hour	1.00	61.21	–	–	-	**61.21**
RESOURCES – PLANT						
Asphalting						
45 litre portable tar boiler including sprayer (50% of time)		–	0.59	–	-	**0.59**
2 t dumper (50% of time)		–	2.71	–	-	**2.71**
gas oil for ditto		–	2.70	–	-	**2.70**
Total Rate/Hour		–	6.00	–	-	**6.00**
Damp-proofing						
2 t dumper (50% of time)		–	2.71	–	-	**2.71**
gas oil for ditto		–	2.70	–	-	**2.70**
Total Rate/Hour		–	5.41	–	-	**5.41**
WATERPROOFING						
Mastic asphalt; BS 6925 Type T 1097; 20 mm thick; two coats						
over 300 mm wide; ne 30° to horizontal	0.33	20.00	14.89	19.76	m^2	**54.65**
over 300 mm wide; 30–90° to horizontal	0.50	30.31	9.82	19.76	m^2	**59.90**
ne 300 mm wide; at any inclination	0.60	36.37	9.82	19.76	m^2	**65.96**
to domed surfaces	0.75	45.47	9.82	19.76	m^2	**75.05**

SERIES 2000: WATERPROOFING FOR STRUCTURES

Item	Gang hours	Labour £	Plant £	Material £	Unit	**Total rate £**
Bituthene 1000; lapped joints						
over 300 mm wide; ne 30° to horizontal	0.05	3.06	0.27	11.05	m^2	**14.38**
over 300 mm wide; 30–90° to horizontal	0.06	3.67	0.32	14.09	m^2	**18.09**
ne 300 mm wide; at any inclination	0.08	4.90	0.43	14.09	m^2	**19.42**
Extra; one coat primer on vertical surfaces	0.03	1.84	0.16	0.04	m^2	**2.04**
Bituthene 4000; lapped joints						
over 300 mm wide; ne 30° to horizontal	0.05	3.06	0.27	12.58	m^2	**15.91**
over 300 mm wide; 30–90° to horizontal	0.06	3.67	1.78	16.77	m^2	**22.23**
ne 300 mm wide; at any inclination	0.08	4.90	2.05	16.77	m^2	**23.72**
Famguard (hot applied) with Fam-primer						
over 300 mm wide; ne 30° to horizontal	0.32	19.59	1.73	319.56	m^2	**340.88**
over 300 mm wide; 30–90° to horizontal	0.35	21.43	1.89	325.81	m^2	**349.13**
ne 300 mm wide; at any inclination	0.40	24.49	2.16	367.79	m^2	**394.44**
Famflex (hot applied) with Fam-primer						
over 300 mm wide; ne 30° to horizontal	0.32	19.59	1.73	244.89	m^2	**266.21**
over 300 mm wide; 30–90° to horizontal	0.34	20.81	1.84	249.58	m^2	**272.23**
ne 300 mm wide; at any inclination	0.38	23.26	2.05	278.96	m^2	**304.28**
Two coats of RIW liquid asphaltic composition sprayed or brushed on						
over 300 mm wide; ne 30° to horizontal	0.03	0.80	0.16	68.69	m^2	**69.65**
over 300 mm wide; 30–90° to horizontal	0.03	0.80	0.16	68.69	m^2	**69.65**
ne 300 mm wide; at any inclination	0.04	1.06	0.22	68.69	m^2	**69.97**
Two coats of Mulseal sprayed or brushed on any inclination	0.07	1.86	0.38	30.73	m^2	**32.97**
20 mm thick red tinted sand asphalt layer onto bridge deck	0.02	1.22	0.65	13.09	m^2	**14.97**
SURFACE IMPREGNATION OF CONCRETE						
Silane waterproofing						
Surface impregnation to plain surfaces		–	–	–	m^2	**5.44**
REMOVAL OF EXISTING ASPHALT WATERPROOFING						
over 300 mm wide; ne 30° to horizontal	0.13	7.88	–	–	m^2	**7.88**
over 300 mm wide; 30–90° to horizontal	0.18	10.91	–	–	m^2	**10.91**
over 300 mm wide; at any inclination	0.20	12.12	–	–	m^2	**12.12**
to domed surfaces	0.25	15.16	–	–	m^2	**15.16**

SERIES 2100: BRIDGE BEARINGS

Item	Gang hours	Labour £	Plant £	Material £	Unit	Total rate £
NOTES						
Bridge bearings are manufactured and installed to individual specifications. The following guide prices are for different sizes of simple bridge bearings. If requirements are known, then advice ought to be obtained from specialist manufacturers such as CCL.						
RESOURCES – LABOUR						
Bridge bearing gang						
1 ganger/chargehand (skill rate 4)	2.00	21.42	–	–	-	**21.42**
2 unskilled operatives (general)	2.00	39.20	–	–	-	**39.20**
Total Gang Rate/Hour	1.00	59.40	–	–	-	**59.40**
BEARINGS						
Supply plain rubber bearings (3 m and 5 m lengths)						
150 × 20 mm	0.35	20.79	–	55.99	m	**76.78**
150 × 25 mm	0.35	20.79	–	67.21	m	**88.00**
Supply and place in position laminated elastomeric rubber bearing						
250 × 150 × 19 mm	0.25	14.85	–	41.61	nr	**56.46**
300 × 200 × 19 mm	0.25	14.85	–	45.61	nr	**60.46**
300 × 200 × 30 mm	0.27	16.04	–	57.60	nr	**73.63**
300 × 200 × 41 mm	0.27	16.04	–	86.40	nr	**102.44**
300 × 250 × 41 mm	0.30	17.82	–	86.40	nr	**104.22**
300 × 250 × 63 mm	0.30	17.82	–	152.00	nr	**169.82**
400 × 250 × 19 mm	0.32	19.01	–	80.00	nr	**99.01**
400 × 250 × 52 mm	0.32	19.01	–	176.41	nr	**195.42**
400 × 300 × 19 mm	0.32	19.01	–	92.79	nr	**111.80**
600 × 450 × 24 mm	0.35	20.79	–	191.98	nr	**212.77**
Adhesive fixings to laminated elastomeric rubber bearings						
2 mm thick epoxy adhesive	1.00	59.40	–	64.21	m^2	**123.61**
15 mm thick epoxy mortar	1.50	89.10	–	399.61	m^2	**488.71**
15 mm thick epoxy pourable grout	2.00	118.80	–	408.90	m^2	**527.70**
Supply and install mechanical guides for laminated elastomeric rubber bearings						
500kN SLS design load; FP50 fixed pin Type 1	2.00	118.80	–	1235.45	nr	**1354.25**
500kN SLS design load; FP50 fixed pin Type 2	2.00	118.80	–	1292.72	nr	**1411.52**
750kN SLS design load; FP75 fixed pin Type 1	2.10	124.74	–	1466.34	nr	**1591.08**
750kN SLS design load; FP75 fixed pin Type 2	2.10	124.74	–	1652.02	nr	**1776.76**
300kN SLS design load; UG300 Uniguide Type 1	2.00	118.80	–	1503.18	nr	**1621.98**
300kN SLS design load; UG300 Uniguide Type 2	2.00	118.80	–	1726.41	nr	**1845.21**
Supply and install fixed pot bearings						
355 × 355 mm; PF200	2.00	118.80	–	1558.68	nr	**1677.48**
425 × 425 mm; PF300	2.10	124.74	–	1724.77	nr	**1849.51**

SERIES 2100: BRIDGE BEARINGS

Item	Gang hours	Labour £	Plant £	Material £	Unit	**Total rate £**
Supply and install free sliding pot bearings						
445 × 345 mm; PS200	2.10	124.74	–	1807.81	nr	**1932.55**
520 × 415 mm; PS300	2.20	130.68	–	2303.52	nr	**2434.20**
Supply and install guided sliding pot bearings						
455 × 375 mm; PG200	2.20	130.68	–	2399.62	nr	**2530.30**
545 × 435 mm; PG300	2.30	136.62	–	2857.53	nr	**2994.15**
TESTING BEARINGS						
If there is a requirement for testing bridge bearings prior to their being installed then the tests should be enumerated separately. Specialist advice should be sought once details are known.						
Compression test for laminated elastomeric bearings						
generally		–	–	–	nr	**94.61**
Shear test for laminated elastomeric bearings						
generally		–	–	–	nr	**110.38**

SERIES 2300: BRIDGE EXPANSION JOINTS AND SEALING OF GAPS

Item	Gang hours	Labour £	Plant £	Material £	Unit	Total rate £
NOTES						
Major movement joints to bridge and viaduct decks are manufactured and installed to individual specifications determined by the type of structure location in the deck, amount of movement to be expected, and many other variables. The following are unit rates for other types of movement joints found in structures.						
RESOURCES – LABOUR						
Bridge jointing gang						
1 ganger/chargehand (skill rate 4)	2.00	21.42	–	–	-	**21.42**
1 skilled operative (skill rate 4)	1.00	20.20	–	–	-	**20.20**
1 unskilled operative	1.00	19.60	–	–	-	**19.60**
Total Gang Rate/Hour	1.00	61.21	–	–	-	**61.21**
SEALING OF GAPS						
Flexcell joint filler board						
10 mm thick	0.10	6.12	–	4.47	m^2	**10.59**
19 mm thick	0.16	9.79	–	7.66	m^2	**17.45**
25 mm thick	0.16	9.79	–	9.56	m^2	**19.35**
Building paper slip joint to abutment toe	0.01	0.67	–	22.57	m^2	**23.24**
Bond breaking agent	0.03	1.53	–	4.32	m^2	**5.85**
Hot poured rubber bitumen joint sealant						
PVC flat dumbbell waterstop						
10 × 20 mm	0.03	2.02	–	11.39	m	**13.41**
20 × 20 mm	0.04	2.45	–	4.73	m	**7.18**
25 × 15 mm	0.07	3.98	–	4.53	m	**8.50**
25 × 25 mm	0.07	4.47	–	7.40	m	**11.87**
Cold applied polysulphide joint sealant						
20 × 20 mm	0.07	3.98	–	7.10	m	**11.08**
Gun grade cold applied elastomeric joint sealant						
25 × 25 mm on 3 mm foam strip	0.07	3.98	–	10.50	m	**14.47**
50 × 25 mm on 3 mm foam strip	0.09	5.51	–	21.56	m	**27.07**
PVC centre bulb waterstop						
150 mm wide	0.08	4.90	–	5.13	m	**10.03**
230 mm wide	0.09	5.51	–	7.32	m	**12.83**
305 mm wide	0.11	6.73	–	8.80	m	**15.53**
PVC flat dumbell waterstop						
150 mm wide	0.08	4.90	–	30.46	m	**35.36**
230 mm wide	0.10	6.12	–	46.80	m	**52.92**
305 mm wide	0.12	7.35	–	76.42	m	**83.76**
Dowels, plain or greased						
12 mm mild steel 450 mm long	0.04	2.45	–	2.35	nr	**4.80**
16 mm mild steel 750 mm long	0.05	2.75	–	3.66	nr	**6.41**
16 mm mild steel 750 mm long with debonding agent for 375 mm	0.05	3.24	–	4.49	nr	**7.74**
Dowels, sleeved or capped						
12 mm mild steel 450 mm long with debonding agent for 225 mm and PVC dowel cap	0.05	2.75	–	2.58	nr	**5.34**

SERIES 2400: BRICKWORK BLOCKWORK AND STONEWORK

Item	Gang hours	Labour £	Plant £	Material £	Unit	**Total rate £**
RESOURCES – LABOUR						
Masonry gang						
1 foreman bricklayer (craftsman)	2.00	26.14	–	–	-	**26.14**
4 bricklayers (craftsman)	4.00	91.20	–	–	-	**91.20**
1 unskilled operative (general)	1.00	19.60	–	–	-	**19.60**
Total Gang Rate/Hour	1.00	136.93	–	–	-	**136.93**
RESOURCES – PLANT						
Masonry						
dumper 2 t (50% of time)	–		2.71	–	-	**2.71**
gas oil for ditto	–		2.70	–	-	**2.70**
cement mixer 4/3 (50% of time)	–		0.86	–	-	**0.86**
petrol for ditto	–		1.20	–	-	**1.20**
small power tools (masonry)	–		0.97	–	-	**0.97**
minor scaffolding (masonry)	–		1.48	–	-	**1.48**
Total Rate/Hour	–		9.92	–	-	**9.92**
RESOURCES – MATERIALS						
Half brick thick walls are in stretcher bond, thicker than this in english bond (3 stretchers: 1 header) unless otherwise stated.						
DfT Table 24/1: Mortar Proportions by Volume:						

Mortar Type	Cement: Lime: sand	Masonry Cement: sand	Cement: sand
(i)	1.0:0.25:4.0	–	–
(ii)	1: ½:4 to 4½	1:2½ to 3½	1:3 to 4
(iii)	1:1:5 to 6	1:4½	1:5 to 6

Item	Gang hours	Labour £	Plant £	Material £	Unit	**Total rate £**
BRICKWORK						
Common bricks; in stretcher bond; in cement mortar designation (ii)						
Walls						
half brick thick	0.30	40.53	2.90	49.58	m^2	**93.02**
one brick thick	0.44	60.52	5.89	99.80	m^2	**166.22**
one and a half bricks thick	0.64	88.19	6.38	141.01	m^2	**235.58**
two bricks thick	1.00	136.93	8.27	204.82	m^2	**350.03**
Walls, curved on plan						
half brick thick	0.30	40.53	2.90	49.58	m^2	**93.02**
one brick thick	0.57	78.05	5.65	99.80	m^2	**183.50**
one and a half bricks thick	0.82	112.29	8.16	150.46	m^2	**270.91**
two bricks thick	1.06	144.88	10.49	200.95	m^2	**356.31**
Walls, with a battered face						
half brick thick	0.33	45.05	3.26	49.58	m^2	**97.90**
one brick thick	0.63	86.27	6.25	99.80	m^2	**192.32**
one and a half bricks thick	0.91	124.20	8.99	150.46	m^2	**283.65**
two bricks thick	1.16	159.12	11.52	200.95	m^2	**371.59**

SERIES 2400: BRICKWORK BLOCKWORK AND STONEWORK

Item	Gang hours	Labour £	Plant £	Material £	Unit	Total rate £
BRICKWORK – cont						
Common bricks – cont						
Facework to concrete						
half brick thick	0.25	33.82	2.45	55.86	m^2	**92.13**
one brick thick	0.48	65.45	4.74	106.07	m^2	**176.27**
one and a half bricks thick	0.69	95.03	6.88	156.74	m^2	**258.65**
two bricks thick	0.90	122.97	8.90	207.15	m^2	**339.02**
In alteration work						
half brick thick	0.30	40.53	2.94	55.86	m^2	**99.33**
one brick thick	0.57	78.05	5.65	106.07	m^2	**189.78**
one and a half bricks thick	0.82	112.70	8.16	156.74	m^2	**277.60**
two bricks thick	1.06	144.88	10.49	207.15	m^2	**362.52**
ADD or DEDUCT to materials costs for variation of £10.00/1000 in PC of common bricks						
half brick thick		–	–	0.76	m^2	**0.76**
one brick thick		–	–	1.52	m^2	**1.52**
one and a half bricks thick		–	–	2.28	m^2	**2.28**
two bricks thick		–	–	3.04	m^2	**3.04**
Copings; standard header-on-edge; 215 mm wide × 103 mm high	0.11	14.38	1.04	11.14	m	**26.55**
ADD or DEDUCT to copings for variation of £1.00/100 in PC of common bricks		–	–	0.16	m	**0.16**
Class A engineering bricks,perforated; in cement mortar designation (ii)						
Walls						
half brick thick	0.50	68.47	4.61	41.58	m^2	**114.65**
one brick thick	0.52	71.21	5.16	83.79	m^2	**160.15**
one and a half bricks thick	0.75	103.11	7.47	126.45	m^2	**237.02**
two bricks thick	0.97	132.96	9.63	168.92	m^2	**311.51**
Walls, curved on plan						
half brick thick	0.60	82.16	3.67	41.58	m^2	**127.41**
one brick thick	1.00	136.93	6.99	83.79	m^2	**227.71**
one and a half bricks thick	1.40	191.71	10.01	126.45	m^2	**328.17**
two bricks thick	1.70	232.79	12.77	168.92	m^2	**414.48**
Walls, with a battered face						
half brick thick	0.70	95.85	3.67	41.58	m^2	**141.10**
one brick thick	1.00	136.93	6.99	83.79	m^2	**227.71**
one and a half bricks thick	1.50	205.40	10.01	126.45	m^2	**341.86**
two bricks thick	1.80	246.48	12.77	168.92	m^2	**428.18**
Facework to concrete						
half brick thick	0.60	82.16	3.12	47.85	m^2	**133.13**
one brick thick	0.90	123.24	6.00	90.06	m^2	**219.30**
ADD or DEDUCT to materials costs for variation of £10.00/1000 in PC of engineering bricks						
half brick thick		–	–	0.76	m^2	**0.76**
one brick thick		–	–	1.52	m^2	**1.52**
one and a half bricks thick		–	–	2.28	m^2	**2.28**
two bricks thick		–	–	3.04	m^2	**3.04**

SERIES 2400: BRICKWORK BLOCKWORK AND STONEWORK

Item	Gang hours	Labour £	Plant £	Material £	Unit	**Total rate £**
Brick coping in standard bricks in headers on edge;						
215 mm wide × 103 mm high	0.29	39.71	2.87	11.37	m	**53.96**
ADD or DEDUCT to copings for variation of £1.00/ 100 in PC of Class A enginreering bricks		–	–	0.16	m	**0.16**
Class B engineering bricks, perforated; in cement mortar designation (ii)						
Walls						
half brick thick	0.29	39.71	2.87	30.36	m^2	**72.95**
one brick thick	0.49	66.41	4.81	62.43	m^2	**133.66**
one and a half bricks thick	0.70	95.17	6.89	90.61	m^2	**192.67**
two bricks thick	0.91	123.92	8.97	116.66	m^2	**249.56**
Walls, curved on plan						
half brick thick	0.42	56.96	4.13	30.36	m^2	**91.45**
one brick thick	0.78	106.81	7.74	62.43	m^2	**176.98**
one and a half bricks thick	1.04	142.41	10.31	90.61	m^2	**243.33**
two bricks thick	1.30	178.01	12.90	116.66	m^2	**307.57**
Walls, with battered face						
half brick thick	0.42	56.96	4.13	30.36	m^2	**91.45**
one brick thick	0.78	106.81	7.74	62.43	m^2	**176.98**
one and a half bricks thick	1.04	142.41	10.31	90.61	m^2	**243.33**
two bricks thick	1.30	178.01	12.90	116.66	m^2	**307.57**
Facework to concrete						
half brick thick	0.32	43.54	3.16	36.56	m^2	**83.26**
one brick thick	0.56	76.68	5.55	68.64	m^2	**150.87**
ADD or DEDUCT to materials costs for variation of £10.00/1000 in PC of engineering bricks						
half brick thick		–	–	0.76	m^2	**0.76**
one brick thick		–	–	1.52	m^2	**1.52**
one and a half bricks thick		–	–	2.28	m^2	**2.28**
two bricks thick		–	–	3.04	m^2	**3.04**
Brick coping in standard bricks in headers on edge;						
215 mm wide × 103 mm high	0.13	18.35	1.33	11.46	m	**31.14**
ADD or DEDUCT to copings for variation of £1.00/ 100 in PC of Class B engineering bricks		–	–	0.16	m	**0.16**
Facing bricks; in lime mortar designation (ii)						
Walls						
half brick thick	0.34	46.56	3.38	51.30	m^2	**101.23**
one brick thick	0.57	78.60	8.09	103.22	m^2	**189.92**
one and a half bricks thick	0.83	113.24	8.20	155.60	m^2	**277.04**
two bricks thick	1.08	147.89	10.71	207.80	m^2	**366.40**
Walls, curved on plan						
half brick thick	0.45	61.62	4.46	51.30	m^2	**117.38**
one brick thick	0.84	115.02	8.33	103.22	m^2	**226.58**
one and a half bricks thick	1.12	152.68	11.05	155.60	m^2	**319.34**
two bricks thick	1.39	190.34	13.79	207.80	m^2	**411.92**
Walls, with a battered face						
half brick thick	0.45	61.62	4.46	51.30	m^2	**117.38**
one brick thick	0.84	115.02	8.33	103.22	m^2	**226.58**
one and a half bricks thick	1.12	152.68	11.05	155.60	m^2	**319.34**
two bricks thick	1.39	190.34	13.79	207.80	m^2	**411.92**

SERIES 2400: BRICKWORK BLOCKWORK AND STONEWORK

Item	Gang hours	Labour £	Plant £	Material £	Unit	Total rate £
BRICKWORK – cont						
Facing bricks – cont						
Facework to concrete						
half brick thick	0.37	50.39	3.48	57.57	m^2	**111.44**
one brick thick	0.66	90.38	6.55	109.50	m^2	**206.42**
Extra over common brickwork in English bond for facing with facing bricks in lime mortar designation (ii)	0.11	15.34	1.18	34.48	m^2	**50.99**
ADD or DEDUCT to materials costs for variation of £10.00/1000 in PC of facing bricks						
half brick thick		–	–	0.76	m^2	**0.76**
one brick thick		–	–	1.52	m^2	**1.52**
one and a half bricks thick		–	–	2.28	m^2	**2.28**
two bricks thick		–	–	3.04	m^2	**3.04**
Brick coping in standard bricks in headers on edge;						
215 mm wide × 103 mm high	0.13	18.35	1.33	11.07	m	**30.75**
Flat arches in standard stretchers on end;						
103 mm wide × 215 mm high	0.21	28.76	2.08	10.80	m	**41.63**
Flat arches in bullnose stretchers on end;						
103 mm × 215 mm high	0.22	30.13	2.19	6.08	m	**38.39**
Segmental arches in single ring stretchers on end;						
103 mm wide × 215 mm high	0.37	50.67	3.67	10.79	m	**65.13**
Segmental arches in double ring stretchers on end;						
103 mm wide × 440 mm high	0.49	67.10	4.86	43.57	m	**115.52**
Segmental arches; cut voussoirs						
103 mm wide × 215 mm high	0.70	95.85	3.87	24.53	m	**124.26**
ADD or DEDUCT to copings and arches for variation of £1.00/100 in PC of facing bricks						
header-on-edge		–	–	0.16	m	**0.16**
stretcher-on-end		–	–	0.16	m	**0.16**
stretcher-on-end bullnose specials		–	–	0.16	m	**0.16**
single ring		–	–	0.16	m	**0.16**
two ring		–	–	0.32	m	**0.32**
BLOCKWORK AND STONEWORK						
Lightweight concrete blocks; solid; 3.5 N/mm²; in cement-lime mortar						
Walls						
100 mm thick;	0.17	23.83	1.73	18.54	m^2	**44.10**
140 mm thick;	0.23	30.81	2.23	24.20	m^2	**57.24**
215 mm thick;	0.28	37.66	2.72	32.01	m^2	**72.40**
Walls, curved on plan						
100 mm thick;	0.23	31.63	2.29	18.54	m^2	**52.46**
140 mm thick;	0.30	40.94	2.96	24.20	m^2	**68.11**
215 mm thick;	0.37	50.12	3.63	32.01	m^2	**85.76**

SERIES 2400: BRICKWORK BLOCKWORK AND STONEWORK

Item	Gang hours	Labour £	Plant £	Material £	Unit	**Total rate £**
Facework to concrete						
100 mm thick;	0.18	24.51	1.77	24.81	m^2	**51.10**
140 mm thick;	0.23	31.77	2.30	30.47	m^2	**64.54**
215 mm thick;	0.28	38.75	2.80	38.29	m^2	**79.84**
In alteration work						
100 mm thick;	0.17	23.83	1.73	18.54	m^2	**44.10**
140 mm thick;	0.23	30.81	2.23	24.20	m^2	**57.24**
215 mm thick;	0.28	37.66	2.72	32.01	m^2	**72.40**
Dense concrete blocks; solid; 3.5 or 7 N/mm²; in cement-lime mortar						
Walls						
100 mm thick;	0.17	23.28	1.68	9.82	m^2	**34.78**
140 mm thick;	0.20	27.39	1.98	15.44	m^2	**44.81**
215 mm thick;	0.24	32.86	2.38	20.73	m^2	**55.98**
Walls, curved on plan						
100 mm thick;	0.23	30.95	2.24	9.82	m^2	**43.01**
140 mm thick;	0.23	30.95	2.64	15.44	m^2	**49.02**
215 mm thick;	0.32	43.68	3.16	19.94	m^2	**66.78**
Facework to concrete						
100 mm thick;	0.18	23.96	1.74	16.10	m^2	**41.79**
140 mm thick;	0.21	28.21	2.04	21.71	m^2	**51.96**
215 mm thick;	0.35	47.93	2.70	26.21	m^2	**76.84**
In alteration work						
100 mm thick;	0.27	36.42	2.24	9.82	m^2	**48.49**
140 mm thick;	0.26	35.60	2.64	15.44	m^2	**53.67**
215 mm thick;	0.32	43.68	3.16	19.94	m^2	**66.78**
Reconstituted stone; Bradstone 100 bed weathered Cotswold or North Cerney masonry blocks; rough hewn rockfaced blocks; in coloured cement-lime mortar designation (1:2:9) (iii)						
Walls, thickness 100 mm						
vertical and straight	0.30	41.08	2.97	68.90	m^2	**112.95**
curved on plan	0.39	53.40	3.87	68.90	m^2	**126.17**
with a battered face	0.35	47.24	3.42	68.90	m^2	**119.56**
in arches	0.58	78.74	5.70	68.90	m^2	**153.34**
Facing to concrete; wall ties						
vertical and straight	0.24	32.45	2.35	100.61	m^2	**135.42**
curved on plan	0.32	43.13	3.12	100.61	m^2	**146.86**
with a battered face	0.36	48.75	3.53	100.61	m^2	**152.88**
Reconstituted stone; Bradstone Architectural dressings in weathered Cotswold or North Cerney shades; in coloured cement-lime mortar designation (1:2:9) (iii)						
Copings; twice weathered and throated						
152 × 76 mm ;	0.08	10.95	0.79	18.28	m	**30.03**
152 × 76 mm; curved on plan;	0.11	14.51	1.05	64.54	m	**80.11**
305 × 76 mm;	0.10	13.69	1.00	39.17	m	**53.86**
305 × 76 mm; curved on plan;	0.13	18.21	1.32	81.95	m	**101.49**

SERIES 2400: BRICKWORK BLOCKWORK AND STONEWORK

Item	Gang hours	Labour £	Plant £	Material £	Unit	Total rate £
BLOCKWORK AND STONEWORK – cont						
Reconstituted stone – cont						
Corbels						
479 × 100 × 215 mm, splayed	0.49	67.10	4.86	353.61	nr	**425.56**
665 × 100 × 215 mm, splayed	0.55	75.31	5.46	352.70	nr	**433.47**
Pier caps						
305 × 305 mm	0.09	12.32	0.89	24.92	nr	**38.14**
381 × 381 mm	0.11	15.06	1.09	36.18	nr	**52.34**
457 × 457 mm	0.13	17.80	1.29	48.29	nr	**67.38**
533 × 533 mm	0.15	20.54	1.49	67.11	nr	**89.14**
Lintels						
100 × 140 mm	0.11	15.06	1.09	52.19	m	**68.35**
100 × 215 mm	0.16	21.91	1.59	69.86	m	**93.36**
Natural stone ashlar; Portland Whitbed limestone; in cement-lime mortar designation (iii)						
Walls						
vertical and straight	12.70	1739.05	119.78	3169.80	m^3	**5028.63**
curved on plan	19.30	2642.81	191.40	4732.30	m^3	**7566.51**
with a battered face	19.30	2642.81	191.40	4537.73	m^3	**7371.95**
Facing to concrete; wall ties						
vertical and straight	17.00	2327.86	168.59	3074.55	m^3	**5571.00**
curved on plan	25.80	3532.88	255.86	4795.69	m^3	**8584.43**
with a battered face	25.80	3532.88	255.86	4795.69	m^3	**8584.43**
Copings; twice weathered and throated						
250 × 150 mm	0.45	61.62	4.46	170.61	m	**236.69**
250 × 150 mm; curved on plan	0.45	61.62	4.46	204.64	m	**270.72**
400 × 150 mm	0.49	67.10	4.86	250.69	m	**322.65**
400 × 150 mm; curved on plan	0.49	67.10	4.86	300.75	m	**372.71**
Shaped and dressed string courses						
75 mm projection × 150 mm high	0.45	61.62	4.46	149.18	m	**215.26**
Corbel						
500 × 450 × 300 mm	0.55	75.31	5.46	228.56	nr	**309.33**
Keystone						
750 × 900 × 300 mm (extreme)	1.30	178.01	12.90	718.13	nr	**909.04**
Random rubble uncoursed , weighing 2.0 t/m^3 of wall; in cement-lime mortar designation (iii)						
Walls						
vertical and straight	4.17	571.01	41.35	339.37	m^3	**951.73**
curved on plan	4.67	639.48	46.31	427.01	m^3	**1112.80**
with a battered face	4.67	639.48	46.31	427.01	m^3	**1112.80**
in arches	8.53	1168.04	84.59	427.01	m^3	**1679.64**
Facework to concrete						
vertical and straight	4.17	571.01	41.35	433.28	m^3	**1045.65**
curved on plan	4.67	639.48	46.31	433.28	m^3	**1119.07**
with a battered face	4.67	639.48	46.31	433.28	m^3	**1119.07**
in arches	8.53	1168.04	84.59	433.28	m^3	**1685.91**
Copings						
500 × 125 mm	0.49	67.10	4.86	280.48	m	**352.43**

SERIES 2400: BRICKWORK BLOCKWORK AND STONEWORK

Item	Gang hours	Labour £	Plant £	Material £	Unit	**Total rate £**
Squared random rubble uncoursed, weighing 2.0 t/m³ of wall; in cement-lime mortar designation (iii)						
Walls						
vertical and straight	4.17	571.01	41.35	580.82	m^3	**1193.19**
curved on plan	4.67	639.48	46.31	580.82	m^3	**1266.62**
with a battered face	4.67	639.48	46.31	580.82	m^3	**1266.62**
in arches	8.53	1168.04	84.59	580.82	m^3	**1833.45**
Facework to concrete						
vertical and straight	4.17	571.01	41.35	587.10	m^3	**1199.46**
curved on plan	4.67	639.48	46.31	587.10	m^3	**1272.89**
with a battered face	4.67	639.48	46.31	587.10	m^3	**1272.89**
in arches	8.53	1168.04	84.59	587.10	m^3	**1839.73**
Copings						
500 × 125 mm	0.49	67.10	4.86	280.46	m	**352.42**
Dry rubble , weighing 2.0 t/m³ of wall						
Walls						
vertical and straight	3.83	524.45	37.98	442.01	m^3	**1004.45**
curved on plan	4.33	592.92	42.94	442.01	m^3	**1077.87**
with a battered face	4.33	592.92	42.94	442.01	m^3	**1077.87**
Copings formed of rough stones						
275 × 200 mm (average) high	0.45	61.62	4.46	37.46	m	**103.54**
500 × 200 mm	0.55	75.31	5.46	64.12	m	**144.89**

SERIES 2500: SPECIAL STRUCTURES

Item	Gang hours	Labour £	Plant £	Material £	Unit	**Total rate £**
SPECIAL STRUCTURES DESIGNED BY CONTRACTOR						
Notes This section envisages the following types of structure which may be required to be designed by the Contractor based on stipulated performance criteria: * Buried structures * Earth retaining structures * Environmental barriers * Underbridges up to 8 m span * Footbridges * Piped culverts * Box culverts * Drainage exceeding 900 mm dia. * Other structures Naturally, this work cannot be catered for directly in this section and will require the preparation of a sketch solution and approximate quantities to allow pricing using the various other Unit Costs sections as well as the Approximate Estimates section. An allowance must be added to such an estimate to cover the Contractor's design fee(s) and expenses.						

SERIES 2700: ACCOMMODATION WORKS, WORKS FOR STATUTORY UNDERTAKERS

Item	Gang hours	Labour £	Plant £	Material £	Unit	**Total rate £**
ACCOMMODATION WORKS, WORKS FOR STATUTORY UNDERTAKERS						
Note Cost items in this series will be specific to individual contract agreements and it is felt that inclusion of prices in this publication would not provide useful guidance.						

SERIES 3000 LANDSCAPE INFRASTRUCTURE

Item	Gang hours	Labour £	Plant £	Material £	Unit	Total rate £
GROUND PREPARATION AND CULTIVATION						
Cultivation by tractor; Agripower Ltd; Specialist subcontract prepared turf and sports area preparation						
Ripping up subsoil; using approved subsoiling machine; minimum depth 250 mm below topsoil; at 1.20 m centres; in						
gravel or sandy clay	–	–	4.10	–	100 m²	**4.10**
soil compacted by machines	–	–	7.66	–	100 m²	**7.66**
clay	–	–	11.72	–	100 m²	**11.72**
chalk or other soft rock	–	–	13.67	–	100 m²	**13.67**
Extra for subsoiling at 1 m centres	–	–	2.34	–	100 m²	**2.34**
Breaking up existing ground; using tractor drawn tine cultivator or rotavator with stone burier						
Single pass						
100 mm deep	–	–	1.64	–	100 m²	**1.64**
150 mm deep	–	–	2.05	–	100 m²	**2.05**
200 mm deep	–	–	2.73	–	100 m²	**2.73**
Cultivating ploughed ground; using disc, drag or chain harrow						
4 passes	–	–	0.82	–	100 m²	**0.82**
Cultivation; pedestrian operated rotavator						
Breaking up existing ground; tine cultivator or rotavator						
100 mm deep	0.22	9.90	1.23	–	100 m²	**11.13**
150 mm deep	0.28	12.38	1.53	–	100 m²	**13.91**
200 mm deep	0.37	16.50	2.04	–	100 m²	**18.54**
As above but in heavy clay or wet soils						
100 mm deep	0.44	19.80	2.45	–	100 m²	**22.25**
150 mm deep	0.66	29.70	3.68	–	100 m²	**33.38**
200 mm deep	0.82	37.13	4.60	–	100 m²	**41.73**
Clearing stones; disposing off site						
by hand; stones not exceeding 50 mm in any direction; loading to skip 6.1 m³	0.01	0.45	0.04	–	m²	**0.49**
by mechanical stone rake; stones not exceeding 50 mm in any direction; loading to 15 m³ truck by mechanical loader	–	0.09	0.33	0.01	m²	**0.43**
Lightly cultivating; weeding; to fallow areas; disposing debris off site						
by hand	0.01	0.64	–	0.14	m²	**0.78**
Pre-seeding pre-planting or pre turfing general purpose fertilizers (9+5+5); PC £1.00/kg; to soil surface; by machine						
35 g/m²	–	–	1.28	3.53	100 m²	**4.81**
50 g/m²	–	–	1.28	5.04	100 m²	**6.32**

SERIES 3000 LANDSCAPE INFRASTRUCTURE

Item	Gang hours	Labour £	Plant £	Material £	Unit	**Total rate £**
Pre-seeding pre-planting or pre turfing general purpose fertilizers (9+5+5); PC £1.00/kg; to soil surface; by hand						
35 g/m²	0.17	7.50	–	3.53	100 m²	**11.03**
50 g/m²	0.17	7.50	–	5.04	100 m²	**12.54**
AHS Ltd; peat free tree and shrub compost; Organic green; PAS 100 (69 m³ loads); placing on beds by mechanical loader; spreading and rotavating into topsoil by tractor drawn machine						
50 mm thick	–	–	18.59	67.50	100 m²	**86.09**
100 mm thick	–	–	29.14	141.75	100 m²	**170.89**
150 mm thick	–	–	34.24	212.63	100 m²	**246.87**
200 mm thick	–	–	42.06	283.50	100 m²	**325.56**
Composted fine bark compost; AHS Ltd; (25 m³ loads); from not further than 25 m from location; cultivating into topsoil by pedestrian drawn machine						
50 mm thick	2.86	128.57	1.02	259.50	100 m²	**389.09**
100 mm thick	6.05	272.20	1.02	519.00	100 m²	**792.22**
150 mm thick	8.90	400.72	1.02	778.50	100 m²	**1180.24**
SEEDING						
Seeding						
Seeding labours only in two operations; by machine						
35 g/m²	–	–	2.07	–	100 m²	**2.07**
Grass seed; spreading in two operations; PC £5.90/kg						
35 g/m²	–	–	2.07	20.65	100 m²	**22.72**
50 g/m²	–	–	2.07	29.50	100 m²	**31.57**
350 kg/ha	–	–	206.53	2065.00	ha	**2271.53**
500 kg/ha	–	–	206.53	2950.00	ha	**3156.53**
Extra over seeding by machine for slopes over 30° (allowing for the actual area but measured in plan)						
35 g/m²	–	–	0.31	3.10	100 m²	**3.41**
50 g/m²	–	–	0.31	4.42	100 m²	**4.73**
350 kg/ha	–	–	30.98	309.75	ha	**340.73**
500 kg/ha	–	–	30.98	442.50	ha	**473.48**
Seeding labours only in two operations; by machine (for seed prices see above)						
35 g/m²	0.17	7.50	–	–	100 m²	**7.50**
Grass seed; spreading in two operations; PC £5.90/kg (for changes in material prices please refer to table above); by hand						
35 g/m²	0.17	7.50	–	20.65	100 m²	**28.15**
50 g/m²	0.17	7.50	–	29.50	100 m²	**37.00**
Extra over seeding by hand for slopes over 30° (allowing for the actual area but measured in plan)						
35 g/m²	–	0.17	–	3.07	100 m²	**3.24**
50 g/m²	–	0.17	–	4.42	100 m²	**4.59**
Harrowing seeded areas; light chain harrow	–	–	0.64	–	100 m²	**0.64**

SERIES 3000 LANDSCAPE INFRASTRUCTURE

Item	Gang hours	Labour £	Plant £	Material £	Unit	Total rate £
SEEDING – cont						
Seeding labours only in two operations – cont						
Raking over seeded areas						
by mechanical stone rake	–	–	8.85	–	100 m²	**8.85**
by hand	0.80	36.00	–	–	100 m²	**36.00**
Rolling seeded areas; light roller						
by tractor drawn roller	–	–	3.90	–	100 m²	**3.90**
by pedestrian operated mechanical roller	0.08	3.75	0.44	–	100 m²	**4.19**
by hand drawn roller	0.17	7.50	–	–	100 m²	**7.50**
Extra over harrowing, raking or rolling seeded areas for slopes over 30°; by machine or hand	–	–	–	–	25%	**–**
Turf edging; to seeded areas; 300 mm wide						
part load	0.05	2.14	–	3.70	m²	**5.84**
full load	0.05	2.14	–	2.80	m²	**4.94**
TURFING						
Preparation for turfing						
Bringing existing topsoil to a fine tilth for turfing by raking or harrowing; stones not to exceed 6 mm; by machine	–	0.18	0.20	–	m²	**0.38**
Bringing existing topsoil to a fine tilth for turfing by raking; stones not to exceed 6 mm; by hand						
general commercial or amenity turf	0.01	0.56	–	–	m²	**0.56**
higher grade fine culivation	0.02	0.90	–	–	m²	**0.90**
Hand load and transport and unload turf from delivery area to laying area						
by small site dumper up to 50 m	0.01	0.90	0.04	–	m²	**0.94**
by hand 25 m	0.03	1.13	–	–	m²	**1.13**
by hand 50 m	0.03	1.29	–	–	m²	**1.29**
Turfing; laying only; to stretcher bond; butt joints; including providing and working from barrow plank runs where necessary to surfaces not exceeding 30° from horizontal						
specially selected lawn turves from previously lifted stockpile	0.08	3.38	–	–	m²	**3.38**
cultivated lawn turves; to large open areas	0.06	2.62	–	–	m²	**2.62**
cultivated lawn turves; to domestic or garden areas	0.08	3.50	–	–	m²	**3.50**
road verge quality turf	0.04	1.80	–	–	m²	**1.80**
Industrially grown turf; PC prices listed represent the general range of industrial turf prices for sportsfields and amenity purposes; prices will vary with quantity and site location						
Rolawn						
RB Medallion; sports fields, domestic lawns, general landscape; full loads 1720 m²	0.08	3.75	–	2.80	m²	**6.55**
RB Medallion; sports fields, domestic lawns, general landscape; part loads	0.08	3.75	–	3.70	m²	**7.45**

SERIES 3000 LANDSCAPE INFRASTRUCTURE

Item	Gang hours	Labour £	Plant £	Material £	Unit	Total rate £
TREE PLANTING						
Generally						
The cost of planting semi-mature trees will depend on the size and species, and on the access to the site for tree handling machines. Prices should be obtained for individual trees and planting.						
Tree planting additional notes regardeing Quercus (Oak) species						
Designers and organizations requiring oak tree prices should exercise additional checks as there are severe restrictions on sizes imported and quarantine requirements due to processionary moth regulations. This impacts on the price and availability of oak.						
Additional charges for plants imported from the European union						
Phytosanitary certificate	–	–	–	120.00	/load	**120.00**
Customs Clearance charges	–	–	–	50.00	/load	**50.00**
Tree planting; containerized trees; nursery stock; James Coles & Sons (Nurseries) Ltd						
Acer platanoides 'Emerald Queen'; including backfillling with excavated material (other operations not included)						
standard; 8–10 cm girth	0.48	21.60	–	88.20	nr	**109.80**
selected standard; 10–12 cm girth	0.56	25.23	–	147.00	nr	**172.23**
heavy standard; 12–14 cm girth	0.76	34.40	–	166.60	nr	**201.00**
extra heavy standard; 14–16 cm girth	1.20	54.00	–	196.00	nr	**250.00**
Quercus robur; including backfillling with excavated material (other operations not included)						
standard; 8–10 cm girth	0.48	21.60	–	88.20	nr	**109.80**
selected standard; 10–12 cm girth	0.56	25.23	–	127.40	nr	**152.63**
heavy standard; 12–14 cm girth	0.76	34.40	–	166.60	nr	**201.00**
extra heavy standard; 14–16 cm girth	1.20	54.00	–	196.00	nr	**250.00**
Betula pendula; multistemmed; including backfillling with excavated material (other operations not included)						
300/350 mm high	0.75	101.25	8.20	210.00	nr	**319.45**
350/400 mm high	1.00	135.00	10.94	280.00	nr	**425.94**
400/450 mm high	1.00	135.00	10.94	420.00	nr	**565.94**
450/500 mm high	1.05	141.75	11.48	560.00	nr	**713.23**
Tree planting; root balled trees; advanced nursery stock and semi-mature – General						
Preamble: The cost of planting semi-mature trees will depend on the size and species and on the access to the site for tree handling machines. Prices should be obtained for individual trees and planting.						

SERIES 3000 LANDSCAPE INFRASTRUCTURE

Item	Gang hours	Labour £	Plant £	Material £	Unit	Total rate £
TREE PLANTING – cont						
Tree planting; bare root trees; nursery stock; James Coles & Sons (Nurseries) Ltd						
Acer platanoides; including backfillling with excavated material (other operations not included)						
standard; 8–10 cm girth	0.40	18.00	–	29.40	nr	**47.40**
selected standard; 10–12 cm girth	0.58	26.10	–	39.20	nr	**65.30**
heavy standard; 12–14 cm girth	0.83	37.50	–	63.70	nr	**101.20**
extra heavy standard; 14–16 cm girth	1.00	45.00	–	78.40	nr	**123.40**
Quercus robur; including backfillling with excavated material (other operations not included)						
light standard; 6–8 cm girth	0.35	15.75	–	27.44	nr	**43.19**
standard; 8–10 cm girth	0.40	18.00	–	43.05	nr	**61.05**
selected standard; 10–12 cm girth	0.58	26.10	–	58.80	nr	**84.90**
heavy standard; 12–14 cm girth	0.83	37.50	–	98.00	nr	**135.50**
Robinia pseudoacacia; including backfillling with excavated material (other operations not included)						
light standard; 6–8 cm girth	0.35	15.75	–	24.50	nr	**40.25**
standard; 8–10 cm girth	0.40	18.00	–	29.40	nr	**47.40**
selected standard; 10–12 cm girth	0.58	26.10	–	39.20	nr	**65.30**
Tree planting; root balled trees; nursery stock; James Coles & Sons (Nurseries) Ltd						
Acer platanoides; including backfillling with excavated material (other operations not included)						
standard; 8–10 cm girth	0.48	21.60	–	38.67	nr	**60.27**
selected standard; 10–12 cm girth	0.56	25.23	–	52.08	nr	**77.31**
heavy standard; 12–14 cm girth	0.76	34.40	–	79.66	nr	**114.06**
extra heavy standard; 14–16 cm girth	1.20	54.00	–	131.41	nr	**185.41**
Quercus robur; including backfillling with excavated material (other operations not included)						
standard; 8–10 cm girth	0.48	21.60	–	69.12	nr	**90.72**
selected standard; 10–12 cm girth	0.56	25.23	–	96.18	nr	**121.41**
heavy standard; 12–14 cm girth	0.76	34.40	–	139.51	nr	**173.91**
extra heavy standard; 14–16 cm girth	1.20	54.00	–	184.26	nr	**238.26**
Tree planting; Airpot container grown trees; advanced nursery stock and semi-mature; Hillier Trees Ltd						
Acer platanoides 'Emerald Queen'; including backfilling with excavated material (other operations not included)						
16–18 cm girth	1.98	89.10	74.04	165.00	nr	**328.14**
18–20 cm girth	2.18	98.01	79.58	195.00	nr	**372.59**
20–25 cm girth	2.38	106.92	88.85	255.00	nr	**450.77**
25–30 cm girth	2.97	133.65	136.19	315.00	nr	**584.84**
30–35 cm girth	3.96	178.20	148.08	440.00	nr	**766.28**

SERIES 3000 LANDSCAPE INFRASTRUCTURE

Item	Gang hours	Labour £	Plant £	Material £	Unit	**Total rate £**
Quercus palustris 'Pin Oak'; including backfilling with excavated material (other operations not included)						
16–18 cm girth	1.98	89.10	74.04	205.00	nr	**368.14**
18–20 cm girth	1.60	72.00	79.58	285.00	nr	**436.58**
20–25 cm girth	2.38	106.92	88.85	317.00	nr	**512.77**
25–30 cm girth	2.97	133.65	136.19	525.00	nr	**794.84**
30–35 cm girth	3.96	178.20	148.08	900.00	nr	**1226.28**
Tree planting; Airpot container grown trees; semi-mature and mature trees; Hillier Trees Ltd; planting and back filling; planted by telehandler or by crane; delivery included; all other operations priced separately						
Semi-mature trees; indicative prices						
40–45 cm girth	4.00	180.00	83.42	775.00	nr	**1038.42**
45–50 cm girth	4.00	180.00	83.42	1100.00	nr	**1363.42**
55–60 cm girth	6.00	270.00	83.42	1950.00	nr	**2303.42**
60–70 cm girth	7.00	315.00	111.16	3300.00	nr	**3726.16**
70–80 cm girth	7.50	337.50	138.89	3800.00	nr	**4276.39**
80–90 cm girth	8.00	360.00	166.85	6000.00	nr	**6526.85**
Tree planting; rootballed trees; advanced nursery stock and semi-mature; Lorenz von Ehren						
Acer platanoides 'Emerald Queen'; including backfilling with excavated material (other operations not included)						
16–18 cm girth	1.30	58.32	4.06	150.00	nr	**212.38**
18–20 cm girth	1.60	72.00	4.06	200.00	nr	**276.06**
20–25 cm girth	4.50	202.50	24.77	250.00	nr	**477.27**
25–30 cm girth	6.00	270.00	30.54	350.00	nr	**650.54**
30–35 cm girth	11.00	495.00	40.20	450.00	nr	**985.20**
Quercus palustris 'Pin Oak'; including backfilling with excavated material (other operations not included)						
16–18 cm girth	1.30	58.32	4.06	200.00	nr	**262.38**
18–20 cm girth	1.60	72.00	4.06	250.00	nr	**326.06**
20–25 cm girth	4.50	202.50	24.77	300.00	nr	**527.27**
25–30 cm girth	6.00	270.00	30.54	450.00	nr	**750.54**
30–35 cm girth	11.00	495.00	40.20	550.00	nr	**1085.20**
Betula pendula (3 stems); including backfilling with excavated material (other operations not included)						
3.0–3.5 m high	1.98	89.10	74.04	250.00	nr	**413.14**
3.5–4.0 m high	1.60	72.00	79.58	300.00	nr	**451.58**
4.0–4.5 m high	2.38	106.92	88.85	400.00	nr	**595.77**
4.5–5.0 m high	2.97	133.65	136.19	500.00	nr	**769.84**
5.0–6.0 m high	3.96	178.20	148.08	750.00	nr	**1076.28**
6.0–7.0 m high	4.50	202.50	173.47	950.00	nr	**1325.97**

SERIES 3000 LANDSCAPE INFRASTRUCTURE

Item	Gang hours	Labour £	Plant £	Material £	Unit	Total rate £
TREE PLANTING – cont						
Tree planting – cont						
Pinus sylvestris; including backfilling with excavated material (other operations not included)						
3.0–3.5 m high	1.98	89.10	74.04	950.00	nr	**1113.14**
3.5–4.0 m high	1.60	72.00	79.58	1200.00	nr	**1351.58**
4.0–4.5 m high	2.38	106.92	88.85	840.00	nr	**1035.77**
4.5–5.0 m high	2.97	133.65	136.19	1500.00	nr	**1769.84**
5.0–6.0 m high	3.96	178.20	148.08	2500.00	nr	**2826.28**
6.0–7.0 m high	4.50	202.50	179.06	3500.00	nr	**3881.56**
Tree planting; containerized trees; Lorenz von Ehren; to the tree prices above add for Airpot containerization only						
Tree size						
20–25 cm	–	–	–	100.00	nr	**100.00**
25–30 cm	–	–	–	125.00	nr	**125.00**
30–35 cm	–	–	–	150.00	nr	**150.00**
35–40 cm	–	–	–	200.00	nr	**200.00**
40–45 cm	–	–	–	300.00	nr	**300.00**
45–50 cm	–	–	–	400.00	nr	**400.00**
50–60 cm	–	–	–	500.00	nr	**500.00**
60–70 cm	–	–	–	700.00	nr	**700.00**
70–80 cm	–	–	–	900.00	nr	**900.00**
80–90 cm	–	–	–	1200.00	nr	**1200.00**
Tree planting; rootballed trees; semi-mature and mature trees; Lorenz von Ehren; planting and back filling; planted by telehandler or by crane; delivery included; all other operations priced separately						
Semi-mature trees						
40–45 cm girth	8.00	360.00	87.42	1250.00	nr	**1697.42**
45–50 cm girth	8.00	360.00	111.47	2500.00	nr	**2971.47**
50–60 cm girth	10.00	450.00	136.60	3500.00	nr	**4086.60**
60–70 cm girth	15.00	675.00	326.64	5000.00	nr	**6001.64**
70–80 cm girth	18.00	810.00	296.09	7500.00	nr	**8606.09**
80–90 cm girth	18.00	810.00	296.09	9500.00	nr	**10606.09**
90–100 cm girth	18.00	810.00	296.09	11500.00	nr	**12606.09**

SERIES 3000 LANDSCAPE INFRASTRUCTURE

Item	Gang hours	Labour £	Plant £	Material £	Unit	Total rate £
SHRUB PLANTING						
Setting out of shrubs						
Setting out; selecting planting from holding area; loading to wheelbarrows; planting as plan or as directed; distance from holding area maximum 50 m; plants 2–3 litre containers						
single plants not grouped	0.04	1.80	–	–	nr	**1.80**
plants in groups of 3–5 nr	0.03	1.13	–	–	nr	**1.13**
plants in groups of 10–100 nr	0.02	0.75	–	–	nr	**0.75**
plants in groups of 100 nr minimum	0.01	0.52	–	–	nr	**0.52**
Forming planting holes; in cultivated ground (cultivating not included); by mechanical auger; trimming holes by hand; depositing excavated material alongside holes						
250 mm dia.	0.03	1.50	0.07	–	nr	**1.57**
250 × 250 mm	0.04	1.80	0.13	–	nr	**1.93**
300 × 300 mm	0.08	3.38	0.16	–	nr	**3.54**
Planting shrubs; hand excavation; forming planting holes; in cultivated ground (cultivating not included); depositing excavated material alongside holes; Placing plants previously set out alongside and backfilling (plants not included)						
100 × 100 × 100 mm deep; with mattock or hoe	0.01	0.50	–	–	nr	**0.50**
250 × 250 × 300 mm deep	0.01	0.60	–	–	nr	**0.60**
300 × 300 × 300 mm deep	0.02	0.75	–	–	nr	**0.75**
400 × 400 × 400 mm deep	0.08	3.75	–	–	nr	**3.75**
500 × 500 × 500 mm deep	0.17	7.50	–	–	nr	**7.50**
600 × 600 × 600 mm deep	0.25	11.25	–	–	nr	**11.25**
900 × 900 × 600 mm deep	0.50	22.50	–	–	nr	**22.50**
1.00 × 1.00 × 600 mm deep	1.00	45.00	–	–	nr	**45.00**
1.25 × 1.25 × 600 mm deep	1.33	60.00	–	–	nr	**60.00**
Hand excavation; forming planting holes; in uncultivated ground; depositing excavated material alongside holes						
300 × 300 × 300 mm deep	0.06	2.81	–	–	nr	**2.81**
400 × 400 × 400 mm deep	0.25	11.25	–	–	nr	**11.25**
500 × 500 × 500 mm deep	0.33	14.65	–	–	nr	**14.65**
600 × 600 × 600 mm deep	0.55	24.75	–	–	nr	**24.75**
900 × 900 × 600 mm deep	1.25	56.25	–	–	nr	**56.25**
1.00 × 1.00 × 600 mm deep	1.54	69.19	–	–	nr	**69.19**
1.25 × 1.25 × 600 mm deep	2.41	108.56	–	–	nr	**108.56**
Bare root planting; to planting holes (forming holes not included); including backfilling with excavated material (bare root plants not included)						
bare root 1+1; 30–90 mm high	0.02	0.75	–	–	nr	**0.75**
bare root 1+2; 90–120 mm high	0.02	0.75	–	–	nr	**0.75**

SERIES 3000 LANDSCAPE INFRASTRUCTURE

Item	Gang hours	Labour £	Plant £	Material £	Unit	**Total rate £**
SHRUB PLANTING – cont						
Shrub planting where plants are set out at equal off set centres where the spacings are always the speciifed distance between the plant centres. This introduces more uniform but slightly higher than rectangular centres						
2 litre containerized plants; in cultivated ground (cultivating not included); PC £3.05/nr						
1.00 m centres; 1.43 plants/m^2	0.02	0.85	–	4.36	m^2	**5.21**
750 mm centres; 2.6 plants/m^2	0.04	1.60	–	7.93	m^2	**9.53**
600 mm centres; 3.04 plants/m^2	0.04	1.82	–	9.27	m^2	**11.09**
500 mm cenrtes; 4.5 plants/m^2	0.06	2.70	–	13.72	m^2	**16.42**
450 mm centres; 6.75 plants/m^2	0.09	4.05	–	20.59	m^2	**24.64**
Shrub planting; 3 litre containerized plants; in cultivated ground (cultivating not included); PC £3.20/nr						
1.00 m centres; 1.43 plants/m^2	0.02	0.85	–	5.58	m^2	**6.43**
750 mm centres; 2.6 plants/m^2	0.04	1.60	–	10.14	m^2	**11.74**
600 mm centres; 3.04 plants/m^2	0.04	1.82	–	11.86	m^2	**13.68**
500 mm cenrtes; 4.5 plants/m^2	0.06	2.70	–	17.55	m^2	**20.25**
450 mm centres; 6.75 plants/m^2	0.09	4.05	–	26.32	m^2	**30.37**
Extra over shrubs for stakes	0.02	0.94	–	1.08	nr	**2.02**
HEDGES						
Hedges						
Excavating trench for hedges; depositing soil alongside trench; by machine						
300 mm deep × 300 mm wide	0.03	2.73	1.70	–	m	**4.43**
300 mm deep × 600 mm wide	0.05	4.09	2.58	–	m	**6.67**
500 mm deep × 500 mm wide	0.06	5.64	3.55	–	m	**9.19**
Excavating trench for hedges; depositing soil alongside trench; by hand						
300 mm deep × 300 mm wide	0.08	11.25	–	–	m	**11.25**
300 mm deep × 450 mm wide	0.13	16.88	–	–	m	**16.88**
300 mm deep × 600 mm wide	0.17	22.50	–	–	m	**22.50**
500 mm deep × 500 mm wide	0.23	30.52	–	–	m	**30.52**
Setting out; notching out; excavating trench; breaking up subsoil to minimum depth 300 mm						
minimum 400 mm deep	0.25	11.25	–	–	m	**11.25**
Disposal of excavated soil to stockpile						
Up to 25 m distant						
By hand	2.00	90.00	–	–	m^3	**90.00**
By machine	–	–	3.87	–	m^3	**3.87**

SERIES 3000 LANDSCAPE INFRASTRUCTURE

Item	Gang hours	Labour £	Plant £	Material £	Unit	Total rate £
Native species and bare root hedge planting						
Hedge planting operations only (excavation of trenches priced separately) including backfill with excavated topsoil; PC £0.65/nr						
single row; 200 mm centres	0.06	2.81	–	3.25	m	**6.06**
single row; 300 mm centres	0.06	2.50	–	2.17	m	**4.67**
single row; 400 mm centres	0.04	1.88	–	1.63	m	**3.51**
single row; 500 mm centres	0.03	1.50	–	1.30	m	**2.80**
double row; 200 mm centres	0.17	7.50	–	6.50	m	**14.00**
double row; 300 mm centres	0.13	6.00	–	4.33	m	**10.33**
double row; 400 mm centres	0.08	3.75	–	3.25	m	**7.00**
double row; 500 mm centres	0.07	3.00	–	2.60	m	**5.60**
Extra over hedges for incorporating manure; at 1 m^3 per 30 m	0.03	1.13	–	1.17	m	**2.30**
MULCHING						
Mulch to ornamental planting areas; for highways and civil works mulch will only normally be applied where ornamental planting is employed at plant spacings of less than 400 mm centres. For bare root and tree planting mulch will usually be mulchmat or to individual tree pits. Mulch will not normally be used on slopes						
Mulch; AHS Ltd; Amenity Bark Mulch FSC; to plant beds; delivered in 80 m^3 loads; maximum distance 25 m						
50 mm thick	0.04	2.00	–	1.65	m^2	**3.65**
75 mm thick	0.07	3.00	–	2.47	m^2	**5.47**
100 mm thick	0.09	4.00	–	3.30	m^2	**7.30**
Mulch; AHS Ltd; Amenity Bark Mulch FSC; to plant beds; delivered in 25 m^3 loads; maximum distance 25 m						
50 mm thick	0.04	2.00	–	2.07	m^2	**4.07**
75 mm thick	0.07	3.00	–	3.11	m^2	**6.11**
100 mm thick	0.09	4.00	–	4.15	m^2	**8.15**
TREE AND SHRUB PROTECTION						
Tree planting; tree protection – General						
Preamble: Care must be taken to ensure that tree grids and guards are removed when trees grow beyond the specified dia. of guard.						
Tree planting; tree protection						
Tree tube; olive green						
1200 mm high × 80 × 80 mm	0.07	3.00	–	2.42	nr	**5.42**
stakes; 1200 mm high for tree tube; driving into ground	0.05	2.25	–	1.08	nr	**3.33**

SERIES 3000 LANDSCAPE INFRASTRUCTURE

Item	Gang hours	Labour £	Plant £	Material £	Unit	Total rate £
TREE AND SHRUB PROTECTION – cont						
Tree planting – cont						
Expandable plastic tree guards; including 25 mm softwood stakes						
500 mm high	0.17	7.50	–	1.32	nr	**8.82**
1.00 m high	0.17	7.50	–	1.56	nr	**9.06**
English Woodlands; Weldmesh tree guards; nailing to tree stakes (tree stakes not included)						
1800 mm high × 300 mm dia.	0.25	11.25	–	28.00	nr	**39.25**
J. Toms Ltd; spiral rabbit guards; clear or brown						
610 × 38 mm	0.03	1.50	–	0.29	nr	**1.79**
English Woodlands; Plastic Mesh Tree Guards; black; supplied in 50 m rolls						
13 × 13 mm small mesh; roll width 60 cm	0.04	1.61	–	2.50	nr	**4.11**
13 × 13 mm small mesh; roll width 120 cm	0.06	2.65	–	4.48	nr	**7.13**
Tree guards of 3 nr 2.40 m × 100 mm stakes; driving 600 mm into firm ground; bracing with timber braces at top and bottom.	1.00	45.00	–	–	nr	**45.00**
English Woodlands; strimmer guard in heavy duty black plastic; 225 mm high	0.07	3.00	–	5.94	nr	**8.94**
Tubex Ltd; Standard Treeshelter inclusive of 25 mm stake; prices shown for quantities of 500 nr						
0.6 m high	0.05	2.25	–	1.99	nr	**4.24**
0.75 m high	0.05	2.25	–	2.41	nr	**4.66**
1.2 m high	0.05	2.25	–	3.23	nr	**5.48**
1.5 m high	0.05	2.25	–	4.09	nr	**6.34**
Tubex Ltd; Shrubshelter inclusive of 25 mm stake; prices shown for quantities of 500 nr						
Ecostart shelter for forestry transplants and seedlings	0.07	3.00	–	2.04	nr	**5.04**
0.6 m high	0.07	3.00	–	2.79	nr	**5.79**
0.75 m high	0.07	3.00	–	3.14	nr	**6.14**

SERIES 5000: MAINTENANCE PAINTING OF STEELWORK

Item	Gang hours	Labour £	Plant £	Material £	Unit	**Total rate £**
SURFACE PREPARATION						
Surface preparation to general surfaces by dry blast cleaning to DfT Clause 5003 to remove unsound paint down to sound paint		–	–	–	–	**14.23**
Protective System						
Protective system Type I (M) to DfT Table 50/2 to general surfaces prepared down to sound paint		–	–	–	–	**34.50**

PART 8

Daywork

This part of the book contains the following section:

Malcolm Copson, Peter Kendrick
and Steve Beresford
Roadwork:
Theory and Practice
SIXTH EDITION

INTRODUCTION

'Dayworks' relates to work which is carried out incidental to a contract but where no other rates have been agreed.

The most recent Schedule is dated April 2022 and is published by the Civil Engineering Contractors Association. Copies may be obtained from:

CIVIL ENGINEERING CONTRACTORS ASSOCIATION
1 Birdcage Walk
London
SW1H 9JJ
Tel: 020 7340 0450
Email: enquiries@ceca.co.uk

These schedules identify a number of items that are excluded from the rates and percentages quoted, but should be recovered by the Contractor when valuing his Daywork account. Under normal circumstances it is impractical to accurately value these items for each Daywork event, although certain items can be allowed for by means of a percentage addition. Suggested methods of calculating such additions are included in this section.

In Summary the Daywork schedule allows for an addition:

of 148% to wages paid to workmen.
of 88% to labour only subcontractors and hired plant drivers.
of 12.5% to subsistence allowances and travel paid to workmen.
of 12.5% to materials used on dayworks.
of 12.5% to full cost of plant hired for dayworks.
of 12.5% to cost of operating welfare facilities.
of 12.5% to cost of additional insurance premiums for abnormal contract work or special conditions.

Part 4 of this book (Resources – Plant) includes detailed references to the plant section of the 2011 CECA Daywork Schedule.

The text of the document is as follows:

SCHEDULES OF DAYWORKS CARRIED OUT INCIDENTAL TO CONTRACT WORK

The General clauses in the Schedule are repeated in full as follows:

Labour

Add to the amount of wages paid to operatives – 148%.

1 'Amount of wages' means:

Actual wages and bonus paid, daily travelling allowances (fare and/or time), tool allowance and all prescribed payments including those in respect of time lost due to inclement weather paid to operatives at plain time rates and/or at overtime rates.

2 The percentage addition provides for all statutory charges at the date of publication and other charges including:

- National Insurance and Surcharge.
- Normal Contract Works, Third Party & Employer's Liability Insurances
- Annual and Public Holidays with pay.
- Statutory and industry sick pay.
- Welfare benefits.
- Industrial Training Levy.
- Redundancy Payments.

- Employment Rights Act 1996
- Employment Relations Act 1999
- Employment Act 2002
- Site Supervision and staff including foremen and walking gangers, but the time of the gangers or charge hands' working with their gangs is to be paid for as for operatives.
- Small tools – such as picks, shovels, barrows, trowels, hand saws, buckets, trestles, hammers, chisels and all items of a like nature.
- Protective clothing.
- Head Office charges and profit.

3 The time spent in training, mobilization, demobilization etc. for the Dayworks operation is chargeable.
4 All hired plant drivers and labour subcontractor's accounts to be charged in full (without deduction of any cash discounts not exceeding 2.5%) plus 88%.
5 Subsistence or lodging allowances and periodic travel allowances (fare and/or time) paid to or incurred on behalf of operatives are chargeable at cost plus 12.5%.

Materials

Add to the cost of materials – 12.5%.

1 The percentage addition provides for Head Office charges and profit.
2 The cost of materials means the invoiced price of materials including delivery to site without deduction of any cash discounts not exceeding 2.5%.
3 Unloading of materials:
The percentage added to the cost of materials excludes the cost of handling which shall be charged in addition. An allowance for unloading into site stock or storage including wastage should be added where materials are taken from existing stock.

Supplementary Charges

1 Transport provided by contractors for operatives to, from, in and around the site to be charged at the appropriate Schedule rates.
2 Any other charges incurred in respect of any Dayworks operation including, tipping charges, professional fees, subcontractor's accounts and the like shall be paid for in full plus 12.5% (without deduction of cash discounts not exceeding 2.5%). Labour subcontractors being dealt with in Section 1.
3 The cost of operating welfare facilities to be charged by the contractor at cost plus 12.5%.
4 The cost of additional insurance premiums for abnormal contract work or special site conditions to be charged at cost plus 12.5%.
5 The cost of watching and lighting specially necessitated by Dayworks is to be paid for separately at Schedule rates.

Plant

1 These rates apply only to plant already on site, exclusive of drivers and attendants, but inclusive of fuel and consumable stores unless stated to be charged in addition, repairs and maintenance, insurance of plant but excluding time spent on general servicing.
2 Where plant is hired specifically for Dayworks: plant hire (exclusive of drivers and attendants), fuel, oil and grease, insurance, transport etc., to be charged at full amount of invoice (without deduction of any cash discount not exceeding 2.5%) to which should be added consumables where supplied by the contractor, all plus 12.5%.
3 Fuel distribution, mobilization and demobilization are not included in the rates quoted which shall be an additional charge.
4 Metric capacities are adopted and these are not necessarily exact conversions from their imperial equivalents, but cater for the variations arising from comparison of plant manufacturing firms' ratings.
5 SAE rated capacities of plant means rated in accordance with the standards specified by the Society of Automotive Engineers.
6 Minimum hire charge will be for the period quoted.
7 Hire rates for plant not included below shall be settled at prices reasonably related to the rates quoted.
8 The rates provide for Head Office charges and profit.

The Schedule then gives 21 pages of hire rates for a wide range of plant and equipment.

APPLICATION OF DAYWORKS

Generally

A check should be made on the accuracy of the recorded resources and times.

Tender documents generally allow the contractor to tender percentage variations to the figure calculated using the published percentage additions. These vary widely but a reasonable average indication can be along the lines of:

Labour	20% less
Materials	10% less
Plant	30% less

Labour

The Contractor should provide substantiation of the hourly rates he wishes to be paid for the various classes of labour and should demonstrate that the basic 'amount of wages' does not include any of the items actually covered by the percentage addition.

The wage bill is intended to reflect the cost to the Contractor. The time involved is not restricted to the duration of the task, but also includes mobilization and demobilization, together with any training needed – which would include induction courses required for Health & Safety requirements. The rate paid is the actual value of wages and bonuses paid (not simply the basic rate promulgated for the labour grade involved, and includes overtime rates if applicable, tool money, time lost due to inclement weather. In addition, daily travelling allowances are included, as are periodic travel allowances and also subsistence or lodging allowances.

Care should be taken that the matters deemed included in the percentage addition are not duplicated in the amount of wages. For example, it should be noted that foremen, gangers and other supervisory staff are covered by the percentage addition, unless they work, in which case they are paid for at the correct rate for the task involved. The proportion of the time they spend working rather than supervising must be agreed. Refer to the amplification of labour categories in Part 4.

Hired or subcontracted labour is paid at invoiced cost, adjusted only where any cash discount exceeded 2½%, in which case the excess percentage is deducted.

Materials

The cost of materials delivered to the site is simply the invoiced price of the materials plus any delivery charges.

Should the cash discount exceed 2½%, the excess percentage is deducted from the amount to be paid.

The percentage addition simply covers the cost of Head Office charges and profit. It does not include for unloading or temporarily storing the materials, nor for distributing them on site to the work place. Such cost can be charged, even in cases where the materials may already be in the site stock.

Material waste should be added direct to the cost of materials used for each particular Daywork items as an appropriate percentage.

Handling and offloading materials

Schedule 2.3 states that an allowance for handling materials, and an allowance for unloading or storage including wastage should be an additional charge.

Example

For a 12 month, £10.0 m Civil Engineering scheme where the total cost of materials that require handling (excluding Ready Mix concrete, imported fills, fuels and similar items) is £2,500,000.

The following gang is employed (part time) throughout the contract for offloading and handling of materials.

		Net cost (£)
2 labourers		27.18
Lorry (8T) with Hiab lift		23.94
	Rate per hour (£)	51.12
Allow an average of 5 hours per week over 50 weeks		
250 hours @ £51.12 / hour =		£ 12,780
This cost as a percentage of the materials element of the contract.		
(£12,780/£2,500,000) × 100		= 0.51 %

Supplementary charges

Transport

Schedule 3.1 states that transport provided by contractors to take operatives to and from the site as well as in and around the site shall be charged at the appropriate Schedule Rate. This would entail the driver and vehicle being included with the labour and plant parts of the Daywork calculation.

Any other charges

This relates to any other charges incurred in respect of any Dayworks operation and includes tipping charges, professional fees, subcontractor's accounts and the like. Schedule 3.2 provides for full payment of such charges in full plus the addition of 12½% for Head Office charges and profit.

Should the cash discount exceed 2½%, the excess percentage is deducted from the amount to be paid.

Welfare facilities

Schedule 3.3 allows for the net cost of operating these facilities plus 12.5%.

Example

How the costs of operating welfare facilities may be charged to the Daywork account on a 12 month, £10.0 m Civil Engineering scheme, where the total labour element is £1,400,000.

Facility		Weekly Cost
		£
Toilet unit (4 nr)		160.00
Jack leg hutments 24' (2 nr)		120.00
Jack leg hutment 12' (1 nr)		42.00
Labour to clean, maintain and make tea, etc. = 1 man, 2 hours per day, 6 days × 24.88		149.28
Consumables (heat, light, soap, disinfectant, etc.) say		50.00
Rates, insurance, taxes, etc. (add 2%)		10.43
	Total weekly cost	531.71
Multiply by 50 weeks (construction period) plus 6 weeks (maintenance period)		
Total cost to contract = £ 531.71 × 56		29,775.76
Thus cost as a percentage of the Labour element of the contract =(£29,776/£1,400,000) × 100 = £1,400,000		2.13 %
Add, as schedule 3.3 12.5%		0.27 %
Percentage addition for facilities		2.40 %

Insurances

Schedule 3.4 allows for the cost of additional insurance premiums for abnormal work or special site conditions to be charged at cost plus 12.5%.

Watching and lighting

Schedule 3.5 allows for all such costs necessitated by Dayworks to be paid for separately at Schedule rates.

Plant

The cost of the driver(s) and any required attendants such as banksmen should be covered in the labour section of the dayworks calculation.

The Schedule rates include fuel and consumable stores, repairs and maintenance (but not the time spent on general servicing) and insurance.

The Schedule rates only apply to machinery which is on site at the time of the work – where plant is specifically hired for the task, the Contractor is entitled to be paid the invoiced value. If the invoice excludes consumables used (fuel, oil and grease) then the Contractor is entitled to add the cost – together with insurance and any transport costs incurred in getting the equipment on site all subject to a 12½% addition for Head Office costs and profit.

Head Office charges and profit allowances are included in the Schedule rates, 12½% being added to the charged value of hired plant.

General servicing of plant

Schedule 4.1 specifically excludes time spent on general servicing from the Hire Rates.

General servicing in this context can be assumed to mean:

Checking, replenishing (or changing, if applicable). i.e.:

- engine lubrication
- transmission lubrications
- general greasing
- coolants
- hydraulic oils and brake systems
- filters
- tyres

Inspecting special items, e.g.:

- buckets
- hoses/airlines
- shank protectors
- cables/ropes/hawsers
- rippers
- blades, steels, etc.

Example

Assuming an 8 hour working day these operations could take a plant operator on average:

large machine	20 minutes per day	(equating to 1 hour for each 24 worked)
medium machine	10 minutes per day	(equating to 1 hour for each 48 worked)
small machine	5 minutes per day	(equating to 1 hour for each 96 worked)

The cost of the servicing labour would be as follows

large machine	£26.54/hr / 24hrs	= £1.11
medium machine	£17.58/hr / 48 hrs	= £0.37
small machine	£14.68/hr / 96 hrs	= £0.15

These labour costs can be expressed as a percentage of the schedule hire rates:

D8 Dozer	£1.11 / £121.54/hr × 100	= 0.9 %
JCB 3CX	£0.37 / £18.16/hr × 100	= 2.0%
2 tonne dumper	£0.15 / £6.62/hr × 100	= 2.3 %

Taking into account the range of these sizes of plant which are normally deployed on site, the following would provide a reasonable average percentage addition:

0.9 % × 2 = 1.8 %

2.0 % × 3 = 6.0 %

2.3 % × 6 = 13.8%

Total = 21.6 % for 11 items of plant i.e. average = 2.0 %

Fuel distribution

Schedule 4.3 allows for charging for fuel distribution.

Assuming this is done with a towed fuel bowser behind a farm type tractor with driver/labourer in attendance, the operation cycle would involve visiting, service, and return or continue on to the next machine.

The attendance cost based on the hourly rate tractor/bowser/driver would be:

= £15.96 + £1.38 + £13.59 = £30.93

Example

A heavy item of plant, for example a Cat D8R Tractor Bulldozer with a 212 kW engine

Fuel consumption is 30.4 litres/hr (38 litres/hr × 80% site utilization factor)

Fuel capacity is 200 litres

Requires filling after 6.5 working hours operation (200 litres divided by 30.40 l/hr)

Tractor/bowser service taking 30 minutes

The machine cost during this 6.5 hr period would be £121.54 × 6.5, i.e. £ 790.01

The attendance cost for the 30 min cycle would be £30.93 × 0.5, i.e. £15.47

The percentage addition for fuelling the machine would be:

(£15.47/£790.01) × 100 = 2.0%

Example

A medium sized item of plant, for example a JCB 3CX

Fuel consumption is 5.6 litres/hr (7.5 litres/hr × 75% site utilization factor)

Fuel capacity is 90 litres

Requires filling after 16 working hours operation (90 litres divided by 5.60 l/hr)

Tractor/bowser service taking 10 minutes

The machine cost during this 16 hr period would be £18.16 × 16 hrs, i.e. £290.56

The attendance cost for the 10 min cycle would be £30.93 × 0.17 hrs, i.e. £5.26

The percentage addition for fuelling the machine would be:

(£5.26/£290.56) × 100 = 1.8%

Example

A medium sized item of plant, for example a 2 tonne dumper

Fuel consumption is 2.4 litres/hr (3 litres/hr × 80% site utilization factor)

Fuel capacity is 35 litres

Requires filling after 14.5 working hours operation (35 litres divided by 2.40 l/hr)

Tractor/bowser service taking 5 minutes

The machine cost during this 14.5 hr period would be £6.62 × 14.5 hrs, i.e. £95.99

The attendance cost for the 5 min cycle would be £30.93 × 0.08 hrs, i.e. £2.47

The percentage addition for fuelling the machine would be:

(£2.47/£95.99) × 100 = 2.6%

Considering the range of these categories of plant which are normally deployed on site, the following would provide a reasonable average percentage addition for the above:

2.0% × 2 = 4.0 %

1.8% × 3 = 5.4 %

2.6% × 6 = 15.6 %

Total = 25.0% for 11 items of plant – average = 2.3%

Mobilization

Schedule 4.3 allows for charging for mobilization and demobilization.

PART 9

Professional Fees

This part of the book contains the following sections:

CONSULTING ENGINEERS' CONDITIONS OF APPOINTMENT

Introduction

A scale of professional charges for consulting engineering services in connection with civil engineering works is published by the Association for Consultancy and Engineering (ACE).

Copies of the document can be obtained direct from:

Association for Consultancy and Engineering
Alliance House
12 Caxton Street
London SW1H OQL
Tel 020 7222 6557
Fax 020 7222 0750

Comparisons

Instead of the previous arrangement of having different agreements designed for each major discipline of engineering, the current agreements have been developed primarily to suit the different roles that Consulting Engineers may be required to perform, with variants of some of them for different disciplines. The agreements have been standardized as far as possible whilst retaining essential differences.

Greater attention is required than with previous agreements to ensure the documents are completed properly. This is because of the perceived need to allow for a wider choice of arrangements, particularly of methods of payment.

The agreements are not intended to be used as unsigned reference material with the details of an engagement being covered in an exchange of letters, although much of their content could be used as a basis for drafting such correspondence.

In 2019 the ACE published a new suite of Agreements with a broader set of services, these replaced the 2009 publications and are listed below.

Forms of Agreement

ACE Agreement 1: Design
ACE Agreement 2: Advisory, Investigatory and other Services
ACE Agreement 3: Design and Construct
ACE Agreement 4: Sub-Consultancy
ACE Agreement 5: Homeowner
ACE Agreement 6: Expert Witness (Sole Practitioner)
ACE Agreement 7: Expert Witness (Firm)
ACE Agreement 8: Adjudicator

To a number of the above ACE Agreements, Schedules of Services are appended and currently these are:

For use with ACE Agreement 1: Design

- ACE Schedule of Services – Part G(a):
  Civil and Structural Engineer – single consultant or non-lead consultant
- ACE Schedule of Services – Part G(b):
  Mechanical and Electrical Engineering (detailed design in buildings)
- ACE Schedule of Services – Part G(c):
  Mechanical and Electrical Engineering (performance design in buildings)
- ACE Schedule of Services – Part G(d):
  Civil and Structural Engineer – Lead consultant
- ACE Schedule of Services – Part G(e):
  Mechanical and Electrical Engineering Design in buildings – Lead consultant

For use with ACE Agreement 3: Design and Construct

- ACE Schedule of Services – Part G(f):
  Civil and Structural Engineer
- ACE Schedule of Services – Part G(g):
  Mechanical and Electrical Engineering (detailed design in buildings)
- ACE Schedule of Services – Part G(h):
  Mechanical and Electrical Engineering (performance design in buildings)

ACE Agreement 1: Design

Design for the appointment of a consultant by a client to undertake detailed design and/or specification of permanent works to be undertaken or installed by a contractor including any studies, appraisals, investigations, contract administration or construction monitoring leading to or resulting from such detailed design and/or specification.

ACE Agreement 2: Advisory, Investigatory and Other Services

Advise and report for the appointment of a consultant by a client to provide any type of advisory, research, checking, reviewing, investigatory, monitoring, reporting or technical services in the built and natural environments where such services do not consist of detailed design or specification of permanent works to be constructed or installed by a contractor.

ACE Agreement 3: Design and Construct

For the appointment of a single entity e.g. design and build contractor to provide design and construction services. The contract can then appoint the design consultant to provide design services.

ACE Agreement 4: Sub-consultancy

For the appointment of a sub-consultant by a consultant in circumstances where the consultant is appointed on the terms of an ACE Agreement by its client

ACE Agreement 5: Homeowner

Model letter for the appointment of a consultant by a homeowner

ACE Agreement 6: Expert Witness (Sole Practitioner)

For the appointment of an individual to act as an expert witness

ACE Agreement 7: Expert Witness (Firm)

For the appointment of a firm to provide an expert witness

ACE Agreement 8: Adjudicator

For the appointment of an adjudicator

Collateral Warranties

The association is convinced that collateral warranties are generally unnecessary and should only be used in exceptional circumstances. The interests of clients, employers and others are better protected by taking out project or BUILD type latent defects insurance. Nevertheless, in response to observations raised when the pilot editions excluded any mention of warranties, references and arrangements have been included in the Memorandum and elsewhere by which Consulting Engineers may agree to enter into collateral warranty agreements; these should however only be given when the format and requirements thereof have been properly defined and recorded in advance of undertaking the commission.

Requirements for the provision of collateral warranties will be justified even less with commissions under Agreement D than with those under the other ACE agreements. Occasional calls may be made for them, such as when a client intends to dispose of property and needs evidence of a duty of care being owed to specific third parties, but these will be few and far between.

Remuneration

Guidance on appropriate levels of fees to be charged is given at the end of each agreement. Firms and their clients may use this or other sources, including their own records, to determine suitable fee arrangements.

Need for formal documentation

The Association for Consultancy and Engineering recommends that formal written documentation should be executed to record the details of each commission awarded to a Consulting Engineer. These Conditions are published as model forms of agreement suitable for the purpose. However, even if these particular Conditions are not used, it is strongly recommended that, whenever a Consulting Engineer is appointed, there should be at least an exchange of letters defining the duties to be performed and the terms of payment.

Appointments outside the United Kingdom

These conditions of Engagement are designed for use within the UK. For work overseas it is impracticable to give definite recommendations; circumstances differ too widely between countries. There are added complications in documentation relating to local legislation, import customs, conditions of payment, insurance, freight, etc. Furthermore, it is often necessary to arrange for visits to be made by principals and senior staff whose absence abroad during such periods represents a serious reduction of their earning power. The additional duties, responsibilities and non-recoverable costs involved, and the extra work on general co-ordination, should be reflected in the levels of fees. Special arrangements are also necessary to cover travelling and other out-of-pocket expenses in excess of those normally incurred on similar work in the UK, including such matters as local cost-of-living allowances and the cost of providing home-leave facilities for expatriate staff.

CONDITIONS OF ENGAGEMENT

Obligations of the Consulting Engineer

The responsibilities of the Consultant Engineer for the works are as set out in the actual agreement The various standard clauses in the Conditions relate to such matters as differentiating between Normal and Additional services, the duty to exercise skill and care, the need for Client's written consent to the assignment or transfer of any benefit or obligation of the agreement, the rendering of advice if requested on the appointment of other consultants and specialist sub- consultants, any recommendations for design of any part of the Works by Contractors or Subcontractors (with the proviso that the Consulting Engineer is not responsible for detailed design of contractors or for defects or omissions in such design), the designation of a Project Leader, the need for timeliness in requests to the Client for information etc., freezing the design once it has been given Client approval and the specific exclusion of any duty to advise on the actual or possible presence of pollution or contamination or its consequences.

Obligations of the Client

The Consultant Engineer shall be supplied with all necessary data and information in good time. The Client shall designate a Representative authorized to make decisions on his behalf and ensure that all decisions, instructions, and approvals are given in time so as not to delay or disrupt the Consultant Engineer.

Site Staff

The Consulting Engineer may employ site staff he feels are required to perform the task, subject to the prior written agreement of the Client. The Client shall bear the cost of local office accommodation, equipment and running costs.

Commencement, Determination, Postponement, Disruption and Delay

The Consulting Engineer's appointment commences at the date of the execution of the Memorandum of Agreement or such earlier date when the Consulting Engineer first commenced the performance of the Services, subject to the right of the Client to determine or postpone all or any of the Services at any time by Notice.

The Client or the Consulting Engineer may determine the appointment in the event of a breach of the Agreement by the other party after two weeks' notice. In addition, the Consulting Engineer may determine his appointment after two weeks' notice in the event of the Client failing to make proper payment.

The Consulting Engineer may suspend the performance of all or any of the Services for up to twenty-six weeks if he is prevented or significantly impeded from performance by circumstances outside his control. The appointment may be determined by either party in the event of insolvency subject to the issue of notice of determination.

Payments

The Client shall pay fees for the performance of the agreed service(s) together with all fees and charges to the local or other authorities for seeking and obtaining statutory permissions, for all site staff on a time basis, together with additional payments for any variation or the disruption of the Consulting Engineer's work due to the Client varying the task list or brief or to delay caused by the Client, others or unforeseeable events.

If any part of any invoice submitted by the Consulting Engineer is contested, payment shall be made in full of all that is not contested.

Payments shall be made within 28 days of the date of the Consulting Engineer's invoice; interest shall be added to all amounts remaining unpaid thereafter.

Ownership of Documents and Copyright

The Consulting Engineer retains the copyright in all drawings, reports, specifications, calculations etc. prepared in connection with the Task; with the agreement of the Consulting Engineer and subject to certain conditions, the Client may have a licence to copy and use such intellectual property solely for his own purpose on the Task in hand, subject to reservations.

The Consulting Engineer must obtain the client's permission before he publishes any articles, photographs or other illustrations relating to the Task, nor shall he disclose to any person any information provided by the Client as private and confidential unless so authorized by the Client.

Liability, Insurance and Warranties

The liability of the Consulting Engineer is defined, together with the duty of the Client to indemnify the Consulting Engineer against all claims etc. in excess of the agreed liability limit.

The Consulting Engineer shall maintain Professional Indemnity Insurance for an agreed amount and period at commercially reasonable rates, together with Public Liability Insurance and shall produce the brokers' certificates for inspection to show that the required cover is being maintained as and when requested by the Client.

The Consulting Engineer shall enter into and provide collateral warranties for the benefit of other parties if so agreed.

Disputes and Differences

Provision is made for mediation to solve disputes, subject to a time limit of six weeks of the appointment of the mediator at which point it should be referred to an independent adjudicator. Further action could be by referring the dispute to an arbitrator.

QUANTITY SURVEYORS' CONDITIONS OF APPOINTMENT

Introduction

Authors' Note:

The Royal Institution of Chartered Surveyors formally abolished standard Quantity Surveyor's fee scales with effect from 31st December 1998. However, in the absence of any alternative guidance and for the benefit of the user, extracts from relevant fee scales have been reproduced in part with the permission of the Royal Institution of Chartered Surveyors, which owns the copyright.

Summary of Scale of Professional Charges

Scale No 38. issued by The Royal Institution of Chartered Surveyors provides an itemized scale of professional charges for Quantity Surveying Services for Civil Engineering Works which is summarized as follows: -

1.1 Generally

1.2 The Scale of professional charges is applicable where the contract provides for the bills of quantities and final account to be based on measurements prepared in accordance with or based on the principles of the Standard Method of Measurement of Civil Engineering Quantities issued by the Institution of Civil Engineers.

1.3 The fees are in all cases exclusive of travelling and other expenses (for which the actual disbursement is recoverable unless there is some prior arrangement for such charges) and of the cost of reproduction of bills of quantities and other documents, which are chargeable in addition at net cost.

1.4 The fees are in all cases exclusive of services in connection with the allocation of the cost of the works for purposes of calculating value added tax for which there shall be an additional fee based on the time involved.

1.5 If any of the materials used in the works are supplied by the employer or charged at a preferential rate, then the actual or estimated market value thereof shall be included in the amounts upon which fees are to be calculated.

1.6 The fees are in all cases exclusive of preparing a specification of the materials to be used and the works to be done.

1.7 If the quantity surveyor incurs additional costs due to exceptional delays in construction operations or any other cause beyond the control of the quantity surveyor then the fees may be adjusted by agreement between the employer and the quantity surveyor.

1.8 If the works are substantially varied at any stage or if the quantity surveyor is involved in abortive work there shall be an additional fee based on the time involved.

1.9 The fees and charges are in all cases exclusive of value added tax which will be applied in accordance with legislation.

1.10 The scale is not intended to apply to works of a civil engineering nature which form a subsidiary part of a building contractor to buildings which are ancillary to a civil engineering contract. In these cases the fees to be charged for quantity surveying services shall be in accordance with the scales applicable to building works.

1.11 When works of both categories I* and II** are included in one contract the fee to be charged shall be calculated by taking the total value of the sections of work in each of the categories and applying the appropriate scale from the beginning in each case. General items such as preliminaries (and in the case of post contract fees contract price fluctuations) or sections of works which cannot be specifically allocated to either category shall be apportioned pro-rata to the values of the other sections of the works and added thereto in order to ascertain the total value of works in each category.

1.12 When a project is the subject of a number of contracts then, for the purpose of calculating fees, the value of such contracts shall not be aggregated but each contract shall be taken separately and the scale of charges applied as appropriate.

1.13 Roads, railways, earthworks and dredging which are ancillary only to any Category II** work shall be regarded as Category II** work. Works or sections of works of Category I* which incorporate piled construction shall be regarded as Category II** works.

1.14 No addition to the fees given hereunder shall be made in respect of works of alteration or repair where such works are incidental to the new works. If the work covered by a single contract is mainly one of alteration or repair then an additional fee shall be negotiated.

1.15 In the absence of agreement to the contrary, payments to the quantity surveyor shall be made by instalments by arrangement between the employer and the quantity surveyor.

1.16 Copyright in bills of quantities and other documents prepared by the quantity surveyor is reserved to the quantity surveyor.

*Category I	Works or sections of works such as monolithic walls for quays, jetties dams and reservoirs; caissons; tunnels; airport runways and tracks roads; railways; and earthworks and dredging
** Category II	Works or sections of works such as piled quay walls; suspended jetties and quays; bridges and their abutments; culverts; sewers; pipe-lines; electric mains; storage and treatment tanks; water cooling towers and structures for housing heavy industrial and public utility furnace houses and rolling mills to steel works; and boiler houses, plant, e.g. reactor blocks and turbine halls to electricity generating stations.

Scale of charges

For the full Scale of Fees for Professional Charges for Quantity Surveying Services together with a detailed description of the full service provided the appropriate RICS Fee Scale should be consulted.

PART 10

Outputs

This part lists a selection of OUTPUT CONSTANTS for use within various areas of Civil Engineering Work

This part lists a selection of OUTPUT CONSTANTS for use within various areas of Civil Engineering Work.

DISPOSAL OF EXCAVATED MATERIALS

Outputs Per Hundred Cubic Metres

Tipper Capacity and Length of Haul	Driver (hours)	Attendant Labour (hours)	Number of cycles	Average Speed (km/hr)	Cycle time (minutes)				
					Loading	Haul	Discharge	Return	Total
$3\,m^3$ tipper:									
1 km haul	8.8	0.0	43.3	10	0.70	6.0	0.7	5.5	12.2
5 km haul	26.3	0.0	43.3	16	0.70	18.8	0.7	17.0	36.5
9 km haul	41.9	0.0	43.3	18	0.70	30.0	0.7	27.3	58.0
12 km haul	52.7	0.0	43.3	19	0.70	37.9	0.7	34.4	73.0
15 km haul	65.8	0.0	43.3	19	0.70	47.4	0.7	43.1	91.2
$8\,m^3$ tipper:									
1 km haul	10.6	0.8	16.3	3	2.0	20.0	0.9	18.2	39.1
5 km haul	19.7	0.8	16.3	8	2.0	37.5	0.9	34.1	72.5
9 km haul	25.7	0.8	16.3	11	2.0	49.1	0.9	44.6	94.6
12 km haul	26.9	0.8	16.3	14	2.0	51.4	0.9	46.8	99.1
15 km haul	31.4	0.8	16.3	15	2.0	60.0	0.9	54.5	115.4
$12\,m^3$ tipper:									
1 km haul	7.1	0.7	10.8	3	2.90	20.0	1.1	18.2	39.3
5 km haul	13.1	0.7	10.8	8	2.90	37.5	1.1	34.1	72.7
9 km haul	17.1	0.7	10.8	11	2.90	49.1	1.1	44.6	94.8
12 km haul	17.9	0.7	10.8	14	2.90	51.4	1.1	46.8	99.3
15 km haul	20.8	0.7	10.8	15	2.90	60.0	1.1	54.5	115.6
$15\,m^3$ tipper:									
1 km haul	5.8	0.8	8.7	3	3.70	20.0	1.5	18.2	39.7
5 km haul	10.6	0.8	8.7	8	3.70	37.5	1.5	34.1	73.1
9 km haul	13.8	0.8	8.7	11	3.70	49.1	1.5	44.6	95.2
12 km haul	14.5	0.8	8.7	14	3.70	51.4	1.5	46.8	99.7
15 km haul	16.8	0.8	8.7	15	3.70	60.0	1.5	54.5	116.0

Man hours include round trip for tipper and driver together with attendant labour for positioning during loading and unloading.

The number of cycles is based on the stated heaped capacity of the tipper divided into the total volume of $100\,m^3$ to be moved, being multiplied by a bulking factor of × 1.30 to the loose soil volume.

The average speeds are calculated assuming that the vehicles run on roads or on reasonably level firm surfaces and allow for acceleration and deceleration with the return empty journey being say 10% faster than the haul.

The cycle time shows in detail the time spent being loaded (calculated using a $1.5\,m^3$ loader with a cycle time of 22 seconds), haul journey, discharge (turning/manoeuvring/tipping) and return journey.

BREAKING OUT OBSTRUCTIONS BY HAND

Breaking out pavements, brickwork, concrete and masonry by hand and pneumatic breaker

		By hand using picks, shovels & points	Using Compressor			
			7 m^3 Compressor (2 Tool)		10 m^3 Compressor (3 Tool)	
Description	Unit	Labour	Compressor	Labour	Compressor	Labour
Break out bitmac surfaces on subbase or hardcore						
75 mm thick	m^2/hr	1.00	25	13	50	17
100 mm thick	m^2/hr	1.00	20	10	33	11
Break out asphalt roads on hardcore						
150 mm thick	m^2/hr	0.60	8	4	11	4
225 mm thick	m^2/hr	0.50	6	3	8	3
300 mm thick	m^2/hr	0.40	4	2	7	2
Remove existing set paving	m^2/hr	0.80	10	5	17	6
Break out brickwork in cement mortar 215 mm thick	m^2/hr	0.20	3	2	5	2
Break out concrete in areas						
100 mm thick	m^2/hr	0.90	9	5	14	5
150 mm thick	m^2/hr	0.50	7	4	10	4
225 mm thick	m^2/hr	0.30	4	2	6	2
300 mm thick	m^2/hr	0.20	3	1	4	1
Break out reinforced concrete	m^3/hr	0.02	0.40	0.20	0.60	0.20
Break out sandstone	m^3/hr	0.03	0.50	0.30	0.80	0.30

Loading loose materials and items by hand

Material	Unit	Loading into Vehicles	
		Tonne	m^3
Bricks	hr	1.7	2.9
Concrete, batches	hr	1.4	1.2
Gulley grates and frames	hr	1.0	
Kerb	hr	1.1	
Paving slabs	hr	0.9	
Pipes, concrete and clayware	hr	0.9	
Precast concrete items	hr	0.9	
Soil	hr	1.4	2.0
Steel reinforcement	hr	0.8	
Steel sections, etc.	hr	1.0	
Stone and aggregates:			
bedding material	hr	1.5	1.8
filter/subbase	hr	1.3	1.4
rock fill (6" down)	hr	1.3	1.3
Trench planking and shoring	hr	0.9	2.9

CONCRETE WORK

Placing ready mixed concrete in the works

Description	Labour Gang (m^3 per hour)
MASS CONCRETE	
Blinding	
150 mm thick	5.50
150–300 mm thick	6.25
300–500 mm thick	7.00
Bases and oversite concrete	
not exceeding 150 mm thick	5.00
not exceeding 300 mm thick	5.75
not exceeding 500 mm thick	6.75
exceeding 500 mm thick	7.00
REINFORCED CONCRETE	
Bases	
not exceeding 300 mm thick	5.50
not exceeding 500 mm thick	6.25
exceeding 500 mm thick	6.75
Suspended slabs (not exceeding 3 m above pavement level)	
not exceeding 150 mm thick	3.75
not exceeding 300 mm thick	4.75
exceeding 300 mm thick	5.75
Walls and stems (not exceeding 3 m above pavement):	
not exceeding 150 mm thick	3.50
not exceeding 300 mm thick	4.50
exceeding 300 mm thick	5.00
Beams, columns and piers (not exceeding 3 m above pavement):	
sectional area not exceeding 0.03 m^2	2.00
sectional area not exceeding 0.03–1.0 m^2	2.50
sectional area exceeding 1.0 m^2	3.50

Fixing bar reinforcement

All bars delivered to site cut and bent and marked, including craneage and hoisting (maximum height 5 m).

Description		up to 6 mm		7 to 12 mm		13 to 19 mm		over 19 mm	
(fix only)	Unit	steelfixer	labourer	steelfixer	labourer	steelfixer	labourer	steelfixer	labourer
Straight round bars									
to beams, floors, roofs and walls	t/hr	0.03	0.03	0.04	0.04	0.06	0.06	0.08	0.08
to braces, columns, sloping roofs and battered walls	t/hr	0.01	0.01	0.02	0.02	0.03	0.03	0.05	0.05
Bent round bars									
to beams, floors, roofs and walls	t/hr	0.02	0.02	0.03	0.03	0.03	0.03	0.04	0.04
to braces, columns, sloping roofs and battered walls	t/hr	0.01	0.01	0.01	0.01	0.02	0.02	0.03	0.03
Straight, indented or square bars									
to beams, floors, roofs and walls	t/hr	0.02	0.02	0.04	0.04	0.05	0.05	0.07	0.07
to braces, columns, sloping roofs and battered walls	t/hr	0.01	0.01	0.02	0.02	0.02	0.02	0.04	0.04
Bent, indented or square bars									
to beams, floors, roofs and walls	t/hr	0.02	0.02	0.02	0.02	0.03	0.03	0.03	0.03
to braces, columns, sloping roofs and battered walls	t/hr	0.01	0.01	0.01	0.01	0.02	0.02	0.02	0.02
(Average based on Gang D)		0.125		0.148		0.163		0.225	

Erecting formwork to beams and walls

Erect and strike formwork	Unit	Joiner	Labourer
Walls – vertical face (first fix)			
up to 1.5 m	m²/hr	1.7	0.8
1.5 to 3.0 m	m²/hr	1.4	0.7
3.0 to 4.5 m	m²/hr	1.2	0.6
4.5 to 6.0 m	m²/hr	1.0	0.5

Erecting formwork to slabs

Erect and strike formwork	Unit	Joiner	Labourer
Horizontal flat formwork at heights (first fix)			
up to 3.0 m	m²/hr	1.11	1.11
3.0 to 3.6 m	m²/hr	1.05	1.05
3.6 to 4.2 m	m²/hr	1.00	1.00
4.2 to 4.8 m	m²/hr	0.95	0.95
4.8 to 5.4 m	m²/hr	0.90	0.90
5.4 to 6.0 m	m²/hr	0.83	0.83

Multipliers for formwork

Description	Multiplier
Walls built to batter	1.20
Walls built circular to large radius	1.70
Walls built circular to small radius	2.10
Formwork used once	1.00
Formwork used twice, per use	0.85
Formwork used three times, per use	0.75
Formwork used four times, per use	0.72
Formwork used five times, per use	0.68
Formwork used six times, per use	0.66
Formwork used seven or more times, per use	0.63
Formwork to slope not exceeding 45° from horizontal	1.25

DRAINAGE

DRAINAGE

Laying and jointing flexible-jointed clayware pipes

Dia. of pipe in mm	Drainage Gang	In trench not exceeding 1.5 m	In trench not exceeding 3 m	In trench 3–4.5 m
Pipework				
100 mm	m/hr	10	8	7
150 mm	m/hr	7	6	5
225 mm	m/hr	5	4	3
300 mm	m/hr	4	3	2
375 mm	m/hr	3	2	2
450 mm	m/hr	2	2	1
Bends				
100 mm	nr/hr	20	17	14
150 mm	nr/hr	17	14	13
225 mm	nr/hr	13	10	8
300 mm	nr/hr	10	8	7
375 mm	nr/hr	7	6	5
450 mm	nr/hr	3	3	2
Single junctions				
100 mm	nr/hr	13	10	8
150 mm	nr/hr	7	6	5
225 mm	nr/hr	6	5	4
300 mm	nr/hr	4	3	3
375 mm	nr/hr	3	2	2
450 mm	nr/hr	2	2	1

Precast concrete manholes in sections

Description	Unit	Pipelayer	Labourer
Place 675 mm dia. shaft rings	m/hr	1.00	0.30
Place 900 mm manhole rings	m/hr	0.60	0.20
Place 1200 mm manhole rings	m/hr	0.40	0.10
Place 1500 mm manhole rings	m/hr	0.30	0.10
Place 900/675 mm tapers	nr/hr	1.00	0.30
Place 1200/675 mm tapers	nr/hr	0.60	0.20
Place 1500/675 mm tapers	nr/hr	0.50	0.20
Place cover slabs to 675 mm rings	nr/hr	2.50	0.80
Place cover slabs to 900 mm rings	nr/hr	2.00	0.70
Place cover slabs to 1200 mm rings	nr/hr	1.40	0.50
Place cover slabs to 1500 mm rings	nr/hr	0.90	0.30
Build in pipes and make good base:			
150 mm dia.	nr/hr	4.00	-
300 mm dia.	nr/hr	2.00	-
450 mm dia.	nr/hr	1.00	-
Benching 150 mm thick	m²/hr	1.20	1.20
Benching 300 mm thick	m²/hr	0.60	0.60
Render benching 25 mm thick	m²/hr	1.10	1.10
Fix manhole covers frame	nr/hr	1.30	1.30

Tunnelling, Shafts, Diaphragm Walls, Major Earthworks & Fencing

Activity	**Item Description**	**Average Output**
Bore Tunnelling	Strong impervious <8 m dia. (24/7)	8 m/day
	Strong impervious > 8 m dia. (24/7)	15–18 m/day
	Strong impervious & limestone bedrock < 8 m dia. (24/7)	13–15 m/day
	Strong impervious & limestone bedrock > 8 m dia. (24/7)	15–17 m/day
	Cross passage excavation including primary lining (24/7)	1–2.2 m per day
Spray Concrete Lining (SCL) Tunnel	SCL drive >5 m dia. <10 m dia.	1.2–2.3 m a day
	1st stage concrete	130 m a day
	Secondary lining	15–35 m a day
Tunnel Ventilation Shafts	SCL Shaft excavation (average)	347 m^3/day
	SCL main adit excavation – (24/7)	1.5 m/day
	Internal structures (average double shift)	25–30 m/week
Ground Engineering – Diaphragm Wall	(1.5 m width; excavation of panel)	50–80 m^2 per day/rig
Ground Engineering – Piling	Bore piles (dependent upon dia. & length)	2–4 no. per day/rig
	Secant piles (M&F dependent upon dia. & length)	2–6 no. per day/rig
	Contagious piles (CFA – continuous flight auger)	5–10 no per day/rig
	Sheet pile (dependent upon type and depth)	275 m^2 per week/rig
Earthworks	Shafts/box excavation depending on the depth of the shaft or the box	280 m^3–550 m^3 per day/gang
	Bulk excavation – cuttings etc.	2300 m^3–2750 m^3 per day/gang
	Structural fill e.g. Embankments	320 m^3–575 m^3 per day/gang
Hoarding and Perimeter Fence	Hoarding	8 m–12 m per day/gang
	Permanent fence	40 m–60 m per day/gang
	Temporary fence e.g. site boundary	85 m–130 m per day/gang

PART 11

Useful Addresses for Further Information

ACOUSTICAL INVESTIGATION & RESEARCH ORGANISATION LTD (AIRO)
Tel: 01442 247 146
Website: www.airo.co.uk
Email: airo@airo.co.uk

AGGREGATE INDUSTRIES
Tel: 01530 510066
Website: www.aggregate.com

AINSCOUGH
Tel: 0800 272 637
Website: www.ainscough.co.uk
Email: info@ainscough.co.uk

ALUMINIUM FEDERATION LTD (ALFED)
Tel: 0121 601 6363
Website: www.alfed.org.uk
Email: m_lane@alfed.org.uk

ALUMINIUM STOCKHOLDERS ASSOCIATION
Tel: 0121 555 5111
Website: https://www.asauk.co.uk/
Email: info@asauk.co.uk

AMERICAN HARDWOOD EXPORT COUNCIL (AHEC)
Tel: 020 7626 4111
Website: https://www.americanhardwood.org/
Email: Europe@americanhardwood.org

AMIBLU
Tel: 0783486919
Website: https://www.amiblu.com
Email: unitend.kingdom@amiblu.com

ANCIENT MONUMENTS SOCIETY (AMS)
Tel: 020 7236 3934
Website: http://www.ancientmonumentssociety.org.uk/
Email: office@ancientmonumentssociety.org.uk

ARBORICULTURAL ASSOCIATION
Tel: 01242 522152
Website: https://www.trees.org.uk/
Email: admin@trees.org.uk

ARCHITECTURAL ASSOCIATION (AA)
Tel: 020 7887 4000
Website: https://www.aaschool.ac.uk/

ASSOCIATION FOR CONSULTANCY AND ENGINEERING
Tel: 020 7222 6557
Website: https://www.acenet.co.uk/
Email: consult@acenet.co.uk

ASSOCIATION OF FENCING INDUSTRIES
DL6 2NJ
Tel: 020 8253 4516
Website: http://www.afiorg.uk/
Email: info@afiorg.uk

ASSOCIATION OF LOADING AND ELEVATING EQUIPMENT MANUFACTURERS
Tel: 020 8253 4501
Website: https://www.alem.org.uk/
Email: alem@admin.co.uk

ASSOCIATION OF PROJECT MANAGEMENT
Tel: 0845 458 1944
Website: https://www.apm.org.uk/

BATHROOM MANUFACTURERS ASSOCIATION
Tel: 01782 631 619
Website: https://bathroom-association.org.uk/
Email: info@bathroom-association.org.uk

BEAMA Limited (Formerly UNDERFLOOR HEATING MANUFACTURERS ASSOCIATION)
Tel: 020 7793 3000
Website: http://www.beama.org.uk/
Email: info@beama.org.uk

BOX CULVERT ASSOCIATION (BCA)
Tel: 0116 232 5170
Website: https://www.precastdrainage.co.uk/
Email: email@precastdrainage.co.uk

BRETT MARTIN LTD
Tel: +44 (0) 28 9084 9999
Website: http://www.brettmartin.com/
Email: mail@brettmartin.com

BRITISH ADHESIVES AND SEALANTS ASSOCIATION
Tel: 03302 233290
Website: https://www.basa.uk.com/
Email: secretary@basa.uk.com

BRITISH APPROVALS FOR FIRE EQUIPMENT (BAFE)
Tel: 0844 355 0897 / 01608 653 350
Website: https://www.bafe.org.uk/
Email: info@bafe.org.uk

BRITISH APPROVALS SERVICE FOR CABLES (BASEC)
Tel: 01908 267 300
Website: https://www.basec.org.uk/
Email: mail@basec.org.uk

BRITISH ARCHITECTURAL LIBRARY (BAL)
Tel: 020 7307 3882
Website: https://www.architecture.com/
Email: info@riba.org

BRITISH ASSOCIATION OF REINFORCEMENT
Tel: 01276 607140
Website: https://www.uk-bar.org/

BRITISH ASSOCIATION OF LANDSCAPE INDUSTRIES (BALI)
Website: https://www.bali.org.uk/home/
Email: contact@bali.org.uk

BRITISH BOARD OF AGREMENT (BBA)
Tel: 01923 665 300
Website: https://www.bbacerts.co.uk/
Email: clientservices@bbacerts.co.uk

BRITISH CABLES ASSOCIATION (BCA)
Website: https://www.bcauk.org/
Email: secgen@bcauk.org

BRITISH CERAMIC CONFEDERATION (BCC)
Website: https://www.ceramfed.co.uk/
Email: bcc@ceramfed.co.uk

BRITISH CERAMIC TILE COUNCIL (BCTC TILE ASSOCIATION)
Tel: 01782 747 147
Website: www.tpb.org.uk/

BRITISH COATINGS FEDERATION LTD
Website: https://www.coatings.org.uk/
Email: info@bcf.co.uk

BRITISH CONSTRUCTIONAL STEELWORK ASSOCIATION LTD (BCSA)
Tel: 020 7839 8566
Website: https://www.steelconstruction.org/

BRITISH CONTRACT FURNISHING ASSOCIATION (BCFA)
Tel: 01494 896790
Website: https://www.thebcfa.com/

BRITISH FURNITURE MANUFACTURER'S ASSOCIATION (BFM)
Tel: 01494 523021
Website: http://www.bfm.org.uk/
Email: info@bfm.org.uk

BRITISH GEOLOGICAL SURVEY (BGS)
Tel: 0115 936 3100
Website: https://www.bgs.ac.uk/
Email: enquiries@bgs.ac.uk

CHARTERED INSTITUTE OF ARCHITECTURAL TECHNOLOGISTS (CIAT)
Tel: 020 7278 2206
Website: https://ciat.org.uk/
Email: info@ciat.org.uk

BRITISH LAMINATE FABRICATORS ASSOCIATION (BLFA)
Tel: 01159 213889
Website: http://www.blfa.co.uk/
Email: info@blfa.co.uk

BRITISH LIBRARY BIBLIOGRAPHIC SERVICE AND DOCUMENT SUPPLY
Tel: 01937 546 060
Website: https://www.bl.uk/
Email: customer-services@bl.uk

BRITISH LIBRARY ENVIRONMENTAL INFORMATION SERVICE
Tel: 01937 546 060
Website: https://www.bl.uk/
Email: customer-services@bl.uk

BRITISH PLASTICS FEDERATION (BPF)
Tel: 020 7457 5000
Website: https://www.bpf.co.uk/
Email: reception@bpf.co.uk

BRITISH PLASTICS FEDERATION PIPES GROUP
Tel: 020 7457 5000
Website: https://www.bpfpipesgroup.com/
Email: reception@bpf.co.uk

BRITISH PRECAST CONCRETE FEDERATION LTD
Tel: 0116 232 5170
Website: https://www.britishprecast.org/
Email: info@britishprecast.org

BRITISH PRECAST DRAINAGE ASSOCIATION
Tel: 0116 232 5170
Website: https://www.precastdrainage.co.uk/
Email: email@precastdrainage.co.uk

BRITISH PROPERTY FEDERATION (BPF)
Tel: 020 7802 0110
Website: https://www.bpf.org.uk/
Email: info@bpf.org.uk

BRITISH STANDARDS INSTITUTION (BSI)
Tel: 0345 086 9001
Website: https://www.bsigroup.com/en-GB/
Email: cservices@bsigroup.com

BRITISH STAINLESS STEEL ASSOCIATION (BSSA)
Tel: 0114 292 2636
Website: https://www.bssa.org.uk/
Email: enquiry@bssa.org.uk

BRITISH WATER
Tel: 0203 567 0950
Website: https://www.britishwater.co.uk/
Email: info@britishwater.co.uk

BRITISH WOODWORKING FEDERATION
Tel: 0844 209 2610
Website: https://www.bwf.org.uk/
Email: bwf@bwf.org.uk

BUILDING & ENGINEERING SERVICES ASSOCIATION
Tel: 020 7313 4900
Website: https://www.thebesa.com/
Email: contact@thebesa.com

BUILDERS MERCHANTS FEDERATION
Tel: 02476 854980
Website: https://www.bmf.org.uk/:
Email: info@bmf.org.uk

BUILDING CENTRE
Tel: 020 7692 4000
Website: https://www.buildingcentre.co.uk/

BUILDING COST INFORMATION SERVICE LTD (BCIS)
Tel: 024 7686 8555
Website: www.bcis.co.uk
Email: contactrics@rics.org

BUILDING MAINTENANCE INFORMATION (BMI)
Tel: 024 7686 8555
Website: www.bcis.co.uk
Email: contactrics@rics.org

BUILDING RESEARCH ESTABLISHMENT
Tel: 0333 321 8811
Website: https://www.bregroup.com/
Email: enquiries@bregroup.com

BUILDING RESEARCH ESTABLISHMENT: SCOTLAND
Tel: 01414 474999
Website: https://www.bregroup.com/
Email: enquiries@bregroup.com

BUILDING SERVICES RESEARCH AND INFORMATION ASSOCIATION LTD
Tel: 01344 465 600 / 0800 254 5700
Website: https://www.bsria.co.uk/
Email: bsria@bsria.co.uk

BUILD UK (Formerly National Specialist Contractors' Council)
Tel: 0844 249 5351
Website: https://builduk.org/
Email: info@BuildUK.org

CARPET FOUNDATION
LTD (BCMA)
Tel: 01562 755 568
Website: https://www.carpetfoundation.com/
Email: info@carpetfoundation.com

CASTINGS TECHNOLOGY INTERNATIONAL
Advanced Manufacturing Park
Tel: 0114 215 8300
Website: http://castingstechnology.com/
Email: info@castingstechnology.com

CATERING EQUIPMENT MANUFACTURERS ASSOCIATION (CESA)
Tel: 020 7793 3030
Website: http://www.cesa.org.uk/
Email: enquiries@cesa.org.uk

CEMENT ADMIXTURES ASSOCIATION
Tel: 01564 776 362
Website: http://www.admixtures.org.uk/
Email: info@admixtures.org.uk

CHARTERED INSTITUTE OF BUILDING (CIOB)
Tel: 01344 630 700
Website: https://www.ciob.org/
Email: reception@ciob.org.uk

CHARTERED INSTITUTE OF ARBITRATORS
Tel: 020 7421 7444
Website: https://www.ciarb.org/

CHARTERED INSTITUTE OF ARCHITECTURAL TECHNOLOGISTS
Tel: 020 7278 2206
Website: https://ciat.org.uk/
Email: info@ciat.org.uk

CHARTERED INSTITUTION OF BUILDING SERVICES ENGINEERS (CIBSE)
Tel: 020 8675 5211
Website: https://www.cibse.org/
Email: websupport@cibse.org

CHARTERED INSTITUTE OF PLUMBING AND HEATING ENGINEERING
Tel: 01708 472 791
Website: www.ciphe.org.uk
Email: info@ciphe.org.uk

CHARTERED INSTITUTE OF WASTE MANAGEMENT
Tel: 01604 620 426
Website: https://www.ciwm.co.uk/
Email: ciwm@ciwm.co.uk

CIVIL ENGINEERING CONTRACTORS ASSOCIATION
Tel: 020 7340 0450
Website: https://www.ceca.co.uk/
Email: lauraellis@ceca.co.uk

COLD ROLLED SECTIONS ASSOCIATION (CRSA)
Tel: 0121 601 6350
Website: http://www.crsauk.com/

COMBUSTION ENGINEERING ASSOCIATION
Tel: 01740 625 538
Website: https://cea.org.uk/
Email: info@cea.org.uk

CONCRETE BRIDGE DEVELOPMENT GROUP
Tel: 01276 33777
Website: http://www.cbdg.org.uk/
Email: enquiries@cbdg.org.uk

CONCRETE REPAIR ASSOCIATION (CRA)
Tel: 01420 471615
Website: https://www.cra.org.uk/
Email: admin@cra.org.uk

CONFEDERATION OF BRITISH INDUSTRY (CBI)
Tel: 020 7379 7400
Website: www.cbi.org.uk
Email: enquiries@cbi.org.uk

CONSTRUCT – CONCRETE STRUCTURES GROUP LTD
Tel: 0844 249 1355
Website: https://www.construct.org.uk/
Email: enquiries@construct.org.uk

CONSTRUCTION EMPLOYERS FEDERATION LTD (CEF)
Tel: 028 9087 7143
Website: https://www.cefni.co.uk/
Email: mail@cefni.co.uk

CONSTRUCTION INDUSTRY FEDERATION
Tel: 01 406 6000
Website:
https://cif.ie/
Email: info@cif.ie

CONSTRUCTION INDUSTRY RESEARCH AND INFORMATION ASSOCIATION (CIRIA)
Tel: 020 7549 3300
Website: https://www.ciria.org/
Email: enquiries@ciria.org

CONSTRUCTION PLANT-HIRE ASSOCIATION (CPA)
Tel: 020 7796 3366
Website: https://www.cpa.uk.net/
Email: enquiries@cpa.uk.net

CONTRACT FLOORING ASSOCIATION (CFA)
Tel: 0115 941 1126
Website: https://www.cfa.org.uk/

CONTRACTORS MECHANICAL PLANT ENGINEERS (CMPE)
Tel: 0330 838 2557
Website: https://www.cmpe.co.uk/
Email: info@cmpe.co.uk

COPPER DEVELOPMENT ASSOCIATION
Tel: 01442 275 705
Website: https://copperalliance.org.uk/
Email: cda@copperalliance.org.uk

COUNCIL FOR ALUMINIUM IN BUILDING (CAB)
Tel: 01453 828851
Website: http://c-a-b.org.uk/

DANNY SULLIVAN GROUP LTD
Tel: 020 8961 1900
Email: enquiries@dannysullivan.co.uk

DEPARTMENT FOR BUSINESS, ENERGY AND INDUSTRIAL STRATEGY
Website: https://www.gov.uk/government/organizations/department-for-business-energy-and-industrial-strategy
Email: enquiries@beis.gov.uk

DEPARTMENT FOR TRANSPORT
Tel: 0300 330 3000
Website: https://www.gov.uk/government/organizations/department-for-transport

DOORS & HARDWARE FEDERATION
Tel: 01827 52337
Website: https://www.dhfonline.org.uk/
Email: info@dhfonline.org.uk

DRAINAGE SUPERSTORE (CMO LTD)
Tel: 01752 692 221
Website: https://www.drainagesuperstore.co.uk/
Email: sales@drainagesuperstore.co.uk

DRY STONE WALLING ASSOCIATION OF GREAT BRITAIN (DSWA)
Tel: 01539 567 953
Website: https://www.dswa.org.uk/
Email: information@dswa.org.uk

ELECTRICAL CONTRACTORS ASSOCIATION (ECA)
Tel: 020 7313 4800
Website: https://eca.co.uk/

ELECTRICAL CONTRACTORS ASSOCIATION OF SCOTLAND (SELECT)
Tel: 0131 445 5577
Website: https://www.select.org.uk/
Email: admin@select.org.uk

FEDERATION OF ENVIRONMENTAL TRADE ASSOCIATIONS
Tel: 0118 940 3416
Website: https://www.feta.co.uk/
Email: info@feta.co.uk

FEDERATION OF MASTER BUILDERS (FMB)
Tel: 0330 333 7777
Website: https://www.fmb.org.uk/
Email: reception@fmb.org.uk

FEDERATION OF PILING SPECIALISTS
Tel: 020 8663 0947
Website: https://www.fps.org.uk/
Email: fps@fps.org.uk

FENCING CONTRACTORS ASSOCIATION
Tel: 020 8253 4516
Website: http://fencingcontractors.org/

FINISHES & INTERIORS SECTOR
Tel: 0121 707 0077
Website: https://www.thefis.org/
Email: info@thefis.org

FIRSTFENCE
Tel: 01283512111
Website: https://www.firstfence.co.uk
Email: sales@firsfence.co.uk

FURNITURE INDUSTRY RESEARCH ASSOCIATION (FIRA INTERNATIONAL LTD)
Tel: 01438 777 700
Website: https://www.fira.co.uk/
Email: info@fira.co.uk

GLASS & GLAZING FEDERATION (GGF)
40 Rushworth Street
London
SE1 0RB
Tel: 020 7939 9100
Website: https://www.ggf.org.uk/

GRAFTON MERCHANTING
Tel: 03456 854467
Website: http://www.graftonmerchantinggb.co.uk/

GUTTER SUPPLIES
Tel: 01206 931470
Website: https://www.guttersupplies.co.uk/
Email: sales@guttersupplies.co.uk

HOT WATER ASSOCIATION (HWA)
Tel: 01926 513 777
Website: https://www.hotwater.org.uk/
Email: info@hotwater.org.uk

HYDRO CLEANSING LTD
Tel: 020 8689 3339
Website: https://hydro-cleansing.com/
Email: sales@hydro-cleansing.com

ICOM ENERGY ASSOCIATION
Tel: 01926 513748
Website: http://icom.org.uk/

INSTITUTE OF ACOUSTICS
Tel: 0300 999 9675
Website: https://www.ioa.org.uk/
Email: ioa@ioa.org.uk

INSTITUTE OF ASPHALT TECHNOLOGY
Tel: 01360 440 682
Website: https://www.instituteofasphalt.org/
Email: info@instituteofasphalt.org

INSTITUTION OF CIVIL ENGINEERS (ICE)
Tel: 020 7222 7722
Website: https://www.ice.org.uk/

INSTITUTE OF MATERIALS, MINERALS AND MINING
London NW1 3AD
Tel: 020 7451 7300
Website: https://www.iom^3.org/

INSTITUTION OF ENGINEERING AND TECHNOLOGY (IET)
Tel: 01438 313 311
Website: https://www.theiet.org/
Email: postmaster@theiet.org

INSTITUTION OF MECHANICAL ENGINEERS
Tel: 020 7222 7899
Website: http://www.imeche.org/
Email: enquiries@imeche.org

INSTITUTION OF STRUCTURAL ENGINEERS (ISE)
Tel: 020 7235 4535
Website: https://www.istructe.org/
Email: mail@istructe.org

INSULATION EXPRESS
Tel: 0330 123 1761
Website: https://www.insulationexpress.co.uk/
Email: info@insulationexpress.co.uk

INTERNATIONAL LEAD ASSOCIATION
Tel: 020 7833 8090
Website: https://www.ila-lead.org/
Email: enq@ila-lead.org

INTERPAVE (THE PRECAST CONCRETE PAVING & KERB ASSOCIATION)
Tel: 0116 232 5170
Website: http://www.paving.org.uk/
Email: info@paving.org.uk

JACKSONS FENCING
Tel: 08004082234
Website: http://www.jackson-fencing.co.uk
Email: sales@jacksons-fencing.co.uk

JOINT CONTRACTS TRIBUNAL LTD (JCT)
Tel: 0345 0821080
Website: https://www.jctltd.co.uk/

KITCHEN SPECIALISTS ASSOCIATION
Tel: 01623 818808
Website: https://www.kbsa.org.uk/

LIGHTING ASSOCIATION LTD
Tel: 01952 290 905
Website: https://www.thelia.org.uk/
Email: enquiries@thelia.org.uk

LUCIDEON
Tel: 01782 764 444
Website: https://www.lucideon.com/
Email: enquiries@lucideon.com

MASTIC ASPHALT COUNCIL LTD
Tel: 01273 242 778
Website: http://masticasphaltcouncil.co.uk/
Email: info@masticasphaltcouncil.co.uk

METAL CLADDING & ROOFING MANUFACTURERS ASSOCIATION
Tel: 01633 895 633
Website: https://mcrma.co.uk/
Email: info@mcrma.co.uk

METALS4U
Tel: 01937534318
Website: http://www.metals4u.co.uk
Email: info@metals4u.uk

MET OFFICE
Tel: 0370 900 0100
Website: https://www.metoffice.gov.uk/
Email: enquiries@metoffice.gov.uk

MINERAL PRODUCTS ASSOCIATION
Tel: 020 7963 8000
Website: https://mineralproducts.org/
Email: info@mineralproducts.org

MOVERIGHT INTERNATIONAL LTD
Tel: 01675 475590
Website: http://www.moverightinternational.com/
Email: andrew@moverightinternational.com

MYTUB LTD
Tel: 0844 556 1818
Website: https://www.mytub.co.uk/

NATIONAL ASSOCIATION OF STEEL STOCKHOLDERS
Tel: 0121 200 2288
Website: https://www.nass.org.uk/

NATIONAL HOUSE-BUILDING COUNCIL (NHBC)
Tel: 0344 633 1000
Website: http://www.nhbc.co.uk/

NATIONAL SOIL RESOURCES INSTITUTE
Tel: 01234 750 111
Website: https://www.cranfield.ac.uk/centres/soil-and-agrifood-institute/research-groups/national-soil-resources-institute

NHS PROPERTY (Formerly NHS ESTATES)
Tel: 0800 085 3015
Website: https://www.property.nhs.uk/
Email: customer.service@property.nhs.uk

ORDNANCE SURVEY
Tel: 03454 56 04 20 / 0345 050505
Website: https://www.ordnancesurvey.co.uk/

PAINTING AND DECORATING ASSOCIATION
Tel: 024 7635 3776
Website: https://paintingdecoratingassociation.co.uk/
Email: info@paintingdecoratingassociation.co.uk

PIPELINE INDUSTRIES GUILD
Tel: 020 7235 7938
Website: https://pipeguild.com/
Email: enquiries@pipeguild.com

PROPERTY CARE ASSOCIATION (FORMERLY BRITISH WOOD PRESERVING & DAMPPROOFING ASSOCIATION (BWPDA))
Tel: 0844 375 4301
Website: https://www.property-care.org/
Email: pca@property-care.org

PROPERTY CONSULTANTS SOCIETY LTD
Tel: 01903 889590
Website: http://www.propertyconsultantssociety.org/
Email: info@propertyconsultantssociety.org

Pryce's
Tel: 01502574141
Website: http://www.rjpryce.com
Email: Sales@rjpryce.co.uk

ROMTECH LTD
Tel: 0870 011 3609
Website: https://www.rom.co.uk/
Email: ireland.sales@rom-tech.co.uk

ROYAL INCORPORATION OF ARCHITECTS IN SCOTLAND (RIAS)
Tel: 0131 229 7545
Website: https://www.rias.org.uk/
Email: info@rias.org.uk

ROYAL INSTITUTE OF BRITISH ARCHITECTS (RIBA)
Tel: 020 7580 5533
Website: https://www.architecture.com/
Email: info@riba.org

ROYAL INSTITUTION OF CHARTERED SURVEYORS
Tel: 024 7686 8555
Website: https://www.rics.org/uk/
Email: RICSSouth@rics.org

ROYAL TOWN PLANNING INSTITUTE (RTPI)
Tel: 020 7929 9494
Website: https://www.rtpi.org.uk/
Email: contact@rtpi.org.uk

RURAL AND INDUSTRIAL DESIGN AND BUILDING ASSOCIATION
Tel: 0844 249 0043
Website: https://www.ridba.org.uk/
Email: admin@ridba.org.uk

SAVIY PIPING INC
Website: https://www.savoypipinginc.com/
Email: sales@savoypipinginc.com

SCOTTISH BUILDING FEDERATION
Tel: 0131 556 8866
Website: https://www.scottish-building.co.uk/
Email: info@scottish-building.co.uk

SCOTTISH GOVERNEMENT
Tel: 0300 244 4000
Website: https://www.gov.scot
Email: ceu@gov.scot

SCOTTISH NATURAL HERITAGE
Tel: 01463 725 000
Website: https://www.nature.scot/
Email: enquiries@nature.scot

SINGLE PLY ROOFING ASSOCIATION
Tel: 0845 1547188
Website: www.spra.co.uk/
Email: enquiries@spra.co.uk

SMOKE CONTROL ASSOCIATION
Tel: 0118 940 3416
Website: www.feta.co.uk
Email: smokecontrol@feta.co.uk

SOCIETY FOR THE PROTECTION OF ANCIENT BUILDINGS (SPAB)
Tel: 020 7377 1644
Website: https://www.spab.org.uk/
Email: info@spab.org.uk

SOCIETY OF GLASS TECHNOLOGY
Tel: 0114 263 4455
Website: https://www.sgt.org/

SOLAR ENERGY SOCIETY
Tel: 07760 163 559
Website: http://uk-ises.org/
Email: info@uk-ises.org

SPON'S PRICE BOOK EDITORS
AECOM Ltd
Tel: 020 7061 7000
Website: www.aecom.com

SPORT ENGLAND
Website: https://www.sportengland.org/
Email: funding@sportengland.org

SPORT SCOTLAND
Tel: 0141 534 6500
Website: https://sportscotland.org.uk/
Email: sportscotland.enquiries@sportscotland.org.uk

SPORT WALES
Tel: 0300 300 3111
Website: http://sport.wales/
Email: info@sportwales.org.uk

SPORTS TURF RESEARCH INSTITUTE (STRI)
Tel: 01274 565 131
Website: https://strigroup.com/stri-uk/
Email: enquiries@strigroup.com

SPRAYED CONCRETE ASSOCIATION
Tel: 01420 471 622
Website: http://sca.org.uk/
Email: admin@sca.org.uk

STEEL CONSTRUCTION INSTITUTE
Tel: (0)1344 636525
Website: https://steel-sci.com/
Email: reception@steel-sci.com

STEEL WINDOW ASSOCIATION
Tel: 020 8543 2841
Website: https://www.steel-window-association.co.uk/
Email: info@steel-window-association.co.uk

STONE FEDERATION GREAT BRITAIN
Channel Business Centre
Tel: 01303 856123
Website: http://www.stonefed.org.uk/
Email: enquiries@stonefed.org.uk

SPECIALIST ACCESS ENGINEERING & MAINTENANCE ASSOCIATION
Tel: 07909 333 581 / 07854 226251
Website: https://www.saema.org/
Email: info@saema.org

SWIMMING POOL & ALLIED TRADES ASSOCIATION (SPATA)
Tel: 01264 356210
Website: https://www.spata.co.uk/
Email: admin@spata.co.uk

THE CONCRETE SOCIETY
Tel: 01276 607 140
Website: http://www.concrete.org.uk/default.asp

THERMAL INSULATION CONTRACTORS ASSOCIATION
Tel: 01325 466 704
Website: https://tica-acad.co.uk/tica-home/
Email: enquiries@tica.uk.com

TIMBER RESEARCH & DEVELOPMENT ASSOCIATION (TRADA)
Tel: 01494 569 603
Website: https://www.trada.co.uk/
Email: advisory@trada.co.uk

TIMBER TRADE FEDERATION
Tel: 020 3205 0067
Website: https://ttf.co.uk/
Email: ttf@ttf.co.uk

TOWN & COUNTRY PLANNING ASSOCIATION (TCPA)
Tel: 020 7930 8903
Website: https://www.tcpa.org.uk/
Email: tcpa@tcpa.org.uk

THE TREE COUNCIL
Tel: 020 7407 9992
Website: https://www.treecouncil.org.uk/
Email: info@treecouncil.org.uk

TRACEY CONCRETE LIMITED
Tel: 028 6632 6437
Website: https://www.traceyconcrete.com/site/
Email: headoffice@traceyconcrete.com

TRAVIS PERKINS PLC
Tel: 01604 752424
Website: https://www.travisperkinsplc.co.uk/
Email: customerservice@travisperkins.co.uk

TRUSSED RAFTER ASSOCIATION
Tel: 020 3205 0032
Website: https://www.tra.org.uk/
Email: info@tra.org.uk

TWI
Tel: 01223 899 000
Website: https://www.twi-global.com/

WATER RESEARCH CENTRE
Tel: 01793 865000
Website: http://www.wrcplc.co.uk/
Email: solutions@wrcplc.co.uk

WATER UK
Tel: 020 7344 1844
Website: https://www.water.org.uk/
Email: webadmin@water.org.uk

WOOD PANEL INDUSTRIES FEDERATION
Tel: 01476 512 381
Website: http://wpif.org.uk/
Email: enquiries@wpif.org.uk

WRAP
Tel: 01295 819900
Website: http://www.wrap.org.uk/

WYDELS
Tel: 01536264064
Website: http://www.wydels.co.uk
Email: enquireies@wydels.co.uk

ZINC INFORMATION CENTRE
Tel: 0121 362 1201
Website: http://www.zincinfocentre.org/
Email: zincinfocentre@hdg.org.uk

PART 12

Tables and Memoranda

This part contains the following sections:

CONVERSION TABLES

CONVERSION TABLES

Length	Unit	Conversion factors			
Millimetre	mm	1 in	= 25.4 mm	1 mm	= 0.0394 in
Centimetre	cm	1 in	= 2.54 cm	1 cm	= 0.3937 in
Metre	m	1 ft	= 0.3048 m	1 m	= 3.2808 ft
		1 yd	= 0.9144 m		= 1.0936 yd
Kilometre	km	1 mile	= 1.6093 km	1 km	= 0.6214 mile

Note:

1 cm	= 10 mm	1 ft	= 12 in
1 m	= 1 000 mm	1 yd	= 3 ft
1 km	= 1 000 m	1 mile	= 1 760 yd

Area	Unit	Conversion factors			
Square Millimetre	mm^2	1 in^2	= 645.2 mm^2	1 mm^2	= 0.0016 in^2
Square Centimetre	cm^2	1 in^2	= 6.4516 cm^2	1 cm^2	= 1.1550 in^2
Square Metre	m^2	1 ft^2	= 0.0929 m^2	1 m^2	= 10.764 ft^2
		1 yd^2	= 0.8361 m^2	1 m^2	= 1.1960 yd^2
Square Kilometre	km^2	1 $mile^2$	= 2.590 km^2	1 km^2	= 0.3861 $mile^2$

Note:

1 cm^2	= 100 mm^2	1 ft^2	= 144 in^2
1 m^2	= 10 000 cm^2	1 yd^2	= 9 ft^2
1 km^2	= 100 hectares	1 acre	= 4 840 yd^2
		1 $mile^2$	= 640 acres

Volume	Unit	Conversion factors			
Cubic Centimetre	cm^3	1 cm^3	= 0.0610 in^3	1 in^3	= 16.387 cm^3
Cubic Decimetre	dm^3	1 dm^3	= 0.0353 ft^3	1 ft^3	= 28.329 dm^3
Cubic Metre	m^3	1 m^3	= 35.3147 ft^3	1 ft^3	= 0.0283 m^3
		1 m^3	= 1.3080 yd^3	1 yd^3	= 0.7646 m^3
Litre	l	1 l	= 1.76 pint	1 pint	= 0.5683 l
			= 2.113 US pt		= 0.4733 US l

Note:

1 dm^3	= 1 000 cm^3	1 ft^3	= 1 728 in^3	1 pint	= 20 fl oz
1 m^3	= 1 000 dm^3	1 yd^3	= 27 ft^3	1 gal	= 8 pints
1 l	= 1 dm^3				

Neither the Centimetre nor Decimetre are SI units, and as such their use, particularly that of the Decimetre, is not widespread outside educational circles.

Mass	Unit	Conversion factors			
Milligram	mg	1 mg	= 0.0154 grain	1 grain	= 64.935 mg
Gram	g	1 g	= 0.0353 oz	1 oz	= 28.35 g
Kilogram	kg	1 kg	= 2.2046 lb	1 lb	= 0.4536 kg
Tonne	t	1 t	= 0.9842 ton	1 ton	= 1.016 t

Note:

1 g	= 1000 mg	1 oz	= 437.5 grains	1 cwt	= 112 lb
1 kg	= 1000 g	1 lb	= 16 oz	1 ton	= 20 cwt
1 t	= 1000 kg	1 stone	= 14 lb		

Force	Unit	Conversion factors			
Newton	N	1 lbf	= 4.448 N	1 kgf	= 9.807 N
Kilonewton	kN	1 lbf	= 0.004448 kN	1 ton f	= 9.964 kN
Meganewton	MN	100 tonf	= 0.9964 MN		

CONVERSION TABLES

Pressure and stress	Unit	Conversion factors	
Kilonewton per square metre	kN/m^2	1 lbf/in^2	= 6.895 kN/m^2
		1 bar	= 100 kN/m^2
Meganewton per square metre	MN/m^2	1 $tonf/ft^2$	= 107.3 kN/m^2 = 0.1073 MN/m^2
		1 kgf/cm^2	= 98.07 kN/m^2
		1 lbf/ft^2	= 0.04788 kN/m^2

Coefficient of consolidation (Cv) or swelling	Unit	Conversion factors	
Square metre per year	m^2/year	1 cm^2/s	= 3 154 m^2/year
		1 ft^2/year	= 0.0929 m^2/year

Coefficient of permeability	Unit	Conversion factors	
Metre per second	m/s	1 cm/s	= 0.01 m/s
Metre per year	m/year	1 ft/year	= 0.3048 m/year
			= 0.9651 × $(10)^8$m/s

Temperature	Unit	Conversion factors	
Degree Celsius	°C	°C = 5/9 × (°F − 32)	°F = (9 × °C)/ 5 + 32

CONVERSION TABLES

SPEED CONVERSION

km/h	m/min	mph	fpm
1	16.7	0.6	54.7
2	33.3	1.2	109.4
3	50.0	1.9	164.0
4	66.7	2.5	218.7
5	83.3	3.1	273.4
6	100.0	3.7	328.1
7	116.7	4.3	382.8
8	133.3	5.0	437.4
9	150.0	5.6	492.1
10	166.7	6.2	546.8
11	183.3	6.8	601.5
12	200.0	7.5	656.2
13	216.7	8.1	710.8
14	233.3	8.7	765.5
15	250.0	9.3	820.2
16	266.7	9.9	874.9
17	283.3	10.6	929.6
18	300.0	11.2	984.3
19	316.7	11.8	1038.9
20	333.3	12.4	1093.6
21	350.0	13.0	1148.3
22	366.7	13.7	1203.0
23	383.3	14.3	1257.7
24	400.0	14.9	1312.3
25	416.7	15.5	1367.0
26	433.3	16.2	1421.7
27	450.0	16.8	1476.4
28	466.7	17.4	1531.1
29	483.3	18.0	1585.7
30	500.0	18.6	1640.4
31	516.7	19.3	1695.1
32	533.3	19.9	1749.8
33	550.0	20.5	1804.5
34	566.7	21.1	1859.1
35	583.3	21.7	1913.8
36	600.0	22.4	1968.5
37	616.7	23.0	2023.2
38	633.3	23.6	2077.9
39	650.0	24.2	2132.5
40	666.7	24.9	2187.2
41	683.3	25.5	2241.9
42	700.0	26.1	2296.6
43	716.7	26.7	2351.3
44	733.3	27.3	2405.9
45	750.0	28.0	2460.6

CONVERSION TABLES

km/h	m/min	mph	fpm
46	766.7	28.6	2515.3
47	783.3	29.2	2570.0
48	800.0	29.8	2624.7
49	816.7	30.4	2679.4
50	833.3	31.1	2734.0

GEOMETRY

GEOMETRY

Two dimensional figures

Figure	Diagram of figure	Surface area	Perimeter
Square	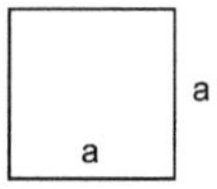	a^2	$4a$
Rectangle	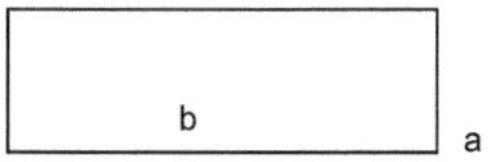	ab	$2(a+b)$
Triangle	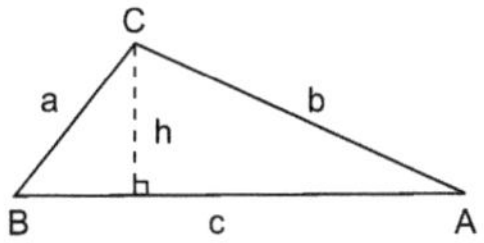	$\frac{1}{2}ch$	$a+b+c$
Circle	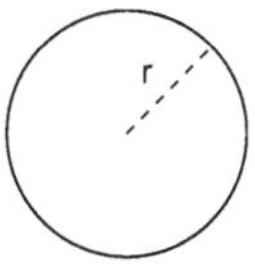	πr^2 $\frac{1}{4}\pi d^2$ where $2r = d$	$2\pi r$ πd
Parallelogram	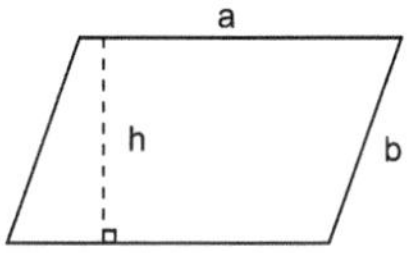	ah	$2(a+b)$
Trapezium	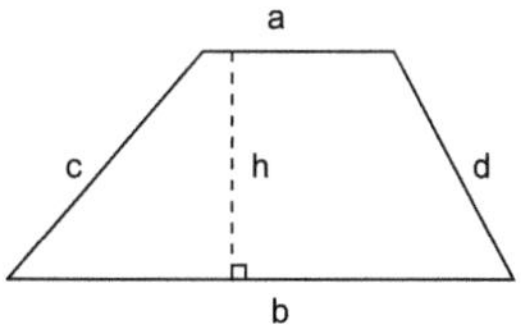	$\frac{1}{2}h(a+b)$	$a+b+c+d$
Ellipse	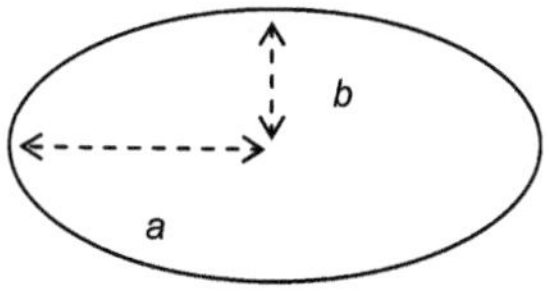	Approximately πab	$\pi(a+b)$
Hexagon	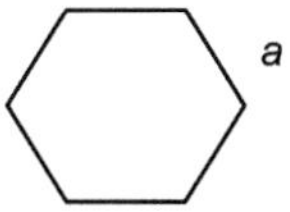	$2.6 \times a^2$	

GEOMETRY

Figure	Diagram of figure	Surface area	Perimeter
Octagon		$4.83 \times a^2$	6a
Sector of a circle	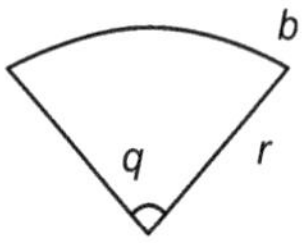	$\frac{1}{2}rb$ or $\frac{q}{360}\pi r^2$ note $b =$ angle $\frac{q}{360} \times \pi 2r$	
Segment of a circle	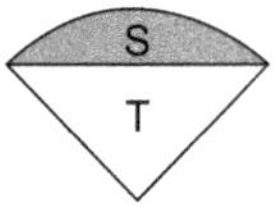	S – T where S = area of sector, T = area of triangle	
Bellmouth	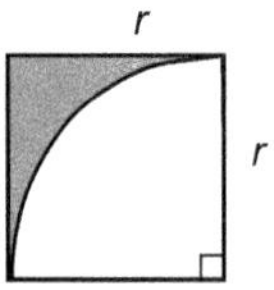	$\frac{3}{14} \times r^2$	

GEOMETRY

Three dimensional figures

Figure	Diagram of figure	Surface area	Volume
Cube	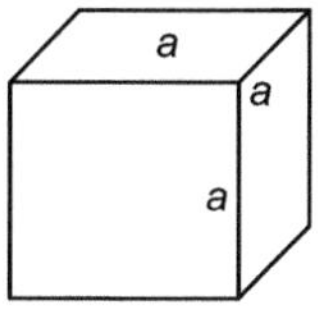	$6a^2$	a^3
Cuboid/ rectangular block	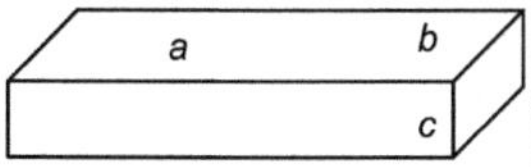	$2(ab + ac + bc)$	abc
Prism/ triangular block	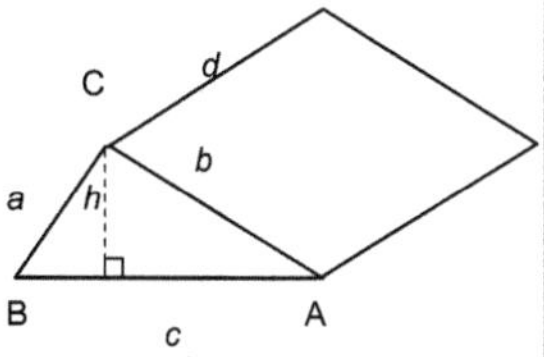	$bd + hc + dc + ad$	$\frac{1}{2}hcd$
Cylinder	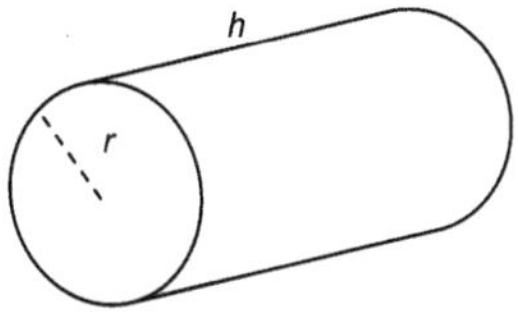	$2\pi r^2 + 2\pi h$	$\pi r^2 h$
Sphere	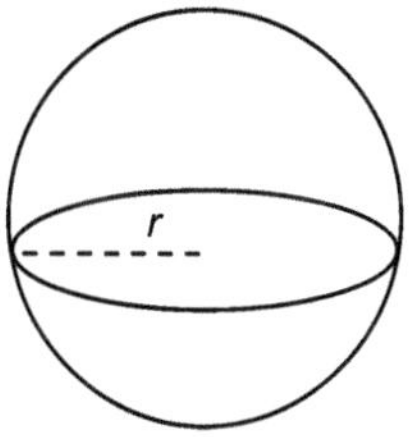	$4\pi r^2$	$\frac{4}{3}\pi r^3$
Segment of sphere	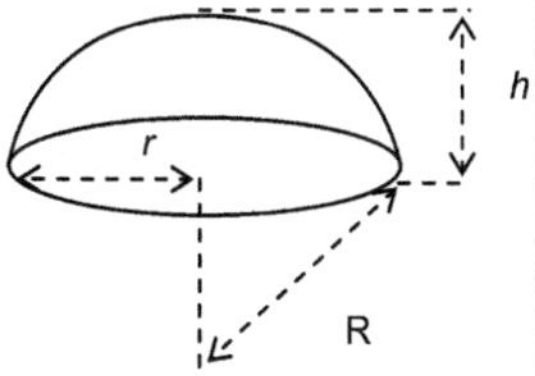	$2\pi Rh$	$\frac{1}{6}\pi h(3r^2 + h^2)$ $\frac{1}{3}\pi h^2(3R - H)$
Pyramid	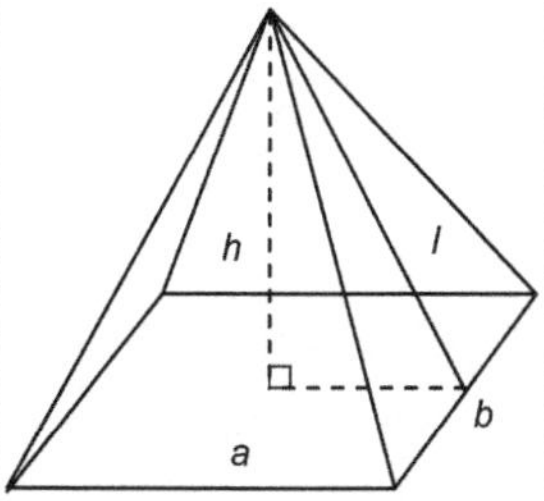	$(a + b)l + ab$	$\frac{1}{3}abh$

GEOMETRY

Figure	Diagram of figure	Surface area	Volume
Frustum of a pyramid	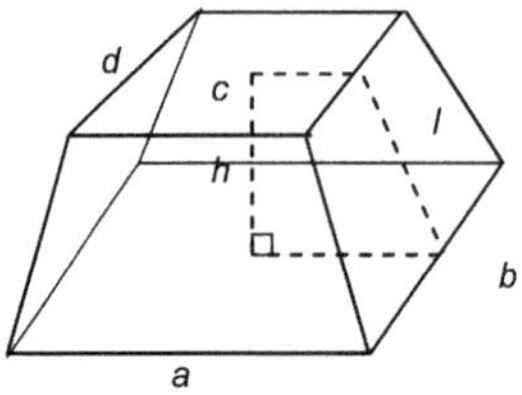	$l(a + b + c + d) + \sqrt{(ab + cd)}$ [rectangular figure only]	$\frac{h}{3}(ab + cd + \sqrt{abcd})$
Cone	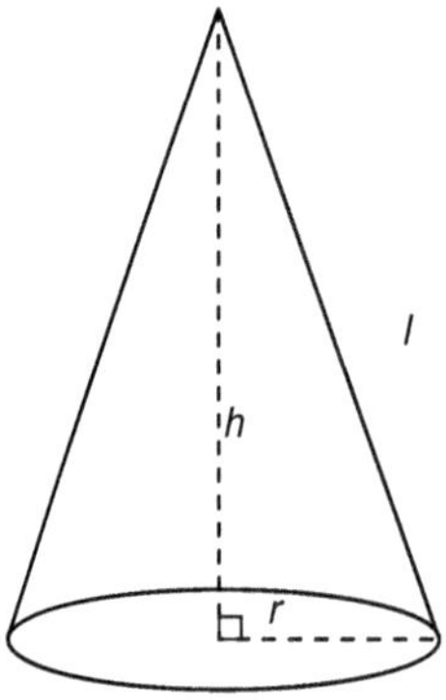	πrl (excluding base) $\pi rl + \pi r^2$ (including base)	$\frac{1}{3}\pi r^2 h$ $\frac{1}{12}\pi d^2 h$
Frustum of a cone	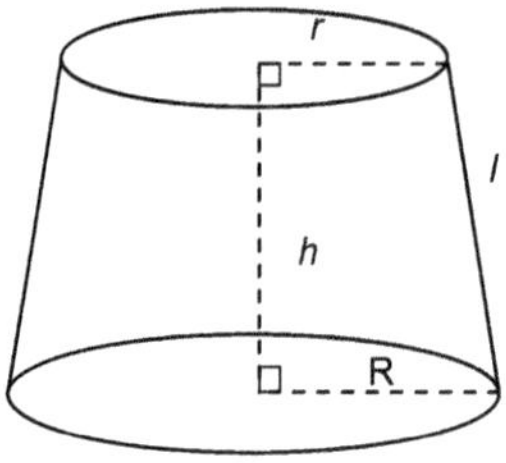	$\pi r^2 + \pi R^2 + \pi l(R + r)$	$\frac{1}{3}\pi(R^2 + Rr + r^2)$

FORMULAE

Formulae

Formula	Description
Pythagoras Theorem	$A^2 = B^2 + C^2$ where A is the hypotenuse of a right-angled triangle and B and C are the two adjacent sides
Simpsons Rule	The Area is divided into an even number of strips of equal width, and therefore has an odd number of ordinates at the division points area = $\frac{S(A + 2B + 4C)}{3}$ where S = common interval (strip width) A = sum of first and last ordinates B = sum of remaining odd ordinates C = sum of the even ordinates The Volume can be calculated by the same formula, but by substituting the area of each coordinate rather than its length
Trapezoidal Rule	A given trench is divided into two equal sections, giving three ordinates, the first, the middle and the last volume = $\frac{S \times (A + B + 2C)}{2}$ where S = width of the strips A = area of the first section B = area of the last section C = area of the rest of the sections
Prismoidal Rule	A given trench is divided into two equal sections, giving three ordinates, the first, the middle and the last volume = $\frac{L \times (A + 4B + C)}{6}$ where L = total length of trench A = area of the first section B = area of the middle section C = area of the last section

TYPICAL THERMAL CONDUCTIVITY OF BUILDING MATERIALS

(Always check manufacturer's details – variation will occur depending on product and nature of materials)

	Thermal conductivity (W/mK)		Thermal conductivity (W/mK)
Acoustic plasterboard	0.25	Oriented strand board	0.13
Aerated concrete slab (500 kg/m^3)	0.16	Outer leaf brick	0.77
Aluminium	237	Plasterboard	0.22
Asphalt (1700 kg/m^3)	0.5	Plaster dense (1300 kg/m^3)	0.5
Bitumen-impregnated fibreboard	0.05	Plaster lightweight (600 kg/m^3)	0.16
Blocks (standard grade 600 kg/m^3)	0.15	Plywood (950 kg/m^3)	0.16
Blocks (solar grade 460 kg/m^3)	0.11	Prefabricated timber wall panels (check manufacturer)	0.12
Brickwork (outer leaf 1700 kg/m^3)	0.84	Screed (1200 kg/m^3)	0.41
Brickwork (inner leaf 1700 kg/m^3)	0.62	Stone chippings (1800 kg/m^3)	0.96
Dense aggregate concrete block 1800 kg/m^3 (exposed)	1.21	Tile hanging (1900 kg/m^3)	0.84
Dense aggregate concrete block 1800 kg/m^3 (protected)	1.13	Timber (650 kg/m^3)	0.14
Calcium silicate board (600 kg/m^3)	0.17	Timber flooring (650 kg/m^3)	0.14
Concrete general	1.28	Timber rafters	0.13
Concrete (heavyweight 2300 kg/m^3)	1.63	Timber roof or floor joists	0.13
Concrete (dense 2100 kg/m^3 typical floor)	1.4	Roof tile (1900 kg/m^3)	0.84
Concrete (dense 2000 kg/m^3 typical floor)	1.13	Timber blocks (650 kg/m^3)	0.14
Concrete (medium 1400 kg/m^3)	0.51	Cellular glass	0.045
Concrete (lightweight 1200 kg/m^3)	0.38	Expanded polystyrene	0.034
Concrete (lightweight 600 kg/m^3)	0.19	Expanded polystyrene slab (25 kg/m^3)	0.035
Concrete slab (aerated 500 kg/m^3)	0.16	Extruded polystyrene	0.035
Copper	390	Glass mineral wool	0.04
External render sand/cement finish	1	Mineral quilt (12 kg/m^3)	0.04
External render (1300 kg/m^3)	0.5	Mineral wool slab (25 kg/m^3)	0.035
Felt – Bitumen layers (1700 kg/m^3)	0.5	Phenolic foam	0.022
Fibreboard (300 kg/m^3)	0.06	Polyisocyanurate	0.025
Glass	0.93	Polyurethane	0.025
Marble	3	Rigid polyurethane	0.025
Metal tray used in wriggly tin concrete floors (7800 kg/m^3)	50	Rock mineral wool	0.038
Mortar (1750 kg/m^3)	0.8		

EARTHWORK

Weights of Typical Materials Handled by Excavators

The weight of the material is that of the state in its natural bed and includes moisture. Adjustments should be made to allow for loose or compacted states

Material	Mass (kg/m^3)	Mass (lb/cu yd)
Ashes, dry	610	1028
Ashes, wet	810	1365
Basalt, broken	1954	3293
Basalt, solid	2933	4943
Bauxite, crushed	1281	2159
Borax, fine	849	1431
Caliche	1440	2427
Cement, clinker	1415	2385
Chalk, fine	1221	2058
Chalk, solid	2406	4055
Cinders, coal, ash	641	1080
Cinders, furnace	913	1538
Clay, compacted	1746	2942
Clay, dry	1073	1808
Clay, wet	1602	2700
Coal, anthracite, solid	1506	2538
Coal, bituminous	1351	2277
Coke	610	1028
Dolomite, lumpy	1522	2565
Dolomite, solid	2886	4864
Earth, dense	2002	3374
Earth, dry, loam	1249	2105
Earth, Fullers, raw	673	1134
Earth, moist	1442	2430
Earth, wet	1602	2700
Felsite	2495	4205
Fieldspar, solid	2613	4404
Fluorite	3093	5213
Gabbro	3093	5213
Gneiss	2696	4544
Granite	2690	4534
Gravel, dry ¼ to 2 inch	1682	2835
Gravel, dry, loose	1522	2565
Gravel, wet ¼ to 2 inch	2002	3374
Gypsum, broken	1450	2444
Gypsum, solid	2787	4697
Hardcore (consolidated)	1928	3249
Lignite, dry	801	1350
Limestone, broken	1554	2619
Limestone, solid	2596	4375
Magnesite, magnesium ore	2993	5044
Marble	2679	4515
Marl, wet	2216	3735
Mica, broken	1602	2700
Mica, solid	2883	4859
Peat, dry	400	674
Peat, moist	700	1179

EARTHWORK

Material	Mass (kg/m^3)	Mass (lb/cu yd)
Peat, wet	1121	1889
Potash	1281	2159
Pumice, stone	640	1078
Quarry waste	1438	2423
Quartz sand	1201	2024
Quartz, solid	2584	4355
Rhyolite	2400	4045
Sand and gravel, dry	1650	2781
Sand and gravel, wet	2020	3404
Sand, dry	1602	2700
Sand, wet	1831	3086
Sandstone, solid	2412	4065
Shale, solid	2637	4444
Slag, broken	2114	3563
Slag, furnace granulated	961	1619
Slate, broken	1370	2309
Slate, solid	2667	4495
Snow, compacted	481	810
Snow, freshly fallen	160	269
Taconite	2803	4724
Trachyte	2400	4045
Trap rock, solid	2791	4704
Turf	400	674
Water	1000	1685

Transport Capacities

Type of vehicle	Capacity of vehicle	
	Payload	Heaped capacity
Wheelbarrow	150	0.10
1 tonne dumper	1250	1.00
2.5 tonne dumper	4000	2.50
Articulated dump truck (Volvo A20 6 × 4)	18500	11.00
Articulated dump truck (Volvo A35 6 × 6)	32000	19.00
Large capacity rear dumper (Euclid R35)	35000	22.00
Large capacity rear dumper (Euclid R85)	85000	50.00

EARTHWORK

Machine Volumes for Excavating and Filling

Machine type	Cycles per minute	Volume per minute (m^3)
1.5 tonne excavator	1	0.04
	2	0.08
	3	0.12
3 tonne excavator	1	0.13
	2	0.26
	3	0.39
5 tonne excavator	1	0.28
	2	0.56
	3	0.84
7 tonne excavator	1	0.28
	2	0.56
	3	0.84
21 tonne excavator	1	1.21
	2	2.42
	3	3.63
Backhoe loader JCB3CX excavator	1	0.28
Rear bucket capacity 0.28 m^3	2	0.56
	3	0.84
Backhoe loader JCB3CX loading	1	1.00
Front bucket capacity 1.00 m^3	2	2.00

Machine Volumes for Excavating and Filling

Machine type	Loads per hour	Volume per hour (m^3)
1 tonne high tip skip loader	5	2.43
Volume 0.485 m^3	7	3.40
	10	4.85
3 tonne dumper	4	7.60
Max volume 2.40 m^3	5	9.50
Available volume 1.9 m^3	7	13.30
	10	19.00
6 tonne dumper	4	15.08
Max volume 3.40 m^3	5	18.85
Available volume 3.77 m^3	7	26.39
	10	37.70

EARTHWORK

Bulkage of Soils (after excavation)

Type of soil	Approximate bulking of 1 m^3 after excavation
Vegetable soil and loam	25–30%
Soft clay	30–40%
Stiff clay	10–20%
Gravel	20–25%
Sand	40–50%
Chalk	40–50%
Rock, weathered	30–40%
Rock, unweathered	50–60%

Shrinkage of Materials (on being deposited)

Type of soil	Approximate bulking of 1 m^3 after excavation
Clay	10%
Gravel	8%
Gravel and sand	9%
Loam and light sandy soils	12%
Loose vegetable soils	15%

Voids in Material Used as Subbases or Beddings

Material	m^3 of voids/m^3
Alluvium	0.37
River grit	0.29
Quarry sand	0.24
Shingle	0.37
Gravel	0.39
Broken stone	0.45
Broken bricks	0.42

Angles of Repose

Type of soil		Degrees
Clay	– dry	30
	– damp, well drained	45
	– wet	15–20
Earth	– dry	30
	– damp	45
Gravel	– moist	48
Sand	– dry or moist	35
	– wet	25
Loam		40

EARTHWORK

Slopes and Angles

Ratio of base to height	Angle in degrees
5:1	11
4:1	14
3:1	18
2:1	27
1½:1	34
1:1	45
1:1½	56
1:2	63
1:3	72
1:4	76
1:5	79

Grades (in Degrees and Percents)

Degrees	Percent	Degrees	Percent
1	1.8	24	44.5
2	3.5	25	46.6
3	5.2	26	48.8
4	7.0	27	51.0
5	8.8	28	53.2
6	10.5	29	55.4
7	12.3	30	57.7
8	14.0	31	60.0
9	15.8	32	62.5
10	17.6	33	64.9
11	19.4	34	67.4
12	21.3	35	70.0
13	23.1	36	72.7
14	24.9	37	75.4
15	26.8	38	78.1
16	28.7	39	81.0
17	30.6	40	83.9
18	32.5	41	86.9
19	34.4	42	90.0
20	36.4	43	93.3
21	38.4	44	96.6
22	40.4	45	100.0
23	42.4		

EARTHWORK

Bearing Powers

Ground conditions		Bearing power		
		kg/m²	lb/in²	Metric t/m²
Rock,	broken	483	70	50
	solid	2415	350	240
Clay,	dry or hard	380	55	40
	medium dry	190	27	20
	soft or wet	100	14	10
Gravel,	cemented	760	110	80
Sand,	compacted	380	55	40
	clean dry	190	27	20
Swamp and alluvial soils		48	7	5

CONCRETE WORK

CONCRETE WORK

Weights of Concrete and Concrete Elements

Type of material		kg/m³	lb/cu ft
Ordinary concrete (dense aggregates)			
Non-reinforced plain or mass concrete			
Nominal weight		2305	144
Aggregate	– limestone	2162 to 2407	135 to 150
	– gravel	2244 to 2407	140 to 150
	– broken brick	2000 (av)	125 (av)
	– other crushed stone	2326 to 2489	145 to 155
Reinforced concrete			
Nominal weight		2407	150
Reinforcement	– 1%	2305 to 2468	144 to 154
	– 2%	2356 to 2519	147 to 157
	– 4%	2448 to 2703	153 to 163
Special concretes			
Heavy concrete			
Aggregates	– barytes, magnetite	3210 (min)	200 (min)
	– steel shot, punchings	5280	330
Lean mixes			
Dry-lean (gravel aggregate)		2244	140
Soil-cement (normal mix)		1601	100

CONCRETE WORK

Type of material		kg/m² per mm thick	lb/sq ft per inch thick
Ordinary concrete (dense aggregates)			
Solid slabs (floors, walls etc.)			
Thickness:	75 mm or 3 in	184	37.5
	100 mm or 4 in	245	50
	150 mm or 6 in	378	75
	250 mm or 10 in	612	125
	300 mm or 12 in	734	150
Ribbed slabs			
Thickness:	125 mm or 5 in	204	42
	150 mm or 6 in	219	45
	225 mm or 9 in	281	57
	300 mm or 12 in	342	70
Special concretes			
Finishes etc.			
	Rendering, screed etc. Granolithic, terrazzo	1928 to 2401	10 to 12.5
	Glass-block (hollow) concrete	1734 (approx)	9 (approx)
Prestressed concrete		Weights as for reinforced concrete (upper limits)	
Air-entrained concrete		Weights as for plain or reinforced concrete	

CONCRETE WORK

Average Weight of Aggregates

Materials	Voids %	Weight kg/m³
Sand	39	1660
Gravel 10–20 mm	45	1440
Gravel 35–75 mm	42	1555
Crushed stone	50	1330
Crushed granite (over 15 mm)	50	1345
(n.e. 15 mm)	47	1440
'All-in' ballast	32	1800–2000

Material	kg/m³	lb/cu yd
Vermiculite (aggregate)	64–80	108–135
All-in aggregate	1999	125

Applications and Mix Design

Site mixed concrete

Recommended mix	Class of work suitable for	Cement (kg)	Sand (kg)	Coarse aggregate (kg)	Nr 25 kg bags cement per m³ of combined aggregate
1:3:6	Roughest type of mass concrete such as footings, road haunching over 300 mm thick	208	905	1509	8.30
1:2.5:5	Mass concrete of better class than 1:3:6 such as bases for machinery, walls below ground etc.	249	881	1474	10.00
1:2:4	Most ordinary uses of concrete, such as mass walls above ground, road slabs etc. and general reinforced concrete work	304	889	1431	12.20
1:1.5:3	Watertight floors, pavements and walls, tanks, pits, steps, paths, surface of 2 course roads, reinforced concrete where extra strength is required	371	801	1336	14.90
1:1:2	Works of thin section such as fence posts and small precast work	511	720	1206	20.40

CONCRETE WORK

Ready mixed concrete

Application	Designated concrete	Standardized prescribed concrete	Recommended consistence (nominal slump class)
Foundations			
Mass concrete fill or blinding	GEN 1	ST2	S3
Strip footings	GEN 1	ST2	S3
Mass concrete foundations			
Single storey buildings	GEN 1	ST2	S3
Double storey buildings	GEN 3	ST4	S3
Trench fill foundations			
Single storey buildings	GEN 1	ST2	S4
Double storey buildings	GEN 3	ST4	S4
General applications			
Kerb bedding and haunching	GEN 0	ST1	S1
Drainage works – immediate support	GEN 1	ST2	S1
Other drainage works	GEN 1	ST2	S3
Oversite below suspended slabs	GEN 1	ST2	S3
Floors			
Garage and house floors with no embedded steel	GEN 3	ST4	S2
Wearing surface: Light foot and trolley traffic	RC30	ST4	S2
Wearing surface: General industrial	RC40	N/A	S2
Wearing surface: Heavy industrial	RC50	N/A	S2
Paving			
House drives, domestic parking and external parking	PAV 1	N/A	S2
Heavy-duty external paving	PAV 2	N/A	S2

CONCRETE WORK

Prescribed Mixes for Ordinary Structural Concrete

Weights of cement and total dry aggregates in kg to produce approximately one cubic metre of fully compacted concrete together with the percentages by weight of fine aggregate in total dry aggregates

Conc. grade	Nominal max size of aggregate (mm)	40		20		14		10	
	Workability	**Med.**	**High**	**Med.**	**High**	**Med.**	**High**	**Med.**	**High**
	Limits to slump that may be expected (mm)	**50–100**	**100–150**	**25–75**	**75–125**	**10–50**	**50–100**	**10–25**	**25–50**
7	Cement (kg)	180	200	210	230	–	–	–	–
	Total aggregate (kg)	1950	1850	1900	1800	–	–	–	–
	Fine aggregate (%)	30–45	30–45	35–50	35–50	–	–	–	–
10	Cement (kg)	210	230	240	260	–	–	–	–
	Total aggregate (kg)	1900	1850	1850	1800	–	–	–	–
	Fine aggregate (%)	30–45	30–45	35–50	35–50	–	–	–	–
15	Cement (kg)	250	270	280	310	–	–	–	–
	Total aggregate (kg)	1850	1800	1800	1750	–	–	–	–
	Fine aggregate (%)	30–45	30–45	35–50	35–50	–	–	–	–
20	Cement (kg)	300	320	320	350	340	380	360	410
	Total aggregate (kg)	1850	1750	1800	1750	1750	1700	1750	1650
	Sand								
	Zone 1 (%)	35	40	40	45	45	50	50	55
	Zone 2 (%)	30	35	35	40	40	45	45	50
	Zone 3 (%)	30	30	30	35	35	40	40	45
25	Cement (kg)	340	360	360	390	380	420	400	450
	Total aggregate (kg)	1800	1750	1750	1700	1700	1650	1700	1600
	Sand								
	Zone 1 (%)	35	40	40	45	45	50	50	55
	Zone 2 (%)	30	35	35	40	40	45	45	50
	Zone 3 (%)	30	30	30	35	35	40	40	45
30	Cement (kg)	370	390	400	430	430	470	460	510
	Total aggregate (kg)	1750	1700	1700	1650	1700	1600	1650	1550
	Sand								
	Zone 1 (%)	35	40	40	45	45	50	50	55
	Zone 2 (%)	30	35	35	40	40	45	45	50
	Zone 3 (%)	30	30	30	35	35	40	40	45

REINFORCEMENT

Weights of Bar Reinforcement

Nominal sizes (mm)	Cross-sectional area (mm^2)	Mass (kg/m)	Length of bar (m/tonne)
6	28.27	0.222	4505
8	50.27	0.395	2534
10	78.54	0.617	1622
12	113.10	0.888	1126
16	201.06	1.578	634
20	314.16	2.466	405
25	490.87	3.853	260
32	804.25	6.313	158
40	1265.64	9.865	101
50	1963.50	15.413	65

Weights of Bars (at specific spacings)

Weights of metric bars in kilogrammes per square metre

Size	Spacing of bars in millimetres									
(mm)	75	100	125	150	175	200	225	250	275	300
6	2.96	2.220	1.776	1.480	1.27	1.110	0.99	0.89	0.81	0.74
8	5.26	3.95	3.16	2.63	2.26	1.97	1.75	1.58	1.44	1.32
10	8.22	6.17	4.93	4.11	3.52	3.08	2.74	2.47	2.24	2.06
12	11.84	8.88	7.10	5.92	5.07	4.44	3.95	3.55	3.23	2.96
16	21.04	15.78	12.63	10.52	9.02	7.89	7.02	6.31	5.74	5.26
20	32.88	24.66	19.73	16.44	14.09	12.33	10.96	9.87	8.97	8.22
25	51.38	38.53	30.83	25.69	22.02	19.27	17.13	15.41	14.01	12.84
32	84.18	63.13	50.51	42.09	36.08	31.57	28.06	25.25	22.96	21.04
40	131.53	98.65	78.92	65.76	56.37	49.32	43.84	39.46	35.87	32.88
50	205.51	154.13	123.31	102.76	88.08	77.07	68.50	61.65	56.05	51.38

Basic weight of steelwork taken as 7850 kg/m^3
Basic weight of bar reinforcement per metre run = 0.00785 kg/mm^2
The value of π has been taken as 3.141592654

REINFORCEMENT

Fabric Reinforcement

Preferred range of designated fabric types and stock sheet sizes

Fabric reference	Longitudinal wires			Cross wires			
	Nominal wire size (mm)	Pitch (mm)	Area (mm^2/m)	Nominal wire size (mm)	Pitch (mm)	Area (mm^2/m)	Mass (kg/m^2)
Square mesh							
A393	10	200	393	10	200	393	6.16
A252	8	200	252	8	200	252	3.95
A193	7	200	193	7	200	193	3.02
A142	6	200	142	6	200	142	2.22
A98	5	200	98	5	200	98	1.54
Structural mesh							
B1131	12	100	1131	8	200	252	10.90
B785	10	100	785	8	200	252	8.14
B503	8	100	503	8	200	252	5.93
B385	7	100	385	7	200	193	4.53
B283	6	100	283	7	200	193	3.73
B196	5	100	196	7	200	193	3.05
Long mesh							
C785	10	100	785	6	400	70.8	6.72
C636	9	100	636	6	400	70.8	5.55
C503	8	100	503	5	400	49.0	4.34
C385	7	100	385	5	400	49.0	3.41
C283	6	100	283	5	400	49.0	2.61
Wrapping mesh							
D98	5	200	98	5	200	98	1.54
D49	2.5	100	49	2.5	100	49	0.77

Stock sheet size 4.8 m × 2.4 m, Area 11.52 m^2

Average weight kg/m^3 of steelwork reinforcement in concrete for various building elements

Substructure	kg/m^3 concrete	Substructure	kg/m^3 concrete
Pile caps	110–150	Plate slab	150–220
Tie beams	130–170	Cant slab	145–210
Ground beams	230–330	Ribbed floors	130–200
Bases	125–180	Topping to block floor	30–40
Footings	100–150	Columns	210–310
Retaining walls	150–210	Beams	250–350
Raft	60–70	Stairs	130–170
Slabs – one way	120–200	Walls – normal	40–100
Slabs – two way	110–220	Walls – wind	70–125

Note: For exposed elements add the following %:
Walls 50%, Beams 100%, Columns 15%

FORMWORK

Formwork Stripping Times – Normal Curing Periods

Conditions under which concrete is maturing	Minimum periods of protection for different types of cement					
	Number of days (where the average surface temperature of the concrete exceeds 10°C during the whole period)			Equivalent maturity (degree hours) calculated as the age of the concrete in hours multiplied by the number of degrees Celsius by which the average surface temperature of the concrete exceeds 10°C		
	Other	SRPC	OPC or RHPC	Other	SRPC	OPC or RHPC
1. Hot weather or drying winds	7	4	3	3500	2000	1500
2. Conditions not covered by 1	4	3	2	2000	1500	1000

KEY
OPC – Ordinary Portland Cement
RHPC – Rapid-hardening Portland Cement
SRPC – Sulphate-resisting Portland Cement

Minimum Period before Striking Formwork

	Minimum period before striking		
	Surface temperature of concrete		
	16°C	17°C	t°C (0–25)
Vertical formwork to columns, walls and large beams	12 hours	18 hours	300 hours t+10
Soffit formwork to slabs	4 days	6 days	100 days t+10
Props to slabs	10 days	15 days	250 days t+10
Soffit formwork to beams	9 days	14 days	230 days t+10
Props to beams	14 days	21 days	360 days t+10

MASONRY

MASONRY

Number of Bricks Required for Various Types of Work per m² of Walling

Description	Brick size	
	215 × 102.5 × 50 mm	215 × 102.5 × 65 mm
Half brick thick		
Stretcher bond	74	59
English bond	108	86
English garden wall bond	90	72
Flemish bond	96	79
Flemish garden wall bond	83	66
One brick thick and cavity wall of two half brick skins		
Stretcher bond	148	119

Quantities of Bricks and Mortar Required per m² of Walling

	Unit	No of bricks required	Mortar required (cubic metres)		
Standard bricks			No frogs	Single frogs	Double frogs
Brick size 215 × 102.5 × 50 mm					
half brick wall (103 mm)	m²	72	0.022	0.027	0.032
2 × half brick cavity wall (270 mm)	m²	144	0.044	0.054	0.064
one brick wall (215 mm)	m²	144	0.052	0.064	0.076
one and a half brick wall (322 mm)	m²	216	0.073	0.091	0.108
Mass brickwork	m³	576	0.347	0.413	0.480
Brick size 215 × 102.5 × 65 mm					
half brick wall (103 mm)	m²	58	0.019	0.022	0.026
2 × half brick cavity wall (270 mm)	m²	116	0.038	0.045	0.055
one brick wall (215 mm)	m²	116	0.046	0.055	0.064
one and a half brick wall (322 mm)	m²	174	0.063	0.074	0.088
Mass brickwork	m³	464	0.307	0.360	0.413
Metric modular bricks			Perforated		
Brick size 200 × 100 × 75 mm					
90 mm thick	m²	67	0.016	0.019	
190 mm thick	m²	133	0.042	0.048	
290 mm thick	m²	200	0.068	0.078	
Brick size 200 × 100 × 100 mm					
90 mm thick	m²	50	0.013	0.016	
190 mm thick	m²	100	0.036	0.041	
290 mm thick	m²	150	0.059	0.067	
Brick size 300 × 100 × 75 mm					
90 mm thick	m²	33	–	0.015	
Brick size 300 × 100 × 100 mm					
90 mm thick	m²	44	0.015	0.018	

Note: Assuming 10 mm thick joints

MASONRY

Mortar Required per m² Blockwork (9.88 blocks/m²)

Wall thickness	75	90	100	125	140	190	215
Mortar m^3/m^2	0.005	0.006	0.007	0.008	0.009	0.013	0.014

Mortar Group	Cement: lime: sand	Masonry cement: sand	Cement: sand with plasticizer
1	1:0–0.25:3		
2	1:0.5:4–4.5	1:2.5-3.5	1:3–4
3	1:1:5–6	1:4–5	1:5–6
4	1:2:8–9	1:5.5–6.5	1:7–8
5	1:3:10–12	1:6.5–7	1:8

Group 1: strong inflexible mortar
Group 5: weak but flexible

All mixes within a group are of approximately similar strength
Frost resistance increases with the use of plasticizers
Cement: lime: sand mixes give the strongest bond and greatest resistance to rain penetration
Masonry cement equals ordinary Portland cement plus a fine neutral mineral filler and an air entraining agent

Calcium Silicate Bricks

Type	Strength	Location
Class 2 crushing strength	14.0 N/mm²	not suitable for walls
Class 3	20.5 N/mm²	walls above dpc
Class 4	27.5 N/mm²	cappings and copings
Class 5	34.5 N/mm²	retaining walls
Class 6	41.5 N/mm²	walls below ground
Class 7	48.5 N/mm²	walls below ground

The Class 7 calcium silicate bricks are therefore equal in strength to Class B bricks
Calcium silicate bricks are not suitable for DPCs

Durability of Bricks	
FL	Frost resistant with low salt content
FN	Frost resistant with normal salt content
ML	Moderately frost resistant with low salt content
MN	Moderately frost resistant with normal salt content

MASONRY

Brickwork Dimensions

No. of horizontal bricks	Dimensions (mm)	No. of vertical courses	Height of vertical courses (mm)
½	112.5	1	75
1	225.0	2	150
1½	337.5	3	225
2	450.0	4	300
2½	562.5	5	375
3	675.0	6	450
3½	787.5	7	525
4	900.0	8	600
4½	1012.5	9	675
5	1125.0	10	750
5½	1237.5	11	825
6	1350.0	12	900
6½	1462.5	13	975
7	1575.0	14	1050
7½	1687.5	15	1125
8	1800.0	16	1200
8½	1912.5	17	1275
9	2025.0	18	1350
9½	2137.5	19	1425
10	2250.0	20	1500
20	4500.0	24	1575
40	9000.0	28	2100
50	11250.0	32	2400
60	13500.0	36	2700
75	16875.0	40	3000

TIMBER

TIMBER

Weights of Timber

Material	kg/m^3	lb/cu ft
General	806 (avg)	50 (avg)
Douglas fir	479	30
Yellow pine, spruce	479	30
Pitch pine	673	42
Larch, elm	561	35
Oak (English)	724 to 959	45 to 60
Teak	643 to 877	40 to 55
Jarrah	959	60
Greenheart	1040 to 1204	65 to 75
Quebracho	1285	80
Material	**kg/m^2 per mm thickness**	**lb/sq ft per inch thickness**
Wooden boarding and blocks		
Softwood	0.48	2.5
Hardwood	0.76	4
Hardboard	1.06	5.5
Chipboard	0.76	4
Plywood	0.62	3.25
Blockboard	0.48	2.5
Fibreboard	0.29	1.5
Wood-wool	0.58	3
Plasterboard	0.96	5
Weather boarding	0.35	1.8

TIMBER

Conversion Tables (for timber only)

Inches	Millimetres	Feet	Metres
1	25	1	0.300
2	50	2	0.600
3	75	3	0.900
4	100	4	1.200
5	125	5	1.500
6	150	6	1.800
7	175	7	2.100
8	200	8	2.400
9	225	9	2.700
10	250	10	3.000
11	275	11	3.300
12	300	12	3.600
13	325	13	3.900
14	350	14	4.200
15	375	15	4.500
16	400	16	4.800
17	425	17	5.100
18	450	18	5.400
19	475	19	5.700
20	500	20	6.000
21	525	21	6.300
22	550	22	6.600
23	575	23	6.900
24	600	24	7.200

Planed Softwood

The finished end section size of planed timber is usually 3/16" less than the original size from which it is produced. This however varies slightly depending upon availability of material and origin of the species used.

Standards (timber) to cubic metres and cubic metres to standards (timber)

Cubic metres	Cubic metres standards	Standards
4.672	1	0.214
9.344	2	0.428
14.017	3	0.642
18.689	4	0.856
23.361	5	1.070
28.033	6	1.284
32.706	7	1.498
37.378	8	1.712
42.050	9	1.926
46.722	10	2.140
93.445	20	4.281
140.167	30	6.421
186.890	40	8.561
233.612	50	10.702
280.335	60	12.842
327.057	70	14.982
373.779	80	17.122

TIMBER

1 cu metre = 35.3148 cu ft = 0.21403 std

1 cu ft = 0.028317 cu metres

1 std = 4.67227 cu metres

Basic sizes of sawn softwood available (cross-sectional areas)

Thickness (mm)	Width (mm)								
	75	100	125	150	175	200	225	250	300
16	X	X	X	X					
19	X	X	X	X					
22	X	X	X	X					
25	X	X	X	X	X	X	X	X	X
32	X	X	X	X	X	X	X	X	X
36	X	X	X	X					
38	X	X	X	X	X	X	X		
44	X	X	X	X	X	X	X	X	X
47*	X	X	X	X	X	X	X	X	X
50	X	X	X	X	X	X	X	X	X
63	X	X	X	X	X	X	X		
75	X	X	X	X	X	X	X	X	
100		X		X		X		X	X
150				X		X			X
200						X			
250								X	
300									X

* This range of widths for 47 mm thickness will usually be found to be available in construction quality only

Note: The smaller sizes below 100 mm thick and 250 mm width are normally but not exclusively of European origin. Sizes beyond this are usually of North and South American origin

Basic lengths of sawn softwood available (metres)

1.80	2.10	3.00	4.20	5.10	6.00	7.20
	2.40	3.30	4.50	5.40	6.30	
	2.70	3.60	4.80	5.70	6.60	
		3.90			6.90	

Note: Lengths of 6.00 m and over will generally only be available from North American species and may have to be recut from larger sizes

TIMBER

Reductions from basic size to finished size by planning of two opposed faces

Purpose	Reductions from basic sizes for timber			
	15–35 mm	36–100 mm	101–150 mm	over 150 mm
a) Constructional timber	3 mm	3 mm	5 mm	6 mm
b) Matching interlocking boards	4 mm	4 mm	6 mm	6 mm
c) Wood trim not specified in BS 584	5 mm	7 mm	7 mm	9 mm
d) Joinery and cabinet work	7 mm	9 mm	11 mm	13 mm

Note: The reduction of width or depth is overall the extreme size and is exclusive of any reduction of the face by the machining of a tongue or lap joints

Maximum Spans for Various Roof Trusses

Maximum permissible spans for rafters for Fink trussed rafters

Basic size (mm)	Actual size (mm)	Pitch (degrees)								
		15 (m)	17.5 (m)	20 (m)	22.5 (m)	25 (m)	27.5 (m)	30 (m)	32.5 (m)	35 (m)
38 × 75	35 × 72	6.03	6.16	6.29	6.41	6.51	6.60	6.70	6.80	6.90
38 × 100	35 × 97	7.48	7.67	7.83	7.97	8.10	8.22	8.34	8.47	8.61
38 × 125	35 × 120	8.80	9.00	9.20	9.37	9.54	9.68	9.82	9.98	10.16
44 × 75	41 × 72	6.45	6.59	6.71	6.83	6.93	7.03	7.14	7.24	7.35
44 × 100	41 × 97	8.05	8.23	8.40	8.55	8.68	8.81	8.93	9.09	9.22
44 × 125	41 × 120	9.38	9.60	9.81	9.99	10.15	10.31	10.45	10.64	10.81
50 × 75	47 × 72	6.87	7.01	7.13	7.25	7.35	7.45	7.53	7.67	7.78
50 × 100	47 × 97	8.62	8.80	8.97	9.12	9.25	9.38	9.50	9.66	9.80
50 × 125	47 × 120	10.01	10.24	10.44	10.62	10.77	10.94	11.00	11.00	11.00

TIMBER

Sizes of Internal and External Doorsets

Description	Internal size (mm)	Permissible deviation	External size (mm)	Permissible deviation
Coordinating dimension:				
height of door leaf			2100	
height sets	2100			
Coordinating dimension:			2300	
height of ceiling	2300		2350	
height set	2350		2400	
	2400		2700	
	2700		3000	
	3000			
Coordinating dimension:			900 S	
width of all doorsets	600 S		1000 S	
	700 S		1200 D	
S = Single leaf set	800 S&D			
D = Double leaf set	900 S&D		1800 D	
	1000 S&D		2100 D	
	1200 D			
	1500 D			
	1800 D			
	2100 D			
Work size: height of				
door leaf height set	2090	± 2.0	2095	± 2.0
Work size: height of	2285		2295	
ceiling height set	2335		2345	
	2385	± 2.0	2395	± 2.0
	2685		2695	
	2985		2995	
Work size: width of	590 S		895 S	
all doorsets	690 S		995 S	
	790 S&D		1195 D	
S = Single leaf set	890 S&D		1495 D	± 2.0
D = Double leaf set	990 S&D	± 2.0	1795 D	
	1190 D		2095 D	
	1490 D			
	1790 D			
	2090 D			
Width of door leaf	526 F		806 F&P	± 1.5
in single leaf sets	626 F		906 F&P	
F = Flush leaf	726 F&P	± 1.5		
P = Panel leaf	826 F&P			
	926 F&P			
Width of door leaf	362 F		552 F&P	
in double leaf sets	412 F		702 F&P	± 1.5
F = Flush leaf	426 F		852 F&P	
P = Panel leaf	562 F&P	± 1.5	1002 F&P	
	712 F&P			
	826 F&P			
	1012 F&P			
Door leaf height for all				
doorsets	2040	± 1.5	1994	± 1.5

ROOFING

Total Roof Loadings for Various Types of Tiles/Slates

		Roof load (slope) kg/m²		
		Slate/Tile	**Roofing underlay and battens²**	**Total dead load kg/m**
Asbestos cement slate (600 × 300)		21.50	3.14	24.64
Clay tile	interlocking	67.00	5.50	72.50
	plain	43.50	2.87	46.37
Concrete tile	interlocking	47.20	2.69	49.89
	plain	78.20	5.50	83.70
Natural slate (18" × 10")		35.40	3.40	38.80
		Roof load (plan) kg/m²		
Asbestos cement slate (600 × 300)		28.45	76.50	104.95
Clay tile	interlocking	53.54	76.50	130.04
	plain	83.71	76.50	60.21
Concrete tile	interlocking	57.60	76.50	134.10
	plain	96.64	76.50	173.14

ROOFING

Tiling Data

Product		Lap (mm)	Gauge of battens	No. slates per m²	Battens (m/m²)	Weight as laid (kg/m²)
CEMENT SLATES						
Eternit slates	600 × 300 mm	100	250	13.4	4.00	19.50
(Duracem)		90	255	13.1	3.92	19.20
		80	260	12.9	3.85	19.00
		70	265	12.7	3.77	18.60
	600 × 350 mm	100	250	11.5	4.00	19.50
		90	255	11.2	3.92	19.20
	500 × 250 mm	100	200	20.0	5.00	20.00
		90	205	19.5	4.88	19.50
		80	210	19.1	4.76	19.00
		70	215	18.6	4.65	18.60
	400 × 200 mm	90	155	32.3	6.45	20.80
		80	160	31.3	6.25	20.20
		70	165	30.3	6.06	19.60
CONCRETE TILES/SLATES						
Redland Roofing						
Stonewold slate	430 × 380 mm	75	355	8.2	2.82	51.20
Double Roman tile	418 × 330 mm	75	355	8.2	2.91	45.50
Grovebury pantile	418 × 332 mm	75	343	9.7	2.91	47.90
Norfolk pantile	381 × 227 mm	75	306	16.3	3.26	44.01
		100	281	17.8	3.56	48.06
Renown interlocking tile	418 × 330 mm	75	343	9.7	2.91	46.40
'49' tile	381 × 227 mm	75	306	16.3	3.26	44.80
		100	281	17.8	3.56	48.95
Plain, vertical tiling	265 × 165 mm	35	115	52.7	8.70	62.20
Marley Roofing						
Bold roll tile	420 × 330 mm	75	344	9.7	2.90	47.00
		100	–	10.5	3.20	51.00
Modern roof tile	420 × 330 mm	75	338	10.2	3.00	54.00
		100	–	11.0	3.20	58.00
Ludlow major	420 × 330 mm	75	338	10.2	3.00	45.00
		100	–	11.0	3.20	49.00
Ludlow plus	387 × 229 mm	75	305	16.1	3.30	47.00
		100	–	17.5	3.60	51.00
Mendip tile	420 × 330 mm	75	338	10.2	3.00	47.00
		100	–	11.0	3.20	51.00
Wessex	413 × 330 mm	75	338	10.2	3.00	54.00
		100	–	11.0	3.20	58.00
Plain tile	267 × 165 mm	65	100	60.0	10.00	76.00
		75	95	64.0	10.50	81.00
		85	90	68.0	11.30	86.00
Plain vertical tiles (feature)	267 × 165 mm	35	110	53.0	8.70	67.00
		34	115	56.0	9.10	71.00

ROOFING

Slate Nails, Quantity per Kilogram

Length	Type			
	Plain wire	Galvanized wire	Copper nail	Zinc nail
28.5 mm	325	305	325	415
34.4 mm	286	256	254	292
50.8 mm	242	224	194	200

Metal Sheet Coverings

Thicknesses and weights of sheet metal coverings								
Lead to BS 1178								
BS Code No	3	4	5	6	7	8		
Colour code	Green	Blue	Red	Black	White	Orange		
Thickness (mm)	1.25	1.80	2.24	2.50	3.15	3.55		
Density kg/m^2	14.18	20.41	25.40	30.05	35.72	40.26		
Copper to BS 2870								
Thickness (mm)		0.60	0.70					
Bay width								
Roll (mm)		500	650					
Seam (mm)		525	600					
Standard width to form bay	600	750						
Normal length of sheet	1.80	1.80						
Zinc to BS 849								
Zinc Gauge (Nr)	9	10	11	12	13	14	15	16
Thickness (mm)	0.43	0.48	0.56	0.64	0.71	0.79	0.91	1.04
Density (kg/m^2)	3.1	3.2	3.8	4.3	4.8	5.3	6.2	7.0
Aluminium to BS 4868								
Thickness (mm)	0.5	0.6	0.7	0.8	0.9	1.0	1.2	
Density (kg/m^2)	12.8	15.4	17.9	20.5	23.0	25.6	30.7	

ROOFING

Type of felt	Nominal mass per unit area (kg/10 m)	Nominal mass per unit area of fibre base (g/m^2)	Nominal length of roll (m)
Class 1			
1B fine granule surfaced bitumen	14	220	10 or 20
	18	330	10 or 20
	25	470	10
1E mineral surfaced bitumen	38	470	10
1F reinforced bitumen	15	160 (fibre) 110 (hessian)	15
1F reinforced bitumen, aluminium faced	13	160 (fibre) 110 (hessian)	15
Class 2			
2B fine granule surfaced bitumen asbestos	18	500	10 or 20
2E mineral surfaced bitumen asbestos	38	600	10
Class 3			
3B fine granule surfaced bitumen glass fibre	18	60	20
3E mineral surfaced bitumen glass fibre	28	60	10
3E venting base layer bitumen glass fibre	32	60*	10
3H venting base layer bitumen glass fibre	17	60*	20

* Excluding effect of perforations

GLAZING

GLAZING

Nominal thickness (mm)	Tolerance on thickness (mm)	Approximate weight (kg/m²)	Normal maximum size (mm)
Float and polished plate glass			
3	+ 0.2	7.50	2140 × 1220
4	+ 0.2	10.00	2760 × 1220
5	+ 0.2	12.50	3180 × 2100
6	+ 0.2	15.00	4600 × 3180
10	+ 0.3	25.00)	6000 × 3300
12	+ 0.3	30.00)	
15	+ 0.5	37.50	3050 × 3000
19	+ 1.0	47.50)	3000 × 2900
25	+ 1.0	63.50)	
Clear sheet glass			
2 *	+ 0.2	5.00	1920 × 1220
3	+ 0.3	7.50	2130 × 1320
4	+ 0.3	10.00	2760 × 1220
5 *	+ 0.3	12.50)	2130 × 2400
6 *	+ 0.3	15.00)	
Cast glass			
3	+ 0.4		
	− 0.2	6.00)	2140 × 1280
4	+ 0.5	7.50)	
5	+ 0.5	9.50	2140 × 1320
6	+ 0.5	11.50)	3700 × 1280
10	+ 0.8	21.50)	
Wired glass (Cast wired glass)			
6	+ 0.3	−)	
	− 0.7	)	3700 × 1840
7	+ 0.7	−)	
(Polished wire glass)			
6	+ 1.0	–	330 × 1830

* The 5 mm and 6 mm thickness are known as *thick drawn sheet*. Although 2 mm sheet glass is available it is not recommended for general glazing purposes

METAL

Weights of Metals

Material	kg/m³	lb/cu ft
Metals, steel construction, etc.		
Iron		
– cast	7207	450
– wrought	7687	480
– ore – general	2407	150
– (crushed) Swedish	3682	230
Steel	7854	490
Copper		
– cast	8731	545
– wrought	8945	558
Brass	8497	530
Bronze	8945	558
Aluminium	2774	173
Lead	11322	707
Zinc (rolled)	7140	446
	g/mm^2 per metre	**lb/sq ft per foot**
Steel bars	7.85	3.4
Structural steelwork	Net weight of member @ 7854 kg/m^3	
riveted	+ 10% for cleats, rivets, bolts, etc.	
welded	+ 1.25% to 2.5% for welds, etc.	
Rolled sections		
beams	+ 2.5%	
stanchions	+ 5% (extra for caps and bases)	
Plate		
web girders	+ 10% for rivets or welds, stiffeners, etc.	
	kg/m	**lb/ft**
Steel stairs: industrial type		
1 m or 3 ft wide	84	56
Steel tubes		
50 mm or 2 in bore	5 to 6	3 to 4
Gas piping		
20 mm or ¾ in	2	1¼

METAL

Universal Beams BS 4: Part 1: 2005

Designation	Mass (kg/m)	Depth of section (mm)	Width of section (mm)	Thickness		Surface area (m²/m)
				Web (mm)	Flange (mm)	
1016 × 305 × 487	487.0	1036.1	308.5	30.0	54.1	3.20
1016 × 305 × 438	438.0	1025.9	305.4	26.9	49.0	3.17
1016 × 305 × 393	393.0	1016.0	303.0	24.4	43.9	3.15
1016 × 305 × 349	349.0	1008.1	302.0	21.1	40.0	3.13
1016 × 305 × 314	314.0	1000.0	300.0	19.1	35.9	3.11
1016 × 305 × 272	272.0	990.1	300.0	16.5	31.0	3.10
1016 × 305 × 249	249.0	980.2	300.0	16.5	26.0	3.08
1016 × 305 × 222	222.0	970.3	300.0	16.0	21.1	3.06
914 × 419 × 388	388.0	921.0	420.5	21.4	36.6	3.44
914 × 419 × 343	343.3	911.8	418.5	19.4	32.0	3.42
914 × 305 × 289	289.1	926.6	307.7	19.5	32.0	3.01
914 × 305 × 253	253.4	918.4	305.5	17.3	27.9	2.99
914 × 305 × 224	224.2	910.4	304.1	15.9	23.9	2.97
914 × 305 × 201	200.9	903.0	303.3	15.1	20.2	2.96
838 × 292 × 226	226.5	850.9	293.8	16.1	26.8	2.81
838 × 292 × 194	193.8	840.7	292.4	14.7	21.7	2.79
838 × 292 × 176	175.9	834.9	291.7	14.0	18.8	2.78
762 × 267 × 197	196.8	769.8	268.0	15.6	25.4	2.55
762 × 267 × 173	173.0	762.2	266.7	14.3	21.6	2.53
762 × 267 × 147	146.9	754.0	265.2	12.8	17.5	2.51
762 × 267 × 134	133.9	750.0	264.4	12.0	15.5	2.51
686 × 254 × 170	170.2	692.9	255.8	14.5	23.7	2.35
686 × 254 × 152	152.4	687.5	254.5	13.2	21.0	2.34
686 × 254 × 140	140.1	383.5	253.7	12.4	19.0	2.33
686 × 254 × 125	125.2	677.9	253.0	11.7	16.2	2.32
610 × 305 × 238	238.1	635.8	311.4	18.4	31.4	2.45
610 × 305 × 179	179.0	620.2	307.1	14.1	23.6	2.41
610 × 305 × 149	149.1	612.4	304.8	11.8	19.7	2.39
610 × 229 × 140	139.9	617.2	230.2	13.1	22.1	2.11
610 × 229 × 125	125.1	612.2	229.0	11.9	19.6	2.09
610 × 229 × 113	113.0	607.6	228.2	11.1	17.3	2.08
610 × 229 × 101	101.2	602.6	227.6	10.5	14.8	2.07
533 × 210 × 122	122.0	544.5	211.9	12.7	21.3	1.89
533 × 210 × 109	109.0	539.5	210.8	11.6	18.8	1.88
533 × 210 × 101	101.0	536.7	210.0	10.8	17.4	1.87
533 × 210 × 92	92.1	533.1	209.3	10.1	15.6	1.86
533 × 210 × 82	82.2	528.3	208.8	9.6	13.2	1.85
457 × 191 × 98	98.3	467.2	192.8	11.4	19.6	1.67
457 × 191 × 89	89.3	463.4	191.9	10.5	17.7	1.66
457 × 191 × 82	82.0	460.0	191.3	9.9	16.0	1.65
457 × 191 × 74	74.3	457.0	190.4	9.0	14.5	1.64
457 × 191 × 67	67.1	453.4	189.9	8.5	12.7	1.63
457 × 152 × 82	82.1	465.8	155.3	10.5	18.9	1.51
457 × 152 × 74	74.2	462.0	154.4	9.6	17.0	1.50
457 × 152 × 67	67.2	458.0	153.8	9.0	15.0	1.50
457 × 152 × 60	59.8	454.6	152.9	8.1	13.3	1.50
457 × 152 × 52	52.3	449.8	152.4	7.6	10.9	1.48
406 × 178 × 74	74.2	412.8	179.5	9.5	16.0	1.51
406 × 178 × 67	67.1	409.4	178.8	8.8	14.3	1.50
406 × 178 × 60	60.1	406.4	177.9	7.9	12.8	1.49

METAL

Designation	Mass (kg/m)	Depth of section (mm)	Width of section (mm)	Thickness		Surface area (m^2/m)
				Web (mm)	Flange (mm)	
406 × 178 × 50	54.1	402.6	177.7	7.7	10.9	1.48
406 × 140 × 46	46.0	403.2	142.2	6.8	11.2	1.34
406 × 140 × 39	39.0	398.0	141.8	6.4	8.6	1.33
356 × 171 × 67	67.1	363.4	173.2	9.1	15.7	1.38
356 × 171 × 57	57.0	358.0	172.2	8.1	13.0	1.37
356 × 171 × 51	51.0	355.0	171.5	7.4	11.5	1.36
356 × 171 × 45	45.0	351.4	171.1	7.0	9.7	1.36
356 × 127 × 39	39.1	353.4	126.0	6.6	10.7	1.18
356 × 127 × 33	33.1	349.0	125.4	6.0	8.5	1.17
305 × 165 × 54	54.0	310.4	166.9	7.9	13.7	1.26
305 × 165 × 46	46.1	306.6	165.7	6.7	11.8	1.25
305 × 165 × 40	40.3	303.4	165.0	6.0	10.2	1.24
305 × 127 × 48	48.1	311.0	125.3	9.0	14.0	1.09
305 × 127 × 42	41.9	307.2	124.3	8.0	12.1	1.08
305 × 127 × 37	37.0	304.4	123.3	7.1	10.7	1.07
305 × 102 × 33	32.8	312.7	102.4	6.6	10.8	1.01
305 × 102 × 28	28.2	308.7	101.8	6.0	8.8	1.00
305 × 102 × 25	24.8	305.1	101.6	5.8	7.0	0.992
254 × 146 × 43	43.0	259.6	147.3	7.2	12.7	1.08
254 × 146 × 37	37.0	256.0	146.4	6.3	10.9	1.07
254 × 146 × 31	31.1	251.4	146.1	6.0	8.6	1.06
254 × 102 × 28	28.3	260.4	102.2	6.3	10.0	0.904
254 × 102 × 25	25.2	257.2	101.9	6.0	8.4	0.897
254 × 102 × 22	22.0	254.0	101.6	5.7	6.8	0.890
203 × 133 × 30	30.0	206.8	133.9	6.4	9.6	0.923
203 × 133 × 25	25.1	203.2	133.2	5.7	7.8	0.915
203 × 102 × 23	23.1	203.2	101.8	5.4	9.3	0.790
178 × 102 × 19	19.0	177.8	101.2	4.8	7.9	0.738
152 × 89 × 16	16.0	152.4	88.7	4.5	7.7	0.638
127 × 76 × 13	13.0	127.0	76.0	4.0	7.6	0.537

METAL

Universal Columns BS 4: Part 1: 2005

Designation	Mass (kg/m)	Depth of section (mm)	Width of section (mm)	Thickness		Surface area (m^2/m)
				Web (mm)	Flange (mm)	
356 × 406 × 634	633.9	474.7	424.0	47.6	77.0	2.52
356 × 406 × 551	551.0	455.6	418.5	42.1	67.5	2.47
356 × 406 × 467	467.0	436.6	412.2	35.8	58.0	2.42
356 × 406 × 393	393.0	419.0	407.0	30.6	49.2	2.38
356 × 406 × 340	339.9	406.4	403.0	26.6	42.9	2.35
356 × 406 × 287	287.1	393.6	399.0	22.6	36.5	2.31
356 × 406 × 235	235.1	381.0	384.8	18.4	30.2	2.28
356 × 368 × 202	201.9	374.6	374.7	16.5	27.0	2.19
356 × 368 × 177	177.0	368.2	372.6	14.4	23.8	2.17
356 × 368 × 153	152.9	362.0	370.5	12.3	20.7	2.16
356 × 368 × 129	129.0	355.6	368.6	10.4	17.5	2.14
305 × 305 × 283	282.9	365.3	322.2	26.8	44.1	1.94
305 × 305 × 240	240.0	352.5	318.4	23.0	37.7	1.91
305 × 305 × 198	198.1	339.9	314.5	19.1	31.4	1.87
305 × 305 × 158	158.1	327.1	311.2	15.8	25.0	1.84
305 × 305 × 137	136.9	320.5	309.2	13.8	21.7	1.82
305 × 305 × 118	117.9	314.5	307.4	12.0	18.7	1.81
305 × 305 × 97	96.9	307.9	305.3	9.9	15.4	1.79
254 × 254 × 167	167.1	289.1	265.2	19.2	31.7	1.58
254 × 254 × 132	132.0	276.3	261.3	15.3	25.3	1.55
254 × 254 × 107	107.1	266.7	258.8	12.8	20.5	1.52
254 × 254 × 89	88.9	260.3	256.3	10.3	17.3	1.50
254 × 254 × 73	73.1	254.1	254.6	8.6	14.2	1.49
203 × 203 × 86	86.1	222.2	209.1	12.7	20.5	1.24
203 × 203 × 71	71.0	215.8	206.4	10.0	17.3	1.22
203 × 203 × 60	60.0	209.6	205.8	9.4	14.2	1.21
203 × 203 × 52	52.0	206.2	204.3	7.9	12.5	1.20
203 × 203 × 46	46.1	203.2	203.6	7.2	11.0	1.19
152 × 152 × 37	37.0	161.8	154.4	8.0	11.5	0.912
152 × 152 × 30	30.0	157.6	152.9	6.5	9.4	0.901
152 × 152 × 23	23.0	152.4	152.2	5.8	6.8	0.889

METAL

Joists BS 4: Part 1: 2005 (retained for reference, Corus have ceased manufacture in UK)

Designation	Mass (kg/m)	Depth of section (mm)	Width of section (mm)	Thickness		Surface area (m^2/m)
				Web (mm)	Flange (mm)	
254 × 203 × 82	82.0	254.0	203.2	10.2	19.9	1.210
203 × 152 × 52	52.3	203.2	152.4	8.9	16.5	0.932
152 × 127 × 37	37.3	152.4	127.0	10.4	13.2	0.737
127 × 114 × 29	29.3	127.0	114.3	10.2	11.5	0.646
127 × 114 × 27	26.9	127.0	114.3	7.4	11.4	0.650
102 × 102 × 23	23.0	101.6	101.6	9.5	10.3	0.549
102 × 44 × 7	7.5	101.6	44.5	4.3	6.1	0.350
89 × 89 × 19	19.5	88.9	88.9	9.5	9.9	0.476
76 × 76 × 13	12.8	76.2	76.2	5.1	8.4	0.411

Parallel Flange Channels

Designation	Mass (kg/m)	Depth of section (mm)	Width of section (mm)	Thickness		Surface area (m^2/m)
				Web (mm)	Flange (mm)	
430 × 100 × 64	64.4	430	100	11.0	19.0	1.23
380 × 100 × 54	54.0	380	100	9.5	17.5	1.13
300 × 100 × 46	45.5	300	100	9.0	16.5	0.969
300 × 90 × 41	41.4	300	90	9.0	15.5	0.932
260 × 90 × 35	34.8	260	90	8.0	14.0	0.854
260 × 75 × 28	27.6	260	75	7.0	12.0	0.79
230 × 90 × 32	32.2	230	90	7.5	14.0	0.795
230 × 75 × 26	25.7	230	75	6.5	12.5	0.737
200 × 90 × 30	29.7	200	90	7.0	14.0	0.736
200 × 75 × 23	23.4	200	75	6.0	12.5	0.678
180 × 90 × 26	26.1	180	90	6.5	12.5	0.697
180 × 75 × 20	20.3	180	75	6.0	10.5	0.638
150 × 90 × 24	23.9	150	90	6.5	12.0	0.637
150 × 75 × 18	17.9	150	75	5.5	10.0	0.579
125 × 65 × 15	14.8	125	65	5.5	9.5	0.489
100 × 50 × 10	10.2	100	50	5.0	8.5	0.382

METAL

Equal Angles BS EN 10056-1

Designation	Mass (kg/m)	Surface area (m²/m)
200 × 200 × 24	71.1	0.790
200 × 200 × 20	59.9	0.790
200 × 200 × 18	54.2	0.790
200 × 200 × 16	48.5	0.790
150 × 150 × 18	40.1	0.59
150 × 150 × 15	33.8	0.59
150 × 150 × 12	27.3	0.59
150 × 150 × 10	23.0	0.59
120 × 120 × 15	26.6	0.47
120 × 120 × 12	21.6	0.47
120 × 120 × 10	18.2	0.47
120 × 120 × 8	14.7	0.47
100 × 100 × 15	21.9	0.39
100 × 100 × 12	17.8	0.39
100 × 100 × 10	15.0	0.39
100 × 100 × 8	12.2	0.39
90 × 90 × 12	15.9	0.35
90 × 90 × 10	13.4	0.35
90 × 90 × 8	10.9	0.35
90 × 90 × 7	9.61	0.35
90 × 90 × 6	8.30	0.35

Unequal Angles BS EN 10056-1

Designation	Mass (kg/m)	Surface area (m²/m)
200 × 150 × 18	47.1	0.69
200 × 150 × 15	39.6	0.69
200 × 150 × 12	32.0	0.69
200 × 100 × 15	33.7	0.59
200 × 100 × 12	27.3	0.59
200 × 100 × 10	23.0	0.59
150 × 90 × 15	26.6	0.47
150 × 90 × 12	21.6	0.47
150 × 90 × 10	18.2	0.47
150 × 75 × 15	24.8	0.44
150 × 75 × 12	20.2	0.44
150 × 75 × 10	17.0	0.44
125 × 75 × 12	17.8	0.40
125 × 75 × 10	15.0	0.40
125 × 75 × 8	12.2	0.40
100 × 75 × 12	15.4	0.34
100 × 75 × 10	13.0	0.34
100 × 75 × 8	10.6	0.34
100 × 65 × 10	12.3	0.32
100 × 65 × 8	9.94	0.32
100 × 65 × 7	8.77	0.32

METAL

Structural Tees Split from Universal Beams BS 4: Part 1: 2005

Designation	Mass (kg/m)	Surface area (m^2/m)
305 × 305 × 90	89.5	1.22
305 × 305 × 75	74.6	1.22
254 × 343 × 63	62.6	1.19
229 × 305 × 70	69.9	1.07
229 × 305 × 63	62.5	1.07
229 × 305 × 57	56.5	1.07
229 × 305 × 51	50.6	1.07
210 × 267 × 61	61.0	0.95
210 × 267 × 55	54.5	0.95
210 × 267 × 51	50.5	0.95
210 × 267 × 46	46.1	0.95
210 × 267 × 41	41.1	0.95
191 × 229 × 49	49.2	0.84
191 × 229 × 45	44.6	0.84
191 × 229 × 41	41.0	0.84
191 × 229 × 37	37.1	0.84
191 × 229 × 34	33.6	0.84
152 × 229 × 41	41.0	0.76
152 × 229 × 37	37.1	0.76
152 × 229 × 34	33.6	0.76
152 × 229 × 30	29.9	0.76
152 × 229 × 26	26.2	0.76

Universal Bearing Piles BS 4: Part 1: 2005

Designation	Mass (kg/m)	Depth of Section (mm)	Width of section (mm)	Thickness	
				Web (mm)	Flange (mm)
356 × 368 × 174	173.9	361.4	378.5	20.3	20.4
356 × 368 × 152	152.0	356.4	376.0	17.8	17.9
356 × 368 × 133	133.0	352.0	373.8	15.6	15.7
356 × 368 × 109	108.9	346.4	371.0	12.8	12.9
305 × 305 × 223	222.9	337.9	325.7	30.3	30.4
305 × 305 × 186	186.0	328.3	320.9	25.5	25.6
305 × 305 × 149	149.1	318.5	316.0	20.6	20.7
305 × 305 × 126	126.1	312.3	312.9	17.5	17.6
305 × 305 × 110	110.0	307.9	310.7	15.3	15.4
305 × 305 × 95	94.9	303.7	308.7	13.3	13.3
305 × 305 × 88	88.0	301.7	307.8	12.4	12.3
305 × 305 × 79	78.9	299.3	306.4	11.0	11.1
254 × 254 × 85	85.1	254.3	260.4	14.4	14.3
254 × 254 × 71	71.0	249.7	258.0	12.0	12.0
254 × 254 × 63	63.0	247.1	256.6	10.6	10.7
203 × 203 × 54	53.9	204.0	207.7	11.3	11.4
203 × 203 × 45	44.9	200.2	205.9	9.5	9.5

METAL

Hot Formed Square Hollow Sections EN 10210 S275J2H & S355J2H

Size (mm)	Wall thickness (mm)	Mass (kg/m)	Superficial area (m^2/m)
40 × 40	2.5	2.89	0.154
	3.0	3.41	0.152
	3.2	3.61	0.152
	3.6	4.01	0.151
	4.0	4.39	0.150
	5.0	5.28	0.147
50 × 50	2.5	3.68	0.194
	3.0	4.35	0.192
	3.2	4.62	0.192
	3.6	5.14	0.191
	4.0	5.64	0.190
	5.0	6.85	0.187
	6.0	7.99	0.185
	6.3	8.31	0.184
60 × 60	3.0	5.29	0.232
	3.2	5.62	0.232
	3.6	6.27	0.231
	4.0	6.90	0.230
	5.0	8.42	0.227
	6.0	9.87	0.225
	6.3	10.30	0.224
	8.0	12.50	0.219
70 × 70	3.0	6.24	0.272
	3.2	6.63	0.272
	3.6	7.40	0.271
	4.0	8.15	0.270
	5.0	9.99	0.267
	6.0	11.80	0.265
	6.3	12.30	0.264
	8.0	15.00	0.259
80 × 80	3.2	7.63	0.312
	3.6	8.53	0.311
	4.0	9.41	0.310
	5.0	11.60	0.307
	6.0	13.60	0.305
	6.3	14.20	0.304
	8.0	17.50	0.299
90 × 90	3.6	9.66	0.351
	4.0	10.70	0.350
	5.0	13.10	0.347
	6.0	15.50	0.345
	6.3	16.20	0.344
	8.0	20.10	0.339
100 × 100	3.6	10.80	0.391
	4.0	11.90	0.390
	5.0	14.70	0.387
	6.0	17.40	0.385
	6.3	18.20	0.384
	8.0	22.60	0.379
	10.0	27.40	0.374
120 × 120	4.0	14.40	0.470
	5.0	17.80	0.467
	6.0	21.20	0.465

METAL

Size (mm)	Wall thickness (mm)	Mass (kg/m)	Superficial area (m²/m)
	6.3	22.20	0.464
	8.0	27.60	0.459
	10.0	33.70	0.454
	12.0	39.50	0.449
	12.5	40.90	0.448
140 × 140	5.0	21.00	0.547
	6.0	24.90	0.545
	6.3	26.10	0.544
	8.0	32.60	0.539
	10.0	40.00	0.534
	12.0	47.00	0.529
	12.5	48.70	0.528
150 × 150	5.0	22.60	0.587
	6.0	26.80	0.585
	6.3	28.10	0.584
	8.0	35.10	0.579
	10.0	43.10	0.574
	12.0	50.80	0.569
	12.5	52.70	0.568
Hot formed from seamless hollow	16.0	65.2	0.559
160 × 160	5.0	24.10	0.627
	6.0	28.70	0.625
	6.3	30.10	0.624
	8.0	37.60	0.619
	10.0	46.30	0.614
	12.0	54.60	0.609
	12.5	56.60	0.608
	16.0	70.20	0.599
180 × 180	5.0	27.30	0.707
	6.0	32.50	0.705
	6.3	34.00	0.704
	8.0	42.70	0.699
	10.0	52.50	0.694
	12.0	62.10	0.689
	12.5	64.40	0.688
	16.0	80.20	0.679
200 × 200	5.0	30.40	0.787
	6.0	36.20	0.785
	6.3	38.00	0.784
	8.0	47.70	0.779
	10.0	58.80	0.774
	12.0	69.60	0.769
	12.5	72.30	0.768
	16.0	90.30	0.759
250 × 250	5.0	38.30	0.987
	6.0	45.70	0.985
	6.3	47.90	0.984
	8.0	60.30	0.979
	10.0	74.50	0.974
	12.0	88.50	0.969
	12.5	91.90	0.968
	16.0	115.00	0.959

METAL

Size (mm)	Wall thickness (mm)	Mass (kg/m)	Superficial area (m^2/m)
300 × 300	6.0	55.10	1.18
	6.3	57.80	1.18
	8.0	72.80	1.18
	10.0	90.20	1.17
	12.0	107.00	1.17
	12.5	112.00	1.17
	16.0	141.00	1.16
350 × 350	8.0	85.40	1.38
	10.0	106.00	1.37
	12.0	126.00	1.37
	12.5	131.00	1.37
	16.0	166.00	1.36
400 × 400	8.0	97.90	1.58
	10.0	122.00	1.57
	12.0	145.00	1.57
	12.5	151.00	1.57
	16.0	191.00	1.56
(Grade S355J2H only)	20.00*	235.00	1.55

Note: * SAW process

METAL

Hot Formed Square Hollow Sections JUMBO RHS: JIS G3136

Size (mm)	Wall thickness (mm)	Mass (kg/m)	Superficial area (m^2/m)
350 × 350	19.0	190.00	1.33
	22.0	217.00	1.32
	25.0	242.00	1.31
400 × 400	22.0	251.00	1.52
	25.0	282.00	1.51
450 × 450	12.0	162.00	1.76
	16.0	213.00	1.75
	19.0	250.00	1.73
	22.0	286.00	1.72
	25.0	321.00	1.71
	28.0 *	355.00	1.70
	32.0 *	399.00	1.69
500 × 500	12.0	181.00	1.96
	16.0	238.00	1.95
	19.0	280.00	1.93
	22.0	320.00	1.92
	25.0	360.00	1.91
	28.0 *	399.00	1.90
	32.0 *	450.00	1.89
	36.0 *	498.00	1.88
550 × 550	16.0	263.00	2.15
	19.0	309.00	2.13
	22.0	355.00	2.12
	25.0	399.00	2.11
	28.0 *	443.00	2.10
	32.0 *	500.00	2.09
	36.0 *	555.00	2.08
	40.0 *	608.00	2.06
600 × 600	25.0 *	439.00	2.31
	28.0 *	487.00	2.30
	32.0 *	550.00	2.29
	36.0 *	611.00	2.28
	40.0 *	671.00	2.26
700 × 700	25.0 *	517.00	2.71
	28.0 *	575.00	2.70
	32.0 *	651.00	2.69
	36.0 *	724.00	2.68
	40.0 *	797.00	2.68

Note: * SAW process

METAL

Hot Formed Rectangular Hollow Sections: EN10210 S275J2h & S355J2H

Size (mm)	Wall thickness (mm)	Mass (kg/m)	Superficial area (m²/m)
50 × 30	2.5	2.89	0.154
	3.0	3.41	0.152
	3.2	3.61	0.152
	3.6	4.01	0.151
	4.0	4.39	0.150
	5.0	5.28	0.147
60 × 40	2.5	3.68	0.194
	3.0	4.35	0.192
	3.2	4.62	0.192
	3.6	5.14	0.191
	4.0	5.64	0.190
	5.0	6.85	0.187
	6.0	7.99	0.185
	6.3	8.31	0.184
80 × 40	3.0	5.29	0.232
	3.2	5.62	0.232
	3.6	6.27	0.231
	4.0	6.90	0.230
	5.0	8.42	0.227
	6.0	9.87	0.225
	6.3	10.30	0.224
	8.0	12.50	0.219
76.2 × 50.8	3.0	5.62	0.246
	3.2	5.97	0.246
	3.6	6.66	0.245
	4.0	7.34	0.244
	5.0	8.97	0.241
	6.0	10.50	0.239
	6.3	11.00	0.238
	8.0	13.40	0.233
90 × 50	3.0	6.24	0.272
	3.2	6.63	0.272
	3.6	7.40	0.271
	4.0	8.15	0.270
	5.0	9.99	0.267
	6.0	11.80	0.265
	6.3	12.30	0.264
	8.0	15.00	0.259
100 × 50	3.0	6.71	0.292
	3.2	7.13	0.292
	3.6	7.96	0.291
	4.0	8.78	0.290
	5.0	10.80	0.287
	6.0	12.70	0.285
	6.3	13.30	0.284
	8.0	16.30	0.279

METAL

Size (mm)	Wall thickness (mm)	Mass (kg/m)	Superficial area (m²/m)
100 × 60	3.0	7.18	0.312
	3.2	7.63	0.312
	3.6	8.53	0.311
	4.0	9.41	0.310
	5.0	11.60	0.307
	6.0	13.60	0.305
	6.3	14.20	0.304
	8.0	17.50	0.299
120 × 60	3.6	9.70	0.351
	4.0	10.70	0.350
	5.0	13.10	0.347
	6.0	15.50	0.345
	6.3	16.20	0.344
	8.0	20.10	0.339
120 × 80	3.6	10.80	0.391
	4.0	11.90	0.390
	5.0	14.70	0.387
	6.0	17.40	0.385
	6.3	18.20	0.384
	8.0	22.60	0.379
	10.0	27.40	0.374
150 × 100	4.0	15.10	0.490
	5.0	18.60	0.487
	6.0	22.10	0.485
	6.3	23.10	0.484
	8.0	28.90	0.479
	10.0	35.30	0.474
	12.0	41.40	0.469
	12.5	42.80	0.468
160 × 80	4.0	14.40	0.470
	5.0	17.80	0.467
	6.0	21.20	0.465
	6.3	22.20	0.464
	8.0	27.60	0.459
	10.0	33.70	0.454
	12.0	39.50	0.449
	12.5	40.90	0.448
200 × 100	5.0	22.60	0.587
	6.0	26.80	0.585
	6.3	28.10	0.584
	8.0	35.10	0.579
	10.0	43.10	0.574
	12.0	50.80	0.569
	12.5	52.70	0.568
	16.0	65.20	0.559
250 × 150	5.0	30.40	0.787
	6.0	36.20	0.785
	6.3	38.00	0.784
	8.0	47.70	0.779
	10.0	58.80	0.774
	12.0	69.60	0.769
	12.5	72.30	0.768
	16.0	90.30	0.759

METAL

Size (mm)	Wall thickness (mm)	Mass (kg/m)	Superficial area (m^2/m)
300 × 200	5.0	38.30	0.987
	6.0	45.70	0.985
	6.3	47.90	0.984
	8.0	60.30	0.979
	10.0	74.50	0.974
	12.0	88.50	0.969
	12.5	91.90	0.968
	16.0	115.00	0.959
400 × 200	6.0	55.10	1.18
	6.3	57.80	1.18
	8.0	72.80	1.18
	10.0	90.20	1.17
	12.0	107.00	1.17
	12.5	112.00	1.17
	16.0	141.00	1.16
450 × 250	8.0	85.40	1.38
	10.0	106.00	1.37
	12.0	126.00	1.37
	12.5	131.00	1.37
	16.0	166.00	1.36
500 × 300	8.0	98.00	1.58
	10.0	122.00	1.57
	12.0	145.00	1.57
	12.5	151.00	1.57
	16.0	191.00	1.56
	20.0	235.00	1.55

METAL

Hot Formed Circular Hollow Sections EN 10210 S275J2H & S355J2H

Outside diameter (mm)	Wall thickness (mm)	Mass (kg/m)	Superficial area (m²/m)
21.3	3.2	1.43	0.067
26.9	3.2	1.87	0.085
33.7	3.0	2.27	0.106
	3.2	2.41	0.106
	3.6	2.67	0.106
	4.0	2.93	0.106
42.4	3.0	2.91	0.133
	3.2	3.09	0.133
	3.6	3.44	0.133
	4.0	3.79	0.133
48.3	2.5	2.82	0.152
	3.0	3.35	0.152
	3.2	3.56	0.152
	3.6	3.97	0.152
	4.0	4.37	0.152
	5.0	5.34	0.152
60.3	2.5	3.56	0.189
	3.0	4.24	0.189
	3.2	4.51	0.189
	3.6	5.03	0.189
	4.0	5.55	0.189
	5.0	6.82	0.189
76.1	2.5	4.54	0.239
	3.0	5.41	0.239
	3.2	5.75	0.239
	3.6	6.44	0.239
	4.0	7.11	0.239
	5.0	8.77	0.239
	6.0	10.40	0.239
	6.3	10.80	0.239
88.9	2.5	5.33	0.279
	3.0	6.36	0.279
	3.2	6.76	0.27
	3.6	7.57	0.279
	4.0	8.38	0.279
	5.0	10.30	0.279
	6.0	12.30	0.279
	6.3	12.80	0.279
114.3	3.0	8.23	0.359
	3.2	8.77	0.359
	3.6	9.83	0.359
	4.0	10.09	0.359
	5.0	13.50	0.359
	6.0	16.00	0.359
	6.3	16.80	0.359

METAL

Outside diameter (mm)	Wall thickness (mm)	Mass (kg/m)	Superficial area (m²/m)
139.7	3.2	10.80	0.439
	3.6	12.10	0.439
	4.0	13.40	0.439
	5.0	16.60	0.439
	6.0	19.80	0.439
	6.3	20.70	0.439
	8.0	26.00	0.439
	10.0	32.00	0.439
168.3	3.2	13.00	0.529
	3.6	14.60	0.529
	4.0	16.20	0.529
	5.0	20.10	0.529
	6.0	24.00	0.529
	6.3	25.20	0.529
	8.0	31.60	0.529
	10.0	39.00	0.529
	12.0	46.30	0.529
	12.5	48.00	0.529
193.7	5.0	23.30	0.609
	6.0	27.80	0.609
	6.3	29.10	0.609
	8.0	36.60	0.609
	10.0	45.30	0.609
	12.0	53.80	0.609
	12.5	55.90	0.609
219.1	5.0	26.40	0.688
	6.0	31.50	0.688
	6.3	33.10	0.688
	8.0	41.60	0.688
	10.0	51.60	0.688
	12.0	61.30	0.688
	12.5	63.70	0.688
	16.0	80.10	0.688
244.5	5.0	29.50	0.768
	6.0	35.30	0.768
	6.3	37.00	0.768
	8.0	46.70	0.768
	10.0	57.80	0.768
	12.0	68.80	0.768
	12.5	71.50	0.768
	16.0	90.20	0.768
273.0	5.0	33.00	0.858
	6.0	39.50	0.858
	6.3	41.40	0.858
	8.0	52.30	0.858
	10.0	64.90	0.858
	12.0	77.20	0.858
	12.5	80.30	0.858
	16.0	101.00	0.858
323.9	5.0	39.30	1.02
	6.0	47.00	1.02
	6.3	49.30	1.02
	8.0	62.30	1.02
	10.0	77.40	1.02

METAL

Outside diameter (mm)	Wall thickness (mm)	Mass (kg/m)	Superficial area (m²/m)
	12.0	92.30	1.02
	12.5	96.00	1.02
	16.0	121.00	1.02
355.6	6.3	54.30	1.12
	8.0	68.60	1.12
	10.0	85.30	1.12
	12.0	102.00	1.12
	12.5	106.00	1.12
	16.0	134.00	1.12
406.4	6.3	62.20	1.28
	8.0	79.60	1.28
	10.0	97.80	1.28
	12.0	117.00	1.28
	12.5	121.00	1.28
	16.0	154.00	1.28
457.0	6.3	70.00	1.44
	8.0	88.60	1.44
	10.0	110.00	1.44
	12.0	132.00	1.44
	12.5	137.00	1.44
	16.0	174.00	1.44
508.0	6.3	77.90	1.60
	8.0	98.60	1.60
	10.0	123.00	1.60
	12.0	147.00	1.60
	12.5	153.00	1.60
	16.0	194.00	1.60

METAL

Spacing of Holes in Angles

Nominal leg length (mm)	Spacing of holes						Maximum diameter of bolt or rivet		
	A	B	C	D	E	F	A	B and C	D, E and F
200		75	75	55	55	55		30	20
150		55	55					20	
125		45	60					20	
120									
100	55						24		
90	50						24		
80	45						20		
75	45						20		
70	40						20		
65	35						20		
60	35						16		
50	28						12		
45	25								
40	23								
30	20								
25	15								

KERBS, PAVING, ETC.

KERBS/EDGINGS/CHANNELS

Precast Concrete Kerbs to BS 7263

Straight kerb units: length from 450 to 915 mm

150 mm high × 125 mm thick		
bullnosed	type BN	
half battered	type HB3	
255 mm high × 125 mm thick		
45° splayed	type SP	
half battered	type HB2	
305 mm high × 150 mm thick		
half battered	type HB1	
Quadrant kerb units		
150 mm high × 305 and 455 mm radius to match	type BN	type QBN
150 mm high × 305 and 455 mm radius to match	type HB2, HB3	type QHB
150 mm high × 305 and 455 mm radius to match	type SP	type QSP
255 mm high × 305 and 455 mm radius to match	type BN	type QBN
255 mm high × 305 and 455 mm radius to match	type HB2, HB3	type QHB
225 mm high × 305 and 455 mm radius to match	type SP	type QSP
Angle kerb units		
305 × 305 × 225 mm high × 125 mm thick		
bullnosed external angle	type XA	
splayed external angle to match type SP	type XA	
bullnosed internal angle	type IA	
splayed internal angle to match type SP	type IA	
Channels		
255 mm wide × 125 mm high flat	type CS1	
150 mm wide × 125 mm high flat type	CS2	
255 mm wide × 125 mm high dished	type CD	

KERBS, PAVING, ETC.

Transition kerb units			
from kerb type SP to HB	left handed	type TL	
	right handed	type TR	
from kerb type BN to HB	left handed	type DL1	
	right handed	type DR1	
from kerb type BN to SP	left handed	type DL2	
	right handed	type DR2	

Number of kerbs required per quarter circle (780 mm kerb lengths)

Radius (m)	Number in quarter circle
12	24
10	20
8	16
6	12
5	10
4	8
3	6
2	4
1	2

Precast Concrete Edgings

Round top type ER	Flat top type EF	Bullnosed top type EBN
150 × 50 mm	150 × 50 mm	150 × 50 mm
200 × 50 mm	200 × 50 mm	200 × 50 mm
250 × 50 mm	250 × 50 mm	250 × 50 mm

KERBS, PAVING, ETC.

BASES

Cement Bound Material for Bases and Subbases

CBM1:	very carefully graded aggregate from 37.5–75 mm, with a 7-day strength of 4.5 N/mm^2
CBM2:	same range of aggregate as CBM1 but with more tolerance in each size of aggregate with a 7-day strength of 7.0 N/mm^2
CBM3:	crushed natural aggregate or blast furnace slag, graded from 37.5–150 mm for 40 mm aggregate, and from 20–75 mm for 20 mm aggregate, with a 7-day strength of 10 N/mm^2
CBM4:	crushed natural aggregate or blast furnace slag, graded from 37.5–150 mm for 40 mm aggregate, and from 20–75 mm for 20 mm aggregate, with a 7-day strength of 15 N/mm^2

INTERLOCKING BRICK/BLOCK ROADS/PAVINGS

Sizes of Precast Concrete Paving Blocks

Type R blocks
200 × 100 × 60 mm
200 × 100 × 65 mm
200 × 100 × 80 mm
200 × 100 × 100 mm

Type S
Any shape within a 295 mm space

Sizes of clay brick pavers
200 × 100 × 50 mm
200 × 100 × 65 mm
210 × 105 × 50 mm
210 × 105 × 65 mm
215 × 102.5 × 50 mm
215 × 102.5 × 65 mm

Type PA: 3 kN
Footpaths and pedestrian areas, private driveways, car parks, light vehicle traffic and over-run

Type PB: 7 kN
Residential roads, lorry parks, factory yards, docks, petrol station forecourts, hardstandings, bus stations

KERBS, PAVING, ETC.

PAVING AND SURFACING

Weights and Sizes of Paving and Surfacing

Description of item	Size	Quantity per tonne
Paving 50 mm thick	900 × 600 mm	15
Paving 50 mm thick	750 × 600 mm	18
Paving 50 mm thick	600 × 600 mm	23
Paving 50 mm thick	450 × 600 mm	30
Paving 38 mm thick	600 × 600 mm	30
Path edging	914 × 50 × 150 mm	60
Kerb (including radius and tapers)	125 × 254 × 914 mm	15
Kerb (including radius and tapers)	125 × 150 × 914 mm	25
Square channel	125 × 254 × 914 mm	15
Dished channel	125 × 254 × 914 mm	15
Quadrants	300 × 300 × 254 mm	19
Quadrants	450 × 450 × 254 mm	12
Quadrants	300 × 300 × 150 mm	30
Internal angles	300 × 300 × 254 mm	30
Fluted pavement channel	255 × 75 × 914 mm	25
Corner stones	300 × 300 mm	80
Corner stones	360 × 360 mm	60
Cable covers	914 × 175 mm	55
Gulley kerbs	220 × 220 × 150 mm	60
Gulley kerbs	220 × 200 × 75 mm	120

KERBS, PAVING, ETC.

Weights and Sizes of Paving and Surfacing

Material	kg/m³	lb/cu yd
Tarmacadam	2306	3891
Macadam (waterbound)	2563	4325
Vermiculite (aggregate)	64–80	108–135
Terracotta	2114	3568
Cork – compressed	388	24
	kg/m²	**lb/sq ft**
Clay floor tiles, 12.7 mm	27.3	5.6
Pavement lights	122	25
Damp-proof course	5	1
	kg/m² per mm thickness	**lb/sq ft per inch thickness**
Paving slabs (stone)	2.3	12
Granite setts	2.88	15
Asphalt	2.30	12
Rubber flooring	1.68	9
Polyvinyl chloride	1.94 (avg)	10 (avg)

Coverage (m^2) Per Cubic Metre of Materials Used as Subbases or Capping Layers

Consolidated thickness laid in (mm)	Square metre coverage		
	Gravel	Sand	Hardcore
50	15.80	16.50	–
75	10.50	11.00	–
100	7.92	8.20	7.42
125	6.34	6.60	5.90
150	5.28	5.50	4.95
175	–	–	4.23
200	–	–	3.71
225	–	–	3.30
300	–	–	2.47

KERBS, PAVING, ETC.

Approximate Rate of Spreads

Average thickness of course (mm)	Description	Approximate rate of spread			
		Open Textured		Dense, Medium & Fine Textured	
		(kg/m^2)	(m^2/t)	(kg/m^2)	(m^2/t)
35	14 mm open textured or dense wearing course	60–75	13–17	70–85	12–14
40	20 mm open textured or dense base course	70–85	12–14	80–100	10–12
45	20 mm open textured or dense base course	80–100	10–12	95–100	9–10
50	20 mm open textured or dense, or 28 mm dense base course	85–110	9–12	110–120	8–9
60	28 mm dense base course, 40 mm open textured of dense base course or 40 mm single course as base course		8–10	130–150	7–8
65	28 mm dense base course, 40 mm open textured or dense base course or 40 mm single course	100–135	7–10	140–160	6–7
75	40 mm single course, 40 mm open textured or dense base course, 40 mm dense roadbase	120–150	7–8	165–185	5–6
100	40 mm dense base course or roadbase	–	–	220–240	4–4.5

KERBS, PAVING, ETC.

Surface Dressing Roads: Coverage (m^2) per Tonne of Material

Size in mm	Sand	Granite chips	Gravel	Limestone chips
Sand	168	–	–	–
3	–	148	152	165
6	–	130	133	144
9	–	111	114	123
13	–	85	87	95
19	–	68	71	78

Sizes of Flags

Reference	Nominal size (mm)	Thickness (mm)
A	600 × 450	50 and 63
B	600 × 600	50 and 63
C	600 × 750	50 and 63
D	600 × 900	50 and 63
E	450 × 450	50 and 70 chamfered top surface
F	400 × 400	50 and 65 chamfered top surface
G	300 × 300	50 and 60 chamfered top surface

Sizes of Natural Stone Setts

Width (mm)		Length (mm)		Depth (mm)
100	×	100	×	100
75	×	150 to 250	×	125
75	×	150 to 250	×	150
100	×	150 to 250	×	100
100	×	150 to 250	×	150

SEEDING/TURFING AND PLANTING

Topsoil Quality

Topsoil grade	Properties
Premium	Natural topsoil, high fertility, loamy texture, good soil structure, suitable for intensive cultivation.
General purpose	Natural or manufactured topsoil of lesser quality than Premium, suitable for agriculture or amenity landscape, may need fertilizer or soil structure improvement.
Economy	Selected subsoil, natural mineral deposit such as river silt or greensand. The grade comprises two subgrades; 'Low clay' and 'High clay' which is more liable to compaction in handling. This grade is suitable for low-production agricultural land and amenity woodland or conservation planting areas.

Forms of Trees

Standards:	Shall be clear with substantially straight stems. Grafted and budded trees shall have no more than a slight bend at the union. Standards shall be designated as Half, Extra light, Light, Standard, Selected standard, Heavy, and Extra heavy.
Sizes of Standards	
Heavy standard	12–14 cm girth × 3.50 to 5.00 m high
Extra Heavy standard	14–16 cm girth × 4.25 to 5.00 m high
Extra Heavy standard	16–18 cm girth × 4.25 to 6.00 m high
Extra Heavy standard	18–20 cm girth × 5.00 to 6.00 m high
Semi-mature trees:	Between 6.0 m and 12.0 m tall with a girth of 20 to 75 cm at 1.0 m above ground.
Feathered trees:	Shall have a defined upright central leader, with stem furnished with evenly spread and balanced lateral shoots down to or near the ground.
Whips:	Shall be without significant feather growth as determined by visual inspection.
Multi-stemmed trees:	Shall have two or more main stems at, near, above or below ground.

Seedlings grown from seed and not transplanted shall be specified when ordered for sale as:

1+0	one year old seedling
2+0	two year old seedling
1+1	one year seed bed, one year transplanted = two year old seedling
1+2	one year seed bed, two years transplanted = three year old seedling
2+1	two years seed bed, one year transplanted = three year old seedling
1u1	two years seed bed, undercut after 1 year = two year old seedling
2u2	four years seed bed, undercut after 2 years = four year old seedling

SEEDING/TURFING AND PLANTING

Cuttings

The age of cuttings (plants grown from shoots, stems, or roots of the mother plant) shall be specified when ordered for sale. The height of transplants and undercut seedlings/cuttings (which have been transplanted or undercut at least once) shall be stated in centimetres. The number of growing seasons before and after transplanting or undercutting shall be stated.

0 + 1	one year cutting
0 + 2	two year cutting
0 + 1 + 1	one year cutting bed, one year transplanted = two year old seedling
0 + 1 + 2	one year cutting bed, two years transplanted = three year old seedling

Grass Cutting Capacities in m^2 per hour

Speed mph	Width of cut in metres												
	0.5	0.7	1.0	1.2	1.5	1.7	2.0	2.0	2.1	2.5	2.8	3.0	3.4
1.0	724	1127	1529	1931	2334	2736	3138	3219	3380	4023	4506	4828	5472
1.5	1086	1690	2293	2897	3500	4104	4707	4828	5069	6035	6759	7242	8208
2.0	1448	2253	3058	3862	4667	5472	6276	6437	6759	8047	9012	9656	10944
2.5	1811	2816	3822	4828	5834	6840	7846	8047	8449	10058	11265	12070	13679
3.0	2173	3380	4587	5794	7001	8208	9415	9656	10139	12070	13518	14484	16415
3.5	2535	3943	5351	6759	8167	9576	10984	11265	11829	14082	15772	16898	19151
4.0	2897	4506	6115	7725	9334	10944	12553	12875	13518	16093	18025	19312	21887
4.5	3259	5069	6880	8690	10501	12311	14122	14484	15208	18105	20278	21726	24623
5.0	3621	5633	7644	9656	11668	13679	15691	16093	16898	20117	22531	24140	27359
5.5	3983	6196	8409	10622	12834	15047	17260	17703	18588	22128	24784	26554	30095
6.0	4345	6759	9173	11587	14001	16415	18829	19312	20278	24140	27037	28968	32831
6.5	4707	7322	9938	12553	15168	17783	20398	20921	21967	26152	29290	31382	35566
7.0	5069	7886	10702	13518	16335	19151	21967	22531	23657	28163	31543	33796	38302

Number of Plants per m^2 at the following offset spacings

(All plants equidistant horizontally and vertically)

Distance mm	Nr of Plants
100	114.43
200	28.22
250	18.17
300	12.35
400	6.72
500	4.29
600	3.04
700	2.16
750	1.88
900	1.26
1000	1.05
1200	0.68
1500	0.42
2000	0.23

SEEDING/TURFING AND PLANTING

Grass Clippings Wet: Based on 3.5 m^3/tonne

Annual kg/100 m^2	Average 20 cuts kg/100 m^2	m^2/tonne	m^2/m^3
32.0	1.6	61162.1	214067.3

Nr of cuts	22	20	18	16	12	4
kg/cut	1.45	1.60	1.78	2.00	2.67	8.00
Area capacity of 3 tonne vehicle per load						
m^2	206250	187500	168750	150000	112500	37500
Load m^3	**100 m^2 units/m^3 of vehicle space**					
1	196.4	178.6	160.7	142.9	107.1	35.7
2	392.9	357.1	321.4	285.7	214.3	71.4
3	589.3	535.7	482.1	428.6	321.4	107.1
4	785.7	714.3	642.9	571.4	428.6	142.9
5	982.1	892.9	803.6	714.3	535.7	178.6

Transportation of Trees

To unload large trees a machine with the necessary lifting strength is required. The weight of the trees must therefore be known in advance. The following table gives a rough overview. The additional columns with rootball dimensions and the number of plants per trailer provide additional information, for example about preparing planting holes and calculating unloading times.

Girth in cm	Rootball diameter in cm	Ball height in cm	Weight in kg	Numbers of trees per trailer
16–18	50–60	40	150	100–120
18–20	60–70	40–50	200	80–100
20–25	60–70	40–50	270	50–70
25–30	80	50–60	350	50
30–35	90–100	60–70	500	12–18
35–40	100–110	60–70	650	10–15
40–45	110–120	60–70	850	8–12
45–50	110–120	60–70	1100	5–7
50–60	130–140	60–70	1600	1–3
60–70	150–160	60–70	2500	1
70–80	180–200	70	4000	1
80–90	200–220	70–80	5500	1
90–100	230–250	80–90	7500	1
100–120	250–270	80–90	9500	1

Data supplied by Lorenz von Ehren GmbH
The information in the table is approximate; deviations depend on soil type, genus and weather

FENCING AND GATES

Types of Preservative

Creosote (tar oil) can be 'factory' applied	by pressure to BS 144: pts 1&2 by immersion to BS 144: pt 1 by hot and cold open tank to BS 144: pts 1&2
Copper/chromium/arsenic (CCA)	by full cell process to BS 4072 pts 1&2
Organic solvent (OS)	by double vacuum (vacvac) to BS 5707 pts 1&3 by immersion to BS 5057 pts 1&3
Pentachlorophenol (PCP)	by heavy oil double vacuum to BS 5705 pts 2&3
Boron diffusion process (treated with disodium octaborate to BWPA Manual 1986)	

Note: Boron is used on green timber at source and the timber is supplied dry

Cleft Chestnut Pale Fences

Pales	Pale spacing	Wire lines	
900 mm	75 mm	2	temporary protection
1050 mm	75 or 100 mm	2	light protective fences
1200 mm	75 mm	3	perimeter fences
1350 mm	75 mm	3	perimeter fences
1500 mm	50 mm	3	narrow perimeter fences
1800 mm	50 mm	3	light security fences

Close-Boarded Fences

Close-boarded fences 1.05 to 1.8 m high
Type BCR (recessed) or BCM (morticed) with concrete posts 140 × 115 mm tapered and Type BW with timber posts

Palisade Fences

Wooden palisade fences
Type WPC with concrete posts 140 × 115 mm tapered and Type WPW with timber posts

For both types of fence:
Height of fence 1050 mm: two rails
Height of fence 1200 mm: two rails
Height of fence 1500 mm: three rails
Height of fence 1650 mm: three rails
Height of fence 1800 mm: three rails

FENCING AND GATES

Post and Rail Fences

Wooden post and rail fences Type MPR 11/3 morticed rails and Type SPR 11/3 nailed rails Height to top of rail 1100 mm Rails: three rails 87 mm, 38 mm
Type MPR 11/4 morticed rails and Type SPR 11/4 nailed rails Height to top of rail 1100 mm Rails: four rails 87 mm, 38 mm
Type MPR 13/4 morticed rails and Type SPR 13/4 nailed rails Height to top of rail 1300 mm Rail spacing 250 mm, 250 mm, and 225 mm from top Rails: four rails 87 mm, 38 mm

Steel Posts

Rolled steel angle iron posts for chain link fencing

Posts	Fence height	Strut	Straining post
1500 × 40 × 40 × 5 mm	900 mm	1500 × 40 × 40 × 5 mm	1500 × 50 × 50 × 6 mm
1800 × 40 × 40 × 5 mm	1200 mm	1800 × 40 × 40 × 5 mm	1800 × 50 × 50 × 6 mm
2000 × 45 × 45 × 5 mm	1400 mm	2000 × 45 × 45 × 5 mm	2000 × 60 × 60 × 6 mm
2600 × 45 × 45 × 5 mm	1800 mm	2600 × 45 × 45 × 5 mm	2600 × 60 × 60 × 6 mm
3000 × 50 × 50 × 6 mm with arms	1800 mm	2600 × 45 × 45 × 5 mm	3000 × 60 × 60 × 6 mm

Concrete Posts

Concrete posts for chain link fencing

Posts and straining posts	Fence height	Strut
1570 mm 100 × 100 mm	900 mm	1500 mm × 75 × 75 mm
1870 mm 125 × 125 mm	1200 mm	1830 mm × 100 × 75 mm
2070 mm 125 × 125 mm	1400 mm	1980 mm × 100 × 75 mm
2620 mm 125 × 125 mm	1800 mm	2590 mm × 100 × 85 mm
3040 mm 125 × 125 mm	1800 mm	2590 mm × 100 × 85 mm (with arms)

FENCING AND GATES

Rolled Steel Angle Posts

Rolled steel angle posts for rectangular wire mesh (field) fencing

Posts	Fence height	Strut	Straining post
1200 × 40 × 40 × 5mm	600mm	1200 × 75 × 75mm	1350 × 100 × 100mm
1400 × 40 × 40 × 5mm	800mm	1400 × 75 × 75mm	1550 × 100 × 100mm
1500 × 40 × 40 × 5mm	900mm	1500 × 75 × 75mm	1650 × 100 × 100mm
1600 × 40 × 40 × 5mm	1000mm	1600 × 75 × 75mm	1750 × 100 × 100mm
1750 × 40 × 40 × 5mm	1150mm	1750 × 75 × 100mm	1900 × 125 × 125mm

Concrete Posts

Concrete posts for rectangular wire mesh (field) fencing

Posts	Fence height	Strut	Straining post
1270 × 100 × 100mm	600mm	1200 × 75 × 75mm	1420 × 100 × 100mm
1470 × 100 × 100mm	800mm	1350 × 75 × 75mm	1620 × 100 × 100mm
1570 × 100 × 100mm	900mm	1500 × 75 × 75mm	1720 × 100 × 100mm
1670 × 100 × 100mm	600mm	1650 × 75 × 75mm	1820 × 100 × 100mm
1820 × 125 × 125mm	1150mm	1830 × 75 × 100mm	1970 × 125 × 125mm

Cleft Chestnut Pale Fences

Timber Posts

Timber posts for wire mesh and hexagonal wire netting fences

Round timber for general fences

Posts	Fence height	Strut	Straining post
1300 × 65mm dia.	600mm	1200 × 80mm dia.	1450 × 100mm dia.
1500 × 65mm dia.	800mm	1400 × 80mm dia.	1650 × 100mm dia.
1600 × 65mm dia.	900mm	1500 × 80mm dia.	1750 × 100mm dia.
1700 × 65mm dia.	1050mm	1600 × 80mm dia.	1850 × 100mm dia.
1800 × 65mm dia.	1150mm	1750 × 80mm dia.	2000 × 120mm dia.

Squared timber for general fences

Posts	Fence height	Strut	Straining post
1300 × 75 × 75mm	600mm	1200 × 75 × 75mm	1450 × 100 × 100mm
1500 × 75 × 75mm	800mm	1400 × 75 × 75mm	1650 × 100 × 100mm
1600 × 75 × 75mm	900mm	1500 × 75 × 75mm	1750 × 100 × 100mm
1700 × 75 × 75mm	1050mm	1600 × 75 × 75mm	1850 × 100 × 100mm
1800 × 75 × 75mm	1150mm	1750 × 75 × 75mm	2000 × 125 × 100mm

FENCING AND GATES

Steel Fences to BS 1722: Part 9: 1992

	Fence height	Top/bottom rails and flat posts	Vertical bars
Light	1000 mm	40 × 10 mm 450 mm in ground	12 mm dia. at 115 mm cs
	1200 mm	40 × 10 mm 550 mm in ground	12 mm dia. at 115 mm cs
	1400 mm	40 × 10 mm 550 mm in ground	12 mm dia. at 115 mm cs
Light	1000 mm	40 × 10 mm 450 mm in ground	16 mm dia. at 120 mm cs
	1200 mm	40 × 10 mm 550 mm in ground	16 mm dia. at 120 mm cs
	1400 mm	40 × 10 mm 550 mm in ground	16 mm dia. at 120 mm cs
Medium	1200 mm	50 × 10 mm 550 mm in ground	20 mm dia. at 125 mm cs
	1400 mm	50 × 10 mm 550 mm in ground	20 mm dia. at 125 mm cs
	1600 mm	50 × 10 mm 600 mm in ground	22 mm dia. at 145 mm cs
	1800 mm	50 × 10 mm 600 mm in ground	22 mm dia. at 145 mm cs
Heavy	1600 mm	50 × 10 mm 600 mm in ground	22 mm dia. at 145 mm cs
	1800 mm	50 × 10 mm 600 mm in ground	22 mm dia. at 145 mm cs
	2000 mm	50 × 10 mm 600 mm in ground	22 mm dia. at 145 mm cs
	2200 mm	50 × 10 mm 600 mm in ground	22 mm dia. at 145 mm cs

Notes: Mild steel fences: round or square verticals; flat standards and horizontals. Tops of vertical bars may be bow-top, blunt, or pointed. Round or square bar railings

Timber Field Gates to BS 3470: 1975

Gates made to this standard are designed to open one way only
All timber gates are 1100 mm high
Width over stiles 2400, 2700, 3000, 3300, 3600, and 4200 mm
Gates over 4200 mm should be made in two leaves

Steel Field Gates to BS 3470: 1975

All steel gates are 1100 mm high	
Heavy duty:	width over stiles 2400, 3000, 3600 and 4500 mm
Light duty:	width over stiles 2400, 3000, and 3600 mm

FENCING AND GATES

Domestic Front Entrance Gates to BS 4092: Part 1: 1966

Metal gates:	Single gates are 900 mm high minimum, 900 mm, 1000 mm and 1100 mm wide

Domestic Front Entrance Gates to BS 4092: Part 2: 1966

Wooden gates:	All rails shall be tenoned into the stiles Single gates are 840 mm high minimum, 801 mm and 1020 mm wide Double gates are 840 mm high minimum, 2130, 2340 and 2640 mm wide

Timber Bridle Gates to BS 5709:1979 (Horse or Hunting Gates)

Gates open one way only	
Minimum width between posts	1525 mm
Minimum height	1100 mm

Timber Kissing Gates to BS 5709:1979

Minimum width	700 mm
Minimum height	1000 mm
Minimum distance between shutting posts	600 mm
Minimum clearance at mid-point	600 mm

Metal Kissing Gates to BS 5709:1979

Sizes are the same as those for timber kissing gates
Maximum gaps between rails 120 mm

Categories of Pedestrian Guard Rail to BS 3049:1976

Class A for normal use
Class B where vandalism is expected
Class C where crowd pressure is likely

DRAINAGE

DRAINAGE

Width Required for Trenches for Various Diameters of Pipes

Pipe diameter (mm)	Trench n.e. 1.50 m deep	Trench over 1.50 m deep
n.e. 100 mm	450 mm	600 mm
100–150 mm	500 mm	650 mm
150–225 mm	600 mm	750 mm
225–300 mm	650 mm	800 mm
300–400 mm	750 mm	900 mm
400–450 mm	900 mm	1050 mm
450–600 mm	1100 mm	1300 mm

Weights and Dimensions – Vitrified Clay Pipes

Product	Nominal diameter	Effective length	BS 65 limits of tolerance		Crushing strength	Weight	
			min	max			
	(mm)	(mm)	(mm)	(mm)	(kN/m)	(kg/pipe)	(kg/m)
Supersleve	100	1600	96	105	35.00	14.71	9.19
	150	1750	146	158	35.00	29.24	16.71
Hepsleve	225	1850	221	236	28.00	84.03	45.42
	300	2500	295	313	34.00	193.05	77.22
	150	1500	146	158	22.00	37.04	24.69
Hepseal	225	1750	221	236	28.00	85.47	48.84
	300	2500	295	313	34.00	204.08	81.63
	400	2500	394	414	44.00	357.14	142.86
	450	2500	444	464	44.00	454.55	181.63
	500	2500	494	514	48.00	555.56	222.22
	600	2500	591	615	57.00	796.23	307.69
	700	3000	689	719	67.00	1111.11	370.45
	800	3000	788	822	72.00	1351.35	450.45
Hepline	100	1600	95	107	22.00	14.71	9.19
	150	1750	145	160	22.00	29.24	16.71
	225	1850	219	239	28.00	84.03	45.42
	300	1850	292	317	34.00	142.86	77.22
Hepduct (conduit)	90	1500	–	–	28.00	12.05	8.03
	100	1600	–	–	28.00	14.71	9.19
	125	1750	–	–	28.00	20.73	11.84
	150	1750	–	–	28.00	29.24	16.71
	225	1850	–	–	28.00	84.03	45.42
	300	1850	–	–	34.00	142.86	77.22

DRAINAGE

Weights and Dimensions – Vitrified Clay Pipes

Nominal internal diameter (mm)	Nominal wall thickness (mm)	Approximate weight (kg/m)
150	25	45
225	29	71
300	32	122
375	35	162
450	38	191
600	48	317
750	54	454
900	60	616
1200	76	912
1500	89	1458
1800	102	1884
2100	127	2619

Wall thickness, weights and pipe lengths vary, depending on type of pipe required
The particulars shown above represent a selection of available diameters and are applicable to strength class 1 pipes with flexible rubber ring joints
Tubes with Ogee joints are also available

DRAINAGE

Weights and Dimensions – PVC-u Pipes

	Nominal size	Mean outside diameter (mm)		Wall thickness	Weight
		min	max	(mm)	(kg/m)
Standard pipes	82.4	82.4	82.7	3.2	1.2
	110.0	110.0	110.4	3.2	1.6
	160.0	160.0	160.6	4.1	3.0
	200.0	200.0	200.6	4.9	4.6
	250.0	250.0	250.7	6.1	7.2
Perforated pipes heavy grade	As above	As above	As above	As above	As above
thin wall	82.4	82.4	82.7	1.7	–
	110.0	110.0	110.4	2.2	–
	160.0	160.0	160.6	3.2	–

Width of Trenches Required for Various Diameters of Pipes

Pipe diameter (mm)	Trench n.e. 1.5 m deep (mm)	Trench over 1.5 m deep (mm)
n.e. 100	450	600
100–150	500	650
150–225	600	750
225–300	650	800
300–400	750	900
400–450	900	1050
450–600	1100	1300

DRAINAGE

DRAINAGE BELOW GROUND AND LAND DRAINAGE

Flow of Water Which Can Be Carried by Various Sizes of Pipe

Clay or concrete pipes

	Gradient of pipeline							
	1:10	1:20	1:30	1:40	1:50	1:60	1:80	1:100
Pipe size	**Flow in litres per second**							
DN 100 15.0	8.5	6.8	5.8	5.2	4.7	4.0	3.5	
DN 150 28.0	19.0	16.0	14.0	12.0	11.0	9.1	8.0	
DN 225 140.0	95.0	76.0	66.0	58.0	53.0	46.0	40.0	

Plastic pipes

	Gradient of pipeline							
	1:10	1:20	1:30	1:40	1:50	1:60	1:80	1:100
Pipe size	**Flow in litres per second**							
82.4 mm i/dia.	12.0	8.5	6.8	5.8	5.2	4.7	4.0	3.5
110 mm i/dia.	28.0	19.0	16.0	14.0	12.0	11.0	9.1	8.0
160 mm i/dia.	76.0	53.0	43.0	37.0	33.0	29.0	25.0	22.0
200 mm i/dia.	140.0	95.0	76.0	66.0	58.0	53.0	46.0	40.0

Vitrified (Perforated) Clay Pipes and Fittings to BS En 295-5 1994

Length not specified		
75 mm bore	**250 mm bore**	**600 mm bore**
100	300	700
125	350	800
150	400	1000
200	450	1200
225	500	

Precast Concrete Pipes: Prestressed Non-pressure Pipes and Fittings: Flexible Joints to BS 5911: Pt. 103: 1994

Rationalized metric nominal sizes: 450, 500	
Length:	500–1000 by 100 increments 1000–2200 by 200 increments 2200–2800 by 300 increments
Angles: length:	450–600 angles 45, 22.5,11.25° 600 or more angles 22.5, 11.25°

DRAINAGE

Precast Concrete Pipes: Unreinforced and Circular Manholes and Soakaways to BS 5911: Pt. 200: 1994

Nominal sizes:	
Shafts:	675, 900 mm
Chambers:	900, 1050, 1200, 1350, 1500, 1800, 2100, 2400, 2700, 3000 mm
Large chambers:	To have either tapered reducing rings or a flat reducing slab in order to accept the standard cover
Ring depths:	1. 300–1200 mm by 300 mm increments except for bottom slab and rings below cover slab, these are by 150 mm increments
	2. 250–1000 mm by 250 mm increments except for bottom slab and rings below cover slab, these are by 125 mm increments
Access hole:	750 × 750 mm for DN 1050 chamber 1200 × 675 mm for DN 1350 chamber

Calculation of Soakaway Depth

The following formula determines the depth of concrete ring soakaway that would be required for draining given amounts of water.

$$h = \frac{4ar}{3\pi D^2}$$

h = depth of the chamber below the invert pipe
a = the area to be drained
r = the hourly rate of rainfall (50 mm per hour)
π = pi
D = internal diameter of the soakaway

This table shows the depth of chambers in each ring size which would be required to contain the volume of water specified. These allow a recommended storage capacity of ⅓ (one third of the hourly rainfall figure).

Table Showing Required Depth of Concrete Ring Chambers in Metres

Area m²	**50**	**100**	**150**	**200**	**300**	**400**	**500**
Ring size							
0.9	1.31	2.62	3.93	5.24	7.86	10.48	13.10
1.1	0.96	1.92	2.89	3.85	5.77	7.70	9.62
1.2	0.74	1.47	2.21	2.95	4.42	5.89	7.37
1.4	0.58	1.16	1.75	2.33	3.49	4.66	5.82
1.5	0.47	0.94	1.41	1.89	2.83	3.77	4.72
1.8	0.33	0.65	0.98	1.31	1.96	2.62	3.27
2.1	0.24	0.48	0.72	0.96	1.44	1.92	2.41
2.4	0.18	0.37	0.55	0.74	1.11	1.47	1.84
2.7	0.15	0.29	0.44	0.58	0.87	1.16	1.46
3.0	0.12	0.24	0.35	0.47	0.71	0.94	1.18

DRAINAGE

Precast Concrete Inspection Chambers and Gullies to BS 5911: Part 230: 1994

Nominal sizes:	375 diameter, 750, 900 mm deep 450 diameter, 750, 900, 1050, 1200 mm deep
Depths:	from the top for trapped or untrapped units: centre of outlet 300 mm invert (bottom) of the outlet pipe 400 mm
Depth of water seal for trapped gullies:	85 mm, rodding eye int. dia. 100 mm
Cover slab:	65 mm min

Bedding Flexible Pipes: PVCu Or Ductile Iron

Type 1 =	100 mm fill below pipe, 300 mm above pipe: single size material
Type 2 =	100 mm fill below pipe, 300 mm above pipe: single size or graded material
Type 3 =	100 mm fill below pipe, 75 mm above pipe with concrete protective slab over
Type 4 =	100 mm fill below pipe, fill laid level with top of pipe
Type 5 =	200 mm fill below pipe, fill laid level with top of pipe
Concrete =	25 mm sand blinding to bottom of trench, pipe supported on chocks, 100 mm concrete under the pipe, 150 mm concrete over the pipe

DRAINAGE

Bedding Rigid Pipes: Clay or Concrete

(for vitrified clay pipes the manufacturer should be consulted)

Class D:	Pipe laid on natural ground with cut-outs for joints, soil screened to remove stones over 40 mm and returned over pipe to 150 m min depth. Suitable for firm ground with trenches trimmed by hand.
Class N:	Pipe laid on 50 mm granular material of graded aggregate to Table 4 of BS 882, or 10 mm aggregate to Table 6 of BS 882, or as dug light soil (not clay) screened to remove stones over 10 mm. Suitable for machine dug trenches.
Class B:	As Class N, but with granular bedding extending half way up the pipe diameter.
Class F:	Pipe laid on 100 mm granular fill to BS 882 below pipe, minimum 150 mm granular fill above pipe: single size material. Suitable for machine dug trenches.
Class A:	Concrete 100 mm thick under the pipe extending half way up the pipe, backfilled with the appropriate class of fill. Used where there is only a very shallow fall to the drain. Class A bedding allows the pipes to be laid to an exact gradient.
Concrete surround:	25 mm sand blinding to bottom of trench, pipe supported on chocks, 100 mm concrete under the pipe, 150 mm concrete over the pipe. It is preferable to bed pipes under slabs or wall in granular material.

PIPED SUPPLY SYSTEMS

Identification of Service Tubes From Utility to Dwellings

Utility	Colour	Size	Depth
British Telecom	grey	54 mm od	450 mm
Electricity	black	38 mm od	450 mm
Gas	yellow	42 mm od rigid 60 mm od convoluted	450 mm
Water	may be blue	(normally untubed)	750 mm

ELECTRICAL SUPPLY/POWER/LIGHTING SYSTEMS

Electrical Insulation Class En 60.598 BS 4533

Class 1:	luminaires comply with class 1 (I) earthed electrical requirements
Class 2:	luminaires comply with class 2 (II) double insulated electrical requirements
Class 3:	luminaires comply with class 3 (III) electrical requirements

Protection to Light Fittings

BS EN 60529:1992 Classification for degrees of protection provided by enclosures.
(IP Code – International or ingress Protection)

1st characteristic: against ingress of solid foreign objects		
The figure	2	indicates that fingers cannot enter
	3	that a 2.5 mm diameter probe cannot enter
	4	that a 1.0 mm diameter probe cannot enter
	5	the fitting is dust proof (no dust around live parts)
	6	the fitting is dust tight (no dust entry)
2nd characteristic: ingress of water with harmful effects		
The figure	0	indicates unprotected
	1	vertically dripping water cannot enter
	2	water dripping 15° (tilt) cannot enter
	3	spraying water cannot enter
	4	splashing water cannot enter
	5	jetting water cannot enter
	6	powerful jetting water cannot enter
	7	proof against temporary immersion
	8	proof against continuous immersion
Optional additional codes:		A–D protects against access to hazardous parts
	H	high voltage apparatus
	M	fitting was in motion during water test
	S	fitting was static during water test
	W	protects against weather
Marking code arrangement:		(example) IPX5S = IP (International or Ingress Protection) X (denotes omission of first characteristic) 5 = jetting S = static during water test

RAIL TRACKS

RAIL TRACKS

	kg/m of track	lb/ft of track
Standard gauge		
Bull-head rails, chairs, transverse timber (softwood) sleepers etc.	245	165
Main lines		
Flat-bottom rails, transverse prestressed concrete sleepers, etc.	418	280
Add for electric third rail	51	35
Add for crushed stone ballast	2600	1750
	kg/m²	**lb/sq ft**
Overall average weight – rails connections, sleepers, ballast, etc.	733	150
	kg/m of track	**lb/ft of track**
Bridge rails, longitudinal timber sleepers, etc.	112	75

RAIL TRACKS

Heavy Rails

British Standard Section No.	Rail height (mm)	Foot width (mm)	Head width (mm)	Min web thickness (mm)	Section weight (kg/m)
			Flat Bottom Rails		
60 A	114.30	109.54	57.15	11.11	30.62
70 A	123.82	111.12	60.32	12.30	34.81
75 A	128.59	114.30	61.91	12.70	37.45
80 A	133.35	117.47	63.50	13.10	39.76
90 A	142.88	127.00	66.67	13.89	45.10
95 A	147.64	130.17	69.85	14.68	47.31
100 A	152.40	133.35	69.85	15.08	50.18
110 A	158.75	139.70	69.85	15.87	54.52
113 A	158.75	139.70	69.85	20.00	56.22
50 'O'	100.01	100.01	52.39	10.32	24.82
80 'O'	127.00	127.00	63.50	13.89	39.74
60R	114.30	109.54	57.15	11.11	29.85
75R	128.59	122.24	61.91	13.10	37.09
80R	133.35	127.00	63.50	13.49	39.72
90R	142.88	136.53	66.67	13.89	44.58
95R	147.64	141.29	68.26	14.29	47.21
100R	152.40	146.05	69.85	14.29	49.60
95N	147.64	139.70	69.85	13.89	47.27
			Bull Head Rails		
95R BH	145.26	69.85	69.85	19.05	47.07

Light Rails

British Standard Section No.	Rail height (mm)	Foot width (mm)	Head width (mm)	Min web thickness (mm)	Section weight (kg/m)
			Flat Bottom Rails		
20M	65.09	55.56	30.96	6.75	9.88
30M	75.41	69.85	38.10	9.13	14.79
35M	80.96	76.20	42.86	9.13	17.39
35R	85.73	82.55	44.45	8.33	17.40
40	88.11	80.57	45.64	12.3	19.89
			Bridge Rails		
13	48.00	92	36.00	18.0	13.31
16	54.00	108	44.50	16.0	16.06
20	55.50	127	50.00	20.5	19.86
28	67.00	152	50.00	31.0	28.62
35	76.00	160	58.00	34.5	35.38
50	76.00	165	58.50	–	50.18
			Crane Rails		
A65	75.00	175.00	65.00	38.0	43.10
A75	85.00	200.00	75.00	45.0	56.20
A100	95.00	200.00	100.00	60.0	74.30
A120	105.00	220.00	120.00	72.0	100.00
175CR	152.40	152.40	107.95	38.1	86.92

RAIL TRACKS

Fish Plates

British Standard Section No.	Overall plate length		Hole diameter	Finished weight per pair	
	4 Hole (mm)	6 Hole (mm)	(mm)	4 Hole (kg/pair)	6 Hole (kg/pair)
For British Standard Heavy Rails: Flat Bottom Rails					
60 A	406.40	609.60	20.64	9.87	14.76
70 A	406.40	609.60	22.22	11.15	16.65
75 A	406.40	–	23.81	11.82	17.73
80 A	406.40	609.60	23.81	13.15	19.72
90 A	457.20	685.80	25.40	17.49	26.23
100 A	508.00	–	pear	25.02	–
110 A (shallow)	507.00	–	27.00	30.11	54.64
113 A (heavy)	507.00	–	27.00	30.11	54.64
50 'O' (shallow)	406.40	–	–	6.68	10.14
80 'O' (shallow)	495.30	–	23.81	14.72	22.69
60R (shallow)	406.40	609.60	20.64	8.76	13.13
60R (angled)	406.40	609.60	20.64	11.27	16.90
75R (shallow)	406.40	–	23.81	10.94	16.42
75R (angled)	406.40	–	23.81	13.67	–
80R (shallow)	406.40	609.60	23.81	11.93	17.89
80R (angled)	406.40	609.60	23.81	14.90	22.33
For British Standard Heavy Rails: Bull head rails					
95R BH (shallow)	–	457.20	27.00	14.59	14.61
For British Standard Light Rails: Flat Bottom Rails					
30M	355.6	–	–	–	2.72
35M	355.6	–	–	–	2.83
40	355.6	–	–	3.76	–

FRACTIONS, DECIMALS AND MILLIMETRE EQUIVALENTS

FRACTIONS, DECIMALS AND MILLIMETRE EQUIVALENTS

Fractions	Decimals	(mm)	Fractions	Decimals	(mm)
1/64	0.015625	0.396875	33/64	0.515625	13.096875
1/32	0.03125	0.79375	17/32	0.53125	13.49375
3/64	0.046875	1.190625	35/64	0.546875	13.890625
1/16	0.0625	1.5875	9/16	0.5625	14.2875
5/64	0.078125	1.984375	37/64	0.578125	14.684375
3/32	0.09375	2.38125	19/32	0.59375	15.08125
7/64	0.109375	2.778125	39/64	0.609375	15.478125
1/8	0.125	3.175	5/8	0.625	15.875
9/64	0.140625	3.571875	41/64	0.640625	16.271875
5/32	0.15625	3.96875	21/32	0.65625	16.66875
11/64	0.171875	4.365625	43/64	0.671875	17.065625
3/16	0.1875	4.7625	11/16	0.6875	17.4625
13/64	0.203125	5.159375	45/64	0.703125	17.859375
7/32	0.21875	5.55625	23/32	0.71875	18.25625
15/64	0.234375	5.953125	47/64	0.734375	18.653125
1/4	0.25	6.35	3/4	0.75	19.05
17/64	0.265625	6.746875	49/64	0.765625	19.446875
9/32	0.28125	7.14375	25/32	0.78125	19.84375
19/64	0.296875	7.540625	51/64	0.796875	20.240625
5/16	0.3125	7.9375	13/16	0.8125	20.6375
21/64	0.328125	8.334375	53/64	0.828125	21.034375
11/32	0.34375	8.73125	27/32	0.84375	21.43125
23/64	0.359375	9.128125	55/64	0.859375	21.828125
3/8	0.375	9.525	7/8	0.875	22.225
25/64	0.390625	9.921875	57/64	0.890625	22.621875
13/32	0.40625	10.31875	29/32	0.90625	23.01875
27/64	0.421875	10.71563	59/64	0.921875	23.415625
7/16	0.4375	11.1125	15/16	0.9375	23.8125
29/64	0.453125	11.50938	61/64	0.953125	24.209375
15/32	0.46875	11.90625	31/32	0.96875	24.60625
31/64	0.484375	12.30313	63/64	0.984375	25.003125
1/2	0.5	12.7	1.0	1	25.4

IMPERIAL STANDARD WIRE GAUGE (SWG)

IMPERIAL STANDARD WIRE GAUGE (SWG)

SWG	Diameter		SWG	Diameter	
No.	(inches)	(mm)	No.	(inches)	(mm)
7/0	0.5	12.7	23	0.024	0.61
6/0	0.464	11.79	24	0.022	0.559
5/0	0.432	10.97	25	0.02	0.508
4/0	0.4	10.16	26	0.018	0.457
3/0	0.372	9.45	27	0.0164	0.417
2/0	0.348	8.84	28	0.0148	0.376
1/0	0.324	8.23	29	0.0136	0.345
1	0.3	7.62	30	0.0124	0.315
2	0.276	7.01	31	0.0116	0.295
3	0.252	6.4	32	0.0108	0.274
4	0.232	5.89	33	0.01	0.254
5	0.212	5.38	34	0.009	0.234
6	0.192	4.88	35	0.008	0.213
7	0.176	4.47	36	0.008	0.193
8	0.16	4.06	37	0.007	0.173
9	0.144	3.66	38	0.006	0.152
10	0.128	3.25	39	0.005	0.132
11	0.116	2.95	40	0.005	0.122
12	0.104	2.64	41	0.004	0.112
13	0.092	2.34	42	0.004	0.102
14	0.08	2.03	43	0.004	0.091
15	0.072	1.83	44	0.003	0.081
16	0.064	1.63	45	0.003	0.071
17	0.056	1.42	46	0.002	0.061
18	0.048	1.22	47	0.002	0.051
19	0.04	1.016	48	0.002	0.041
20	0.036	0.914	49	0.001	0.031
21	0.032	0.813	50	0.001	0.025
22	0.028	0.711			

PIPES, WATER, STORAGE, INSULATION

WATER PRESSURE DUE TO HEIGHT

Imperial

Head (Feet)	Pressure (lb/in^2)		Head (Feet)	Pressure (lb/in^2)
1	0.43		70	30.35
5	2.17		75	32.51
10	4.34		80	34.68
15	6.5		85	36.85
20	8.67		90	39.02
25	10.84		95	41.18
30	13.01		100	43.35
35	15.17		105	45.52
40	17.34		110	47.69
45	19.51		120	52.02
50	21.68		130	56.36
55	23.84		140	60.69
60	26.01		150	65.03
65	28.18			

Metric

Head (m)	Pressure (bar)		Head (m)	Pressure (bar)
0.5	0.049		18.0	1.766
1.0	0.098		19.0	1.864
1.5	0.147		20.0	1.962
2.0	0.196		21.0	2.06
3.0	0.294		22.0	2.158
4.0	0.392		23.0	2.256
5.0	0.491		24.0	2.354
6.0	0.589		25.0	2.453
7.0	0.687		26.0	2.551
8.0	0.785		27.0	2.649
9.0	0.883		28.0	2.747
10.0	0.981		29.0	2.845
11.0	1.079		30.0	2.943
12.0	1.177		32.5	3.188
13.0	1.275		35.0	3.434
14.0	1.373		37.5	3.679
15.0	1.472		40.0	3.924
16.0	1.57		42.5	4.169
17.0	1.668		45.0	4.415

1 bar	=	14.5038 lbf/in^2
1 lbf/in^2	=	0.06895 bar
1 metre	=	3.2808 ft or 39.3701 in
1 foot	=	0.3048 metres
1 in wg	=	2.5 mbar (249.1 N/m^2)

PIPES, WATER, STORAGE, INSULATION

Dimensions and Weights of Copper Pipes to BSEN 1057, BSEN 12499, BSEN 14251

Outside diameter (mm)	Internal diameter (mm)	Weight per metre (kg)	Internal diameter (mm)	Weight per metre (kg)	Internal siameter (mm)	Weight per metre (kg)
	Formerly Table X		Formerly Table Y		Formerly Table Z	
6	4.80	0.0911	4.40	0.1170	5.00	0.0774
8	6.80	0.1246	6.40	0.1617	7.00	0.1054
10	8.80	0.1580	8.40	0.2064	9.00	0.1334
12	10.80	0.1914	10.40	0.2511	11.00	0.1612
15	13.60	0.2796	13.00	0.3923	14.00	0.2031
18	16.40	0.3852	16.00	0.4760	16.80	0.2918
22	20.22	0.5308	19.62	0.6974	20.82	0.3589
28	26.22	0.6814	25.62	0.8985	26.82	0.4594
35	32.63	1.1334	32.03	1.4085	33.63	0.6701
42	39.63	1.3675	39.03	1.6996	40.43	0.9216
54	51.63	1.7691	50.03	2.9052	52.23	1.3343
76.1	73.22	3.1287	72.22	4.1437	73.82	2.5131
108	105.12	4.4666	103.12	7.3745	105.72	3.5834
133	130.38	5.5151	–	–	130.38	5.5151
159	155.38	8.7795	–	–	156.38	6.6056

Dimensions of Stainless Steel Pipes to BS 4127

Outside siameter (mm)	Maximum outside siameter (mm)	Minimum outside diameter (mm)	Wall thickness (mm)	Working pressure (bar)
6	6.045	5.940	0.6	330
8	8.045	7.940	0.6	260
10	10.045	9.940	0.6	210
12	12.045	11.940	0.6	170
15	15.045	14.940	0.6	140
18	18.045	17.940	0.7	135
22	22.055	21.950	0.7	110
28	28.055	27.950	0.8	121
35	35.070	34.965	1.0	100
42	42.070	41.965	1.1	91
54	54.090	53.940	1.2	77

PIPES, WATER, STORAGE, INSULATION

Dimensions of Steel Pipes to BS 1387

Nominal Size	Approx. Outside Diameter	Outside diameter				Thickness		
		Light		Medium & Heavy		Light	Medium	Heavy
		Max	Min	Max	Min			
(mm)	(mm)	(mm)	(mm)	(mm)	(mm)	(mm)	(mm)	(mm)
6	10.20	10.10	9.70	10.40	9.80	1.80	2.00	2.65
8	13.50	13.60	13.20	13.90	13.30	1.80	2.35	2.90
10	17.20	17.10	16.70	17.40	16.80	1.80	2.35	2.90
15	21.30	21.40	21.00	21.70	21.10	2.00	2.65	3.25
20	26.90	26.90	26.40	27.20	26.60	2.35	2.65	3.25
25	33.70	33.80	33.20	34.20	33.40	2.65	3.25	4.05
32	42.40	42.50	41.90	42.90	42.10	2.65	3.25	4.05
40	48.30	48.40	47.80	48.80	48.00	2.90	3.25	4.05
50	60.30	60.20	59.60	60.80	59.80	2.90	3.65	4.50
65	76.10	76.00	75.20	76.60	75.40	3.25	3.65	4.50
80	88.90	88.70	87.90	89.50	88.10	3.25	4.05	4.85
100	114.30	113.90	113.00	114.90	113.30	3.65	4.50	5.40
125	139.70	–	–	140.60	138.70	–	4.85	5.40
150	165.1*	–	–	166.10	164.10	–	4.85	5.40

* 165.1 mm (6.5in) outside diameter is not generally recommended except where screwing to BS 21 is necessary
All dimensions are in accordance with ISO R65 except approximate outside diameters which are in accordance with ISO R64
Light quality is equivalent to ISO R65 Light Series II

Approximate Metres Per Tonne of Tubes to BS 1387

Nom. size	BLACK						GALVANIZED					
	Plain/screwed ends			Screwed & socketed			Plain/screwed ends			Screwed & socketed		
	L	M	H	L	M	H	L	M	H	L	M	H
(mm)	(m)	(m)	(m)	(m)	(m)	(m)	(m)	(m)	(m)	(m)	(m)	(m)
6	2765	2461	2030	2743	2443	2018	2604	2333	1948	2584	2317	1937
8	1936	1538	1300	1920	1527	1292	1826	1467	1254	1811	1458	1247
10	1483	1173	979	1471	1165	974	1400	1120	944	1386	1113	939
15	1050	817	688	1040	811	684	996	785	665	987	779	661
20	712	634	529	704	628	525	679	609	512	673	603	508
25	498	410	336	494	407	334	478	396	327	474	394	325
32	388	319	260	384	316	259	373	308	254	369	305	252
40	307	277	226	303	273	223	296	268	220	292	264	217
50	244	196	162	239	194	160	235	191	158	231	188	157
65	172	153	127	169	151	125	167	149	124	163	146	122
80	147	118	99	143	116	98	142	115	97	139	113	96
100	101	82	69	98	81	68	98	81	68	95	79	67
125	–	62	56	–	60	55	–	60	55	–	59	54
150	–	52	47	–	50	46	–	51	46	–	49	45

The figures for 'plain or screwed ends' apply also to tubes to BS 1775 of equivalent size and thickness
Key:
L – Light
M – Medium
H – Heavy

PIPES, WATER, STORAGE, INSULATION

Flange Dimension Chart to BS 4504 & BS 10

Normal Pressure Rating (PN 6) 6 Bar

Nom. size	Flange outside dia.	Table 6/2 Forged Welding Neck	Table 6/3 Plate Slip on	Table 6/4 Forged Bossed Screwed	Table 6/5 Forged Bossed Slip on	Table 6/8 Plate Blank	Raised face		Nr. bolt hole	Size of bolt
							Dia.	T'ness		
15	80	12	12	12	12	12	40	2	4	M10 × 40
20	90	14	14	14	14	14	50	2	4	M10 × 45
25	100	14	14	14	14	14	60	2	4	M10 × 45
32	120	14	16	14	14	14	70	2	4	M12 × 45
40	130	14	16	14	14	14	80	3	4	M12 × 45
50	140	14	16	14	14	14	90	3	4	M12 × 45
65	160	14	16	14	14	14	110	3	4	M12 × 45
80	190	16	18	16	16	16	128	3	4	M16 × 55
100	210	16	18	16	16	16	148	3	4	M16 × 55
125	240	18	20	18	18	18	178	3	8	M16 × 60
150	265	18	20	18	18	18	202	3	8	M16 × 60
200	320	20	22	–	20	20	258	3	8	M16 × 60
250	375	22	24	–	22	22	312	3	12	M16 × 65
300	440	22	24	–	22	22	365	4	12	M20 × 70

Normal Pressure Rating (PN 16) 16 Bar

Nom. size	Flange outside dia.	Table 6/2 Forged Welding Neck	Table 6/3 Plate Slip on	Table 6/4 Forged Bossed Screwed	Table 6/5 Forged Bossed Slip on	Table 6/8 Plate Blank	Raised face		Nr. bolt hole	Size of bolt
							Dia.	T'ness		
15	95	14	14	14	14	14	45	2	4	M12 × 45
20	105	16	16	16	16	16	58	2	4	M12 × 50
25	115	16	16	16	16	16	68	2	4	M12 × 50
32	140	16	16	16	16	16	78	2	4	M16 × 55
40	150	16	16	16	16	16	88	3	4	M16 × 55
50	165	18	18	18	18	18	102	3	4	M16 × 60
65	185	18	18	18	18	18	122	3	4	M16 × 60
80	200	20	20	20	20	20	138	3	8	M16 × 60
100	220	20	20	20	20	20	158	3	8	M16 × 65
125	250	22	22	22	22	22	188	3	8	M16 × 70
150	285	22	22	22	22	22	212	3	8	M20 × 70
200	340	24	24	–	24	24	268	3	12	M20 × 75
250	405	26	26	–	26	26	320	3	12	M24 × 90
300	460	28	28	–	28	28	378	4	12	M24 × 90

PIPES, WATER, STORAGE, INSULATION

Minimum Distances Between Supports/Fixings

Material	BS Nominal pipe size		Pipes – Vertical	Pipes – Horizontal on to low gradients
	(inch)	(mm)	Support distance in metres	Support distance in metres
Copper	0.50	15.00	1.90	1.30
	0.75	22.00	2.50	1.90
	1.00	28.00	2.50	1.90
	1.25	35.00	2.80	2.50
	1.50	42.00	2.80	2.50
	2.00	54.00	3.90	2.50
	2.50	67.00	3.90	2.80
	3.00	76.10	3.90	2.80
	4.00	108.00	3.90	2.80
	5.00	133.00	3.90	2.80
	6.00	159.00	3.90	2.80
muPVC	1.25	32.00	1.20	0.50
	1.50	40.00	1.20	0.50
	2.00	50.00	1.20	0.60
Polypropylene	1.25	32.00	1.20	0.50
	1.50	40.00	1.20	0.50
uPVC	–	82.40	1.20	0.50
	–	110.00	1.80	0.90
	–	160.00	1.80	1.20
Steel	0.50	15.00	2.40	1.80
	0.75	20.00	3.00	2.40
	1.00	25.00	3.00	2.40
	1.25	32.00	3.00	2.40
	1.50	40.00	3.70	2.40
	2.00	50.00	3.70	2.40
	2.50	65.00	4.60	3.00
	3.00	80.40	4.60	3.00
	4.00	100.00	4.60	3.00
	5.00	125.00	5.50	3.70
	6.00	150.00	5.50	4.50
	8.00	200.00	8.50	6.00
	10.00	250.00	9.00	6.50
	12.00	300.00	10.00	7.00
	16.00	400.00	10.00	8.25

PIPES, WATER, STORAGE, INSULATION

Litres of Water Storage Required Per Person Per Building Type

Type of building	Storage (litres)
Houses and flats (up to 4 bedrooms)	120/bedroom
Houses and flats (more than 4 bedrooms)	100/bedroom
Hostels	90/bed
Hotels	200/bed
Nurses homes and medical quarters	120/bed
Offices with canteen	45/person
Offices without canteen	40/person
Restaurants	7/meal
Boarding schools	90/person
Day schools – Primary	15/person
Day schools – Secondary	20/person

Recommended Air Conditioning Design Loads

Building type	Design loading
Computer rooms	500 W/m^2 of floor area
Restaurants	150 W/m^2 of floor area
Banks (main area)	100 W/m^2 of floor area
Supermarkets	25 W/m^2 of floor area
Large office block (exterior zone)	100 W/m^2 of floor area
Large office block (interior zone)	80 W/m^2 of floor area
Small office block (interior zone)	80 W/m^2 of floor area

PIPES, WATER, STORAGE, INSULATION

Capacity and Dimensions of Galvanized Mild Steel Cisterns – BS 417

Capacity (litres)	BS type (SCM)	Dimensions		
		Length (mm)	Width (mm)	Depth (mm)
18	45	457	305	305
36	70	610	305	371
54	90	610	406	371
68	110	610	432	432
86	135	610	457	482
114	180	686	508	508
159	230	736	559	559
191	270	762	584	610
227	320	914	610	584
264	360	914	660	610
327	450/1	1220	610	610
336	450/2	965	686	686
423	570	965	762	787
491	680	1090	864	736
709	910	1070	889	889

Capacity of Cold Water Polypropylene Storage Cisterns – BS 4213

Capacity (litres)	BS type (PC)	Maximum height (mm)
18	4	310
36	8	380
68	15	430
91	20	510
114	25	530
182	40	610
227	50	660
273	60	660
318	70	660
455	100	760

PIPES, WATER, STORAGE, INSULATION

Minimum Insulation Thickness to Protect Against Freezing for Domestic Cold Water Systems (8 Hour Evaluation Period)

Pipe size (mm)	Insulation thickness (mm)					
	Condition 1			Condition 2		
	λ = 0.020	λ = 0.030	λ = 0.040	λ = 0.020	λ = 0.030	λ = 0.040
Copper pipes						
15	11	20	34	12	23	41
22	6	9	13	6	10	15
28	4	6	9	4	7	10
35	3	5	7	4	5	7
42	3	4	5	8	4	6
54	2	3	4	2	3	4
76	2	2	3	2	2	3
Steel pipes						
15	9	15	24	10	18	29
20	6	9	13	6	10	15
25	4	7	9	5	7	10
32	3	5	6	3	5	7
40	3	4	5	3	4	6
50	2	3	4	2	3	4
65	2	2	3	2	3	3

Condition 1: water temperature 7°C; ambient temperature –6°C; evaluation period 8 h; permitted ice formation 50%; normal installation, i.e. inside the building and inside the envelope of the structural insulation
Condition 2: water temperature 2°C; ambient temperature –6°C; evaluation period 8 h; permitted ice formation 50%; extreme installation, i.e. inside the building but outside the envelope of the structural insulation
λ = thermal conductivity [W/(mK)]

Insulation Thickness for Chilled And Cold Water Supplies to Prevent Condensation

On a Low Emissivity Outer Surface (0.05, i.e. Bright Reinforced Aluminium Foil) with an Ambient Temperature of +25°C and a Relative Humidity of 80%

Steel pipe size (mm)	*t* = +10			*t* = +5			*t* = 0		
	Insulation thickness (mm)			Insulation thickness (mm)			Insulation thickness (mm)		
	λ = 0.030	λ = 0.040	λ = 0.050	λ = 0.030	λ = 0.040	λ = 0.050	λ = 0.030	λ = 0.040	λ = 0.050
15	16	20	25	22	28	34	28	36	43
25	18	24	29	25	32	39	32	41	50
50	22	28	34	30	39	47	38	49	60
100	26	34	41	36	47	57	46	60	73
150	29	38	46	40	52	64	51	67	82
250	33	43	53	46	60	74	59	77	94
Flat surfaces	39	52	65	56	75	93	73	97	122

t = temperature of contents (°C)
λ = thermal conductivity at mean temperature of insulation [W/(mK)]

PIPES, WATER, STORAGE, INSULATION

Insulation Thickness for Non-domestic Heating Installations to Control Heat Loss

Steel pipe size (mm)	*t* = 75			*t* = 100			*t* = 150		
	Insulation thickness (mm)			Insulation thickness (mm)			Insulation thickness (mm)		
	λ = 0.030	λ = 0.040	λ = 0.050	λ = 0.030	λ = 0.040	λ = 0.050	λ = 0.030	λ = 0.040	λ = 0.050
10	18	32	55	20	36	62	23	44	77
15	19	34	56	21	38	64	26	47	80
20	21	36	57	23	40	65	28	50	83
25	23	38	58	26	43	68	31	53	85
32	24	39	59	28	45	69	33	55	87
40	25	40	60	29	47	70	35	57	88
50	27	42	61	31	49	72	37	59	90
65	29	43	62	33	51	74	40	63	92
80	30	44	62	35	52	75	42	65	94
100	31	46	63	37	54	76	45	68	96
150	33	48	64	40	57	77	50	73	100
200	35	49	65	42	59	79	53	76	103
250	36	50	66	43	61	80	55	78	105

t = hot face temperature (°C)

λ = thermal conductivity at mean temperature of insulation [W/(mK)]

Index

Ebook Single-User Licence Agreement

We welcome you as a user of this Spon Price Book ebook and hope that you find it a useful and valuable tool. Please read this document carefully. **This is a legal agreement** between you (hereinafter referred to as the "Licensee") and Taylor and Francis Books Ltd. (the "Publisher"), which defines the terms under which you may use the Product. **By accessing and retrieving the access code on the label inside the front cover of this book you agree to these terms and conditions outlined herein. If you do not agree to these terms you must return the Product to your supplier intact, with the seal on the label unbroken and with the access code not accessed.**

1. **Definition of the Product**
The product which is the subject of this Agreement, (the "Product") consists of online and offline access to the VitalSource ebook edition of *Spon's Civil Engineering & Highway Works Price Book 2023.*

2. **Commencement and licence**
2.1 This Agreement commences upon the breaking open of the document containing the access code by the Licensee (the "Commencement Date").
2.2 This is a licence agreement (the "Agreement") for the use of the Product by the Licensee, and not an agreement for sale.
2.3 The Publisher licenses the Licensee on a non-exclusive and non-transferable basis to use the Product on condition that the Licensee complies with this Agreement. The Licensee acknowledges that it is only permitted to use the Product in accordance with this Agreement.

3. **Multiple use**
Use of the Product is not provided or allowed for more than one user or for a wide area network or consortium.

4. **Installation and Use**
4.1 The Licensee may provide access to the Product for individual study in the following manner: The Licensee may install the Product on a secure local area network on a single site for use by one user.
4.2 The Licensee shall be responsible for installing the Product and for the effectiveness of such installation.
4.3 Text from the Product may be incorporated in a coursepack. Such use is only permissible with the express permission of the Publisher in writing and requires the payment of the appropriate fee as specified by the Publisher and signature of a separate licence agreement.
4.4 The Product is a free addition to the book and the Publisher is under no obligation to provide any technical support.

5. **Permitted Activities**
5.1 The Licensee shall be entitled to use the Product for its own internal purposes;
5.2 The Licensee acknowledges that its rights to use the Product are strictly set out in this Agreement, and all other uses (whether expressly mentioned in Clause 6 below or not) are prohibited.

6. **Prohibited Activities**
The following are prohibited without the express permission of the Publisher:
6.1 The commercial exploitation of any part of the Product.
6.2 The rental, loan, (free or for money or money's worth) or hire purchase of this product, save with the express consent of the Publisher.
6.3 Any activity which raises the reasonable prospect of impeding the Publisher's ability or opportunities to market the Product.
6.4 Any networking, physical or electronic distribution or dissemination of the product save as expressly permitted by this Agreement.
6.5 Any reverse engineering, decompilation, disassembly or other alteration of the Product save in accordance with applicable national laws.
6.6 The right to create any derivative product or service from the Product save as expressly provided for in this Agreement.
6.7 Any alteration, amendment, modification or deletion from the Product, whether for the purposes of error correction or otherwise.

7. General Responsibilities of the License

7.1 The Licensee will take all reasonable steps to ensure that the Product is used in accordance with the terms and conditions of this Agreement.

7.2 The Licensee acknowledges that damages may not be a sufficient remedy for the Publisher in the event of breach of this Agreement by the Licensee, and that an injunction may be appropriate.

7.3 The Licensee undertakes to keep the Product safe and to use its best endeavours to ensure that the product does not fall into the hands of third parties, whether as a result of theft or otherwise.

7.4 Where information of a confidential nature relating to the product of the business affairs of the Publisher comes into the possession of the Licensee pursuant to this Agreement (or otherwise), the Licensee agrees to use such information solely for the purposes of this Agreement, and under no circumstances to disclose any element of the information to any third party save strictly as permitted under this Agreement. For the avoidance of doubt, the Licensee's obligations under this sub-clause 7.4 shall survive the termination of this Agreement.

8. Warrant and Liability

8.1 The Publisher warrants that it has the authority to enter into this agreement and that it has secured all rights and permissions necessary to enable the Licensee to use the Product in accordance with this Agreement.

8.2 The Publisher warrants that the Product as supplied on the Commencement Date shall be free of defects in materials and workmanship, and undertakes to replace any defective Product within 28 days of notice of such defect being received provided such notice is received within 30 days of such supply. As an alternative to replacement, the Publisher agrees fully to refund the Licensee in such circumstances, if the Licensee so requests, provided that the Licensee returns this copy of *Spon's Civil Engineering & Highway Works Price Book 2023* to the Publisher. The provisions of this sub-clause 8.2 do not apply where the defect results from an accident or from misuse of the product by the Licensee.

8.3 Sub-clause 8.2 sets out the sole and exclusive remedy of the Licensee in relation to defects in the Product.

8.4 The Publisher and the Licensee acknowledge that the Publisher supplies the Product on an "as is" basis. The Publisher gives no warranties:

8.4.1 that the Product satisfies the individual requirements of the Licensee; or

8.4.2 that the Product is otherwise fit for the Licensee's purpose; or

8.4.3 that the Product is compatible with the Licensee's hardware equipment and software operating environment.

8.5 The Publisher hereby disclaims all warranties and conditions, express or implied, which are not stated above.

8.6 Nothing in this Clause 8 limits the Publisher's liability to the Licensee in the event of death or personal injury resulting from the Publisher's negligence.

8.7 The Publisher hereby excludes liability for loss of revenue, reputation, business, profits, or for indirect or consequential losses, irrespective of whether the Publisher was advised by the Licensee of the potential of such losses.

8.8 The Licensee acknowledges the merit of independently verifying the price book data prior to taking any decisions of material significance (commercial or otherwise) based on such data. It is agreed that the Publisher shall not be liable for any losses which result from the Licensee placing reliance on the data under any circumstances.

8.9 Subject to sub-clause 8.6 above, the Publisher's liability under this Agreement shall be limited to the purchase price.

9. Intellectual Property Rights

9.1 Nothing in this Agreement affects the ownership of copyright or other intellectual property rights in the Product.

9.2 The Licensee agrees to display the Publishers' copyright notice in the manner described in the Product.

9.3 The Licensee hereby agrees to abide by copyright and similar notice requirements required by the Publisher, details of which are as follows:

"© 2023 Taylor & Francis. All rights reserved. All materials in *Spon's Civil Engineering & Highway Works Price Book 2023* are copyright protected. All rights reserved. No such materials may be used, displayed, modified, adapted, distributed, transmitted, transferred, published or otherwise reproduced in any form or by any means now or hereafter developed other than strictly in accordance with the terms of the licence agreement enclosed with *Spon's Civil Engineering & Highway Works Price Book 2023*. However, text and images may be printed and copied for research and private study within the preset program limitations. Please note the copyright notice above, and that any text or images printed or copied must credit the source."

9.4 This Product contains material proprietary to and copyedited by the Publisher and others. Except for the licence granted herein, all rights, title and interest in the Product, in all languages, formats and media throughout the world, including copyrights therein, are and remain the property of the Publisher or other copyright holders identified in the Product.

10. Non-assignment

This Agreement and the licence contained within it may not be assigned to any other person or entity without the written consent of the Publisher.

11. Termination and Consequences of Termination.

11.1 The Publisher shall have the right to terminate this Agreement if:

11.1.1 the Licensee is in material breach of this Agreement and fails to remedy such breach (where capable of remedy) within 14 days of a written notice from the Publisher requiring it to do so; or

11.1.2 the Licensee becomes insolvent, becomes subject to receivership, liquidation or similar external administration; or

11.1.3 the Licensee ceases to operate in business.

11.2 The Licensee shall have the right to terminate this Agreement for any reason upon two month's written notice. The Licensee shall not be entitled to any refund for payments made under this Agreement prior to termination under this sub-clause 11.2.

11.3 Termination by either of the parties is without prejudice to any other rights or remedies under the general law to which they may be entitled, or which survive such termination (including rights of the Publisher under sub-clause 7.4 above).

11.4 Upon termination of this Agreement, or expiry of its terms, the Licensee must destroy all copies and any back up copies of the product or part thereof.

12. General

12.1 *Compliance with export provisions*

The Publisher hereby agrees to comply fully with all relevant export laws and regulations of the United Kingdom to ensure that the Product is not exported, directly or indirectly, in violation of English law.

12.2 *Force majeure*

The parties accept no responsibility for breaches of this Agreement occurring as a result of circumstances beyond their control.

12.3 *No waiver*

Any failure or delay by either party to exercise or enforce any right conferred by this Agreement shall not be deemed to be a waiver of such right.

12.4 *Entire agreement*

This Agreement represents the entire agreement between the Publisher and the Licensee concerning the Product. The terms of this Agreement supersede all prior purchase orders, written terms and conditions, written or verbal representations, advertising or statements relating in any way to the Product.

12.5 *Severability*

If any provision of this Agreement is found to be invalid or unenforceable by a court of law of competent jurisdiction, such a finding shall not affect the other provisions of this Agreement and all provisions of this Agreement unaffected by such a finding shall remain in full force and effect.

12.6 *Variations*

This agreement may only be varied in writing by means of variation signed in writing by both parties.

12.7 *Notices*

All notices to be delivered to: Spon's Price Books, Taylor & Francis Books Ltd., 4 Park Square, Milton Park, Abingdon, Oxfordshire, OX14 4RN, UK.

12.8 *Governing law*

This Agreement is governed by English law and the parties hereby agree that any dispute arising under this Agreement shall be subject to the jurisdiction of the English courts.

If you have any queries about the terms of this licence, please contact:

Spon's Price Books
Taylor & Francis Books Ltd.
4 Park Square, Milton Park, Abingdon, Oxfordshire, OX14 4RN
Tel: +44 (0) 20 7017 6000
www.sponpress.com

Ebook options

Your print copy comes with a free ebook on the VitalSource® Bookshelf platform. Further copies of the ebook are available from https://www.crcpress.com/search/results?kw=spon+2023
Pick your price book and select the ebook under the 'Select Format' option.

To buy Spon's ebooks for five or more users of in your organisation please contact:

Spon's Price Books
eBooks & Online Sales
Taylor & Francis Books Ltd.
4 Park Square, Milton Park, Oxfordshire, OX14 4RN
Tel: (020) 337 73480
onlinesales@informa.com